Signals and Systems
Using MATLAB®

Luis F. Chaparro

Department of Electrical and Computer Engineering
University of Pittsburgh

AMSTERDAM • BOSTON • HEIDELBERG • LONDON
NEW YORK • OXFORD • PARIS • SAN DIEGO
SAN FRANCISCO • SINGAPORE • SYDNEY • TOKYO
Academic Press is an imprint of Elsevier

ELSEVIER

Academic Press is an imprint of Elsevier
30 Corporate Drive, Suite 400, Burlington, MA 01803, USA
Elsevier, The Boulevard, Langford Lane, Kidlington, Oxford, OX5 1GB, UK

Library of Congress Cataloging-in-Publication Data
Chaparro, Luis F.
    Signals and systems using MATLAB® / Luis F. Chaparro.
        p. cm.
    ISBN 978-0-12-374716-7
1. Signal processing–Digital techniques. 2. System analysis. 3. MATLAB. I. Title.
    TK5102.9.C472 2010
    621.382′2–dc22

                                            2010023436

British Library Cataloguing-in-Publication Data
A catalogue record for this book is available from the British Library.

For information on all Academic Press publications
visit our Web site at *www.elsevierdirect.com*

Printed in the United States of America
11 12 13 9 8 7 6 5 4 3 2

Working together to grow
libraries in developing countries

www.elsevier.com | www.bookaid.org | www.sabre.org

ELSEVIER   BOOK AID
            International   Sabre Foundation

To my family, with much love.

Contents

Preface

In this book I have only made up a bunch
of other men's flowers, providing of my own
only the string that ties them together.
M. de Montaigne (1533–1592)
French essayist

Although it is hardly possible to keep up with advances in technology, it is reassuring to know that in science and engineering, development and innovation are possible through a solid understanding of basic principles. The theory of signals and systems is one of those fundamentals, and it will be the foundation of much research and development in engineering for years to come. Not only engineers will need to know about signals and systems—to some degree everybody will. The pervasiveness of computers, cell phones, digital recording, and digital communications will require it.

Learning as well as teaching signals and systems is complicated by the combination of mathematical abstraction and concrete engineering applications. Mathematical sophistication and maturity in engineering are needed. Thus, a course in signals and systems needs to be designed to nurture the students' interest in applications, but also to make them appreciate the significance of the mathematical tools. In writing this textbook, as in teaching this material for many years, the author has found it practical to follow Einstein's recommendation that "Everything should be made as simple as possible, but not simpler," and Melzak's [47] dictum that "It is downright sinful to teach the abstract before the concrete." The aim of this textbook is to serve the students' needs in learning signals and systems theory as well as to facilitate the teaching of the material for faculty by proposing an approach that the author has found effective in his own teaching.

We consider the use of MATLAB, an essential tool in the practice of engineering, of great significance in the learning process. It not only helps to illustrate the theoretical results but makes students aware of the computational issues that engineers face in implementing them. Some familiarity with MATLAB is beneficial but not required.

LEVEL

The material in this textbook is intended for courses in signals and systems at the junior level in electrical and computer engineering, but it could also be used in teaching this material to mechanical engineering and bioengineering students and it might be of interest to students in applied mathematics. The "student-friendly" nature of the text also makes it useful to practicing engineers interested in learning or reviewing the basic principles of signals and systems on their own. The material is organized so that students not only get a solid understanding of the theory—through analytic examples as well as software examples using MATLAB—and learn about applications, but also develop confidence and proficiency in the material by working on problems.

The organization of the material in the book follows the assumption that the student has been exposed to the theory of linear circuits, differential equations, and linear algebra, and that this material will be followed by courses in control, communications, or digital signal processing. The content is guided by the goal of nurturing the interest of students in applications, and of assisting them in becoming more sophisticated mathematically. In teaching signals and systems, the author has found that students typically lack basic skills in manipulating complex variables, in understanding differential equations, and are not yet comfortable with basic concepts in calculus. Introducing discrete-time signals and systems makes students face new concepts that were not explored in their calculus courses, such as summations, finite differences, and difference equations. This text attempts to fill the gap and nurture interest in the mathematical tools.

APPROACH

In writing this text, we have taken the following approach:

1. The material is divided into three parts: introduction, theory and applications of continuous-time signals and systems, and theory and applications of discrete-time signals and systems. To help students understand the connection between continuous- and discrete-time signals and systems, the connection between infinitesimal and finite calculus is made in the introduction part, together with a motivation as to why complex numbers and functions are used in the study of signals and systems. The treatment of continuous- and discrete-time signals and systems is then done separately in the next two parts; combining them is found to be confusing to students. Likewise, the author believes it is important for students to understand the connections and relevance of each of the transformations used in the analysis of signals and systems so that these transformations are seen as a progression rather than as disconnected methods. Thus, the author advocates the presentation of the Laplace analysis followed by the Fourier analysis, and the Z-transform followed by the discrete Fourier, and capping each of these topics with applications to communications, control, and filtering. The mathematical abstraction and the applications become more sophisticated as the material unfolds, taking advantage as needed of the background on circuits that students have.

2. An overview of the topics to be discussed in the book and how each connects with some basic mathematical concepts—needed in the rest of the book—is given in Chapter 0 (analogous to the ground floor of a building). The emphasis is in relating summations, differences, difference equations, and sequence of numbers with the calculus concepts that the students are familiar with, and in doing so providing a new interpretation to integrals, derivatives, differential equations, and functions of time. This chapter also links the theory of complex numbers and functions to vectors and to phasors learned in circuit theory. Because we strongly believe that the material in this chapter should be covered before beginning the discussion of signals and systems, it is not relegated to an appendix but placed at the front of the book where it cannot be ignored. A soft introduction to MATLAB is also provided in this chapter.

3. A great deal of effort has been put into making the text "student friendly." To make sure that the student does not miss some of the important issues presented in a section, we have inserted well-thought-out remarks—we want to minimize the common misunderstandings we have observed from our students in the past. Plenty of analytic examples with different levels of complexity are given to illustrate issues. Each chapter has a set of examples in MATLAB, illustrating topics presented in the text or special issues that the student should know. The MATLAB code is given so that students can learn by example from it. To help students follow the mathematical derivations, we provide extra steps whenever necessary and do not skip steps that are necessary in the understanding of a derivation. Summaries of important issues are boxed and concepts and terms are emphasized to help students grasp the main points and terminology.

4. Without any doubt, learning the material in signals and systems requires working analytical as well as computational problems. It is important to provide problems of different levels of complexity to exercise not only basic problem-solving skills, but to achieve a level of proficiency and mathematical sophistication. The problems at the end of the chapter are of different types, some to be done analytically, others using

MATLAB, and some both. The repetitive type of problem was avoided. Some of the problems explore issues not covered in the text but related to it. The MATLAB problems were designed so that a better understanding of the theoretical concepts is attained by the student working them out.

5. We feel two additional features would be beneficial to students. One is the inclusion of quotations and footnotes to present interesting ideas or historical comments, and the other is the inclusion of sidebars that attempt to teach historical or technical information that students should be aware of. The theory of signals and systems clearly connects with mathematics and a great number of mathematicians have contributed to it. Likewise, there is a large number of engineers who have contributed significantly to the development and application of signals and systems. All of them need to be recognized for their contributions, and we should learn from their experiences.

6. Finally, other features are: (1) the design of the index of the book so that it can be used by students to find definitions, symbols, and MATLAB functions used in the text; and (2) a list of references to the material.

CONTENT

The core of the material is presented in the second and third part of the book. The second part of the book covers the basics of continuous-time signals and systems and illustrates their application. Because the concepts of signals and systems are relatively new to students, we provide an extensive and complete presentation of these topics in Chapters 1 and 2. The presentation in Chapter 1 goes from a very general characterization of signals to very specific classes that will be used in the rest of the book. One of the aims is to familiarize students with continuous-time as well as discrete-time signals so as to avoid confusion in their processing later on—a common difficulty encountered by students. Chapter 1 initiates the representation of signals in terms of basic signals that will be easily processed later with the transform methods. Chapter 2 introduces the general concept of systems, in particular continuous-time systems. The concepts of linearity, time invariance, causality, and stability are introduced in this chapter, trying as much as possible to use the students' background in circuit theory. Using linearity and time invariance, the computation of the output of a continuous-time system using the convolution integral is introduced and illustrated with relatively simple examples. More complex examples are treated with the Laplace transform in the following chapter.

Chapter 3 covers the basics of the Laplace transform and its application in the analysis of continuous-time signals and systems. It introduces the student to the concept of poles and zeros, damping and frequency, and their connection with the signal as a function of time. This chapter emphasizes the solution of differential equations representing linear time-invariant (LTI) systems, paying special attention to transient solutions due to their importance in control, as well as to steady-state solutions due to their importance in filtering and in communications. The convolution integral is dealt with in time and using the Laplace transform to emphasize the operational power of the transform. The important concept of transfer function for LTI systems and the significance of its poles and zeros are studied in detail. Different approaches are considered in computing the inverse Laplace transform, including MATLAB methods.

Fourier analysis of continuous-time signals and systems is covered in detail in Chapters 4 and 5. The Fourier series analysis of periodic signals, covered in Chapter 4, is extended to the analysis of aperiodic signals resulting in the Fourier transform of Chapter 5. The Fourier transform is useful in representing both periodic and aperiodic signals. Special attention is given to the connection of these methods with the Laplace transform so that, whenever possible, known Laplace transforms can be used to compute the Fourier series coefficients and the Fourier transform—thus avoiding integration but using the concept of the region of convergence. The concept of frequency, the response of the system (connected to the location of poles and zeros of the transfer function), and the steady-state response are emphasized in these chapters.

The ordering of the presentation of the Laplace and the Fourier transformations (similar to the Z-transform and the Fourier representation of discrete-time signals) is significant for learning and teaching of the material.

Our approach of presenting first the Laplace transform and then the Fourier series and Fourier transform is justified by several reasons. For one, students coming into a signals and systems course have been familiarized with the Laplace transform in their previous circuits or differential equations courses, and will continue using it in control courses. So expertise in this topic is important and the learned material will stay with them longer. Another is that a common difficulty students have in applying the Fourier series and the Fourier transform is connected with the required integration. The Laplace transform can be used not only to sidestep the integration but to provide a more comprehensive understanding of the frequency representation. By asking students to consider the two-sided Laplace transform and the significance of its region of convergence, they will appreciate better the Fourier representation as a special case of Laplace's in many cases. More importantly, these transforms can be seen as a continuum rather than as different transforms. It also makes theoretical sense to deal with the Laplace representation of systems first to justify the existence of the steady-state solution considered in the Fourier representations, which would not exist unless stability of the system is guaranteed, and stability can only be tested using the Laplace transform. The paradigm of interest is the connection of transient and steady-state responses that must be understood by students before they can understand the connections between Fourier and Laplace analyses.

Chapter 6 presents applications of the Laplace and the Fourier transforms to control, communications, and filtering. The intent of the chapter is to motivate interest in these areas. The chapter illustrates the significance of the concepts of transfer function, response of systems, and stability in control, and of modulation in communications. An introduction to analog filtering is provided. Analytic as well as MATLAB examples illustrate different applications to control, communications, and filter design.

Using the sampling theory as a bridge, the third part of the book covers the theory and illustrates the application of discrete-time signals and systems. Chapter 7 presents the theory of sampling: the conditions under which the signal does not lose information in the sampling process and the recovery of the analog signal from the sampled signal. Once the basic concepts are given, the analog-to-digital and digital-to-analog converters are considered to provide a practical understanding of the conversion of analog-to-digital and digital-to-analog signals.

Discrete-time signals and systems are discussed in Chapter 8, while Chapter 9 introduces the Z-transform. Although the treatment of discrete-time signals and systems in Chapter 8 mirrors that of continuous-time signals and systems, special emphasis is given in this chapter to issues that are different in the two domains. Issues such as the discrete nature of the time, the periodicity of the discrete frequency, the possible lack of periodicity of discrete sinusoids, etc. are considered. Chapter 9 provides the basic theory of the Z-transform and how it relates to the Laplace transform. The material in this chapter bears similarity to the one on the Laplace transform in terms of operational solution of difference equations, transfer function, and the significance of poles and zeros.

Chapter 10 presents the Fourier analysis of discrete signals and systems. Given the accumulated experience of the students with continuous-time signals and systems, we build the discrete-time Fourier transform (DTFT) on the Z-transform and consider special cases where the Z-transform cannot be used. The discrete Fourier transform (DFT) is obtained from the Fourier series of discrete-time signals and sampling in frequency. The DFT will be of great significance in digital signal processing. The computation of the DFT of periodic and aperiodic discrete-time signals using the fast Fourier transform (FFT) is illustrated. The FFT is an efficient algorithm for computing the DFT, and some of the basics of this algorithm are discussed in Chapter 12.

Chapter 11 introduces students to discrete filtering, thus extending the analog filtering in Chapter 6. In this chapter we show how to use the theory of analog filters to design recursive discrete low-pass filters. Frequency transformations are then presented to show how to obtain different types of filters from low-pass prototype filters. The design of finite-impulse filters using the window method is considered next. Finally, the implementation of recursive and nonrecursive filters is shown using some basic techniques. By using MATLAB for the design of recursive and nonrecursive discrete filters, it is expected that students will be motivated to pursue on their own the use of more sophisticated filter designs.

Finally, Chapter 12 explores topics of interest in digital communications, computer control, and digital signal processing. The aim of this chapter is to provide a brief presentation of topics that students could pursue after the basic courses in signals and systems.

TEACHING USING THIS TEXT

The material in this text is intended for a two-term sequence in signals and systems: one on continuous-time signals and systems, followed by a term in discrete-time signals and systems with a lab component using MAT-LAB. These two courses would cover most of the chapters in the text with various degrees of depth, depending on the emphasis the faculty would like to give to the course. As indicated, Chapter 0 was written as a necessary introduction to the rest of the material, but does not need to be covered in great detail—students can refer to it as needed. Chapters 6 and 11 need to be considered together if the emphasis on applications is in filter design. The control, communications, and digital signal processing material in Chapters 6 and 12 can be used to motivate students toward those areas.

TO THE STUDENT

It is important for you to understand the features of this book, so you can take advantage of them to learn the material:

1. Refer as often as necessary to the material in Chapter 0 to review or to learn the mathematical background; to understand the overall structure of the material; or to review or learn MATLAB as it applies to signal processing.

2. As you will see, the complexity of the material grows as it develops. The material in part three has been written assuming good understanding of the material in the first two. See also the connection of the material with applications in your own areas of interest.

3. To help you learn the material, clear and concise results are emphasized by putting them in boxes. Justification of these results is then given, complemented with remarks regarding issues that need a bit more clarification, and illustrated with plenty of analytic and computational examples. Important terms are emphasized throughout the text. Tables provide a good summary of properties and formulas.

4. A heading is used in each of the problems at the end of the chapters, indicating how it relates to specific topics and if it requires to use MATLAB to solve it.

5. One of the objectives of this text is to help you learn MATLAB, as it applies to signal and systems, on your own. This is done by providing the soft introduction to MATLAB in Chapter 0, and then by showing examples using simple code in each of the chapters. You will notice that in the first two parts basic components of MATLAB (scripts, functions, plotting, etc.) are given in more detail than in part three. It is assumed you are very proficient by then to supply that on your own.

6. Finally, notice the footnotes, the vignettes, and the historical sidebars that have been included to provide a glance at the background in which the theory and practice of signals and systems have developed.

Acknowledgments

I would like to acknowledge with gratitude the support and efforts of many people who made the writing of this text possible. First, to my family—my wife Cathy, my children William, Camila, and Juan, and their own families—many thanks for their support and encouragement despite being deprived of my attention. To my academic mentor, Professor Eliahu I. Jury, a deep sense of gratitude for his teachings and for having inculcated in me the love for a scholarly career and for the theory and practice of signals and systems. Thanks to Professor William Stanchina, chair of the Department of Electrical and Computer Engineering at the University of Pittsburgh, for his encouragement and support that made it possible to dedicate time to the project. Sincere thanks to Seda Senay and Mircea Lupus, graduate students in my department. Their contribution to the painful editing and proofreading of the manuscript, and the generation of the solution manual (especially from Ms. Senay) are much appreciated. Equally, thanks to the publisher and its editors, in particular to Joe Hayton and Steve Merken, for their patience, advising, and help with the publishing issues. Thanks also to Sarah Binns for her help with the final editing of the manuscript. Equally, I would like to thank Professor James Rowland from the University of Kansas and the following reviewers for providing significant input and changes to the manuscript: Dimitrie Popescu, Old Dominion University; Hossein Hakim, Worcester Polytechnic Institute; Mark Budnik, Valparaiso University; Periasamy Rajan, Tennessee Tech University; and Mohamed Zohdy, Oakland University. Thanks to my colleagues Amro El-Jaroudi and Juan Manfredi for their early comments and suggestions.

Lastly, I feel indebted to the many students I have had in my courses in signals and systems over the years I have been teaching this material in the Department of Electrical and Computer Engineering at the University of Pittsburgh. Unknown to them, they contributed to my impetus to write a book that I felt would make the teaching of signals and systems more accessible and fun to future students in and outside the university.

RESOURCES THAT ACCOMPANY THIS BOOK

A companion website containing downloadable MATLAB code for the worked examples in the book is available at:

http://booksite.academicpress.com/chaparro

For instructors, a solutions manual and image bank containing electronic versions of figures from the book are available by registering at:

www.textbooks.elsevier.com

Also Available for Use with This Book – Elsevier Online Testing

Web-based testing and assessment feature that allows instructors to create online tests and assignments which automatically assess student responses and performance, providing them with immediate feedback. Elsevier's online testing includes a selection of algorithmic questions, giving instructors the ability to create virtually unlimited variations of the same problem. Contact your local sales representative for additional information, or visit *http://booksite.academicpress.com/chaparro/* to view a demo chapter.

PART

1

Introduction

From the Ground Up!

In theory there is no difference
between theory and practice.
In practice there is.
Lawrence "Yogi" Berra, 1925
New York Yankees baseball player

This chapter provides an overview of the material in the book and highlights the mathematical background needed to understand the analysis of signals and systems. We consider a signal a function of time (or space if it is an image, or of time and space if it is a video signal), just like the voltages or currents encountered in circuits. A system is any device described by a mathematical model, just like the differential equations obtained for a circuit composed of resistors, capacitors, and inductors.

By means of practical applications, we illustrate in this chapter the importance of the theory of signals and systems and then proceed to connect some of the concepts of integro-differential Calculus with more concrete mathematics (from the computational point of view, i.e., using computers). A brief review of complex variables and their connection with the dynamics of systems follows. We end this chapter with a soft introduction to MATLAB, a widely used high-level computational tool for analysis and design.

Significantly, we have called this Chapter 0, because it is the ground floor for the rest of the material in the book. Not everything in this chapter has to be understood in a first reading, but we hope that as you go through the rest of the chapters in the book you will get to appreciate that the material in this chapter is the foundation of the book, and as such you should revisit it as often as needed.

0.1 SIGNALS AND SYSTEMS AND DIGITAL TECHNOLOGIES

In our modern world, signals of all kinds emanate from different types of devices—radios and TVs, cell phones, global positioning systems (GPSs), radars, and sonars. These systems allow us to communicate messages, to control processes, and to sense or measure signals. In the last 60 years, with the advent of the transistor, the digital computer, and the theoretical fundamentals of digital signal

Signals and Systems Using MATLAB®. DOI: 10.1016/B978-0-12-374716-7.00002-8

processing, the trend has been toward digital representation and processing of data, most of which are in analog form. Such a trend highlights the importance of learning how to represent signals in analog as well as in digital forms and how to model and design systems capable of dealing with different types of signals.

1948

The year 1948 is considered the birth year of technologies and theories responsible for the spectacular advances in communications, control, and biomedical engineering since then. In June 1948, Bell Telephone Laboratories announced the invention of the transistor. Later that month, a prototype computer built at Manchester University in the United Kingdom became the first operational stored-program computer. Also in that year, many fundamental theoretical results were published: Claude Shannon's mathematical theory of communications, Richard W. Hamming's theory on error-correcting codes, and Norbert Wiener's *Cybernetics* comparing biological systems with communication and control systems [51].

Digital signal processing advances have gone hand-in-hand with progress in electronics and computers. In 1965, Gordon Moore, one of the founders of Intel, envisioned that the number of transistors on a chip would double about every two years [35]. Intel, the largest chip manufacturer in the world, has kept that pace for 40 years. But at the same time, the speed of the central processing unit (CPU) chips in desktop personal computers has dramatically increased. Consider the well-known Pentium group of chips (the Pentium Pro and the Pentium I to IV) introduced in the 1990s [34]. Figure 0.1 shows the range of speeds of these chips at the time of their introduction into the market, as well as the number of transistors on each of these chips. In five years, 1995 to 2000, the speed increased by a factor of 10 while the number of transistors went from 5.5 million to 42 million.

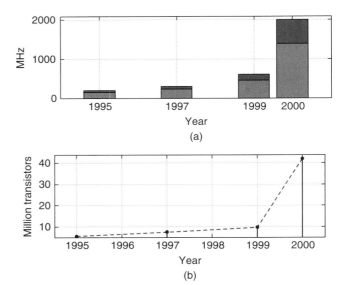

FIGURE 0.1

The Intel Pentium CPU chips. (a) Range of CPU speeds in MHz for the Pentium Pro (1995), Pentium II (1997), Pentium III (1999), and Pentium IV (2000). (b) Number of transistors (in millions) on each of the above chips. (Pentium data taken from [34].)

Advances in digital electronics and in computer engineering in the past 60 years have permitted the proliferation of digital technologies. Digital hardware and software process signals from cell phones, high-definition television (HDTV) receivers, radars, and sonars. The use of digital signal processors (DSPs) and more recently of field-programmable gate arrays (FPGAs) have been replacing the use of application-specific integrated circuits (ASICs) in industrial, medical, and military applications.

It is clear that digital technologies are here to stay. Today, digital transmission of voice, data, and video is common, and so is computer control. The abundance of algorithms for processing signals, and the pervasive presence of DSPs and FPGAs in thousands of applications make digital signal processing theory a necessary tool not only for engineers but for anybody who would be dealing with digital data; soon, that will be everybody! This book serves as an introduction to the theory of signals and systems—a necessary first step in the road toward understanding digital signal processing.

DSPs and FPGAs

A digital signal processor (DSP) is an optimized microprocessor used in real-time signal processing applications [67]. DSPs are typically embedded in larger systems (e.g., a desktop computer) handling general-purpose tasks. A DSP system typically consists of a processor, memory, analog-to-digital converters (ADCs), and digital-to-analog converters (DACs). The main difference with typical microprocessors is they are faster. A field-programmable gate array (FPGA) [77] is a semiconductor device containing programmable logic blocks that can be programmed to perform certain functions, and programmable interconnects. Although FPGAs are slower than their application-specific integrated circuits (ASICs) counterparts and use more power, their advantages include a shorter time to design and the ability to be reprogrammed.

0.2 EXAMPLES OF SIGNAL PROCESSING APPLICATIONS

The theory of signals and systems connects directly, among others, with communications, control, and biomedical engineering, and indirectly with mathematics and computer engineering. With the availability of digital technologies for processing signals, it is tempting to believe there is no need to understand their connection with analog technologies. It is precisely the opposite is illustrated by considering the following three interesting applications: the compact-disc (CD) player, software-defined radio and cognitive radio, and computer-controlled systems.

0.2.1 Compact-Disc Player

Compact discs [9] were first produced in Germany in 1982. Recorded voltage variations over time due to an acoustic sound is called an *analog signal* given its similarity with the differences in air pressure generated by the sound waves over time. Audio CDs and CD players illustrate best the conversion of a binary signal—unintelligible—into an intelligible analog signal. Moreover, the player is a very interesting control system.

To store an analog audio signal (e.g., voice or music) on a CD the signal must be first sampled and converted into a sequence of binary digits—a digital signal—by an ADC and then especially encoded to compress the information and to avoid errors when playing the CD. In the manufacturing of a CD,

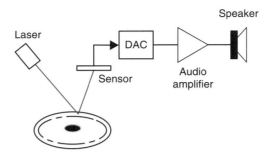

FIGURE 0.2
When playing a CD, the CD player follows the tracks in the disc, focusing a laser on them, as the CD is spun. The laser shines a light that is reflected by the pits and bumps put on the surface of the disc and corresponding to the coded digital signal from an acoustic signal. A sensor detects the reflected light and converts it into a digital signal, which is then converted into an analog signal by the DAC. When amplified and fed to the speakers such a signal sounds like the originally recorded acoustic signal.

pits and bumps corresponding to the ones and zeros from the quantization and encoding processes are impressed on the surface of the disc. Such pits and bumps will be detected by the CD player and converted back into an analog signal that approximates the original signal when the CD is played. The transformation into an analog signal uses a DAC.

As we will see in Chapter 7, an audio signal is sampled at a rate of about 44,000 samples/second (sec) (corresponding to a maximum frequency around 22 KHz for a typical audio signal) and each of these samples is represented by a certain number of bits (typically 8 bits/sample). The need for stereo sound requires that two channels be recorded. Overall, the number of bits representing the signal is very large and needs to be compressed and especially encoded. The resulting data, in the form of pits and bumps impressed on the CD surface, are put into a spiral track that goes from the inside to the outside of the disc.

Besides the binary-to-analog conversion, the CD player exemplifies a very interesting control system (see Figure 0.2). Indeed, the player must: (1) rotate the disc at different speeds depending on the location of the track within the CD being read, (2) focus a laser and a lens system to read the pits and bumps on the disc, and (3) move the laser to follow the track being read. To understand the exactness required, consider that the width of the track and the high of the bumps is typically less than a micrometer (10^{-6} meters or 3.937×10^{-5} inches) and a nanometer (10^{-9} meters or 3.937×10^{-8} inches), respectively.

0.2.2 Software-Defined Radio and Cognitive Radio

Software-defined radio and cognitive radio are important emerging technologies in wireless communications [43]. In software-defined radio (SDR), some of the radio functions typically implemented in hardware are converted into software [64]. By providing smart processing to SDRs, cognitive radio (CR) will provide the flexibility needed to more efficiently use the radio frequency spectrum and to make available new services to users. In the United States the Federal Communication Commission (FCC), and likewise in other parts of the world the corresponding agencies, allocates the bands for different users of the radio spectrum (commercial radio and TV, amateur radio, police, etc.). Although most bands have been allocated, implying a scarcity of spectrum for new users, it has been found that locally at certain times of the day the allocated spectrum is not being fully utilized. Cognitive radio takes advantage of this.

Conventional radio systems are composed mostly of hardware, and as such cannot be easily reconfigured. The basic premise in SDR as a wireless communication system is its ability to reconfigure by changing the software used to implement functions typically done by hardware in a conventional radio. In an SDR transmitter, software is used to implement different types of modulation procedures, while ADCs and DACs are used to change from one type of signal to another. Antennas, audio amplifiers, and conventional radio hardware are used to process analog signals. Typically, an SDR receiver uses an ADC to change the analog signals from the antenna into digital signals that are processed using software on a general-purpose processor. See Figure 0.3.

Given the need for more efficient use of the radio spectrum, cognitive radio (CR) uses SDR technology while attempting to dynamically manage the radio spectrum. A cognitive radio monitors locally the radio spectrum to determine regions that are not occupied by their assigned users and transmits in those bands. If the primary user of a frequency band recommences transmission, the CR either moves to another frequency band, or stays in the same band but decreases its transmission power level or modulation scheme to avoid interference with the assigned user. Moreover, a CR will search

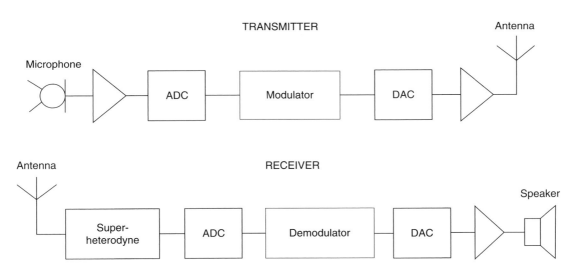

FIGURE 0.3

Schematics of a voice SDR mobile two-way radio. *Transmitter:* The voice signal is inputted by means of a microphone, amplified by an audio amplifier, converted into a digital signal by an ADC, and then modulated using software, before being converted into analog by an DAC, amplified, and sent as a radio frequency signal via an antenna. *Receiver:* The signal received by the antenna is processed by a superheterodyne front-end, converted into a digital signal by an ADC before being demodulated and converted into an analog signal by a DAC, amplified, and fed to a speaker. The modulator and demodulator blocks indicate software processing.

for network services that it can offer to its users. Thus, SDR and CR are bound to change the way we communicate and use network services.

0.2.3 Computer-Controlled Systems

The application of computer control ranges from controlling simple systems such as a heater (e.g., keeping a room temperature comfortable while reducing energy consumption) or cars (e.g., controlling their speed), to that of controlling rather sophisticated machines such as airplanes (e.g., providing automatic flight control) or chemical processes in very large systems such as oil refineries. A significant advantage of computer control is the flexibility computers provide—rather sophisticated control schemes can be implemented in software and adapted for different control modes.

Typically, control systems are feedback systems where the dynamic response of a system is changed to make it follow a desirable behavior. As indicated in Figure 0.4, the plant is a system, such as a heater, car, or airplane, or a chemical process in need of some control action so that its output (it is also possible for a system to have several outputs) follows a reference signal (or signals). For instance, one could think of a cruise-control system in a car that attempts to keep the speed of the car at a certain value by controlling the gas pedal mechanism. The control action will attempt to have the output of the system follow the desired response, despite the presence of disturbances either in the plant (e.g., errors in the model used for the plant) or in the sensor (e.g., measurement error). By comparing the reference signal with the output of the sensor, and using a control law implemented in the computer, a control action is generated to change the state of the plant and attain the desired output.

To use a computer in a control application it is necessary to transform analog signals into digital signals so that they can be inputted into the computer, while it is also necessary that the output of the computer be converted into an analog signal to drive an actuator (e.g., an electrical motor) to provide an action capable of changing the state of the plant. This can be done by means of ADCs and DACs. The sensor should also be able to act as a transducer whenever the output of the plant is

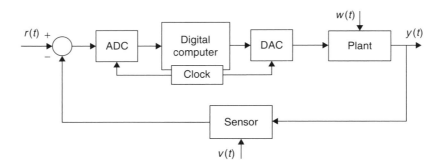

FIGURE 0.4

Computer-controlled system for an analog plant (e.g., cruise control for a car). The reference signal is $r(t)$ (e.g., desired speed) and the output is $y(t)$ (e.g., car speed). The analog signals are converted to digital signals by an ADC, while the digital signal from the computer is converted into an analog signal (an actuator is probably needed to control the car) by a DAC. The signals $w(t)$ and $v(t)$ are disturbances or noise in the plant and the sensor (e.g., electronic noise in the sensor and undesirable vibration in the car).

of a different type than the reference. Such would be the case, for instance, if the plant output is a temperature while the reference signal is a voltage.

0.3 ANALOG OR DISCRETE?

Infinitesimal calculus, or just plain *calculus*, deals with functions of one or more continuously changing variables. Based on the representation of these functions, the concepts of *derivative* and *integral* are developed to measure the rate of change of functions and the areas under the graphs of these functions, or their volumes. Differential equations are then introduced to characterize dynamic systems.

Finite calculus, on the other hand, deals with sequences. Thus, derivatives and integrals are replaced by differences and summations, while differential equations are replaced by difference equations. Finite calculus makes possible the computations of calculus by means of a combination of digital computers and numerical methods—thus, finite calculus becomes the more concrete mathematics.[1] Numerical methods applied to sequences permit us to approximate derivatives, integrals, and the solution of differential equations.

In engineering, as in many areas of science, the inputs and outputs of electrical, mechanical, chemical, and biological processes are measured as functions of time with amplitudes expressed in terms of voltage, current, torque, pressure, etc. These functions are called *analog or continuous-time signals*, and to process them with a computer they must be converted into binary sequences—or a string of ones and zeros that is understood by the computer. Such a conversion is done in a way as to preserve as much as possible the information contained in the original signal. Once in binary form, signals can be processed using algorithms (coded procedures understood by computers and designed to obtain certain desired information from the signals or to change them) in a computer or in a dedicated piece of hardware.

In a digital computer, differentiation and integration can be done only approximately, and the solution of differential equations requires a discretization process as we will illustrate later in this chapter. Not all signals are functions of a continuous parameter—there exist inherently discrete-time signals that can be represented as sequences, converted into binary form, and processed by computers. For these signals the finite calculus is the natural way of representing and processing them.

Analog or continuous-time signals are converted into binary sequences by means of an ADC, which, as we will see, compresses the data by converting the continuous-time signal into a discrete-time signal or a sequence of samples, each sample being represented by a string of ones and zeros giving a binary signal. Both time and signal amplitude are made discrete in this process. Likewise, digital signals can be transformed into analog signals by means of a DAC that uses the reverse process of the ADC. These converters are commercially available, and it is important to learn how they work so that digital representation of analog signals is obtained

---

[1] The use of *concrete*, rather than abstract, mathematics was coined by Graham, Knuth, and Patashnik in *Concrete Mathematics: A Foundation for Computer Science* [26]. Professor Donald Knuth from Stanford University is the the inventor of the Tex and Metafont typesetting systems that are the precursors of Latex, the document layout system in which the original manuscript of this book was done.

with minimal information loss. Chapters 1, 7, and 8 will provide the necessary information about continuous-time and discrete-time signals, and show how to convert one into the other and back. The sampling theory presented in Chapter 7 is the backbone of digital signal processing.

0.3.1 Continuous-Time and Discrete-Time Representations

There are significant differences between continuous-time and discrete-time signals as well as in their processing. A discrete-time signal is a sequence of measurements typically made at uniform times, while the analog signal depends continuously on time. Thus, a discrete-time signal $x[n]$ and the corresponding analog signal $x(t)$ are related by a sampling process:

$$x[n] = x(nT_s) = x(t)_{|t=nT_s} \qquad (0.1)$$

That is, the signal $x[n]$ is obtained by sampling $x(t)$ at times $t = nT_s$, where n is an integer and T_s is the *sampling period* or the time between samples. This results in a sequence,

$$\{\cdots x(-T_s)\ \ x(0)\ \ x(T_s)\ \ x(2T_s) \cdots\}$$

according to the sampling times, or equivalently

$$\{\cdots x[-1]\ \ x[0]\ \ x[1]\ \ x[2] \cdots\}$$

according to the ordering of the samples (as referenced to time 0). This process is called *sampling* or *discretization* of an analog signal.

Clearly, by choosing a small value for T_s we could make the analog and the discrete-time signals look very similar—almost indistinguishable—which is good, but this is at the expense of memory space required to keep the numerous samples. If we make the value of T_s large, we improve the memory requirements, but at the risk of losing information contained in the original signal. For instance, consider a sinusoid obtained from a signal generator:

$$x(t) = 2\cos(2\pi t)$$

for $0 \le t \le 10$ sec. If we sample it every $T_{s1} = 0.1$ sec, the analog signal becomes the following sequence:

$$x_1[n] = x(t)\,|_{t=0.1n} = 2\cos(2\pi n/10) \quad 0 \le n \le 100$$

providing a very good approximation to the original signal. If, on the other hand, we let $T_{s2} = 1$ sec, then the discrete-time signal becomes

$$x_2[n] = x(t)\,|_{t=n} = 2\cos(2\pi n) = 2 \quad 0 \le n \le 10$$

See Figure 0.5. Although for T_{s2} the number of samples is considerably reduced, the representation of the original signal is very poor—it appears as if we had sampled a constant signal, and we have thus lost information! This indicates that it is necessary to come up with a way to choose T_s so that sampling provides not only a reasonable number of samples, but, more importantly, guarantees that the information in the analog and the discrete-time signals remains the same.

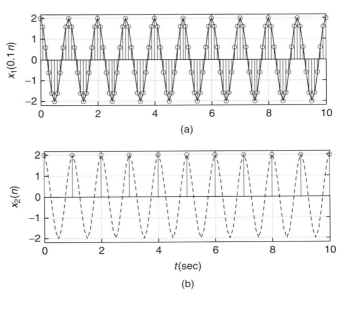

FIGURE 0.5

Sampling an analog sinusoid
$x(t) = 2\cos(2\pi t)$, $0 \le t \le 10$, with two
different sampling periods,
(a) $T_{s1} = 0.1$ sec and (b) $T_{s2} = 1$ sec, giving
$x_1(0.1n)$ and $x_2(n)$. The sinusoid is shown
by dashed lines. Notice the similarity
between the discrete-time signal and the
analog signal when $T_{s1} = 0.1$ sec, while
they are very different when $T_{s2} = 1$ sec,
indicating loss of information.

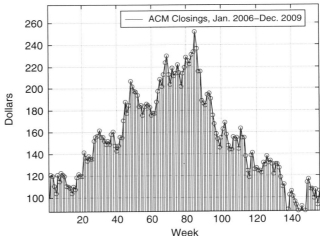

FIGURE 0.6

Weekly closings of ACM stock for 160
weeks in 2006 to 2009. ACM is the trading
name of the stock of the imaginary
company, ACME Inc., makers of everything
you can imagine.

As indicated before, not all signals are analog; there are some that are naturally discrete. Figure 0.6 displays the weekly average of the stock price of a fictitious company, ACME. Thinking of it as a signal, it is naturally discrete-time as it does not come from the discretization of an analog signal.

We have shown in this section the significance of the sampling period T_s in the transformation of an analog signal into a discrete-time signal without losing information. Choosing the sampling period requires knowledge of the frequency content of the signal—this is an example of the relation between time and frequency to be presented in great detail in Chapters 4 and 5, where the Fourier representation of periodic and nonperiodic signals is given. In Chapter 7, where we consider the problem of sampling, we will use this relation to determine appropriate values for the sampling period.

0.3.2 Derivatives and Finite Differences

Differentiation is an operation that is approximated in finite calculus. The derivative operator

$$D[x(t)] = \frac{dx(t)}{dt} = \lim_{h \to 0} \frac{x(t+h) - x(t)}{h} \qquad (0.2)$$

measures the rate of change of an analog signal $x(t)$. In finite calculus the *forward finite-difference operator*

$$\Delta[x(nT_s)] = x((n+1)T_s) - x(nT_s) \qquad (0.3)$$

measures the change in the signal from one sample to the next. If we let $x[n] = x(nT_s)$, for a known T_s, the forward finite-difference operator becomes a function of n:

$$\Delta[x[n]] = x[n+1] - x[n] \qquad (0.4)$$

The forward finite-difference operator measures the difference between two consecutive samples: one in the future $x((n+1)T_s)$ and the other in the present $x(nT_s)$. (See Problem 0.4 for a definition of the *backward finite-difference operator*.) The symbols D and Δ are called operators as they operate on functions to give other functions. The derivative and the finite-difference operators are clearly not the same. In the limit, we have that

$$\frac{dx(t)}{dt} \Big|_{t=nT_s} = \lim_{T_s \to 0} \frac{\Delta[x(nT_s)]}{T_s} \qquad (0.5)$$

Depending on the signal and the chosen value of T_s, the finite-difference operation can be a crude or an accurate approximation to the derivative multiplied by T_s.

Intuitively, if a signal does not change very fast with respect to time, the finite-difference approximates well the derivative for relatively large values of T_s, but if the signal changes very fast one needs very small values of T_s. The concept of frequency of a signal can help us understand this. We will learn that the frequency content of a signal depends on how fast the signal varies with time; thus a constant signal has zero frequency while a noisy signal that changes rapidly has high frequencies. Consider a constant signal $x_0(t) = 2$ having a derivative of zero (i.e., such a signal does not change at all with respect to time or it is a zero-frequency signal). If we convert this signal into a discrete-time signal using a sampling period $T_s = 1$ (or any other positive value), then $x_0[n] = 2$ and so

$$\Delta[x_0[n]] = 2 - 2 = 0$$

coincides with the derivative. Consider then a signal $x_1(t) = t$ with derivative 1 (this signal changes faster than $x(t)$ so it has frequencies larger than zero). If we sample it using $T_s = 1$, then $x_1[n] = n$ and the finite difference is

$$\Delta[x_1[n]] = \Delta[n] = (n+1) - n = 1$$

which again coincides with the derivative. Finally, we consider a signal that changes faster than $x(t)$ and $x_1(t)$ such as $x_2(t) = t^2$. Sampling $x_2(t)$ with $T_s = 1$, we have $x_2[n] = n^2$ and its forward finite difference is given by

$$\Delta[x_2[n]] = \Delta[n^2] = (n+1)^2 - n^2 = 2n + 1$$

which gives as an approximation to the derivative $\Delta[x_2[n]]/T_s = 2n + 1$. The derivative of $x_2(t)$ is $2t$, which at 0 equals 0, and at 1 equals 2. On the other hand, $\Delta[n^2]/T_s$ equals 1 and 3 at $n = 0$ and $n = 1$, respectively, which are different values from those of the derivative. Suppose $T_s = 0.01$, so that $x_2[n] = x_2(nT_s) = (0.01n)^2 = 0.0001n^2$. If we compute the difference for this signal we get

$$\Delta[x_2(0.01n)] = \Delta[(0.01n)^2] = (0.01n + 0.01)^2 - 0.0001n^2 = 10^{-4}(2n + 1)$$

which gives as an approximation to the derivative $\Delta[x_2(0.01n)]/T_s = 10^{-2}(2n + 1)$, or 0.01 when $n = 0$ and 0.03 when $n = 1$ which are a lot closer to the actual values of

$$\frac{dx_2(t)}{dt}\Big|_{t=0.01n} = 2t\,|_{t=0.01n} = 0.02n$$

The error now is 0.01 for each case instead of 1 as in the case when $T_s = 1$. Thus, whenever the rate of change of the signal is faster, the difference gets closer to the derivative by making T_s smaller.

It becomes clear that the faster the signal changes, the smaller the sampling period T_s should be in order to get a better approximation of the signal and its derivative. As we will learn in Chapters 4 and 5 the frequency content of a signal depends on the signal variation over time. A constant signal has frequency zero, while a signal that changes very fast over time would have high frequencies. The higher the frequencies in a signal, the more samples would be needed to represent it with no loss of information, thus requiring that T_s be smaller.

0.3.3 Integrals and Summations

Integration is the opposite of differentiation. To see this, suppose $I(t)$ is the integration of a continuous signal $x(t)$ from some time t_0 to t ($t_0 < t$),

$$I(t) = \int_{t_0}^{t} x(\tau)d\tau \tag{0.6}$$

or the sum of the area under $x(t)$ from t_0 to t. Notice that the upper bound of the integral is t so the integrand depends on a dummy variable.[2] The derivative of $I(t)$ is

$$\frac{dI(t)}{dt} = \lim_{h \to 0} \frac{I(t) - I(t-h)}{h} = \lim_{h \to 0} \frac{1}{h} \int_{t-h}^{t} x(\tau)d\tau$$

$$\approx \lim_{h \to 0} \frac{x(t) + x(t-h)}{2} = x(t)$$

where the integral is approximated as the area of a trapezoid with sides $x(t)$ and $x(t-h)$ and height h. Thus, for a continuous signal $x(t)$,

$$\frac{d}{dt} \int_{t_0}^{t} x(\tau)d\tau = x(t) \tag{0.7}$$

or if using the derivative operator $D[.]$, then its inverse $D^{-1}[.]$ should be the integration operator. That is, the above equation can be written

$$D[D^{-1}[x(t)]] = x(t). \tag{0.8}$$

We will see in Chapter 3 a similar relation between the derivative and the integral. The Laplace transform operators s and $1/s$ (just like D and $1/D$) imply differentiation and integration in the time domain.

Computationally, integration is implemented by sums. Consider, for instance, the integral of $x(t) = t$ from 0 to 10, which we know is equal to

$$\int_{0}^{10} t \; dt = \frac{t^2}{2} \Big|_{t=0}^{10} = 50.$$

That is, the area of a triangle with a base of 10 and a height of 10. For $T_s = 1$, suppose we approximate the signal $x(t)$ by pulses $p[n]$ of width $T_s = 1$ and height $nT_s = n$, or pulses of area n for $n = 0, \ldots, 9$. This can be seen as a lower-bound approximation to the integral, as the total area of these pulses gives a result smaller than the integral. In fact, the sum of the areas of the pulses is given by

$$\sum_{n=0}^{9} p[n] = \sum_{n=0}^{9} n = 0 + 1 + 2 + \cdots 9 = 0.5 \left[\sum_{n=0}^{9} n + \sum_{k=9}^{0} k \right]$$

$$= 0.5 \left[\sum_{n=0}^{9} n + \sum_{n=0}^{9} (9-n) \right] = \frac{9}{2} \sum_{n=0}^{9} 1 = \frac{10 \times 9}{2} = 45$$

---

[2]The integral $I(t)$ is a function of t and as such the integrand needs to be expressed in terms of a so-called *dummy variable* τ that takes values from t_0 to t in the integration. It would be confusing to let the integration variable be t. The variable τ is called a *dummy variable* because it is not crucial to the integration; any other variable could be used with no effect on the integration.

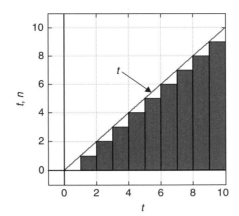

FIGURE 0.7

Approximation of area under
$x(t) = t, t \geq 0$, 0 otherwise, by pulses of
width 1 and height nT_s, where $T_s = 1$ and
$n = 0, 1, \ldots$

The approximation of the area using $T_s = 1$ is very poor (see Figure 0.7). In the above, we used the fact that the sum is not changed whether we add the numbers from 0 to 9 or backwards from 9 to 0, and that doubling the sum and dividing by 2 would not change the final answer. The above sum can thus be generalized to

$$\sum_{n=0}^{N-1} n = \frac{1}{2} \left[\sum_{n=0}^{N-1} n + \sum_{n=0}^{N-1} (N-1-n) \right] = \frac{1}{2} \sum_{n=0}^{N-1} (N-1)$$

$$= \frac{N \times (N-1)}{2} \tag{0.9}$$

a result that Gauss found out when he was a preschooler![3]

To improve the approximation of the integral we use $T_s = 10^{-3}$, which gives a discretized signal nT_s for $0 \leq nT_s < 10$ or $0 \leq n \leq (10/T_s) - 1$. The area of the pulses is nT_s^2 and the approximation to the integral is then

$$\sum_{n=0}^{10^4-1} p[n] = \sum_{n=0}^{10^4-1} n 10^{-6}$$

$$= \frac{10^4 \times (10^4 - 1)}{10^6 \times 2}$$

$$= 49.995$$

---

[3]Carl Friedrich Gauss (1777–1855) was a German mathematician. He was seven years old when he amazed his teachers with his trick for adding the numbers from 1 to 100 [7]. Gauss is one of the most accomplished mathematicians of all times [2]. He is in a group of selected mathematicians and scientists whose pictures appear in the currency of a country. His picture was on the Mark, the previous currency of Germany [6].

which is a lot better result. In general, we have that the integral can be computed quite accurately using a very small value of T_s, indeed

$$\sum_{n=0}^{(10/T_s)-1} p[n] = \sum_{n=0}^{(10/T_s)-1} nT_s^2$$

$$= T_s^2 \frac{(10/T_s) \times ((10/T_s) - 1)}{2}$$

$$= \frac{10 \times (10 - T_s)}{2}$$

which for very small values of T_s (so that $10 - T_s \approx 10$) gives $100/2 = 50$, as desired.

Derivatives and integrals take us into the processing of signals by systems. Once a mathematical model for a dynamic system is obtained, typically differential equations characterize the relation between the input and output variable or variables of the system. A significant subclass of systems (used as a valid approximation in some way to actual systems) is given by linear differential equations with constant coefficients. The solution of these equations can be efficiently found by means of the Laplace transform, which converts them into algebraic equations that are much easier to solve. The Laplace transform is covered in Chapter 3, in part to facilitate the analysis of analog signals and systems early in the learning process, but also so that it can be related to the Fourier theory of Chapters 4 and 5. Likewise for the analysis of discrete-time signals and systems we present in Chapter 9 the Z-transform, having analogous properties to those from the Laplace transform, before the Fourier analysis of those signals and systems.

0.3.4 Differential and Difference Equations

A differential equation characterizes the dynamics of a continuous-time system, or the way the system responds to inputs over time. There are different types of differential equations, corresponding to different systems. Most systems are characterized by nonlinear, time-dependent coefficient differential equations. The analytic solution of these equations is rather complicated. To simplify the analysis, these equations are locally approximated as linear constant-coefficient differential equations.

Solution of differential equations can be obtained by means of analog and digital computers. An electronic *analog computer* consists of operational amplifiers (op-amps), resistors, capacitors, voltage sources, and relays. Using the linearized model of the op-amps, resistors, and capacitors it is possible to realize integrators to solve a differential equation. Relays are used to set the initial conditions on the capacitors, and the voltage source gives the input signal. Although this arrangement permits the solution of differential equations, its drawback is the storage of the solution, which can be seen with an oscilloscope but is difficult to record. Hybrid computers were suggested as a solution—the analog computer is assisted by a digital component that stores the data. Both analog and hybrid computers have gone the way of the dinosaurs, and it is digital computers aided by numerical methods that are used now to solve differential equations.

Before going into the numerical solution provided by digital computers, let us consider why integrators are needed in the solution of differential equations. A first-order (the highest derivative present in the equation); linear (no nonlinear functions of the input or the output are present) with

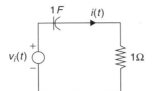

FIGURE 0.8

RC circuit.

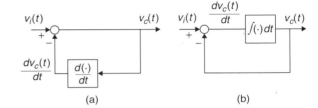

FIGURE 0.9

Realization of first-order differential equation using
(a) a differentiator and (b) an integrator.

(a)   (b)

constant-coefficient differential equations obtained from a simple RC circuit (Figure 0.8) with a constant voltage source $v_i(t)$ as input and with resistor $R = 1\Omega$; and capacitor $C = 1$ F (with huge plates!)
connected in series is given by

$$v_i(t) = v_c(t) + \frac{dv_c(t)}{dt} \qquad (0.10)$$

with an initial voltage $v_c(0)$ across the capacitor.

Intuitively, in this circuit the capacitor starts with an initial charge of $v_c(0)$, and will continue charging
until it reaches saturation, at which point no more charge will flow (the current across the resistor and
the capacitor is zero). Therefore, the voltage across the capacitor is equal to the voltage source–that
is, the capacitor is acting as an open circuit given that the source is constant.

Suppose, ideally, that we have available devices that can perform differentiation. There is then the
tendency to propose that the differential equation (Eq. 0.10) be solved following the block diagram
shown in Figure (0.9). Although nothing is wrong analytically, the problem with this approach is that
in practice most signals are noisy (each device produces electronic noise) and the noise present in the
signal may cause large derivative values given its rapidly changing amplitudes. Thus, the realization
of the differential equation using differentiators is prone to being very noisy (i.e., not good). Instead
of, as proposed years ago by Lord Kelvin,[4] using differentiators we need to smooth out the process by
using integrators, so that the voltage across the capacitor $v_c(t)$ is obtained by integrating both sides of
Equation (0.10). Assuming that the source is switched on at time $t = 0$ and that the capacitor has an
initial voltage $v_c(0)$, using the inverse relation between derivatives and integrals gives

$$v_c(t) = \int_0^t [v_i(\tau) - v_c(\tau)]d\tau + v_c(0) \qquad t \geq 0 \qquad (0.11)$$

---

[4]William Thomson, Lord Kelvin, proposed in 1876 the *differential analyzer,* a type of analog computer capable of solving differential
equations of order 2 and higher. His brother James designed one of the first differential analyzers [78].

which is represented by the block diagram in Figure 0.9(b). Notice that the integrator also provides a way to include the initial condition, which in this case is the initial voltage across the capacitor, $v_c(0)$. Different from accentuating the effect of differentiators on noise, integrators average the noise, thus reducing its effects.

> Block diagrams like the ones shown in Figure 0.9 allow us to visualize the system much better, and are commonly used. Integrators can be efficiently implemented using operational amplifiers with resistors and capacitors.

How to Obtain Difference Equations

Let us then show how Equation (0.10) can be solved using integration and its approximation, resulting in a difference equation. Using Equation (0.11) at $t = t_1$ and $t = t_0$ for $t_1 > t_0$, we have that

$$v_c(t_1) - v_c(t_0) = \int_{t_0}^{t_1} v_i(\tau)d\tau - \int_{t_0}^{t_1} v_c(\tau)d\tau$$

If we let $t_1 - t_0 = \Delta t$ where $\Delta t \to 0$ (i.e., a very small time interval), the integrals can be seen as the area of small trapezoids of height Δt and bases $v_i(t_1)$ and $v_i(t_0)$ for the input source and $v_c(t_1)$ and $v_c(t_0)$ for the voltage across the capacitor (see Figure 0.10). Using the formula for the area of a trapezoid we get an approximation for the above integrals so that

$$v_c(t_1) - v_c(t_0) = [v_i(t_1) + v_i(t_0)]\frac{\Delta t}{2} - [v_c(t_1) + v_c(t_0)]\frac{\Delta t}{2}$$

from which we obtain

$$v_c(t_1)\left[1 + \frac{\Delta t}{2}\right] = [v_i(t_1) + v_i(t_0)]\frac{\Delta t}{2} + v_c(t_0)\left[1 - \frac{\Delta t}{2}\right]$$

Assuming $\Delta t = T$, we then let $t_1 = nT$ and $t_0 = (n-1)T$. The above equation can be written as

$$v_c(nT) = \frac{T}{2+T}[v_i(nT) + v_i((n-1)T)] + \frac{2-T}{2+T}v_c((n-1)T) \qquad n \geq 1 \qquad (0.12)$$

and initial condition $v_c(0) = 0$. This is a first-order linear difference equation with constant coefficients approximating the differential equation characterizing the RC circuit. Letting the input

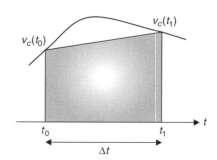

FIGURE 0.10

Approximation of area under the curve by a trapezoid.

be $v_i(t) = 1$ for $t \geq 0$, we have

$$v_c(nT) = \begin{cases} 0 & n = 0 \\ \frac{2T}{2+T} + \frac{2-T}{2+T}v_c((n-1)T) & n \geq 1 \end{cases} \qquad (0.13)$$

The advantage of the difference equation is that it can be solved for increasing values of n using previously computed values of $v_c(nT)$, which is called a *recursive solution*. For instance, letting $T = 10^{-3}$, $v_i(t) = 1$, and defining $M = 2T/(2+T)$, $K = (2-T)/(2+T)$, we obtain

$$n = 0 \qquad v_c(0) = 0$$

$$n = 1 \qquad v_c(T) = M$$

$$n = 2 \qquad v_c(2T) = M + KM = M(1 + K)$$

$$n = 3 \qquad v_c(3T) = M + K(M + KM) = M(1 + K + K^2)$$

$$n = 4 \qquad v_c(4T) = M + KM(1 + K + K^2) = M(1 + K + K^2 + K^3)$$

$$\cdots$$

The values are $M = 2T/(2+T) \approx T = 10^{-3}$, $K = (2-T)/(2+T) < 1$, and $1 - K = M$. The response increases from the zero initial condition to a constant value, which is the effect of the dc source—the capacitor eventually acts as an open circuit, so that the voltage across the capacitor equals that of the input. Extrapolating from the above results it seems that in the steady-state (i.e., when $nT \to \infty$) we have[5]

$$v_c(nT) = M \sum_{m=0}^{\infty} K^m = \frac{M}{1 - K} = 1$$

Even though this is a very simple example, it clearly illustrates that very good approximations to the solution of differential equations can be obtained using numerical methods that are appropriate for implementation in digital computers.

The above example shows how to solve a differential equation using integration and approximation of the integrals to obtain a difference equation that a computer can easily solve. The integral approximation used above is the *trapezoidal rule* method, which is one among many numerical methods used to solve differential equations. Also we will see later that the above results in the *bilinear transformation*, which connects the Laplace s variable with the z variable of the Z-transform, and that will be used in Chapter 11 in the design of discrete filters.

---

[5]The infinite sum converges if $|K| < 1$, which is satisfied in this case. If we multiply the sum by $(1 - K)$ we get

$$(1 - K)\sum_{m=0}^{\infty} K^m = \sum_{m=0}^{\infty} K^m - \sum_{m=0}^{\infty} K^{m+1}$$

$$= 1 + \sum_{m=1}^{\infty} K^m - \sum_{\ell=1}^{\infty} K^\ell = 1$$

where we changed the variable in the second equation to $\ell = m + 1$. This explains why the sum is equal to $1/(1 - K)$.

0.4 COMPLEX OR REAL?

Most of the theory of signals and systems is based on functions of a complex variable. Clearly, signals are functions of a real variable corresponding to time or space (if the signal is two-dimensional, like an image) so why would one need complex numbers in processing signals? As we will see later, time-dependent signals can be characterized by means of frequency and damping. These two characteristics are given by complex variables such as $s = \sigma + j\Omega$ (where σ is the damping factor and Ω is the frequency) in the representation of analog signals in the Laplace transform, or $z = re^{j\omega}$ (where r is the damping factor and ω is the discrete frequency) in the representation of discrete-time signals in the Z-transform. Both of these transformations will be considered in detail in Chapters 3 and 9. The other reason for using complex variables is due to the response of systems to pure tones or sinusoids. We will see that such response is fundamental in the analysis and synthesis of signals and systems. We thus need a solid grasp of what is meant by complex variables and what a function of these is all about. In this section, complex variables will be connected to vectors and phasors (which are commonly used in the sinusoidal steady-state analysis of linear circuits).

0.4.1 Complex Numbers and Vectors

A complex number z represents any point (x, y) in a two-dimensional plane by $z = x + jy$, where $x = \mathcal{R}e[z]$ (real part of z) is the coordinate in the x axis and $y = \mathcal{I}m[z]$ (imaginary part of z) is the coordinate in the y axis. The symbol $j = \sqrt{-1}$ just indicates that z needs to have two components to represent a point in the two-dimensional plane. Interestingly, a vector $\vec{z}$ that emanates from the origin of the complex plane $(0, 0)$ to the point (x, y) with a length

$$|\vec{z}| = \sqrt{x^2 + y^2} = |z| \tag{0.14}$$

and an angle

$$\theta = \angle\vec{z} = \angle z \tag{0.15}$$

also represents the point (x, y) in the plane and has the same attributes as the complex number z. The couple (x, y) is therefore equally representable by the vector $\vec{z}$ or by a complex number z that can be written in a rectangular or in a polar form,

$$z = x + jy = |z|e^{j\theta} \tag{0.16}$$

where the magnitude $|z|$ and the phase θ are defined in Equations (0.14) and (0.15).

It is important to understand that a rectangular plane or a polar complex plane are identical despite the different representation of each point in the plane. Furthermore, when adding or subtracting complex numbers the rectangular form is the appropriate one, while when multiplying or dividing complex numbers the polar form is more advantageous. Thus, if complex numbers $z = x + jy = |z|e^{j\angle z}$ and $v = p + jq = |v|e^{j\angle v}$ are added analytically, we obtain

$$z + v = (x + p) + j(y + q)$$

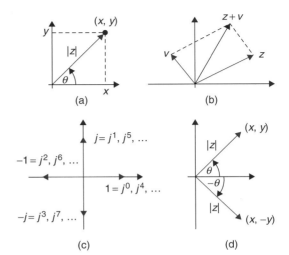

FIGURE 0.11

(a) Representation of a complex number z by a vector (b) addition of complex numbers z and v; (c) integer powers of j; and (d) complex conjugate.

Using their polar representations requires a geometric interpretation: the addition of vectors (see Figure 0.11). On the other hand, the multiplication of z and v is easily done using their polar forms as

$$zv = |z|e^{j\angle z}|v|e^{j\angle v} = |z||v|e^{j(\angle z + \angle v)}$$

but it requires more operations if done in the rectangular form—that is,

$$zv = (x + jy)(p + jq) = (xp - yq) + j(xq + yp)$$

It is even more difficult to obtain a geometric interpretation. Such an interpretation will be seen later on. Addition and subtraction as well as multiplication and division can thus be done more efficiently by choosing the rectangular and the polar representations, respectively. Moreover, the polar representation is also useful when finding powers of complex numbers. For the complex variable $z = |z|e^{\angle z}$, we have that

$$z^n = |z|^n e^{jn\angle z}$$

for n integer or rational. For instance, if $n = 10$, then $z^{10} = |z|^{10} e^{j10\angle z}$, and if $n = 3/2$, then $z^{1.5} = (\sqrt{|z|})^3 e^{j1.5\angle z}$. The powers of j are of special interest. Given that $j = \sqrt{-1}$ then, we have

$$j^n = (-1)^{n/2} = \begin{cases} (-1)^m & n = 2m, \quad n \text{ even} \\ (-1)^m j & n = 2m + 1, \quad n \text{ odd} \end{cases}$$

so that $j^0 = 1$, $j^1 = j$, $j^2 = -1$, $j^3 = -j$, and so on. Letting $j = 1e^{j\pi/2}$, we can see that the increasing powers of $j^n = 1e^{jn\pi/2}$ are vectors with angles of 0 when $n = 0$, $\pi/2$ when $n = 1$, π when $n = 2$, and $3\pi/2$ when $n = 3$. The angles repeat for the next four values, the four after that, and so on. See Figure 0.11.

One operation possible with complex numbers that is not possible with real numbers is *complex conjugation*. Given a complex number $z = x + jy = |z|e^{j\angle z}$ its complex conjugate is $z^* = x - jy = |z|e^{-j\angle z}$—that is, we negate the imaginary part of z or reflect its angle. This operation gives that

$$
\begin{aligned}
&(i) \quad z + z^* = 2x \quad \text{or} \quad \mathcal{Re}[z] = 0.5[z + z^*] \\
&(ii) \quad z - z^* = 2jy \quad \text{or} \quad \mathcal{Im}[z] = -0.5j[z - z^*] \\
&(iii) \quad zz^* = |z|^2 \quad \text{or} \quad |z| = \sqrt{zz^*} \\
&(iv) \quad \frac{z}{z^*} = e^{j2\angle z} \quad \text{or} \quad \angle z = -j0.5[\log(z) - \log(z^*)]
\end{aligned}
\tag{0.17}
$$

The complex conjugation provides a different approach to the division of complex numbers in rectangular form. This is done by making the denominator a positive real number by multiplying both numerator and denominator by the complex conjugate of the denominator. For instance,

$$
z = \frac{1 + j1}{3 + j4} = \frac{(1 + j1)(3 - j4)}{(3 + j4)(3 - j4)} = \frac{7 - j}{9 + 16} = \frac{7 - j}{25}
$$

Finally, the conversion of complex numbers from rectangular to polar needs to be done with care, especially when computing the angles. For instance, $z = 1 + j$ has a vector representing in the first quadrant of the complex plane, and its magnitude is $|z| = \sqrt{2}$ while the tangent of its angle θ is $\tan(\theta) = 1$ or $\theta = \pi/4$ radians. If $z = -1 + j$, the vector representing it is now in the second quadrant with the same magnitude as before, but its angle is now

$$
\theta = \pi - \tan^{-1}(1) = 3\pi/4
$$

That is, we find the angle with respect to the negative real axis and subtract it from π. Likewise, if $z = -1 - j$, the magnitude does not change but the phase is now

$$
\theta = \pi + \tan^{-1}(1) = 5\pi/4
$$

which can also be expressed as $-3\pi/4$. Finally, when $z = 1 - j$, the angle is $-\pi/4$ and the magnitude remains the same. The conversion from polar to rectangular form is much easier. Indeed, given a complex number in polar form $z = |z|e^{j\theta}$ its real part is $x = |z|\cos(\theta)$ (i.e., the projection of the vector corresponding to z onto the real axis) and the imaginary part is $y = |z|\sin(\theta)$, so that $z = x + jy$. For instance, $z = \sqrt{2}e^{3\pi/4}$ can be written as

$$
z = \sqrt{2}\cos(3\pi/4) + j\sqrt{2}\sin(3\pi/4) = -1 + j
$$

0.4.2 Functions of a Complex Variable

Just like real-valued functions, functions of a complex variable can be defined. For instance, the logarithm of a complex number can be written as

$$v = \log(z) = \log(|z|e^{j\theta}) = \log(|z|) + j\theta$$

by using the inverse connection between the exponential and the logarithmic functions. Of particular interest in the theory of signals and systems is the exponential of complex variable z defined as

$$v = e^z = e^{x+jy} = e^x e^{jy}$$

It is important to mention that complex variables as well as functions of complex variables are more general than real variables and real-valued functions. The above definition of the logarithmic function is valid when $z = x$, with x a real value, and also when $z = jy$, a purely imaginary value. Likewise, the exponential function for $z = x$ is a real-valued function.

Euler's Identity

One of the most famous equations of all times[6] is

$$1 + e^{j\pi} = 1 + e^{-j\pi} = 0$$

due to one of the most prolific mathematicians of all times, Leonard Euler.[7] The above equation can be easily understood by establishing Euler's identity, which connects the complex exponential and sinusoids:

$$e^{j\theta} = \cos(\theta) + j\sin(\theta) \tag{0.18}$$

One way to verify this identity is to consider the polar representation of the complex number $\cos(\theta) + j\sin(\theta)$, which has a unit magnitude since $\sqrt{\cos^2(\theta) + \sin^2(\theta)} = 1$ given the trigonometric identity $\cos^2(\theta) + \sin^2(\theta) = 1$. The angle of this complex number is

$$\psi = \tan^{-1}\left[\frac{\sin(\theta)}{\cos(\theta)}\right] = \theta$$

Thus, the complex number

$$\cos(\theta) + j\sin(\theta) = 1e^{j\theta}$$

which is Euler's identity. Now in the case where $\theta = \pm\pi$ the identity implies that $e^{\pm j\pi} = -1$, explaining the famous Euler's equation.

---

[6]A reader's poll done by *Mathematical Intelligencer* named Euler's identity the most beautiful equation in mathematics. Another poll by *Physics World* in 2004 named Euler's identity the greatest equation ever, together with Maxwell's equations. Paul Nahin's book *Dr. Euler's Fabulous Formula* (2006) is devoted to Euler's identity. It states that the identity sets "the gold standard for mathematical beauty" [73].
[7]Leonard Euler (1707–1783) was a Swiss mathematician and physicist, student of John Bernoulli, and advisor of Joseph Lagrange. We owe Euler the notation $f(x)$ for functions, e for the base of natural logs, $i = \sqrt{-1}$, π for pi, Σ for sum, the finite difference notation Δ, and many more!

The relation between the complex exponentials and the sinusoidal functions is of great importance in signals and systems analysis. Using Euler's identity the cosine can be expressed as

$$\cos(\theta) = \mathcal{R}e[e^{j\theta}] = \frac{e^{j\theta} + e^{-j\theta}}{2} \tag{0.19}$$

while the sine is given by

$$\sin(\theta) = \mathcal{I}m[e^{j\theta}] = \frac{e^{j\theta} - e^{-j\theta}}{2j} \tag{0.20}$$

Indeed, we have

$$e^{j\theta} = \cos(\theta) + j\sin(\theta)$$

$$e^{-j\theta} = \cos(\theta) - j\sin(\theta)$$

Adding them we get the above expression for the cosine, and subtracting the second from the first we get the given expression for the sine. The variable θ is in radians, or in the corresponding angle in degrees (recall that 2π radians equals 360 degrees).

These relations can be used to define the hyperbolic sinusoids as

$$\cos(j\alpha) = \frac{e^{-\alpha} + e^{\alpha}}{2} = \cosh(\alpha) \tag{0.21}$$

$$j\sin(j\alpha) = \frac{e^{-\alpha} - e^{\alpha}}{2} = -\sinh(\alpha) \tag{0.22}$$

from which the other hyperbolic functions are defined. Also, we obtain the following expression for the real-valued exponential:

$$e^{-\alpha} = \cosh(\alpha) - \sinh(\alpha) \tag{0.23}$$

Euler's identity can also be used to find different trigonometric identities. For instance,

$$\cos^2(\theta) = \left[\frac{e^{j\theta} + e^{-j\theta}}{2}\right]^2 = \frac{1}{4}[2 + e^{j2\theta} + e^{-j2\theta}] = \frac{1}{2} + \frac{1}{2}\cos(2\theta)$$

$$\sin^2(\theta) = 1 - \cos^2(\theta) = \frac{1}{2} - \frac{1}{2}\cos(2\theta)$$

$$\sin(\theta)\cos(\theta) = \frac{e^{j\theta} - e^{-j\theta}}{2j} \frac{e^{j\theta} + e^{-j\theta}}{2} = \frac{e^{j2\theta} - e^{-j2\theta}}{4j} = \frac{1}{2}\sin(2\theta)$$

0.4.3 Phasors and Sinusoidal Steady State

A sinusoid $x(t)$ is a periodic signal represented by

$$x(t) = A\cos(\Omega_0 t + \psi) \qquad -\infty < t < \infty \tag{0.24}$$

where A is the amplitude, $\Omega_0 = 2\pi f_0$ is the frequency in rad/sec, and ψ is the phase in radians. The signal $x(t)$ is defined for all values of t, and it repeats periodically with a period $T_0 = 1/f_0$ (sec), so that f_0 is the frequency in cycles/sec or in Hertz (Hz) (in honor of H. R. Hertz[8]). Given that the units of Ω_0 is rad/sec, then $\Omega_0 t$ has as units (rad/sec) × (sec) = (rad), which coincides with the units of the phase ψ, and permits the computation of the cosine. If $\psi = 0$, then $x(t)$ is a cosine, and if $\psi = -\pi/2$, then $x(t)$ is a sine.

If one knows the frequency Ω_0 (rad/sec) in Equation (0.24), the cosine is characterized by its amplitude and phase. This permits us to define *phasors*[9] as complex numbers characterized by the amplitude and the phase of a cosine signal of a certain frequency Ω_0. That is, for a voltage signal $v(t) = A\cos(\Omega_0 t + \psi)$ the corresponding phasor is

$$V = Ae^{j\psi} = A\cos(\psi) + jA\sin(\psi) = A\angle\psi \tag{0.25}$$

and such that

$$v(t) = \mathcal{Re}[Ve^{j\Omega_0 t}] = \mathcal{Re}[Ae^{j(\Omega_0 t + \psi)}] = A\cos(\Omega_0 t + \psi) \tag{0.26}$$

One can thus think of the voltage signal $v(t)$ as the projection of the phasor V onto the real axis and turning counterclockwise at a rate of Ω_0 rad/sec. At time $t = 0$ the angle of the phasor is ψ. Clearly the phasor definition is true for only one frequency, in this case Ω_0, and it is always connected to a cosine function.

Interestingly enough, the angle ψ can be used to differentiate cosines and sines. For instance, when $\psi = 0$, the phasor V moving around at a rate of Ω_0 generates as a projection on the real axis the voltage signal $A\cos(\Omega_0 t)$, while when $\psi = -\pi/2$, the phasor V moving around again at a rate of Ω_0 generates a sinusoid $A\sin(\Omega_0 t) = A\cos(\Omega_0 t - \pi/2)$ as it is projected onto the real axis. This establishes the well-known fact that the sine lags the cosine by $\pi/2$ radians or 90 degrees, or that the cosine leads the sine by $\pi/2$ radians or 90 degrees. Thus, the generation and relation of sines and cosines can be easily obtained using the plot in Figure 0.12.

Phasors can be related to vectors. A current source, for instance,

$$i(t) = A\cos(\Omega_0 t) + B\sin(\Omega_0 t)$$

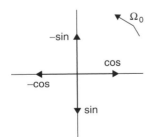

FIGURE 0.12
Generation of sinusoids from phasors of a frequency Ω_0.

---

[8] Heinrich Rudolf Hertz was a German physicist known for being the first to demonstrate the existence of electromagnetic radiation in 1888.
[9] In 1883, Charles Proteus Steinmetz (1885–1923), German-American mathematician and engineer, introduced the concept of phasors for alternating current analysis. In 1902, Steinmetz became a professor of electrophysics at Union College in Schenectady, New York.

can be expressed as

$$i(t) = C\cos(\Omega_0 t + \gamma)$$

where C and γ are to be determined (the sinusoidal components of $i(t)$ must depend on a unique frequency Ω_0; if that was not the case the concept of phasors would not apply). To obtain the equivalent representation, we first obtain the phasor corresponding to $A\cos(\Omega_0 t)$, which is $I_1 = Ae^{j0} = A$, and for $B\sin(\Omega_0 t)$ the corresponding phasor is $I_2 = Be^{-j\pi/2}$, so that

$$i(t) = \mathcal{R}e[(I_1 + I_2)e^{j\Omega_0 t}]$$

Thus, the problem has been transformed into the addition of two vectors I_1 and I_2, which gives a vector

$$I = \sqrt{A^2 + B^2}\,e^{-j\tan^{-1}(B/A)}$$

so that

$$i(t) = \mathcal{R}e[Ie^{j\Omega_0 t}]$$
$$= \mathcal{R}e[\sqrt{A^2 + B^2}\;e^{-j\tan^{-1}(B/A)}e^{j\Omega_0 t}]$$
$$= \sqrt{A^2 + B^2}\cos(\Omega_0 t - \tan^{-1}(B/A))$$

Or, an equivalent source with amplitude $C = \sqrt{A^2 + B^2}$, phase $\gamma = -\tan^{-1}(B/A)$, and frequency Ω_0 – that is, an equivalent phasor that generates $i(t)$ and has the magnitude C, the angle γ, and rotates at frequency Ω_0.

In Figure 0.13 we display the result of adding two phasors (frequency $f_0 = 20$ Hz) and the sinusoid that is generated by the phasor $I = I_1 + I_2 = 27.98e^{j30.4°}$.

0.4.4 Phasor Connection

The fundamental property of a circuit made up of constant resistors, capacitors, and inductors is that its response to a sinusoid is also a sinusoid of the same frequency in steady state. The effect of the circuit on the input sinusoid is on its magnitude and phase and depends on the frequency of the input sinusoid. This is due to the linear and time-invariant nature of the circuit, and can be generalized to more complex continuous-time as well as discrete-time systems as we will see in Chapters 3, 4, 5, 9 and 10.

To illustrate the connection of phasors with dynamic systems consider a simple RC circuit ($R = 1\ \Omega$ and $C = 1$F). If the input to the circuit is a sinusoidal voltage source $v_i(t) = A\cos(\Omega_0 t)$ and the voltage across the capacitor $v_c(t)$ is the output of interest, the circuit can be easily represented by the first-order differential equation

$$\frac{dv_c(t)}{dt} + v_c(t) = v_i(t)$$

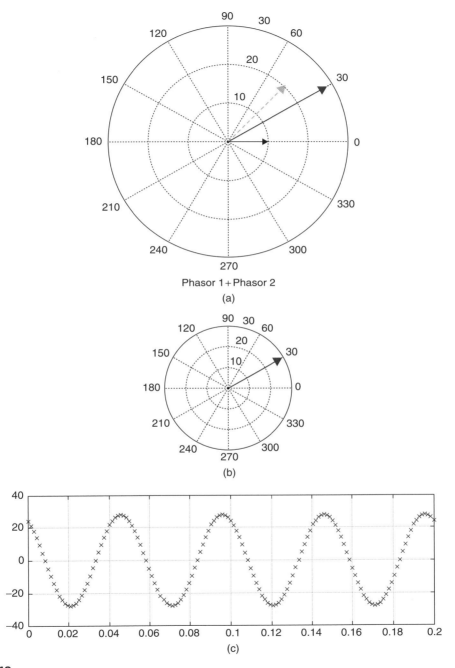

FIGURE 0.13

(a) Sum of phasors $I_1 = 10e^{j0}$ (solid arrow) and $I_2 = 20e^{j\pi/4}$ (dashed arrow) with the result in blue; (c) sinusoid generated by phasor $I = I_1 + I_2$ (b).

Assume that the steady-state response of this circuit (i.e., $v_c(t)$ as $t \to \infty$) is also a sinusoid

$$v_c(t) = C \cos(\Omega_0 t + \psi)$$

of the same frequency as the input, with amplitude C and phase ψ to be determined. This response must satisfy the differential equation, or

$$v_i(t) = \frac{dv_c(t)}{dt} + v_c(t)$$

$$A \cos(\Omega_0 t) = -C\Omega_0 \sin(\Omega_0 t + \psi) + C \cos(\Omega_0 t + \psi)$$

$$= C\Omega_0 \cos(\Omega_0 t + \psi + \pi/2) + C \cos(\Omega_0 t + \psi)$$

$$= C\sqrt{1 + \Omega_0^2} \cos(\Omega_0 t + \psi + \tan^{-1}(C\Omega_0/C))$$

Comparing the two sides of the above equation gives

$$C = \frac{A}{\sqrt{1 + \Omega_0^2}}$$

$$\psi = -\tan^{-1}(\Omega_0)$$

for a steady-state response

$$v_c(t) = \frac{A}{\sqrt{1 + \Omega_0^2}} \cos(\Omega_0 t - \tan^{-1}(\Omega_0)).$$

Comparing the steady-state response $v_c(t)$ with the input sinusoid $v_i(t)$, we see that they both have the same frequency Ω_0, but the amplitude and phase of the input are changed by the circuit depending on the frequency Ω_0. Since at each frequency the circuit responds differently, obtaining the frequency response of the circuit will be useful not only in analysis but in the design of circuits.

The sinusoidal steady-state is obtained using phasors. Expressing the steady-state response of the circuit as

$$v_c(t) = \mathcal{R}e\left[V_c e^{j\Omega_0 t}\right]$$

where $V_c = C e^{j\psi}$ is the corresponding phasor for $v_c(t)$, we find that

$$\frac{dv_c(t)}{dt} = \frac{d\mathcal{R}e[V_c e^{j\Omega_0 t}]}{dt} = \mathcal{R}e\left[V_c \frac{de^{j\Omega_0 t}}{dt}\right] = \mathcal{R}e\left[j\Omega_0 V_c e^{j\Omega_0 t}\right]$$

By replacing $v_c(t)$, $dv_c(t)/dt$, obtained above, and

$$v_i(t) = \mathcal{R}e\left[V_i e^{j\Omega_0 t}\right] \quad \text{where} \quad V_i = A e^{j0}$$

in the differential equation, we obtain

$$\mathcal{R}e\left[V_c(1+j\Omega_0)e^{j\Omega_0 t}\right] = \mathcal{R}e\left[Ae^{j\Omega_0 t}\right]$$

so that

$$V_c = \frac{A}{1+j\Omega_0} = \frac{A}{\sqrt{1+\Omega_0^2}}e^{-j\tan^{-1}(\Omega_0)}$$

$$= Ce^{j\psi}$$

and the sinusoidal steady-state response is

$$v_c(t) = \mathcal{R}e\left[V_c e^{j\Omega_0 t}\right]$$

$$= \frac{A}{\sqrt{1+\Omega_0^2}}\cos(\Omega_0 t - \tan^{-1}(\Omega_0))$$

which coincides with the response obtained above. The ratio of the output phasor V_c to the input phasor V_i,

$$\frac{V_c}{V_i} = \frac{1}{1+j\Omega_0}$$

gives the response of the circuit at frequency Ω_0. If the frequency of the input is a generic Ω, changing Ω_0 above for Ω gives the frequency response for all possible frequencies.

The concepts of *linearity* and *time invariance* will be used in both continuous-time as well as discrete-time systems, along with the Fourier representation of signals in terms of sinusoids or complex exponentials, to simplify the analysis and to allow the design of systems. Thus, transform methods such as Laplace and the Z-transform will be used to solve differential and difference equations in an algebraic setup. Fourier representations will provide the frequency perspective. This is a general approach for both continuous-time and discrete-time signals and systems. The introduction of the concept of the transfer function will provide tools for the analysis as well as the design of linear time-invariant systems. The design of analog and discrete filters is the most important application of these concepts. We will look into this topic in Chapters 5, 6, and 11.

0.5 SOFT INTRODUCTION TO MATLAB

MATLAB is a computing language based on vectorial computations.[10] In this section, we will introduce you to the use of MATLAB for numerical and symbolic computations.

---

[10] MATLAB stands for matrix laboratory. MatWorks, the developer of MATLAB, was founded in 1984 by Jack Little, Steve Bangert, and Cleve Moler. Moler, a math professor at the University of New Mexico, developed the first version of MATLAB in Fortran in the late 1970s. It only had 80 functions and no M-files or toolboxes. Little and Bangert reprogrammed it in C and added M-files, toolboxes, and more powerful graphics [49].

0.5.1 Numerical Computations

The following instructions are intended for users who have no background in MATLAB but are interested in using it in signal processing. Once you get the basic information on how to use the language you will be able to progress on your own.

1. Create a directory where you will put your work, and from where you will start MATLAB. This is important because when executing a program, MATLAB will look at the current directory, and if the file is not present in the current directory, and if it is not a MATLAB function, MATLAB gives an error indicating that it cannot find the desired program.
2. There are two types of programs in MATLAB: the script, which consists in a list of commands using MATLAB functions or your own functions, and the functions, which are programs that can be called with different inputs and provide the corresponding outputs. We will show examples of both.
3. Once you start MATLAB, you will see three windows: the command window, where you will type commands; the command history, which keeps a list of commands that have been used; and the workspace, where the variables used are kept.
4. Your first command on the command window should be to change to your data directory where you will keep your work. You can do this in the command window by using the command CD (change directory) followed by the desired directory. It is also important to use the command clear all and clf to clear all previous variables in memory and all figures.
5. Help is available in several forms in MATLAB. Just type helpwin, helpdesk, or demo to get started. If you know the name of the function, help will give you the necessary information on the particular function, and it will also give you information on help itself. Use help to find more about the functions used in this introduction to MATLAB.
6. To type your scripts or functions you can use the editor provided by MATLAB; simply type edit. You can also use any text editor to create scripts or functions, which need to be saved with the .m extension.

Creating Vectors and Matrices

Comments are preceded by percent, and to begin a script, as the following, it is always a good idea to clear all previous variables and all previous figures.

```
% matlab primer
clear all               % clear all variables
clf                     % clear all figures
% row and column vectors
x = [ 1 2 3 4]          % row vector
y = x'                  % column vector
```

The corresponding output is as follows (notice that there is no semicolon (;) at the end of the lines to stop MATLAB from providing an output when the above script is executed).

```
x =
   1   2   3   4
```

```
y =
    1
    2
    3
    4
```

To see the dimension of *x* and *y* variables, type

```
whos        % provides information on existing variables
```

to which MATLAB responds

```
Name    Size            Bytes  Class
x       1x4                32  double array
y       4x1                32  double array
Grand total is 8 elements using 64 bytes
```

Notice that a vector is thought of as a matrix; for instance, vector *x* is a matrix of one row and four columns. Another way to express the column vector *y* is the following, where each of the row terms is separated by a semicolon (;)

```
y = [1;2;3;4]    % another way to write a column
```

To give as before:

```
y =
    1
    2
    3
    4
```

MATLAB does not allow arguments of vectors or matrices to be zero or negative. For instance, if we want the first entry of the vector *y* we need to type

```
y(1)     % first entry of vector y
```

giving as output

```
ans =
        1
```

If we type

```
y(0)
```

it will give us an error, to which we get the following warning:

```
??? Subscript indices must either be real positive integers or logicals.
```

MATLAB also has a peculiar way to provide information in a vector, for instance:

```
y(1:3)          % first to third entry of column vector y
```

giving as expected the first to the third entries of the column vector y:

```
ans =
     1
     2
     3
```

The following will give the third to the first entry in the row vector x (notice the difference in the two outputs; as expected the values of y are given in a column, while the requested entries of x are given in a row).

```
x(3:-1:1)          % displays entries x(3) x(2) x(1)
```

Thus,

```
ans =
     3   2   1
```

Matrices are constructed as an concatenation of rows (or columns):

```
A = [ 1 2; 3 4; 5 6]    % matrix A with rows [1 2], [3 4] and [5 6]

A =
     1   2
     3   4
     5   6
```

To create a vector corresponding to a sequence of numbers (in this case integers) there are different approaches, as follows:

```
n = 0:10     % vector with entries 0 to 10 increased by 1
```

This approach gives the following as output:

```
n =
   Columns 1 through 10
     0   1   2   3   4   5   6   7   8   9
   Column 11
     10
```

which is the same as the command

```
n = [0:10]
```

If we wish the increment different from 1 (default value), then we indicate it as in the following:

```
n1 = 0:2:10 % vector with entries from 0 to 10 increased by 2
```

which gives

```
n1 =
     0   2   4   6   8   10
```

We can combine the above vectors into one as follows:

```
nn1 = [n n1]    % combination of vectors
```

to get

```
nn1 =
    Columns 1 through 10
       0    1    2    3    4    5    6    7    8    9
    Columns 11 through 17
      10    0    2    4    6    8   10
```

Vectorial Operations

MATLAB allows the conventional vectorial operations as well as facilitates others. For instance, if we wish to multiply by 3 every entry of the row vector x given above, the command

```
z = 3*x          % multiplication by a constant
```

would give

```
z =
    3    6    9   12
```

Besides the conventional multiplication of vectors with the correct dimensions, MATLAB allows two types of multiplications of one vector by another. The first one is where the entries of one vector are multiplied by the corresponding entries of the other. To effect this the two vectors should have the same dimension (i.e., both should be columns or rows with the same number of entries) and it is necessary to put a dot before the multiplication operator—that is, as shown here:

```
v = x.*x   % multiplication of entries of two vectors

v =
    1    4    9   16
```

The other type of multiplication is the conventional multiplication allowed in linear algebra. For instance, with that of a row vector by a column vector,

```
w = x*x'   % multiplication of x (row vector) by x'(column vector)

w = 30
```

the result is a constant—in this case, the length of the row vector should coincide with that of the column vector. If you multiply a column (say x') of dimension 4×1 by a row (say x) of dimension 1×4 (notice that the 1s coincide at the end of the first dimension and at the beginning of the second), the multiplication $z = x' * x$ results in a 4×4 matrix.

The solution of a set of linear equations is very simple in MATLAB. To guarantee that a unique solution exists, the determinant of the matrix should be computed before inverting the matrix. If the determinant is zero MATLAB will indicate the solution is not possible.

```
% Solution of linear set of equations Ax = b
A = [1 0 0; 2 2 0; 3 3 3];     % 3x3 matrix
t = det(A);                    % MATLAB function that calculates determinant
b = [2 2 2]';          % column vector
x = inv(A)*b;          % MATLAB function that inverts a matrix
```

The results of these operations are not given because of the semicolons at the end of the commands. The following script could be used to display them:

```
disp(' Ax = b')          % MATLAB function that displays the text in ' '
A
b
x
t
```

which gives

```
Ax = b
A =
      1   0   0
      2   2   0
      3   3   3
b =
      2
      2
      2
x =
       2.0000
      -1.0000
      -0.3333
t =
      6
```

Another way to solve this set of equations is

```
x = b'/A'
```

Try it!

MATLAB provides a fast way to obtain certain vectors/matrices; for instance,

```
% special vectors and matrices
x = ones(1, 10)    % row of ten 1s

x =
      1   1   1   1   1   1   1   1   1   1

A = ones(5, 5)     % matrix of 5 x 5 1s

A =
      1   1   1   1   1
      1   1   1   1   1
      1   1   1   1   1
      1   1   1   1   1
      1   1   1   1   1

x1 = [x zeros(1, 5)]    % vector with previous x and 5 0s
```

```
x1 =
  Columns 1 through 10
     1   1   1   1   1   1   1   1   1   1
  Columns 11 through 15
     0   0   0   0   0
A(2:5, 2:5) = zeros(4, 4)   % zeros in rows 2—5, columns 2—5

A =
     1   1   1   1   1
     1   0   0   0   0
     1   0   0   0   0
     1   0   0   0   0
     1   0   0   0   0
y = rand(1,10)      % row vector with 10 random values (uniformly
                            % distributed in [0,1]

y =
  Columns 1 through 6
    0.9501   0.2311   0.6068   0.4860   0.8913   0.7621
  Columns 7 through 10
    0.4565   0.0185   0.8214   0.4447
```

Notice that these values are between 0 and 1. When using the normal or Gaussian-distributed noise the values can be positive or negative reals.

```
y1 = randn(1,10)      % row vector with 10 random values
                                    % (Gaussian distribution)

y1 =
  Columns 1 through 6
   −0.4326  −1.6656   0.1253   0.2877  −1.1465   1.1909
  Columns 7 through 10
    1.1892  −0.0376   0.3273   0.1746
```

Using Built-In Functions and Creating Your Own

MATLAB provides a large number of built-in functions. The following script uses some of them.

```
% using built-in functions
t = 0:0.01:1;   % time vector from 0 to 1 with interval of 0.01
x = cos(2*pi*t/0.1);      % cos processes each of the entries in
      % vector t to get the corresponding value in vector x
% plotting the function x
figure(1)   % numbers the figure
plot(t, x)   % interpolated continuous plot
xlabel('t (sec)')   % label of x-axis
ylabel('x(t)')   % label of y-axis
```

```
% let's hear it
sound(1000∗x, 10000)
```

The results are given in Figure 0.14.

To learn about any of these functions use *help*. In particular, use *help* to learn about MATLAB routines for plotting *plot* and *stem*. Use *help sound* and *help waveplay* to learn about the sound routines available in MATLAB. Additional related functions are put at the end of these help files. Explore all of these and become aware of the capabilities of MATLAB. To illustrate the plotting and the sound routines, let us create a chirp that is a sinusoid for which the frequency is varying with time.

```
y = sin(2∗pi∗t.^2/.1);   % notice the dot in the squaring
                         % t was defined before
sound(1000∗y, 10000)  % to listen to the sinusoid
figure(2) % numbering of the figure
plot(t(1:100), y(1:100))  % plotting of 100 values of y
figure(3)
plot(t(1:100), x(1:100), 'k', t(1:100), y(1:100), 'r')  % plotting x and y on same plot
```

Let us hope you were able to hear the chirp, unless you thought it was your neighbor grunting. In this case, we plotted the first 100 values of *t* and *y* and let MATLAB choose the color for them. In the second plot we chose the colors: black (dashed lines) for *x* and blue (continuous line) for the second signal *y(t)* (see Figure 0.15).

Other built-in functions are sin, tan, acos, asin, atan, atan2, log, log10, exp, etc. Find out what each does using help and obtain a listing of all the functions in the signal processing toolbox.

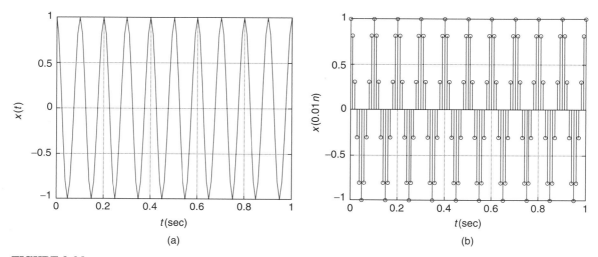

FIGURE 0.14

(a) Plotting of a sinusoid using *plot*, which gives a continuous plot, and (b) *stem*, which gives a discrete plot.

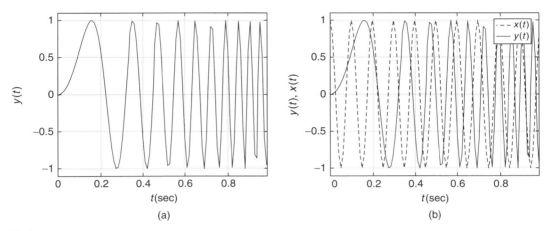

FIGURE 0.15
(a) Plotting chirp (MATLAB chooses color), (b) sinusoid and chirp (the sinusoid is plotted with dashed lines and the chirp with solid lines).

You do not need to define π, as it is already done in MATLAB. For complex numbers also you do not need to define the square root of -1, which for engineers is 'j' and for mathematicians 'i' (they have no current to worry about).

```
% pi and j
pi
j
i

ans =
    3.1416
ans =
        0 + 1.0000i
ans =
        0 + 1.0000i
```

Creating Your Own Functions
MATLAB has created a lot of functions to make our lives easier, and it allows us also to create—in the same way—our own. The following file is for a function f with an input of a scalar x and output of a scalar y related by a mathematical function:

```
function y = f(x)
y = x*exp(-sin(x))/(1 + x^2);
```

Functions cannot be executed on their own—they need to be part of a script. If you try to execute the above function MATLAB will give the following:

```
??? format compact;function y = f(x)
                    |
Error: A function declaration cannot appear within a script M-file.
```

A function is created using the word "function" and then defining the output (y), the name of the function (f), and the input of the function (x), followed by lines of code defining the function, which in this case is given by the second line. In our function the input and the output are scalars. If you want vectors as input/output you need to do the computation in vectorial form—more later.

Once the function is created and saved (the name of the function followed by the extension .m), MAT-LAB will include it as a possible function that can be executed within a script. If we wish to compute the value of the function for $x = 2$ ($f.m$ should be in the working directory) we proceed as follows:

$$y = f(2)$$

gives

$$y = 0.1611$$

To compute the value of the function for a vector as input, we compute for each of the values in the vector the corresponding output using a for loop as shown in the following.

```
x = 0:0.1:100;          % create an input vector x
N = length(x);          % find the length of x
y = zeros(1,N);          % initialize the output y to zeros
for n = 1:N,               % for the variable n from 1 to N, compute
    y(n) = f(x(n));            % the function
end
figure(3)
plot(x, y)
grid                        % put a grid on the figure
title('Function f(x)')
xlabel('x')
ylabel('y')
```

This is not very efficient. A general rule in MATLAB is: Loops are to be avoided, and vectorial computations are encouraged. The results are shown in Figure 0.16.

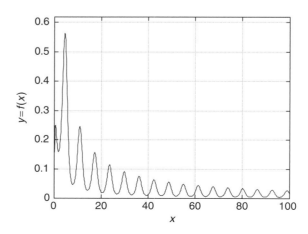

FIGURE 0.16

Result of using the function $f(.)$

The function working on a vector x, rather than one value, takes the following form (to make it different from the above function we let the denominator be $1 + x$ instead of $1 + x^2$):

```
function yy = ff(x)
% vectorial function
yy = x.*exp(−sin(x))./(1 + x);
```

Again, this function must be in the working directory. Notice that the computation of yy is done considering x a vector; the .* and ./ are indicative of this. Thus, this function will accept a vector x and will give as output a vector yy, computed as indicated in the last line. When we use a function, the names of the variables used in the script that calls the function do not need to coincide with the ones in the definition of the function. Consider the following script:

```
z = ff(x);     % x defined before,
                    % z instead of yy is the output of the function ff
figure(4)
plot(x, z); grid
title('Function ff(x)') % MATLAB function that puts title in plot
xlabel('x')  % MATLAB function to label x-axis
ylabel('z')   % MATLAB function to label y-axis
```

The difference between plot and stem is important. The function plot interpolates the vector to be plotted and so the plot appears continuous, while stem simply plots the entries of the vector, separating them uniformly. The input x and the output of the function are discrete time and we wish to plot them as such, so we use stem.

```
stem(x(1:30),  z(1:30))
grid
title('Function ff(x)')
xlabel('x')
ylabel('z')
```

The results are shown in Figure 0.17.

More on Plotting

There are situations where we want to plot several plots together. One can superpose two or more plots by using hold on and hold off. To put several figures in the same plot, we can use the function subplot. Suppose we wish to plot four figures in one plot and they could be arranged as two rows of two figures each. We do the following:

```
subplot(221)
plot(x, y)
subplot(222)
plot(x, z)
subplot(223)
stem(x, y)
subplot(224)
stem(x, z)
```

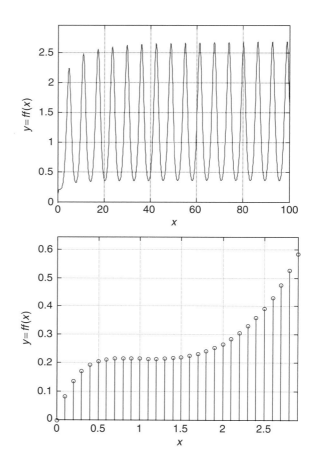

FIGURE 0.17
Results of using the function ff(.)
(notice the difference in scale in the
x axis).

In the subplot function the first two numbers indicate the number of rows and the number of columns, and the last digit refers to the order of the graph that is, 1, 2, 3, and 4 (see Figure 0.18).

There is also a way to control the values in the axis, by using the function (you guessed!) **axis**. This function is especially useful after we have a graph and want to improve its looks. For instance, suppose that the professor would like the above graphs to have the same scales in the y-axis (picky professor). You notice that there are two scales in the y-axis, one 0-0.8 and another 0-3. To have both with the same scale, we choose the one 0-3, and modify the above code to the following

```
subplot(221)
plot(x, y)
axis([0 100 0 3])
subplot(222)
plot(x, z)
axis([0 100 0 3])
subplot(223)
stem(x, y)
```

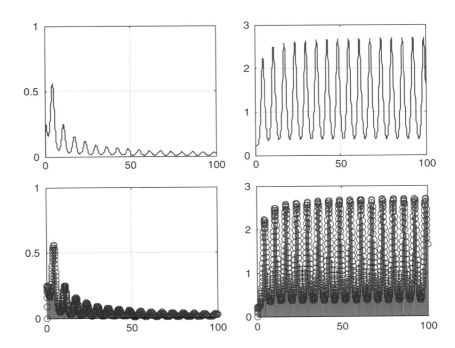

FIGURE 0.18
Plotting four figures in one.

```
axis([0 100 0 3])
subplot(224)
stem(x, z)
axis([0 100 0 3])
```

Saving and Loading Data

In many situations you would like to either save some data or load some data. The following is one
way to do it. Suppose you want to build and save a table of sine values for angles between 0 and
360 degrees in intervals of 3 degrees. This can be done as follows:

```
x = 0:3:360;
y = sin(x*pi/180);  % sine computes the argument in radians
xy = [x' y'];  % vector with 2 columns one for x'
               % and  another for y'
```

Let's now save these values in a file "sine.mat" by using the function save (use help save to learn more):

```
save sine.mat xy
```

To load the table, we use the function load with the name given to the saved table "sine" (the extension
*.mat is not needed). The following script illustrates this:

```
clear all
load sine
whos
```

where we use whos to check its size:

```
Name    Size            Bytes   Class
xy      121x2           1936    double array
Grand total is 242 elements using 1936 bytes
```

This indicates that the array *xy* has 121 rows and 2 columns, the first colum corresponding to *x*, the degree values, and the second column corresponding to the sine values, *y*. Verify this and plot the values by using

```
x = xy(:, 1);
y = xy(:, 2);
stem(x, y)
```

Finally, MATLAB provides some data files for experimentation and you only need to load them. The following "train.mat" is the recording of a train whistle, sampled at the rate of F_s samples/sec, which accompanies the sampled signal $y(n)$ (see Figure 0.19).

```
clear all
load train
whos
```

```
Name        Size            Bytes       Class

Fs          1x1             8           double array
y           12880x1         103040      double array
```

Grand total is 12881 elements using 103048 bytes

```
sound(y, Fs)
plot(y)
```

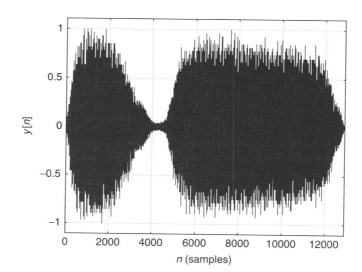

FIGURE 0.19
Train signal.

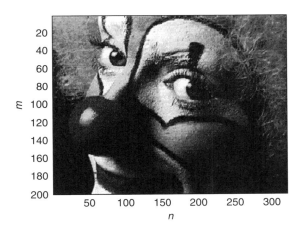

FIGURE 0.20
Clown in gray scale.

MATLAB also provides two-dimensional signals, or images, such as "clown.mat," a 200×320 pixels image.

```
clear all
load clown
whos
```

Name	Size	Bytes	Class
X	200x320	512000	double array
caption	2x1	4	char array
map	81x3	1944	double array

Grand total is 64245 elements using 513948 bytes

We can display this image in gray levels by using the following script (see Figure 0.20):

```
colormap('gray')
imagesc(X)
```

Or in color using

```
colormap('hot')
imagesc(X)
```

0.5.2 Symbolic Computations

We have considered the numerical capabilities of MATLAB, by which numerical data are transformed into numerical data. There will be many situations when we would like to do algebraic or calculus operations resulting in terms of variables rather than numerical data. For instance, we might want to find a formula to solve quadratic algebraic equations, to find a difficult integral, or to obtain the Laplace or the Fourier transform of a signal. For those cases MATLAB provides the Symbolic Math Toolbox a symbolic computing system. In this section, we provide you with an introduction to symbolic computations by means of examples, and hope to get you interested in learning more on your own.

Derivatives and Differences

The following script compares symbolic with numeric computations of the derivative of a chirp signal (a sinusoid with changing frequency) $y(t) = \cos(t^2)$, which is

$$z(t) = \frac{dy(t)}{dt} = -2t \sin(t^2)$$

```
clf; clear all
% symbolic
    syms t y z  % define the symbolic variables
    y = cos(t^2)  % chirp signal -- notice no . before ^ since t is no vector
    z = diff(y)  % derivative
    figure(1)
    subplot(211)
    ezplot(y, [0, 2*pi]);grid  % plotting for symbolic y between 0 and 2*pi
    hold on
    subplot(212)
    ezplot(z, [0, 2*pi]);grid
    hold on
%numeric
    Ts = 0.1;  % sampling period
    t1 = 0:Ts:2*pi;  % sampled time
    y1 = cos(t1.^2);  % sampled signal --notice difference with y above
    z1 = diff(y1)./diff(t1);  % difference -- approximation to derivative
    figure(1)
    subplot(211)
    stem(t1, y1, 'r');axis([0 2*pi 1.1*min(y1) 1.1*max(y1)])
    subplot(212)
    stem(t1(1:length(y1) - 1), z1, 'r');axis([0 2*pi 1.1*min(z1) 1.1*max(z1)])
    legend('Derivative (black)','Difference (blue)')
    hold off
```

The symbolic function syms defines the symbolic variables (use help syms to learn more). The signal $y(t)$ is written differently than $y_1(t)$ in the numeric computation. Since t_1 is a vector, squaring it requires a dot before the symbol. That is not the case for t, which is not a vector but a variable. The results of using diff to compute the derivative of $y(t)$ is given in the same form as you would have obtained doing the derivative by hand—that is,

```
y = cos(t^2)
z = -2*t*sin(t^2)
```

The symbolic toolbox provides its own graphic routines (use help to learn about the different ez-routines). For plotting $y(t)$ and $z(t)$, we use the function ezplot, which plots the above two functions for $t \in [0, 2\pi]$ and titles the plots with these functions.

The numeric computations differ from the symbolic in that vectors are being processed, and we are obtaining an approximation to the derivative $z(t)$. We sample the signal with $T_s = 0.1$ and use again

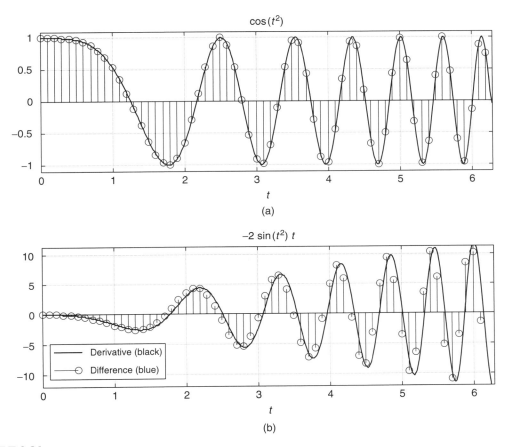

FIGURE 0.21

Symbolic and numeric computation of the derivative of the chirp $y(t) = \cos(t^2)$. (a) $y(t)$ and the sampled signal $y(nT_s)$, $T_s = 0.1$ sec. (b) Displays the exact derivative (continuous line) and the approximation of the derivative at samples nT_s. Better approximation to the derivative can be obtained by using a smaller value of T_s.

the function diff to approximate the derivative (the denominator *diff(t1)* is the same as T_s). Plotting the exact derivative (continuous line) with the approximated one (samples) using stem clarifies that the numeric computation is an approximation at nT_s values of time. See Figure 0.21.

The Sinc Function and Integration

The sinc function is very significant in the theory of signals and systems. It is defined as

$$y(t) = \frac{\sin \pi t}{\pi t} \qquad -\infty < t < \infty$$

It is symmetric with respect to the origin, and defined from $-\infty$ to ∞. The value of $y(0)$ can be found using L'Hôpital's rule. We will see later (Parseval's result in Chapter 5) that the integral of $y^2(t)$ is equal to 1. In the following script we are combining numeric and symbolic computations to show this. First, after defining the variables, we use the symbolic function int to compute the integral of the squared sinc function, with respect to t, from 0 to integer values $1 \leq k \leq 10$. We then use the function subs to convert the symbolic results into a numerical array zz. The numeric part of the script defines a vector y to have the values of the sinc function for 100 time values equally spaced between $[-4, 4]$, obtained using the function linspace. We then use plot and stem to plot the sinc and the values of the integrals, which as seen in Figure 0.22 reach a value close to unity in less than 10 steps. Please use help to learn more about each of these functions.

```
clf; clear all
% symbolic
syms t z
for k = 1:10,
```

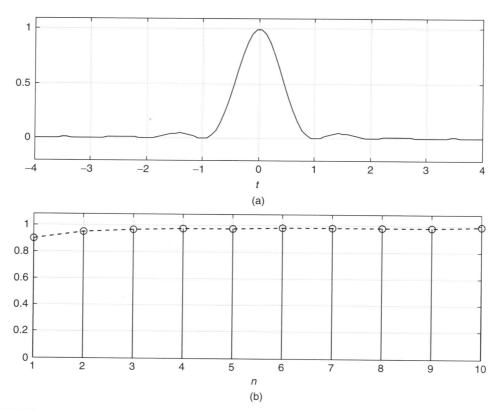

(a)

(b)

FIGURE 0.22
(a) Computation of the integral of the squared sinc function (b) Illustrates that the area under the curve of this function, or its integral, is unity. Using the symmetry of the function only the integral for $t \geq 0$ needs to be computed.

```
    z = int(sinc(t)^2, t, 0, k);    % integral of sinc^2 from 0 to k
    zz(k) = subs(2*z);          % substitution to numeric value zz
end
% numeric
t1 = linspace(-4, 4);        % 100 equally spaced points in [-4,4]
y = sinc(t1).^2;    % numeric definition of the squared sinc function
n = 1:10;
figure(1)
subplot(211)
plot(t1, y);grid;axis([-4 4 -0.2 1.1*max(y)]);title('y(t)=sinc^2(t)');
xlabel('t')
subplot(212)
stem(n(1:10), zz(1:10)); hold on
plot(n(1:10), zz(1:10), 'r');grid;title('∫ y(τ) dτ'); hold off
axis([1 10 0 1.1*max(zz)]); xlabel('n')
```

Figure 0.22 shows the squared sinc function and the values of the integral

$$2 \int_0^k sinc^2(t)dt = 2 \int_0^k \left[\frac{\sin(\pi t)}{\pi t} \right]^2 dt \qquad k = 1, \ldots, 10$$

which quickly reaches the final value of unity. In computing the integral from $(-\infty, \infty)$ we are using the symmetry of the function and thus the multiplication by 2.

Chebyshev Polynomials and Lissajous Figures

The Chebyshev polynomials are used in the design of filters. They can be obtained by plotting two cosine functions as they change with time t, one of fix frequency and the other with increasing frequency:

$$x(t) = \cos(2\pi t)$$
$$y(t) = \cos(2\pi kt) \qquad k = 1, \ldots, N$$

The $x(t)$ gives the x axis coordinate and $y(t)$ the y axis coordinate at each value of t. If we solve for t in the top equation, we get

$$t = \frac{1}{2\pi} \cos^{-1}(x(t))$$

which then replaced in the bottom equation gives

$$y(t) = \cos \left[k \cos^{-1}(x(t)) \right] \qquad k = 1, \ldots, N$$

as an expression for the Chebyshev polynomials (we will see in Chapter 6 that these equations can be expressed as regular polynomials). Figure 0.23 shows the Chebyshev polynomials for $N = 4$. The following script is used to compute and plot these polynomials.

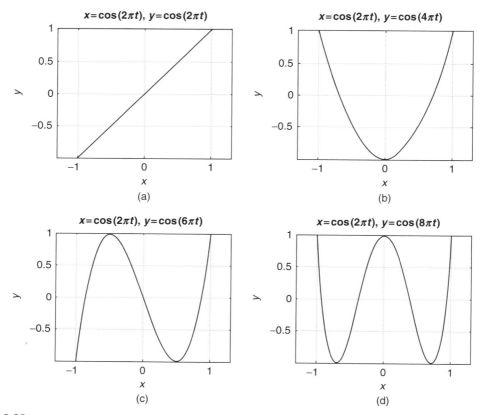

FIGURE 0.23

The Chebyshev polynomials for $n = 1, 2, 3, 4$. First (a) to fourth (d) polynomials. Notice that these polynomials are defined between $[-1, 1]$ in the x axis.

```
clear all;clf
syms x y t
x = cos(2*pi*t); theta=0;
figure(1)
for k = 1:4,
   y = cos(2*pi*k*t + theta);
   if k == 1, subplot(221)
     elseif k == 2, subplot(222)
     elseif k == 3, subplot(223)
     else subplot(224)
   end
   ezplot(x, y);grid;hold on
end
hold off
```

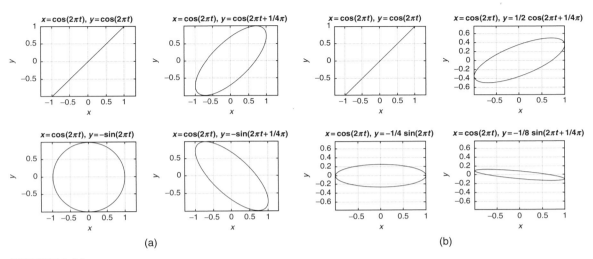

FIGURE 0.24

Lissajous figures: (a) (four left plots) case 1 input and output of same amplitude ($A = 1$) but phase differences of $0, \pi/4, \pi/2$, and $3\pi/4$; (b) (four right plots) case 2 input has unit amplitude but output has decreasing amplitudes and same phase differences as in case 1.

The Lissajous figures we consider next are a very useful extension of the above plotting of sinusoids in the x and y axes. These figures are used to determine the difference between a sinusoidal input and its corresponding sinusoidal steady state. In the case of linear systems, which we will formally define in Chapter 2, for a sinusoidal input the outputs of the system are also sinusoids of the same frequency, but they differ with the input in the amplitude and phase.

The differences in amplitude and phase can be measured using an oscilloscope for which we put the input in the horizontal sweep and the output in the vertical sweep, giving figures from which we can find the differences in amplitude and phase. Two situations are simulated in the following script, one where there is no change in amplitude but the phase changes from zero to $3\pi/4$, while in the other case the amplitude decreases as indicated and the phase changes in the same way as before. The plots, or Lissajous figures, indicate such changes. The difference between the maximum and the minimum of each of the figures in the x axis gives the amplitude of the input, while the difference between the maximum and the minimum in the y axis gives the amplitude of the output. The orientation of the ellipse provides the difference in phase with respect to that of the input. The following script is used to obtain the Lissajous figures in these cases. Figure 0.24 displays the results.

```
clear all;clf
syms x y t
x = cos(2*pi*t); % input of unit amplitude and frequency 2*pi
A = 1;figure(1)   % amplitude of output in case 1
for i = 1:2,
for k = 0:3,
```

```
      theta = k*pi/4;   % phase of output
      y = A^k*cos(2*pi*t + theta);
    if k == 0,subplot(221)
      elseif k == 1,subplot(222)
      elseif k == 2,subplot(223)
      else subplot(224)
    end
    ezplot(x, y);grid;hold on
  end
  A = 0.5; figure(2) % amplitude of output in case 2
  end
```

Ramp, Unit-Step, and Impulse Responses

To close this introduction to symbolic computations we illustrate the response of a linear system represented by a differential equation,

$$\frac{d^2 y(t)}{dt^2} + 5\frac{dy(t)}{dt} + 6y(t) = x(t)$$

where $y(t)$ is the output and $x(t)$ the input. The input is a constant $x(t) = 1$ for $t \geq 0$ and zero otherwise (MATLAB calls this function *heaviside*, but we will call it the *unit-step signal*). We then let the input be the derivative of $x(t)$, which is a signal that we will call *impulse*, and finally we let the input be the integral of $x(t)$, which is what we will call the *ramp* signal. The following script is used to find the responses, which are displayed in Figure 0.25.

```
clear all; clf
syms y t x z
% input a unit-step  (heaviside) response
y = dsolve('D2y + 5*Dy + 6*y = heaviside(t)','y(0) = 0','Dy(0) = 0','t');
x = diff(y); % impulse response
z = int(y); % ramp response
figure(1)
subplot(311)
ezplot(y, [0,5]);title('Unit-step response')
subplot(312)
ezplot(x, [0,5]);title('Impulse response')
subplot(313)
ezplot(z, [0,5]);title('Ramp response')
```

This example illustrates the intuitive appeal of linear systems. When the input is a constant value (or a unit-step signal or a heaviside signal) the output tries to follow the input after some initial inertia and it ends up being constant. The impulse signal (obtained as the derivative of the unit-step signal) is a signal of very short duration equivalent to shocking the system with a signal that disappears very fast, different from the unit-step signal that is like a dc source. Again the output tries to follow the input, eventually disappearing as t increases (no energy from the input!), and the ramp that is

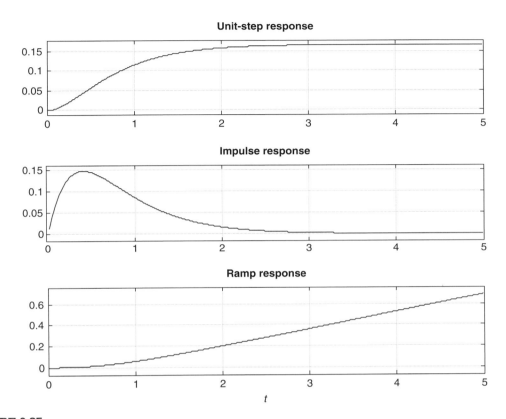

FIGURE 0.25

Response of a second order system represented by a differential equation for input of the unit-step signal, its derivative, or the impulse signal and the ramp signal that is the integral of the unit-step input.

the integral of the unit-step signal grows with time, providing more and more energy to the system as time increases, thus the response we obtained. The function dsolve solves differential equations explicitly given (D stands for the derivative operator, so D is the first derivative and D2 is the second derivative). A second-order system requires two initial conditions, the output and its derivative at $t = 0$.

We hope this introduction to MATLAB has provided you with the necessary background to understand the basic way MATLAB operates, and shown you how to continue increasing your knowledge of it. Your best source of information is the *help* command. Explore the different modules that MATLAB has and you will become quickly convinced that these modules provide a great number of computational tools for many areas of engineering and mathematics. Try it—you will like it! Tables 0.1 and 0.2 provide a listing of the numeric and symbolic variables and operations.

Table 0.1 Basic Numeric Matlab

Special variables	ans	Default name for result
	pi	π value
	inf, NaN	infinity, not-a-number error (e.g., $0/0$)
	i, j	$i = j = \sqrt{-1}$
	Function(s)	**Operation**
Mathematical	abs, angle	magnitude, angle of complex number
	acos, asine, atan	inverse cosine, sine, tangent
	acosh, asinh, atanh	inverse cosh, sinh, tanh
	cos, sin, tan	cosine, sine, tangent
	cosh, sinh, tanh	hyperbolic cosine, sine, tangent
	conj, imag, real	complex conjugate, imaginary, real parts
	exp, log, log10	exponential, natural and base 10 logarithms
Special operations	ceil, floor	round up, round down to integer
	fix, round	round toward zero, to nearest integer
	.*, ./	entry-by-entry multiplication, division
	.^	entry-by-entry power
	x', A'	transpose of vector x, matrix A
Array operations	x=*first:increment:last*	row vector x from *first* to *last* by *increment*
	x=linspace(first,last,n)	row vector x with n elements from *first* to *last*
	A=[x1;x2]	matrix A with rows $x1, x2$
	ones(N,M), zeros(N,M)	$N \times M$ ones and zeros arrays
	A(i,j)	(i, j) entry of matrix A
	A(i,:), A(:,j)	i row (j-column) and all columns (rows) of matrix A
	whos	display variables in workspace
	size(A)	(number rows, number of colums) of matrix A
	length(x)	number rows (colums) of vector x
Control flow	for, if, elseif	for loop, if, else-if loop
	while	while loop
	pause, pause(n)	pause and pause n seconds
Plotting	plot, stem	continuous, discrete plots
	figure	figure for plotting
	subplot	subplots
	hold on, hold off	hold plot on or off
	axis, grid	axis, grid of plots
	xlabel, ylabel, title, legend	labeling of axes, plots, and subplots
Saving and loading	save, load	saving and loading data
Information and managing	help	help
	clear, clf	clear variables from memory, clear figures
Operating system	cd, pwd	change directory, current working directory

Table 0.2 Basic Symbolic Matlab Functions

	Function	Operation
Calculus	diff	differentiate
	int	integrate
	limit	limit
	taylor	Taylor series
	symsum	summation
Simplification	simplify	simplify
	expand	expand
	factor	factor
	simple	find shortest form
	subs	symbolic substitution
Solving equations	solve	solve algebraic equations
	dsolve	solve differential equations
Transforms	fourier	Fourier transform
	ifourier	inverse Fourier transform
	laplace	Laplace transform
	ilaplace	inverse Laplace transform
	ztrans	Z-transform
	iztrans	inverse Z-transform
Symbolic operations	sym	create symbolic objects
	syms	create symbolic objects
	pretty	make pretty expression
Special functions	dirac	Dirac or delta function
	heaviside	unit-step function
Plotting	ezplot	function plotter
	ezpolar	polar coordinate plotter
	ezcontour	contour plotter
	ezsurf	surface plotter
	ezmesh	mesh (surface) plotter

PROBLEMS

For the problems requiring implementation in MATLAB, write scripts or functions to solve them numerically or symbolically. Label the axes of the plots, give a title, and use legend to identify different signals in a plot. To save space use subplot to put several plots into one. To do the problem numerically, sample analog signals with a small T_s.

0.1. Bits or bytes

Just to get an idea of the number of bits or bytes generated and processed by a digital system consider the following applications:

(a) A compact disc is capable of storing 75 minutes of "CD-quality" stereo (left and right channels are recorded) music. Calculate the number of bytes and the number of bits that are stored in the CD.
Hint: Find out what "CD quality" means in the binary representation of each sample, and what is the sampling rate your CD player uses.

(b) Find out what the vocoder in your cell phone is used for. Assume then that in attaining "telephone quality" you use a sampling rate of 10,000 samples/sec to achieve that type of voice quality. Each sample is represented by 8 bits. With this information, calculate the number of bits that your cell phone has to process every second that you talk. Why would you then need a vocoder?

(c) Find out whether text messaging is cheaper or more expensive than voice. Explain how text messaging works.

(d) Find out how an audio CD and an audio DVD compare. Find out why it is said that a vinyl long play record reproduces sounds much better. Are we going backwards with digital technology in music recording? Explain.

(e) To understand why video streaming in the Internet is many times of low quality, consider the amount of data that need to be processed by a video compressor every second. Assume the size of a video frame, in picture elements or pixels, is 352×240, and that an acceptable quality for the image is obtained by allocating 8 bits/pixel, and to avoid jerking effects we use 60 frames/sec.

- How many pixels would have to be processed every second?

- How many bits would be available for transmission every second?

- The above are raw data. Compression changes the whole picture (literally); find out what some of the compression methods are.

0.2. Sampling—MATLAB

Consider an analog signal $x(t) = 4\cos(2\pi t)$ defined for $-\infty < t < \infty$. For the following values of the sampling period T_s, generate a discrete-time signal $x[n] = x(nT_s) = x(t)|_{t=nT_s}$.

- $T_s = 0.1$ sec
- $T_s = 0.5$ sec
- $T_s = 1$ sec

Determine for which values of T_s the discrete-time signal has lost the information in the analog signal. Use MATLAB to plot the analog signal (use the plot function) and the resulting discrete-time signals (use the stem function). Superimpose the analog and the discrete-time signals for $0 \le t \le 3$; use subplot to plot the four figures as one figure. For plotting the analog signal use $T_s = 10^{-4}$. You also need to figure out how to label the different axes and have the same scales and units. In Chapter 7 on sampling we will show how to reconstruct sampled signals.

0.3. Derivative and finite difference—MATLAB

Let $y(t) = dx(t)/dt$, where $x(t)$ is the signal in Problem 0.2. Find $y(t)$ analytically and determine a value of T_s for which $\Delta[x(nT_s)]/T_s = y(nT_s)$ (consider $T_s = 0.01$ and $T_s = 0.1$). Use the MATLAB function diff or create your own to compute the finite difference. Plot the finite difference in the range [0,1] and compare it with the actual derivative $y(t)$ in that range. Explain your results for the given values of T_s.

0.4. Backward difference—MATLAB

Another definition for the finite difference is the backward difference:

$$\Delta[x(nT_s)] = x(nT_s) - x((n-1)T_s)$$

($\Delta[x(nT_s)]/T_s$ approximates the derivative of $x(t)$.)

(a) Indicate how this new definition connects with the finite difference defined earlier in this chapter.

(b) Solve Problem 0.3 with MATLAB using this new finite difference and compare your results with the ones obtained there.

(c) For the value of $T_s = 0.1$, use the average of the two finite differences to approximate the derivative of the analog signal $x(t)$. Compare this result with the previous ones. Provide an expression for calculating this new finite difference directly.

0.5. Differential and difference equations—MATLAB

Find the differential equation relating a current source $i_s(t) = \cos(\Omega_0 t)$ with the current $i_L(t)$ in an inductor, with inductance $L = 1$ H, connected in parallel with a resistor of $R = 1\Omega$ (see Figure 0.26). Assume a zero initial current in the inductor.

(a) Obtain a discrete equation from the differential equation using the trapezoidal approximation of an integral.

(b) Create a MATLAB script to solve the difference equation for $T_s = 0.01$ and three frequencies for $i_s(t)$, $\Omega_0 = 0.005\pi, 0.05\pi$, and 0.5π. Plot the input current source $i_s(t)$ and the approximate solution $i_L(nT_s)$ in the same figure. Use the MATLAB function plot. Use the MATLAB function filter to solve the difference equation (use help to learn about filter).

(c) Solve the differential equation using symbolic MATLAB when the input frequency is $\Omega_0 = 0.5\pi$.

(d) Use phasors to find the amplitude of $i_L(t)$ when the input is $i_s(t)$ with the given three frequencies.

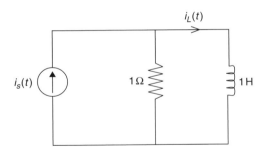

FIGURE 0.26

Problem 0.5. RL circuit: input $i_s(t)$ and output $i_L(t)$.

0.6. Sums and Gauss—MATLAB

Three rules in the computation of sums are

■ Distributive law:

$$\sum_k ca_k = c \sum_k a_k$$

■ Associative law:

$$\sum_k (a_k + b_k) = \sum_k a_k + \sum_k b_k$$

■ Commutative law:

$$\sum_k a_k = \sum_{p(k)} a_{p(k)}$$

for any permutation $p(k)$ of the set of integers k in the summation.

(a) Explain why the above rules make sense when computing sums. To do that consider

$$\sum_k a_k = \sum_{k=0}^{2} a_k$$

and similarly for $\sum_k b_k$. Let c be a constant, and choose any permutation of the values [0,1,2] for instance [2,1,0] or [1,0,2].

(b) The trick that Gauss played when he was a preschooler can be explained by using the above rules. Suppose you want to find the sum of the integers from 0 to 10000 (Gauss did it for integers between 0 and 100 but he was then just a little boy, and we can do better!). That is, we want to find S where

$$S = \sum_{k=0}^{10000} k = 0 + 1 + 2 + \cdots + 10000$$

To do so, consider

$$2S = \sum_{k=0}^{10000} k + \sum_{k=10000}^{0} k$$

and apply the above rules to find S.

(c) Find the sum of an arithmetic progression

$$S = \sum_{k=0}^{N} (\alpha + \beta k)$$

for constants α and β, using the given three rules.

(d) Find out if MATLAB can do these sums symbolically (i.e., without having numerical values).

0.7. Integrals and sums—MATLAB

Suppose you wish to find the area under a signal using sums. You will need the following result found above:

$$\sum_{n=0}^{N} n = \frac{N(N+1)}{2}$$

(a) Consider first $x(t) = t, 0 \leq t \leq 1$, and zero otherwise. The area under this signal is 0.5. The integral can be approximated from above and below as

$$\sum_{n=1}^{N-1} (nT_s)T_s < \int_0^1 t\, dt < \sum_{n=1}^{N} (nT_s)T_s$$

where $NT_s = 1$ (i.e., we segment the interval [0,1] into N intervals of width T_s). Graphically show that the above equation makes sense by showing the right and left bounds as approximations for the area under $x(t)$.

(b) Let $T_s = 0.001$. Use the symbolic function symsum to compute the left and right bounds for the above integral. Find the average of these results and compare it with the actual value of the integral.

(c) Verify the symbolic results by finding the sums on the left and the right of the above inequality using the summation given at the beginning of the problem. You need to change the dummy variables.

(d) Write a similar MATLAB script to compute the area under the signal $y(t) = t^2$ from $0 \le t \le 1$. Let $T_s = 0.001$. Compare the average of the lower and upper bounds to the value of the integral.

0.8. Integrals and sums—MATLAB

Although sums behave like integrals, because of the discrete nature of sums one needs to be careful with the upper and lower limits more than in the integral case. To illustrate this, consider the separation of an integral into two integrals and compare them with the separation of a sum into two sums. For the integral we have that

$$\int_0^1 t\,dt = \int_0^{0.5} t\,dt + \int_{0.5}^1 t\,dt$$

Show that this is true by computing the three integrals. Then consider the sum

$$S = \sum_{n=0}^{100} n$$

Find this sum and determine which of the following is equal to this sum:

$$S_1 = \sum_{n=0}^{50} n + \sum_{n=50}^{100} n$$

$$S_2 = \sum_{n=0}^{50} n + \sum_{n=51}^{100} n$$

Use symbolic MATLAB function symsum to verify your answers.

0.9. Sum of geometric series

The geometric series

$$S = \sum_{n=0}^{N-1} \alpha^n$$

will be used quite frequently in the next chapters, so let us look at some of its properties:

(a) Suppose $\alpha = 1$; what is S equal to?

(b) Suppose $\alpha \ne 1$; show that

$$S = \frac{1 - \alpha^N}{1 - \alpha}$$

This can be done by showing that $(1 - \alpha)S = (1 - \alpha^N)$. Why do you need the constraint that $\alpha \ne 1$? Would this sum exist if $\alpha > 1$? Explain.

(c) Give an expression of the above sum for all possible values of α.

(d) Suppose now that $N = \infty$; under what conditions will S exist? If it does, what would S be equal to? Explain.

(e) Suppose the derivative of S with respect to α is

$$S_1 = \frac{dS}{d\alpha} = \sum_{n=0}^{\infty} n\alpha^{n-1}$$

Obtain an expression to find S_1.

0.10. Exponentials—MATLAB

The exponential $x(t) = e^{at}$ for $t \geq 0$ and zero otherwise is a very common analog signal. Likewise, $y[n] = \alpha^n$ for integers $n \geq 0$ and zero otherwise is a very common discrete-time signal. Let us see how they are related. Do the following using MATLAB:

(a) Let $a = -0.5$; plot $x(t)$.

(b) Let $a = -1$; plot the corresponding signal $x(t)$. Does this signal go to zero faster than the exponential for $a = -0.5$?

(c) Suppose we sample the signal $x(t)$ using $T_s = 1$; what would be $x(nT_s)$ and how can it be related to $y(n)$ (i.e., what is the value of α that would make the two equal)?

(d) Suppose that a current $x(t) = e^{-0.5t}$ for $t \geq 0$ and zero otherwise is applied to a discharged capacitor of capacitance $C = 1$ F at $t = 0$. What would be the voltage in the capacitor at $t = 1$ sec?

(e) How would you obtain an approximate result to the above problem using a computer? Explain.

0.11. Algebra of complex numbers

Consider complex numbers $z = 1 + j1$, $w = -1 + j1$, $v = -1 - j1$, and $u = 1 - j1$.

(a) In the complex plane, indicate the point (x, y) that corresponds to z and then show a vector $\vec{z}$ that joins the point (x, y) to the origin. What is the magnitude and the angle corresponding to z or $\vec{z}$?

(b) Do the same for the complex numbers w, v, and u. Plot the four complex numbers and find their sum $z + w + v + u$ analytically and graphically.

(c) Find the ratios z/w, w/v, and u/z. Determine the real and imaginary parts of each, as well as their magnitudes and phases. Using the ratios find u/w.

(d) The phase of a complex number is only significant when the magnitude of the complex number is significant. Consider z and $y = 10^{-16}z$; compare their magnitudes and phases. What would you say about the phase of y?

0.12. Algebra of complex numbers

Consider a function of $z = 1 + j1$,

$$w = e^z$$

(a) Find $\log(w)$.

(b) Find the real and the imaginary parts of w.

(c) What is $w + w^*$, where w^* is the complex conjugate of w?

(d) Determine $|w|$, $\angle w$.

(e) What is $|\log(w)|^2$?

(f) Express $\cos(1)$ in terms of w using Euler's equation.

0.13. Euler's identity and trigonometric identities

Use Euler's identity to obtain an expression for $e^{j(\alpha+\beta)} = e^{j\alpha}e^{j\beta}$; obtain its real and imaginary components and show the following identities:

- $\cos(\alpha + \beta) = \cos(\alpha)\cos(\beta) - \sin(\alpha)\sin(\beta)$
- $\sin(\alpha + \beta) = \sin(\alpha)\cos(\beta) + \sin(\beta)\cos(\alpha)$

Hint: Find real and imaginary parts of $e^{j\alpha}e^{j\beta}$ and of $e^{j(\alpha+\beta)}$.

0.14. Euler's identity and trigonometric identities

Use Euler's identity to find an expression for $\cos(\alpha)\cos(\beta)$, and from the relation between cosines and sines obtain an expression for $\sin(\alpha)\sin(\beta)$.

0.15. Algebra of complex numbers

(a) The complex conjugate of $z = x + jy$ is $z^* = x - jy$. Using these rectangular representations, show that

$$zz^* = x^2 + y^2$$

$$\frac{1}{z} = \frac{z^*}{zz^*}$$

(b) Show that it is easier to find the above results by using the polar representation $z = |z|e^{j\theta}$ of z where

$$|z| = \sqrt{x^2 + y^2}$$

is the magnitude of z and

$$\theta = \tan^{-1}\left(\frac{y}{x}\right)$$

is the angle or phase of z. Thus, whenever we are multiplying or dividing complex numbers the polar form is more appropriate.

(c) Whenever we are adding or subtracting complex numbers the rectangular representation is more appropriate. Show that for two complex numbers $z = x + jy$ and $w = v + jq$; then,

$$(z + w)^* = z^* + w^*$$

On the other hand, when showing that $(zw)^* = z^*w^*$ the polar form is more appropriate.

(d) If the above conclusions still do not convince you, consider then the case of multiplying two complex numbers:

$$z = r\cos(\theta) + jr\sin(\theta)$$
$$w = \rho\cos(\phi) + j\rho\sin(\phi)$$

Find the polar forms of z and w and then find zw by using the rectangular and then the polar forms and decide which is easier. As a bonus you should get the trigonometric identities for $\cos(\theta + \phi)$ and $\sin(\theta + \phi)$. What are they?

0.16. Vectors and complex numbers

Using the vectorial representation of complex numbers it is possible to get some interesting inequalities:

(a) Is it true that for a complex number $z = x + jy$:

$$|x| \leq |z|?$$

Show it geometrically by representing z as a vector.

(b) The so-called *triangle inequality* says that for any complex (or real) numbers z and v we have that

$$|z + v| \leq |z| + |v|$$

Show a geometric example that verifies this.

0.17. Complex functions of time—MATLAB

Consider the complex function $x(t) = (1 + jt)^2$ for $-\infty < t < \infty$.

(a) Find the real and the imaginary parts of $x(t)$ and carefully plot them with MATLAB. Try to make MATLAB plot $x(t)$ directly. What do you get? Does MATLAB warn you? Does it make sense?

(b) Compute the derivative $y(t) = dx(t)/dt$ and plot its real and imaginary parts. How do these relate to the real and the imaginary parts of $x(t)$?

(c) Compute the integral

$$\int_0^1 x(t)\,dt$$

(d) Would the following statement be true (remember * indicates complex conjugate)?

$$\left(\int_0^1 x(t)\,dt\right)^* = \int_0^1 x^*(t)\,dt$$

0.18. Euler's equation and orthogonality of sinusoids

Euler's equation,

$$e^{j\theta} = \cos(\theta) + j\sin(\theta)$$

is very useful not only in obtaining the rectangular and polar forms of complex numbers, but in many other respects as we will explore in this problem.

(a) Carefully plot $x[n] = e^{j\pi n}$ for $-\infty < n < \infty$. Is this a real or a complex signal?

(b) Suppose you want to find the trigonometric identity corresponding to

$$\sin(\alpha)\sin(\beta)$$

Use Euler's equation to express the sines in terms of exponentials, multiply the resulting exponentials, and use Euler's equation to regroup the expression in terms of sinusoids.

(c) As we will see later on, two periodic signals $x(t)$ and $y(t)$ of period T_0 are said to be orthogonal if the integral over a period T_0 is

$$\int_{T_0} x(t)y(t)\,dt = 0$$

For instance, consider $x(t) = \cos(\pi t)$ and $y(t) = \sin(\pi t)$. Check first that these functions repeat every $T_0 = 2$ (i.e., show that $x(t + 2) = x(t)$ and that $y(t + 2) = y(t)$). Thus, $T_0 = 2$ can be seen as their period. Then use the representation of a cosine in terms of complex exponentials,

$$\cos(\theta t) = \frac{e^{j\theta} + e^{-j\theta}}{2}$$

to express the integrand in terms of exponentials and calculate the integral.

0.19. Euler's equation and trigonometric expressions

Obtain using Euler's equation an expression for $\sin(\theta)$ in terms of exponentials and then

(a) Use it to obtain the trigonometric identity for $\sin^2(\theta)$.

(b) Compute the integral

$$\int_0^1 \sin^2(2\pi t)\,dt$$

0.20. **De Moivre's theorem for roots**
Consider the calculation of roots of an equation,

$$z^N = \alpha$$

where $N \geq 1$ is an integer and $\alpha = |\alpha|e^{j\phi}$ a nonzero complex number.
(a) First verify that there are exactly N roots of this equation and that they are given by

$$z_k = re^{j\theta_k}$$

where $r = |\alpha|^{1/N}$ and $\theta_k = (\phi + 2\pi k)/N$ for $k = 0, 1, \ldots, N-1$.
(b) Use the above result to find the roots of the following equations:

$$z^2 = 1$$
$$z^2 = -1$$
$$z^3 = 1$$
$$z^3 = -1$$

and plot them in a polar plane (i.e., indicating their magnitude and phase).
(c) Explain how the roots are distributed around a circle of radius r in the complex polar plane.

0.21. **Natural log of complex numbers**
Suppose you want to find the log of a complex number $z = |z|e^{j\theta}$. Its logarithm can be found to be

$$\log(z) = \log(|z|e^{j\theta}) = \log(|z|) + \log(e^{j\theta}) = \log(|z|) + j\theta$$

If z is negative it can be written as $z = |z|e^{j\pi}$ and we can find $\log(z)$ by using the above derivation. The log of any complex number can be obtained this way also.
(a) Justify each one of the steps in the above equation.
(b) Find

$$\log(-2)$$
$$\log(1 + j1)$$
$$\log(2e^{j\pi/4})$$

0.22. **Hyperbolic sinusoids—MATLAB**
In filter design you will be asked to use hyperbolic functions. In this problem we relate these functions to sinusoids and obtain a definition of these functions so that we can actually plot them.
(a) Consider computing the cosine of an imaginary number—that is, use

$$\cos(x) = \frac{e^{jx} + e^{-jx}}{2}$$

Let $x = j\theta$ and find $\cos(x)$. The resulting function is called the hyperbolic cosine or

$$\cos(j\theta) = \cosh(\theta)$$

(b) Consider then the computation of the hyperbolic sine $\sinh(\theta)$; how would you do it? Carefully plot it as a function of θ.

    (c) Show that the hyperbolic cosine is always positive and bigger than 1 for all values of θ.

    (d) Show that $\sinh(\theta) = -\sinh(-\theta)$.

    (e) Write a MATLAB script to compute and plot these functions between -10 and 10.

0.23. Phasors!

A phasor can be thought of as a vector, representing a complex number, rotating around the polar plane at a certain frequency expressed in radians/sec. The projection of such a vector onto the real axis gives a cosine. This problem will show the algebra of phasors, which would help you with some of the trigonometric identities that are hard to remember.

    (a) When you plot a sine signal $y(t) = A \sin(\Omega_0 t)$, you notice that it is a cosine $x(t) = A \cos(\Omega_0 t)$ shifted in time—that is,

$$y(t) = A \sin(\Omega_0 t) = A \cos(\Omega_0(t - \Delta_t)) = x(t - \Delta_t)$$

How much is this shift Δ_t? Better yet, what is $\Delta_\theta = \Omega_0 \Delta_t$ or the shift in phase? One thus only need to consider cosine functions with different phase shifts instead of sines and cosines.

    (b) You should have found the answer above is $\Delta_\theta = \pi/2$ (if not, go back and try it and see if it works). Thus, the phasor that generates $x(t) = A \cos(\Omega_0 t)$ is Ae^{j0} so that $x(t) = \mathcal{R}e[Ae^{j0}e^{j\Omega_0 t}]$. The phasor corresponding to the sine $y(t)$ should then be $Ae^{-j\pi/2}$. Obtain an expression for $y(t)$ similar to the one for $x(t)$ in terms of this phasor.

    (c) According to the above results, give the phasors corresponding to $-x(t) = -A \cos(\Omega_0 t)$ and $-y(t) = -\sin(\Omega_0 t)$. Plot the phasors that generate $\cos, \sin, -\cos,$ and $-\sin$ for a given frequency. Do you see now how these functions are connected? How many radians do you need to shift in a positive or negative direction to get a sine from a cosine, etc.

    (d) Suppose then you have the sum of two sinusoids, for instance $z(t) = x(t) + y(t)$, adding the corresponding phasors for $x(t)$ and $y(t)$ at some time (e.g., $t = 0$), which is just a sum of two vectors, you should get a vector and the corresponding phasor. Get the phasor for $z(t)$ and the expression for it in terms of a cosine.

PART

2

Theory and Application of Continuous-Time Signals and Systems

CHAPTER 1

Continuous-Time Signals

A journey of a thousand miles
begins with a single step.
Lao Tzu (604–531 BCE)
Chinese philosopher

1.1 INTRODUCTION

In this second part of the book, we will concentrate on the representation and processing of continuous-time signals. Such signals are familiar to us. Voice, music, as well as images and video coming from radios, cell phones, IPods, and MP3 players exemplify these signals. Clearly each of these signals has some type of information, but what is not clear is how we could capture, represent, and perhaps modify these signals and their information content.

To process signals we need to understand their nature—to classify them—so as to clarify the limitations of our analysis and our expectations. Several realizations could then come to mind. One could be that almost all signals vary randomly and continuously with time. Consider a voice signal. If you are able to capture such a signal, by connecting a microphone to your computer and using the hardware and software necessary to display it, you realize that when you speak into the microphone a rather complicated signal that changes in unpredictable ways is displayed. You would ask yourself how is it that your spoken words are converted into this signal, and how could it be represented mathematically to allow you to develop algorithms to change it. In this book we consider the representation of deterministic—rather than random—signals, clearly a first step in the long process of answering these significant questions.

A second realization could be that to input signals into a computer the signals must be in binary form. How do we convert the voltage signal generated by the microphone into a binary form? This requires that we compress the information in a way that permits us to get it back, as when we wish to listen to the voice signal stored in the computer.

Signals and Systems Using MATLAB®. DOI: 10.1016/B978-0-12-374716-7.00004-1

One more realization could be that the processing of signals requires us to consider systems. In our example, one could think of the human vocal system and of a microphone as a system that converts differences in air pressure into a voltage signal. Signals and systems go together. We will consider the interaction of signals and systems in the next chapter.

Specifically in this chapter we will discuss the following issues:

- *The mathematical representation of signals*—Generally, how to think of a signal as a function of either time (e.g., music and voice signals), space (e.g., images), or of time and space (e.g., videos). In this book we will concentrate on time-dependent signals.

- *Classification of signals*—Using practical characteristics of signals we offer a classification of signals indicating the way a signal is stored, processed, or both. As indicated, this second part of the book will concentrate on the representation and analysis of continuous-time signals and systems, while the next part will discuss the representation and analysis of discrete-time signals and systems.

- *Signal manipulation*—What it means to delay or advance a signal, to reflect it, or to find its odd or even components. These are signal operations that will help us in their representation and processing.

- *Basic signal representation*—We show that any signal can be represented using basic signals. This will permit us to highlight certain characteristics of the signal and to simplify finding the corresponding outputs of systems. In particular, the representation in terms of sinusoids is of great interest as it allows the development of the so-called Fourier representation, which is essential in the development of the theory of linear systems.

1.2 CLASSIFICATION OF TIME-DEPENDENT SIGNALS

Considering signals as functions of time-carrying information, there are many ways in which they can be classified:

(a) According to the predictability of their behavior, signals can be *random* or *deterministic*. While a deterministic signal can be represented by a formula or a table of values, random signals can only be approached probabilistically. In this book we will only consider deterministic signals.

(b) According to the variation of their time variable and their amplitude, signals can be either *continuous-time* or *discrete-time*, *analog* or *discrete* amplitude, or *digital*. This classification relates to the way signals are either processed, stored, or both.

(c) According to their energy content, signals can be characterized as *finite-* or *infinite-energy* signals.

(d) According to whether the signals exhibit repetitive behavior or not as *periodic* or *aperiodic* signals.

(e) According to the symmetry with respect to the time origin, signals can be *even* or *odd*.

(f) According to the dimension of their support, signals can be of *finite* or of *infinite* support. Support can be understood as the time interval of the signal outside of which the signal is always zero.

1.3 CONTINUOUS-TIME SIGNALS

That signals are functions of time-carrying information is easily illustrated with a recorded voice signal. Such a signal can be thought of as a continuously varying voltage, generated by a microphone, that can be transformed into an audible acoustic signal—providing the voice information—by means of an amplifier and speakers. Thus, the speech signal is represented by a function of time

$$v(t), \quad t_b \leq t \leq t_f \tag{1.1}$$

where t_b is the time at which this signal starts, and t_f the time at which it ends. The function $v(t)$ varies continuously with time, and its amplitude can take any possible value (as long as the speakers are not too loud!). This signal obviously carries the information provided by the voice message.

Not all signals are functions of time alone. A digital image stored in a computer provides visual information. The intensity of the illumination of the image depends on its location within the image. Thus, a digital image can be represented as a function of two space variables (m, n) that vary discretely, creating an array of values called *picture elements* or *pixels*. The visual information in the image is thus provided by the signal $p(m, n)$ where $0 \leq m \leq M - 1$ and $0 \leq n \leq N - 1$ for an image of size $M \times N$ pixels. Each of the pixel values can be represented, for instance, by 256 gray scale values or 8 bits/pixel. Thus, the signal $p(m, n)$ varies discretely in space and in amplitude. A video, as a sequence of images in time, is accordingly a function of time and of two space variables. How their time or space variables and their amplitudes vary characterizes signals.

For a time-dependent signal, time and amplitude vary continuously or discretely. Thus, according to the independent variable, signals are *continuous-time* or *discrete-time* signals—that is, t takes an innumerable or a finite set of values. Likewise, the amplitude of either a continuous-time or a discrete-time signal can vary continuously or discretely. Thus, continuous-time signals can be continuous-amplitude as well as discrete-amplitude signals. Continuous-amplitude, continuous-time signals are called *analog signals* given that they resemble the pressure variations caused by an acoustic signal. A continuous-amplitude, discrete-time signal is called a *discrete-time signal*. A *digital signal* has discrete time and discrete amplitude. If the samples of a digital signal are given as binary codes the signal is called a *binary signal*.

A good way to illustrate the signal classification is to consider the steps needed to process the voice signal $v(t)$ in Equation (1.1) with a computer. As indicated above, in $v(t)$ time varies continuously between t_b and t_f, and the amplitude also varies continuously, and we assume it could take any possible real value (i.e., $v(t)$ is an analog signal). As such, $v(t)$ cannot be processed with a computer. It would require to store an innumerable number of signal values (even when t_b is very close to t_f) and for an accurate representation of the amplitude values $v(t)$, we might need a large number of bits. Thus, it is necessary to reduce the amount of data without losing the information provided by the signal. To accomplish that, we sample the signal by taking signal values at equally spaced times nT_s, where n is an integer and T_s is the *sampling period*, which is appropriately chosen for this signal (in Chapter 7 we will learn how to chose T_s).

As a result of the sampling, we obtain the discrete-time signal

$$v(nT_s) = v(t)|_{t=nT_s} \qquad 0 \leq n \leq N-1 \tag{1.2}$$

where $T_s = (t_f - t_b)/N$ and we have taken samples at times $t_b + T_s n$. Clearly, this discretization of the time variable reduces the number of values to enter into the computer, but the amplitudes of these samples still can take possibly innumerable values. Now, to represent each of the $v(nT_s)$ values with a certain number of bits, we also discretize the amplitude of the samples. To do so, the dynamic range (the difference between the maximum and the minimum amplitude) of the analog signal is equally divided into a certain number of levels. A sample value falling within one of these levels is allocated a unique binary code. For instance, if we want each sample to be represented by 8 bits we have 2^8 or 256 possible levels. These operations are called *quantization and coding*. The resulting signal is digital, where each sample is represented as a binary number.

Given that many of the signals we encounter in practical applications are analog, if it is desirable to process such signals with a computer, the above procedure is commonly done. The device that converts an analog signal into a digital signal is called an analog-to-digital converter (ADC) and it is characterized by the number of samples it takes per second (sampling rate $1/T_s$) and by the number of bits that it allocates to each sample. To convert a digital signal into an analog signal a digital-to-analog converter (DAC) is used. Such a device inverts the ADC process: binary values are converted into pulses with amplitudes approximating those of the original samples, which are then smoothed out resulting in an analog signal. We will discuss in Chapter 7 how the sampling, binary representation, and reconstruction of an analog signal is done.

Figure 1.1 shows how the discretization of an analog signal in time and amplitude can be understood, while Figure 1.2 illustrates the sampling and quantization of a segment of speech.

A *continuous-time* signal can be thought of as a real-(or complex-) valued function of time:

$$x(.) : \mathcal{R} \to \mathcal{R} \ (\mathcal{C})$$

$$t \qquad x(t) \tag{1.3}$$

FIGURE 1.1

Discretization in time and amplitude of an analog signal. The parameters are the sampling period T_s and the quantization level Δ. In time, samples are taken at uniform times $\{nT_s\}$, and in amplitude the range of amplitudes is divided into a finite number of levels so that each sample value is approximated by them.

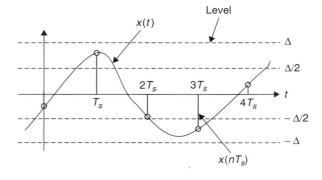

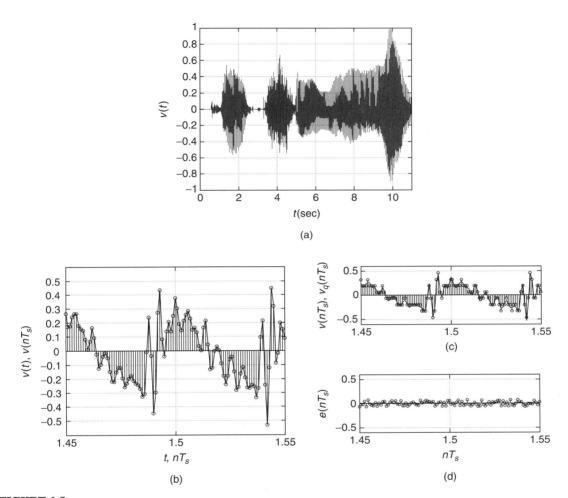

FIGURE 1.2

(a) A segment of this speech signal is sampled and quantized. (b) The speech segment (continuous line) and the sampled signal (vertical samples) using a sampling period $T_s = 10^{-3}$ sec. (c) The sampled and the quantized signal. (d) The quantization error (that is, the difference between the sampled and the quantized signals) is shown.

Thus, the independent variable is time t, and the value of the function at some time t_0, $x(t_0)$, is a real (or a complex) value. (Although in practice signals are real, it is useful in theory to have the option of complex-valued signals.) It is assumed that both time t and signal amplitude $x(t)$ can vary continuously, if needed, from $-\infty$ to ∞.

The term *analog* used for continuous-time signals derives from the similarity of acoustic signals to the pressure variations generated by voice, music, or any other acoustic signal. The terms *continuous-time* and *analog* are used interchangeably for these signals.

■ Example 1.1

Characterize the sinusoidal signal

$$x(t) = \sqrt{2}\cos(\pi t/2 + \pi/4) \qquad -\infty < t < \infty$$

Solution

The signal $x(t)$ is

- Deterministic, as the value of the signal can be obtained for any possible value of t.
- Analog, as there is a continuous variation of the time variable t from $-\infty$ to ∞, and of the amplitude of the signal between $-\sqrt{2}$ to $\sqrt{2}$.
- Of infinite support, as the signal does not become zero outside any finite interval.

The amplitude of the sinusoid is $\sqrt{2}$, its frequency is $\Omega = \pi/2$ (rad/sec), and its phase is $\pi/4$ rad (notice that Ωt has radians as units so that it can be added to the phase). Because of the infinite support, this signal cannot exist in practice, but we will see that sinusoids are extremely important in the representation and processing of signals. ■

■ Example 1.2

A complex signal $y(t)$ is defined as

$$y(t) = (1 + j)e^{j\pi t/2} \qquad 0 \le t \le 10$$

and zero otherwise. Express $y(t)$ in terms of the signal $x(t)$ from Example 1.1. Characterize $y(t)$.

Solution

Since $1 + j = \sqrt{2}e^{j\pi/4}$, then using Euler's identity:

$$y(t) = \sqrt{2}e^{j(\pi t/2 + \pi/4)} = \sqrt{2}\left[\cos(\pi t/2 + \pi/4) + j\sin(\pi t/2 + \pi/4)\right] \qquad 0 \le t \le 10$$

Thus, the real and imaginary parts of this signal are

$$\mathcal{R}e[y(t)] = \sqrt{2}\cos(\pi t/2 + \pi/4)$$
$$\mathcal{I}m[y(t)] = \sqrt{2}\sin(\pi t/2 + \pi/4)$$

for $0 \le t \le 10$ and zero otherwise. The signal $y(t)$ can be written as

$$y(t) = x(t) + jx(t-1) \qquad 0 \le t \le 10$$

and zero otherwise. Notice that

$$x(t-1) = \sqrt{2}\cos(\pi(t-1)/2 + \pi/4) = \sqrt{2}\cos(\pi t/2 - \pi/2 + \pi/4) = \sqrt{2}\sin(\pi t/2 + \pi/4)$$

The signal $y(t)$ is

- Analog of finite support—that is, the signal is zero outside the interval $0 \le t \le 10$.

■ Complex, composed of two sinusoids of frequency $\Omega = \pi/2$ rad/sec, phase $\pi/4$ in rad, and amplitude $\sqrt{2}$ in $0 \le t \le 10$, and it is zero outside that time interval. ■

■ Example 1.3

Consider the pulse signal

$$p(t) = 1 \qquad 0 \le t \le 10$$

and zero elsewhere. Characterize this signal, and use it along with $x(t)$ in Example 1.1, to represent $y(t)$ in the above example.

Solution

The analog signal $p(t)$ is of finite support and real-valued. We have that

$$\mathcal{R}e[y(t)] = x(t)p(t)$$
$$\mathcal{I}m[y(t)] = x(t-1)p(t)$$

so that

$$y(t) = [x(t) + jx(t-1)]p(t)$$

The multiplication by $p(t)$ makes $x(t)p(t)$ and $x(t-1)p(t)$ finite-support signals. This operation is called time windowing as the signal $p(t)$ only allows us to see the values of $x(t)$ wherever $p(t) = 1$, while ignoring the values of $x(t)$ wherever $p(t) = 0$. It acts like a window. ■

Examples 1.1–1.3 not only illustrate how different types of signal can be related to each other, but also how signals can be be defined in shorter or more precise forms. Although the representations for $y(t)$ in Example 1.2 and in this example are equivalent, the one here is shorter and easier to visualize by the use of the pulse $p(t)$.

1.3.1 Basic Signal Operations—Time Shifting and Reversal

The following are basic signal operations used in the representation and processing of signals (for some of these operations we indicate the system that is used to realize the operation):

■ *Signal addition*—Two signals $x(t)$ and $y(t)$ are added to obtain their sum $z(t)$. An *adder* is used.
■ *Constant multiplication*—A signal $x(t)$ is multiplied by a constant α. A *constant multiplier* is used.
■ *Time and frequency shifting*—The signal $x(t)$ is delayed τ seconds to get $x(t-\tau)$, and advanced by τ to get $x(t+\tau)$. A signal can be shifted in frequency or frequency modulated by multiplying it by a complex exponential or a sinusoid. A *delay* shifts right a time signal, while a *modulator* shifts the signal in frequency.
■ *Time scaling*—The time variable of a signal $x(t)$ is scaled by a constant α to give $x(\alpha t)$. If $\alpha = -1$, the signal is reversed in time (i.e., $x(-t)$), or reflected. Only the delay can be implemented in practice.
■ *Time windowing*—A signal $x(t)$ is multiplied by a window signal $w(t)$ so that $x(t)$ is available in the support of $w(t)$.

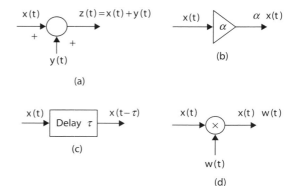

FIGURE 1.3

Diagrams of basic signal operations: (a) adder, (b) constant multiplier, (c) delay, and (d) time windowing or modulation.

Given the simplicity of the first two operations we will only discuss the others. In this section we consider time shifting and reflection (a special case of the time scaling) and leave the rest for a later section.

In Figure 1.3 we show the diagrams used for the implementation of the addition of two signals, the multiplication of a signal by a constant, the delay of a signal, and the time windowing or modulation of a signal. These will be used in the block diagrams for systems in the next chapters.

It is important to understand that advancing or reflecting cannot be implemented in *real time*—that is as the signal is being processed. Delays can be implemented in real time. Advancing and reflection require that the signal be saved or recorded. Thus, an acoustic signal recorded on magnetic tape can be delayed or advanced with respect to an initial time, or played back, faster or slower, but it can only be delayed if we have the signal coming from a live microphone.

We will see later in this chapter that shifting in frequency results in the process of *signal modulation*, which is of great significance in communications. Scaling of the time variable results in a contracted and expanded version of the original signal and causes changes in the frequency content of the signal.

- For a positive value τ, a signal $x(t - \tau)$ is the original signal $x(t)$ shifted right or delayed τ seconds, as illustrated in Figure 1.4(b). That the original signal has been shifted to the right can be verified by finding that the $x(0)$ value of the original signal appears in the delayed signal at $t = \tau$ (which results from making $t - \tau = 0$).
- Likewise, a signal $x(t + \tau)$ is the original signal $x(t)$ shifted left or advanced by τ seconds as illustrated in Figure 1.4(c). The original signal is now shifted to the left—that is, the value $x(0)$ of the original signal occurs now earlier (i.e., it has been advanced) at time $t = -\tau$.
- *Reflection* consists in negating the time variable. Thus, the reflection of $x(t)$ is $x(-t)$. This operation can be visualized as flipping the signal about the origin. See Figure 1.4(d).

> Given an analog signal $x(t)$ and $\tau > 0$ we have that with respect to $x(t)$:
> **(a)** $x(t - \tau)$ is *delayed* or *shifted right* τ seconds.
> **(b)** $x(t + \tau)$ is *advanced* or *shifted left* τ seconds.

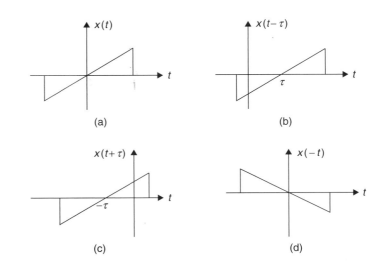

FIGURE 1.4
(a) Continuous-time signal, and its
(b) delayed, (c) advanced, and
(d) reflected versions.

(c) $x(-t)$ is *reflected*.
(d) $x(-t - \tau)$ is *reflected and shifted left* τ *seconds*, while $x(-t + \tau)$ is *reflected and shifted right* τ *seconds*.

Remarks *Whenever we combine the delaying or advancing with reflection, delaying and advancing are swapped. Thus, $x(-t + 1)$ is $x(t)$ reflected and delayed, or shifted to the right, by 1. Likewise, $x(-t - 1)$ is $x(t)$ reflected and advanced, or shifted to the left by 1. Again, the value $x(0)$ of the original signal is found in $x(-t + 1)$ at $t = 1$, and in $x(-t - 1)$ at $t = -1$.*

■ Example 1.4

Consider an analog pulse

$$x(t) = \begin{cases} 1 & 0 \le t \le 1 \\ 0 & \text{otherwise} \end{cases}$$

Find mathematical expressions for $x(t)$ delayed by 2, advanced by 2, and the reflected signal $x(-t)$.

Solution

The delayed signal $x(t - 2)$ can be found mathematically by replacing the variable t by $t - 2$ so that

$$x(t - 2) = \begin{cases} 1 & 0 \le t - 2 \le 1 \text{ or } 2 \le t \le 3 \\ 0 & \text{otherwise} \end{cases}$$

The value $x(0)$ (which in $x(t)$ occurs at $t = 0$) in $x(t - 2)$ now occurs when $t = 2$, so that the signal $x(t)$ has been shifted to the right two units of time, and since the values are occurring later, the signal $x(t - 2)$ is said to be "delayed" by 2 with respect to $x(t)$.

Likewise, we have that

$$x(t+2) = \begin{cases} 1 & 0 \leq t+2 \leq 1 \text{ or } -2 \leq t \leq -1 \\ 0 & \text{otherwise} \end{cases}$$

The signal $x(t+2)$ can be seen to be the advanced version of $x(t)$, as it is this signal shifted to the left by two units of time. The value $x(0)$ for $x(t+2)$ now occurs at $t = -2$, which is ahead of $t = 0$.

Finally, the signal $x(-t)$ is given by

$$x(-t) = \begin{cases} 1 & 0 \leq -t \leq 1 \text{ or } -1 \leq t \leq 0 \\ 0 & \text{otherwise} \end{cases}$$

This signal is a mirror image of the original: the value $x(0)$ still occurs at the same time, but $x(1)$ occurs when $t = -1$. ∎

■ Example 1.5

When the shifting and reflecting operations are considered together the best approach to visualize the operation is to make a table computing several values of the new signal and comparing these with those from the original signal. Consider the pulse in Example 1.4, and plot the signal $x(-t+2)$.

Solution

Although one can see that this signal is reflected, it is not clear whether it is advanced or delayed by 2. By computing a few values:

t	$x(-t+2)$
2	$x(0) = 1$
1.5	$x(0.5) = 1$
1	$x(1) = 1$
0	$x(2) = 0$
−1	$x(3) = 0$

it becomes clear that $x(-t+2)$ is reflected and "delayed" by 2. In fact, as indicated above, whenever the signal is a function of $-t$ (i.e., reflected), the $-t+\tau$ operation becomes reflection and "delay," and $-t-\tau$ becomes reflection and "advancing." ∎

Remarks *When computing the convolution integral later on, we will consider the signal $x(t-\tau)$ as a function of τ for different values of t. As indicated from Example 1.5, this signal is a reflected version of $x(\tau)$ being shifted to the right t seconds. To see this, consider $t = 0$ then $x(t-\tau)|_{t=0} = x(-\tau)$, the reflected version, and $x(0)$ occurs at $\tau = 0$. When $t = 1$, then $x(t-\tau)|_{t=1} = x(1-\tau)$ and $x(0)$ occurs at $\tau = 1$, so that $x(1-\tau)$ is $x(-\tau)$ shifted to the right by 1, and so on.*

1.3.2 Even and Odd Signals

Symmetry with respect to the origin differentiates signals and will be useful in their Fourier analysis. We have that an analog signal $x(t)$ is called

- *Even* whenever $x(t)$ coincides with its reflection $x(-t)$. Such a signal is symmetric with respect to the time origin.
- *Odd* whenever $x(t)$ coincides with $-x(-t)$—that is, the negative of its reflection. Such a signal is asymmetric with respect to the time origin.

Even and odd signals are defined as follows:

$$x(t) \quad \text{even}: \quad x(t) = x(-t) \tag{1.4}$$

$$x(t) \quad \text{odd}: \quad x(t) = -x(-t) \tag{1.5}$$

Even and odd decomposition: Any signal $y(t)$ is representable as a sum of an even component $y_e(t)$ and an odd component $y_o(t)$:

$$y(t) = y_e(t) + y_o(t) \tag{1.6}$$

where

$$y_e(t) = 0.5 \left[y(t) + y(-t) \right] \tag{1.7}$$

$$y_o(t) = 0.5 \left[y(t) - y(-t) \right] \tag{1.8}$$

Using the definitions of even and odd signals, any signal $y(t)$ can be decomposed into the sum of an even and an odd function. Indeed, the following is an identity:

$$y(t) = \frac{1}{2} \left[y(t) + y(-t) \right] + \frac{1}{2} \left[y(t) - y(-t) \right]$$

where the first term is the even and the second is the odd components of $y(t)$. It can be easily verified that $y_e(t)$ is even and that $y_o(t)$ is odd.

■ Example 1.6

Consider the analog signal

$$x(t) = \cos(2\pi t + \theta) \qquad -\infty < t < \infty$$

Determine the value of θ for which $x(t)$ is even and odd. If $\theta = \pi/4$, is $x(t) = \cos(2\pi t + \pi/4)$, $-\infty < t < \infty$, even or odd?

Solution

The reflection of $x(t)$ is $x(-t) = \cos(-2\pi t + \theta)$. Then:

1. $x(t)$ is even if $x(t) = x(-t)$ or

$$\cos(2\pi t + \theta) = \cos(-2\pi t + \theta)$$
$$= \cos(2\pi t - \theta)$$

or $\theta = -\theta$ or $\theta = 0, \pi$. Thus, $x_1(t) = \cos(2\pi t)$ as well as $x_2(t) = \cos(2\pi t + \pi) = -\cos(2\pi t)$ are even.

2. for $x(t)$ to be odd, we need that $x(t) = -x(-t)$ or

$$\cos(2\pi t + \theta) = -\cos(-2\pi t + \theta) = \cos(-2\pi t + \theta \pm \pi) = \cos(2\pi t - \theta \mp \pi)$$

which can be obtained with $\theta = -\theta \mp \pi$ or $\theta = \mp\pi/2$. Indeed, $\cos(2\pi t - \pi/2) = \sin(2\pi t)$ and $\cos(2\pi t + \pi/2) = -\sin(2\pi t)$ are both odd. Thus, $x_3(t) = \pm \sin(2\pi t)$ is odd.

When $\theta = \pi/4$, $x(t) = \cos(2\pi t + \pi/4)$ is neither even nor odd according to the above.   ■

■ Example 1.7

Consider the signal

$$x(t) = \begin{cases} 2\cos(4t) & t > 0 \\ 0 & \text{otherwise} \end{cases}$$

Find its even and odd decomposition. What would happen if $x(0) = 2$ instead of 0—that is, when we define the sinusoid at $t = 0$? Explain.

Solution

The signal $x(t)$ is neither even nor odd given that its values for $t \leq 0$ are zero. For its even–odd decomposition, the even component is given by

$$x_e(t) = 0.5[x(t) + x(-t)]$$

$$= \begin{cases} \cos(4t) & t > 0 \\ \cos(4t) & t < 0 \\ 0 & t = 0 \end{cases}$$

and the odd component is given by

$$x_o(t) = 0.5[x(t) - x(-t)]$$

$$= \begin{cases} \cos(4t) & t > 0 \\ -\cos(4t) & t < 0 \\ 0 & t = 0 \end{cases}$$

which when added together become the given signal.

If $x(0) = 2$, we have

$$x_e(t) = 0.5[x(t) + x(-t)]$$

$$= \begin{cases} \cos(4t) & t > 0 \\ \cos(4t) & t < 0 \\ 2 & t = 0 \end{cases}$$

while the odd component is the same. The even component has a discontinuity at $t = 0$. ■

1.3.3 Periodic and Aperiodic Signals

A useful characterization of signals is whether they are *periodic* or *aperiodic* (nonperiodic).

An analog signal $x(t)$ is periodic if
- it is defined for all possible values of t, $-\infty < t < \infty$, and
- there is a positive real value T_0, the *period* of $x(t)$, such that

$$x(t + kT_0) = x(t) \tag{1.9}$$

for any integer k.

The period of $x(t)$ is the smallest possible value of $T_0 > 0$ that makes the periodicity possible. Thus, although NT_0 for an integer $N > 1$ is also a period of $x(t)$ it should not be considered the period.

Remarks

- *The infinite support and the unique characteristic of the period make periodic signals nonexistent in practical applications. Despite this, periodic signals are of great significance in the Fourier representation of signals and in their processing, as we will see later. The representation of aperiodic signals is obtained from that of periodic signals, and the response of systems to periodic sinusoids is fundamental in the theory of linear systems.*
- *Although seemingly redundant, the first part of the definition of a periodic signal indicates that it is not possible to have a nonzero periodic signal with a finite support (i.e., the analog signal is zero outside an interval $t \in [t_1, t_2]$). This first part of the definition is needed for the second part to make sense.*
- *It is exasperating to find the period of a constant signal $x(t) = A$; visually $x(t)$ is periodic but its period is not clear. Any positive value could be considered the period, but none will be taken. The reason is that $x(t) = A = A\cos(0t)$ or of zero frequency, and as such its period is not determined since we would have to divide by zero—not permitted. Thus, a constant signal is a periodic signal of nondefinable period!*

■ Example 1.8

Consider the analog sinusoid

$$x(t) = A\cos(\Omega_0 t + \theta) \qquad -\infty < t < \infty$$

Determine the period of this signal, and indicate for what frequency Ω_0 the period of $x(t)$ is not clearly defined.

Solution

The analog frequency is $\Omega_0 = 2\pi/T_0$ so $T_0 = 2\pi/\Omega_0$ is the period. Whenever $T_0 > 0$ (or $\Omega_0 > 0$) these sinusoidals are periodic. For instance, consider

$$x(t) = 2\cos(2t - \pi/2) \qquad -\infty < t < \infty$$

Its period is found by noticing that this signal has an analog frequency $\Omega_0 = 2 = 2\pi f_0$ (rad/sec), or a hertz frequency of $f_0 = 1/\pi = 1/T_0$, so that $T_0 = \pi$ is the period in seconds. That this is the period can be seen for an integer N,

$$x(t + NT_0) = 2\cos(2(t + NT_0) - \pi/2) = 2\cos(2t + 2\pi N - \pi/2)$$
$$= 2\cos(2t - \pi/2) = x(t)$$

since adding $2\pi N$ (a multiple of 2π) to the angle of the cosine gives the original angle. If $\Omega_0 = 0$— that is, dc frequency—the period cannot be defined because of the division by zero when finding $T_0 = 2\pi/\Omega_0$.   ∎

■ Example 1.9

Consider a periodic signal $x(t)$ of period T_0. Determine whether the following signals are periodic, and if so, find their corresponding periods:

(a) $y(t) = A + x(t)$.
(b) $z(t) = x(t) + v(t)$ where $v(t)$ is periodic of period $T_1 = NT_0$, where N is a positive integer.
(c) $w(t) = x(t) + u(t)$ where $u(t)$ is periodic of period T_1, not necessarily a multiple of T_0. Determine under what conditions $w(t)$ could be periodic.

Solution

(a) Adding a constant to a periodic signal does not change the periodicity, so $y(t)$ is periodic of period T_0—that is, for an integer k, $y(t + kT_0) = A + x(t + kT_0) = A + x(t)$ since $x(t)$ is periodic of period T_0.

(b) The period $T_1 = NT_0$ of $v(t)$ is also a period of $x(t)$, and so $z(t)$ is periodic of period T_1 since for any integer k,

$$z(t + kT_1) = x(t + kT_1) + v(t + kT_1) = x(t + kNT_0) + v(t) = x(t) + v(t)$$

given that $v(t + kT_1) = v(t)$, and that kN is an integer so that $x(t + kNT_0) = x(t)$. The periodicity can be visualized by considering that in one period of $v(t)$ we can place N periods of $x(t)$.

(c) The condition for $w(t)$ to be periodic is that the ratio of the periods of $x(t)$ and of $u(t)$ be

$$\frac{T_1}{T_0} = \frac{N}{M}$$

where N and M are positive integers not divisible by each other so that $MT_1 = NT_0$ becomes the period of $w(t)$. That is,

$$w(t + MT_1) = x(t + MT_1) + u(t + MT_1) = x(t + NT_0) + u(t + MT_1) = x(t) + u(t)$$

■

■ Example 1.10

Let $x(t) = e^{j2t}$ and $y(t) = e^{j\pi t}$, and consider their sum $z(t) = x(t) + y(t)$, and their product $w(t) = x(t)y(t)$. Determine if $z(t)$ and $w(t)$ are periodic, and if so, find their periods. Is $p(t) = (1 + x(t))(1 + y(t))$ periodic?

Solution

According to Euler's identity,

$$x(t) = \cos(2t) + j\sin(2t)$$

$$y(t) = \cos(\pi t) + j\sin(\pi t)$$

indicating $x(t)$ is periodic of period $T_0 = \pi$ (the frequency of $x(t)$ is $\Omega_0 = 2 = 2\pi/T_0$) and $y(t)$ is periodic of period $T_1 = 2$ (the frequency of $y(t)$ is $\Omega_1 = \pi = 2\pi/T_1$).

For $z(t)$ to be periodic requires that T_1/T_0 be a rational number, which is not the case as $T_1/T_0 = 2/\pi$. So $z(t)$ is not periodic.

The product is $w(t) = x(t)y(t) = e^{j(2+\pi)t} = \cos(\Omega_2 t) + j\sin(\Omega_2 t)$ where $\Omega_2 = 2 + \pi = 2\pi/T_2$ so that $T_2 = 2\pi/(2 + \pi)$, so $w(t)$ is periodic of period T_2.

The terms $1 + x(t)$ and $1 + y(t)$ are periodic of period $T_0 = \pi$ and $T_1 = 2$, and from the case of the product above, one would hope this product be periodic. But since $p(t) = 1 + x(t) + y(t) + x(t)y(t)$ and $x(t) + y(t)$ is not periodic, then $p(t)$ is not periodic. ■

■ Analog sinusoids of frequency $\Omega_0 > 0$ are periodic of period $T_0 = 2\pi/\Omega_0$. If $\Omega_0 = 0$, the period is not well defined.

■ The sum of two periodic signals $x(t)$ and $y(t)$, of periods T_1 and T_2, is periodic if the ratio of the periods T_1/T_2 is a rational number N/M, with N and M being nondivisible. The period of the sum is $MT_1 = NT_2$.

■ The product of two sinusoids is periodic. The product of two periodic signals is not necessarily periodic.

1.3.4 Finite-Energy and Finite Power Signals

Another possible classification of signals is based on their energy and power. The concepts of energy and power introduced in circuit theory can be extended to any signal. Recall that for a resistor of unit resistance its *instantaneous power* is given by

$$p(t) = v(t)i(t) = i^2(t) = v^2(t)$$

where $i(t)$ and $v(t)$ are the current and voltage in the resistor. The *energy* in the resistor for an interval $[t_0, t_1]$, of duration $T = t_1 - t_0$, is the accumulation of instantaneous power over that time interval,

$$E_T = \int_{t_0}^{t_1} p(t)dt = \int_{t_0}^{t_1} i^2(t)dt = \int_{t_0}^{t_1} v^2(t)dt$$

The *power* in the interval $T = t_1 - t_0$ is the average energy

$$P_T = \frac{E_T}{T} = \frac{1}{T}\int_{t_0}^{t_1} i^2(t)dt = \frac{1}{T}\int_{t_0}^{t_1} v^2(t)dt$$

corresponding to the heat dissipated by the resistor (and for which you pay the electric company). The energy and power concepts can thus be easily generalized.

The *energy* and the *power* of an analog signal $x(t)$ are defined for either finite or infinite-support signals as:

$$E_x = \int_{-\infty}^{\infty} |x(t)|^2 dt \qquad (1.10)$$

$$P_x = \lim_{T \to \infty} \frac{1}{2T} \int_{-T}^{T} |x(t)|^2 dt \qquad (1.11)$$

The signal $x(t)$ is then said to be finite energy, or square integrable, whenever

$$E_x < \infty \qquad (1.12)$$

The signal is said to have finite power if

$$P_x < \infty \qquad (1.13)$$

Remarks

- *The above definitions of energy and power are valid for any signal of finite or infinite support, since a finite-support signal is zero outside its support.*
- *In the formulas for energy and power we are considering the possibility that the signals might be complex and so we are squaring its magnitude: If the signal being considered is real, this simply is equivalent to squaring the signal.*
- *According to the above definitions, a finite-energy signal has zero power. Indeed, if the energy of the signal is some constant $E_x < \infty$, then*

$$P_x = \lim_{T \to \infty} \frac{E_x}{2T} = 0$$

■ *An analog signal $x(t)$ is said to be* absolutely integrable *if $x(t)$ satisfies the condition*

$$\int_{-\infty}^{\infty} |x(t)|dt < \infty \tag{1.14}$$

■ Example 1.11

Find the energy and the power of the following:

(a) The periodic signal $x(t) = \cos(\pi t/2 + \pi/4)$.
(b) The complex signal $y(t) = (1 + j)e^{j\pi t/2}$, for $0 \le t \le 10$ and zero otherwise.
(c) The pulse $z(t) = 1$, for $0 \le t \le 10$ and zero otherwise.

Determine whether these signals are finite energy, finite power, or both.

Solution

The energy in these signals is computed as follows:

$$E_x = \int_{-\infty}^{\infty} \cos^2(\pi t/2 + \pi/4)dt \to \infty$$

$$E_y = \int_{0}^{10} |(1 + j)e^{j\pi t/2}|^2 dt = 2\int_{0}^{10} dt = 20$$

$$E_z = \int_{0}^{10} dt = 10$$

where we used $|(1 + j)e^{j\pi t/2}|^2 = |1 + j|^2 |e^{j\pi t/2}|^2 = |1 + j|^2 = 2$. Thus, $x(t)$ is an infinite-energy signal while $y(t)$ and $z(t)$ are finite-energy signals. The power of $y(t)$ and $z(t)$ are zero because they have finite energy. The power of $x(t)$ can be calculated by using the symmetry of the signal squared and letting $T = NT_0$:

$$P_x = \lim_{T\to\infty} \frac{2}{2T} \int_{0}^{T} \cos^2(\pi t/2 + \pi/4)dt = \lim_{N\to\infty} \frac{1}{NT_0} \int_{0}^{NT_0} \cos^2(\pi t/2 + \pi/4)dt$$

$$= \lim_{N\to\infty} \frac{1}{NT_0} \left[N \int_{0}^{T_0} \cos^2(\pi t/2 + \pi/4)dt \right] = \frac{1}{T_0} \int_{0}^{T_0} \cos^2(\pi t/2 + \pi/4)dt$$

Using the trigonometric identity

$$\cos^2(\pi t/2 + \pi/4) = \frac{1}{2}[\cos(\pi t + \pi/2) + 1]$$

we have that

$$P_x = \frac{1}{8} \int_0^4 \cos(\pi t + \pi/2)dt + \frac{1}{8} \int_0^4 dt = 0 + 0.5 = 0.5$$

The first integral is the area of the sinusoid over two of its periods, thus zero. So we have that $x(t)$ is a finite-power but infinite-energy signal, while $y(t)$ and $z(t)$ are finite-power and finite-energy signals. ∎

■ Example 1.12

Consider an aperiodic signal $x(t) = e^{-at}$, $a > 0$, for $t \geq 0$ and zero otherwise. Find the energy and the power of this signal and determine whether the signal is finite energy, finite power, or both.

Solution

The energy of $x(t)$ is given by

$$E_x = \int_0^\infty e^{-2at}dt = \frac{1}{2a} < \infty$$

for any value of $a > 0$. The power of $x(t)$ is then zero. Thus, $x(t)$ is a finite-energy and finite-power signal. ∎

■ Example 1.13

Consider the following analog signal, which we call a *causal* sinusoid because it is zero for $t < 0$:

$$x(t) = \begin{cases} 2\cos(4t - \pi/4) & t \geq 0 \\ 0 & \text{otherwise} \end{cases}$$

This is the kind of signal that you would get from a signal generator that is started at a certain initial time (in this case 0) and that continues until the signal generator is switched off (in this case possibly infinity). Determine if this signal is finite energy, finite power or both.

Solution

Clearly, the analog signal $x(t)$ has infinite energy:

$$E_x = \int_{-\infty}^\infty x^2(t)dt$$

$$= \int_0^\infty 4\cos^2(4t - \pi/4)dt \to \infty$$

Although this signal has infinite energy, it has finite power. Letting $T = NT_0$ where T_0 is the period of $2\cos(4t - \pi/4)$ (or $T_0 = 2\pi/4$), then its power is

$$P_x = \lim_{T\to\infty} \frac{1}{2T} \int_{-T}^{T} x^2(t)dt = \lim_{T\to\infty} \frac{1}{2T} \int_{0}^{T} x^2(t)dt$$

$$= \lim_{N\to\infty} \frac{N}{2NT_0} \int_{0}^{T_0} x^2(t)dt = \frac{1}{2T_0} \int_{0}^{T_0} 4\cos^2(4t - \pi/4)dt$$

which is a finite value and therefore the signal has finite power but infinite energy. ■

As we will see later in the Fourier series representation, any periodic signal is representable as a possibly infinite sum of sinusoids of frequencies multiples of the fundamental frequency of the periodic signal being represented. These frequencies are said to be *harmonically related*, and for this case the power of the signal is shown to be the sum of the power of each of the sinusoidal components—that is, there is superposition of the power. This superposition is still possible when a sum of sinusoids creates a nonperiodic signal. This is illustrated in Example 1.14.

■ Example 1.14

Consider the signals $x(t) = \cos(2\pi t) + \cos(4\pi t)$ and $y(t) = \cos(2\pi t) + \cos(2t)$, $-\infty < t < \infty$. Determine if these signals are periodic, and if so, find their periods. Compute the power of these signals.

Solution

The sinusoids $\cos(2\pi t)$ and $\cos(4\pi t)$ periods $T_1 = 1$ and $T_2 = 1/2$, so $x(t)$ is periodic since $T_1/T_2 = 2$ with period $T_1 = 2T_2 = 1$. The two frequencies are harmonically related. The sinusoid $\cos(2t)$ has as period $T_3 = \pi$. Therefore, the ratio of the periods of the sinusoidal components of $y(t)$ is $T_1/T_3 = 1/\pi$, which is not rational, and so $y(t)$ is not periodic and the frequencies 2π and 2 are not harmonically related.

Using the trigonometric identities

$$\cos^2(\theta) = \frac{1}{2}(1 + \cos(2\theta))$$

$$\cos(\alpha)\cos(\beta) = \frac{1}{2}(\cos(\alpha + \beta) + \cos(\alpha - \beta))$$

we have that

$$x^2(t) = \cos^2(2\pi t) + \cos^2(4\pi t) + 2\cos(2\pi t)\cos(4\pi t)$$

$$= 1 + \frac{1}{2}\cos(4\pi t) + \frac{1}{2}\cos(8\pi t) + \cos(6\pi t) + \cos(2\pi t)$$

which is again a sum of harmonically related frequency sinusoids, so that $x^2(t)$ is periodic of period $T_0 = 1$. As in the previous examples, we have

$$P_x = \frac{1}{T_0} \int_0^{T_0} x^2(t)dt = 1$$

which is the integral of the constant since the other integrals are zero. In this case, we used the periodicity of $x(t)$ and $x^2(t)$ to calculate the power directly. That is not possible when computing the power of $y(t)$ because it is not periodic, so we have to consider each of its components. We have that

$$y^2(t) = \cos^2(2\pi t) + \cos^2(2t) + 2\cos(2\pi t)\cos(2t)$$

$$= 1 + \frac{1}{2}\cos(4\pi t) + \frac{1}{2}\cos(4t) + \cos(2(\pi + 1)t) + \cos(2(\pi - 1)t)$$

and the power of $y(t)$ is

$$P_y = \lim_{T \to \infty} \frac{1}{2T} \int_{-T}^{T} y^2(t)dt$$

$$= 1 + \frac{1}{2T_4} \int_0^{T_4} \cos(4\pi t)dt + \frac{1}{2T_5} \int_0^{T_5} \cos(4t)dt$$

$$+ \frac{1}{T_6} \int_0^{T_6} \cos(2(\pi + 1)t)dt + \frac{1}{T_7} \int_0^{T_7} \cos(2(\pi - 1)t)dt = 1$$

where T_4, T_5, T_6, and T_7 are the periods of the sinusoidal components of $y^2(t)$. Fortunately, only the first integral is not zero and the others are zero (the average over a period of the sinusoidal components of $y^2(t)$). Fortunately, too, we have that the power of $x(t)$ and the power of $y(t)$ are the sum of the powers of its components. That is if

$$x(t) = \cos(2\pi t) + \cos(4\pi t) = x_1(t) + x_2(t)$$

$$y(t) = \cos(2\pi t) + \cos(2t) = y_1(t) + y_2(t)$$

then as in previous examples $P_{x_1} = P_{x_2} = P_{y_1} = P_{y_2} = 0.5$, so that

$$P_x = P_{x_1} + P_{x_2} = 1$$

$$P_y = P_{y_1} + P_{y_2} = 1$$

∎

The power of a sum of sinusoids,

$$x(t) = \sum_k A_k \cos(\Omega_k t) = \sum_k x_k(t) \qquad (1.15)$$

with harmonically or nonharmonically related frequencies $\{\Omega_k\}$, is the sum of the power of each of the ~sinusoidal components,

$$P_X = \sum_k P_{x_k} \qquad (1.16)$$

1.4 REPRESENTATION USING BASIC SIGNALS

A fundamental idea in signal processing is to attempt to represent signals in terms of basic signals, which we know how to process. In this section we consider some of these basic signals (complex exponentials, sinusoids, impulse, unit-step, and ramp) that will be used to represent signals and for which we will obtain their responses in a simple way in the next chapter.

1.4.1 Complex Exponentials

A complex exponential is a signal of the form

$$x(t) = Ae^{at}$$
$$= |A|e^{rt}\left[\cos(\Omega_0 t + \theta) + j\sin(\Omega_0 t + \theta)\right] \quad -\infty < t < \infty \qquad (1.17)$$

where $A = |A|e^{j\theta}$, and $a = r + j\Omega_0$ are complex numbers.

Using Euler's identity, $e^{j\phi} = \cos(\phi) + j\sin(\phi)$, and from the definitions of A and a as complex numbers, we have that

$$x(t) = |A|e^{j\theta}e^{(r+j\Omega_0)t} = |A|e^{rt}e^{(j\Omega_0 t + \theta)}$$
$$= |A|e^{rt}\left[\cos(\Omega_0 t + \theta) + j\sin(\Omega_0 t + \theta)\right]$$

We will see later that complex exponentials are fundamental in the Fourier representation of signals.

Remarks

- *Suppose that A and a are real, then*

$$x(t) = Ae^{at} \qquad -\infty < t < \infty$$

is a decaying exponential if $a < 0$, and a growing exponential if $a > 0$. See Figure 1.5.

- If A is real, but $a = j\Omega_0$, then we have

$$x(t) = Ae^{j\Omega_0 t}$$

$$= A\cos(\Omega_0 t) + jA\sin(\Omega_0 t)$$

where the real part of $x(t)$ is $\mathcal{R}e[x(t)] = A\cos(\Omega_0 t)$ and the imaginary part of $x(t)$ is $\mathcal{I}m[x(t)] = A\sin(\Omega_0 t)$, and $j = \sqrt{-1}$.

- If both A and a are complex, $x(t)$ is a complex signal and we need to consider separately its real and imaginary parts. For instance, the real part function is

$$g(t) = \mathcal{R}e[x(t)]$$

$$= |A|e^{rt}\cos(\Omega_0 t + \theta)$$

The envelope of $g(t)$ can be found by considering that

$$-1 \le \cos(\Omega_0 t + \theta) \le 1$$

and that when multiplied by $|A|e^{rt} > 0$, we have

$$-|A|e^{rt} \le |A|e^{rt}\cos(\Omega_0 t + \theta) \le |A|e^{rt}$$

so that

$$-|A|e^{rt} \le g(t) \le |A|e^{rt}$$

Whenever $r < 0$ the $g(t)$ signal is a damped sinusoid, and when $r > 0$ then $g(t)$ grows, as illustrated in Figure 1.5.

- According to the above, several signals can be obtained from the complex exponential.

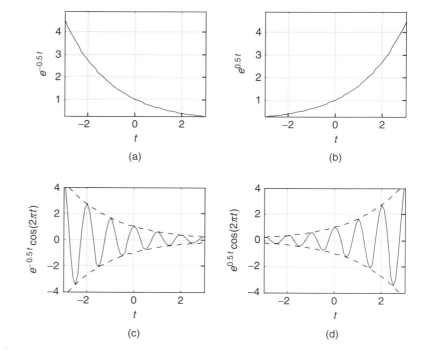

FIGURE 1.5

Analog exponentials:
(a) decaying exponential,
(b) growing exponential, and
(c–d) modulated exponential
(c) decaying and (d) growing.

Sinusoids

Sinusoids are of the general form

$$A\cos(\Omega_0 t + \theta) = A\sin(\Omega_0 t + \theta + \pi/2) \qquad -\infty < t < \infty \qquad (1.18)$$

where A is the amplitude of the sinusoid, $\Omega_0 = 2\pi f_0 (\text{rad/sec})$ is the frequency, and θ is a phase shift. The frequency and time variables are inversely related, as follows:

$$\Omega_0 = 2\pi f_0 = \frac{2\pi}{T_0}$$

The cosine and the sine signals, as indicated above, are out of phase by $\pi/2$ radians. The frequency can also be expressed in hertz or 1/sec units, and in that case $\Omega_0 = 2\pi f_0$, and the period is found by the relation $f_0 = 1/T_0$ (it is important to point out the inverse relation between time and frequency shown here, which will be important in the representation of signals later on).

Recall from Chapter 0, that the Euler's identity provides the relation of the sinusoids with the complex exponential

$$e^{j\Omega_0 t} = \cos(\Omega_0 t) + j\sin(\Omega_0 t) \qquad (1.19)$$

that will allow us to represent in terms of sines and cosines any signal that is represented in terms of complex exponentials. Likewise, the Euler's identity also permits us to represent sines and cosines in terms of complex exponentials, since

$$\cos(\Omega_0 t) = \frac{1}{2}\left(e^{j\Omega_0 t} + e^{-j\Omega_0 t}\right) \qquad (1.20)$$

$$\sin(\Omega_0 t) = \frac{1}{2j}\left(e^{j\Omega_0 t} - e^{-j\Omega_0 t}\right) \qquad (1.21)$$

Remarks *A sinusoid is characterized by its amplitude, frequency, and phase. When we allow these three parameters to be functions of time, or*

$$A(t)\cos(\Omega(t)t + \theta(t))$$

the following different types of modulation systems in communications are obtained:

- Amplitude modulation (AM)—*The amplitude $A(t)$ changes according to the message, while the frequency and the phase are constant,*
- Frequency modulation (FM)—*The frequency $\Omega(t)$ changes according to the message, while the amplitude and phase are constant,*
- Phase modulation (PM)—*The phase $\theta(t)$ varies according to the message and the other parameters are kept constant.*

1.4.2 Unit-Step, Unit-Impulse, and Ram Signals
Unit-Step and Unit-Impulse Signals

Consider a rectangular pulse of duration Δ and unit area

$$p_\Delta(t) = \begin{cases} \dfrac{1}{\Delta} & -\Delta/2 \leq t \leq \Delta/2 \\ 0 & t < -\Delta/2 \text{ and } t > \Delta/2 \end{cases} \tag{1.22}$$

Its integral is

$$u_\Delta(t) = \int_{-\infty}^{t} p_\Delta(\tau)d\tau = \begin{cases} 1 & t > \Delta/2 \\ \dfrac{1}{\Delta}\left(t + \dfrac{\Delta}{2}\right) & -\Delta/2 \leq t \leq \Delta/2 \\ 0 & t < -\Delta/2 \end{cases} \tag{1.23}$$

The pulse $p_\Delta(t)$ and its integral $u_\Delta(t)$ are shown in Figure 1.6.

Suppose that $\Delta \to 0$, then

- The pulse $p_\Delta(t)$ still has a unit area but is an extremely narrow pulse. We will call the limit the *unit-impulse* signal,

$$\delta(t) = \lim_{\Delta \to 0} p_\Delta(t) \tag{1.24}$$

  which is zero for all values of t except at $t = 0$ when its value is not defined.
- The integral $u_\Delta(t)$, as $\Delta \to 0$ has a left-side limit of $u_\Delta(-\epsilon) \to 0$ and a right-side limit of $u_\Delta(\epsilon) \to 1$, for some infinitesimal $\epsilon > 0$, and at $t = 0$ it is $1/2$. Thus, the limit is

$$\lim_{\Delta \to 0} u_\Delta(t) = \begin{cases} 1 & t > 0 \\ 1/2 & t = 0 \\ 0 & t < 0 \end{cases} \tag{1.25}$$

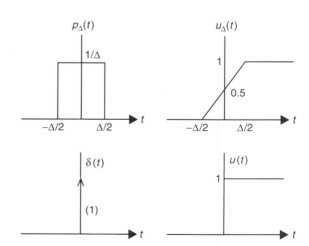

FIGURE 1.6
Generation of $\delta(t)$ and $u(t)$ from limit as $\Delta \to 0$ of a pulse $p_\Delta(t)$ and its integral $u_\Delta(t)$.

Ignoring the value at $t = 0$ we define the *unit-step* signal as

$$u(t) = \begin{cases} 1 & t > 0 \\ 0 & t < 0 \end{cases}$$

You can think of the $u(t)$ as the switching of a dc signal generator from off to on, while $\delta(t)$ is a very strong pulse of very short duration.

The impulse *signal* $\delta(t)$ is:

■ Zero everywhere except at the origin where its value is not well defined (i.e., $\delta(t) = 0, t \neq 0$, and undefined at $t = 0$).

■ its area is unity, i.e.,

$$\int_{-\infty}^{t} \delta(\tau)d\tau = \begin{cases} 1 & t > 0 \\ 0 & t < 0. \end{cases} \tag{1.26}$$

The unit-step signal is

$$u(t) = \begin{cases} 1 & t > 0 \\ 0 & t < 0 \end{cases}$$

The $\delta(t)$ and $u(t)$ are related as follows:

$$u(t) = \int_{-\infty}^{t} \delta(\tau)d\tau \tag{1.27}$$

$$\delta(t) = \frac{du(t)}{dt} \tag{1.28}$$

According to calculus we have

$$u_\Delta(t) = \int_{-\infty}^{t} p_\Delta(\tau)d\tau$$

$$p_\Delta(t) = \frac{du_\Delta(t)}{dt}$$

and so letting $\Delta \to 0$, we obtain the relation between $u(t)$ and $\delta(t)$.

Remarks

■ *Since $u(t)$ is not a continuous function, it jumps from 0 to 1 instantaneously around $t = 0$, from the calculus point of view it should not have a derivative. That $\delta(t)$ is its derivative must be taken with suspicion, which makes the $\delta(t)$ signal also suspicious. Such signals can, however, be formally defined using the theory of distributions.*

■   *The impulse $\delta(t)$ is impossible to generate physically, but characterizes very brief pulses of any shape. It can be derived using pulses or functions different from the rectangular pulse (see Eq. 1.22). In Problem 1.7 at the end of the chapter it is indicated how it can be derived from either a triangular pulse or a sinc function of unit area.*

■   *Signals with jump discontinuities can be represented as the sum of a continuous signal and unit-step signals at the discontinuities. This is useful in computing the derivative of these signals.*

Ramp Signal

The ramp signal is defined as

$$r(t) = t\, u(t) \tag{1.29}$$

Its relation to the unit-step and the unit-impulse signals is

$$\frac{dr(t)}{dt} = u(t) \tag{1.30}$$

$$\frac{d^2 r(t)}{dt^2} = \delta(t) \tag{1.31}$$

The ramp is a continuous function and its derivative is given by

$$\frac{dr(t)}{dt} = \frac{dt u(t)}{dt} = u(t) + t\frac{du(t)}{dt} = u(t) + t\,\delta(t)$$

$$= u(t) + 0\,\delta(t) = u(t)$$

■ Example 1.15

Consider the discontinuous signals

$$x_1(t) = \cos(2\pi t)[u(t) - u(t-1)]$$

$$x_2(t) = u(t) - 2u(t-1) + u(t-2)$$

Represent each of these signals as the sum of a continuous signal and unit-step signals, and find their derivatives.

Solution

The signal $x_1(t)$ is a period of a cosine of period $T_0 = 1$, $0 \le t \le 1$, with a discontinuity of 1 at $t = 0$ and $t = 1$. Subtracting $u(t) - u(t-1)$ from $x_1(t)$ we obtain a continuous signal, but to compensate we must add a unit pulse between $t = 0$ and $t = 1$, giving

$$x_1(t) = (\cos(2\pi t) - 1)[u(t) - u(t-1)] + [u(t) - u(t-1)] = x_{1a}(t) + x_{1b}(t)$$

where the first term $x_{1a}(t)$ is continuous and the second $x_{1b}(t)$ is discontinuous. The derivative is

$$\frac{dx_1(t)}{dt} = -2\pi \sin(2\pi t)[u(t) - u(t-1)] + (\cos(2\pi t) - 1)[\delta(t) - \delta(t-1)] + \delta(t) - \delta(t-1)$$

$$= -2\pi \sin(2\pi t)[u(t) - u(t-1)] + \delta(t) - \delta(t-1)$$

since

$$(\cos(2\pi t) - 1)[\delta(t) - \delta(t-1)] = (\cos(2\pi t) - 1)\delta(t) - (\cos(2\pi t) - 1)\delta(t-1)$$

$$= (\cos(0) - 1)\delta(t) - (\cos(2\pi) - 1)\delta(t-1)$$

$$= 0\delta(t) + 0\delta(t-1) = 0$$

The term $\delta(t)$ in the derivative indicates that there is a jump from 0 to 1 in $x_1(t)$ at $t = 0$ and that in $-\delta(t-1)$ there is a jump of -1 (from 1 to 0) at $t = 1$. See Figure 1.7.

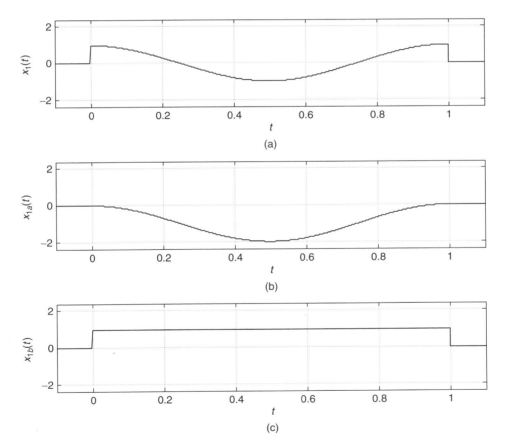

FIGURE 1.7

(a) Decomposition of $x_1(t) = \cos(2\pi t)[u(t) - u(t-1)]$ into (b) a continuous and (c) a discontinuous signal (a pulse).

The signal $x_2(t)$ has jump discontinuities at $t = 0$, $t = 1$, and $t = 2$, and we can think of it as completely discontinuous so that its continuous component is 0. The derivative is

$$\frac{dx_2(t)}{dt} = \delta(t) - 2\delta(t-1) + \delta(t-2)$$

The area of each of the deltas coincides with the jump in the discontinuities.   ■

Signal Generation with MATLAB

In the following examples we illustrate how to generate analog signals using MATLAB. This is done by either approximating continuous-time signals by discrete-time signals or by using the symbolic toolbox. The function plot uses an interpolation algorithm that makes the plots of discrete-time signals look like analog signals.

■ Example 1.16

Write a script and the necessary functions to generate a signal,

$$y(t) = 3r(t+3) - 6r(t+1) + 3r(t) - 3u(t-3)$$

Then plot it and verify analytically that the obtained figure is correct.

Solution

We wrote functions ramp and ustep to generate ramp and unit-step signals for obtaining a numeric approximation of the signal $y(t)$. The following script shows how these functions are used to generate $y(t)$. The arguments of ramp determine the support of the signal, the slope, and the shift (for advance, a positive number, and for delay, a negative number). For ustep we need to provide the support and the shift.

```
%%%%%%%%%%%%%%%%%%%%
% Example 1.16
%%%%%%%%%%%%%%%%%%%%
clear all; clf
Ts = 0.01; t = -5:Ts:5; % support of signal
% ramp with support [-5, 5], slope of 3 and advanced
% (shifted left) with respect to the origin by 3
y1 = ramp(t,3,3);
y2 = ramp(t,-6,1);
y3 = ramp(t,3,0);
% unit-step function with support [-5,5], delayed by 3
y4 = -3 * ustep(t,-3);
y = y1 + y2 + y3 + y4;
plot(t,y,'k'); axis([-5 5 -1 7]); grid
```

Our functions ramp and ustep are as follows.

```
function y = ramp(t,m,ad)
% ramp generation
```

```
% t: time support
% m: slope of ramp
% ad : advance (positive), delay (negative) factor
% USE: y = ramp(t,m,ad)
N = length(t);
y = zeros(1,N);
 for i = 1:N,
   if t(i) >= -ad,
   y(i) = m * (t(i)+ad);
 end
 end

function y = ustep(t,ad)
% generation of unit step
% t: time
% ad : advance (positive), delay (negative)
% USE y = ustep(t,ad)
N = length(t);
y = zeros(1,N);
 for i = 1:N,
   if t(i) >= -ad,
   y(i) = 1;
   end
 end
```

Analytically,

- $y(t) = 0$ for $t < -3$ and for $t > 3$, so the chosen support $-5 \leq t \leq 5$ displays the signal in a region where the signal occurs.
- For $-3 \leq t \leq -1$, $y(t)$ is $3r(t+3) = 3(t+3)$, which is 0 at $t = -3$ and 6 at $t = -1$.
- For $-1 \leq t \leq 0$, $y(t)$ is $3r(t+3) - 6r(t+1) = 3(t+3) - 6(t+1) = -3t + 3$, which is 6 at $t = -1$ and 3 at $t = 0$.
- For $0 \leq t \leq 3$, $y(t)$ is $3r(t+3) - 6r(t+1) + 3r(t) = -3t + 3 + 3t = 3$.
- For $t \geq 3$ the signal is $3r(t+3) - 6r(t+1) + 3r(t) - 3u(t-3) = 3 - 3 = 0$.

These coincide with the signal shown in Figure 1.8. ■

■ Example 1.17

Consider the following script that uses the functions ramp and ustep to generate a signal $y(t)$. Obtain analytically the formula for the signal $y(t)$. Write a function to compute and plot the even and odd components of $y(t)$.

```
clear all; clf
t = -5:0.01:5;
y1 = ramp(t,2,2.5);
y2 = ramp(t,-5,0);
```

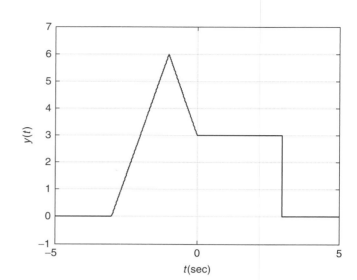

FIGURE 1.8
Generation of
$y(t) = 3r(t + 3) - 6r(t + 1) + 3r(t) - 3u(t - 3), -5 \le t \le 5$, and zero otherwise.

```
y3 = ramp(t,3,-2);
y4 = ustep(t,-4);
y = y1 + y2 + y3 + y4;
plot(t,y,'k'); axis([-5 5 -3 5]); grid
```

The signal $y(t) = 0$ for $t < -5$ and $t > 5$.

Solution

The signal $y(t)$ displayed on Figure 1.9(a) is given analytically by

$$y(t) = 2r(t + 2.5) - 5r(t) + 3r(t - 2) + u(t - 4)$$

Clearly, $y(t)$ is neither even nor odd. To find its even and odd components we use the function evenodd, shown in the following code with inputs as the signal and its support and outputs as the even and odd components. The results are shown on the bottom plots of Figure 1.9. Adding these two signals gives back the original signal $y(t)$. The script used is as follows.

```
%%%%%%%%%%%%%%%%%%
% Example 1.17
%%%%%%%%%%%%%%%%%%%%
[ye, yo] = evenodd(t,y);
subplot(211)
plot(t,ye,'r')
grid
axis([min(t) max(t) -2 5])
subplot(212)
plot(t,yo,'r')
```

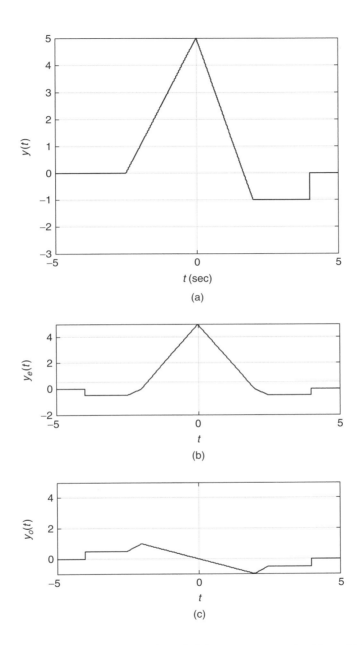

FIGURE 1.9

(a) Signal $y(t) = 2r(t + 2.5) - 5r(t) + 3r(t - 2) + u(t - 4)$, (b) even component $y_e(t)$, and (c) odd component $y_o(t)$.

```
grid
axis([min(t) max(t) -1 5])

function [ye,yo] = evenodd(t,y)
% even/odd decomposition
% t: time
% y: analog signal
% ye, yo: even and odd components
% USE [ye,yo] = evenodd(t,y)
%
yr = fliplr(t,y);
ye = 0.5 * (y + yr);
yo = 0.5 * (y - yr);
```

The MATLAB function fliplr reverses the values of the vector y giving the reflected signal. ∎

■ Example 1.18

Use symbolic MATLAB to generate the following analog signals.

(a) For the damped sinusoid signal

$$x(t) = e^{-t} \cos(2\pi t)$$

obtain a script to generate $x(t)$ and its envelope.

(b) For a rough approximation of a periodic pulse generated by adding three cosines of frequencies multiples of $\Omega_0 = \pi/10$—that is

$$x_1(t) = 1 + 1.5 \cos(2\Omega_0 t) - 0.6 \cos(4\Omega_0 t)$$

write a script to generate $x_1(t)$.

Solution

The following script generates the damped sinusoid signal, and its envelope $y(t) = \pm e^{-t}$.

```
%%%%%%%%%%%%%%%%%%%%
% Example 1.18 --- damped sinusoid
%%%%%%%%%%%%%%%%%%%%
t = sym('t');
x = exp(-t) * cos(2 * pi * t);
y = exp(-t);
ezplot(x,[-2,4])
grid
hold on
ezplot(y,[-2,4])
```

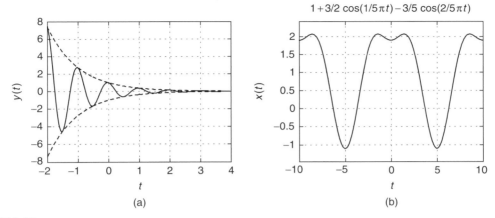

FIGURE 1.10
(a) Damped sinusoid, and (b) sum of weigthed cosines approximating a pulse.

```
hold on
ezplot(-y,[-2,4])
axis([-2 4 -8 8])
hold off
```

The approximate pulse signal is generated by the following script.

```
clear; clf
t = sym('t');
% sum of constant and cosines
x = 1 + 1.5 * cos(2 * pi * t/10)-.6 * cos(4 * pi * t/10);
ezplot(x,[-10,10]); grid
```

The plots of the damped sinusoid and the approximate pulse are given in Figure 1.10. ∎

■ Example 1.19

Consider the generation of a triangular signal,

$$\Lambda(t) = \begin{cases} t & 0 \le t \le 1 \\ -t + 2 & 1 < t \le 2 \\ 0 & \text{otherwise} \end{cases}$$

using ramp signals $r(t)$. Use the unit-step signal to represent the derivative of $d\Lambda(t)/dt$.

Solution

The triangular pulse can be represented as

$$\Lambda(t) = r(t) - 2r(t-1) + r(t-2) \tag{1.32}$$

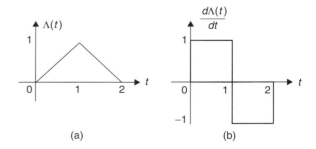

FIGURE 1.11
(a) The triangular signal $\Lambda(t)$ and (b) its derivative.

(a)   (b)

In fact, since $r(t-1)$ and $r(t-2)$ have values different from 0 for $t \geq 1$ and $t \geq 2$, respectively, then

$$\Lambda(t) = r(t) = t \quad \text{for } 0 \leq t \leq 1$$

and that for $1 \leq t \leq 2$,

$$\Lambda(t) = r(t) - 2r(t-1) = t - 2(t-1) = -t + 2$$

Finally, for $t > 2$ the three ramp signals are different from zero, so

$$\begin{aligned}
\Lambda(t) &= r(t) - 2r(t-1) + r(t-2) \\
&= t - 2(t-1) + (t-2) \\
&= 0 \quad t > 2
\end{aligned}$$

and by definition $\Lambda(t)$ is zero for $t < 0$. So the given expression for $\Lambda(t)$ in terms of the ramp functions is identical to its given mathematical definition.

Using the mathematical definition of the triangular function, its derivative is given by

$$\frac{d\Lambda(t)}{dt} = \begin{cases} 1 & 0 \leq t \leq 1 \\ -1 & 1 < t \leq 2 \\ 0 & \text{otherwise} \end{cases}$$

Using the representation in Equation (1.32) this derivative is also given by

$$\frac{d\Lambda(t)}{dt} = u(t) - 2u(t-1) + u(t-2)$$

which are two unit pulses, as shown in Figure 1.11. ∎

■ Example 1.20

Consider a full-wave rectified signal,

$$x(t) = |\cos(2\pi t)| \quad -\infty < t < \infty$$

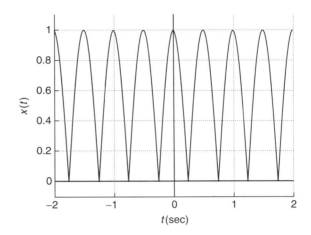

FIGURE 1.12

Eight periods of full-wave rectified signal $x(t) = |\cos(2\pi t)|, \quad -\infty < t < \infty.$

of period $T_0 = 0.5$. Obtain a representation for a period between 0 and 0.5, and represent $x(t)$ in terms of shifted versions of it. A full-wave rectified signal is used in designing dc sources. It is a first step in converting an alternating voltage into a dc voltage. See Figure 1.12.

Solution

The period between 0 and 0.5 can be expressed as

$$p(t) = x(t)[u(t) - u(t - 0.5)] = |\cos(2\pi t)|[u(t) - u(t - 0.5)]$$

Since $x(t)$ is a periodic signal of period $T_0 = 0.5$, we have then that

$$x(t) = \sum_{k=-\infty}^{\infty} p(t - kT_0)$$

■

■ Example 1.21

Generate a causal train of pulses that repeats every two units of time using as first period

$$s(t) = u(t) - 2u(t - 1) + u(t - 2)$$

Find the derivative of the train of pulses.

Solution

Considering that $s(t)$ is the first period of the train of pulses of period two, then

$$\rho(t) = \sum_{k=0}^{\infty} s(t - 2k)$$

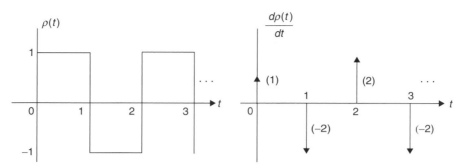

FIGURE 1.13

Causal train of pulses $\rho(t)$ and its derivative. The number enclosed in () is the area of the corresponding delta function and it indicates the jump at the particular discontinuity—positive when increasing and negative when decreasing.

is the desired signal. Notice that $\rho(t)$ equals zero for $t < 0$, thus it is causal. Given that the derivative of a sum of signals is the sum of the derivative of each of the signals, the derivative of $\rho(t)$ is

$$\frac{d\rho(t)}{dt} = \sum_{k=0}^{\infty} \frac{ds(t - 2k)}{dt}$$

$$= \sum_{k=0}^{\infty} [\delta(t - 2k) - 2\delta(t - 1 - 2k) + \delta(t - 2 - 2k)]$$

which can be simplified to

$$\frac{d\rho(t)}{dt} = [\delta(t) - 2\delta(t - 1) + \delta(t - 2)] + [\delta(t - 2) - 2\delta(t - 3) + \delta(t - 4)] + [\delta(t - 4) \cdots$$

$$= \delta(t) + 2\sum_{k=1}^{\infty} \delta(t - 2k) - 2\sum_{k=1}^{\infty} \delta(t - 2k + 1)$$

where $\delta(t)$, $2\delta(t - 2k)$, and $-2\delta(t - 2k + 1)$ for $k \geq 1$ occur at $t = 0$, $t = 2k$, and $t = 2k - 1$ for $k \geq 1$, or the times at which the discontinuities of $\rho(t)$ occur. The value associated with the $\delta(t)$ corresponds to the jump of the signal from the left to the right. Thus, $\delta(t)$ indicates there is a discontinuity in $\rho(t)$ at zero as it jumps from 0 to 1, while the discontinuities at 2, 4, ... have a jump of 2 from -1 to 1, increasing. The discontinuities indicated by $\delta(t - 2k - 1)$ occurring at 1, 3, 5, ... are from 1 to -1 (i.e., decreasing, so the value of -2). See Figure 1.13.   ■

1.4.3 Special Signals—the Sampling Signal and the Sinc

Two signals of great significance in the sampling of continuous-time signals and their reconstruction are the sampling signal and the sinc. Sampling a continuous-time signal consists in taking samples of the signal at uniform times. One can think of this process as the multiplication of a continuous-time signal $x(t)$ by a train of very narrow pulses of the sampling period T_s. For simplicity, considering that the width of the pulses is much smaller than T_s, the train of pulses can be approximated by a train of impulses that is periodic of period T_s—that is, the *sampling signal* $\delta_{T_s}(t)$ is

$$\delta_{T_s}(t) = \sum_{n=-\infty}^{\infty} \delta(t - nT_s) \tag{1.33}$$

The sampled signal $x_s(t)$ is then

$$x_s(t) = x(t)\delta_{T_s}(t)$$

$$= \sum_{n=-\infty}^{\infty} x(nT_s)\delta(t - nT_s) \tag{1.34}$$

or a sequence of uniformly shifted impulses with amplitude the value of the signal $x(t)$ at the time when the impulse occurs.

A fundamental result in sampling theory is the recovery of the original signal, under certain constrains, by means of an interpolation using *sinc signals*. Moreover, we will see that the sinc is connected with ideal low-pass filters. The sinc function is defined as

$$S(t) = \frac{\sin \pi t}{\pi t} \qquad -\infty < t < \infty \tag{1.35}$$

This signal has the following characteristics:

- The time support of this signal is infinite.
- It is an even function of t, as

$$S(-t) = \frac{\sin(-\pi t)}{-\pi t} = \frac{-\sin(\pi t)}{-\pi t} = S(t) \tag{1.36}$$

- At $t = 0$ the numerator and the denominator of the sinc are zero; thus the limit as $t \to 0$ is found using L'Hôpital's rule—that is,

$$\lim_{t \to 0} S(t) = \lim_{t \to 0} \frac{d \sin(\pi t)/dt}{d\pi t/dt}$$

$$= \lim_{t \to 0} \frac{\pi \cos(\pi t)}{\pi} = 1 \tag{1.37}$$

- $S(t)$ is bounded—that is, since $-1 \leq \sin(\pi t) \leq 1$, then for $t \geq 0$,

$$\frac{-1}{\pi t} \leq \frac{\sin(\pi t)}{\pi t} = S(t) \leq \frac{1}{\pi t} \tag{1.38}$$

and given that $S(t)$ is even, it is equally bounded for $t < 0$. As $t \to \pm\infty$, $S(t) \to 0$.

- The zero-crossing time of $S(t)$ are found by letting the numerator equal zero—that is, when $\sin(\pi t) = 0$, the zero-crossing times are such that $\pi t = k\pi$, or $t = k$ for a nonzero integer k or $k = \pm 1, \pm 2, \ldots$.

■ A property that is not obvious and that requires the frequency representation of $S(t)$ is that the integral

$$\int_{-\infty}^{\infty} |S(t)|^2 dt = 1 \tag{1.39}$$

Recall that we showed this in Chapter 0 using numeric and symbolic MATLAB.

The sinc signal will appear in several places in the rest of the book.

1.4.4 Basic Signal Operations—Time Scaling, Frequency Shifting, and Windowing

Given a signal $x(t)$, and real values $\alpha \neq 0$ or 1, and $\phi > 0$:
■ $x(\alpha t)$ is said to be contracted if $|\alpha| > 1$, and if $\alpha < 0$ it is also reflected.
■ $x(\alpha t)$ is said to be expanded if $|\alpha| < 1$, and if $\alpha < 0$ it is also reflected.
■ $x(t)e^{j\phi t}$ is said to be shifted in frequency by ϕ radians.
■ For a window signal $w(t)$, $x(t)w(t)$ displays $x(t)$ within the support of $w(t)$.

To illustrate the time scaling, consider a signal $x(t)$ with a finite support $t_0 \leq t \leq t_1$. Assume that $\alpha > 1$, then $x(\alpha t)$ is defined in $t_0 \leq \alpha t \leq t_1$ or $t_0/\alpha \leq t \leq t_1/\alpha$, a smaller support than the original one. For instance, for $\alpha = 2$, $t_0 = 2$, and $t_1 = 4$, then the support of $x(2t)$ is $1 \leq t \leq 2$, while the support of $x(t)$ is $2 \leq t \leq 4$. If $\alpha = -2$, then $x(-2t)$ is not only contracted but also reflected. Similarly, $x(0.5t)$ would have a support of $2t_0 \leq t \leq 2t_1$, which is larger than the original support.

Multiplication by an exponential shifts the frequency of the original signal. To illustrate this consider the case of an exponential $x(t) = e^{j\Omega_0 t}$ of frequency Ω_0. If we multiply $x(t)$ by an exponential $e^{j\phi t}$, then

$$x(t)e^{j\phi t} = e^{j(\Omega_0+\phi)t} = \cos((\Omega_0 + \phi)t) + j\sin((\Omega_0 + \phi)t)$$

so that the frequency of the new exponential is greater than Ω_0 if $\phi > 0$ or smaller if $\phi < 0$. So we have shifted the frequency of $x(t)$. If we have a sum of exponentials (they do not need to be harmonically related as in the Fourier series we will consider later),

$$x(t) = \sum_k A_k e^{j\Omega_k t}$$

then

$$x(t)e^{j\phi t} = \sum_k A_k e^{j(\Omega_k+\phi)t}$$

so that each of the frequencies of the signal $x(t)$ has been shifted. This shifting of the frequency is significant in the development of amplitude modulation, and as such this frequency shift process is called *modulation*—that is, the signal $x(t)$ modulates the exponential and $x(t)e^{j\phi t}$ is the modulated signal.

Notice that time scaling also changes the frequency content of the signal. For instance, a signal $x(t) = e^{j\Omega_0 t}$ is periodic of period $T_0 = 2\pi/\Omega_0$, while $x(\alpha t) = e^{j\alpha\Omega_0 t}$ has a period T_0/α or a frequency $\alpha\Omega_0$, which is larger than the original frequency of Ω_0 when $\alpha > 1$ and smaller than Ω_0 when $0 < \alpha < 1$.

Remarks *We can thus summarize the above as follows:*

- *If $x(t)$ is periodic of period T_0 then the time-scaled signal $x(\alpha t)$, $\alpha \neq 0$, is also periodic of period $T_0/|\alpha|$.*
- *The support in time of a periodic or nonperiodic signal is inversely proportional to the support in frequency for that signal.*
- *The frequencies present in a signal can be changed by modulation—that is, multiplying the signal by a complex exponential or, equivalently, by sines and cosines. The frequency change is also possible by expansion and compression of the signal.*
- *Reflection is a special case of time scaling with $\alpha = -1$.*

■ Example 1.22

Let $x_1(t)$, for $0 \leq t \leq T_0$, be one period of a periodic signal $x(t)$ of period T_0. Represent $x(t)$ in terms of advanced and delayed versions of $x_1(t)$. What would be $x(2t)$?

Solution

The periodic signal $x(t)$ can be written as

$$x(t) = \cdots + x_1(t + 2T_0) + x_1(t + T_0) + x_1(t) + x_1(t - T_0) + x_1(t - 2T_0) + \cdots$$

$$= \sum_{k=-\infty}^{\infty} x_1(t - kT_0)$$

and the contracted signal $x(2t)$ is then

$$x(2t) = \sum_{k=-\infty}^{\infty} x_1(2t - kT_0)$$

and periodic of period $T_0/2$. ■

■ Example 1.23

An acoustic signal $x(t)$ has a duration of 3.3 minutes and a radio station would like to use the signal for a three-minute segment. Indicate how to make it possible.

Solution

We need to contract the signal by a factor of $\alpha = 3.3/3 = 1.1$, so that $x(1.1t)$ can be used in the three-minute piece. If the signal is recorded on tape, the tape player can be run 1.1 times faster than the recording speed. This would change the voice or music on the tape, as the frequencies $x(1.1t)$ are increased with respect to the original frequencies in $x(t)$. ■

■ Example 1.24

One way of transmitting a message over the airwaves is to multiply it by a sinusoid of frequency higher than those in the message, thus changing the frequency content of the signal. The resulting signal is called an amplitude-modulated (AM) signal: The message changes the amplitude of the sinusoid. To recover the message from the transmitted signal, one can make the envelope of the modulated signal be related to the message. Use again the ramp and ustep functions to generate a signal $y(t) = 2r(t + 2) - 4r(t) + r(t - 2) + r(t - 3) + u(t - 3)$ to modulate a so-called carrier signal $x(t) = \sin(5\pi t)$ to give the AM signal $z(t)$. Obtain a script to generate and plot the AM signal. Indicate whether the envelope of the AM signal is connected with the message signal $y(t)$.

Solution

The signal $y(t)$ analytically equals

$$y(t) = \begin{cases} 0 & t < -2 \\ 2r(t+2) = 2(t+2) & -2 \leq t < 0 \\ 2r(t+2) - 4r(t) = -2t + 4 & 0 \leq t < 2 \\ 2r(t+2) - 4r(t) + r(t-2) = -t + 2 & 2 \leq t < 3 \\ 2r(t+2) - 4r(t) + r(t-2) + r(t-3) + u(t-3) = 0 & t \geq 3 \end{cases}$$

The following script is used to generate the message signal $y(t)$, the AM signal $z(t)$, and the corresponding plots. The MATLAB function sound is used to produce the sound corresponding to $100z(t)$. In Figure 1.14 we show $z(t)$ and emphasize the envelope (dashed line) that corresponds to $\pm y(t)$.

```
%%%%%%%%%%%%%%
% Example 1.24 --- AM signal
%%%%%%%%%%%%%%%%%%
t = -5:0.01:5;
x = sin(5*pi*t);
y1 = ramp(t,2,2);
y2 = ramp(t,-4,0);
y3 = ramp(t,1,-2);
y4 = ramp(t,1,-3);
y5 = ustep(t,-3);
y = y1 + y2 + y3 + y4 + y5;
z = y.*x;
sound(100*z,1000)
plot(t,z,'k'); hold on
plot(t,y,'r',t,-y,'r'); axis([-5 5 -5 5]); grid
hold off
xlabel('t'); ylabel('z(t)')
```

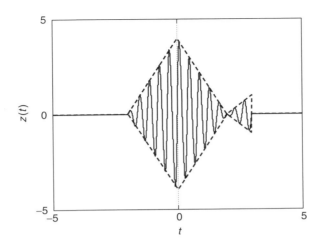

FIGURE 1.14
AM signal.

1.4.5 Generic Representation of Signals

Consider the following integral:

$$\int_{-\infty}^{\infty} f(t)\delta(t)dt$$

The product of $f(t)$ and $\delta(t)$ gives zero everywhere except at the origin where we get an impulse of area $f(0)$—that is, $f(t)\delta(t) = f(0)\delta(t)$ (let $t_0 = 0$ in Figure 1.15). Therefore,

$$\int_{-\infty}^{\infty} f(t)\delta(t)dt = \int_{-\infty}^{\infty} f(0)\delta(t)dt = f(0)\int_{-\infty}^{\infty} \delta(t)dt = f(0) \qquad (1.40)$$

since the area under the curve of the impulse is unity. This property of the impulse function is appropriately called the *sifting property*. The result of this integration is to sift out $f(t)$ for all t except for $t = 0$, where $\delta(t)$ occurs. If we delay or advance the $\delta(t)$ function in the integrand, the result is that all values of $f(t)$ are sifted out except for the value corresponding to the location of the delta function—that is,

$$\int_{-\infty}^{\infty} f(t)\delta(t - \tau)dt = \int_{-\infty}^{\infty} f(\tau)\delta(t - \tau)dt = f(\tau)\int_{-\infty}^{\infty} \delta(t - \tau)dt$$
$$= f(\tau) \qquad \text{for any } \tau$$

since the last integral is still unity. Figure 1.15 illustrates the multiplication of a signal $f(t)$ by an impulse signal $\delta(t - t_0)$, located at $t = t_0$.

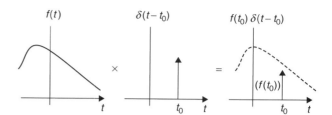

FIGURE 1.15

Multiplication of a signal $f(t)$ by an impulse signal $\delta(t - t_0)$.

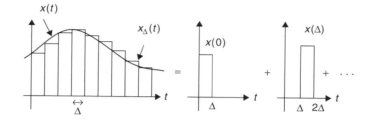

FIGURE 1.16

Generic representation of $x(t)$ as an infinite sum of pulses of height $x(k\Delta)$ and width Δ when $\Delta \to 0$, so that the sum becomes an integral of weighted impulse signals.

By the sifting property of the impulse function $\delta(t)$, any signal $x(t)$ can be represented by the following *generic representation*:

$$x(t) = \int_{-\infty}^{\infty} x(\tau)\delta(t - \tau)d\tau \tag{1.41}$$

Figure 1.16 shows a generic representation. Equation (1.41) basically indicates that any signal can be viewed as a stacking of pulses $x(k\Delta)p_\Delta(t - k\Delta)$, which in the limit as $\Delta \to 0$ become impulses $x(\tau)\delta(t - \tau)$.

Equation (1.41) provides a generic representation of a signal in terms of basic signals, in this case impulse signals. As we will see in the next chapter, once we determine the response of a system to an impulse we will use the generic representation to find the response of the system to any signal.

1.5 WHAT HAVE WE ACCOMPLISHED? WHERE DO WE GO FROM HERE?

We have taken another step in our long journey. In this chapter we discussed the main classification of signals and have started the study of deterministic, continuous-time signals. We have also discussed important characteristics of signals such as periodicity, energy, power, evenness, and oddness, and learned basic signal operations that will be useful as we will see in the next chapters. Interestingly,

Table 1.1 Basic Signals

Signal	Definition
Complex exponential	$\|A\|e^{rt}[\cos(\Omega_0 t + \theta) + j\sin(\Omega_0 t + \theta)] \quad -\infty < t < \infty$
Sinusoid	$A\cos(\Omega_0 t + \theta) = A\sin(\Omega_0 t + \theta + \pi/2) \quad -\infty < t < \infty$
Unit impulse	$\delta(t) = 0 \ \ t \neq 0$, undefined at $t = 0$
	$\displaystyle\int_{-\infty}^{t} \delta(\tau)d\tau = 1, \quad t > 0$
	$\displaystyle\int_{-\infty}^{\infty} f(\tau)\delta(t - \tau)d\tau = f(t)$
Unit step	$u(t) = \begin{cases} 1 & t > 0 \\ 0 & t < 0 \end{cases}$
Ramp	$r(t) = tu(t) = \begin{cases} t & t > 0 \\ 0 & t < 0 \end{cases}$
	$\delta(t) = du(t)/dt$
	$u(t) = \displaystyle\int_{-\infty}^{t} \delta(\tau)d\tau$
	$r(t) = \displaystyle\int_{-\infty}^{t} u(\tau)d\tau$
Rectangular pulse	$p(t) = A[u(t) - u(t-1)] = \begin{cases} A & 0 \leq t \leq 1 \\ 0 & \text{otherwise} \end{cases}$
Triangular pulse	$\Lambda(t) = A[r(t) - 2r(t-1) + r(t-2)] = \begin{cases} At & 0 \leq t \leq 1 \\ A(2-t) & 1 < t \leq 2 \\ 0 & \text{otherwise} \end{cases}$
Sampling	$\delta_{T_s}(t) = \sum_k \delta(t - kT_s)$
Sinc	$S(t) = \sin(\pi t)/(\pi t)$
	$S(0) = 1$
	$S(k) = 0 \ \ k \neq 0$ integer
	$\displaystyle\int_{-\infty}^{\infty} S^2(t)dt = 1$

we began to see how some of these operations lead to practical applications, such as amplitude, frequency, and phase modulations, which are basic in the theory of communications. Very importantly, we have also begun to represent signals in terms of basic signals, which in later chapters will allow us to simplify the analysis and will give us flexibility in the synthesis of systems. These basic signals are used as test signals in control systems. Table 1.1 displays basic signals.

Our next step is to connect signals with systems. We are particularly interested in developing a theory that can be used to approximate, to some degree, the behavior of most systems of interest in engineering. After that we consider the analysis of signals and systems time and frequency domains.

PROBLEMS

1.1. Signal energy and RC circuit—MATLAB

The signal $x(t) = e^{-|t|}$ is defined for all values of t.

(a) Plot the signal $x(t)$ and determine if this signal is finite energy. That is, compute the integral

$$\int_{-\infty}^{\infty} |x(t)|^2 dt$$

and determine if it is finite.

(b) If you determine that $x(t)$ is absolutely integrable, or that the integral

$$\int_{-\infty}^{\infty} |x(t)| dt$$

is finite, could you say that $x(t)$ has finite energy? Explain why or why not. *Hint:* Plot $|x(t)|$ and $|x(t)|^2$ as functions of time.

(c) From your results above, is it true the energy of the signal

$$y(t) = e^{-t} \cos(2\pi t) u(t)$$

is less than half the energy of $x(t)$? Explain. To verify your result, use symbolic MATLAB to plot $y(t)$ and to compute its energy.

(d) To discharge a capacitor of 1 mF charged with a voltage of 1 volt we connect it, at time $t = 0$, with a resistor of $R\ \Omega$. When we measure the voltage in the resistor we find it to be $v_R(t) = e^{-t}u(t)$. Determine the resistance R. If the capacitor has a capacitance of $1\ \mu F$, what would be R? In general, how are R and C related?

1.2. Power in RL circuits

Consider a circuit consisting of a sinusoidal source $v_s(t) = \cos(t)u(t)$ volts connected in series to a resistor R and an inductor L and assume they have been connected for a very long time.

(a) Let $R = 0$ and $L = 1$ H. Compute the instantaneous and the average powers delivered to the inductor.

(b) Let $R = 1\ \Omega$ and $L = 1$ H. Compute the instantaneous and the average powers delivered to the resistor and the inductor.

(c) Let $R = 1\ \Omega$ and $L = 0$ H. Compute the instantaneous and the average powers delivered to the resistor. *Hint:* In the above parts of the problem use phasors or the trigonometric formula

$$\cos(\alpha) \cos(\beta) = 0.5[\cos(\alpha - \beta) + \cos(\alpha + \beta)]$$

(d) The average power used by the resistor is what you pay to the electric company, but there is also a reactive power for which you do not. The complex power supplied to the circuit is defined as

$$P = \frac{1}{2} V_s I^*$$

where V_s and I are the phasors corresponding to the source and the current in the circuit, and I^* is the complex conjugate of I. Consider the values of the resistor and the inductor given above, and compute the complex power and relate it to the average power computed in each case.

1.3. Power in periodic and nonperiodic sum of sinusoids

Consider the periodic signal $x(t) = \cos(2\Omega_0 t) + 2 \cos(\Omega_0 t)$, $-\infty < t < \infty$, and $\Omega_0 = \pi$. The frequencies of the two sinusoids are said to be harmonically related (one is a multiple of the other).

(a) Determine the period T_0 of $x(t)$.

(b) Compute the power P_x of $x(t)$.

(c) Verify that the power P_x is the sum of the power P_1 of $x_1(t) = \cos(2\pi t)$ and the power P_2 of $x_2(t) = 2\cos(\pi t)$.

(d) In the above case you are able to show that there is superposition of the powers because the frequencies are harmonically related. Suppose that $y(t) = \cos(t) + \cos(\pi t)$ where the frequencies are not harmonically related. Find out whether $y(t)$ is periodic or not. Indicate how you would find the power P_y of $y(t)$. Would $P_y = P_1 + P_2$ where P_1 is the power of $\cos(t)$ and P_2 is the power of $\cos(\pi t)$? Explain what is the difference with respect to the case of harmonic frequencies.

1.4. Periodicity of sum of sinusoids—MATLAB

Consider the periodic signals $x_1(t) = 4\cos(\pi t)$ and $x_2(t) = -\sin(3\pi t + \pi/2)$.

(a) Find the periods of $x_1(t)$ and $x_2(t)$.

(b) Is the sum $x(t) = x_1(t) + x_2(t)$ periodic? If so, what is its period?

(c) In general, two periodic signals $x_1(t)$ and $x_2(t)$ having periods T_1 and T_2 such that their ratio $T_1/T_2 = M/K$ is a rational number (i.e., M and K are positive integers), then the sum $x(t) = x_1(t) + x_2(t)$ is periodic. Suppose the rationality condition is satisfied and $M = 3$ and $K = 12$. Determine the period of $x(t)$.

(d) Determine whether $x(t) = x_1(t) + x_2(t)$ is periodic when

 ■ $x_1(t) = 4\cos(2\pi t)$ and $x_2(t) = -\sin(3\pi t + \pi/2)$
 ■ $x_1(t) = 4\cos(2t)$ and $x_2(t) = -\sin(3\pi t + \pi/2)$

 Use symbolic MATLAB to plot $x(t)$ in the above two cases and confirm your analytic results about the periodicity or lack of periodicity of $x(t)$.

1.5. Time shifting

Consider a finite-support signal

$$x(t) = t \qquad 0 \leq t \leq 1$$

and zero elsewhere.

(a) Carefully plot $x(t + 1)$.

(b) Carefully plot $x(-t + 1)$.

(c) Add the above two signals to get a new signal $y(t)$. To verify your results, represent each of the above signals analytically and show that the resulting signal is correct.

(d) How does $y(t)$ compare to the signal $\Lambda(t) = (1 - |t|)(u(t + 1) - u(t - 1))$? Plot them. Compute the integrals of $y(t)$ and $\Lambda(t)$ for all values of t and compare them. Explain.

1.6. Even and odd hyperbolic functions—MATLAB

According to Euler's identity the sine and the cosine are defined in terms of complex exponentials. You would then ask what if instead of complex exponentials you were to use real exponentials. Well, using Euler's identity we obtain the hyperbolic functions defined in $-\infty < t < \infty$:

$$\cosh(\Omega_0 t) = \frac{e^{\Omega_0 t} + e^{-\Omega_0 t}}{2}$$

$$\sinh(\Omega_0 t) = \frac{e^{\Omega_0 t} - e^{-\Omega_0 t}}{2}$$

(a) Let $\Omega_0 = 1$ rad/sec. Use the definition of the real exponentials to plot $\cosh(t)$ and $\sinh(t)$.

(b) Is $\cosh(t)$ even or odd?

(c) Is sinh(t) even or odd?

(d) Obtain an expression for $x(t) = e^{-t}u(t)$ in terms of the hyperbolic functions. Use symbolic MATLAB to plot $x(t) = e^{-t}u(t)$ and to plot your expression in terms of the hyperbolic functions. Compare them.

1.7. Impulse signal generation—MATLAB

When defining the impulse or $\delta(t)$ signal, the shape of the signal used to do so is not important. Whether we use the rectangular pulse we considered in this chapter or another pulse, or even a signal that is not a pulse, in the limit we obtain the same impulse signal. Consider the following cases:

(a) The triangular pulse,

$$\Lambda_\Delta(t) = \frac{1}{\Delta}\left(1 - \left|\frac{t}{\Delta}\right|\right)(u(t + \Delta) - u(t - \Delta))$$

Carefully plot it, compute its area, and find its limit as $\Delta \to 0$. What do you obtain in the limit? Explain.

(b) Consider the signal

$$S_\Delta(t) = \frac{\sin(\pi t/\Delta)}{\pi t}$$

Use the properties of the sinc signal $S(t) = \sin(\pi t)/(\pi t)$ to express $S_\Delta(t)$ in terms of $S(t)$. Then find its area, and the limit as $\Delta \to 0$. Use symbolic MATLAB to show that for decreasing values of Δ the $S_\Delta(t)$ becomes like the impulse signal.

1.8. Impulse and unit-step signals

By introducing the impulse $\delta(t)$ and the unit-step $u(t)$ signals, we expand the conventional calculus. One of the advantages of having the $\delta(t)$ function is that we are now able to find the derivative of discontinuous signals. Let us illustrate this advantage. Consider a periodic sinusoid defined for all times,

$$x(t) = \cos(\Omega_0 t) \quad -\infty < t < \infty$$

and a causal sinusoid defined as

$$x_1(t) = \cos(\Omega_0 t)u(t)$$

where the unit-step function indicates that the function has a discontinuity at zero, since for $t = 0+$ the function is close to 1, and for $t = 0-$ the function is zero.

(a) Find the derivative $y(t) = dx(t)/dt$ and plot it.

(b) Find the derivative $z(t) = dx_1(t)/dt$ (treat $x_1(t)$ as the product of two functions $\cos(\Omega_0 t)$ and $u(t)$) and plot it. Express $z(t)$ in terms of $y(t)$.

(c) Verify that the integral of $z(t)$ gives you back $x_1(t)$.

1.9. Series RC circuit response to a unit-step signal

A unit-step function $u(t)$ can be considered a causal constant source (e.g., a battery in a circuit if the units of $u(t)$ is volts).

(a) From basic principles consider the response of an RC circuit to $u(t)$—that is, a battery connected in series with the resistor and the capacitor. Remember that the voltage across the capacitor results from an accumulation of charge, and that the presence of the resistor simply means that the charge is slowly accumulated. Therefore, plot what would be the voltage across the capacitor for $t > 0$ (assume the capacitor has no initial voltage at $t = 0$).

(b) What would be the voltage across the capacitor in the steady state? Explain.

(c) Finally, suppose that the capacitor is disconnected from the circuit at some time $t_0 \gg 0$. Ideally, what would be the voltage across the capacitor from then on?

(d) If you disconnect the capacitor, again at $t_0 \gg 0$, but somehow it is left connected to the resistor, so they are in parallel, what would happen to the voltage across the capacitor? Plot approximately the voltage across the capacitor for all times and explain the reason for your plot.

1.10. Ramp in terms of unit-step signals

A ramp, $r(t) = tu(t)$, can be expressed as

$$r(t) = \int_0^\infty u(\tau)u(t-\tau)d\tau$$

(a) Show that the above expression for $r(t)$ is equivalent to

$$r(t) = \int_0^t d\tau = tu(t)$$

(b) Compute the derivative of

$$r(t) = \int_0^\infty u(\tau)u(t-\tau)d\tau$$

to show that

$$u(t) = \int_0^\infty u(\tau)\delta(t-\tau)d\tau$$

1.11. Sampling signal and impulse signal—MATLAB

Consider the sampling signal

$$\delta_T(t) = \sum_{k=0}^\infty \delta(t-kT)$$

which we will use in the sampling of analog signals later on.

(a) Plot $\delta_T(t)$. Find

$$ss_T(t) = \int_{-\infty}^t \delta_T(\tau)d\tau$$

and carefully plot it for all t. What does the resulting signal $ss(t)$ look like? In reference 17, Craig calls it the "stairway to the stars." Explain.

(b) Use MATLAB function stairs to plot $ss_T(t)$ for $T = 0.1$. Determine what signal would be the limit as $T \to 0$.

(c) A sampled signal is

$$x_s(t) = x(t)\delta_T(t) = \sum_{k=0}^\infty x(kT_s)\delta(t-kT_s)$$

Let $x(t) = \cos(2\pi t)u(t)$ and $T_s = 0.1$. Find the integral

$$\int_{-\infty}^t x_s(t)dt$$

and use MATLAB to plot it for $0 \leq t \leq 10$. In a simple way this problem illustrates the operation of a discrete-to-analog converter, which converts a discrete-time into a continuous-time signal (its cousin is the digital-to-analog converter or DAC).

1.12. Reflection and time shifting

Do the reflection and the time-shifting operations commute? That is, do the two block diagrams in Figure 1.17 provide identical signals (i.e., is $y(t)$ equal to $z(t)$)? To provide an answer to this consider the signal $x(t)$ shown in Figure 1.17. Reflect $x(t)$ to get $v(t) = x(-t)$ and then shift it to get $y(t) = v(t - 2)$. Then consider delaying $x(t)$ to get $w(t) = x(t - 2)$, and reflecting it to get $z(t) = w(-t)$. Perform each of these operations on $x(t)$ to get $y(t)$ and $z(t)$; plot them and compare these plots. What is your conclusion? Explain

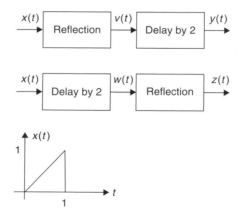

FIGURE 1.17
Problem 1.12.

1.13. Contraction and expansion of signals

Let $x(t)$ be the analog signal considered in Problem 1.12 (see Figure 1.17). In this problem we would like to consider expanded and compressed versions of that signal.

(a) Plot $x(2t)$ and determine if it is a compressed or expanded version of $x(t)$.

(b) Plot $x(t/2)$ and determine if it is a compressed or expanded version of $x(t)$.

(c) Suppose $x(t)$ is an acoustic signal—let's say it is a music signal recorded in a magnetic tape. What would be a possible application of the expanding and compression operations? Explain.

1.14. Even and odd decomposition and power

Consider the analog signal $x(t)$ in Figure 1.18.

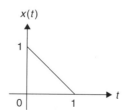

FIGURE 1.18
Problem 1.14.

(a) Plot the even–odd decomposition of $x(t)$ (i.e., find and plot the even $x_e(t)$ and the odd $x_o(t)$ components of $x(t)$).

(b) Show that the energy of the signal $x(t)$ can be expressed as the sum of the energies of its even and odd components—that is, that

$$\int_{-\infty}^{\infty} x^2(t)dt = \int_{-\infty}^{\infty} x_e^2(t)dt + \int_{-\infty}^{\infty} x_o^2(t)dt$$

(c) Verify that the energy of $x(t)$ is equal to the sum of the energies of $x_e(t)$ and $x_o(t)$.

1.15. Generation of periodic signals

A periodic signal can be generated by repeating a period.
   (a) Find the function $g(t)$, defined in $0 \leq t \leq 2$ only, in terms of basic signals and such, that when repeated using a period of 2, generates the periodic signal $x(t)$, as shown in Figure 1.19.
   (b) Obtain an expression for $x(t)$ in terms of $g(t)$ and shifted versions of it.
   (c) Suppose we shift and multiply by a constant the periodic signal $x(t)$ to get new signals $y(t) = 2x(t-2)$, $z(t) = x(t+2)$, and $v(t) = 3x(t)$. Are these signals periodic?
   (d) Let then $w(t) = dx(t)/dt$, and plot it. Is $w(t)$ periodic? If so, determine its period.

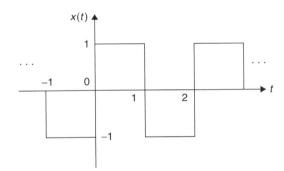

FIGURE 1.19
Problem 1.15.

1.16. Contraction and expansion and periodicity—MATLAB

Consider the periodic signal $x(t) = \cos(\pi t)$ of period $T_0 = 2$ sec.
   (a) Is the expanded signal $x(t/2)$ periodic? If so, indicate its period.
   (b) Is the compressed signal $x(2t)$ periodic? If so, indicate its period.
   (c) Use MATLAB to plot the above two signals and verify your analytic results.

1.17. Derivatives and integrals of periodic signals

Consider the triangular train of pulses $x(t)$ in Figure 1.20.

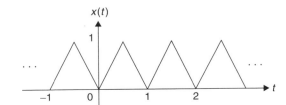

FIGURE 1.20
Problem 1.17.

(a) Carefully plot the signal $y(t) = dx(t)/dt$, the derivative of $x(t)$.

(b) Can you compute

$$z(t) = \int_{-\infty}^{\infty} [x(t) - 0.5] dt$$

If so, what is it equal to? If not, explain why.

(c) Is $x(t)$ a finite-energy signal? How about $y(t)$?

1.18. Complex exponentials

For a complex exponential signal $x(t) = 2e^{j2\pi t}$:

(a) Determine its analog frequency Ω_0 in rad/sec and its analog frequency f in hertz. Then find the signal's period.

(b) Suppose $y(t) = e^{j\pi t}$. Would the sum of these signals $z(t) = x(t) + y(t)$ also be periodic? If so, what is the period of $z(t)$?

(c) Suppose we then generate a signal $v(t) = x(t)y(t)$, with the $x(t)$ and $y(t)$ signals given before. Is $v(t)$ periodic? If so, what is its period?

1.19. Full-wave rectified signal—MATLAB

Consider the full-wave rectified signal

$$y(t) = |\sin(\pi t)| \qquad -\infty < t < \infty$$

part of which is shown in Figure 1.21.

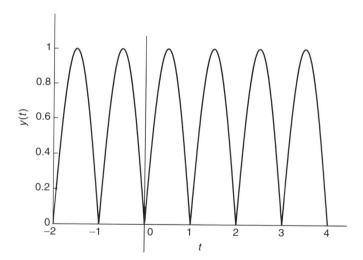

FIGURE 1.21
Problem 1.19.

(a) As a periodic signal, $y(t)$ does not have finite energy, but it has a finite power P_y. Find it.

(b) It is always useful to get a quick estimate of the power of a periodic signal by finding a bound for the signal squared. Find a bound for $|y(t)|^2$ and show that $P_y < 1$.

(c) Use symbolic MATLAB to check if the full-wave rectified signal has finite power and if that value coincides with the P_y you found above. Plot the signal and provide the script for the computation of the power. How does it coincide with the analytical result?

1.20. Multipath effects, first part—MATLAB

In wireless communications, the effects of *multipath* significantly affect the quality of the received signal. Due to the presence of buildings, cars, etc. between the transmitter and the receiver, the sent signal does not typically go from the transmitter to the receiver in a straight path (called *line of sight*). Several copies of the signal, shifted in time and frequency as well as attenuated, are received—that is, the transmission is done over multiple paths each attenuating and shifting the sent signal. The sum of these versions of the signal appears quite different from the original signal given that constructive as well as destructive effects may occur. In this problem we consider the time-shift of an actual signal to illustrate the effects of attenuation and time shift. In the next problem we consider the effects of time and frequency shifting and attenuation.

Assume that the MATLAB "handel.mat" signal is an analog signal $x(t)$ that it is transmitted over three paths, so that the received signal is

$$y(t) = x(t) + 0.8x(t - \tau) + 0.5x(t - 2\tau)$$

and let $\tau = 0.5$ seconds. Determine the number of samples corresponding to a delay of τ seconds by using the sampling rate F_s (samples per second) given when the file "handel.mat" is loaded.

To simplify matters, just work with a signal of duration 1 second—that is, generate a signal from "handel.mat" with the appropriate number of samples. Plot the segment of the original "handel.mat" signal $x(t)$ and the signal $y(t)$ to see the effect of multipath. Use the MATLAB function sound to listen to the original and the received signals.

1.21. Multipath effects, second part—MATLAB

Consider now the Doppler effect in wireless communications. The difference in velocity between the transmitter and the receiver causes a shift in frequency in the signal, which is called the *Doppler effect* (e.g., this is just like the acoustic effect of a train whistle as a train goes by).

To illustrate the frequency-shift effect, consider a complex exponential $x(t) = e^{j\Omega_0 t}$. Assume two paths: One that does not change the signal, while the other causes the frequency shift and attenuation, resulting in the signal

$$y(t) = e^{j\Omega_0 t} + \alpha e^{j\Omega_0 t} e^{j\phi t}$$
$$= e^{j\Omega_0 t} \left[1 + \alpha e^{j\phi t} \right]$$

where α is the attenuation and ϕ is the Doppler frequency shift, which is typically much smaller than the signal frequency. Let $\Omega_0 = \pi$, $\phi = \pi/100$, and $\alpha = 0.7$. This is analogous to the case where the received signal is the sum of the line-of-sight signal and an attenuated signal affected by Doppler.

(a) Consider the term $\alpha e^{j\phi t}$, a phasor with frequency $\phi = \pi/100$ to which we add 1. Use the MATLAB plotting function compass to plot the addition $1 + 0.7e^{j\phi t}$ for times from 0 to 256 sec, changing in increments of $T = 0.5$ sec.

(b) If we write $y(t) = A(t)e^{j(\Omega_0 t + \theta(t))}$, give analytical expressions for $A(t)$ and $\theta(t)$, and compute and plot them using MATLAB for the times indicated above.

(c) Compute the real part of the signal

$$y_1(t) = x(t) + 0.7x(t - 100)e^{j\phi(t - 100)}$$

That is, the effects of time and frequency delays, put together with attenuation, for the times indicated in part (a). Use the function sound (let $F_s = 2000$ in this function) to listen to the different signals.

1.22. Beating or pulsation—MATLAB

An interesting phenomenon in the generation of musical sounds is beating or pulsation. Suppose *NP* different players try to play a pure tone, a sinusoid of frequency 160 Hz, and that the signal recorded is the sum of these sinusoids. Assume the NP players while trying to play the pure tone end up playing tones separated by 0.02 Hz, so that the recorded signal is

$$y(t) = \sum_{i=1}^{NP} 10 \cos(2\pi f_i t)$$

where the f_i are frequencies from 159 to 161 separated by Δ Hz.

(a) Generate the signal $y(t)$ $0 \le t \le 200$ sec in MATLAB. Let each musician play a unique frequency. Consider an increasing number of players, letting NP be first 51 players with $\Delta = 0.04$ Hz, and then 101 players with $\Delta = 0.02$ Hz. Plot $y(t)$ for each of the different number of players.

(b) Explain how this is related with multipath and Doppler effects discussed in the previous problems.

1.23. **Chirps—MATLAB**

Pure tones or sinusoids are not very interesting to listen to. Modulation and other techniques are used to generate more interesting sounds. Chirps, which are sinusoids with time-varying frequency, are some of those more interesting sounds. For instance, the following is a chirp signal:

$$y(t) = A \cos(\Omega_c t + s(t))$$

(a) Let $A = 1$, $\Omega_c = 2$, and $s(t) = t^2/4$. Use MATLAB to plot this signal for $0 \le t \le 40$ sec in steps of 0.05 sec. Use the sound function to listen to the signal.

(b) Let $A = 1$, $\Omega_c = 2$, and $s(t) = -2 \sin(t)$. Use MATLAB to plot this signal for $0 \le t \le 40$ sec in steps of 0.05 sec. Use the sound function to listen to the signal.

(c) The frequency of these chirps is not clear. The instantaneous frequency $IF(t)$ is the derivative with respect to t of the argument of the cosine. For instance, for a cosine $\cos(\Omega_0 t)$, the $IF(t) = d\Omega_0 t/dt = \Omega_0$, so that the instantaneous frequency coincides with the conventional frequency. Determine the instantaneous frequencies of the two chirps and plot them. Do they make sense as frequencies? Explain.

CHAPTER 2

Continuous-Time Systems

Things should be made as simple as possible,
but not any simpler.
Albert Einstein (1879–1955)
physicist

2.1 INTRODUCTION

In this chapter we will consider the following topics:

- *Systems and their classification*—The concept of *system* is useful in dealing with actual devices or processes for purposes of analysis and synthesis. A transmission line, for instance, carrying information from one point to another is a system, even though physically it is just wires connecting two terminals. Voltages and currents in this system are not just functions of time but also of space. It takes time for a voltage signal to "travel" from one point to another separated by miles—Kirchhoff's laws do not apply. Resistance, capacitance, and inductance of the line are distributed over the length of the line—that is, the line is modeled as a concatenation of circuits characterized by values of resistance, capacitance, and inductance per unit length. A less complicated system could be one consisting of resistors, capacitors, and inductors where ideal models are used to represent these elements and to perform analysis and synthesis. The word "ideal" indicates that the models only approximate the real behavior of resistors, capacitors, and inductors. A more realistic model for a resistor would need to consider possible changes in the resistance due to temperature, and perhaps other marginal effects present in the resistor. Although this would result in a better model, for most practical applications it would be unnecessarily complicated.

- *Linear time-invariant systems*—We initiate the characterization of systems, and propose the linear time-invariant (LTI) model as a mathematical idealization of the behavior of systems—a good starting point. It will be seen that most practical systems deviate from it, but despite that, the behavior of many devices is approximated as linear and time invariant. A transistor, which is a nonlinear device, is analyzed using linear models around an operating point. Although the vocal system is hardly time invariant or linear, or even represented by a differential equation, in speech

Signals and Systems Using MATLAB®. DOI: 10.1016/B978-0-12-374716-7.00005-3

synthesis short intervals of speech are modeled as the output of linear time-invariant models. Finally, it will be seen that the LTI model is not appropriate to represent communication systems; rather, nonlinear or time-varying systems are more appropriate.

■ *Convolution integral, causality, and stability*— The output of a LTI system due to any signal is obtained by means of the generic signal representation obtained in Chapter 1. The response due to an impulse, together with the linearity and time-invariance of the system, gives the output as an integral. This convolution integral, although difficult to compute, even in simple cases, has significant theoretical value. It allows us not only to determine the response of the system for very general cases, but also provides a way to characterize causal and stable systems. Causality relates to the cause and effect of the input and the output of the system, giving us the conditions for real-time processing while stability characterizes useful systems. These two conditions are of great practical significance.

2.2 SYSTEM CONCEPT

Although we view a *system* as a mathematical transformation of an input signal (or signals) into an output signal (or signals), it is important to understand that such transformation results from an idealized model of the physical device or process we are interested in.

For instance, in the interconnection of physical resistors, capacitors, and inductors, the model idealizes how to deal with the resistors, capacitors, and inductors. In this simple RLC circuit, we would ignore, for instance, stray inductive and capacitive effects and the effect of temperature on the resistors. The resistance, capacitance, and impedance would be assumed localized in the physical devices and the wires would not have resistance, inductance, or capacitance. We would then use the circuits laws to obtain a differential equation to characterize the interconnection. A wire that in the RLC circuit model connects two elements, in a transmission line a similar wire is modeled as having capacitance, inductance, and resistance distributed over the line to realize the way the voltages travel over it. In practice, the model and the mathematical representation are not unique.

A system can be considered a connection of subsystems. Thinking of the RLC circuit as a system, for instance, the resistor, the capacitor, the inductor, and the source are the subsystems.

In engineering, the models are typically developed in the different areas. There will be, however, analogs as it is the case between mechanical and electrical systems. In such cases, the mathematical equations are similar, or even identical, but their significance is very different.

2.2.1 System Classification

According to general characteristics attributed to systems, they can be classified as follows:

■ *Static or dynamic systems*—A dynamic system has the capability of storing energy, or remembering its state, while a static system does not. A battery connected to resistors is a static system, while the same battery connected to resistors, capacitors, and inductors constitutes a dynamic system. The main difference is the capability of capacitors and inductors to store energy, to remember the state of the device, that resistors do not have.

- *Lumped- or distributed-parameter systems*—This classification relates as to how the elements of the system are viewed. In the case of the RLC circuit, the resistance, capacitance, and inductance are localized so that these physical elements are modeled as lumped elements. In the case of a transmission line resistance, capacitance and inductance are modeled as distributed over the length of the line.
- *Passive or active systems*—A system is passive if it is not able to deliver energy to the outside world. Constant resistors, capacitors, and inductors are passive elements. An operational amplifier is an active system.

Dynamic systems with lumped parameters, such as the RLC circuit, are typically represented by ordinary differential equations, while distributed-parameter dynamic systems like the transmission line are represented by partial differential equations. In the case of lumped systems only the time variation is of interest, while in the case of distributed systems we are interested in both time and space variations of the signals. In this book we consider only dynamic systems with lumped parameters, possibly changing with time, with a single input and a single output.

A further classification of systems is obtained by considering the types of signals present at the input and the output of the system.

> Whenever the input(s) and output(s) are both continuous time, discrete time, or digital, the corresponding systems are *continuous time*, *discrete time*, or *digital*, respectively. It is also possible to have *hybrid* systems when the input(s) and output(s) are not of the same type.

Of the systems presented in Chapter 0, the CD player is a hybrid system as it has a digital input (the bits stored on the disc) and an analog output (the acoustic signal put out by the player). The SDR system, on the other hand, can be considered to have an analog input (in the transmitter) and an analog output (at the receiver), making it an analog system, but having hybrid subsystems.

2.3 LTI CONTINUOUS-TIME SYSTEMS

> A continuous-time system is a system in which the signals at its input and output are continuous-time signals. Mathematically we represent it as a transformation S that converts an input signal $x(t)$ into an output signal $y(t) = S[x(t)]$ (see Figure 2.1):
>
> $$x(t) \quad \Rightarrow \quad y(t) = S[x(t)] \qquad (2.1)$$
>
> Input $\qquad$ Output

FIGURE 2.1

System S with input $x(t)$ and output $y(t)$.

Input
$x(t)$

S

Output
$y(t) = S[x(t)]$

When developing a mathematical model for a continuous-time system it is important to contrast the accuracy of the model with its simplicity and practicality. The following are some of the characteristics of the model being considered:

- Linearity
- Time invariance
- Causality
- Stability

The linearity between the input and the output, as well as the constancy of the system parameters, simplify the mathematical model. Causality, or nonanticipatory behavior of the system, relates to the cause–effect relationship between the input and the output. It is essential when the system is working under real-time situations—that is, when there is limited time for the system to process signals coming into the system. Stability is needed in practical systems. A stable system behaves well under reasonable inputs. Unstable systems are useless.

2.3.1 Linearity

A system represented by S is said to be *linear* if for inputs $x(t)$ and $v(t)$, and any constants α and β, superposition holds—that is,

$$S[\alpha x(t) + \beta v(t)] = S[\alpha x(t)] + S[\beta v(t)]$$
$$= \alpha S[x(t)] + \beta S[v(t)] \tag{2.2}$$

When checking the linearity of a system we first need to check the scaling—that is, if the output $y(t) = S[x(t)]$ for some input $x(t)$ is known, then for a scaled input $\alpha x(t)$ the output should be $\alpha y(t) = \alpha S[x(t)]$. If this condition is not satisfied, the system is nonlinear. If the condition is satisfied, you would then test the additivity or that the response to the sum of weighted inputs, $S[\alpha x(t) + \beta v(t)]$, is the sum of the corresponding responses $\alpha S[x(t)] + \beta S[v(t)]$.

The scaling property of linear systems indicates that whenever the input of a linear system is zero the output is zero. Thus, if the output corresponding to an input $x(t)$ is $y(t)$, then the output corresponding to $\alpha x(t)$ is $\alpha y(t)$; and if, in particular, $\alpha = 0$, then both input and output are zero.

■ Example 2.1

Consider a biased averager—that is, the output $y(t)$ of such a system is given by

$$y(t) = \frac{1}{T} \int_{t-T}^{t} x(\tau)d\tau + B$$

for an input $x(t)$. The system finds the average over an interval T and adds a constant value B. Is this system linear? If not, is there a way to make it linear? Explain.

Solution

Let $y(t)$ be the system response corresponding to $x(t)$. Assume then that we scale the input by a factor α so that the input is $\alpha x(t)$. The corresponding output is then

$$\frac{1}{T} \int_{t-T}^{t} \alpha x(\tau)d\tau + B = \frac{\alpha}{T} \int_{t-T}^{t} x(\tau)d\tau + B$$

which is not equal to

$$\alpha y(t) = \frac{\alpha}{T} \int_{t-T}^{t} x(\tau)d\tau + \alpha B$$

so the system is not linear. Notice that the difference is due to the term associated with B, which is not affected at all by the scaling of the input. So to make the system linear we let $B = 0$.

The constant B is the response due to zero input, and as such, the response can be seen as the sum of a linear system and a zero-input response. This type of system is called *incrementally linear* given that if

$$S[x_1(t)] = y_1(t) - B$$
$$S[x_2(t)] = y_2(t) - B$$

then

$$S[x_1(t) - x_2(t)] = S[x_1(t)] - S[x_2(t)] = y_1(t) - y_2(t)$$

$$= \frac{1}{T} \int_{t-T}^{t} [x_1(\tau) - x_2(\tau)]d\tau$$

That is, the difference of the responses to two inputs is linear. ∎

■ Example 2.2

Whenever the explicit relation between the input and the output of a system is represented by a nonlinear expression the system is nonlinear. Consider the following input–output relations that show the corresponding systems are nonlinear:

$$(i)\ y(t) = |x(t)|$$
$$(ii)\ z(t) = \cos(x(t))\ \text{assuming}\ |x(t)| \leq 1$$
$$(iii)\ v(t) = x^2(t)$$

where $x(t)$ is the input and $y(t)$, $z(t)$, and $v(t)$ are the outputs.

Solution

Superposition is not satisfied for the first system. If the outputs for $x_1(t)$ and $x_2(t)$ are $y_1(t) = |x_1(t)|$ and $y_2(t) = |x_2(t)|$, respectively, the output for $x_1(t) + x_2(t)$ is

$$y_{12}(t) = |x_1(t) + x_2(t)| \le |x_1(t)| + |x_2(t)| = y_1(t) + y_2(t)$$

For the second system, if the response for $x(t)$ is $z(t) = \cos(x(t))$, the response for $-x(t)$ is not $-z(t)$ because the cosine is an even function of its argument. Thus,

$$-x(t) \to \cos(-x(t)) = \cos(x(t)) = z(t)$$

For the third system, if $x_1(t) \to v_1(t) = (x_1(t))^2$ and $x_2(t) \to v_2(t) = (x_2(t))^2$ are corresponding input–output pairs, then

$$x_1(t) + x_2(t) \to (x_1(t) + x_2(t))^2 = (x_1(t))^2 + (x_2(t))^2 + 2x_1(t)x_2(t) \ne v_1(t) + v_2(t)$$

Thus, it is nonlinear.   ∎

■ Example 2.3

Consider each of the components of an RLC circuit and determine under what conditions they are linear.

Solution

Because a resistor, a capacitor, and an inductor are one-port or two-terminal elements, input and output variables are not obvious. However, from physics the cause and effect are well understood.

A resistor R has a voltage–current relation

$$v(t) = Ri(t) \tag{2.3}$$

If this relation is a straight line through the origin the resistor is linear; otherwise it is non-linear. A diode is an example of a nonlinear resistor: its voltage–current relation is nonlinear.

If the voltage-current relation is a straight line of constant slope R, considering the current is the input, superposition is satisfied. Indeed, if we apply to the resistor a current $i_1(t)$ to get $Ri_1(t) = v_1(t)$ and get $Ri_2(t) = v_2(t)$ when we apply a current $i_2(t)$, then when a current $ai_1(t) + bi_2(t)$, for any constants a and b, is applied, the voltage across the resistor is $v(t) = R(ai_1(t) + bi_2(t)) = av_1(t) + bv_2(t)$—that is, the resistor R is a linear system.

A capacitor is characterized by the *charge–voltage relation*

$$q(t) = Cv_c(t) \tag{2.4}$$

If this relation is not a straight line, the capacitor is nonlinear. A varactor is a diode for which its capacitance depends nonlinearly on the voltage applied to its terminals, and thus it is a nonlinear capacitor.

When the relation is a straight line through the origin with a constant slope C, using the current–charge relation $i(t) = dq(t)/dt$, we get the differential equation

$$i(t) = Cdv_c(t)/dt$$

characterizing the capacitor. Letting $i(t)$ be the input, solving this differential equation gives as output the voltage

$$v_c(t) = \frac{1}{C} \int_0^t i(\tau)d\tau + v_c(0) \qquad (2.5)$$

which explains the way the capacitor works. For time $t > 0$, the capacitor accumulates charge on its plates beyond the original charge due to an initial voltage $v_c(0)$. The capacitor is seen to be a linear system if $v_c(0) = 0$; otherwise it is not. In fact, when $v_c(0) = 0$, the outputs corresponding to $i_1(t)$ and $i_2(t)$ are

$$v_{c1}(t) = \frac{1}{C} \int_0^t i_1(\tau)d\tau$$

$$v_{c2}(t) = \frac{1}{C} \int_0^t i_2(\tau)d\tau$$

respectively, and the output due to a combination $ai_1(t) + bi_2(t)$ is

$$\frac{1}{C} \int_0^t [ai_1(\tau) + bi_2(\tau)]d\tau = av_{c1}(t) + bv_{c2}(t)$$

Thus, a linear capacitor is a linear system if it is not initially charged. When the initial condition is not zero, the capacitor is affected by the current input $i(t)$ as well as by the initial condition $v_c(0)$, and as such it is not possible to satisfy linearity, as only the current input can be changed. The capacitor is thus an incrementally linear system.

The inductor L is the dual of the capacitor (replacing currents by voltages and C by L in the above equations, we obtain the equations for the inductor). A *linear inductor* is characterized by the *magnetic flux–current relation*

$$\phi(t) = Li_L(t) \qquad (2.6)$$

being a straight line of slope $L > 0$. If the plot of the magnetic flux $\phi(t)$ and the current $i_L(t)$ is not a line, the inductor is nonlinear. The voltage across the inductor is

$$v(t) = \frac{d\phi(t)}{dt} = L\frac{di_L(t)}{dt}$$

according to Faraday's induction law. Solving this differential equation for the current we obtain

$$i_L(t) = \frac{1}{L} \int_0^t v(\tau)d\tau + i_L(0) \tag{2.7}$$

Like the capacitor, the inductor is not a linear system unless the initial current in the inductor is zero. The inductor can be considered an incrementally linear system.

Notice that an explicit relation between the input and the output was necessary to determine linearity. ■

Op-Amps and Feedback

Operational amplifiers, or op-amps, are high-gain amplifiers typically used with feedback. In the 1930s, Harold S. Black developed the principles of feedback amplifiers—that is the application of a portion of the output back to the input to reduce the overall gain. By doing so, the characteristics of the amplifier are greatly enhanced. In the late 1930s, George A. Philbrick developed a vacuum-tube circuit that performed some of the op-amp functions. Professor John Ragazzini, from Columbia University, coined the name of "operational amplifier" in 1947. Early op-amps were vacuum-tube based, and thus bulky and expensive. The trend to cheaper and smaller op-amps began in the 1960s [50, 72].

The Op-Amp

An excellent example of a device that can be used as either a nonlinear or a linear system is the operational amplifier or *op-amp*. It is a two-port device (see Figure 2.2) with two voltage inputs: $v_-(t)$, in the *inverting terminal*, and $v_+(t)$, in the *noninverting terminal*. The output voltage $v_0(t)$ is a nonlinear function of the difference between the two inputs—that is,

$$v_o(t) = f[v_+(t) - v_-(t)] = f(v_d(t))$$

The function $f(v_d(t))$ is approximately linear for small values $\pm\Delta V$ of $v_d(t)$, in the order of millivolts, and it becomes constant beyond $\pm\Delta V$. The output voltage $v_0(t)$ is, however, in the order of volts, so that letting

$$v_o(t) = A v_d(t) \qquad -\Delta V \leq v_d(t) \leq \Delta V$$

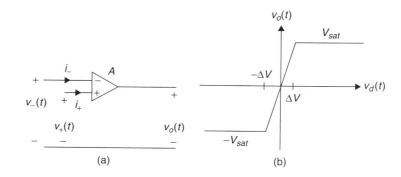

FIGURE 2.2
Operational amplifier: (a) circuit diagram, and (b) input–output voltage relation.

be a line through the origin, its slope A is very large. If $|v_d(t)| > \Delta V$ the output voltage is a constant V_{sat}. That is, the gain of the amplifier saturates. Furthermore, the input resistance of the op-amp is large so that the currents into the negative and the positive terminals are very small. The op-amp output resistance is relatively small.

Thus, depending on the dynamic range of the input signals, the op-amp operates in either a *linear* region or a *nonlinear* region. Restricting the operational amplifier to operate in the linear region simplifies the model. Assuming that $A \to \infty$, and that $R_{in} \to \infty$, then we obtain the following equations defining an *ideal operational amplifier*:

$$i_- = i_+ = 0$$
$$v_d(t) = v_+(t) - v_-(t) = 0 \qquad (2.8)$$

These equations are called the *virtual short* and are valid only if the output voltage of the operational amplifier is limited by the saturation voltage V_{sat}—that is, when

$$-V_{sat} \leq v_o(t) \leq V_{sat}$$

Later in the chapter we will consider ways to use the op-amp to get inverters, integrators, adders, and buffers.

2.3.2 Time Invariance

A continuous-time system S is *time invariant* if whenever for an input $x(t)$ with a corresponding output $S[x(t)]$, the output corresponding to a shifted input $x(t \mp \tau)$ (delayed or advanced) is the original output shifted in time $S[x(t \mp \tau)]$ (delayed or advanced). Thus,

$$x(t) \quad \Rightarrow \quad y(t) = S[x(t)]$$
$$x(t \mp \tau) \quad \Rightarrow \quad y(t \mp \tau) = S[x(t \pm \tau)] \qquad (2.9)$$

That is, the system does not age—its parameters are constant.

A system that satisfies both the linearity and the time invariance is called a *linear time-invariant* or LTI system.

Remarks

- *It should be clear that linearity and time invariance are independent of each other. Thus, one can have linear time-varying or nonlinear time-invariant systems.*
- *Although most actual systems are, according to the above definitions, nonlinear and time varying, linear models are used to approximate around an operating point the nonlinear behavior, and time-invariant models are used to approximate in short segments the system's time-varying behavior. For instance, in speech synthesis the vocal system is typically modeled as a linear time-invariant system for intervals of about 20 msec, attempting to approximate the continuous variation in shape in the different parts of the vocal system (mouth, cheeks, nose, etc.). A better model for such a system is clearly a linear time-varying model.*

■ *In many cases time invariance can be determined by identifying—if possible—the input and the output, and letting the rest represent the parameters of the system. If these parameters change with time, the system is time varying. For instance, if the input $x(t)$ and the output $y(t)$ of a system are related by the equation*

$$y(t) = f(t)x(t)$$

the parameter of the system is the function $f(t)$, and if it is not constant, the system is time varying. Thus, the system $y(t) = tx(t)$ is time varying as can be easily verified. Likewise, the AM modulation system given by $y(t) = \cos(\Omega_0 t)x(t)$ is time varying as the function $f(t) = \cos(\Omega_0 t)$.

AM Communication Systems

Amplitude modulation (AM) communication systems arose from the need to send an acoustic signal, the "message," over the airwaves using a reasonably sized antenna to radiate it. The size of the antenna depends inversely on the frequencies present in the message, and voice and music have relatively low frequencies. A voice signal typically has frequencies in the range of 100 Hz to about 5 KHz (the frequencies needed to make a telephone conversation intelligible), while music typically displays frequencies up to about 22 KHz. The transmission of such signals with a practical antenna is impossible. To make the transmission possible, *modulation* was introduced—that is, multiplying the message $m(t)$ by a periodic signal such as a cosine $\cos(\Omega_0 t)$, the carrier, with a frequency Ω_0 much larger than those in the acoustic signal. Amplitude modulation provided the larger frequencies needed to reduce the size of the antenna. Thus, $y(t) = m(t)\cos(\Omega_0 t)$ is the signal to be transmitted, and we will see later that the effect of this multiplication is to change the frequency content of the input. Such a system is clearly linear, but time-varying. Indeed, if the input is $m(t - \tau)$ the output would be $m(t - \tau)\cos(\Omega_0 t)$, which is not $y(t - \tau) = \cos(\Omega_0(t - \tau))m(t - \tau)$, as a time-invariant system would give. Figure 2.3 illustrates the AM transmitter and receiver. In Chapter 6, we will discuss AM and other modulation systems and will illustrate them with MATLAB simulations.

In comparison with the AM system, a frequency modulation (FM) system is represented by the following input–output equation, where $m(t)$ is the input message and $z(t)$ the output:

$$z(t) = \cos\left(\Omega_c t + \int_{-\infty}^{t} m(\tau)d\tau\right)$$

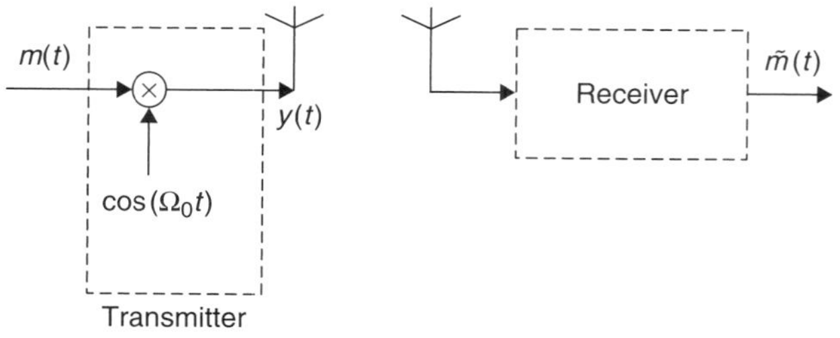

FIGURE 2.3

AM modulation: transmitter and receiver.

The FM system is nonlinear. Suppose that we scale the message to $\gamma m(t)$, for some constant γ, the corresponding output is given by

$$\cos\left(\Omega_c t + \gamma \int_{-\infty}^{t} m(\tau)d\tau\right)$$

which is not the previous output scaled (i.e., $\gamma z(t)$); thus FM is a nonlinear system.

The Beginnings of Radio

The names of Nikola Tesla (1856–1943) and Reginald Fessenden (1866–1932) are linked to the invention of radio and amplitude modulation [3, 58, 75]. Radio was initially called "wireless telegraphy" and then "wireless." Tesla was a mechanical as well as an electrical engineer, but mostly an inventor. He has been credited with significant contributions to electricity and magnetism in the late 19th and early 20th centuries. His work is the basis of the alternating-current (AC) power system and the induction motor. His work on wireless communications using the "Tesla coils" was capable of transmitting and receiving radio signals. Although Tesla submitted a patent application for his basic radio before Guglielmo Marconi, it was Marconi who was initially given the patent for the invention of the radio (1904). The Supreme Court in 1943 reversed the decision in favor of Tesla [45].

Fessenden has been called the "father of radio broadcasting." His early work on radio led to demonstrations in December 1906 of the capability of point-to-point wireless telephony, and what appears to be the first radio broadcasts of entertainment and music ever made to an audience (in this case, shipboard radio operators in the Atlantic). Fessenden was a professor of electrical engineering at Purdue University and the first chairman of the electrical engineering department of the University of Pittsburgh in 1893.

Vocal System

A remarkable system that we all have is the vocal system (see Figure 2.4). The air pushed out from the lungs in this system is directed by the trachea through the vocal cords, making them vibrate and create resonances similar to those from a wind musical instrument. The generated sounds are then muffled by the mouth and the nasal cavities, resulting in an acoustic signal carrying a message. Given the length of the typical vocal system, it is modeled as a distributed system and represented by partial differential equations. Due to the complexity of this model, it is the speech signal along with the understanding of the speech production that is used to obtain models of the vocal system. Speech processing is one of the most fascinating areas of electrical engineering.

A typical linear time-invariant model for speech production considers segments of speech of about 20 msec, and for each develops a low-order LTI system. The input is either a periodic pulse for the generation of voiced sounds (e.g., vowels) or a noiselike signal for unvoiced sounds (e.g., the /sh/ sound). Processing these inputs gives speechlike signals. A linear time-varying model would take into consideration the variations of the vocal system with time and it would thus be more appropriate.

■ Example 2.4

Characterize time-varying resistors, capacitors, and inductors. Assume zero initial conditions in the capacitors and inductors.

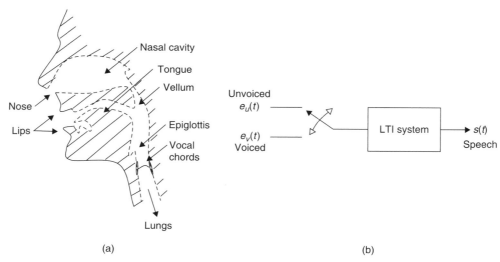

(a)   (b)

FIGURE 2.4

(a) Vocal system: principal organs of articulation. (b) Model for speech production.

Solution

If we generalize the characteristic equations for the resistor, capacitor, and inductor to be

$$v(t) = R(t)i(t)$$

$$q(t) = C(t)v_c(t)$$

$$\phi(t) = L(t)i_L(t)$$

as straight lines with time-varying slope, we have linear but time-varying elements. Using $i(t) = dq(t)/dt$ and $v(t) = d\phi(t)/dt$, we obtain the following voltage–current relations:

$$v(t) = R(t)i(t)$$

$$i(t) = C(t)\frac{dv_c(t)}{dt} + \frac{dC(t)}{dt}v_c(t)$$

$$v(t) = L(t)\frac{di_L(t)}{dt} + \frac{dL(t)}{dt}i_L(t)$$

As $R(t)$ is a function of time, the resistor is a time-varying system. The second and third equation are linear differential equations with time-varying coefficients representing time-varying capacitors and inductors. ∎

■ Example 2.5

Consider constant linear capacitors and inductors, represented by differential equations

$$\frac{dv_c(t)}{dt} = \frac{1}{C}i(t)$$

$$\frac{di_L(t)}{dt} = \frac{1}{L}v(t)$$

with initial conditions $v_c(0) = 0$ and $i_L(0) = 0$. Under what conditions are these time-invariant systems?

Solution

Given the duality of the capacitor and the inductor, we only need to consider one of these. Solving the differential equation for the capacitor, we get

$$v_c(t) = \frac{1}{C}\int_0^t i(\tau)d\tau$$

Let us then find out what happens when we delay (or advance) the input current $i(t)$ by λ sec. The corresponding output for $t > \lambda$ is given by

$$\frac{1}{C}\int_0^t i(\tau - \lambda)d\tau = \frac{1}{C}\int_{-\lambda}^0 i(\rho)d\rho + \frac{1}{C}\int_0^{t-\lambda} i(\rho)d\rho \qquad (2.10)$$

by changing the integration variable to $\rho = \tau - \lambda$. For Equation (2.10) to equal the voltage at the capacitor delayed λ sec, given by

$$v_c(t - \lambda) = \frac{1}{C}\int_0^{t-\lambda} i(\rho)d\rho$$

we need that $i(t) = 0$ for $t < 0$, so that the first integral in the right expression in Equation (2.10) is zero. Thus, the system is time invariant if the input current $i(t) = 0$ for $t < 0$. If the initial condition $v(0)$ is not zero, or if the input $i(t)$ is not zero for $t < 0$, then linearity or time invariance, or both, are not satisfied. A similar situation occurs with the inductor.

Thus, an RLC circuit is an LTI system provided that it is not energized for $t < 0$—that is, that the initial conditions as well as the input are zero for $t < 0$.   ■

RLC Circuits

An RLC circuit is represented by an ordinary differential equation of order equal to the number of independent inductors and capacitors (i.e., if two or more capacitors are connected in parallel, or if two or more inductors are connected in series they share the same initial conditions and can be simplified to one capacitor and one inductor), and with constant coefficients (due to the assumption

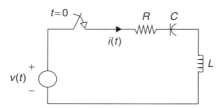

FIGURE 2.5

RLC circuit.

that the R, L, and C values are constant). If the initial conditions of the RLC circuit are zero, and the input is zero for $t < 0$, then the system represented by the linear differential equation with constant coefficients is LTI.

Consider, for instance, the circuit in Figure 2.5 consisting of a series connection of a resistor R, an inductor L, and a capacitor C. The switch has been open for a very long time and it is closed at $t = 0$, so that there is no initial energy stored in either the inductor or the capacitor (the initial current in the inductor is $i_L(0) = 0$ and the initial voltage in the capacitor is $v_C(0) = 0$) and the voltage applied to the elements is zero for $t < 0$. This circuit is represented by a second-order differential equation with constant coefficients. According to Kirchhoff's voltage law,

$$v(t) = Ri(t) + L\frac{di(t)}{dt} + \frac{1}{C}\int_0^t i(\tau)d\tau$$

and taking a derivative of $v(t)$ with respect to t we obtain

$$\frac{dv(t)}{dt} = R\frac{di(t)}{dt} + L\frac{d^2i(t)}{dt^2} + \frac{1}{C}i(t)$$

a second-order differential equation, with input the voltage source $v(t)$ and output the current $i(t)$.

2.3.3 Representation of Systems by Differential Equations

Given a dynamic system represented by a linear differential equation with constant coefficients,

$$a_0y(t) + a_1\frac{dy(t)}{dt} + \cdots + \frac{d^Ny(t)}{dt^N} = b_0x(t) + b_1\frac{dx(t)}{dt} + \cdots + b_M\frac{d^Mx(t)}{dt^M} \qquad t \geq 0$$

with N initial conditions $y(0)$, $d^ky(t)/dt^k|_{t=0}$ for $k = 1, \ldots, N-1$ and input $x(t) = 0$ for $t < 0$, its *complete response* $y(t)$ for $t \geq 0$ has two components:

- The *zero-state response*, $y_{zs}(t)$, due exclusively to the input as the initial conditions are zero.
- The *zero-input response*, $y_{zi}(t)$, due exclusively to the initial conditions as the input is zero. So that

$$y(t) = y_{zs}(t) + y_{zi}(t) \tag{2.11}$$

Thus, when the initial conditions are zero, then $y(t)$ depends exclusively on the input (i.e., $y(t) = y_{zs}(t)$), and the system is linear and time invariant or LTI.

On the other hand, if the initial conditions are different from zero, when checking linearity and time invariance we only change the input and do not change the initial conditions so that $y_{zi}(t)$ remains the same, and thus the system is nonlinear. The Laplace transform will provide the solution of these systems.

Most continuous-time dynamic systems with lumped parameters are represented by linear ordinary differential equations with constant coefficients. By linear it is meant that there are no nonlinear terms such as products of the input and the output, quadratic terms of the input and the output, etc. If the coefficients change with time the system is time varying. The order of the differential equation equals the number of independent elements capable of storing energy.

Consider a dynamic system represented by an Nth-order linear differential equation with constant coefficients, and with $x(t)$ as the input and $y(t)$ as the output:

$$a_0 y(t) + a_1 \frac{dy(t)}{dt} + \cdots + \frac{d^N y(t)}{dt^N} = b_0 x(t) + b_1 \frac{dx(t)}{dt} + \cdots + b_M \frac{d^M x(t)}{dt^M} \qquad t \geq 0 \qquad (2.12)$$

The corresponding N initial conditions are $y(0)$, $d^k y(t)/dt^k |_{t=0}$ for $k = 1, \ldots, N-1$. Defining the derivative operator as

$$D^n[y(t)] = \frac{d^n y(t)}{dt^n} \qquad n > 0, \text{ integer}$$

$$D^0[y(t)] = y(t)$$

we write the differential Equation (2.12) as

$$(a_0 + a_1 D + \cdots + D^N)[y(t)] = (b_0 + b_1 D + \cdots + b_M D^M)[x(t)] \qquad t \geq 0$$

$$D^k[y(t)]|_{t=0}, \qquad k = 0, \ldots, N-1$$

As indicated before, the system represented by this differential equation is LTI if the initial conditions as well as the input are zero for $t < 0$—that is, the system is not energized for $t < 0$. However, many LTI systems represented by differential equations have nonzero initial conditions. Considering that the input signal $x(t)$ is independent of the initial conditions, we can think of these as two different inputs. As such, using superposition we have that the *complete solution* of the differential equation is composed of a *zero-input solution*, due to the initial conditions when the input $x(t)$ is zero, and the *zero-state response* due to the input $x(t)$ with zero initial conditions.

Thus, to find the complete solution we need to solve the following two related differential equations:

$$(a_0 + a_1 D + \cdots + D^N)[y(t)] = 0 \qquad (2.13)$$

with initial conditions $D^k[y(t)]|_{t=0}, k = 0, \ldots, N-1$, and the differential equation

$$(a_0 + a_1 D + \cdots + D^N)[y(t)] = (b_0 + b_1 D + \cdots + b_M D^M)[x(t)] \qquad (2.14)$$

with zero initial conditions. If $y_{zi}(t)$ is the response of the zero-input differential Equation (2.13), and $y_{zs}(t)$ the zero-state (or zero initial conditions) differential Equation (2.14), we have that the complete response is their sum,

$$y(t) = y_{zi}(t) + y_{zs}(t)$$

Indeed, $y_{zi}(t)$ and $y_{zs}(t)$ satisfy

$$(a_0 + a_1 D + \cdots + D^N)[y_{zi}(t)] = 0$$

$$D^k[y_{zi}(t)]|_{t=0}, \qquad k = 0, \ldots, N-1$$

$$(a_0 + a_1 D + \cdots + D^N)[y_{zs}(t)] = (b_0 + b_1 D + \cdots + b_M D^M)[x(t)]$$

Adding these equations gives

$$(a_0 + a_1 D + \cdots + D^N)[y_{zi}(t) + y_{zs}(t)] = (b_0 + b_1 D + \cdots + b_M D^M)[x(t)]$$

$$D^k[y(t)]|_{t=0}, \qquad k = 0, \ldots, N-1$$

indicating that $y_{zi}(t) + y_{zs}(t)$ is the complete solution.

To find the solution of the zero-input and the zero-state equations we need to factor out the derivative operator $a_0 + a_1 D + \cdots + D^N$. We can do so by replacing D by a complex variable s, as the roots will be either real or in complex-conjugate pairs, simple or multiple. The *characteristic polynomial*

$$a_0 + a_1 s + \cdots + s^N = \prod_k (s - p_k)$$

is then obtained. The roots of this polynomial are called the *natural frequencies* or *eigenvalues* and characterize the dynamics of the system as it is being represented by the differential equation. The solution of the zero-state can be obtained from a modified characteristic polynomial.

The solution of differential equations will be efficiently done using the Laplace transform in the next chapter.

■ Example 2.6

Consider a circuit that is a series connection of a resistor $R = 1\ \Omega$ and an inductor $L = 1$ H, with a voltage source $v(t) = Bu(t)$, and I_0 amps is the initial current in the inductor. Find and solve the differential equation for $B = 1$ and $B = 2$ for initial conditions $I_0 = 1$ and $I_0 = 0$, respectively. Determine the zero-input and the zero-output responses. Under what conditions is the system linear and time invariant?

Solution

The first-order differential equation representing this circuit is given by

$$v(t) = i(t) + \frac{di(t)}{dt}$$

$$i(0) = I_0$$

The solution of this differential equation is given by

$$i(t) = [I_0 e^{-t} + B(1 - e^{-t})]u(t) \tag{2.15}$$

which satisfies the initial condition $i(0) = I_0$ and the differential equation. In fact, if $t = 0+$ (slightly larger than 0) we have that the solution gives $i(0+) = I_0$, and that for $t > 0$ when we replace in the differential equation the input voltage by B, $i(t)$, and $di(t)/dt$ (using the above solution), we get

$$\underbrace{B}_{v(t)} = \underbrace{[I_0 e^{-t} + B(1 - e^{-t})]}_{i(t)} + \underbrace{[Be^{-t} - I_0 e^{-t}]}_{di(t)/dt} = B \qquad t > 0$$

or an identity indicating $i(t)$ in Equation (2.15) is the solution of the differential equation.

Initial Condition Different from Zero

When $I_0 = 1$ and $B = 1$, the complete solution given by Equation (2.15) becomes

$$i_1(t) = [e^{-t} + (1 - e^{-t})]u(t)$$
$$= u(t) \tag{2.16}$$

The zero-state response (i.e., the response due to $v(t) = u(t)$ and zero initial condition) is

$$i_{1zs}(t) = (1 - e^{-t})u(t)$$

which is obtained by letting $B = 1$ and $I = 0$ in Equation (2.15). The zero-input response, when $v(t) = 0$ and the initial condition is $I_0 = 1$, is

$$i_{1zi}(t) = e^{-t}u(t)$$

obtained by subtracting the zero-state response from the complete response in Equation (2.16).

If we then consider $B = 2$ (i.e., we double the original input) and keep $I_0 = 1$, the complete solution is given by

$$i_2(t) = [e^{-t} + 2(1 - e^{-t})]u(t)$$
$$= (2 - e^{-t})u(t)$$

which is completely different from the expected $2i_1(t) = 2u(t)$ for a linear system. Thus, the system is not linear (see Figure 2.6). In this case we have that the zero-state response due to $v(t) = 2u(t)$ and zero-initial conditions is doubled so that

$$i_{2zs}(t) = 2(1 - e^{-t})u(t)$$

while the zero-input response remains the same, as the initial condition did not change. So,

$$i_{2zi}(t) = e^{-t}u(t)$$

and we get the complete solution shown above. The output in this case depends on the input $v(t)$ and on the initial condition, and when testing linearity we are only changing $v(t)$.

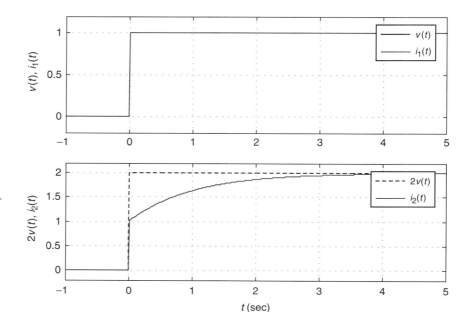

FIGURE 2.6
Nonlinear behavior of RL circuit: (top) $I_0 = 1$, $B = 1$, $v(t) = u(t)$, $i_1(t) = u(t)$, and (bottom) $I_0 = 1$, $B = 2$, $v(t) = 2u(t)$, $i_2(t) = (2 - e^{-t})u(t)$, and $i_2(t) \neq 2i_i(t)$.

Zero initial conditions

Suppose then we perform the above experiments with $I_0 = 0$ when $B = 1$ and when $B = 2$. We get

$$i_1(t) = (1 - e^{-t})u(t)$$

for $B = 1$, and for $B = 2$ we get

$$i_2(t) = 2(1 - e^{-t})u(t)$$
$$= 2i_1(t)$$

which indicates the system is linear. In this case the response only depends on the input $v(t)$.

Time invariance

Suppose now that $B = 1$, $v(t) = u(t - 1)$, and the initial condition is I_0. The complete response is

$$i_3(t) = I_0 e^{-t}u(t) + (1 - e^{-(t-1)})u(t - 1)$$

If $I_0 = 0$, then the above response is $i_3(t) = (1 - e^{-(t-1)})u(t - 1)$, which equals $i(t - 1)$ (Equation (2.15) with $B = 1$ and $I_0 = 0$ delayed by 1) indicating the system is time invariant. On the other hand, when $I_0 = 1$ the complete response is not equal to $i(t - 1)$ because the term with the initial condition is not shifted like the second term. The system in that case is time varying. Thus, if $I_0 = 0$ the system is LTI.

■

Analog mechanical systems

Making the analogy shown in Table 2.1 between the different variables and elements in a circuit and in a mechanical system the differential equations representing mechanical systems are found to be like those for RLC circuits.

Consider the translational mechanical system shown in Figure 2.7, composed of a mass M to which an external force $f(t)$ is being applied, and is moving at a velocity $w(t)$. It is assumed that between the mass and the floor there is a damping with a damping coefficient D. Just as with Kirchhoff's voltage law, the applied force equals the sum of the forces generated by the mass and the damper. Thus,

$$f(t) = M\frac{dw(t)}{dt} + Dw(t)$$

which is analogous to the differential equation of an RL series circuit with a voltage source $v(t)$:

$$v(t) = L\frac{di(t)}{dt} + Ri(t)$$

Exactly the same as with the RL circuit, if the initial velocity and the external force are zero for $t < 0$, the above differential equation represents a LTI mechanical system.

2.3.4 Application of Superposition and Time Invariance

The computation of the output of an LTI system is simplified when the input can be represented as the combination of signals for which we know their response. This is done by applying superposition and time invariance. This property of LTI systems will be of great importance in their analysis as you will soon learn.

Table 2.1 Equivalences in Mechanical and Electrical Systems

Mechanical System	Electrical System
force f(t)	voltage v(t)
velocity w(t)	current i(t)
mass M	inductance L
damping D	resistance R
compliance K	capacitance C

FIGURE 2.7

Analog mechanical and electrical systems. Using the equivalences $R = D$, $L = M$, $v(t) = f(t)$, and $i(t) = w(t)$, the two systems are represented by identical differential equations.

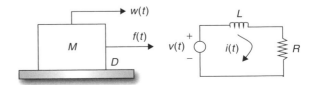

If $\mathcal{S}$ is the transformation corresponding to an LTI system, so that the response of the system is

$$y(t) = \mathcal{S}[x(t)] \text{ for an input } x(t)$$

then we have that

$$\mathcal{S}\left[\sum_k A_k x(t - \tau_k)\right] = \sum_k A_k \mathcal{S}[x(t - \tau_k)] = \sum_k A_k y(t - \tau_k)$$

$$\mathcal{S}\left[\int g(\tau)x(t - \tau)d\tau\right] = \int g(\tau)\mathcal{S}[x(t - \tau)]d\tau = \int g(\tau)y(t - \tau)d\tau$$

In the next section we will see that this property allows us to find the response of a linear time-invariant system due to any signal, if we know the response of the system to an impulse signal.

■ Example 2.7

The response of an RL circuit to a unit-step source $v(t) = u(t)$ is

$$i(t) = (1 - e^{-t})u(t)$$

Find the response to a source $v(t) = u(t) - u(t - 2)$.

Solution

Using superposition and time invariance, the output current due to the pulse $v(t) = u(t) - u(t - 2)$ volts is

$$i(t) - i(t - 2) = 2(1 - e^{-t})u(t) - 2(1 - e^{(t-2)})u(t - 2)$$

Figure 2.8 shows the responses to $u(t)$ and $u(t - 2)$ and the overall response to $v(t) = u(t) - u(t - 2)$. ■

■ Example 2.8

Suppose we know that the response to a rectangular pulse $v_1(t)$ is the current $i_1(t)$ shown in Figure 2.9. If the input voltage is a train of two pulses, $v(t)$, find the corresponding current $i(t)$.

Solution

Graphically the response to $v(t)$ of the LTI system is given by $i(t)$ as shown in Figure 2.9. ■

2.3.5 Convolution Integral

In this section we consider the computation of the output of a continuous-time linear time-invariant (LTI) system due to any continuous-time input signal.

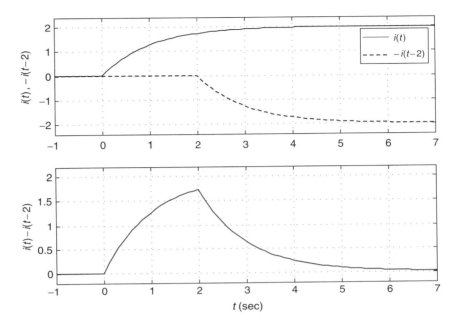

FIGURE 2.8

Response of an RL circuit to a pulse $v(t) = u(t) - u(t - 2)$ using superposition and time invariance.

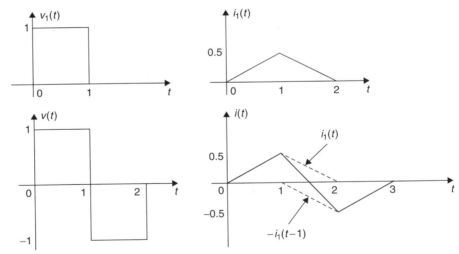

FIGURE 2.9

Application of superposition and time invariance to find the response of an LTI system.

Recall that the generic representation of a signal $x(t)$ in terms of shifted $\delta(t)$ signals found in Chapter 1 is given by

$$x(t) = \int_{-\infty}^{\infty} x(\tau)\delta(t - \tau)d\tau \tag{2.17}$$

Next we define the impulse response of an LTI and find the response due to $x(t)$. The impulse response of an analog LTI system, $h(t)$, is the output of the system corresponding to an impulse $\delta(t)$ as input, and initial conditions equal to zero.

If the input $x(t)$ in Equation (2.17) is seen as an infinite sum of weighted and shifted impulses $x(\tau)\delta(t-\tau)$ then the output of an LTI system is the superposition of the responses to each of these terms.

The response of an LTI system $\mathcal{S}$ represented by its impulse response $h(t) = \mathcal{S}[\delta(t)]$ (i.e., the output of the system to an impulse signal $\delta(t)$ and zero initial conditions) to any signal $x(t)$ is the *convolution integral*

$$y(t) = \int_{-\infty}^{\infty} x(\tau)h(t-\tau)d\tau = \int_{-\infty}^{\infty} x(t-\tau)h(\tau)d\tau$$

$$= [x*h](t) = [h*x](t) \qquad (2.18)$$

where the symbol $*$ stands for the convolution integral of the input signal and the impulse response of the system.

The above can be seen as follows:

- Assuming no energy is initially stored in the system (i.e., initial conditions are zero) the response to $\delta(t)$ is the impulse response $h(t)$.
- Given that the system is time invariant, the response to $\delta(t-\tau)$ is $h(t-\tau)$ and by linearity the response to $x(\tau)\delta(t-\tau)$ is $x(\tau)h(t-\tau)$ since $x(\tau)$ is not a function of time t.
- Thus, the response of the system to the generic representation Equation (2.17)

$$x(t) = \int_{-\infty}^{\infty} x(\tau)\delta(t-\tau)d\tau$$

is by superposition

$$y(t) = \int_{-\infty}^{\infty} x(\tau)h(t-\tau)d\tau$$

or equivalently

$$y(t) = \int_{-\infty}^{\infty} x(t-\sigma)h(\sigma)d\sigma$$

after letting $\sigma = t - \tau$. The two integrals are identical—each gives the response of the LTI system. The impulse response $h(t)$ represents the system. Notice that in the convolution integral the input and the impulse response commute (i.e., are interchangeable).

Remarks

■ *We will see that the impulse response is fundamental in the characterization of linear time-invariant systems.*

■ *Any system characterized by the convolution integral is linear and time invariant by the above construction. The convolution integral is a general representation of LTI systems, given that it was obtained from a generic representation of the input signal.*

■ *We showed before that a system represented by a linear differential equation with constant coefficients and no initial conditions, or input, before $t = 0$ is LTI. Thus, one should be able to represent that system by a convolution integral after finding its impulse response $h(t)$.*

■ Example 2.9

Obtain the impulse response of a capacitor and use it to find its unit-step response by means of the convolution integral. Let $C = 1$ F.

Solution

For a capacitor with a initial voltage $v_c(0) = 0$, we have that

$$v_c(t) = \frac{1}{C} \int_0^t i(\tau) d\tau$$

The impulse response of a capacitor is found by letting the input $i(t) = \delta(t)$ and the output $v_c(t) = h(t)$, which according to the above equation becomes

$$h(t) = \frac{1}{C} \int_0^t \delta(\tau) d\tau = \frac{1}{C} \qquad t > 0$$

and zero if $t < 0$, or $h(t) = (1/C)u(t)$. For $C = 1F$, to compute the unit-step response of the capacitor we let the input $i(t) = u(t)$, and $v_c(0) = 0$. The voltage across the capacitor is

$$v_c(t) = \int_{-\infty}^{\infty} h(t - \tau)i(\tau) d\tau = \int_{-\infty}^{\infty} \frac{1}{C} u(t - \tau)u(\tau) d\tau$$

and since, as a function of τ, $u(t - \tau)u(\tau) = 1$ for $0 \le \tau \le t$ and zero otherwise, we have that

$$v_c(t) = \int_0^t d\tau = t$$

for $t \ge 0$ and zero otherwise (as the input is zero for $t < 0$ and there are no initial conditions), or $v_c(t) = r(t)$. The above result makes physical sense since the capacitor is accumulating charge and the input is providing a constant charge, so that the result is a ramp function. Notice that the impulse response is the derivative of the unit-step response.

The relation between the impulse response and the unit-step and the ramp responses can be generalized for any system as the impulse response $h(t)$, the unit-step response $s(t)$, and the ramp response $\rho(t)$ are related by

$$h(t) = \begin{cases} ds(t)/dt \\ d^2\rho(t)/dt^2 \end{cases} \qquad (2.19)$$

This can be shown by computing first $s(t)$ (the output due to a unit-step input):

$$s(t) = \int_{-\infty}^{\infty} u(t-\tau)h(\tau)d\tau = \int_{-\infty}^{t} h(\tau)d\tau$$

since

$$u(t-\tau) = \begin{cases} 1 & \tau \le t \\ 0 & \tau > t \end{cases}$$

The derivative of $s(t)$ is $h(t)$.

Similarly, the ramp response $\rho(t)$ of a LTI system, represented by the impulse response $h(t)$, is given by

$$\rho(t) = \int_{-\infty}^{\infty} h(\tau)(t-\tau)u(t-\tau)d\tau = \int_{-\infty}^{t} h(\tau)(t-\tau)d\tau = t\int_{-\infty}^{t} h(\tau)d\tau - \int_{-\infty}^{t} h(\tau)\tau d\tau$$

and its derivative is

$$\frac{d\rho(t)}{dt} = \underbrace{\int_{-\infty}^{t} h(\tau)d\tau + th(t)}_{d(t\int_{-\infty}^{t} h(\tau)d\tau)/dt} - \underbrace{th(t)}_{d(\int_{-\infty}^{t} h(\tau)\tau d\tau)/dt} = \int_{-\infty}^{t} h(\tau)d\tau$$

so that the second derivative of $\rho(t)$ is $h(t)$—that is,

$$\frac{d^2\rho(t)}{dt^2} = \frac{d}{dt}\left[\int_{-\infty}^{t} h(\tau)d\tau\right] = h(t)$$

Using the Laplace transform, one is able to obtain the above relations in a much simpler way. ∎

■ Example 2.10

The output $y(t)$ of an analog averager is given by

$$y(t) = \frac{1}{T}\int_{t-T}^{t} x(\tau)d\tau$$

which corresponds to the accumulation of values of $x(t)$ in a segment $[t - T, t]$ divided by its length T, or the average of $x(t)$ in $[t - T, t]$. Use the convolution integral to find the response of the averager to a ramp.

Solution

To find the ramp response using the convolution integral we first need $h(t)$. The impulse response of an averager can be found by letting $x(t) = \delta(t)$ and $y(t) = h(t)$ or

$$h(t) = \frac{1}{T} \int_{t-T}^{t} \delta(\tau) d\tau$$

If $t < 0$ or if $t - T > 0$ this integral is zero as in these two situations $t = 0$, where the delta function occurs, is not included in the integral limits. However, when $t - T < 0$ and $t > 0$, or $0 < t < T$, the integral is 1 as the origin $t = 0$, where $\delta(t)$ occurs, is included in this interval. Thus, the impulse response of the analog averager is

$$h(t) = \begin{cases} \frac{1}{T} & 0 < t < T \\ 0 & \text{otherwise} \end{cases}$$

We then have that the output $y(t)$, for a given input $x(t)$, is given by the convolution integral

$$y(t) = \int_{-\infty}^{\infty} h(\tau) x(t - \tau) d\tau = \int_{0}^{T} \frac{1}{T} x(t - \tau) d\tau$$

which can be shown to equal the definition of the averager by a change of variable. Indeed, let $\sigma = t - \tau$, so when $\tau = 0$ then $\sigma = t$, and when $\tau = T$ then $\sigma = t - T$. Moreover, we have that $d\sigma = -d\tau$. The above integral becomes

$$y(t) = -\frac{1}{T} \int_{t}^{t-T} x(\sigma) d\sigma = \frac{1}{T} \int_{t-T}^{t} x(\sigma) d\sigma$$

Thus, we have that

$$y(t) = \frac{1}{T} \int_{0}^{t} x(t - \tau) d\tau = \frac{1}{T} \int_{t-T}^{t} x(\sigma) d\sigma \qquad (2.20)$$

If the input is a ramp, $x(t) = tu(t)$, the ramp response $\rho(t)$ is

$$\rho(t) = \frac{1}{T} \int_{t-T}^{t} x(\sigma) d\sigma = \frac{1}{T} \int_{t-T}^{t} \sigma u(\sigma) d\sigma$$

If $t - T < 0$ and $t \geq 0$, the above integral becomes

$$\rho(t) = \frac{1}{T} \int_0^t \sigma \, d\sigma = \frac{t^2}{2T} \qquad 0 \leq t < T$$

but if $t - T \geq 0$, we would then get

$$\rho(t) = \frac{1}{T} \int_{t-T}^t \sigma \, d\sigma = \frac{t^2 - (t - T)^2}{2T} = t - \frac{T}{2} \qquad t \geq T$$

So that the ramp response is

$$\rho(t) = \begin{cases} 0 & t < 0 \\ t^2/(2T) & 0 \leq t < T \\ t - T/2 & t \geq T \end{cases}$$

Notice that the second derivative of $\rho(t)$ is

$$\frac{d^2\rho(t)}{dt^2} = \begin{cases} 1/T & 0 \leq t < T \\ 0 & \text{otherwise} \end{cases}$$

which is the impulse response of the averager as found before. ∎

■ Example 2.11

Find the convolution integral $y_T(t)$ of a pulse $x(t) = u(t) - u(t - T_0)$ with a sampling signal

$$\delta_T(t) = \sum_{k=-\infty}^{\infty} \delta(t - kT)$$

Consider $T = T_0$ and $T = 2T_0$. Find and plot the corresponding $y_T(t)$.

Solution

For any value of T the convolution integral is given by

$$y_T(t) = \int_{-\infty}^{\infty} \delta_T(\tau)x(t - \tau)d\tau = \int_{-\infty}^{\infty} \sum_{k=-\infty}^{\infty} \delta(\tau - kT)x(t - \tau)d\tau$$

$$= \sum_{k=-\infty}^{\infty} \int_{-\infty}^{\infty} \delta(\tau - kT)x(t - \tau)d\tau = \sum_{k=-\infty}^{\infty} x(t - kT) \int_{-\infty}^{\infty} \delta(\tau - kT)d\tau$$

$$= \sum_{k=-\infty}^{\infty} x(t - kT)$$

where we used the sifting property of the impulse and that its area is unity. If we let $T = T_0$ and let the unit step be $u(0) = 0.5$, the signal $y_{T_0}(t) = 1$ for $-\infty < t < \infty$. When $T = 2T_0$, the signal $y_{2T_0}(t)$ is a periodic train of rectangular pulses of period $2T_0$. See Figure 2.10.

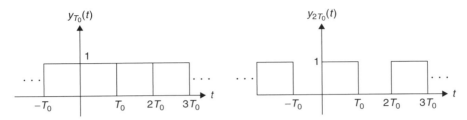

FIGURE 2.10
Convolution with a sequence of unit impulses as input. Notice the result is the superposition of the input signal shifted by the time-shift kT of the impulses. For $T = T_0$, $y_T(t) = 1$ and for $T = 2T_0$ is a sequence of pulses.   ■

2.3.6 Causality

Causality relates to the conditions under which processing of a signal can be performed in real time—when it is necessary to process the signal as it comes into the system. For real-time processing the system needs to be causal. In many situations the data can be stored and processed without the requirements of real-time processing; under such circumstances causality is not necessary.

A continuous-time system S is called *causal* if:
- Whenever the input $x(t) = 0$ and there are no initial conditions, the output is $y(t) = 0$.
- The output $y(t)$ does not depend on future inputs.

For a value $\tau > 0$, when considering causality it is helpful to think of

- The time t (the time at which the output $y(t)$ is being computed) as the *present*.
- Times $t - \tau$ as the *past*.
- Times $t + \tau$ as the *future*.

Remarks

Causality is independent of the linearity and the time-invariance properties of a system. For instance, the system represented by the input–output equation

$$y(t) = x^2(t)$$

where $x(t)$ is the input and $y(t)$ the output, is nonlinear but time invariant, and according to the above definition is a causal system. Likewise, an LTI system can be noncausal. Consider the following averager:

$$y(t) = \frac{1}{2T} \int_{t-T}^{t+T} x(\tau)d\tau$$

which can be written as

$$y(t) = \frac{1}{2T} \int_{t-T}^{t} x(\tau)d\tau + \frac{1}{2T} \int_{t}^{t+T} x(\tau)d\tau$$

At the present time t, y(t) consists of the average of a past and present values in $[t - T, t]$ of the input, and of the average of future values of the signal (i.e., the average of values x(t) for $[t, t + T]$). Thus, this system is not causal.

An LTI system represented by its impulse response $h(t)$ is *causal* if

$$h(t) = 0 \qquad \text{for } t < 0 \tag{2.21}$$

The output of a causal LTI system with a causal input $x(t)$ (i.e., $x(t) = 0$ for $t < 0$) is

$$y(t) = \int_{0}^{t} x(\tau)h(t - \tau)d\tau \tag{2.22}$$

One can understand the above results by considering the following:

- The choice of the starting time as $t = 0$ is for convenience. It is purely arbitrary as the system being considered is time invariant, so that similar results are obtained for any other starting time.
- When computing the impulse response $h(t)$, the input $\delta(t)$ only occurs at $t = 0$ and there are no initial conditions. Thus, $h(t)$ should be zero for $t < 0$ since for $t < 0$ there is no input and there are no initial conditions.
- A causal LTI system is represented by the convolution integral

$$y(t) = \int_{-\infty}^{\infty} x(\tau)h(t - \tau)d\tau$$

$$= \int_{-\infty}^{t} x(\tau)h(t - \tau)d\tau + \int_{t}^{\infty} x(\tau)h(t - \tau)d\tau$$

where the second integral is zero according to the causality of the system ($h(t - \tau) = 0$ when $\tau > t$ since the argument of $h(.)$ becomes negative). Thus, we obtain

$$y(t) = \int_{-\infty}^{t} x(\tau)h(t - \tau)d\tau$$

■ If the input signal $x(t)$ is causal (i.e., $x(t) = 0$ for $t < 0$), we can simplify further the above equation. Indeed

$$y(t) = \int_0^t x(\tau)h(t - \tau)d\tau$$

where the lower limit of the integral is set by the causality of the input signal, and the upper limit is set by the causality of the system. This equation clearly indicates that the system is causal, as the output $y(t)$ depends on present and past values of the input (considering the integral an infinite sum, the integrand depends continuously on $x(\tau)$, from $\tau = 0$ to $\tau = t$, which are past and present input values). Also if $x(t) = 0$ the output is also zero.

2.3.7 Graphical Computation of Convolution Integral

Graphically, the computation of the convolution integral, Equation (2.18), consists in multiplying $x(\tau)$ (as a function of τ) by a reflected (again as function of τ) and shifted to the right t sec impulse response $h(t - \tau)$. Once this product is obtained we integrate it from 0 to t (the time at which we are computing the convolution). The computational cost of this operation is rather high considering that these operations need to be done for each value of t for which we are interested in finding the output $y(t)$. A more efficient way will be by using the Laplace transform as we will see in the next chapter.

■ Example 2.12

Graphically find the unit-step $y(t)$ response of an averager, with $T = 1$ sec, which has an impulse response

$$h(t) = u(t) - u(t - 1)$$

Solution

Plotting the input signal $x(\tau) = u(\tau)$ and the reflected and delayed impulse response $h(t - \tau)$, both as functions of τ, for some value of t (notice that when $t = 0$, $h(-\tau)$ is the reflected version of the impulse response, and for $t > 0$, $h(t - \tau)$ is $h(-\tau)$ shifted by t to the right) are as shown in Figure 2.11. Notice the position of $h(t - \tau)$ with respect to $x(\tau)$ as it moves from left to right as t goes from $-\infty$ to ∞.

We then have the following results for different values of t:

■ If $t < 0$, then $h(t - \tau)$ and $x(\tau)$ do not overlap and so the convolution integral is zero, or $y(t) = 0$ for $t < 0$. That is, the system for $t < 0$ has not yet been affected by the input.
■ For $t \geq 0$ and $t - 1 < 0$, or equivalently $0 \leq t < 1$, $h(t - \tau)$ and $x(\tau)$ increasingly overlap, and as such the integral increases linearly from 0 at $t = 0$ to 1 when $t = 1$. So that $y(t) = t$ for $0 \leq t < 1$. That is, for this period of time the system starts reacting slowly to the input.

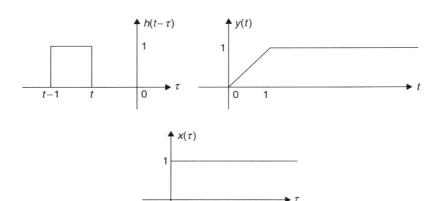

FIGURE 2.11
Graphical convolution for a
unit-step input into an
averager with $T = 1$.

- For $t \geq 1$, the overlap of $h(t - \tau)$ and $x(\tau)$ remains constant, and as such the integral is unity from then on, or $y(t) = 1$ for $t \geq 1$. The response for $t \geq 1$ has attained steady state. Thus, the complete response is given as

$$y(t) = r(t) - r(t - 1)$$

where $r(t) = tu(t)$, the ramp function. ∎

■ Example 2.13

Consider the graphical computation of the convolution integral of two pulses of the same duration (see Figure 2.12).

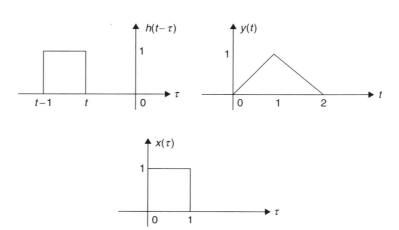

FIGURE 2.12
Graphical convolution of two equal
pulses—that is, a system with input
$x(t) = u(t) - u(t - 1)$ and impulse
response $h(t) = x(t)$.

Solution

In this case, $x(t) = h(t) = u(t) - u(t-1)$. Again we plot $x(\tau)$ and $h(t - \tau)$ both as functions of τ, for $-\infty < t < \infty$.

- It should be noticed that while computing the convolution integral for t increasing from negative to positive values, $h(t - \tau)$ moves from left to right while $x(\tau)$ remains stationary, and that they only overlap on a finite support.
- For $t < 0$, $h(t - \tau)$ and $x(\tau)$ do not overlap, so $y(t) = 0$ for $t < 0$.
- $h(t - \tau)$ and $x(\tau)$ increasingly overlap for $0 \le t < 1$ and decreasingly overlap for $1 \le t < 2$. So that $y(t) = t$ for $0 \le t < 1$, and $y(t) = 2 - t$ for $1 \le t < 2$.
- For $t > 2$, there is no overlap and so $y(t) = 0$ for $t > 2$.

Thus, the complete response is

$$y(t) = r(t) - 2r(t-1) + r(t-2)$$

where $r(t) = tu(t)$ is the ramp signal.

Notice in this example that:

- The result of the convolution of these two pulses, $y(t)$, is smoother than $x(t)$ and $h(t)$. This is because $y(t)$ is the continuous average of $x(t)$, as $h(t)$ is the impulse response of the averager in example 2.12.
- The length of the support of $y(t)$ equals the sum of the lengths of the supports of $x(t)$ and $h(t)$. This is a general result that applies to any two signals $x(t)$ and $h(t)$. ■

The length of the support of $y(t) = [x * h](t)$ is equal to the sum of the lengths of the supports of $x(t)$ and $h(t)$.

2.3.8 Interconnection of Systems—Block Diagrams

Systems can be considered a connection of subsystems. In the case of LTI systems, to visualize the interaction of the different subsystems each of the subsystems is represented by a block with the corresponding impulse response, or equivalently by its Laplace transform as we will see in the next chapter. The flow of the signals is indicated by arrows, and the addition of signals or multiplication of a signal by a constant is indicated by means of circles.

Two possible connections, the *cascade* and the *parallel* connections, result from the properties of the convolution integral, while the *feedback* connection is found in many natural systems and has been replicated in engineering, especially in control. The concept of feedback is one of the greatest achievements of the 20th century. See Figure 2.13.

Cascade Connection

When connecting LTI systems in cascade the impulse response of the overall system can be found using the convolution integral.

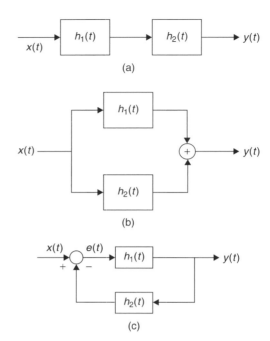

FIGURE 2.13

Block diagrams for connecting two LTI systems with impulse responses $h_1(t)$ and $h_2(t)$ in (a) cascade, (b) parallel, and (c) negative feedback.

Two LTI systems with impulse responses $h_1(t)$ and $h_2(t)$ connected in *cascade* have as an overall impulse response

$$h(t) = [h_1 * h_2](t) = [h_2 * h_1](t)$$

where $h_1(t)$ and $h_2(t)$ commute (i.e., they can be interchanged).

In fact, if the input to the cascade connection is $x(t)$, the output $y(t)$ is found as

$$
\begin{aligned}
y(t) &= [[x * h_1] * h_2](t) \\
     &= [x * [h_1 * h_2]](t) \\
     &= [x * [h_2 * h_1]](t)
\end{aligned}
$$

where the last two equations show the *commutative property* of convolution. The impulse response of the cascade connection indicates that the order in which we connect LTI systems is not important—that we can put the system with impulse response $h_1(t)$ first, or the system with impulse response $h_2(t)$ first with no effect in the overall response of the system (we will see later that this is true provided that the two systems do not load each other). When dealing with linear but time-varying systems, however, the order in which we connect the systems in cascade is important.

Parallel Connection

If we connect in *parallel* two LTI systems with impulse responses $h_1(t)$ and $h_2(t)$, the impulse response of the overall system is

$$h(t) = h_1(t) + h_2(t)$$

In fact, the output of the parallel combination is

$$
\begin{aligned}
y(t) &= [x * h_1](t) + [x * h_2](t) \\
&= [x * (h_1 + h_2)](t)
\end{aligned}
$$

which is the *distributive property* of convolution.

Feedback Connection

In these connections the output of the system is fed back and compared with the input of the system. The fedback output is either added to the input giving a *positive feedback* system or subtracted from the input giving a *negative feedback* system. In most cases, especially in control systems, negative feedback is used. Figure 2.13(c) illustrates the negative feedback connection.

Given two LTI systems with impulse responses $h_1(t)$ and $h_2(t)$, a negative feedback connection (Figure 2.13(c)) is such that the output is

$$y(t) = [h_1 * e](t)$$

where the error signal is

$$e(t) = x(t) - [y * h_2](t)$$

The overall impulse response $h(t)$, or the impulse response of the *closed-loop* system, is given by the implicit expression

$$h(t) = [h_1 - h * h_1 * h_2](t)$$

If $h_2(t) = 0$ (i.e., there is no feedback) the system is called an *open-loop* system and $h(t) = h_1(t)$.

Using the Laplace transform we will obtain later an explicit expression for the Laplace transform of $h(t)$. To obtain the above result we consider the output of the system as the overall impulse response $y(t) = h(t)$ due to an input $x(t) = \delta(t)$. Then $e(t) = \delta(t) - [h * h_2](t)$, and so when replaced in the expression for the output

$$h(t) = [e * h_1](t) = [(\delta - h * h_2) * h_1](t) = [h_1 - h * h_1 * h_2](t)$$

the implicit expression is as given above. When there is no feedback, $h_2(t) = 0$, then $h(t) = h_1(t)$.

■ Example 2.14

Consider the block diagram in Figure 2.14 with input a unit-step signal, $u(t)$. The averager is such that for an input $x(t)$ its output is

$$y(t) = \frac{1}{T} \int_{t-T}^{t} x(\tau)d\tau$$

Determine what the system is doing as we let the delay $\Delta \to 0$. Consider that the averager and the system with input $u(t)$ and output $x(t)$ are LTI.

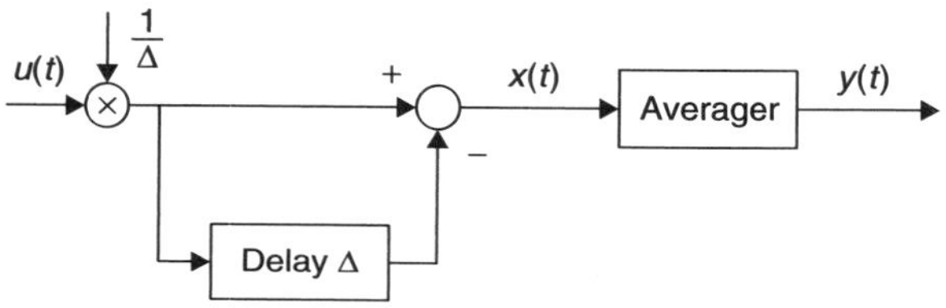

FIGURE 2.14
Block diagram of the cascading of two LTI systems, one of them being an averager.

Solution

Since it is not clear from the given block diagram what the system is doing, using the LTI of the two systems connected in cascade lets us reverse their order so that the averager is first (see Figure 2.15), obtaining an equivalent block diagram.

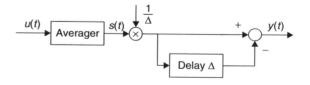

FIGURE 2.15
Equivalent block diagram of the cascading of two LTI systems, one of them being an averager.

The output of the averager is

$$s(t) = \frac{1}{T} \int_{t-T}^{t} u(\tau)d\tau = \begin{cases} 0 & t < 0 \\ t/T & 0 \le t < T \\ 1 & t \ge T \end{cases}$$

as we obtained before in example 2.12. The output $y(t)$ of the other system is given by

$$y(t) = \frac{1}{\Delta}[s(t) - s(t - \Delta)]$$

If we then let $\Delta \to 0$ we have that (recall that $ds(t)/dt = h(t)$ is the relation between the unit-step response $s(t)$ and the impulse response $h(t)$)

$$\lim_{\Delta \to 0} y(t) = \frac{ds(t)}{dt} = h(t) = \frac{1}{T}[u(t) - u(t - T)]$$

That is, this system approximates the impulse response of the averager. ∎

■ Example 2.15

Consider the circuits obtained with an operational amplifier when we feed back its output with a wire, a resistor, and a capacitor (Figure 2.16). Assume the linear model for the op-amp. The circuits in Figure 2.16 are called a *voltage follower*, an *integrator*, and an *adder*.

Solution

Virtual follower circuit. Although the operational amplifier can be made linear, its large gain A makes it not useful. Feedback is needed to make the op-amp useful. The *voltage follower* circuit (Figure 2.16(a)), which is used to isolate cascaded circuits, is a good example of a feedback system. Given that the voltage differential is assumed to be zero, then $v_-(t) = v_i(t)$, and therefore the output voltage is

$$v_o(t) = v_i(t)$$

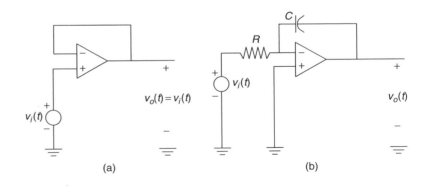

(a)   (b)

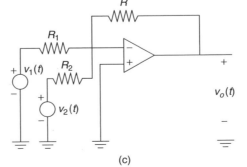

FIGURE 2.16

Operational amplifier circuits:
(a) virtual follower, (b) inverting
integrator, and (c) adder with
inversion.

(c)

The input resistance of this circuit is $R_{in} = \infty$ and the output resistance is $R_{out} = 0$ so that the output behaves as an ideal voltage source. The voltage follower is used to isolate two circuits connected in cascade, as the connected circuit at either the input or the output port does not draw any current from the first—that is, it does not load the other circuit. This is because of the infinite input resistance, or the behavior of the circuit as a voltage source ($R_{out} = 0$). This circuit is very useful in the implementation of analog filters.

Inverting integrator circuit. If we let the feedback element be a capacitor, we obtain the following equation from the virtual short equations. The current through the resistor R is $v_i(t)/R$ given that $v_-(t) = 0$ and it is the current through the capacitor as no current enters the negative terminal. Therefore, the output voltage is

$$v_o(t) = -v_c(t) = -\frac{1}{C} \int_0^t \frac{v_i(\tau)}{R} d\tau - v_c(0)$$

where $v_c(0)$ is the voltage across the capacitor at $t = 0$, when the voltage source is turned on. If we let $v_c(0) = 0$ and $RC = 1$ the above equation is the negative of the integral of the voltage source. Thus, we have a circuit that realizes an integrator with a sign inversion. Again this circuit will be very useful in the implementation of analog filters.

Adder circuit. Since the circuit components are linear, the circuit is linear and we can use superposition. Letting $v_2(t) = 0$ the output voltage due to it is zero, and the output voltage due to $v_1(t)$ is $v_{o1}(t) = -v_1(t)R/R_1$. Similarly, if we let $v_1(t) = 0$, its corresponding output is zero, and the output due to $v_2(t)$ is $v_{o2}(t) = -v_2(t)R/R_2$, so that when both $v_1(t)$ and $v_2(t)$ are considered the output is

$$v_o(t) = v_{o1}(t) + v_{o2}(t) = -v_1(t)\frac{R}{R_1} - v_2(t)\frac{R}{R_2}$$

Using this circuit:

1. When $R_1 = R_2 = R$, we have an adder with a sign inversion:

$$v_o(t) = -[v_1(t) + v_2(t)]$$

2. When $R_2 \to \infty$ and $R_1 = R$, we get an inverter of the input

$$v_o(t) = -v_1(t),$$

3. When $R_2 \to \infty$ and $R_1 = \alpha R$, we get a constant multiplier with sign inversion:

$$v_o(t) = -\frac{1}{\alpha}v_1(t)$$

i.e., the inverted input with a gain $1/\alpha$.

The above three circuits illustrate the realization of a buffer, an integrator, and an adder that can be used to realize analog filters.   ∎

2.3.9 Bounded-Input Bounded-Output Stability

Stability characterizes useful systems. A stable system is such that well-behaved outputs are obtained for well-behaved inputs. Of the many possible definitions of stability, we consider here bounded-input bounded-output (BIBO) stability.

> *Bounded-input bounded-output (BIBO) stability* establishes that for a bounded (i.e., well-behaved) input $x(t)$ the output of a BIBO stable system $y(t)$ is also bounded. *This means that if there is a finite bound $M < \infty$ such that $|x(t)| < M$* (you can think of it as an envelope $[-M, M]$ inside which the input is in) the output is also bounded.
>
> An LTI system with an absolutely integrable impulse response—that is,
>
> $$\int_{-\infty}^{\infty} |h(t)|dt < \infty \tag{2.23}$$
>
> is BIBO stable. A simpler way, using the Laplace transform, to test the BIBO stability of a system is given later.

For a bounded input, the output $y(t)$ of an LTI system is represented by a convolution integral that is bounded as follows:

$$
\begin{aligned}
|y(t)| &= \left| \int_{-\infty}^{\infty} x(t - \tau)h(\tau)d\tau \right| \\
&\leq \int_{-\infty}^{\infty} |x(t - \tau)||h(\tau)|d\tau \\
&\leq M \int_{-\infty}^{\infty} |h(\tau)|d\tau \\
&\leq ML < \infty
\end{aligned}
$$

where L is the bound for $\int_{-\infty}^{\infty} |h(\tau)|d\tau$, or equivalently the impulse response is absolutely integrable.

■ Example 2.16

Consider the BIBO stability and causality of RLC circuits. Consider, for instance, a series RL circuit where $R = 1\,\Omega$ and $L = 1$ H, and a voltage source $v_s(t)$, which is bounded. Discuss why such a system would be causal and stable.

Solution

RLC circuits are naturally stable. As you know, inductors and capacitors simply store energy and so LC circuits simply exchange energy between these elements. Resistors consume energy, which is transformed into heat, and so RLC circuits spend the energy given to them. This characteristic is called *passivity*, indicating that RLC circuits can only use energy, not generate it. Clearly, RLC circuits are also causal systems as one would not expect them to provide any output before they are activated.

According to Kirchhoff's voltage law, the RL circuit is represented by a first-order differential equation

$$v_s(t) = i(t)R + L\frac{di(t)}{dt} = i(t) + \frac{di(t)}{dt}$$

To find its impulse response we would need to solve this equation with input $v_s(t) = \delta(t)$ and zero initial condition, $i(0) = 0$. In the next chapter, the Laplace domain will provide us an algebraic way to solve the differential equation and will confirm our intuitive solution given here. Intuitively, in response to a large and sudden impulse $v_s(t) = \delta(t)$, the inductor tries to follow it by instantaneously increasing its current. But as time goes by and the input is not providing any additional energy, the current in the inductor goes to zero. Thus, we conjecture that the current in the inductor is $i(t) = h(t) = e^{-t}u(t)$ when $v_s(t) = \delta(t)$ and initial conditions are zero, $i(0) = 0$. It is possible to confirm that is the case. Replacing $v_s(t) = \delta(t)$ and $i(t) = e^{-t}u(t)$ in the differential equation, we get

$$\underbrace{\delta(t)}_{v_s(t)} = \underbrace{e^{-t}u(t)}_{i(t)} + \underbrace{[e^{-t}\delta(t) - e^{-t}u(t)]}_{di(t)/dt} = e^0\delta(t) = \delta(t)$$

which is an identity, confirming that indeed our conjectured solution is the solution of the differential equation. The initial condition is also satisfied by remembering that there is no initial current at the source—that is, $\delta(t)$ is zero right before we close the switch—and that physically the inductor remains at that point for an extremely short time before reacting to the strong input.

Thus, the RL circuit where $R = 1\ \Omega$ and $L = 1$ H has an impulse response of

$$h(t) = e^{-t}u(t)$$

indicating that it is causal since $h(t) = 0$ for $t < 0$; that is, the circuit output is zero given that the initial conditions are zero, and that the input $\delta(t)$ is also zero before 0. We can also show that the RL circuit is stable. In fact,

$$\int_{-\infty}^{\infty} |h(t)|dt = \int_{0}^{\infty} e^{-t}dt = 1$$

∎

■ Example 2.17

Consider the causality and BIBO stability of an echo system (or a multipath system). See Figure 2.17. Let the output $y(t)$ be given by

$$y(t) = \alpha_1 x(t - \tau_1) + \alpha_2 x(t - \tau_2)$$

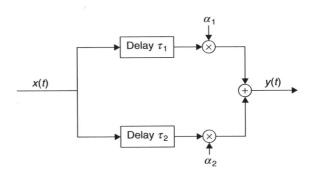

FIGURE 2.17
Echo system with two paths.

where $x(t)$ is the input, and α_i, $\tau_i > 0$, for $i = 1$ and 2, are attenuation factors and delays. Thus, the output is the superposition of attenuated and delayed versions of the input. Typically, the attenuation factors are less than unity. Is this system causal and BIBO stable?

Solution

Since the output depends only on past values of the input, the echo system is causal. To determine if the system is BIBO stable we consider a bounded input signal $x(t)$, and determine if the output is bounded. Suppose $x(t)$ is bounded by a finite value M, or $|x(t)| < M < \infty$, for all times, which means that the value of $x(t)$ cannot exceed an envelope $[-M, M]$ at all times. This would also hold when we shift $x(t)$ in time, so that

$$|y(t)| \leq |\alpha_1||x(t - \tau_1)| + |\alpha_2||x(t - \tau_2)| < [|\alpha_1| + |\alpha_2|]M$$

so the corresponding output is bounded. The system is BIBO stable.

We can also find the impulse response $h(t)$ of the echo system, and show that it satisfies the absolutely integrable condition of BIBO stability. Indeed, if we let the input of the echo system be $x(t) = \delta(t)$ the output is

$$y(t) = h(t) = \alpha_1 \delta(t - \tau_1) + \alpha_2 \delta(t - \tau_2)$$

and the integral is

$$\int_{-\infty}^{\infty} |h(t)|dt = |\alpha_1| \int_{-\infty}^{\infty} \delta(t - \tau_1)dt + |\alpha_2| \int_{-\infty}^{\infty} \delta(t - \tau_2)dt = |\alpha_1| + |\alpha_2| < \infty \qquad \blacksquare$$

■ **Example 2.18**

Consider a positive feedback system created by a microphone close to a set of speakers that are putting out an amplified acoustic signal (see Figure 2.18). The microphone picks up the input signal $x(t)$ as well as the amplified and delayed signal $\beta y(t - \tau)$, $|\beta| \geq 1$. Find the equation that connects the input $x(t)$ and the output $y(t)$ and recursively from it obtain an expression for $y(t)$ in terms of past values of the input. Determine if the system is BIBO stable or not—use $x(t) = u(t)$, $\beta = 2$, and $\tau = 1$ in doing so.

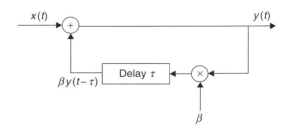

FIGURE 2.18
Positive feedback system: the microphone picks up input signal $x(t)$ and the amplified and delayed signal $\beta y(t - \tau)$, making the system unstable.

Solution

The input–output equation is

$$y(t) = x(t) + \beta y(t - \tau)$$

If we use this expression to obtain $y(t - \tau)$, we get that

$$y(t - \tau) = x(t - \tau) + \beta y(t - 2\tau)$$

and replacing it in the input–output equation, we get

$$y(t) = x(t) + \beta[x(t - \tau) + \beta y(t - 2\tau)] = x(t) + \beta x(t - \tau) + \beta^2 y(t - 2\tau)$$

Repeating the above scheme, we will obtain the following expression for $y(t)$ in terms of the input

$$y(t) = x(t) + \beta x(t - \tau) + \beta^2 x(t - 2\tau) + \beta^3 x(t - 3\tau) + \cdots$$

If we let $x(t) = u(t)$ and $\beta = 2$, the corresponding output is

$$y(t) = u(t) + 2u(t - 1) + 4u(t - 2) + 8u(t - 3) + \cdots$$

which continuously grows as time increases. The output is clearly not a bounded signal, although the input is bounded. Thus, the system is unstable, and the screeching sound from the speakers will prove it—you need to separate the speakers and the microphone to avoid it. ■

2.4 WHAT HAVE WE ACCOMPLISHED? WHERE DO WE GO FROM HERE?

By now you should have begun to see the forest for the trees. In this chapter we connected signals with systems. Especially, we initiated the study of linear time-invariant dynamic systems. As you will learn throughout your studies, this model is of great use in representing systems in many engineering applications. The appeal is its simplicity and mathematical structure. We also indicated some practical properties of systems such as causality and stability. Simple yet significant examples of systems, ranging from the vocal system to simple RLC circuits, illustrate the use of the LTI model and point to its practical application. At the same time, modulators also show that more complicated systems need to be explored to be able to communicate wirelessly. Finally, you were given a system's approach to the theory of differential equations and shown some features that will come back when we apply transforms.

Our next step is to do the analysis of systems with continuous-time signals by means of transforms. In Chapter 3 we discuss the Laplace transform that allows transient as well as steady-state analysis and that will convert the solution of differential equations into an algebraic problem. More important, it will provide the concept of transfer function that connects with the impulse response and the convolution integral covered in this chapter. The Laplace transform is very significant in the area of classic control.

PROBLEMS

2.1. Temperature measuring system—MATLAB

The op-amp circuit shown in Figure 2.19 is used to measure the changes of temperature in a system. The output voltage is given by

$$v_o(t) = -R(t)v_i(t)$$

Suppose that the temperature in the system changes cyclically after $t = 0$, so that

$$R(t) = [1 + 0.5\cos(20\pi t)]\, u(t)$$

Let the input be $v_i(t) = 1$ volt.

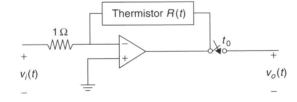

FIGURE 2.19
Problem 2.1.

(a) Assuming that the switch closes at $t_0 = 0$ sec, use MATLAB to plot the output voltage $v_o(t)$ for $0 \le t \le 0.2$ sec in time intervals of 0.01 sec.

(b) If the switch closes at $t_0 = 50$ msec, plot the output voltage $v_0(t)$ for $0 \le t \le 0.2$ sec in time intervals of 0.01 sec.

(c) Use the above results to determine if this system is time invariant. Explain.

2.2. Zener diode—MATLAB

A zener diode circuit is such that the output corresponding to an input $v_s(t) = \cos(\pi t)$ is a "clipped" sinusoid

$$x(t) = \begin{cases} 0.5 & v_s(t) > 0.5 \\ -0.5 & v_s(t) < 0.5 \\ v_s(t) & \text{otherwise} \end{cases}$$

as shown in Figure 2.20 for a few periods. Use MATLAB to generate the input and the output signals and plot them in the same plot for $0 \le t \le 4$ at time intervals of 0.001.

(a) Is this system linear? Compare the output obtained from $v_s(t)$ with that obtained from $0.3v_s(t)$.

(b) Is the system time invariant? Explain.

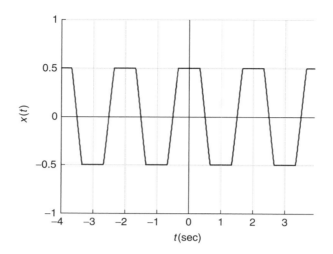

FIGURE 2.20
Problem 2.2.

2.3. Analog averaging system

Consider the analog averager

$$y(t) = \frac{1}{T} \int_{t-T/2}^{t+T/2} x(\tau)d\tau$$

where $x(t)$ is the input and $y(t)$ is the output.

(a) Find the impulse response $h(t)$ of the averager. Is this system causal?

(b) Let $x(t) = u(t)$. Find the output of the averager.

2.4. LTI determination from input–output relation

An analog system has the input–output relation

$$y(t) = \int_{0}^{t} e^{-(t-\tau)}x(\tau)d\tau \qquad t \geq 0$$

and zero otherwise. The input is $x(t)$ and $y(t)$ is the output.

(a) Is this a linear time-invariant system? If so, can you determine without any computation the impulse response of the system? Explain.

(b) Is this system causal? Explain.

(c) Find the unit-step response $s(t)$ and from it find the impulse response $h(t)$. Is this a BIBO-stable system? Explain.

(d) Find the response due to a pulse $x(t) = u(t) - u(t-1)$.

2.5. p-n diode—MATLAB

The voltage–current characterization of a p-n diode is given by (see Figure 2.21)

$$i(t) = I_s(e^{qv(t)/kT} - 1)$$

where $i(t)$ and $v(t)$ are the current and the voltage in the diode (in the direction indicated in the diode), I_s is the reversed saturation current, and kT/q is a constant.

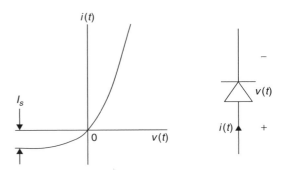

FIGURE 2.21

Problem 2.5: p-n diode and i-v characteristic.

(a) Consider the voltage $v(t)$ as the input and the current $i(t)$ as the output of the diode. Is the p-n diode a linear system? Explain.

(b) An *ideal diode* is such that when the voltage is negative, $v(t) < 0$, the current is zero (i.e., open circuit), and when the current is positive, $i(t) > 0$, the voltage is zero or short circuit. Under what conditions does the p-n diode voltage–current characterization approximate the characterization of the ideal diode? Use MATLAB to plot the current–voltage plot for a diode with $I_s = 0.0001$ and $kT/q = 0.026$, and compare it to the ideal diode current–voltage plot. Determine if the ideal diode is linear.

(c) Consider the circuit using an ideal diode in Figure 2.22, where the source is a sinusoid signal $v(t) = \sin(2\pi t)u(t)$ and the output is the voltage in the resistor $R = 1\ \Omega$ or $v_R(t)$. Plot $v_R(t)$. Is this system linear? Where would you use this circuit?

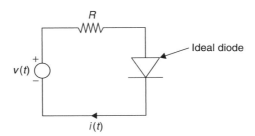

FIGURE 2.22

Problem 2.5: ideal diode circuit.

2.6. Capacitor/inductor circuit

Consider the circuit in Figure 2.23, where the value of the capacitor is $C = 1$ F, and the initial condition is $v_C(0) = -1$ volts. Assume the input is the current source $i_s(t)$ and the voltage in the capacitor $v_C(t)$ the output.

(a) Let the current in the circuit be $i_s(t) = u(t) - u(t-1)$, and the initial voltage in the capacitor be $v_C(0) = -1$ volts. Plot the voltage in the capacitor $v_C(t)$ for all times. Suppose then we double the current, $i_s(t) = 2(u(t) - u(t-1))$, but keep the same initial condition. Plot the voltage in the capacitor $v_C(t)$ for all times, and compare it with the one obtained before. Is the capacitor with non-zero initial conditions a linear system? Explain.

(b) Consider the dual circuit where the value of the inductor is $L = 1$ H, the initial current in the inductor is $i_L(0) = -1$ amps, and the input is the source $v_s(t)$. Let $v_s(t) = u(t) - u(t-1)$. Plot the corresponding current $i_L(t)$ for all times. Double the voltage source $v_s(t) = 2[u(t) - u(t-1)]$, and plot the corresponding current in the inductor $i_L(t)$ for all times. Compare the two currents and determine if the inductor with the initial current is linear. What if $i_L(0) = 0$?

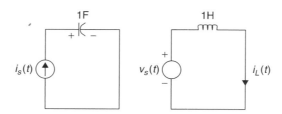

FIGURE 2.23
Problem 2.6.

2.7. Time-varying capacitor

A time-varying capacitor is characterized by the charge–voltage equation

$$q(t) = C(t)v(t)$$

That is, the capacitance is not a constant but a function of time.
(a) Given that $i(t) = dq(t)/dt$, find the voltage–current relation for this time-varying capacitor.
(b) Let $C(t) = 1 + \cos(2\pi t)$ and $v(t) = \cos(2\pi t)$. Determine the current $i_1(t)$ in the capacitor for all t.
(c) Let $C(t)$ be as above, but delay $v(t)$ by 0.25 sec. Determine $i_2(t)$ for all time. Is the system TI?

2.8. Sinusoidal Test for LTI

A fundamental property of linear time-invariant systems is that whenever the input of the system is a sinusoid of a certain frequency, the output will also be a sinusoid of the same frequency but with an amplitude and phase determined by the system. For the following systems let the input be $x(t) = \cos(t)$, $-\infty < t < \infty$, and find the output $y(t)$ and determine if the system is LTI.

$$(a)\ \ y(t) = |x(t)|^2$$

$$(b)\ \ y(t) = 0.5[x(t) + x(t-1)]$$

$$(c)\ \ y(t) = x(t)u(t)$$

$$(d)\ \ y(t) = \frac{1}{2} \int_{t-2}^{t} x(\tau)d\tau$$

2.9. Testing the time invariance of systems

Consider the following systems and find the response to $x_1(t) = u(t)$ and $x_2(t) = u(t-1)$. Determine from the corresponding outputs whether the system is time-varying or not.

$$(a)\ \ y(t) = x(t)\cos(\pi t)$$

$$(b)\ \ y(t) = x(t)[u(t) - u(t-2)]$$

$$(c)\ \ y(t) = 0.5[x(t) + x(t-1)]$$

Plot $y_1(t)$ and $y_2(t)$ for each case.

2.10. Window/modulator

Consider the system where for an input $x(t)$ the output is $y(t) = x(t)f(t)$ for some function $f(t)$.
(a) Let $f(t) = u(t) - u(t-10)$. Determine whether the system with input $x(t)$ and output $y(t)$ is linear, time invariant, causal, and BIBO stable.
(b) Suppose $x(t) = 4\cos(\pi t/2)$, and $f(t) = \cos(6\pi t/7)$ and both are periodic. Is the output $y(t)$ also periodic? What frequencies are present in the output? Is this system linear? Is it time invariant? Explain.

(c) Let $f(t) = u(t) - u(t-2)$ and the input $x(t) = u(t)$. Find the corresponding output $y(t)$. Suppose you shift the input so that it is $x_1(t) = x(t-3)$. What is the corresponding output $y_1(t)$. Is the system time invariant? Explain.

2.11. Initial conditions, LTI, steady state, and stability

The input–output characterization of a system is

$$y(t) = e^{-2t}y(0) + 2 \int_0^t e^{-2(t-\tau)}x(\tau)d\tau \qquad t \geq 0$$

and zero otherwise. In the above equation $x(t)$ is the input and $y(t)$ is the output.

(a) Is this system LTI? Is it possible to determine a value for $y(0)$ that would make this an LTI system? Explain.

(b) Find the differential equation that also characterizes this system.

(c) Suppose for $x(t) = u(t)$ and any value of $y(0)$, we wish to determine the steady-state response of the system. Is the value of $y(0)$ of any significance—that is, do we get the same steady-state response if $y(0) = 0$ or $y(0) = 1$? Explain.

(d) Compute the steady-state response when $y(0) = 0$ and $x(t) = u(t)$ using the convolution integral. To do so, first find the impulse response of the system $h(t)$. Then relate the integral in the equation given above with the convolution integral and graphically compute it.

(e) Suppose the input is zero. Is the system depending on the initial condition BIBO stable? Find the zero-input response $y(t)$ when $y(0) = 1$. Is it bounded?

2.12. Amplifier with saturation

The input–output equation characterizing an amplifier that saturates once the input reaches certain values is

$$y(t) = \begin{cases} 100x(t) & -10 \leq x(t) \leq 10 \\ 1000 & x(t) > 10 \\ -1000 & x(t) < 10 \end{cases}$$

where $x(t)$ is the input and $y(t)$ is the output.

(a) Plot the relation between the input $x(t)$ and the output $y(t)$. Is this a linear system? Explain.

(b) For what range of input values is the system linear, if any?

(c) Suppose the input is a sinusoid $x(t) = 20 \cos(2\pi t)u(t)$. Carefully plot $x(t)$ and $y(t)$ for $t = -2$ to 4.

(d) Let the input be delayed by two units of time (i.e., the input is $x_1(t) = x(t-2)$). Find the corresponding output $y_1(t)$ and indicate how it relates to the output $y(t)$ due to $x(t)$ found above. Is the system time invariant?

2.13. QAM system

A quadrature amplitude modulation (QAM) system is a communication system capable of transmitting two messages $m_1(t)$, $m_2(t)$ at the same time. The transmitted signal $s(t)$ is

$$s(t) = m_1(t) \cos(\Omega_c t) + m_2(t) \sin(\Omega_c t)$$

Carefully draw a block diagram for the QAM system.

(a) Determine if the system is time invariant or not.

(b) Assume $m_1(t) = m_2(t) = m(t)$—that is, we are sending the same message using two different modulators. Express the modulated signal in terms of a cosine with carrier frequency Ω_c, amplitude A, and phase θ. Obtain A and θ. Is the system linear? Explain.

2.14. Steady-state response of averager—MATLAB

An analog averager is given by

$$y(t) = \frac{1}{T} \int_{t-T}^{t} x(\tau)d\tau$$

(a) Let $x(t) = u(t) - u(t-1)$. Find the average signal $y(t)$ using the above integral. Let $T = 1$. Carefully plot $y(t)$. Verify your result by graphically computing the convolution of $x(t)$ and the impulse response $h(t)$ of the averager.

(b) To see the effect of T on the averager, consider the signal to be averaged to be $x(t) = \cos(2\pi t/T_0)u(t)$. Select the smallest possible value of T in the averager so that the steady-state response of the system, $y(t)$ as $t \to \infty$, will be 0.

(c) Use MATLAB to compute the output in part (b). Compute the output $y(t)$ for $0 \le t \le 2$ at intervals $T_s = 0.001$. Approximate the convolution integral using the function conv (use help to find about conv) multiplied by T_s.

2.15. Echo system modeling

An echo system could be modeled as follows:

(a) Using feedback systems is of great interest in control and in the modeling of many systems. An echo is created as the sum of one or more delayed and attenuated output signals that are fed back into the present signal. A possible model for an echo system is

$$y(t) = x(t) + \alpha_1 y(t-\tau) + \cdots + \alpha_N y(t - N\tau)$$

where $x(t)$ is the present input signal, $y(t)$ is the present output, $y(t - k\tau)$ is the previous delayed outputs, and the $|\alpha_k| < 1$ values are attenuation factors. Carefully draw a block diagram for this system.

(b) Consider the echo model for $N = 1$ and parameters $\tau = 1$ and $\alpha_1 = 0.1$. Is the resulting echo system LTI? Explain.

(c) Another possible model is given by a nonrecursive, or without feedback, system,

$$z(t) = x(t) + \beta_1 x(t-\tau) + \cdots + \beta_M x(t - M\tau)$$

where several present and past inputs are delayed and attenuated and added up to form the output. The parameters $|\beta_k| < 1$ are attenuation factors and τ is a delay. Carefully draw a block diagram for the echo system characterized by the above equation. Does the above equation represent an LTI system? Explain.

2.16. An ideal low-pass filter—MATLAB

The impulse response of an ideal low-pass filter is

$$h(t) = \frac{\sin(t)}{t}$$

or a sinc signal.

(a) Given that the impulse response is the response of the system to an input $x(t) = \delta(t)$ with zero initial conditions, can an ideal low-pass filter be used for real-time processing? Explain.

(b) Is the ideal low-pass filtering bounded-input bounded-output stable? Use MATLAB to check if the impulse response satisfies the condition for BIBO stability.

2.17. Response to unbounded inputs versus BIBO stability

The BIBO stability assumes that the input is always bounded, limited in amplitude. If that is not the case, even a stable system would provide an unbounded output. Consider the analog averager, with an input–output relationship of

$$y(t) = \frac{1}{T} \int_{t-T}^{t} x(\tau)d\tau$$

(a) Suppose that the input to the averager is a bounded signal $x(t)$ (i.e., there is a finite value M such that $|x(t)| < M$). Find the value for the bound of the output $y(t)$ and determine whether the averager is BIBO stable or not.

(b) Let the input to the averager be $x(t) = tu(t)$ (i.e., a ramp signal). Compute the output $y(t)$ and determine if it is bounded or not. If $y(t)$ is not bounded, does that mean that the averager is an unstable system? Explain.

2.18. Sampler and hold circuit

In an analog-to-digital converter (ADC), the analog signal is first sampled and then each of its samples is converted into a digital value. Since each of the samples is obtained momentarily, there is the need for a circuit that holds the value long enough for the ADC to convert it into a binary number. The circuit having the sampler and the hold circuit is called the *sampler and hold* circuit, an example of which is shown in Figure 2.24. The input is the sampled signal $x_s(t)$, which we are considering a train of rectangular pulses of duration Δ and periodicity T_s and different magnitudes corresponding to $x(nT_s)$.

The value $rC << \Delta$, where Δ is the duration of the pulse and $RC >> T_s$ where $T_s >> \Delta$, is the sampling period. The first condition allows the capacitor to be charged fast in Δ seconds, and the second condition allows slow discharge in T_s seconds.

(a) Consider the first sample conversion. Let the input to the hold circuit be a pulse of duration Δ and amplitude $x(0)$. The switch has been opened before $t = 0$ so that the capacitor is discharged. The switch closes at $t = 0$ and remains closed until $t = \Delta$ and then it opens. Carefully draw the voltage in the capacitor from $t = 0$ to T_s.

(b) Since the RC circuit is a linear time-invariant system, the output corresponding to the other samples can be found from the result of the first sample. Suppose the analog signal is a ramp $x(t) = tu(t)$, sampled with $T_s = 1$ and $\Delta = 0.1$. Plot the voltage in the capacitor from $t = 0$ to $t = 4$ sec.

2.19. AM envelope detector—MATLAB

Consider an envelope detector that is used to detect the message sent in the AM system shown in the examples. The envelope detector as a system is composed of two cascaded systems: one that computes the absolute value of the input (implemented with ideal diodes), and a second that low-pass filters its input (implemented with an RC circuit). The following is an implementation of these operations in the discrete time so we can use numeric MATLAB.

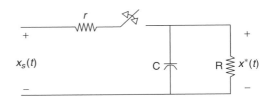

FIGURE 2.24

Problem 2.18: sample and hold circuit.

Let the input to the envelope detector be

$$x(t) = [p(t) + P] \cos(\Omega_0 t)$$

where P is the minimum of $p(t)$ scaled. Use MATLAB to solve numerically this problem.

(a) Consider first

$$p(t) = 20[u(t) - u(t - 40)] - 10[u(t - 40) - u(t - 60)]$$

Let $\Omega_0 = 2\pi$, $P = 1.1|\min(p(t)|$. Generate the signals $p(t)$ and $x(t)$ for $0 \le t \le 100$ with an interval of $T_s = 0.01$.

(b) Consider then the subsystem that computes the absolute value of the input $x(t)$.

(c) Compute the low-pass filtered signal by using an RC circuit with impulse response $h(t) = e^{-0.8t}u(t)$. To implement the convolution use the conv function multiplied by T_s. Plot together the message signal $p(t)$, the modulated signal $x(t)$, the absolute value $y(t)$, and the envelope of $x(t)$. Does this envelope look like $p(t)$?

(d) Consider the message signal $p(t) = 2\cos(0.2\pi t)$, $\Omega_0 = 10\pi$, and $P = |\min(p(t)|$, and repeat the process. Scale the signal to get the original $p(t)$.

2.20. Frequency modulation (FM)—MATLAB

Frequency modulation, or FM, uses a wider bandwidth than amplitude modulation, or AM, but it is not affected as much by noise as AM is. The output of an FM transmitter is of the form

$$y(t) = \cos\left(\Omega_c t + 2\pi\nu \int_0^t m(\tau)d\tau\right)$$

where $m(t)$ is the message and ν is a factor in Hz/volt if the units of the message are in volts.

(a) Create as the message a signal

$$m(t) = \cos(t)$$

Find the FM signal $y(t)$ for $\nu = 10$ and then for $\nu = 1$. Let the carrier frequency $\Omega_c = 2\pi$. Use MATLAB to generate the different signals for times $0 \le t \le 10$ at intervals of $T_s = 0.01$. Plot $m(t)$ and the two FM signals (one for $\nu = 10$ and the other for $\nu = 1$) in the same plot. Is the FM transmitter a linear system? Explain.

(b) Create a message signal

$$m_1(t) = \begin{cases} 1 & \text{when } m(t) \ge 0 \\ -1 & \text{when } m(t) < 0 \end{cases}$$

Find the corresponding FM signal for $\nu = 1$.

CHAPTER 3

The Laplace Transform

What we know is not much.
What we do not know is immense.
Pierre-Simon marquis de Laplace (1749–1827)
French mathematician and astronomer

3.1 INTRODUCTION

The material in this chapter is very significant for the analysis of continuous-time signals and systems. The main issues discussed are:

- *Frequency domain analysis of continuous-time signals and systems*—We begin the frequency domain analysis of continuous-time signals and systems using transforms. The Laplace transform, the most general of these transforms, will be followed by the Fourier transform. Both provide complementary representations of a signal to its own in the time domain, and an algebraic characterization of systems. The Laplace transform depends on a complex variable $s = \sigma + j\Omega$, composed of damping σ and frequency Ω, while the Fourier transform considers only frequency Ω.

- *Damping and frequency characterization of continuous-time signals*—The growth or decay of a signal—damping—as well as its repetitive nature—frequency—in the time domain are characterized in the Laplace domain by the location of the roots of the numerator and denominator, or zeros and poles, of the Laplace transform of the signal.

- *Transfer function characterization of continuous-time LTI systems*—The Laplace transform provides a significant algebraic characterization of continuous-time systems: The ratio of the Laplace transform of the output to that of the input—or the transfer function of the system. It unifies the convolution integral and the differential equations system representations. The concept of transfer function is not only useful in analysis but also in design, as we will see later. The location of the poles and the zeros of the transfer function relates to the dynamic characteristics of the system.

- *Stability, and transient and steady-state responses*—Certain characteristics of continuous-time systems can only be verified or understood via the Laplace transform. Such is the case of stability,

Signals and Systems Using MATLAB®. DOI: 10.1016/B978-0-12-374716-7.00006-5

and of transient and steady-state responses. This is a significant reason to study the Laplace analysis before the Fourier analysis, which deals exclusively with the frequency characterization of continuous-time signals and systems. Stability and transients are important issues in classic control theory, thus the importance of the Laplace transform in this area. The frequency characterization of signals and the frequency response of systems—provided by the Fourier transform—are significant in communications.

- *One- and two-sided Laplace transforms*—Given the prevalence of causal signals (those that are zero for negative time) and of causal systems (having zero impulse responses for negative time) the Laplace transform is typically known as "one-sided," but the "two-sided" transform also exists. The impression is that these are two different transforms, but in reality it is the Laplace transform applied to two different types of signals and systems. We will show that by separating the signal into its causal and its anti-causal components, we only need to apply the one-sided transform. Care should be exercised, however, when dealing with the inverse transform so as to get the correct signal.

- *Region of convergence and the Fourier transform*—Since the Laplace transform requires integration over an infinite domain, it is necessary to consider if and where this integration converges—or the "region of convergence" in the s-plane. Now, if such a region includes the $j\Omega$ axis of the s-plane, then the Laplace transform exists for $s = j\Omega$, and when computed there it coincides with the Fourier transform of the signal. Thus, the Fourier transform for a large class of functions can be obtained directly from their Laplace transforms—a good reason to study first the Laplace transform. In a subtle way, the Laplace transform is also connected with the Fourier series representation of periodic continuous-time signals. Such a connection reduces the computational complexity of the Fourier series by eliminating integration in cases when we can compute the Laplace transform of a period.

- *Eigenfunctions of LTI systems*—LTI systems respond to complex exponentials in a very special way: The output is the exponential with its magnitude and phase changed by the response of the system at the exponent. This provides the characterization of the system by the Laplace transform, in the case of exponents of the complex frequency s, and by the Fourier representation when the exponent is $j\Omega$. The eigenfunction concept is linked to phasors used to compute the steady-state response in circuits (see Figure 3.1).

3.2 THE TWO-SIDED LAPLACE TRANSFORM

Rather than giving the definitions of the Laplace transform and its inverse, let us see how they could be obtained intuitively. As indicated before, a basic idea in characterizing signals—and their response when applied to LTI systems—is to consider them a combination of basic signals for which we can easily obtain a response. In Chapter 2, when considering the time-domain solutions, we represented the input as an infinite combination of impulses occurring at all possible times and weighted by the value of the input signal at those times. The reason we did so is because the response due to an impulse is the impulse response of the LTI system, which is fundamental in our studies. A similar approach will be followed when attempting to obtain the frequency-domain representation of signals and their responses when applied to an LTI system. In this case, the basic functions used are complex exponentials or sinusoids that depend on frequency. The concept of eigenfunction is somewhat

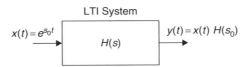

FIGURE 3.1

Eigenfunction property of LTI systems. The input of the system is $x(t) = e^{s_0 t} = e^{\sigma_0 t} e^{j\Omega_0 t}$ and the output of the system is the same input multiplied by the complex value $H(s_0)$ where $H(s) = \mathcal{L}[h(t)]$—that is, the Laplace transform of the impulse response $h(t)$ of the LTI system.

abstract at the beginning, but after you see it applied here and in the Fourier representation later you will think of it as a way to obtain a representation analogous to the impulse representation. You will soon discover the importance of using complex exponentials, and it will then become clear that eigenfunctions are connected with phasors that greatly simplify the sinusoidal steady-state solution of circuits.

3.2.1 Eigenfunctions of LTI Systems

Consider as the input of an LTI system the complex signal

$$x(t) = e^{s_0 t} \qquad s_0 = \sigma_0 + j\Omega_0$$

for $-\infty < t < \infty$, and let $h(t)$ be the impulse response of the system. According to the convolution integral, the output of the system is

$$y(t) = \int_{-\infty}^{\infty} h(\tau) x(t-\tau) d\tau = \int_{-\infty}^{\infty} h(\tau) e^{s_0(t-\tau)} d\tau$$

$$= e^{s_0 t} \int_{-\infty}^{\infty} h(\tau) e^{-\tau s_0} d\tau = x(t) H(s_0) \tag{3.1}$$

Since the same exponential at the input appears at the output, $x(t) = e^{s_0 t}$ is called an *eigenfunction*[1] of the LTI system. The input $x(t)$ is changed at the output by the complex function $H(s_0)$, which is related to the system through the impulse response $h(t)$. In general, for any s, the eigenfunction at the output is modified by a complex function

$$H(s) = \int_{-\infty}^{\infty} h(\tau) e^{-\tau s} d\tau$$

which corresponds to the Laplace transform of $h(t)$!

---

[1] German mathematician David Hilbert (1862–1943) seems to be the first to use the German word *eigen* to denote eigenvalues and eigenvectors in 1904. The word *eigen* means own or proper.

An input $x(t) = e^{s_0 t}$, $s_0 = \sigma_0 + j\Omega_0$, is called an eigenfunction of an LTI system with impulse response $h(t)$ if the corresponding output of the system is

$$y(t) = x(t) \int_{-\infty}^{\infty} h(t)e^{-s_0 t} = x(t)H(s_0)$$

where $H(s_0)$ is the Laplace transform of $h(t)$ computed at $s = s_0$. This property is only valid for LTI systems—it is not satisfied by time-varying or nonlinear systems.

Remarks

■ *You could think of $H(s)$ as an infinite combination of complex exponentials, weighted by the impulse response $h(\tau)$. One can use a similar representation for signals.*

■ *Consider now the significance of applying the eigenfunction result. Suppose a signal $x(t)$ is expressed as a sum of complex exponentials in $s = \sigma + j\Omega$,*

$$x(t) = \frac{1}{2\pi j} \int_{\sigma - j\infty}^{\sigma + j\infty} X(s)e^{st} ds$$

That is, an infinite sum of exponentials in s each weighted by the function $X(s)/(2\pi j)$ (this equation is connected with the inverse Laplace transform as we will see soon). Using the superposition property of LTI systems, and considering that for an LTI system with impulse response $h(t)$ the output due to e^{st} is $H(s)e^{st}$, then the output due to $x(t)$ is

$$y(t) = \frac{1}{2\pi j} \int_{\sigma - j\infty}^{\sigma + j\infty} X(s)\left[H(s)e^{st}\right] ds = \frac{1}{2\pi j} \int_{\sigma - j\infty}^{\sigma + j\infty} Y(s)e^{st} ds$$

where we let $Y(s) = X(s)H(s)$. But from Chapter 2 we have that $y(t)$ is the convolution $y(t) = [x * h](t)$. Thus, these two expressions are connected:

$$y(t) = [x * h](t) \qquad \Leftrightarrow \qquad Y(s) = X(s)H(s)$$

The expression on the left indicates how to compute the output in the time domain, and the one on the right shows how to compute the Laplace transform of the output in the frequency domain. This is the most important property of the Laplace transform: It reduces the complexity of the convolution integral in time to the multiplication of the Laplace transforms of the input $X(s)$ and of the impulse response $H(s)$.

Now we are ready for the proper definition of the direct and inverse Laplace transforms of a signal or of the impulse response of a system.

The two-sided Laplace transform of a continuous-time function $f(t)$ is

$$F(s) = \mathcal{L}[f(t)] = \int_{-\infty}^{\infty} f(t)e^{-st}dt \qquad s \in \text{ROC} \qquad (3.2)$$

where the variable $s = \sigma + j\Omega$, with Ω as the frequency in rad/sec and σ as a damping factor. ROC stands for the region of convergence—that is, where the integral exists.

The inverse Laplace transform is given by

$$f(t) = \mathcal{L}^{-1}[F(s)] = \frac{1}{2\pi j} \int_{\sigma-j\infty}^{\sigma+j\infty} F(s)e^{st}ds \qquad \sigma \in \text{ROC} \qquad (3.3)$$

Remarks

- *The Laplace transform $F(s)$ provides a representation of $f(t)$ in the s-domain, which in turn can be converted back into the original time-domain functon in a one-to-one manner using the region of convergence. Thus,*

$$F(s) \quad ROC \quad \Leftrightarrow \quad f(t)$$

- *If $f(t) = h(t)$, the impulse response of an LTI system, then $H(s)$ is called the* system *or* transfer function *of the system and it characterizes the system in the s-domain just like $h(t)$ does in the time-domain. If $f(t)$ is a signal, then $F(s)$ is its Laplace transform.*

- *The inverse Laplace transform in Equation (3.3) can be understood as the representation of $f(t)$ (whether it is a signal or an impulse response) by an infinite summation of complex exponentials with weights $F(s)$ at each. The computation of the inverse Laplace transform using Equation (3.3) requires complex integration. Algebraic methods will be used later to find the inverse Laplace transform, thus avoiding the complex integration.*

Laplace and Heaviside

The Marquis Pierre-Simon de Laplace (1749–1827) [2, 7] was a French mathematician and astronomer. Although from humble beginnings he became royalty by his political abilities. As an astronomer, he dedicated his life to the work of applying the Newtonian law of gravitation to the entire solar system. He was considered an applied mathematician and, as a member of the Academy of Sciences, knew other great mathematicians of the time such as Legendre, Lagrange, and Fourier. Besides his work on celestial mechanics, Laplace did significant work in the theory of probability from which the Laplace transform probably comes. He felt that "the theory of probabilities is only common sense expressed in number." Early transformations similar to Laplace's had been used by Euler and Lagrange. It was, however, Oliver Heaviside (1850–1925) who used the Laplace transform in the solution of differential equations. Heaviside, an Englishman, was a self-taught electrical engineer, mathematician, and physicist [76].

■ Example 3.1

A problem in wireless communications is the so-called *multipath effect* on the transmitted message. Consider the channel between the transmitter and the receiver as a system like the one depicted in Figure 3.2. The sent message $x(t)$ does not necessarily go from the transmitter to the receiver directly (line of sight) but it may take different paths, each with different length so that the signal in each path is attenuated and delayed differently.[2] At the receiver, these delayed and attenuated signals are added, causing a fading effect—given the different phases of the incoming signals their addition at the receiver results in a weak or a strong signal, thus giving the sensation of the message fading back and forth. If $x(t)$ is the message sent from the transmitter, and the channel has N different paths with attenuation factors $\{\alpha_i\}$ and corresponding delays $\{t_i\}$, $i = 0, \ldots, N$, use the eigenfunction property to find the system function of the channel causing the multipath effect.

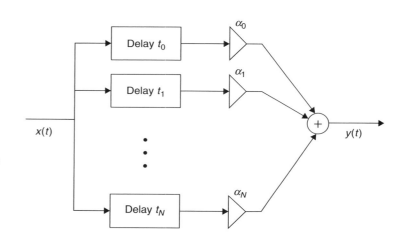

FIGURE 3.2

Block diagram of a wireless communication channel causing a multipath effect on the sent message $x(t)$. The message $x(t)$ is delayed and attenuated when sent over $N + 1$ paths. The effect is similar to that of an echo in acoustic signals.

Solution

The output of the channel or multipath system in Figure 3.2 can be written as

$$y(t) = \alpha_0 x(t - t_0) + \alpha_1 x(t - t_1) + \cdots + \alpha_N x(t - t_N) \tag{3.4}$$

Considering $s = \sigma + j\Omega$ as the variable, the response of the multipath system to $x(t) = e^{st}$ is $y(t) = x(t)H(s)$, so that when replacing them in Equation (3.4), we get

$$x(t)H(s) = x(t)\left[\alpha_0 e^{-st_0} + \cdots + \alpha_N e^{-st_N}\right]$$

---

[2] Typically, there are three effects each path can have on the sent signal: The distance the signal needs to travel (in each path this is due to reflection or refraction on buildings, structures, cars, etc.) determines how much it is attenuated and delayed (the longer the path, the more attenuated and delayed with respect to the time it was sent) and the third effect is a frequency shift—or Doppler effect—that is caused by the relative velocity between the transmitter and the receiver.

giving as the system function for the channel,

$$H(s) = \alpha_0 e^{-st_0} + \cdots + \alpha_N e^{-st_N}$$

Notice that the time shifts in the input–output equation became exponentials in the Laplace domain, a property we will see later. ∎

Let us consider the different types of functions (either continuous-time signals or the impulse responses of continuous-time systems) we might be interested in calculating Laplace transforms of.

- *Finite support functions:* the function $f(t)$ in this case is

$$f(t) = 0 \qquad \text{for } t \notin \text{finite segment } t_1 \leq t \leq t_2$$

for any finite, positive or negative t_1 and t_2, and so that $t_1 < t_2$. We will see that the Laplace transform of these finite support signals is of particular interest in the computation of the coefficients of the Fourier series of periodic signals.
- *Infinite support functions:* In this case, $f(t)$ is defined over an infinite support (e.g., $t_1 < t < t_2$ where either t_1 or t_2 are infinite, or both are infinite as long as $t_1 < t_2$).

A finite, or infinite, support function $f(t)$ is called (see examples in Figure 3.3):

- *Casual* if $f(t) = 0$ $\quad t < 0$,
- *Anti-causal* if $f(t) = 0$ $\quad t \geq 0$,
- *Non causal* if a combination of the above.

In each of these cases we need to consider the region in the s-plane where the transform exists or its region of convergence (ROC). This is obtained by looking at the convergence of the transform.

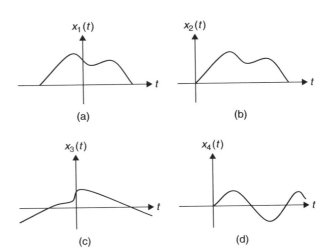

FIGURE 3.3

Examples of different types of signals:
(a) noncausal finite support signal $x_1(t)$, (b) causal finite support signal $x_2(t)$, (c) noncausal infinite support signal $x_3(t)$, and (d) causal infinite support signal $x_4(t)$.

For the Laplace transform of $f(t)$ to exist we need that

$$\left| \int_{-\infty}^{\infty} f(t)e^{-st}dt \right| = \left| \int_{-\infty}^{\infty} f(t)e^{-\sigma t}e^{-j\Omega t}dt \right|$$

$$\leq \int_{-\infty}^{\infty} |f(t)e^{-\sigma t}| \, dt < \infty$$

or that $f(t)e^{-\sigma t}$ be absolutely integrable. This may be possible by choosing an appropriate σ even in the case when $f(t)$ is not absolutely integrable. The value chosen for σ determines the ROC of $F(s)$; the frequency Ω does not affect the ROC.

3.2.2 Poles and Zeros and Region of Convergence

The *region of convergence* (ROC) can be obtained from the conditions for the integral in the Laplace transform to exist. The ROC is related to the *poles* of the transform, which is in general a complex rational function.

For a rational function $F(s) = \mathcal{L}[f(t)] = N(s)/D(s)$, its zeros are the values of s that make the function $F(s) = 0$, and its poles are the values of s that make the function $F(s) \to \infty$. Although only finite zeros and poles are considered, infinite zeros and poles are also possible.

Typically, $F(s)$ is rational, a ratio of two polynomials $N(s)$ and $D(s)$, or $F(s) = N(s)/D(s)$, and as such its zeros are the values of s that make the numerator polynomial $N(s) = 0$, while the poles are the values of s that make the denominator polynomial $D(s) = 0$. For instance, for

$$F(s) = \frac{2(s^2 + 1)}{s^2 + 2s + 5} = \frac{2(s + j)(s - j)}{(s + 1)^2 + 4} = \frac{2(s + j)(s - j)}{(s + 1 + 2j)(s + 1 - 2j)}$$

we have the zeros are at $s = \pm j$, roots of $N(s) = 0$, since $F(\pm j) = 0$, and a pair of complex conjugate poles $-1 \pm 2j$, the roots of the equation $D(s) = 0$ and such that $F(-1 \pm 2j) \to \infty$. Geometrically, zeros can be visualized as those values that make the function go to zero, and poles as those values that make the function approach infinity (looking like the main "pole" of a circus tent). See Figure 3.4.

Not all rational functions have poles or a finite number of zeros. Consider the Laplace transform

$$P(s) = \frac{1}{s} \left(e^s - e^{-s} \right)$$

$P(s)$ seems to have a pole at $s = 0$. Its zeros are obtained by letting $e^s - e^{-s} = 0$, which when multiplied by e^s gives

$$e^{2s} = 1 = e^{j2\pi k}$$

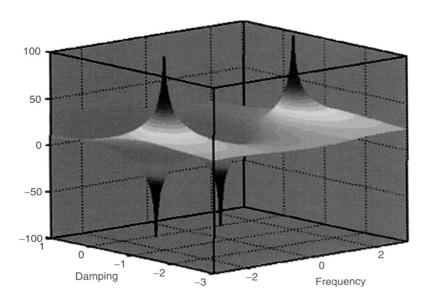

FIGURE 3.4
Three-dimensional plot of the logarithm of the magnitude of $F(s) = 2(s^2 + 1)/(s^2 + 2s + 5)$ as a function of damping σ and frequency Ω. The poles shoot up, while the zeros shoot down. In the logarithmic scale both poles and zeros will have infinite value: When $F(s) = 0$ (zero) its logarithm is $-\infty$, while when $F(s) \to \infty$ (pole) the logarithm is ∞.

for an integer $k = 0, \pm 1, \pm 2, \ldots$. Thus, the zeros are $s_k = j\pi k$, $k = 0, \pm 1, \pm 2, \ldots$. Now, when $k = 0$, the zero at 0 cancels the pole at zero; therefore, $P(s)$ has only zeros, an infinite number of them, $\{j\pi k, k = \pm 1, \pm 2, \ldots\}$.

Poles and ROC
The ROC consists of the values of σ such that

$$\left| \int_{-\infty}^{\infty} x(t)e^{-st}dt \right| \leq \int_{-\infty}^{\infty} |x(t)||e^{-(\sigma+j\Omega)t}|dt = \int_{-\infty}^{\infty} |x(t)|e^{-\sigma t}dt < \infty \qquad (3.5)$$

This is equivalent to choosing values of σ for which $x(t)e^{-\sigma t}$ is absolutely integrable.

Two general comments that apply to all types of signals when finding ROCs are:

- No poles are included in the ROC, which means that for the ROC to be that region where the Laplace transform is defined, the transform cannot become infinite at any point in it. So poles should not be present in the ROC.
- The ROC is a plane parallel to the $j\Omega$ axis, which means that it is the damping σ that defines the ROC, not frequency Ω. This is because when we compute the absolute value of the integrand in

the Laplace transform to test for convergence, we let $s = \sigma + j\Omega$ and the term $|e^{-j\Omega t}| = 1$. Thus, all regions of convergence will contain $-\infty < \Omega < \infty$.

If $\{\sigma_i\}$ are the real parts of the poles of $F(s) = \mathcal{L}[f(t)]$, the region of convergence corresponding to different types of signals or impulse responses is determined from its poles as follows:

■ For a causal $f(t)$, $f(t) = 0$ for $t < 0$, the region of convergence of its Laplace transform $F(s)$ is a plane to the right of the poles,

$$\mathcal{R}_c = \{(\sigma, \Omega) : \sigma > \max\{\sigma_i\}, -\infty < \Omega < \infty\}$$

■ For an anti-causal $f(t)$, $f(t) = 0$ for $t > 0$, the region of convergence of its Laplace transform $F(s)$ is a plane to the left of the poles,

$$\mathcal{R}_{ac} = \{(\sigma, \Omega) : \sigma < \min\{\sigma_i\}, -\infty < \Omega < \infty\}$$

■ For a noncausal $f(t)$ (i.e., $f(t)$ defined for $-\infty < t < \infty$), the region of convergence of its Laplace transform $F(s)$ is the intersection of the regions of convergence corresponding to the causal component, $\mathcal{R}_c$, and $\mathcal{R}_{ac}$ corresponding to the anti-causal component:

$$\mathcal{R}_c \bigcap \mathcal{R}_{ac}$$

See Figure 3.5 for an example illustrating how the ROCs connect with the poles and the type of signal.

Special case: The Laplace transform of a function $f(t)$ of finite support $t_1 \leq t \leq t_2$, has the whole s-plane as ROC.

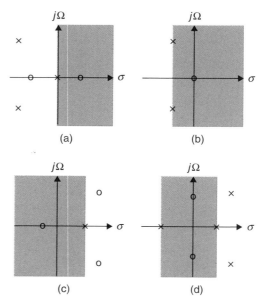

FIGURE 3.5
ROC for (a) causal signal with poles with $\sigma_{max} = 0$; (b) causal signal with poles with $\sigma_{max} < 0$; (c) anti-causal signal with poles with $\sigma_{min} > 0$; (d) noncausal signal where ROC is bounded by poles (poles on the left-hand s-plane give causal component and poles on the right-hand s-plane give the anti-causal component of the signal). The ROCs do not contain poles, but they can contain zeros.

Indeed, the integral defining the Laplace transform is bounded for any value of $\sigma \neq 0$. If $A = \max(|f(t)|)$, then

$$|F(s)| \leq \int_{t_1}^{t_2} |f(t)||e^{-st}|dt \leq A \int_{t_1}^{t_2} e^{-\sigma t}dt = A\frac{e^{-\sigma t_1} - e^{-\sigma t_2}}{\sigma} < \infty \qquad \sigma \neq 0$$

The Laplace transform of a

- Finite support function (i.e., $f(t) = 0$ for $t < t_1$ and $t > t_2$, for $t_1 < t_2$) is

$$\mathcal{L}[f(t)] = \mathcal{L}[f(t)[u(t - t_1) - u(t - t_2)]] \qquad \text{whole } s\text{-plane}$$

- Causal function (i.e., $f(t) = 0$ for $t < 0$) is

$$\mathcal{L}[f(t)u(t)] \qquad \mathcal{R}_c = \{(\sigma, \Omega) : \sigma > \max\{\sigma_i\}, -\infty < \Omega < \infty\}$$

- Anti-causal function (i.e., $f(t) = 0$ for $t > 0$) is

$$\mathcal{L}[f(t)u(-t)] \qquad \mathcal{R}_{ac} = \{(\sigma, \Omega) : \sigma < \min\{\sigma_i\}, -\infty < \Omega < \infty\}$$

- Noncausal function (i.e., $f(t) = f_{ac}(t) + f_c(t) = f(t)u(-t) + f(t)u(t)$) is

$$\mathcal{L}[f(t)] = \mathcal{L}[f_{ac}(-t)u(t)]_{(-s)} + \mathcal{L}[f_c(t)u(t)] \qquad \mathcal{R}_c \bigcap \mathcal{R}_{ac}$$

Although redundant, a causal function $f(t)$ (i.e., $f(t) = 0$ for $t < 0$) is denoted as $f(t)u(t)$. Its Laplace transform is thus

$$\mathcal{L}[f(t)u(t)] = \int_{-\infty}^{\infty} f(t)u(t)e^{-st}dt = \int_{0}^{\infty} f(t)e^{-st}dt$$

which is called the *one-sided Laplace transform*. Likewise, if $f(t)$ is anti-causal (i.e., $f(t) = 0$ for $t > 0$), we will denote it as $f(t)u(-t)$ and its Laplace transform is given by

$$\mathcal{L}[f(t)u(-t)] = \int_{-\infty}^{0} f(t)u(-t)e^{-st}dt = \int_{0}^{\infty} f(-t')u(t')e^{st'}dt'$$

or the one-sided Laplace transform of the causal signal $f(-t)u(t)$, with s changed into $-s$.

A noncausal signal $f(t)$ is defined for all values of t (i.e., for $-\infty < t < \infty$). Such a signal has a causal component $f_c(t)$, which is obtained by multiplying $f(t)$ by the unit-step function, $f_c(t) = f(t)u(t)$, and an anti-causal component $f_{ac}(t)$, which is obtained by multiplying $f(t)$ by $u(-t)$, so that

$$f(t) = f_{ac}(t) + f_c(t)$$
$$= f(t)u(-t) + f(t)u(t) \tag{3.6}$$

At $t = 0$ we assume that $u(0) = 0.5$ to get $f(0)$ from the sum $f_c(0) + f_{ac}(0)$. The Laplace transform of the two-sided signal $f(t)$ can then be computed as

$$F(s) = \int_0^\infty f(-t)u(t)e^{st}dt + \int_0^\infty f(t)u(t)e^{-st}dt$$

$$= \mathcal{L}[f_{ac}(-t)u(t)]_{(-s)} + \mathcal{L}[f_c(t)u(t)] \qquad (3.7)$$

with an ROC the intersection of the ROCs of the causal and the anti-causal Laplace transforms.

3.3 THE ONE-SIDED LAPLACE TRANSFORM

The *one-sided Laplace transform* is defined as

$$F(s) = \mathcal{L}[f(t)u(t)] = \int_{0-}^\infty f(t)u(t)e^{-st}dt \qquad (3.8)$$

where $f(t)$ is either a causal function or made into a causal function by the multiplication by $u(t)$. The one-sided Laplace transform is of significance given that most of the applications deal with causal systems and signals, and that any signal or system can be decomposed into causal and anti-causal components requiring only the computation of one-sided Laplace transforms.

Remarks

- *If $f(t)$ is causal the multiplication by $u(t)$ is redundant but harmless, but if $f(t)$ is not causal the multiplication by $u(t)$ makes $f(t)u(t)$ causal. Notice that when $f(t)$ is causal, the two-sided and the one-sided Laplace transforms of $f(t)$ coincide.*
- *The lower limit of the integral in the one-sided Laplace transform is set to $0- = 0 - \varepsilon$, which corresponds to a value on the left side of 0 for an infinitesimal value ε. The reason for this is to make sure that an impulse function $\delta(t)$, only defined at $t = 0$, is included when we are computing its Laplace transform. For any other signal this limit can be taken as 0 with no effect on the transform.*
- *As we will see, the advantage of the one-sided Laplace transform is that it can be used in the solution of differential equations with initial conditions. In fact, the two-sided Laplace transform by starting at $t = -\infty$ (lower bound of the integral) ignores initial conditions at $t = 0$, and thus it is not useful in solving differential equations unless the initial conditions are zero.*

■ Example 3.2

Find the Laplace transforms of $\delta(t)$, $u(t)$, and a pulse $p(t) = u(t) - u(t - 1)$. Use MATLAB to verify the transforms.

Solution

Even though $\delta(t)$ is not a regular signal, its Laplace transform can be easily obtained:

$$\mathcal{L}[\delta(t)] = \int_{-\infty}^{\infty} \delta(t)e^{-st}dt = \int_{-\infty}^{\infty} \delta(t)e^{-s0}dt = \int_{-\infty}^{\infty} \delta(t)dt = 1$$

Since there are no conditions for the integral to exist, we say that $\mathcal{L}[\delta(t)] = 1$ exists for all values of s, or that its ROC is the whole s-plane. This is also indicated by the fact that $\mathcal{L}[\delta(t)] = 1$ has no poles.

The Laplace transform of $u(t)$ can be found as

$$U(s) = \mathcal{L}[u(t)] = \int_{-\infty}^{\infty} u(t)e^{-st}dt = \int_{0}^{\infty} e^{-st}dt = \int_{0}^{\infty} e^{-\sigma t}e^{-j\Omega t}dt$$

where we replaced the variable $s = \sigma + j\Omega$. Using Euler's equation, the above equation becomes

$$U(s) = \int_{0}^{\infty} e^{-\sigma t}[\cos(\Omega t) - j\sin(\Omega t)]dt$$

and since the sine and the cosine are bound, then we need to find a value for σ so that the exponential $e^{-\sigma t}$ does not grow as t increases. If $\sigma < 0$, the exponential $e^{-\sigma t}$ for $t \geq 0$ will grow and the integral will not converge. On the other hand, if $\sigma > 0$, the integral will converge as $e^{-\sigma t}$ for $t \geq 0$ decays, and it is not clear what happens when $\sigma = 0$. Thus, the integral exists in the region defined by $\sigma > 0$ and all frequencies $-\infty < \Omega < \infty$ (the frequency values do not interfere in the convergence). Such a region is the open right-hand s-plane, and is called the ROC of $U(s)$.

In the region of convergence, the integral is found to be

$$U(s) = \frac{e^{-st}}{-s}\Big|_{t=0}^{\infty} = \frac{1}{s}$$

where the limit for $t = \infty$ is zero since $\sigma > 0$. So the Laplace transform $U(s) = 1/s$ converges in the region defined by $\{(\sigma, \Omega) : \sigma > 0, -\infty < \Omega < \infty\}$, or the open (i.e., the $j\Omega$ axis is not included) right-hand s-plane. This ROC can also be obtained by considering that the pole of $U(s)$ is at $s = 0$ and that $u(t)$ is casual.

We can find the Laplace transform of signals using symbolic computations in MATLAB. For the unit-step and the delta functions, once the symbolic parameters are defined, the MATLAB function laplace computes their Laplace transforms as indicated by the following script.

```
%%%%%%%%%%%%%%%%%
% Example 3.2
%%%%%%%%%%%%%%%%%
syms t s
% Unit-step function
```

```
u = sym('heaviside(t)')
U=laplace(u)
% Delta function
d = sym('dirac(t)')
D = laplace(d)
```

giving

```
u = Heaviside(t)
U = 1/s

d = Dirac(t)si
D = 1
```

where U and D stand for the Laplace transforms of u and d. The naming of $u(t)$ and $\delta(t)$ as Heaviside and Dirac functions is used in MATLAB.[3]

The pulse $p(t) = u(t) - u(t-1)$ is a finite support signal and so its ROC is the whole s-plane. Its Laplace transform is

$$P(s) = \mathcal{L}[u(t+1) - u(t-1)] = \int_{-1}^{1} e^{-st} dt = \frac{-e^{-st}}{s}\Big|_{t=-1}^{1} = \frac{1}{s}[e^s - e^{-s}] = \frac{e^s}{s}[1 - e^{-2s}]$$

which as shown before has an infinite number of zeros, and the one at the origin cancels the pole, so that

$$P(s) = \prod_{k=-\infty, k\neq 0}^{\infty} (s - j\pi k)$$

■ Example 3.3

Let us find and use the Laplace transform of $e^{j(\Omega_0 t + \theta)} u(t)$ to obtain the Laplace transform of $x(t) = \cos(\Omega_0 t + \theta) u(t)$. Consider the special cases for $\theta = 0$ and $\theta = -\pi/2$. Determine the ROCs. Use MATLAB to plot the signals and the corresponding poles/zeros when $\Omega_0 = 2$ and $\theta = 0$ and $\pi/4$.

Solution

The Laplace transform of the complex causal signal $e^{j(\Omega_0 t + \theta)} u(t)$ is found to be

$$\mathcal{L}[e^{j(\Omega_0 t + \theta)} u(t)] = \int_{0}^{\infty} e^{j(\Omega_0 t + \theta)} e^{-st} dt = e^{j\theta} \int_{0}^{\infty} e^{-(s - j\Omega_0)t} dt$$

---

[3] Oliver Heaviside (1850–1925) was an English electrical engineer who adapted the Laplace transform to the solution of differential equations (the so-called operational calculus), while Paul Dirac (1902–1984) was also an English electrical engineer, better known for his work in physics.

$$= \frac{-e^{j\theta}}{s - j\Omega_0} e^{-\sigma t - j(\Omega - \Omega_0)t} \big|_{t=0}^{\infty} = \frac{e^{j\theta}}{s - j\Omega_0} \qquad \text{ROC: } \sigma > 0$$

According to Euler's identity

$$\cos(\Omega_0 t + \theta) = \frac{e^{j(\Omega_0 t + \theta)} + e^{-j(\Omega_0 t + \theta)}}{2}$$

by the linearity of the integral and using the above result, we get that

$$\mathcal{L}[\cos(\Omega_0 t + \theta)u(t)] = 0.5\mathcal{L}[e^{j(\Omega_0 t + \theta)}u(t)] + 0.5\mathcal{L}[e^{-j(\Omega_0 t + \theta)}u(t)]$$

$$= 0.5 \frac{e^{j\theta}(s + j\Omega_0) + e^{-j\theta}(s - j\Omega_0)}{s^2 + \Omega_0^2}$$

$$= \frac{s\cos(\theta) - \Omega_0 \sin(\theta)}{s^2 + \Omega_0^2}$$

and a region of convergence $\{(\sigma, \Omega) : \sigma > 0, -\infty < \Omega < \infty\}$.

Now if we let $\theta = 0, -\pi/2$ in the above equation we have the following Laplace transforms:

$$\mathcal{L}[\cos(\Omega_0 t)u(t)] = \frac{s}{s^2 + \Omega_0^2}$$

$$\mathcal{L}[\sin(\Omega_0 t)u(t)] = \frac{\Omega_0}{s^2 + \Omega_0^2}$$

as $\cos(\Omega_0 t - \pi/2) = \sin(\Omega_0 t)$. The ROC of the above Laplace transforms is $\{(\sigma, \Omega) : \sigma > 0, -\infty < \Omega < \infty\}$, or the open right-hand s-plane (i.e., not including the $j\Omega$ axis). See Figure 3.6 for the pole-zero plots and the corresponding signals for $\theta = 0$, $\theta = \pi/4$, and $\Omega_0 = 2$. ∎

■ Example 3.4

Use MATLAB symbolic computation to find the Laplace transform of a real exponential, $x(t) = e^{-t}u(t)$, and of $x(t)$ modulated by a cosine or $y(t) = e^{-t}\cos(10t)u(t)$. Plot the signals and the poles and zeros of their Laplace transforms.

Solution

The following script is used. The MATLAB function laplace is used for the computation of the Laplace transform and the function ezplot allows us to do the plotting. For the plotting of the poles and zeros we use our function splane. When you run the script you obtain the Laplace transforms

$$X(s) = \frac{1}{s + 1}$$

$$Y(s) = \frac{s + 1}{s^2 + 2s + 101}$$

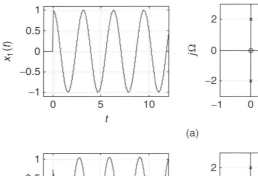

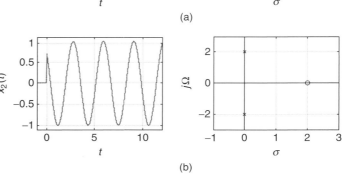

FIGURE 3.6

Location of the poles and zeros of $\cos(2t + \theta)u(t)$ for (a) $\theta = 0$ and for (b) $\theta = \pi/4$. Note that the zero is moved to the right to 2 because the zero of the Laplace transform is $s = \Omega_0 \tan(\theta) = 2 \tan(\pi/4) = 2$.

(a)

(b)

```
%%%%%%%%%%%%%%%%%
% Example 3.4
%%%%%%%%%%%%%%%%%
syms t
x = exp (-t);
y = x * cos(10 * t);
X = laplace(x)
Y = laplace(y)
% plotting of signals and poles/zeros
figure(1)
subplot(221)
ezplot(x,[0,5]);grid
axis([0 5 0 1.1]);title('x(t) = exp(-t)u(t)')
numx = [0 1];denx = [1 1];
subplot(222)
splane(numx,denx)
subplot(223)
ezplot(y,[-1,5]);grid
axis([0 5 -1.1 1.1]);title('y(t) = cos(10t)exp(-t)u(t)')
numy = [0 1 1];deny = [1 2 101];
subplot(224)
splane(numy,deny)
```

The results are shown in Figure 3.7. ∎

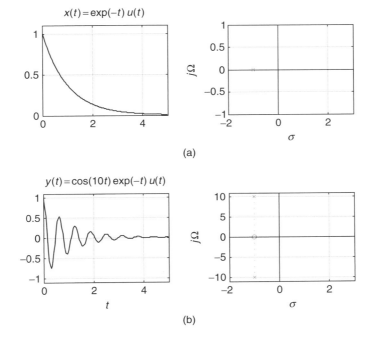

$x(t) = \exp(-t)\, u(t)$

$y(t) = \cos(10t)\, \exp(-t)\, u(t)$

(a)

(b)

FIGURE 3.7

Poles and zeros of the Laplace transform of (a) causal signal $x(t) = e^{-t}u(t)$ and of (b) causal decaying signal $y(t) = e^{-t}\cos(10t)u(t)$.

■ Example 3.5

In statistical signal processing, the autocorrelation function $c(\tau)$ of a random signal describes the correlation that exists between the random signal $x(t)$ and shifted versions of it, $x(t + \tau)$ and $x(t - \tau)$ for shifts $-\infty < \tau < \infty$. Clearly, $c(\tau)$ is two-sided (i.e., nonzero for both positive and negative values of τ) and symmetric. Its two-sided Laplace transform is related to the power spectrum of the signal $x(t)$. Let $c(t) = e^{-a|t|}$, where $a > 0$ (we replaced the τ variable for t for convenience). Find its Laplace transform. Determine if it would be possible to compute $|C(\Omega)|^2$, which is called the power spectrum of the random signal $x(t)$.

Solution

The autocorrelation can be expressed as

$$c(t) = c(t)u(t) + c(t)u(-t)$$
$$= c_c(t) + c_{ac}(t)$$

where $c_c(t)$ is the causal component and $c_{ac}(t)$ the anti-causal component of $c(t)$. The Laplace transform of $c(t)$ is then given by

$$C(s) = \mathcal{L}[c_c(t)u(t)] + \mathcal{L}[c_{ac}(-t)u(t)]_{(-s)}$$

The Laplace transform for $c_c(t) = e^{-at}u(t)$, as seen before, is

$$C_c(s) = \frac{1}{s+a}$$

with a region of convergence $\{(\sigma, \Omega) : \sigma > -a, -\infty < \Omega < \infty\}$. The Laplace transform of the anti-causal part is

$$\mathcal{L}[c_{ac}(-t)u(t)]_{(-s)} = \frac{1}{-s+a}$$

and since it is anti-causal and has a pole at $s = a$, its region of convergence is $\{(\sigma, \Omega) : \sigma < a, -\infty < \Omega < \infty\}$.

We thus have that

$$C(s) = \frac{1}{s+a} + \frac{1}{-s+a}$$

$$= \frac{2a}{a^2 - s^2}$$

with a region of convergence the intersection of $\sigma > -a$ with $\sigma < a$ or $\{(\sigma, \Omega) : -a < \sigma < a, -\infty < \Omega < \infty\}$. This region contains the $j\Omega$ axis, which permits us to compute the distribution of the power over frequencies or the power spectrum of the random signals $|C(\Omega)|^2$ (see in Figure 3.8).

∎

■ Example 3.6

Consider a noncausal LTI system with impulse response

$$h(t) = e^{-t}u(t) + e^{2t}u(-t)$$

$$= h_c(t) + h_{ac}(t)$$

Let us compute the system function $H(s)$ for this system, and find out whether we could compute $H(j\Omega)$ from its Laplace transform.

Solution

As from before, the Laplace transform of the causal component, $h_c(t)$, is

$$H_c(s) = \frac{1}{s+1}$$

provided that $\sigma > -1$. For the anti-causal component

$$\mathcal{L}[h_{ac}(t)] = \mathcal{L}[h_{ac}(-t)u(t)]_{(-s)} = \frac{1}{-s+2}$$

which converges when $\sigma - 2 < 0$ or $\sigma < 2$, or its region of convergence is $\{(\sigma, \Omega) : \sigma < 2, -\infty < \Omega < \infty\}$.

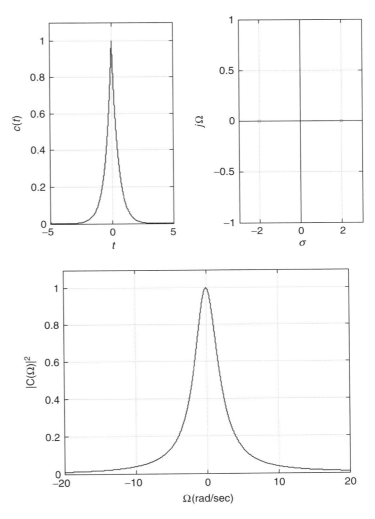

FIGURE 3.8
Poles (top right) of the Laplace transform of the autocorrelation $c(t) = e^{-2|t|}$ (top left), which is noncausal. The ROC of $C(s)$ is the region in between the poles, which includes the $j\Omega$ axis. The spectrum $|C(\Omega)|^2$ corresponding to $c(t)$ is shown in the bottom plot—this is the magnitude square of the Fourier transform of $c(t)$.

Thus, the system function is

$$H(s) = \frac{1}{s+1} + \frac{1}{-s+2} = \frac{-3}{(s+1)(s-2)}$$

with a region of convergence the intersection of the regions of convergence of its components, or the intersection of $\{(\sigma, \Omega) : \sigma > -1, -\infty < \Omega < \infty\}$ and $\{(\sigma, \Omega) : \sigma < 2, -\infty < \Omega < \infty\}$, or

$$\{(\sigma, \Omega) : -1 < \sigma < 2, -\infty < \Omega < \infty\}$$

which is a sector of the s-plane that includes the $j\Omega$ axis. Thus, $H(j\Omega)$ can be computed from its Laplace transform. ∎

■ Example 3.7

Compute the Laplace transform of the ramp function $r(t) = tu(t)$ and use it to find the Laplace of a triangular pulse $\Lambda(t) = r(t+1) - 2r(t) + r(t-1)$.

Solution

Notice that although the ramp is an ever-increasing function of t, we still can obtain its Laplace transform

$$R(s) = \int_0^\infty te^{-st}dt = \frac{e^{-st}}{s^2}(-st-1)\,\big|_{t=0}^\infty = \frac{1}{s^2}$$

where we let $\sigma > 0$ for the integral to exist. Thus, $R(s) = 1/s^2$ with region of convergence $\{(\sigma, \Omega) : \sigma > 0, -\infty < \Omega < \infty\}$. The above integration can be avoided by noticing that if we find the derivative with respect to s of the Laplace transform of $u(t)$, or

$$\frac{d\,U(s)}{ds} = \int_0^\infty \frac{de^{-st}}{ds}dt$$

$$= \int_0^\infty (-t)e^{-st}dt$$

$$= -R(s)$$

where we assumed the derivative and the integral can be interchanged. We then have

$$R(s) = -\frac{d\,U(s)}{ds} = \frac{1}{s^2}$$

The Laplace transform of $\Lambda(t)$ can then be shown to be (try it!)

$$\Lambda(s) = \frac{1}{s^2}[e^s - 2 + e^{-s}]$$

The zeros of $\Lambda(s)$ are the values of s that make $e^s - 2 + e^{-s} = 0$, or multiplying by e^{-s},

$$1 - 2e^{-s} + e^{-2s} = (1 - e^{-s})^2 = 0$$

which is equivalent to $e^{-s} = 1 = e^{j2\pi k}$, for integer k, or double zeros at

$$s_k = j2\pi k \qquad k = 0, \pm 1, \pm 2, \ldots$$

In particular, when $k = 0$ there are two zeros at 0, which cancel the two poles at 0 resulting from the denominator s^2. Thus, $\Lambda(s)$ has an infinite number of zeros but no poles given this pole-zero cancellation (see Figure 3.9). Therefore, $\Lambda(s)$ has the whole s-plane as its region of convergence, and can be calculated at $s = j\Omega$. ■

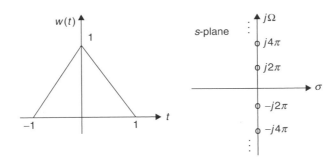

FIGURE 3.9
The Laplace transform of triangular signal $\Lambda(t)$ has as ROC the whole s-plane, since it has no poles but an infinite number of double zeros at $\pm j2\pi k$, for $k = \pm1, \pm2, \ldots$.

We will consider next the basic properties of the one-sided Laplace transform—many of these properties will be encountered in the Fourier analysis, presented in a slightly different form, given the connection between the Laplace and the Fourier transforms. Something to observe is the duality that exists between the time and the frequency domains. The time and the frequency domain representations of continuous-time signals and systems are complementary—that is, certain characteristics of the signal or the system can be seen better in one domain than in the other. In the following, we consider the properties of the Laplace transform of signals but they equally apply to the impulse response of a system.

3.3.1 Linearity

For signals $f(t)$ and $g(t)$, with Laplace transforms $F(s)$ and $G(s)$, and constants a and b, we have the Laplace transform is linear:

$$\mathcal{L}[af(t)u(t) + bg(t)u(t)] = aF(s) + bG(s)$$

The linearity of the Laplace transform is easily verified using integration properties:

$$\mathcal{L}[af(t)u(t) + bg(t)u(t)] = \int_0^\infty [af(t) + bg(t)]u(t)e^{-st}dt$$

$$= a\int_0^\infty f(t)u(t)e^{-st}dt + b\int_0^\infty g(t)u(t)e^{-st}dt$$

$$= a\mathcal{L}[f(t)u(t)] + b\mathcal{L}[g(t)(t)]$$

We will use the linearity property to illustrate the significance of the location of the poles of the Laplace transform of causal signals. As seen before, the Laplace transform of an exponential signal

$f(t) = Ae^{-at}u(t)$ where a in general can be a complex number is

$$F(s) = \frac{A}{s+a} \qquad \text{ROC: } \sigma > -a$$

The location of the pole $s = -a$ closely relates to the signal. For instance, if $a = 5$, $f(t) = Ae^{-5t}u(t)$ is a decaying exponential and the pole of $F(s)$ is at $s = -5$ (in left-hand s-plane); if $a = -5$, we have an increasing exponential and the pole is at $s = 5$ (in right-hand s-plane). The larger the value of $|a|$ the faster the exponential decays (for $a > 0$) or increases (for $a < 0$); thus, $Ae^{-10t}u(t)$ decays a lot faster that $Ae^{-5t}u(t)$, and $Ae^{10t}u(t)$ grows a lot faster than $Ae^{5t}u(t)$.

The Laplace transform $F(s) = 1/(s+a)$ of $f(t) = e^{-at}u(t)$, for any real value of a, has a pole on the real axis σ of the s-plane, and we have the following three cases:

- For $a = 0$, the pole at the origin $s = 0$ corresponds to the signal $f(t) = u(t)$, which is constant for $t \geq 0$ (i.e., it does not decay).
- For $a > 0$, the signal is $f(t) = e^{-at}u(t)$, a decaying exponential, and the pole $s = -a$ of $F(s)$ is in the real axis σ of the left-hand s-plane. As the pole is moved away from the origin toward the left, the faster the exponential decays, and as it moves toward the origin, the slower the exponential decays.
- For $a < 0$, the pole $s = -a$ is on the real axis σ of the right-hand s-plane, and corresponds to a growing exponential. As the pole moves to the right the exponential grows faster, and as it is moved toward the origin it grows at a slower rate—clearly this signal is not useful, as it grows continuously.

The conclusion is that the σ axis of the Laplace plane corresponds to damping. A single pole on this axis and in the left-hand s-plane corresponds to a decaying exponential, and a single pole on this axis and in the right-hand s-plane corresponds to a growing exponential.

Suppose then we consider

$$g(t) = A\cos(\Omega_0 t)u(t) = A\frac{e^{j\Omega_0 t}}{2}u(t) + A\frac{e^{-j\Omega_0 t}}{2}u(t)$$

and let $a = j\Omega_0$ to express $g(t)$ as

$$g(t) = 0.5[Ae^{at}u(t) + Ae^{-at}u(t)]$$

Then, by the linearity of the Laplace transform and the previous result we obtain

$$G(s) = \frac{A}{2}\frac{1}{s - j\Omega_0} + \frac{A}{2}\frac{1}{s + j\Omega_0} = \frac{As}{s^2 + \Omega_0^2} \qquad (3.9)$$

with a zero at $s = 0$, and the poles are values for which

$$s^2 + \Omega_0^2 = 0 \implies s^2 = -\Omega_0^2 \text{ or } s_{1,2} = \pm j\Omega_0$$

which are located on the $j\Omega$ axis. The farther away from the origin of the $j\Omega$ axis the poles are, the higher the frequency Ω_0, and the closer the poles are to the origin, the lower the frequency. Thus, the $j\Omega$ axis corresponds to the frequency axis. Furthermore, notice that to generate the real-valued signal $g(t)$ we need two complex conjugate poles, one at $+j\Omega_0$ and the other at $-j\Omega_0$. Although frequency, as measured by frequency meters, is a positive value, "negative" frequencies are needed to represent "real" signals (if the poles are not complex conjugate pairs, the inverse Laplace transform is complex—rather than real valued).

> The conclusion is that the Laplace transform of a sinusoid has a pair of poles on the $j\Omega$ axis. For these poles to correspond to a real-valued signal they should be complex conjugate pairs, requiring negative as well as positive values of the frequency. Furthermore, when these poles are moved away from the origin of the $j\Omega$ axis, the frequency increases, and the frequency decreases whenever the poles are moved toward the origin.

Finally, consider the case of a signal $d(t) = Ae^{-\alpha t}\cos(\Omega_0 t)u(t)$ or a causal sinusoid multiplied (or modulated) by $e^{-\alpha t}$. According to Euler's identity,

$$d(t) = A\left[\frac{e^{(-\alpha+j\Omega_0)t}}{2}u(t) + \frac{e^{(-\alpha-j\Omega_0)t}}{2}u(t)\right]$$

and as such we can again use linearity to get

$$D(s) = \frac{A(s+\alpha)}{(s+\alpha)^2 + \Omega_0^2} \tag{3.10}$$

Notice the connection between Equations (3.9) and (3.10). Given $G(s)$, then $D(s) = G(s+\alpha)$, with $G(s)$ corresponding to $g(t) = A\cos(\Omega_0 t)$ and $D(s)$ to $d(t) = g(t)e^{-\alpha t}$. Multiplying a function $g(t)$ by an exponential $e^{-\alpha t}$, with α real or imaginary, shifts the transform to $G(s+\alpha)$—that is, it is a *complex frequency-shift* property. The poles of $D(s)$ have as the real part the damping factor $-\alpha$ and as the imaginary part the frequencies $\pm\Omega_0$. The real part of the pole indicates decay (if $\alpha > 0$) or growth (if $\alpha < 0$) in the signal, while the imaginary part indicates the frequency of the cosine in the signal. Again, the poles will be complex conjugate pairs since the signal $d(t)$ is real valued.

> The conclusion is that the location of the poles (and to some degree the zeros), as indicated in the previous two cases, determines the characteristics of the signal. Signals are characterized by their damping and frequency and as such can be described by the poles of its Laplace transform.

If we were to add the different signals considered above, then the Laplace transform of the resulting signal would be the sum of the Laplace transform of each of the signals and the poles would be the aggregation of the poles from each. This observation will be important when finding the inverse Laplace transform, then we would like to do the opposite: To isolate poles or pairs of poles (when they are complex conjugate) and associate with each a general form of the signal with parameters that are found by using the zeros and the other poles of the transform. Figure 3.10 provides an example illustrating the importance of the location of the poles, and the significance of the σ and $j\Omega$ axes.

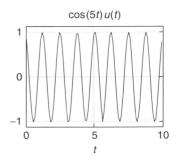

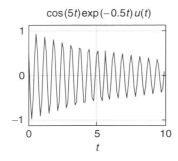

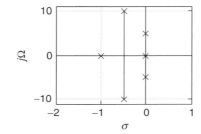

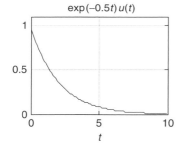

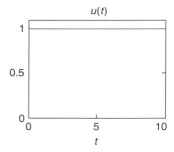

FIGURE 3.10

For poles shown in the middle, possible signals are displayed around them anti–clockwise from bottom right. The pole $s = 0$ corresponds to a unit-step signal; the complex conjugate poles on the $j\Omega$ axis correspond to a sinusoid; the pair of complex conjugate poles with a negative real part provides a sinusoid multiplied by an exponential; and the pole in the negative real axis gives a decaying exponential. The actual amplitudes and phases are determined by the other poles and by the zeros.

3.3.2 Differentiation

For a signal $f(t)$ with Laplace transform $F(s)$ the one-sided Laplace transform of its first-and second-order derivatives are

$$\mathcal{L}\left[\frac{df(t)}{dt}u(t)\right] = sF(s) - f(0-) \tag{3.11}$$

$$\mathcal{L}\left[\frac{d^2f(t)}{dt^2}u(t)\right] = s^2F(s) - sf(0-) - \frac{df(t)}{dt}\Big|_{t=0-} \tag{3.12}$$

In general, if $f^{(N)}(t)$ denotes an Nth-order derivative of a function $f(t)$ that has a Laplace transform $F(s)$, we have

$$\mathcal{L}[f^{(N)}(t)u(t)] = s^N F(s) - \sum_{k=0}^{N-1} f^{(k)}(0-)s^{N-1-k} \tag{3.13}$$

where $f^{(m)}(t) = d^m f(t)/dt^m$ is the mth-order derivative, $m > 0$, and $f^{(0)}(t) \triangleq f(t)$.

The Laplace transform of the derivative of a causal signal is

$$\mathcal{L}\left[\frac{df(t)}{dt}u(t)\right] = \int_{0-}^{\infty}\frac{df(t)}{dt}e^{-st}dt$$

This integral is evaluated by parts. Let $w = e^{-st}$, then $dw = -se^{-st}dt$, and let $v = f(t)$ so that $dv = [df(t)/dt]dt$, and

$$\int wdv = wv - \int vdw$$

We would then have

$$\int_{0-}^{\infty}\frac{df(t)}{dt}e^{-st}dt = e^{-st}f(t)\Big|_{0-}^{\infty} - \int_{0-}^{\infty}f(t)(-se^{-st})dt$$

$$= s\int_{0-}^{\infty}f(t)e^{-st}dt - f(0-)$$

$$= sF(s) - f(0-)$$

where $e^{-st}f(t)|_{t=0-} = f(0-)$ and $e^{-st}f(t)|_{t\to\infty} = 0$ since the region of convergence guarantees that

$$\lim_{t\to\infty}f(t)e^{-\sigma t} = 0$$

For a second-order derivative we have that

$$\mathcal{L}\left[\frac{d^2f(t)}{dt^2}u(t)\right] = \mathcal{L}\left[\frac{df^{(1)}(t)}{dt}u(t)\right]$$

$$= s\mathcal{L}[f^{(1)}(t)] - f^{(1)}(0-)$$

$$= s^2F(s) - sf(0-) - \frac{df(t)}{dt}\Big|_{t=0-}$$

where we used the notation $f^{(1)}(t) = df(t)/dt$. This approach can be extended to any higher order to obtain the general result shown above.

Remarks

■ *The derivative property for a signal $x(t)$ defined for all t is*

$$\int_{-\infty}^{\infty} \frac{dx(t)}{dt} e^{-st} dt = sX(s)$$

This can be seen by computing the derivative of the inverse Laplace transform with respect to t, assuming that the integral and the derivative can be interchanged. Using Equation (3.3):

$$\frac{dx(t)}{dt} = \frac{1}{2\pi j} \int_{\sigma-j\infty}^{\sigma+j\infty} X(s) \frac{de^{st}}{dt} ds$$

$$= \frac{1}{2\pi j} \int_{\sigma-j\infty}^{\sigma+j\infty} (sX(s)) e^{st} ds$$

or that $sX(s)$ is the Laplace transform of the derivative of $x(t)$. Thus, the two-sided transform does not include initial conditions. The above result can be generalized to any order of the derivative as

$$\mathcal{L}[d^N x(t)/dt^N] = s^N X(s)$$

■ *Application of the linearity and the derivative properties of the Laplace transform makes solving differential equations an algebraic problem.*

■ Example 3.8

Find the impulse response of an RL circuit in series with a voltage source $v_s(t)$ (see Figure 3.11). The current $i(t)$ is the output and the input is the voltage source $v_s(t)$.

Solution

To find the impulse response of the RL circuit we let $v_s(t) = \delta(t)$ and set the initial current in the inductor to zero. According to Kirchhoff's voltage law,

$$v_s(t) = L \frac{di(t)}{dt} + Ri(t) \qquad i(0-) = 0$$

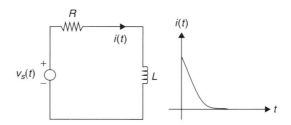

FIGURE 3.11

Impulse response $i(t)$ of an RL circuit with input $v_s(t)$.

which is a first-order linear differential equation with constant coefficients, zero initial condition, and a causal input so that it is a linear time-invariant system, as discussed before.

Letting $v_s(t) = \delta(t)$ and computing the Laplace transform of the above equation (using the linearity and the derivative properties of the transform and remembering the initial condition is zero), we obtain the following equation in the s-domain:

$$\mathcal{L}[\delta(t)] = \mathcal{L}\left[L\frac{di(t)}{dt} + Ri(t) \right]$$

$$1 = sLI(s) + RI(s)$$

where $I(s)$ is the Laplace transform of $i(t)$. Solving for $I(s)$ we have that

$$I(s) = \frac{1/L}{s + R/L}$$

which as we have seen is the Laplace transform of

$$i(t) = \frac{1}{L}e^{-(R/L)t}u(t)$$

Notice that $i(0-) = 0$ and that the response has the form of a decaying exponential trying to follow the input signal, a delta function. ■

■ Example 3.9

In this example we consider the duality between the time and the Laplace domains. The differentiation property indicates that computing the derivative of a function in the time domain corresponds to multiplying by s the Laplace transform of the function (assuming initial conditions are zero). We will illustrate in this example the dual of this—that is, when we differentiate a function in the s-domain its effect in the time domain is to multiply by $-t$. Consider the connection between $\delta(t)$, $u(t)$, and $r(t)$ (i.e., the unit impulse, the unit step, and the ramp, respectively), and relate it to the indicated duality. Explain how this property connects with the existence of multiple poles, real and complex, in general.

Solution

The relation between the signals $\delta(t)$, $u(t)$, and $r(t)$ is seen from

$$\mathcal{L}[r(t)] = \frac{1}{s^2}$$

$$\mathcal{L}\left[u(t) = \frac{dr(t)}{dt} \right] = s\frac{1}{s^2} = \frac{1}{s}$$

$$\mathcal{L}\left[\delta(t) = \frac{du(t)}{dt} \right] = s\frac{1}{s} = 1$$

which also shows that a double pole at the origin, $1/s^2$, corresponds to a ramp function $r(t) = tu(t)$.

The above results can be explained by looking for a dual of the derivative property. Multiplying by $-t$ the signal $x(t)$ corresponds to differentiating $X(s)$ with respect to s. Indeed for an integer $N > 1$,

$$\frac{d^N X(s)}{ds^N} = \int_0^\infty x(t) \frac{d^N e^{-st}}{ds^N} dt$$

$$= \int_0^\infty x(t)(-t)^N e^{-st} dt$$

Thus, if $x(t) = u(t)$, $X(s) = 1/s$, then $-tx(t)$ has Laplace transform $dX(s)/ds = -1/s^2$, or $tu(t)$ and $1/s^2$ are Laplace transform pairs. In general, the Laplace transform of $t^{N-1}u(t)$, for $N \geq 1$, has N poles at the origin.

What about multiple real (different from zero) and multiple complex poles? What are the corresponding inverse Laplace transforms? The inverse Laplace transform of

$$2\Omega_0 s/(s^2 + \Omega_0^2)^2$$

having double complex poles at $\pm j\Omega_0$, is

$$t\sin(\Omega_0 t)u(t)$$

Likewise,

$$te^{-at}u(t)$$

has as Laplace transform $1/(s+a)^2$. So multiple poles correspond to multiplication by t in the time domain. ∎

■ Example 3.10

Obtain from the Laplace transform of $x(t) = \cos(\Omega_0 t)u(t)$ the Laplace transform of $\sin(t)u(t)$ using the derivative property.

Solution

The causal sinusoid

$$x(t) = \cos(\Omega_0 t)u(t)$$

has a Laplace transform

$$X(s) = \frac{s}{s^2 + \Omega_0^2}$$

Then,

$$\frac{dx(t)}{dt} = u(t)\frac{d\cos(\Omega_0 t)}{dt} + \cos(\Omega_0 t)\frac{du(t)}{dt}$$

$$= -\Omega_0 \sin(\Omega_0 t)u(t) + \cos(\Omega_0 t)\delta(t)$$

$$= -\Omega_0 \sin(\Omega_0 t)u(t) + \delta(t)$$

so that the Laplace transform of $dx(t)/dt$ is given by

$$sX(s) - x(0-) = -\Omega_0 \mathcal{L}[\sin(\Omega_0 t)u(t)] + \mathcal{L}[\delta(t)]$$

Thus, the Laplace transform of the sine is

$$\mathcal{L}[\sin(\Omega_0 t)u(t)] = -\frac{sX(s) - x(0-) - 1}{\Omega_0}$$

$$= \frac{1 - sX(s)}{\Omega_0}$$

$$= \frac{\Omega_0}{s^2 + \Omega_0^2}$$

since $x(0-) = 0$ and $X(s) = \mathcal{L}[\cos(\Omega_o T)]$ given above. ∎

Notice that whenever the signal is discontinuous at $t = 0$, as in the case of $x(t) = \cos(\Omega_0 t)u(t)$, its derivative will include a $\delta(t)$ signal due to the discontinuity. On the other hand, whenever the signal is continuous at $t = 0$, for instance $y(t) = \sin(\Omega_0 t)u(t)$, its derivative does not contain $\delta(t)$ signals. In fact,

$$\frac{dy(t)}{dt} = \Omega_0 \cos(\Omega_0 t)u(t) + \sin(\Omega_0 t)\delta(t)$$

$$= \Omega_0 \cos(\Omega_0 t)u(t)$$

since the sine is zero at $t = 0$.

3.3.3 Integration

The Laplace transform of the integral of a causal signal $y(t)$ is given by

$$\mathcal{L}\left[\int_0^t y(\tau)d\tau \; u(t)\right] = \frac{Y(s)}{s} \tag{3.14}$$

This property can be shown by using the derivative property. Call the integral

$$f(t) = \int_0^t y(\tau)d\tau u(t)$$

Using the fundamental theorem of calculus, we then have that

$$\frac{df(t)}{dt} = y(t)u(t)$$

and so

$$\mathcal{L}\left[\frac{df(t)}{dt}\right] = sF(s) - f(0)$$
$$= Y(s)$$

since $f(0) = 0$ (the integral over a point), then

$$F(s) = \mathcal{L}\left[\int_0^t y(\tau)d\tau\right] = \frac{Y(s)}{s}$$

■ **Example 3.11**

Suppose that

$$\int_0^t y(\tau)d\tau = 3u(t) - 2y(t)$$

Find the Laplace transform of $y(t)$, a causal signal.

Solution

Applying the integration property gives

$$\frac{Y(s)}{s} = \frac{3}{s} - 2Y(s)$$

so that solving for $Y(s)$ we obtain

$$Y(s) = \frac{3}{2(s + 0.5)}$$

corresponding to $y(t) = 1.5e^{-0.5t}u(t)$.   ■

3.3.4 Time Shifting

If the Laplace transform of $f(t)u(t)$ is $F(s)$, the Laplace transform of the time-shifted signal $f(t - \tau)u(t - \tau)$ is

$$\mathcal{L}[f(t - \tau)u(t - \tau)] = e^{-\tau s}F(s) \tag{3.15}$$

This indicates that when we delay (advance) the signal to get $f(t - \tau)u(t - \tau)$ ($f(t + \tau)u(t + \tau)$) its corresponding Laplace transform is $F(s)$ multiplied by $e^{-\tau s}$ ($e^{\tau s}$). This property is easily shown by a change of variable when computing the Laplace transform of the shifted signals.

■ **Example 3.12**

Suppose we wish to find the Laplace transform of the causal sequence of pulses $x(t)$ shown in Figure 3.12. Let $x_1(t)$ denote the first pulse (i.e., for $0 \leq t < 1$).

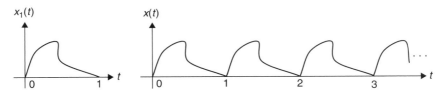

FIGURE 3.12
Generic causal pulse signal.

Solution

We have for $t \geq 0$,

$$x(t) = x_1(t) + x_1(t-1) + x_1(t-2) + \cdots$$

and 0 for $t < 0$. According to the shifting and linearity properties, we have

$$X(s) = X_1(s) \left[1 + e^{-s} + e^{-2s} + \cdots \right]$$
$$= X_1(s) \left[\frac{1}{1 - e^{-s}} \right]$$

Notice that $1 + e^{-s} + e^{-2s} + \cdots = 1/(1 - e^{-s})$, which is verified by cross-mutiplying:

$$[1 + e^{-s} + e^{-2s} + \cdots](1 - e^{-s}) = (1 + e^{-s} + e^{-2s} + \cdots) - (e^{-s} + e^{-2s} + \cdots) = 1$$

The poles of $X(s)$ are the poles of $X_1(s)$ and the roots of $1 - e^{-s} = 0$ (the s values such that $e^{-s} = 1$, or $s_k = \pm j2\pi k$ for any integer $k \geq 0$). Thus, there is an infinite number of poles for $X(s)$, and the partial fraction expansion method that uses poles to invert Laplace transforms, presented later, will not be useful. The reason this example is presented here, ahead of the inverse Laplace, is to illustrate that when we are finding the inverse of this type of Laplace function we need to consider the time-shift property, otherwise we would need to consider an infinite partial fraction expansion.   ■

■ **Example 3.13**

Consider the causal full-wave rectified signal shown in Figure 3.13. Find its Laplace transform.

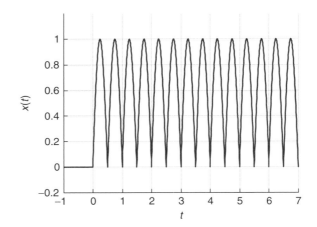

FIGURE 3.13
Full-wave rectified causal signal.

Solution

The first period of the full-wave rectified signal can be expressed as

$$x_1(t) = \sin(2\pi t)u(t) + \sin(2\pi(t - 0.5))u(t - 0.5)$$

and its Laplace transform is

$$X_1(s) = \frac{2\pi(1 + e^{-0.5s})}{s^2 + (2\pi)^2}$$

And the train of these sinusoidal pulses

$$x(t) = \sum_{k=0}^{\infty} x_1(t - 0.5k)$$

will then have the following Laplace transform:

$$X(s) = X_1(s)[1 + e^{-s/2} + e^{-s} + \cdots] = X_1(s)\frac{1}{1 - e^{-s/2}} = \frac{2\pi(1 + e^{-s/2})}{(1 - e^{-s/2})(s^2 + 4\pi^2)} \qquad ■$$

3.3.5 Convolution Integral

Because this is the most important property of the Laplace transform we provide a more extensive coverage later, after considering the inverse Laplace transform.

The Laplace transform of the convolution integral of a causal signal $x(t)$, with Laplace transforms $X(s)$, and a causal impulse response $h(t)$, with Laplace transform $H(s)$, is given by

$$\mathcal{L}[(x * h)(t)] = X(s)H(s) \qquad (3.16)$$

If the input of an LTI system is the causal signal $x(t)$ and the impulse response of the system is $h(t)$, then the output $y(t)$ can be written as

$$y(t) = \int_0^\infty x(\tau)h(t-\tau)d\tau \qquad t \geq 0$$

and zero otherwise. Its Laplace transform is

$$Y(s) = \mathcal{L}\left[\int_0^\infty x(\tau)h(t-\tau)d\tau\right] = \int_0^\infty \left[\int_0^\infty x(\tau)h(t-\tau)d\tau\right] e^{-st}dt$$

$$= \int_0^\infty x(\tau)\left[\int_0^\infty h(t-\tau) \; e^{-s(t-\tau)} \; dt\right] e^{-s\tau} d\tau = X(s)H(s)$$

where the internal integral is shown to be $H(s) = \mathcal{L}[h(t)]$ (change variable to $v = t - \tau$) using the causality of $h(t)$. The remaining integral is the Laplace transform of $x(t)$.

The system function or transfer function $H(s) = \mathcal{L}[h(t)]$, the Laplace transform of the impulse response $h(t)$ of an LTI system, can be expressed as the ratio

$$H(s) = \frac{\mathcal{L}[y(t)]}{\mathcal{L}[x(t)]} = \frac{\mathcal{L}[\text{ output }]}{\mathcal{L}[\text{ input }]} \qquad (3.17)$$

This function is called *transfer function* because it transfers the Laplace transform of the input to the output. Just as with the Laplace transform of signals, $H(s)$ characterizes an LTI system by means of its poles and zeros. Thus, it becomes a very important tool in the analysis and synthesis of systems.

3.4 INVERSE LAPLACE TRANSFORM

Inverting the Laplace transform consists in finding a function (either a signal or an impulse response of a system) that has the given transform with the given region of convergence. We will consider three cases:

- Inverse of one-sided Laplace transforms giving causal functions.
- Inverse of Laplace transforms with exponentials.
- Inverse of two-sided Laplace transforms giving anti-causal or noncausal functions.

The given function $X(s)$ we wish to invert can be the Laplace transform of a signal or a transfer function—that is, the Laplace transform of an impulse response.

3.4.1 Inverse of One-Sided Laplace Transforms

When we consider a causal function $x(t)$, the region of convergence of $X(s)$ is of the form

$$\{(\sigma, \Omega) : \sigma > \sigma_{max}, -\infty < \Omega < \infty\}$$

where σ_{max} is the maximum of the real parts of the poles of $X(s)$. Since in this section we only consider causal signals, the region of convergence will be assumed and will not be shown with the Laplace transform.

The most common inverse Laplace method is the so-called *partial fraction expansion*, which consists in expanding the given function in s into a sum of components of which the inverse Laplace transforms can be found in a table of Laplace transform pairs. Assume the signal we wish to find has a rational Laplace transform—that is,

$$X(s) = \frac{N(s)}{D(s)} \tag{3.18}$$

where $N(s)$ and $D(s)$ are polynomials in s with real-valued coefficients. In order for the partial fraction expansion to be possible, it is required that $X(s)$ be *proper rational*, which means that the degree of the numerator polynomial $N(s)$ is less than that of the denominator polynomial $D(s)$. If $X(s)$ is not proper, then we need to do long division until we obtain a proper rational function—that is,

$$X(s) = g_0 + g_1 s + \cdots + g_m s^m + \frac{B(s)}{D(s)} \tag{3.19}$$

where the degree of $B(s)$ is now less than that of $D(s)$—so that we can perform partial expansion for $B(s)/D(s)$. The inverse of $X(s)$ is then given by

$$x(t) = g_0 \delta(t) + g_1 \frac{d\delta(t)}{dt} + \cdots + g_m \frac{d^m \delta(t)}{dt^m} + \mathcal{L}^{-1}\left[\frac{B(s)}{D(s)} \right] \tag{3.20}$$

The presence of $\delta(t)$ and its derivatives (called doublets, triplets, etc.) are very rare in actual signals, and as such the typical rational function has a numerator polynomial that is of lower degree than the denominator polynomial.

Remarks

- *Things to remember before performing the inversion are:*
  - *The poles of $X(s)$ provide the basic characteristics of the signal $x(t)$.*
  - *If $N(s)$ and $D(s)$ are polynomials in s with real coefficients, then the zeros and poles of $X(s)$ are real and/or complex conjugate pairs, and can be simple or multiple.*
  - *In the inverse, $u(t)$ should be included since the result of the inverse is causal—the function $u(t)$ is an integral part of the inverse.*
- *The basic idea of the partial expansion is to decompose proper rational functions into a sum of rational components of which the inverse transform can be found directly in tables. Table 3.1 displays common one-sided Laplace transform pairs, while Table 3.2 provides properties of the one-sided Laplace transform.*

We will consider now how to obtain a partial fraction expansion when the poles are real, simple and multiple, and in complex conjugate pairs, simple and multiple.

Table 3.1 One-Sided Laplace Transforms

	Function of Time	Function of s, ROC
1.	$\delta(t)$	$1,\quad$ whole s-plane
2.	$u(t)$	$\frac{1}{s},\ \mathcal{R}e[s] > 0$
3.	$r(t)$	$\frac{1}{s^2},\ \mathcal{R}e[s] > 0$
4.	$e^{-at}u(t),\ a > 0$	$\frac{1}{s+a},\ \mathcal{R}e[s] > -a$
5.	$\cos(\Omega_0 t)u(t)$	$\frac{s}{s^2+\Omega_0^2},\ \mathcal{R}e[s] > 0$
6.	$\sin(\Omega_0 t)u(t)$	$\frac{\Omega_0}{s^2+\Omega_0^2},\ \mathcal{R}e[s] > 0$
7.	$e^{-at}\cos(\Omega_0 t)u(t),\ a > 0$	$\frac{s+a}{(s+a)^2+\Omega_0^2},\ \mathcal{R}e[s] > -a$
8.	$e^{-at}\sin(\Omega_0 t)u(t),\ a > 0$	$\frac{\Omega_0}{(s+a)^2+\Omega_0^2},\ \mathcal{R}e[s] > -a$
9.	$2A\,e^{-at}\cos(\Omega_0 t + \theta)u(t),\ a > 0$	$\frac{A\angle\theta}{s+a-j\Omega_0} + \frac{A\angle-\theta}{s+a+j\Omega_0},\ \mathcal{R}e[s] > -a$
10.	$\frac{1}{(N-1)!}\,t^{N-1}u(t)$	$\frac{1}{s^N}$ N an integer, $\mathcal{R}e[s] > 0$
11.	$\frac{1}{(N-1)!}\,t^{N-1}e^{-at}u(t)$	$\frac{1}{(s+a)^N}$ N an integer, $\mathcal{R}e[s] > -a$
12.	$\frac{2A}{(N-1)!}\,t^{N-1}e^{-at}\cos(\Omega_0 t + \theta)u(t)$	$\frac{A\angle\theta}{(s+a-j\Omega_0)^N} + \frac{A\angle-\theta}{(s+a+j\Omega_0)^N},\ \mathcal{R}e[s] > -a$

Table 3.2 Basic Properties of One-Sided Laplace Transforms

Causal functions and constants	$\alpha f(t),\ \beta g(t)$	$\alpha F(s),\ \beta G(s)$
Linearity	$\alpha f(t) + \beta g(t)$	$\alpha F(s) + \beta G(s)$
Time shifting	$f(t - \alpha)$	$e^{-\alpha s}F(s)$
Frequency shifting	$e^{\alpha t}f(t)$	$F(s - \alpha)$
Multiplication by t	$t f(t)$	$-\frac{dF(s)}{ds}$
Derivative	$\frac{df(t)}{dt}$	$sF(s) - f(0-)$
Second derivative	$\frac{d^2 f(t)}{dt^2}$	$s^2 F(s) - sf(0-) - f^{(1)}(0)$
Integral	$\int_{0-}^{t} f(t')dt'$	$\frac{F(s)}{s}$
Expansion/contraction	$f(\alpha t)\ \alpha \neq 0$	$\frac{1}{|\alpha|}F\left(\frac{s}{\alpha}\right)$
Initial value	$f(0+) = \lim_{s\to\infty} sF(s)$	
Final value	$\lim_{t\to\infty} f(t) = \lim_{s\to 0} sF(s)$	

Simple Real Poles

If $X(s)$ is a proper rational function

$$X(s) = \frac{N(s)}{D(s)} = \frac{N(s)}{\prod_k(s - p_k)},$$
(3.21)

where the $\{p_k\}$ are simple real poles of $X(s)$, its partial fraction expansion and its inverse are given by

$$X(s) = \sum_k \frac{A_k}{s - p_k} \quad \Leftrightarrow \quad x(t) = \sum_k A_k e^{p_k t} u(t) \qquad (3.22)$$

where the expansion coefficients are computed as

$$A_k = X(s)(s - p_k) \big|_{s=p_k}$$

According to Laplace transform tables the time function corresponding to $A_k/(s - p_k)$ is $A_k e^{p_k t} u(t)$, thus the form of the inverse $x(t)$. To find the coefficients of the expansion, say A_j, we multiply both sides of the Equation (3.22) by the corresponding denominator $(s - p_j)$ so that

$$X(s)(s - p_j) = A_j + \sum_{k \neq j} \frac{A_k(s - p_j)}{s - p_k}$$

If we let $s = p_j$, or $s - p_j = 0$, in the above expression, all the terms in the sum will be zero and we find that

$$A_j = X(s)(s - p_j) \big|_{s=p_j}$$

■ Example 3.14

Consider the proper rational function

$$X(s) = \frac{3s + 5}{s^2 + 3s + 2} = \frac{3s + 5}{(s + 1)(s + 2)}$$

Find its causal inverse.

Solution

The partial fraction expansion is

$$X(s) = \frac{A_1}{s + 1} + \frac{A_2}{s + 2}$$

Given that the two poles are real, the expected signal $x(t)$ will be a superposition of two decaying exponentials, with damping factors -1 and -2, or

$$x(t) = [A_1 e^{-t} + A_2 e^{-t}] u(t)$$

where as indicated above,

$$A_1 = X(s)(s + 1)|_{s=-1} = \frac{3s + 5}{s + 2} \Big|_{s=-1} = 2$$

and

$$A_2 = X(s)(s+2)|_{s=-2} = \frac{3s+5}{s+1}\Big|_{s=-2} = 1$$

Therefore,

$$X(s) = \frac{2}{s+1} + \frac{1}{s+2}$$

and as such

$$x(t) = [2e^{-t} + e^{-2t}]u(t)$$

To check that the solution is correct one could use the initial or the final value theorems shown in Table 3.2. According to the initial value theorem, $x(0) = 3$ should coincide with

$$\lim_{s\to\infty}\left[sX(s) = \frac{3s^2+5s}{s^2+3s+2}\right] = \lim_{s\to\infty}\frac{3+5/s}{1+3/s+2/s^2} = 3$$

as it does. The final value theorem indicates that $\lim_{t\to\infty} x(t) = 0$ should coincide with

$$\lim_{s\to 0}\left[sX(s) = \frac{3s^2+5s}{s^2+3s+2}\right] = 0$$

as it does. Both of these validations seem to indicate that the result is correct. ■

Remarks *The coefficients A_1 and A_2 can be found using other methods. For instance,*

■ *We can compute*

$$X(s) = \frac{A_1}{s+1} + \frac{A_2}{s+2} \tag{3.23}$$

for two different values of s (as long as we do not divide by zero), such as $s = 0$ and $s = 1$,

$$s = 0 \quad X(0) = \frac{5}{2} = A_1 + \frac{1}{2}A_2$$

$$s = 1 \quad X(1) = \frac{8}{6} = \frac{1}{2}A_1 + \frac{1}{3}A_2$$

which gives a set of two linear equations with two unknowns, and applying Cramer's rule we find that $A_1 = 2$ and $A_2 = 1$.

■ *We cross-multiply the partial expansion given by Equation (3.23) to get*

$$X(s) = \frac{3s+5}{s^2+3s+2} = \frac{s(A_1+A_2)+(2A_1+A_2)}{s^2+3s+2}$$

Comparing the numerators, we have that $A_1 + A_2 = 3$ and $2A_1 + A_2 = 5$, two equations with two unknowns, which can be shown to have as unique solutions $A_1 = 2$ and $A_2 = 1$, as before.

Simple Complex Conjugate Poles

The partial fraction expansion of a proper rational function

$$X(s) = \frac{N(s)}{(s+\alpha)^2 + \Omega_0^2} = \frac{N(s)}{(s+\alpha - j\Omega_0)(s+\alpha + j\Omega_0)} \tag{3.24}$$

with complex conjugate poles $\{s_{1,2} = -\alpha \pm j\Omega_0\}$ is given by

$$X(s) = \frac{A}{s+\alpha - j\Omega_0} + \frac{A^*}{s+\alpha + j\Omega_0}$$

where

$$A = X(s)(s+\alpha - j\Omega_0)|_{s=-\alpha+j\Omega_0} = |A|e^{j\theta}$$

so that the inverse is the function

$$x(t) = 2|A|e^{-\alpha t}\cos(\Omega_0 t + \theta)u(t) \tag{3.25}$$

Because the numerator and the denominator polynomials of $X(s)$ have real coefficients, the zeros and poles whenever complex appear as complex conjugate pairs. One could thus think of the case of a pair of complex conjugate poles as similar to the case of two simple real poles presented above. Notice that the numerator $N(s)$ must be a first-order polynomial for $X(s)$ to be proper rational. The poles of $X(s)$, $s_{1,2} = -\alpha \pm j\Omega_0$, indicate that the signal $x(t)$ will have an exponential $e^{-\alpha t}$, given that the real part of the poles is $-\alpha$, multiplied by a sinusoid of frequency Ω_0, given that the imaginary parts of the poles are $\pm\Omega_0$. We have the expansion

$$X(s) = \frac{A}{s+\alpha - j\Omega_0} + \frac{A^*}{s+\alpha + j\Omega_0}$$

where the expansion coefficients are complex conjugate of each other. From the pole information, the general form of the inverse is

$$x(t) = Ke^{-\alpha t}\cos(\Omega_0 t + \Phi)u(t)$$

for some constants K and Φ. As before, we can find A as

$$A = X(s)(s+\alpha - j\Omega_0)|_{s=-\alpha+j\Omega_0} = |A|e^{j\theta}$$

and that $X(s)(s+\alpha + j\Omega_0)|_{s=-\alpha-j\Omega_0} = A^*$ can be easily verified. Then the inverse transform is given by

$$x(t) = \left[Ae^{-(\alpha - j\Omega_0)t} + A^*e^{-(\alpha + j\Omega_0)t}\right]u(t)$$

$$= |A|e^{-\alpha t}(e^{j(\Omega_0 t+\theta)} + e^{-j(\Omega_0 t+\theta)})u(t)$$

$$= 2|A|e^{-\alpha t}\cos(\Omega_0 t + \theta)u(t).$$

Remarks

- An equivalent partial fraction expansion consists in expressing the numerator $N(s)$ of $X(s)$, for some constants a and b, as $N(s) = a + b(s + \alpha)$, a first-order polynomial, so that

$$X(s) = \frac{a + b(s + \alpha)}{(s + \alpha)^2 + \Omega_0^2} = \frac{a}{\Omega_0} \frac{\Omega_0}{(s + \alpha)^2 + \Omega_0^2} + b \frac{s + \alpha}{(s + \alpha)^2 + \Omega_0^2}$$

so that the inverse is a sum of a sine and a cosine multiplied by a decaying exponential. The inverse Laplace transform is

$$x(t) = \left[\frac{a}{\Omega_0} e^{-\alpha t} \sin(\Omega_0 t) + b e^{-\alpha t} \cos(\Omega_0 t) \right] u(t)$$

which can be simplified, using the sum of phasors corresponding to sine and cosine, to

$$x(t) = \sqrt{\frac{a^2}{\Omega_0^2} + b^2} \; e^{-\alpha t} \cos\left(\Omega_0 t - \tan^{-1}\left(\frac{a}{\Omega_0 b} \right) \right) u(t)$$

- When $\alpha = 0$ the above indicates that the inverse Laplace transform of

$$X(s) = \frac{a + bs}{s^2 + \Omega_0^2}$$

is

$$x(t) = \sqrt{\frac{a^2}{\Omega_0^2} + b^2} \; \cos\left(\Omega_0 t - \tan^{-1}\left(\frac{a}{\Omega_0 b} \right) \right) u(t)$$

which is transform of a cosine with a phase shift not commonly found in tables.
- When the frequency $\Omega_0 = 0$, we get that the inverse Laplace transform of

$$X(s) = \frac{a + b(s + \alpha)}{(s + \alpha)^2} = \frac{a}{(s + \alpha)^2} + \frac{b}{s + \alpha}$$

(corresponds to a double pole at $-\alpha$) is

$$x(t) = \lim_{\Omega_0 \to 0} \left[\frac{a}{\Omega_0} e^{-\alpha t} \sin(\Omega_0 t) + b e^{-\alpha t} \cos(\Omega_0 t) \right] u(t)$$

$$= [ate^{-\alpha t} + be^{-\alpha t}] u(t)$$

where the first limit is found by L'Hôpital's rule. Notice that when computing the partial fraction expansion of the double pole $s = -\alpha$ the expansion is composed of two terms, one with denominator $(s + \alpha)^2$ and the other with denominator $s + \alpha$ of which the sum gives a first-order numerator and a second-order denominator to satisfy the proper rational condition.

■ Example 3.15

Consider the Laplace function

$$X(s) = \frac{2s+3}{s^2+2s+4} = \frac{2s+3}{(s+1)^2+3}$$

Find the corresponding causal signal $x(t)$, then use MATLAB to validate your answer.

Solution

The poles are at $-1 \pm j\sqrt{3}$, so that we expect that $x(t)$ is a decaying exponential with a damping factor of -1 (the real part of the poles) multiplied by a causal cosine of frequency $\sqrt{3}$. The partial fraction expansion is of the form

$$X(s) = \frac{2s+3}{s^2+2s+4} = \frac{a+b(s+1)}{(s+1)^2+3}$$

so that $3+2s = (a+b)+bs$, or $b=2$ and $a+b=3$ or $a=1$. Thus,

$$X(s) = \frac{1}{\sqrt{3}}\frac{\sqrt{3}}{(s+1)^2+3} + 2\frac{s+1}{(s+1)^2+3}$$

which corresponds to

$$x(t) = \left[\frac{1}{\sqrt{3}}\sin(\sqrt{3}t) + 2\cos(\sqrt{3}t)\right]e^{-t}u(t)$$

The value $x(0) = 2$ and according to the initial value theorem the following limit should equal it:

$$\lim_{s\to\infty}\left[sX(s) = \frac{2s^2+3s}{s^2+2s+4}\right] = \lim_{s\to\infty}\frac{2+3/s}{1+2/s+4/s^2} = 2$$

which is the case, indicating the result is probably correct (satisfying the initial value theorem is not enough to indicate the result is correct, but if it does not the result is wrong).

We use the MATLAB function ilaplace to compute symbolically the inverse Laplace transform and plot the response using ezplot, as shown in the following script.

```
%%%%%%%%%%%%%%%%%%
% Example 3.15
%%%%%%%%%%%%%%%%%%
clear all; clf
syms s t w
num = [0 2 3]; den = [1 2 4]; % coefficients of numerator and denominator
subplot(121)
splane(num,den) % plotting poles and zeros
disp('>>>>> Inverse Laplace <<<<<')
x = ilaplace((2*s+3)/(s^2+2*s+4)); % inverse Laplace transform
subplot(122)
```

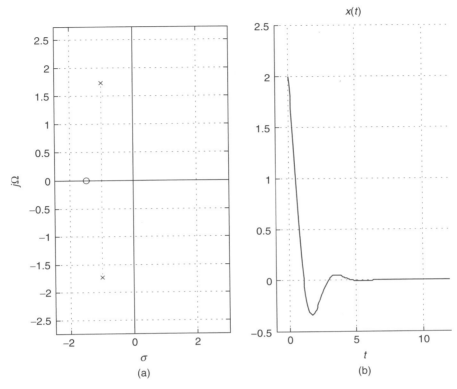

FIGURE 3.14

Inverse Laplace transform of $X(s) = (2s + 3)/(s^2 + 2s + 4)$: (a) poles and zeros and (b) inverse $x(t)$.

```
ezplot(x,[0,12]); title('x(t)')
axis([0 12 -0.5 2.5]); grid
```

The results are shown in Figure 3.14. ∎

Double Real Poles

If a proper rational function has double real poles

$$X(s) = \frac{N(s)}{(s+\alpha)^2} = \frac{a + b(s + \alpha)}{(s + \alpha)^2} = \frac{a}{(s + \alpha)^2} + \frac{b}{s + \alpha} \qquad (3.26)$$

then its inverse is

$$x(t) = [ate^{-\alpha t} + be^{-\alpha t}]u(t) \qquad (3.27)$$

where a can be computed as

$$a = X(s)(s + \alpha)^2 \,|_{s=-\alpha}$$

After replacing it, b is found by computing $X(s_0)$ for a value $s_0 \neq \alpha$.

When we have double real poles we need to express the numerator $N(s)$ as a first-order polynomial, just as in the case of a pair of complex conjugate poles. The values of a and b can be computed in different ways, as we illustrate in the following examples.

■ Example 3.16

Typically, the Laplace transforms appear as combinations of the different terms we have considered, for instance a combination of first- and second-order poles gives

$$X(s) = \frac{4}{s(s+2)^2}$$

which has a pole at $s = 0$ and a double pole at $s = -2$. Find the causal signal $x(t)$. Use MATLAB to plot the poles and zeros of $X(s)$ and to find the inverse Laplace transform $x(t)$.

Solution

The partial fraction expansion is

$$X(s) = \frac{A}{s} + \frac{a + b(s+2)}{(s+2)^2}$$

The value of $A = X(s)s|_{s=0} = 1$, and so

$$X(s) - \frac{1}{s} = \frac{4 - (s+2)^2}{s(s+2)^2} = \frac{-(s+4)}{(s+2)^2}$$

$$= \frac{a + b(s+2)}{(s+2)^2}$$

Comparing the numerators of $X(s) - 1/s$ and the one in the partial fraction expansion gives $b = -1$ and $a + 2b = -4$ or $a = -2$. We then have

$$X(s) = \frac{1}{s} + \frac{-2 - (s+2)}{(s+2)^2}$$

so that

$$x(t) = [1 - 2te^{-2t} - e^{-2t}]u(t)$$

Another way to do this type of problem is to express $X(s)$ as

$$X(s) = \frac{A}{s} + \frac{B}{(s+2)^2} + \frac{C}{s+2}$$

We find the A as before, and then find B by multiplying both sides by $(s+2)^2$ and letting $s = -2$, which gives

$$X(s)(s+2)^2|_{s=-2} = \left[\frac{A(s+2)^2}{s} + B + C(s+2)\right]_{s=-2}$$

so that

$$B = X(s)(s+2)^2|_{s=-2}$$

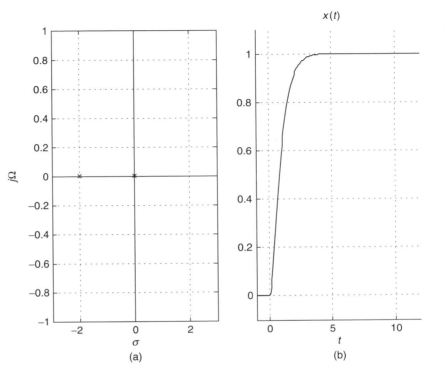

FIGURE 3.15
Inverse Laplace
transform of
$X(s) = 4/(s(s+2)^2)$:
(a) poles and zeros and
(b) $x(t)$.

To find C we compute the partial fraction expansion for a value of s for which no division by zero is possible. For instance, if we let $s = 1$ we can find the value of C, after which we can find the inverse.

The initial value $x(0) = 0$ coincides with

$$\lim_{s \to \infty} \left[sX(s) = \frac{4s}{s(s+2)^2} \right] = \lim_{s \to \infty} \frac{4/s^2}{(1+2/s)^2} = 0$$

To find the inverse Laplace transform with MATLAB we use a similar script to the one used before; only the numerator and denominator description needs to be changed. The plots are shown in Figure 3.15. ∎

■ Example 3.17

Find the inverse Laplace transform of the function

$$X(s) = \frac{4}{s((s+1)^2 + 3)}$$

which has a simple real pole $s = 0$, and complex conjugate poles $s = -1 \pm j\sqrt{3}$.

Solution

The partial fraction expansion is

$$X(s) = \frac{A}{s + 1 - j\sqrt{3}} + \frac{A^*}{s + 1 + j\sqrt{3}} + \frac{B}{s}$$

We then have

$$B = sX(s)|_{s=0} = 1$$

$$A = X(s)(s + 1 - j\sqrt{3})|_{s=-1+j\sqrt{3}} = 0.5\left(-1 + \frac{j}{\sqrt{3}}\right) = \frac{1}{\sqrt{3}}\angle 150°$$

so that

$$x(t) = \frac{2}{\sqrt{3}}e^{-t}\cos(\sqrt{3}t + 150°)u(t) + u(t)$$

$$= -[\cos(\sqrt{3}t) + 0.577\sin(\sqrt{3}t)]e^{-t}u(t) + u(t) \qquad \blacksquare$$

Remarks

■ *Following the above development, when the poles are complex conjugate and double the procedure for the double poles is repeated. Thus, the partial expansion is given as*

$$X(s) = \frac{N(s)}{(s + \alpha - j\Omega_0)^2(s + \alpha + j\Omega_0)^2}$$

$$= \frac{a + b(s + \alpha - j\Omega_0)}{(s + \alpha - j\Omega_0)^2} + \frac{a^* + b^*(s + \alpha + j\Omega_0)}{(s + \alpha + j\Omega_0)^2} \qquad (3.28)$$

so that finding a and b we obtain the inverse.
■ *The partial fraction expansion for second- and higher-order poles should be done with MATLAB.*

■ Example 3.18

In this example we use MATLAB to find the inverse Laplace transform of more complicated functions than the ones considered before. In particular, we want to illustrate some of the additional information that our function pfeLaplace gives. Consider the Laplace transform

$$X(s) = \frac{3s^2 + 2s - 5}{s^3 + 6s^2 + 11s + 6}$$

Find poles and zeros of $X(s)$, and obtain the coefficients of its partial fraction expansion (also called the residues). Use ilaplace to find its inverse and plot it using ezplot.

Solution

The following is the function pfeLaplace.

```
function pfeLaplace(num,den)
%
disp('>>>>>  Zeros <<<<<')
z = roots(num)
[r,p,k]=residue(num,den);
disp('>>>>>  Poles <<<<<')
p
disp('>>>>>  Residues <<<<<')
r
splane(num,den)
```

The function pfeLaplace uses the MATLAB function roots to find the zeros of $X(s)$ defined by the coefficients of its numerator and denominator given in descending order of s. For the partial fraction expansion, pfeLaplace uses the MATLAB function residue, which finds coefficients of the expansion as well as the poles of $X(s)$. (The residue $r(i)$ in the vector r corresponds to the expansion term for the pole $p(i)$; for instance, the residue $r(1) = 8$ corresponds to the expansion term corresponding to the pole $p(1) = -3$.) The symbolic function ilaplace is then used to find the inverse $x(t)$; as input to ilaplace the function $X(s)$ is described in a symbolic way. The MATLAB function ezplot is used for the plotting of the symbolic computations.

The analytic results are shown in the following, and the plot of $x(t)$ is given in Figure 3.16.

```
>>>>>  Zeros <<<<<
z = -1.6667
      1.0000
>>>>>  Poles <<<<<
p = -3.0000
    -2.0000
    -1.0000
>>>>>  Residues <<<<<
r = 8.0000
   -3.0000
   -2.0000
>>>>> Inverse Laplace <<<<<
x = 8 * exp(-3 * t)-3 * exp(-2 * t)-2 * exp(-t)
```

3.4.2 Inverse of Functions Containing $e^{-\rho s}$ Terms

When $X(s)$ has exponentials $e^{-\rho s}$ in the numerator or denominator, ignore these terms and perform partial fraction expansion on the rest, and at the end consider the exponentials to get the correct time shifting. In particular, when

$$X(s) = \frac{N(s)}{D(s)(1 - e^{-\alpha s})} = \frac{N(s)}{D(s)} + \frac{N(s)e^{-\alpha s}}{D(s)} + \frac{N(s)e^{-2\alpha s}}{D(s)} + \cdots$$

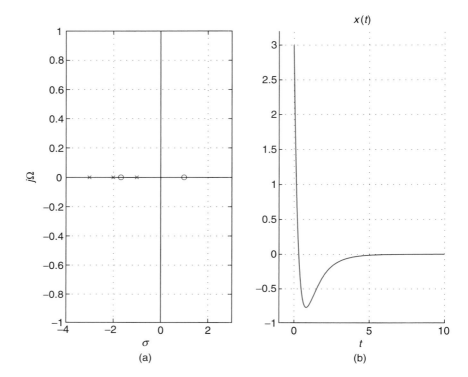

FIGURE 3.16
Inverse Laplace
transform of
$X(s) = (3s^2 + 2s - 5)/$
$(s^3 + 6s^2 + 11s + 6)$.
(a) Poles and zeros of
$X(s)$ are given with (b)
the corresponding
inverse $x(t)$.

if $f(t)$ is the inverse of $N(s)/D(s)$, then

$$x(t) = f(t) + f(t - \alpha) + f(t - 2\alpha) + \cdots$$

Another possibility is when the function is given as

$$X(s) = \frac{N(s)}{D(s)(1 + e^{-\alpha s})} = \frac{N(s)}{D(s)} - \frac{N(s)e^{-\alpha s}}{D(s)} + \frac{N(s)e^{-2\alpha s}}{D(s)} - \cdots$$

If $f(t)$ is the inverse of $N(s)/D(s)$, we then have

$$x(t) = f(t) - f(t - \alpha) + f(t - 2\alpha) - \cdots$$

The time-shifting property of Laplace makes it possible for the numerator $N(s)$ or the denominator $D(s)$ to have $e^{-\sigma s}$ terms. The procedure for inverting such functions is to initially ignore these terms and do the partial fraction expansion on the rest and at the end consider them to do the necessary time shifting. For instance, the inverse of

$$X(s) = \frac{e^s - e^{-s}}{s} = \frac{e^s}{s} - \frac{e^{-s}}{s}$$

is obtained by first considering the term $1/s$, which has $u(t)$ as inverse, and then using the information in the numerator to get the final response,

$$x(t) = u(t + 1) - u(t - 1)$$

The two sums

$$\sum_{k=0}^{\infty} e^{-\alpha sk} = \frac{1}{1 - e^{-\alpha s}}$$

$$\sum_{k=0}^{\infty} (-e^{-\alpha s})^k = \frac{1}{1 + e^{-\alpha s}}$$

can be easily verified by cross-multiplying. So when the function is

$$X_1(s) = \frac{N(s)}{D(s)(1 - e^{-\alpha s})} = \frac{N(s)}{D(s)} \sum_{k=0}^{\infty} e^{-\alpha sk}$$

$$= \frac{N(s)}{D(s)} + \frac{N(s)e^{-\alpha s}}{D(s)} + \frac{N(s)e^{-2\alpha s}}{D(s)} + \cdots$$

and if $f(t)$ is the inverse of $N(s)/D(s)$, we then have

$$x_1(t) = f(t) + f(t - \alpha) + f(t - 2\alpha) + \cdots$$

Likewise, when

$$X_2(s) = \frac{N(s)}{D(s)(1 + e^{-\alpha s})} = \frac{N(s)}{D(s)} \sum_{k=0}^{\infty} (-1)^k e^{-\alpha sk}$$

$$= \frac{N(s)}{D(s)} - \frac{N(s)e^{-\alpha s}}{D(s)} + \frac{N(s)e^{-2\alpha s}}{D(s)} - \cdots$$

if $f(t)$ is the inverse of $N(s)/D(s)$, we then have

$$x_2(t) = f(t) - f(t - \alpha) + f(t - 2\alpha) - \cdots$$

■ Example 3.19

We wish to find the causal inverse of

$$X(s) = \frac{1 - e^{-s}}{(s + 1)(1 - e^{-2s})}$$

Solution

We let

$$X(s) = F(s) \sum_{k=0}^{\infty} (e^{-2s})^k$$

where

$$F(s) = \frac{1 - e^{-s}}{s + 1}$$

The inverse of $F(s)$ is

$$f(t) = e^{-t}u(t) - e^{-(t-1)}u(t - 1)$$

and the inverse of $X(s)$ is thus given by

$$x(t) = f(t) + f(t - 2) + f(t - 4) + \cdots$$ ∎

3.4.3 Inverse of Two-Sided Laplace Transforms

When finding the inverse of a two-sided Laplace transform we need to pay close attention to the region of convergence and to the location of the poles with respect to the $j\Omega$ axis. Three regions of convergence are possible:

- A plane to the right of all the poles, which corresponds to a causal signal.
- A plane to the left of all poles, which corresponds to an anti-causal signal.
- A region that is in between poles on the right and poles on the left (no poles included in it), which corresponds to a two-sided signal.

If the $j\Omega$ axis is included in the region of convergence, bounded-input bounded-output (BIBO) stability of the system, or absolute integrability of the impulse response of the system, is guaranteed. Furthermore, the system with that region of convergence would have a frequency response, and the signal a Fourier transform. The inverses of the causal and the anti-causal components are obtained using the one-sided Laplace transform.

■ Example 3.20

Find the inverse Laplace transform of

$$X(s) = \frac{1}{(s + 2)(s - 2)} \qquad \text{ROC: } -2 < \mathcal{R}e(s) < 2$$

Solution

The ROC $-2 < \mathcal{R}e(s) < 2$ is equivalent to $\{(\sigma, \Omega) : -2 < \sigma < 2, -\infty < \Omega < \infty\}$. The partial fraction expansion is

$$X(s) = \frac{1}{(s + 2)(s - 2)} = \frac{-0.25}{s + 2} + \frac{0.25}{s - 2} \qquad -2 < \mathcal{R}e(s) < 2$$

where the first term with the pole at $s = -2$ corresponds to a causal signal with a region of convergence $\mathcal{R}e(s) > -2$, and the second term corresponds to an anti-causal signal with a region of convergence $\mathcal{R}e(s) < 2$. That this is so is confirmed by the intersection of these two regions of convergence that gives

$$[\mathcal{R}e(s) > -2] \cap [\mathcal{R}e(s) < 2] = -2 < \mathcal{R}e(s) < 2$$

As such, we have

$$x(t) = -0.25e^{-2t}u(t) - 0.25e^{2t}u(-t)$$
■

■ Example 3.21

Consider the transfer function

$$H(s) = \frac{s}{(s+2)(s-1)} = \frac{2/3}{s+2} + \frac{1/3}{s-1}$$

with a zero at $s = 0$, and poles at $s = -2$ and $s = 1$. Find out how many impulse responses can be connected with $H(s)$ by considering different possible regions of convergence and by determining in which cases the system with $H(s)$ as its transfer function is BIBO stable.

Solution

The following are the different possible impulse responses:

■ If ROC: $\mathcal{R}e(s) > 1$, the impulse response

$$h_1(t) = (2/3)e^{-2t}u(t) + (1/3)e^{t}u(t)$$

corresponding to $H(s)$ with this region of convergence is causal. The corresponding system is unstable—due to the pole in the right-hand s-plane, which will make the impulse response grow as t increases.

■ If ROC: $-2 < \mathcal{R}e(s) < 1$, the impulse response corresponding to $H(s)$ with this region of convergence is noncausal, but the system is stable. The impulse response would be

$$h_2(t) = (2/3)e^{-2t}u(t) - (1/3)e^{t}u(-t)$$

Notice that the region of convergence includes the $j\Omega$ axis, and this guarantees the stability (verify that $h_2(t)$ is absolutely integrable), and as we will see later, also the existence of the Fourier transform of $h_2(t)$.

■ If ROC: $\mathcal{R}e(s) < -2$, the impulse response in this case would be anti-causal, and the system is unstable (please verify it), as the impulse response is

$$h_3(t) = -(2/3)e^{-2t}u(-t) - (1/3)e^{t}u(-t)$$
■

Two very important generalizations of the results in this example are:

■ An LTI with a transfer function $H(s)$ and region of convergence $\mathcal{R}$ is BIBO stable if the $j\Omega$ axis is contained in the region of convergence.

- If the system is BIBO stable and causal, then the region of convergence includes the $j\Omega$ axis so that the frequency response $H(j\Omega)$ exists, and all the poles of $H(s)$ are in the open left-hand s-plane (the $j\Omega$ axis is not included).

3.5 ANALYSIS OF LTI SYSTEMS

Dynamic linear time-invariant systems are typically represented by differential equations. Using the derivative property of the one-sided Laplace transform (allowing the inclusion of initial conditions) and the inverse transformation, differential equations are changed into easier-to-solve algebraic equations. The convolution integral is not only a valid alternate representation for systems represented by differential equations, but for other systems. The Laplace transform provides a very efficient computational method for the convolution integral. More important, the convolution property of the Laplace transform introduces the concept of *transfer function*, a very efficient representation of LTI systems whether they are represented by differential equations or not. In Chapter 6, we will present applications of the material in this section to classic control theory.

3.5.1 LTI Systems Represented by Ordinary Differential Equations

Two ways to characterize the response of a causal and stable LTI system are:

- *Zero-state* and *zero-input* responses, which have to do with the effect of the input and the initial conditions of the system.
- *Transient* and *steady-state* responses, which have to do with close and faraway behavior of the response.

The complete response $y(t)$ of a system represented by an Nth-order linear differential equation with constant coefficients,

$$y^{(N)}(t) + \sum_{k=0}^{N-1} a_k y^{(k)}(t) = \sum_{\ell=0}^{M} b_\ell x^{(\ell)}(t) \qquad N > M \tag{3.29}$$

where $x(t)$ is the input and $y(t)$ is the output of the system, and initial conditions

$$\{y^{(k)}(t), \ 0 \le k \le N-1\} \tag{3.30}$$

is obtained by inverting the Laplace transform

$$Y(s) = \frac{B(s)}{A(s)} X(s) + \frac{1}{A(s)} I(s) \tag{3.31}$$

where $Y(s) = \mathcal{L}[y(t)]$, $X(s) = \mathcal{L}[x(t)]$, and

$$A(s) = \sum_{k=0}^{N} a_k s^k \qquad a_N = 1$$

$$B(s) = \sum_{\ell=0}^{M} b_\ell s^\ell$$

$$I(s) = \sum_{k=1}^{N} a_k \left(\sum_{m=0}^{k-1} s^{k-m-1} y^{(m)}(0) \right)$$

That is, $I(s)$ depends on the initial conditions.

The notation $y^{(k)}(t)$ and $x^{(\ell)}(t)$ indicates the kth and the ℓth derivatives of $y(t)$ and of $x(t)$, respectively (it is to be understood that $y^{(0)}(t) = y(t)$ and likewise $x^{(0)}(t) = x(t)$ in this notation). The assumption $N > M$ avoids the presence of $\delta(t)$ and its derivatives in the solution, which are realistically not possible. To obtain the complete response $y(t)$ we compute the Laplace transform of Equation (3.29):

$$\underbrace{\left[\sum_{k=0}^{N} a_k s^k \right]}_{A(s)} Y(s) = \underbrace{\left[\sum_{\ell=0}^{M} b_\ell s^\ell \right]}_{B(s)} X(s) + \underbrace{\sum_{k=1}^{N} a_k \left(\sum_{m=0}^{k-1} s^{(k-1)-m} y^{(m)}(0) \right)}_{I(s)}$$

which can be written as

$$A(s)Y(s) = B(s)X(s) + I(s) \tag{3.32}$$

by defining $A(s)$, $B(s)$, and $I(s)$ as indicated above. Solving for $Y(s)$ in Equation (3.32), we have

$$Y(s) = \frac{B(s)}{A(s)} X(s) + \frac{1}{A(s)} I(s)$$

and finding its inverse we obtain the complete response $y(t)$.

Letting

$$H(s) = \frac{B(s)}{A(s)} \quad \text{and} \quad H_1(s) = \frac{1}{A(s)}$$

the *complete response* $y(t) = \mathcal{L}^{-1}[Y(s)]$ of the system is obtained by the inverse Laplace transform of

$$Y(s) = H(s)X(s) + H_1(s)I(s) \tag{3.33}$$

which gives

$$y(t) = y_{zs}(t) + y_{zi}(t) \tag{3.34}$$

where

zero-state response: $y_{zs}(t) = \mathcal{L}^{-1}[H(s)X(s)]$

zero-input response: $y_{zi}(t) = \mathcal{L}^{-1}[H_1(s)I(s)]$

In terms of convolution integrals,

$$y(t) = \int_0^t x(\tau)h(t-\tau)d\tau + \int_0^t i(\tau)h_1(t-\tau)d\tau \tag{3.35}$$

where $h(t) = \mathcal{L}^{-1}[H(s)]$ and $h_1(t) = \mathcal{L}^{-1}[H_1(s)]$, and

$$i(t) = \mathcal{L}^{-1}[I(s)] = \sum_{k=1}^{N} a_k \left(\sum_{m=0}^{k-1} y^{(m)}(0)\delta^{(k-m-1)}(t) \right)$$

where $\{\delta^{(m)}(t)\}$ are mth derivatives of the impulse signal $\delta(t)$ (as indicated before, $\delta^{(0)}(t) = \delta(t)$).

Zero-State and Zero-Input Responses

Despite the fact that linear differential equations, with constant coefficients, do not represent linear systems unless the initial conditions are zero and the input is causal, linear system theory is based on these representations with initial conditions. Typically, the input is causal so it is the initial conditions not always being zero that causes problems. This can be remedied by a different way of thinking about the initial conditions. In fact, one can think of the input $x(t)$ and the initial conditions as two different inputs to the system, and apply superposition to find the responses to these two different inputs. This defines two responses. One is due completely to the input, with zero initial conditions, called the *zero-state solution*. The other component of the complete response is due exclusively to the initial conditions, assuming that the input is zero, and is called the *zero-input solution*.

Remarks

■ *It is important to recognize that to compute the transfer function of the system*

$$H(s) = \frac{Y(s)}{X(s)}$$

according to Equation (3.33) requires that the initial conditions be zero, or $I(s) = 0$.
■ *If there is no pole-zero cancellation, both $H(s)$ and $H_1(s)$ have the same poles, as both have $A(s)$ as denominator, and as such $h(t)$ and $h_1(t)$ might be similar.*

Transient and Steady-State Responses

Whenever the input of a causal and stable system has poles in the closed left-hand s-plane, poles in the $j\Omega$-axis being simple, the complete response will be bounded. Moreover, whether the response exists as $t \to \infty$ can then be determined without using the inverse Laplace transform.

The complete response $y(t)$ of an LTI system is made up of transient and steady-state components. The transient response can be thought of as the system's reaction to the initial inertia after applying the input, while the steady-state response is how the system reacts to the input away from the initial time when the input starts.

If the poles (simple or multiple, real or complex) of the Laplace transform of the output, $Y(s)$, of an LTI system are in the open left-hand s-plane (i.e., no poles on the $j\Omega$ axis), the steady-state response is

$$y_{ss}(t) = \lim_{t \to \infty} y(t) = 0$$

In fact, for any real pole $s = -\alpha$, $\alpha > 0$, of multiplicity $m \geq 1$, we have that

$$\mathcal{L}^{-1}\left[\frac{N(s)}{(s+\alpha)^m}\right] = \sum_{k=1}^{m} A_k t^{k-1} e^{-\alpha t} u(t)$$

where $N(s)$ is a polynomial of degree less or equal to $m - 1$. Clearly, for any value of $\alpha > 0$ and any order $m \geq 1$, the above inverse will tend to zero as t increases. The rate at which these terms go to zero depends on how close the pole(s) is (are) to the $j\Omega$ axis: The farther away, the faster the term goes to zero. Likewise, complex conjugate pairs of poles with a negative real part also give terms that go to zero as $t \to \infty$, independent of their order. For complex conjugate pairs of poles $s_{1,2} = -\alpha \pm j\Omega_0$ of order $m \geq 1$, we have

$$\mathcal{L}^{-1}\left[\frac{N(s)}{((s+\alpha)^2 + \Omega_0^2)^m}\right] = \sum_{k=1}^{m} 2|A_k| t^{k-1} e^{-\alpha t} \cos(\Omega_0 t + \angle(A_k)) u(t)$$

where again $N(s)$ is a polynomial of degree less or equal to $2m - 1$. Due to the decaying exponentials this type of term will go to zero as t goes to infinity.

Simple complex conjugate poles and a simple real pole at the origin of the s-plane cause a steady-state response. Indeed, if the pole of $Y(s)$ is $s = 0$ we know that its inverse transform is of the form $Au(t)$, and if the poles are complex conjugates $\pm j\Omega_0$ the corresponding inverse transform is a sinusoid— neither of which is transient. *However, multiple poles on the $j\Omega$-axis, or any poles in the right-hand s-plane will give inverses that grow as $t \to \infty$.* This statement is clear for the poles in the right-hand s-plane. For double- or higher-order poles in the $j\Omega$ axis their inverse transform is of the form

$$\mathcal{L}^{-1}\left[\frac{N(s)}{(s^2 + \Omega_0^2)^m}\right] = \sum_{k=1}^{m} 2|A_k| t^{m-1} \cos(\Omega_0 + \angle(A_k)) u(t)$$

which will continuously grow as t increases.

In summary, when solving differential equations—with or without initial conditions—we have

- The steady-state component of the complete solution is given by the inverse Laplace transforms of the partial fraction expansion terms of $Y(s)$ that have simple poles (real or complex conjugate pairs) in the $j\Omega$-axis.
- The transient response is given by the inverse transform of the partial fraction expansion terms with poles in the left-hand s-plane, independent of whether the poles are simple or multiple, real or complex.
- Multiple poles in the $j\Omega$ axis and poles in the right-hand s-plane give terms that will increase as t increases.

■ Example 3.22

Consider a second-order ($N = 2$) differential equation,

$$\frac{d^2 y(t)}{dt^2} + 3\frac{dy(t)}{dt} + 2y(t) = x(t)$$

Assume the above equation represents a system with input $x(t)$ and output $y(t)$. Find the impulse response $h(t)$ and the unit-step response $s(t)$ of the system.

Solution

If the initial conditions are zero, computing the two- or one-sided Laplace transform of the two sides of this equation, after letting $Y(s) = \mathcal{L}[y(t)]$ and $X(s) = \mathcal{L}[x(t)]$, and using the derivative property of Laplace, we get

$$Y(s)[s^2 + 3s + 2] = X(s)$$

To find the impulse response of this system (i.e., the system response $y(t) = h(t)$), we let $x(t) = \delta(t)$ and the initial condition be zero. Since $X(s) = 1$, then $Y(s) = H(s) = \mathcal{L}[h(t)]$ is

$$H(s) = \frac{1}{s^2 + 3s + 2} = \frac{1}{(s+1)(s+2)} = \frac{A}{s+1} + \frac{B}{s+2}$$

We obtain values $A = 1$ and $B = -1$, and the inverse Laplace transform is then

$$h(t) = \left[e^{-t} - e^{-2t}\right]u(t)$$

which is completely transient.

In a similar form we obtain the unit-step response $s(t)$, by letting $x(t) = u(t)$ and the initial conditions be zero. Calling $Y(s) = S(s) = \mathcal{L}[s(t)]$, since $X(s) = 1/s$, we obtain

$$S(s) = \frac{H(s)}{s} = \frac{1}{s(s^2 + 3s + 2)} = \frac{A}{s} + \frac{B}{s+1} + \frac{C}{s+2}$$

It is found that $A = 1/2$, $B = -1$, and $C = 1/2$, so that

$$s(t) = 0.5u(t) - e^{-t}u(t) + 0.5e^{-2t}u(t)$$

The steady state of $s(t)$ is 0.5 as the two exponentials go to zero. Interestingly, the relation $sS(s) = H(s)$ indicates that by computing the derivative of $s(t)$ we obtain $h(t)$. Indeed,

$$\frac{ds(t)}{dt} = 0.5\delta(t) + e^{-t}u(t) - e^{-t}\delta(t) - e^{-2t}u(t) + 0.5e^{-2t}\delta(t)$$

$$= [0.5 - 1 + 0.5]\delta(t) + [e^{-t} - e^{-2t}]u(t)$$

$$= [e^{-t} - e^{-2t}]u(t) = h(t)$$

∎

Remarks

- *Because the existence of the steady-state response depends on the poles of $Y(s)$ it is possible for an unstable causal system (recall that for such a system BIBO stability requires all the poles of the system transfer function be in the open, left-hand s-plane) to have a steady-state response. It all depends on the input. Consider, for instance, an unstable system with $H(s) = 1/(s(s+1))$, being unstable due to the pole at*

$s = 0$; if the system input is $x_1(t) = u(t)$ so that $X_1(s) = 1/s$, then $Y_1(s) = 1/(s^2(s+1))$. There will be no steady state because of the double pole $s = 0$. On the other hand, $X_2(s) = s/(s+2)^2$ will give

$$Y_2(s) = H(s)X_2(s) = \frac{1}{s(s+1)} \frac{s}{(s+2)^2} = \frac{1}{(s+1)(s+2)^2}$$

which will give a zero steady state, even though the system is unstable. This is possible because of the pole-zero cancellation.

■ The steady-state response is the response of the system away from $t = 0$, and it can be found by letting $t \to \infty$ (even though the steady state can be reached at finite times, depending on how fast the transient goes to zero). In Example 3.22, the steady-state response of $h(t) = (e^{-t} - e^{-2t})u(t)$ is zero, while for $s(t) = 0.5u(t) - e^{-t}u(t) + 0.5e^{-2t}u(t)$ it is 0.5. The transient responses are then $h(t) - 0 = h(t)$ and $s(t) - 0.5u(t) = -e^{-t}u(t) + 0.5e^{-2t}u(t)$. These transients eventually disappear.

■ The relation found between the impulse response $h(t)$ and the unit-step response $s(t)$ can be extended to more cases by the definition of the transfer function—that is, $H(s) = Y(s)/X(s)$ so that the response $Y(s)$ is connected with $H(s)$ by $Y(s) = H(s)X(s)$, giving the relation between $y(t)$ and $h(t)$. For instance, if $x(t) = \delta(t)$, then $Y(s) = H(s) \times 1$, with inverse the impulse response. If $x(t) = u(t)$, then $Y(s) = H(s)/s$ is $S(s)$, the Laplace transform of the unit-step response, and so $h(t) = ds(t)/dt$. And if $x(t) = r(t)$, then $Y(s) = H(s)/s^2$ is $\rho(s)$, the Laplace transform of the ramp response, and so $h(t) = d^2\rho(t)/dt^2 = ds(t)/dt$.

■ Example 3.23

Consider again the second-order differential equation in the previous example,

$$\frac{d^2y(t)}{dt^2} + 3\frac{dy(t)}{dt} + 2y(t) = x(t)$$

but now with initial conditions $y(0) = 1$ and $dy(t)/dt|_{t=0} = 0$, and $x(t) = u(t)$. Find the complete response $y(t)$. Could we find the impulse response $h(t)$ from this response? How could we do it?

Solution

The Laplace transform of the differential equation gives

$$[s^2Y(s) - sy(0) - \frac{dy(t)}{dt}\Big|_{t=0}] + 3[sY(s) - y(0)] + 2Y(s) = X(s)$$

$$Y(s)(s^2 + 3s + 2) - (s+3) = X(s)$$

so we have that

$$Y(s) = \frac{X(s)}{(s+1)(s+2)} + \frac{s+3}{(s+1)(s+2)}$$

$$= \frac{1 + 3s + s^2}{s(s+1)(s+2)} = \frac{B_1}{s} + \frac{B_2}{s+1} + \frac{B_3}{s+2}$$

after replacing $X(s) = 1/s$. We find that $B_1 = 1/2$, $B_2 = 1$, and $B_3 = -1/2$, so that the complete response is

$$y(t) = [0.5 + e^{-t} - 0.5e^{-2t}]u(t) \tag{3.36}$$

Again, we can check that this solution satisfies the initial condition $y(0)$ and $dy(0)/dt$ (this is particularly interesting to see, try it!). The steady-state response is 0.5 and the transient $[e^{-t} - 0.5e^{-2t}]u(t)$.

According to Equation (3.36), the complete solution $y(t)$ is composed of the zero-state response, due to the input only, and the response due to the initial conditions only or the zero-input response. Thus, the system considers two different inputs: One that is $x(t) = u(t)$ and the other the initial conditions.

If we are able to find the transfer function $H(s) = Y(s)/X(s)$ its inverse Laplace transform would be $h(t)$. However that is not possible when the initial conditions are nonzero. As shown above, in the case of nonzero initial conditions, we get that the Laplace transform is

$$Y(s) = \frac{X(s)}{A(s)} + \frac{I(s)}{A(s)}$$

where in this case $A(s) = (s+1)(s+2)$ and $I(s) = s+3$, and thus we cannot find the ratio $Y(s)/X(s)$. If we make the second term zero (i.e., $I(s) = 0$), we then have that $Y(s)/X(s) = H(s) = 1/A(s)$ and $h(t) = e^{-t}u(t) - e^{-2t}u(t)$. ∎

■ Example 3.24

Consider an analog averager represented by

$$y(t) = \frac{1}{T} \int_{t-T}^{t} x(\tau)d\tau \tag{3.37}$$

where $x(t)$ is the input and $y(t)$ is the output. The derivative of $y(t)$ gives the first-order differential equation

$$\frac{dy(t)}{dt} = \frac{1}{T}[x(t) - x(t-T)]$$

with a finite difference for the input. Let us find the impulse response of this analog averager.

Solution

The impulse response of the averager is found by letting $x(t) = \delta(t)$ and the initial condition be zero. Computing the Laplace transform of the two sides of the differential equation, we obtain

$$sY(s) = \frac{1}{T}[1 - e^{-sT}]X(s)$$

and substituting $X(s) = 1$, then

$$H(s) = Y(s) = \frac{1}{sT}[1 - e^{-sT}]$$

The impulse response is then

$$h(t) = \frac{1}{T}[u(t) - u(t - T)].$$

■

3.5.2 Computation of the Convolution Integral

From the point of view of signal processing, the convolution property is the most important application of the Laplace transform to systems. The computation of the convolution integral is difficult even for simple signals. In Chapter 2 we showed how to obtain the convolution integral analytically as well as graphically. As we will see in this section, it is not only that the convolution property of the Laplace transform gives an efficient solution to the computation of the convolution integral, but that it introduces an important representation of LTI systems, namely the *transfer function of the system*. A system, like signals, is thus represented by the poles and zeros of the transfer function. But it is not only the pole-zero characterization of the system that can be obtained from the transfer function. The system's impulse response is uniquely obtained from the poles and zeros of the transfer function and the corresponding region of convergence. The way the system responds to different frequencies will be also given by the transfer function. Stability and causality of the system can be equally related to the transfer function. Design of filters depends on the transfer function.

The Laplace transform of the convolution $y(t) = [x * h](t)$ is given by the product

$$Y(s) = X(s)H(s) \tag{3.38}$$

where $X(s) = \mathcal{L}[x(t)]$ and $H(s) = \mathcal{L}[h(t)]$. The transfer function *of the system* $H(s)$ is defined as

$$H(s) = \mathcal{L}[h(t)] = \frac{Y(s)}{X(s)} \tag{3.39}$$

$H(s)$ transfers the Laplace transform $X(s)$ of the input into the Laplace transform of the output $Y(s)$. Once $Y(s)$ is found, $y(t)$ is computed by means of the inverse Laplace transform.

■ Example 3.25

Use the Laplace transform to find the convolution $y(t) = [x * h](t)$ when

(1) the input is $x(t) = u(t)$ and the impulse response is a pulse $h(t) = u(t) - u(t - 1)$, and

(2) the input and the impulse response of the system are $x(t) = h(t) = u(t) - u(t - 1)$.

Solution

■ The Laplace transforms are $X(s) = \mathcal{L}[u(t)] = 1/s$ and $H(s) = \mathcal{L}[h(t)] = (1 - e^{-s})/s$, so that

$$Y(s) = H(s)X(s) = \frac{1 - e^{-s}}{s^2}$$

Its inverse is

$$y(t) = r(t) - r(t - 1)$$

where $r(t)$ is the ramp signal. This result coincides with the one obtained graphically in Example 2.12 in Chapter 2.

■ In the second case, $X(s) = H(s) = \mathcal{L}[u(t) - u(t - 1)] = (1 - e^{-s})/s$, so that

$$Y(s) = H(s)X(s) = \frac{(1 - e^{-s})^2}{s^2} = \frac{1 - 2e^{-s} + e^{-2s}}{s^2}$$

which corresponds to

$$y(t) = r(t) - 2r(t - 1) + r(t - 2)$$

or a triangular pulse as we obtained graphically in Example 2.13 in Chapter 2. ∎

■ Example 3.26

To illustrate the significance of the Laplace approach in computing the output of an LTI system by means of the convolution integral, consider an RLC circuit in series with input a voltage source $x(t)$ and as output the voltage $y(t)$ across the capacitor (see Figure 3.17). Find its impulse response $h(t)$ and its unit-step response $s(t)$. Let $LC = 1$ and $R/L = 2$.

Solution

The RLC circuit is represented by a second-order differential equation given that the inductor and the capacitor are capable of storing energy and their initial conditions are not dependent on each other. To obtain the differential equation we apply Kirchhoff's voltage law (KVL)

$$x(t) = Ri(t) + L\frac{di(t)}{dt} + y(t)$$

where $i(t)$ is the current through the resistor, inductor and capacitor and where the voltage across the capacitor is given by

$$y(t) = \frac{1}{C} \int_0^t i(\sigma)d\sigma + y(0)$$

FIGURE 3.17
RLC circuit with input a voltage source $x(t)$ and output the voltage across the capacitor $y(t)$.

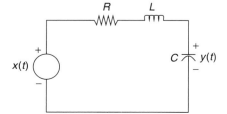

with $y(0)$ the initial voltage in the capacitor and $i(t)$ the current through the resistor, inductor, and capacitor. The above two equations are called *integro-differential* given that they are composed of an integral equation and a differential equation. To obtain a differential equation in terms of the input $x(t)$ and the output $y(t)$, we find the first and second derivative of $y(t)$, which gives

$$\frac{dy(t)}{dt} = \frac{1}{C}i(t) \quad \Rightarrow \quad i(t) = C\frac{dy(t)}{dt}$$

$$\frac{d^2y(t)}{dt^2} = \frac{1}{C}\frac{di(t)}{dt} \quad \Rightarrow \quad L\frac{di(t)}{dt} = LC\frac{d^2y(t)}{dt^2}$$

which when replaced in the KVL equation gives

$$x(t) = RC\frac{dy(t)}{dt} + LC\frac{d^2y(t)}{dt^2} + y(t) \tag{3.40}$$

which, as expected, is a second-order differential equation with two initial conditions: $y(0)$, the initial voltage in the capacitor, and $i(0) = Cdy(t)/dt|_{t=0}$, the initial current in the inductor. To find the impulse response of this circuit, we let $x(t) = \delta(t)$ and the initial conditions be zero. The Laplace transform of Equation (3.40) gives

$$X(s) = [LCs^2 + RCs + 1]Y(s)$$

The impulse response of the system is the inverse Laplace transform of the transfer function

$$H(s) = \frac{Y(s)}{X(s)} = \frac{1/LC}{s^2 + (R/L)s + 1/LC}$$

If $LC = 1$ and $R/L = 2$, then the transfer function is

$$H(s) = \frac{1}{(s+1)^2}$$

which corresponds to the impulse response

$$h(t) = te^{-t}u(t)$$

Now that we have the impulse response of the system, suppose then the input is a unit-step signal, $x(t) = u(t)$. To find its response we consider the convolution integral

$$y(t) = \int_{-\infty}^{\infty} x(\tau)h(t - \tau)d\tau$$

$$= \int_{-\infty}^{\infty} u(\tau)(t - \tau)e^{-(t-\tau)}u(t - \tau)d\tau$$

$$= \int_{0}^{t} (t - \tau)e^{-(t-\tau)}d\tau$$

$$= [1 - e^{-t}(1 + t)]u(t)$$

which satisfies the initial conditions and attempts to follow the input signal. This is the unit-step response.

In the Laplace domain, the above can be easily computed as follows. From the transfer function, we have that

$$Y(s) = H(s)X(s)$$
$$= \frac{1}{(s+1)^2} \frac{1}{s}$$

where we replaced the transfer function and the Laplace transform of $x(t) = u(t)$. The partial fraction expansion of $Y(s)$ is then

$$Y(s) = \frac{A}{s} + \frac{B}{s+1} + \frac{C}{(s+1)^2}$$

and after obtaining that $A = 1$, $C = -1$, and $B = -1$, we get

$$y(t) = s(t) = u(t) - e^{-t}u(t) - te^{-t}u(t)$$

which coincides with the solution of the convolution integral. It has been obtained, however, in a much easier way. ∎

■ Example 3.27

Consider the positive feedback system created by a microphone close to a set of speakers that are putting out an amplified acoustic signal (see Figure 3.18), which we considered in Example 2.18 in Chapter 2. Find the impulse response of the system using the Laplace transform, and use it to express the output in terms of a convolution. Determine the transfer function and show that the system is not BIBO stable. For simplicity, let $\beta = 1$, $\tau = 1$, and $x(t) = u(t)$. Connect the location of the poles of the transfer function with the unstable behavior of the system.

Solution

As we indicated in Example 2.18 in Chapter 2, the impulse response of a feedback system cannot be explicitly obtained in the time domain, but it can be done using the Laplace transform. The input–output equation for the positive feedback is

$$y(t) = x(t) + \beta y(t - \tau)$$

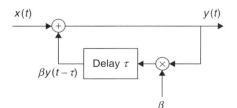

FIGURE 3.18
Positive feedback created by closeness
of a microphone to a set of speakers.

If we let $x(t) = \delta(t)$, the output is $y(t) = h(t)$ or

$$h(t) = \delta(t) + \beta h(t - \tau)$$

and if $H(s) = \mathcal{L}[h(t)]$, then the Laplace transform of the above equation is $H(s) = 1 + \beta H(s)e^{-s\tau}$ or solving for $H(s)$:

$$H(s) = \frac{1}{1 - \beta e^{-s\tau}} = \frac{1}{1 - e^{-s}}$$

$$= \sum_{k=0}^{\infty} e^{-sk} = 1 + e^{-s} + e^{-2s} + e^{-3s} + \cdots$$

after replacing the given values for β and τ. The impulse response $h(t)$ is the inverse Laplace transform of $H(s)$ or

$$h(t) = \delta(t) + \delta(t - 1) + \delta(t - 2) + \cdots = \sum_{k=0}^{\infty} \delta(t - k)$$

If $x(t)$ is the input, the output is given by the convolution integral

$$y(t) = \int_{-\infty}^{\infty} x(t - \tau)h(\tau)d\tau = \int_{-\infty}^{\infty} \sum_{k=0}^{\infty} \delta(\tau - k)x(t - \tau)d\tau$$

$$= \sum_{k=0}^{\infty} \int_{-\infty}^{\infty} \delta(\tau - k)x(t - \tau)d\tau = \sum_{k=0}^{\infty} x(t - k)$$

and replacing $x(t) = u(t)$, we get

$$y(t) = \sum_{k=0}^{\infty} u(t - k)$$

which tends to infinity as t increases.

For this system to be BIBO stable, the impulse response $h(t)$ must be absolutely integrable, which is not the case for this system. Indeed,

$$\int_{-\infty}^{\infty} |h(t)|dt = \int_{-\infty}^{\infty} \sum_{k=0}^{\infty} \delta(t - k)dt$$

$$= \sum_{k=0}^{\infty} \int_{-\infty}^{\infty} \delta(t - k)dt = \sum_{k=0}^{\infty} 1 \to \infty$$

The poles of $H(s)$ are the roots of $1 - e^{-s} = 0$, which are the values of s such that $e^{-s_k} = 1 = e^{j2\pi k}$ or $s_k = \pm j2\pi k$. That is, there is an infinite number of poles on the $j\Omega$ axis, indicating that the system is not BIBO stable. ∎

3.6 WHAT HAVE WE ACCOMPLISHED? WHERE DO WE GO FROM HERE?

In this chapter you have learned the significance of the Laplace transform in the representation of signals as well as of systems. The Laplace transform provides a complementary representation to the time representation of a signal, so that damping and frequency, poles and zeros, together with regions of convergence, conform a new domain for signals. But it is more than that—you will see that these concepts will apply for the rest of this part of the book. When discussing the Fourier analysis of signals and systems we will come back to the Laplace domain for computational tools and for interpretation. The solution of differential equations and the different types of responses are obtained algebraically with the Laplace transform. Likewise, the Laplace transform provides a simple and yet very significant solution to the convolution integral. It also provides the concept of transfer function, which will be fundamental in analysis and synthesis of linear time-invariant systems.

The common thread of the Laplace and the Fourier transforms is the eigenfunction property of LTI systems. You will see that understanding this property will provide you with the needed insight into the Fourier analysis, which we will cover in the next two chapters.

PROBLEMS

3.1. Generic signal representation and the Laplace transform
The generic representation of a signal $x(t)$ in terms of impulses is

$$x(t) = \int_{-\infty}^{\infty} x(\tau)\delta(t-\tau)d\tau$$

Considering the integral an infinite sum of terms $x(\tau)\delta(t-\tau)$ (think of $x(\tau)$ as a constant, as it is not a function of time t), find the Laplace transform of each of these terms and use the linearity property to find $X(s) = \mathcal{L}[x(t)]$. Are you surprised at this result?

3.2. Impulses and the Laplace transform
Given

$$x(t) = 2[\delta(t+1) + \delta(t-1)]$$

  (a) Find the Laplace transform $X(s)$ of $x(t)$ and determine its region of convergence.
  (b) Plot $x(t)$.
  (c) The function $X(s)$ is complex. Let $s = \sigma + j\Omega$ and carefully obtain the magnitude $|X(\sigma + j\Omega)|$ and the phase $\angle X(\sigma + j\Omega)$.

3.3. Sinusoids and the Laplace transform
Consider the following cases involving sinusoids:
  (a) Find the Laplace transform of $y(t) = \sin(2\pi t)u(t) - \sin(2\pi(t-1))u(t-1)$ and its region of convergence. Carefully plot $y(t)$. Determine the region of convergence of $Y(s)$.
  (b) A very smooth pulse, called the raised cosine, $x(t)$ is obtained as

$$x(t) = 1 - \cos(2\pi t) \qquad 0 \leq t \leq 1$$

and zero elsewhere. The raised cosine is used in communications to transmit signals with minimal interference. Find its Laplace transform and its corresponding region of convergence.

(c) Indicate three possible approaches to finding the Laplace transform of $\cos^2(t)u(t)$. Use two of these approaches to find the Laplace transform.

3.4. Unit-step signals and the Laplace transform

Find the Laplace transform of the reflection of the unit-step signal (i.e., $u(-t)$) and its region of convergence. Then use the result together with the Laplace transform of $u(t)$ to see if you can obtain the Laplace transform of a constant or $u(t) + u(-t)$ (assume $u(0) = 0.5$ so there is no discontinuity at $t = 0$).

3.5. Laplace transform of noncausal signal

Consider the noncausal signal

$$x(t) = e^{-|t|}u(t+1)$$

Carefully plot it, and find its Laplace transform $X(s)$ by separating $x(t)$ into a causal signal and an anti-causal signal, $x_c(t)$ and $x_{ac}(t)$, respectively, and plot them separately. Find the ROC of $X(s)$, $X_c(s)$, and $X_{ac}(s)$.

3.6. Transfer function and differential equation

The transfer function of a causal LTI system is

$$H(s) = \frac{1}{s^2 + 4}$$

(a) Find the differential equation that relates the input $x(t)$ and the output $y(t)$ of the system.

(b) Suppose we would like the output $y(t)$ to be identically zero for t greater or equal to zero. If we let $x(t) = \delta(t)$, what would the initial conditions be equal to?

3.7. Transfer function

The input to an LTI system is

$$x(t) = u(t) - 2u(t-1) + u(t-2)$$

If the Laplace transform of the output is given by

$$Y(s) = \frac{(s+2)(1-e^{-s})^2}{s^2(s+1)^2}$$

determine the transfer function of the system.

3.8. Inverse Laplace transform—MATLAB

Consider the following inverse Laplace transform problems for a causal signal $x(t)$:

(a) Given the Laplace transform

$$X(s) = \frac{s^4 + 2s + 1}{s^3 + 4s^2 + 5s + 2}$$

which is not proper, determine the amplitude of the $\delta(t)$ and $d\delta(t)/dt$ terms in the inverse signal $x(t)$.

(b) Find the inverse Laplace transform of

$$X(s) = \frac{s^2 - 3}{(s+1)(s+2)}$$

Can you use the initial-value theorem to check your result? Explain.

(c) The inverse Laplace transform of

$$X(s) = \frac{3s - 4}{s(s + 1)(s + 2)}$$

should give a response of the form

$$x(t) = [Ae^{-t} + B + Ce^{-2t}]u(t)$$

Find the values of A, B, and C. Use the MATLAB function ilaplace to get the inverse.

3.9. Steady state and transient
Consider the following cases where we want to determine either the steady state, transient, or both.
(a) Without computing the inverse of the Laplace transform

$$X(s) = \frac{1}{s(s^2 + 2s + 10)}$$

corresponding to a causal signal $x(t)$, determine its steady-state solution. What is the value of $x(0)$? Show how to obtain this value without computing the inverse Laplace transform.
(b) The Laplace transform of the output of an LTI system is

$$Y(s) = \frac{1}{s((s + 2)^2 + 1)}$$

What would be the steady-state response $y_{ss}(t)$?
(c) The Laplace transform of the output of an LTI system is

$$Y(s) = \frac{e^{-s}}{s((s - 2)^2 + 1)}$$

How would you determine if there is a steady state or not? Explain.
(d) The Laplace transform of the output of an LTI system is

$$Y(s) = \frac{s + 1}{s((s + 1)^2 + 1)}$$

Determine the steady-state and the transient responses corresponding to $Y(s)$.

3.10. Inverse Laplace transformation—MATLAB
Consider the following inverse Laplace problems using MATLAB for causal signal $x(t)$:
(a) Use MATLAB to compute the inverse Laplace transform of

$$X(s) = \frac{s^2 + 2s + 1}{s(s + 1)(s^2 + 10s + 50)}$$

and determine the value of $x(t)$ in the steady state. How would you be able to obtain this value without computing the inverse? Explain
(b) Find the poles and zeros of

$$X(s) = \frac{(1 - se^{-s})}{s(s + 2)}$$

Find the inverse Laplace transform $x(t)$ (use MATLAB to verify your result).

3.11. Convolution integral

Consider the following problems related to the convolution integral:

(a) The impulse response of an LTI system is $h(t) = e^{-2t}u(t)$ and the system input is a pulse $x(t) = u(t) - u(t-3)$. Find the output of the system $y(t)$ by means of the convolution integral graphically and by means of the Laplace transform.

(b) It is known that the impulse response of an analog averager is $h(t) = u(t) - u(t-1)$. Consider the input to the averager $x(t) = u(t) - u(t-1)$, and determine graphically as well as by means of the Laplace transform the corresponding output of the averager $y(t) = [h * x](t)$. Is $y(t)$ smoother than the input signal $x(t)$? Provide an argument for your answer.

(c) Suppose we cascade three analog averagers each with the same impulse response $h(t) = u(t) - u(t-1)$. Determine the transfer function of this system. If the duration of the support of the input to the first averager is M sec, what would be the duration of the support of the output of the third averager?

3.12. Deconvolution

In convolution problems the impulse response $h(t)$ of the system and the input $x(t)$ are given and one is interested in finding the output of the system $y(t)$. The so-called "deconvolution" problem consists in giving two of $x(t)$, $h(t)$, and $y(t)$ to find the other. For instance, given the output $y(t)$ and the impulse response $h(t)$ of the system, one wants to find the input. Consider the following cases:

(a) Suppose the impulse response of the system is $h(t) = e^{-t}\cos(t)u(t)$ and the output has a Laplace transform

$$Y(s) = \frac{4}{s((s+1)^2+1)}$$

What is the input $x(t)$?

(b) The output of an LTI system is $y(t) = r(t) - 2r(t-1) + r(t-2)$, where $r(t)$ is the ramp signal. Determine the impulse response of the system if it is known that the input is $x(t) = u(t) - u(t-1)$.

3.13. Application of superposition

One of the advantages of LTI systems is the superposition property. Suppose that the transfer function of a LTI system is

$$H(s) = \frac{s}{s^2+s+1}$$

Find the unit-step response $s(t)$ of the system, and then use it to find the response due to the following inputs:

$$x_1(t) = u(t) - u(t-1)$$

$$x_2(t) = \delta(t) - \delta(t-1)$$

$$x_3(t) = r(t)$$

$$x_4(t) = r(t) - 2r(t-1) + r(t-2)$$

Express the responses $y_i(t)$ due to $x_i(t)$ for $i = 1, \ldots, 4$ in terms of the unit-step response $s(t)$.

3.14. Properties of the Laplace transform

Consider computing the Laplace transform of a pulse

$$p(t) = \begin{cases} 1 & 0 \le t \le 1 \\ 0 & \text{otherwise} \end{cases}$$

(a) Use the integral formula to find $P(s)$, the Laplace transform of $p(t)$. Determine the region of convergence of $P(s)$.

(b) Represent $p(t)$ in terms of the unit-step function and use its Laplace transform and the time-shift property to find $P(s)$. Find the poles and zeros of $P(s)$ to verify the region of convergence obtained above.

3.15. Frequency-shift property

Duality occurs between time and frequency shifts. As shown, if $\mathcal{L}[x(t)] = X(s)$, then $\mathcal{L}[x(t - t_0)] = X(s)e^{-t_0 s}$. The dual of this would be $\mathcal{L}[x(t)e^{-\alpha t}] = X(s + \alpha)$, which we call the *frequency-shift* property.

(a) Use the integral formula for the Laplace transform to show the frequency-shift property.

(b) Use the above frequency-shift property to find $X(s) = \mathcal{L}[x(t) = \cos(\Omega_0 t)u(t)]$ (represent the cosine using Euler's identity). Find and plot the poles and zeros of $X(s)$.

(c) Recall the definition of the hyperbolic cosine, $\cosh(\Omega_0 t) = 0.5(e^{\Omega_0 t} + e^{-\Omega_0 t})$, and find the Laplace transform $Y(s)$ of $y(t) = \cosh(\Omega_0 t)u(t)$. Find and plot the poles and zeros of $Y(s)$. Explain the relation of the poles of $X(s)$ and $Y(s)$ by connecting $x(t)$ with $y(t)$.

3.16. Poles and zeros

Consider the pulse $x(t) = u(t) - u(t - 1)$.

(a) Find the zeros and poles of $X(s)$ and plot them.

(b) Suppose $x(t)$ is the input of an LTI system with a transfer function $H(s) = 1/(s^2 + 4\pi^2)$. Find and plot the poles and zeros of $Y(s) = \mathcal{L}[y(t)] = H(s)X(s)$ where $y(t)$ is the output of the system.

(c) If the transfer function of the LTI system is

$$G(s) = \frac{Z(s)}{X(s)} = \prod_{k=1}^{\infty} \frac{1}{s^2 + (2k\pi)^2}$$

and the input is the above signal $x(t)$, compute the output $z(t)$.

3.17. Poles and zeros—MATLAB

The poles corresponding to the Laplace transform $X(s)$ of a signal $x(t)$ are

$$p_{1,2} = -3 \pm j\pi/2$$
$$p_3 = 0$$

(a) Within some constants, give a general form of the signal $x(t)$.

(b) Let

$$X(s) = \frac{1}{(s + 3 - j\pi/2)(s + 3 - j\pi/2)s}$$

From the location of the poles, obtain a general form for $x(t)$. Use MATLAB to find $x(t)$ and plot it. How well did you guess the answer?

3.18. Solving differential equations—MATLAB

One of the uses of the Laplace transform is the solution of differential equations.

(a) Suppose you are given the differential equation that represents an LTI system,

$$y^{(2)}(t) + 0.5y^{(1)}(t) + 0.15y(t) = x(t) \qquad t \geq 0$$

where $y(t)$ is the output and $x(t)$ is the input of the system, and $y^{(1)}(t)$ and $y^{(2)}(t)$ are first- and second-order derivatives with respect to t. The input is causal, (i.e., $x(t) = 0$ $t < 0$). What should the initial conditions be for the system to be LTI? Find $Y(s)$ for those initial conditions.

(b) If $y^{(1)}(0) = 1$ and $y(0) = 1$ are the initial conditions for the above differential equation, find $Y(s)$. If the input to the system is doubled—that is, the input is $2x(t)$ with Laplace transform $2X(s)$—is $Y(s)$ doubled so that its inverse Laplace transform $y(t)$ is doubled? Is the system linear?

(c) Use MATLAB to find the poles and zeros and the solutions of the differential equation when the input is $u(t)$ and $2u(t)$ with the initial conditions given above. Compare the solutions and verify your response in (b).

3.19. Differential equation, initial conditions, and impulse response—MATLAB

The following function $Y(s) = \mathcal{L}[y(t)]$ is obtained applying the Laplace transform to a differential equation representing a system with nonzero initial conditions and input $x(t)$, with Laplace transform $X(s)$:

$$Y(s) = \frac{X(s)}{s^2 + 2s + 3} + \frac{s + 1}{s^2 + 2s + 3}$$

(a) Find the differential equation in $y(t)$ and $x(t)$ representing the system.

(b) Find the initial conditions $y'(0)$ and $y(0)$.

(c) Use MATLAB to determine the impulse response $h(t)$ of this system. Find the poles of the transfer function $H(s)$ and determine if the system is BIBO stable.

3.20. Different responses—MATLAB

Let $Y(s) = \mathcal{L}[y(t)]$ be the Laplace transform of the solution of a second-order differential equation representing a system with input $x(t)$ and some initial conditions,

$$Y(s) = \frac{X(s)}{s^2 + 2s + 1} + \frac{s + 1}{s^2 + 2s + 1}$$

(a) Find the zero-state response (response due to the input only with zero initial conditions) for $x(t) = u(t)$.

(b) Find the zero-input response (response due to the initial conditions and zero input).

(c) Find the complete response when $x(t) = u(t)$.

(d) Find the transient and the steady-state response when $x(t) = u(t)$.

(e) Use MATLAB to verify the above responses.

3.21. Poles and stability

The transfer function of a BIBO-stable system has poles only on the open left-hand s-plane (excluding the $j\Omega$ axis).

(a) Let the transfer function of a system be

$$H_1(s) = \frac{Y(s)}{X(s)} = \frac{1}{(s + 1)(s - 2)}$$

and let $X(s)$ be the Laplace transform of signals that are bounded (i.e., the poles of $X(s)$ are on the left-hand s-plane). Find $\lim_{t \to \infty} y(t)$. Determine if the system is BIBO stable. If not, determine what makes the system unstable.

(b) Let the transfer function be

$$H_2(s) = \frac{Y(s)}{X(s)} = \frac{1}{(s + 1)(s + 2)}$$

and $X(s)$ be as indicated above. Find

$$\lim_{t \to \infty} y(t)$$

Can you use this limit to determine if the system is BIBO stable? If not, what would you do to check its stability?

3.22. Poles, stability, and steady-state response

The steady-state solution of stable systems is due to simple poles in the $j\Omega$ axis of the s-plane coming from the input. Suppose the transfer function of the system is

$$H(s) = \frac{Y(s)}{X(s)} = \frac{1}{(s+1)^2 + 4}$$

(a) Find the poles and zeros of $H(s)$ and plot them in the s-plane. Find then the corresponding impulse response $h(t)$. Determine if the impulse response of this system is absolutely integrable so that the system is BIBO stable.

(b) Let the input $x(t) = u(t)$. Find $y(t)$ and from it determine the steady-state solution.

(c) Let the input $x(t) = tu(t)$. Find $y(t)$ and from it determine the steady-state response. What is the difference between this case and the previous one?

(d) To explain the behavior in the case above consider the following: Is the input $x(t) = tu(t)$ bounded? That is, is there some finite value M such that $|x(t)| < M$ for all times? So what would you expect the output to be knowing that the system is stable?

3.23. Responses from an analog averager

The input–output equation for an analog averager is given by

$$y(t) = \frac{1}{T} \int_{t-T}^{t} x(\tau)d\tau$$

where $x(t)$ is the input and $y(t)$ is the output. This equation corresponds to the convolution integral.

(a) Change the above equation so that you can determine from it the impulse response $h(t)$.

(b) Graphically determine the output $y(t)$ corresponding to a pulse input $x(t) = u(t) - u(t-2)$ using the convolution integral (let $T = 1$) relating the input and the output. Carefully plot the input and the output. (The output can also be obtained intuitively from a good understanding of the averager.)

(c) Using the impulse response $h(t)$ found above, use now the Laplace transform to find the output corresponding to $x(t) = u(t) - u(t-2)$. Let again $T = 1$ in the averager.

3.24. Transients for second-order systems—MATLAB

The type of transient you get in a second-order system depends on the location of the poles of the system. The transfer function of the second-order system is

$$H(s) = \frac{Y(s)}{X(s)} = \frac{1}{s^2 + b_1 s + b_0}$$

and let the input be $x(t) = u(t)$.

(a) Let the coefficients of the denominator of $H(s)$ be $b_1 = 5$ and $b_0 = 6$. Find the response $y(t)$. Use MATLAB to verify the response and to plot it.

(b) Suppose then that the denominator coefficients of $H(s)$ are changed to $b_1 = 2$ and $b_0 = 6$. Find the response $y(t)$. Use MATLAB to verify the response and to plot it.

(c) Explain your results above by relating your responses to the location of the poles of $H(s)$.

3.25. Effect of zeros on the sinusoidal steady state

To see the effect of the zeros on the complete response of a system, suppose you have a system with a transfer function

$$H(s) = \frac{Y(s)}{X(s)} = \frac{s^2 + 4}{s((s+1)^2 + 1)}$$

(a) Find and plot the poles and zeros of $H(s)$. Is this BIBO stable?

(b) Find the frequency Ω_0 of the input $x(t) = 2\cos(\Omega_0 t)u(t)$ such that the output of the given system is zero in the steady state. Why do you think this happens?

(c) If the input is a sine instead of a cosine, would you get the same result as above? Explain why or why not.

3.26. Zero steady-state response of analog averager—MATLAB

The analog averager can be represented by the differential equation

$$\frac{dy(t)}{dt} = \frac{1}{T}[x(t) - x(t-T)]$$

where $y(t)$ is the output and $x(t)$ is the input.

(a) If the input–output equation of the averager is

$$y(t) = \frac{1}{T}\int_{t-T}^{t} x(\tau)d\tau$$

show how to obtain the above differential equation and that $y(t)$ is the solution of the differential equation.

(b) If $x(t) = \cos(\pi t)u(t)$, choose the value of T in the averager so that the output is $y(t) = 0$ in the steady state. Graphically show how this is possible for your choice of T. Is there a unique value for T that makes this possible? How does it relate to the frequency $\Omega_0 = \pi$ of the sinusoid?

(c) Use the impulse response $h(t)$ of the averager found before, to show using Laplace that the steady state is zero when $x(t) = \cos(\pi t)u(t)$ and T is the above chosen value. Use MATLAB to solve the differential equation and to plot the response for the value of T you chose. (*Hint:* Consider $x(t)/T$ the input and use superposition and time invariance to find $y(t)$ due to $(x(t) - x(t-T))/T$.)

3.27. Partial fraction expansion—MATLAB

Consider the following functions $Y_i(s) = \mathcal{L}[y_i(t)]$, $i = 1, 2$ and 3:

$$Y_1(s) = \frac{s+1}{s(s^2 + 2s + 4)}$$

$$Y_2(s) = \frac{1}{(s+2)^2}$$

$$Y_3(s) = \frac{s-1}{s^2((s+1)^2 + 9)}$$

where $\{y_i(t), i = 1, 2, 3\}$ are the complete responses of differential equations with zero initial conditions.

(a) For each of these functions, determine the corresponding differential equation, if all of them have as input $x(t) = u(t)$.

(b) Find the general form of the complete response $\{y_i(t), i = 1, 2, 3\}$ for each of the $\{Y_i(s)\ i = 1, 2, 3\}$. Use MATLAB to plot the poles and zeros for each of the $\{Y_i(s)\}$, to find their partial fraction expansions, and the complete responses.

3.28. Iterative convolution integral—MATLAB

Consider the convolution of a pulse $x(t) = u(t+0.5) - u(t-0.5)$ with itself many times. Use MATLAB for the calculations and the plotting.

(a) Consider the result for $N = 2$ of these convolutions—that is,

$$y_2(t) = (x * x)(t)$$

Find $Y_2(s) = \mathcal{L}[y_2(t)]$ using the convolution property of the Laplace transform and find $y_2(t)$.

(b) Consider then the result for $N = 3$ of these convolutions—that is,

$$y_3(t) = (x * x * x)(t)$$

Find $Y_3(s) = \mathcal{L}[y_3(t)]$ using the convolution property of the Laplace transform and find $y_3(t)$.

(c) The signal $x(t)$ can be considered the impulse response of an averager that "smooths" out a signal. Letting $y_1(t) = x(t)$, plot the three functions $y_i(t)$ for $i = 1$, 2, and 3. Compare these signals on their smoothness and indicate their supports in time. (For $y_2(t)$ and $y_3(t)$, how do their supports relate to the supports of the signals convolved?)

3.29. Positive and negative feedback

There are two types of feedback, negative and positive. In this problem we explore their difference.

(a) Consider negative feedback. Suppose you have a system with transfer function $H(s) = Y(s)/E(s)$ where $E(s) = C(s) - Y(s)$, and $C(s)$ and $Y(s)$ are the transforms of the feedback system's reference $c(t)$ and output $y(t)$. Find the transfer function of the overall system $G(s) = Y(s)/C(s)$.

(b) In positive feedback, the only equation that changes is $E(s) = C(s) + Y(s)$; the other equations remain the same. Find the overall feedback system transfer function $G(s) = Y(s)/C(s)$.

(c) Suppose that $C(s) = 1/s$, $H(s) = 1/(s+1)$. Determine $G(s)$ for both negative and positive feedback. Find $y(t) = \mathcal{L}^{-1}[Y(s)]$ for both types of feedback and comment on the difference in these signals.

3.30. Feedback stabilization

An unstable system can be stabilized by using negative feedback with a gain K in the feedback loop. For instance, consider an unstable system with transfer function

$$H(s) = \frac{2}{s-1}$$

which has a pole in the right-hand s-plane, making the impulse response of the system $h(t)$ grow as t increases. Use negative feedback with a gain $K > 0$ in the feedback loop, and put $H(s)$ in the forward loop. Draw a block diagram of the system. Obtain the transfer function $G(s)$ of the feedback system and determine the value of K that makes the overall system BIBO stable (i.e., its poles in the open left-hand s-plane).

3.31. All-pass stabilization

Another stabilization method consists in cascading an all-pass system with the unstable system to cancel the poles in the right-hand s-plane. Consider a system with a transfer function

$$H(s) = \frac{s+1}{(s-1)(s^2 + 2s + 1)}$$

which has a pole in the right-hand s-plane, $s = 1$, so it is unstable.

(a) The poles and zeros of an all-pass filter are such that if $p_{12} = -\sigma \pm j\Omega_0$ are complex conjugate poles of the filter, then $z_{12} = \sigma \pm j\Omega_0$ are the corresponding zeros, and for real poles $p = -\sigma$ there is a corresponding $z = \sigma$. The orders of the numerator and the denominator of the all-pass filter are equal. Write the general transfer function of an all-pass filter $H_{ap}(s) = KN(s)/D(s)$.

(b) Find an all-pass filter $H_{ap}(s)$ so that when cascaded with $H(s)$ the overall transfer function $G(s) = H(s)H_{ap}(s)$ has all its poles in the left-hand s-plane.

(c) Find K of the all-pass filter so that when $s = 0$ the all-pass filter has a gain of unity. What is the relation between the magnitude of the overall system $|G(s)|$ and that of the unstable filter $|H(s)|$.

3.32. Half-wave rectifier—MATLAB

In the generation of DC from AC voltage, the "half-wave" rectified signal is an important part. Suppose the AC voltage is $x(t) = \sin(2\pi t)u(t)$.

(a) Carefully plot the half-wave rectified signal $y(t)$ from $x(t)$.

(b) Let $y_1(t)$ be the period of $y(t)$ between $0 \le t \le 1$. Show that $y_1(t)$ can be written as

$$y_1(t) = \sin(2\pi t)u(t) + \sin(2\pi(t - 0.5))u(t - 0.5)$$

or

$$y_1(t) = \sin(2\pi t)[u(t) - u(t - 0.5)]$$

Use MATLAB to verify this. Find the Laplace transform $X_1(s)$ of $x_1(t)$.

(c) Express $y(t)$ in terms of $y_1(t)$ and find the Laplace transform $Y(s)$ of $y(t)$.

3.33. Polynomial multiplication—MATLAB

When the numerator or denominator is given in a factorized form, we need to multiply polynomials. Although this can be done by hand, MATLAB provides the function conv that computes the coefficients of the polynomial resulting from the product of two polynomials.

(a) Use help in MATLAB to find how conv can be used, and then consider two polynomials

$$P(s) = s^2 + s + 1 \text{ and } Q(s) = 2s^3 + 3s^2 + s + 1$$

Do the multiplication of these polynomials by hand to find $Z(s) = P(s)Q(s)$ and use conv to verify your results.

(b) The output of a system has a Laplace transform

$$Y(s) = \frac{N(s)}{D(s)} = \frac{(s + 2)}{s^2(s + 1)((s + 4)^2) + 9)}$$

Use conv to find the denominator polynomial and then find the inverse Laplace transform using ilaplace.

3.34. Feedback error—MATLAB

Consider a negative feedback system used to control a plant $G(s) = 1/(s(s + 1)(s + 2))$. The output $y(t)$ of the feedback system is connected via a sensor with transfer function $H(s) = 1$ to a differentiator where the reference signal $x(t)$ is also connected. The output of the differentiator is the feedback error $e(t) = x(t) - v(t)$ where $v(t)$ is the output of the feedback sensor.

(a) Carefully draw the feedback system, and find an expression for $E(s)$, the Laplace transform of the feedback error $e(t)$.

(b) Two possible reference test signals for the given plant are $x(t) = u(t)$ and $x(t) = r(t)$. Choose the one that would give a zero steady-state feedback error.

(c) Use MATLAB to do the partial fraction expansions for the two error functions $E_1(s)$, corresponding to when $x(t) = u(t)$ and $E_2(s)$ when $x(t) = r(t)$. Use these partial fraction expansions to find $e_1(t)$ and $e_2(t)$, and thus verify your results obtained before.

CHAPTER 4

Frequency Analysis: The Fourier Series

A Mathematician is a device for
turning coffee into theorems.
Paul Erdos (1913–1996)
mathematician

4.1 INTRODUCTION

In this chapter and the next we consider the frequency analysis of continuous-time signals and systems—the Fourier series for periodic signals in this chapter, and the Fourier transform for both periodic and aperiodic signals as well as for systems in Chapter 5. In these chapters we consider:

- *Spectral representation*—The frequency representation of periodic and aperiodic signals indicates how their power or energy is allocated to different frequencies. Such a distribution over frequency is called the *spectrum of the signal*. For a periodic signal the spectrum is discrete, as its power is concentrated at frequencies multiples of a so-called *fundamental frequency*, directly related to the period of the signal. On the other hand, the spectrum of an aperiodic signal is a continuous function of frequency. The concept of spectrum is similar to the one used in optics for light, or in material science for metals, each indicating the distribution of power or energy over frequency. The Fourier representation is also useful in finding the frequency response of linear time-invariant systems, which is related to the transfer function obtained with the Laplace transform. The frequency response of a system indicates how an LTI system responds to sinusoids of different frequencies. Such a response characterizes the system and permits easy computation of its steady-state response, and will be equally important in the synthesis of systems.
- *Eigenfunctions and Fourier analysis*—It is important to understand the driving force behind the representation of signals in terms of basic signals when applied to LTI systems. For instance, the convolution integral that gives the output of an LTI system resulted from the representation of its input signal in terms of shifted impulses. Along with this result came the concept of the impulse response of an LTI system. Likewise, the Laplace transform can be seen as the representation of signals in terms of general eigenfunctions. In this chapter and the next we will see that complex

Signals and Systems Using MATLAB®. DOI: 10.1016/B978-0-12-374716-7.00007-7

exponentials or sinusoids are used in the Fourier representation of periodic as well as aperiodic signals by taking advantage of the eigenfunction property of LTI systems. The results of the Fourier series in this chapter will be extended to the Fourier transform in Chapter 5.

- *Steady-state analysis*—Fourier analysis is in the steady state, while Laplace analysis considers both transient and steady state. Thus, if one is interested in transients, as in control theory, Laplace is a meaningful transformation. On the other hand, if one is interested in the frequency analysis, or steady state, as in communications theory, the Fourier transform is the one to use. There will be cases, however, where in control and communications both Laplace and Fourier analysis are considered.
- *Application of Fourier analysis*—The frequency representation of signals and systems is extremely important in signal processing and in communications. It explains filtering, modulation of messages in a communication system, the meaning of bandwidth, and how to design filters. Likewise, the frequency representation turns out to be essential in the sampling of analog signals—the bridge between analog and digital signal processing.

4.2 EIGENFUNCTIONS REVISITED

As indicated in Chapter 3, the most important property of stable LTI systems is that when the input is a complex exponential (or a combination of a cosine and a sine) of a certain frequency, the output of the system is the input times a complex constant connected with how the system responds to the frequency at the input. The complex exponential is called an *eigenfunction* of stable LTI systems.

If $x(t) = e^{j\Omega_0 t}$, $-\infty < t < \infty$, is the input to a causal and a stable system with impulse response $h(t)$, the output in the steady state is given by

$$y(t) = e^{j\Omega_0 t} H(j\Omega_0) \tag{4.1}$$

where

$$H(j\Omega_0) = \int_0^\infty h(\tau) e^{-j\Omega_0 \tau} d\tau \tag{4.2}$$

is the frequency response of the system at Ω_0. The signal $x(t) = e^{j\Omega_0 t}$ is said to be an *eigenfunction* of the LTI system as it appears at both input and output.

This can be seen by finding the output corresponding to $x(t) = e^{j\Omega_0 t}$ by means of the convolution integral,

$$y(t) = \int_0^\infty h(\tau) x(t-\tau) d\tau = e^{j\Omega_0 t} \int_0^\infty h(\tau) e^{-j\Omega_0 \tau} d\tau$$

$$= e^{j\Omega_0 t} H(j\Omega_0)$$

where we let $H(j\Omega_0)$ equal the integral in the second equation. The input signal appears in the output modified by the frequency response of the system $H(j\Omega_0)$ at the frequency Ω_0 of the input. Notice that the convolution integral limits indicate that the input started at $-\infty$ and that we are considering the output at finite time t—this means that we are in steady state. The steady-state response of a stable LTI system is attained by either considering that the initial time when the input is applied to the system is $-\infty$ and we reach a finite time t, or by starting at time 0 and going to ∞.

The above result for one frequency can be easily extended to the case of several frequencies present at the input. If the input signal $x(t)$ is a linear combination of complex exponentials, with different amplitudes, frequencies, and phases, or

$$x(t) = \sum_k X_k e^{j\Omega_k t}$$

where X_k are complex values, since the output corresponding to $X_k e^{j\Omega_k t}$ is $X_k e^{j\Omega_k t} H(j\Omega_k)$ by superposition the response to $x(t)$ is

$$y(t) = \sum_k X_k e^{j\Omega_k t} H(j\Omega_k)$$
$$= \sum_k X_k |H(j\Omega_k)| e^{j(\Omega_k t + \angle H(j\Omega_k))} \tag{4.3}$$

The above is valid for any signal that is a combination of exponentials of arbitrary frequencies. As we will see in this chapter, when $x(t)$ is periodic it can be represented by the Fourier series, which is a combination of complex exponentials harmonically related (i.e., the frequencies of the exponentials are multiples of the fundamental frequency of the periodic signal). Thus, when a periodic signal is applied to a causal and stable LTI system its output is computed as in Equation (4.3).

The significance of the eigenfunction property is also seen when the input signal is an integral (a sum, after all) of complex exponentials, with continuously varying frequency, as the integrand. That is, if

$$x(t) = \frac{1}{2\pi} \int_{-\infty}^{\infty} X(\Omega) e^{j\Omega t} d\Omega$$

then using superposition and the eigenfunction property of a stable LTI system, with frequency response $H(j\Omega)$, the output is

$$y(t) = \frac{1}{2\pi} \int_{-\infty}^{\infty} X(\Omega) e^{j\Omega t} H(j\Omega) d\Omega$$
$$= \frac{1}{2\pi} \int_{-\infty}^{\infty} X(\Omega) |H(j\Omega)| e^{(j\Omega t + j\angle H(j\Omega))} d\Omega \tag{4.4}$$

The above representation of $x(t)$ corresponds to the Fourier representation of aperiodic signals, which will be covered in Chapter 5. Again here, the eigenfunction property of LTI systems provides an efficient way to compute the output. Furthermore, we also find that by letting $Y(\Omega) = X(\Omega)H(j\Omega)$ the above equation gives an expression to compute $y(t)$ from $Y(\Omega)$. The product $Y(\Omega) = X(\Omega)H(j\Omega)$ corresponds to the Fourier transform of the convolution integral $y(t) = x(t) * h(t)$, and is connected with the convolution property of the Laplace transform. It is important to start noticing these connections, to understand the link between Laplace and Fourier analysis.

Remarks

- Notice the difference of notation for the frequency representation of signals and systems used above. If $x(t)$ is a periodic signal its frequency representation is given by $\{X_k\}$, and if aperiodic by $X(\Omega)$, while for a system with impulse response $h(t)$ its frequency response is given by $H(j\Omega)$.
- When considering the eigenfunction property, the stability of the LTI system is necessary to ensure that $H(j\Omega)$ exists for all frequencies.
- The eigenfunction property applied to a linear circuit gives the same result as the one obtained from phasors in the sinusoidal steady state. That is, if

$$x(t) = A\cos(\Omega_0 t + \theta) = \frac{Ae^{j\theta}}{2}e^{j\Omega_0 t} + \frac{Ae^{-j\theta}}{2}e^{-j\Omega_0 t} \tag{4.5}$$

is the input of a circuit represented by the transfer function

$$H(s) = \frac{Y(s)}{X(s)} = \frac{\mathcal{L}[y(t)]}{\mathcal{L}[x(t)]}$$

then the corresponding steady-state output is given by

$$y_{ss}(t) = \frac{Ae^{j\theta}}{2}e^{j\Omega_0 t}H(j\Omega_0) + \frac{Ae^{-j\theta}}{2}e^{-j\Omega_0 t}H(-j\Omega_0)$$
$$= A|H(j\Omega_0)|\cos(\Omega_0 t + \theta + \angle H(j\Omega_0)) \tag{4.6}$$

where, very importantly, the frequency of the output coincides with that of the input, and the amplitude and phase of the input are changed by the magnitude and phase of the frequency response of the system for the frequency Ω_0. The frequency response is $H(j\Omega_0) = H(s)|_{s=j\Omega_0}$, and as we will see its magnitude is an even function of frequency, or $|H(j\Omega)| = |H(-j\Omega)|$, and its phase is an odd function of frequency, or $\angle H(j\Omega_0) = -\angle H(-j\Omega_0)$. Using these two conditions we obtain Equation (4.6).
The phasor corresponding to the input

$$x(t) = A\cos(\Omega_0 t + \theta)$$

is defined as a vector,

$$X = A\angle\theta$$

rotating in the polar plane at the frequency of Ω_0. The phasor has a magnitude A and an angle θ with respect to the positive real axis. The projection of the phasor onto the real axis, as it rotates at the given *frequency, with time generates a cosine of the indicated frequency, amplitude, and phase. The transfer function is computed at $s = j\Omega_0$ or*

$$H(s)|_{s=j\Omega_0} = H(j\Omega_0) = \frac{Y}{X}$$

(ratio of the phasors corresponding to the output Y and the input X). The phasor for the output is thus

$$Y = H(j\Omega_0)X = |Y|e^{j\angle Y}$$

Such a phasor is then converted into the sinusoid (which equals Eq. 4.6):

$$y_{ss}(t) = \mathcal{R}e[Ye^{j\Omega_0 t}] = |Y|\cos(\Omega_0 t + \angle Y)$$

■ *A very important application of LTI systems is filtering, where one is interested in preserving desired frequency components of a signal and getting rid of less-desirable components. That an LTI system can be used for filtering is seen in Equations (4.3) and (4.4). In the case of a periodic signal, the magnitude $|H(j\Omega_k)|$ can be set ideally to one for those components we wish to keep and to zero for those we wish to get rid of. Likewise, for an aperiodic signal, the magnitude $|H(j\Omega)|$ could be set ideally to one for those components we wish to keep and zero for those components we wish to get rid of. Depending on the filtering application, an LTI system with the appropriate characteristics can be designed, obtaining the desired transfer function $H(s)$.*

For a stable LTI with transfer function $H(s)$ if the input is

$$x(t) = \mathcal{R}e[Ae^{j(\Omega_0 t + \theta)}] = A\cos(\Omega_0 t + \theta)$$

the steady-state output is given by

$$y(t) = \mathcal{R}e[AH(j\Omega_0)e^{j(\Omega_0 t + \theta)}]$$
$$= A|H(j\Omega_0)|\cos(\Omega_0 t + \theta + \angle H(j\Omega_0)) \qquad (4.7)$$

where

$$H(j\Omega_0) = H(s)|_{s=j\Omega_0}$$

■ Example 4.1

Consider the RC circuit shown in Figure 4.1. Let the voltage source be $v_s(t) = 4\cos(t + \pi/4)$ volts. the resistor be $R = 1\Omega$, and the capacitor $C = 1$ F. Find the steady-state voltage across the capacitor.

Solution

This problem can be approached in two ways.

■ *Phasor approach.* From the phasor circuit in Figure 4.1, by voltage division we have the following phasor ratio, where V_s is the phasor corresponding to the source $v_s(t)$ and V_c the phasor

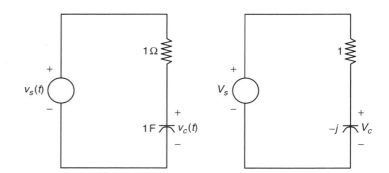

FIGURE 4.1

RC circuit and corresponding phasor circuit.

corresponding to $v_c(t)$:

$$\frac{V_c}{V_s} = \frac{-j}{1-j} = \frac{-j(1+j)}{2} = \frac{\sqrt{2}}{2}\angle-\pi/4$$

Since $V_s = 4\angle\pi/4$, then

$$V_c = 2\sqrt{2}\angle0$$

so that in the steady state,

$$v_c(t) = 2\sqrt{2}\cos(t)$$

■ *Eigenfunction approach.* Considering the output is the voltage across the capacitor and the input is the voltage source, the transfer function is obtained using voltage division as

$$H(s) = \frac{V_c(s)}{V_s(s)} = \frac{1/s}{1+1/s} = \frac{1}{s+1}$$

so that the system frequency response at the input frequency $\Omega_0 = 1$ is

$$H(j1) = \frac{\sqrt{2}}{2}\angle-\pi/4$$

According to the eigenfunction property the steady-state response of the capacitor is

$$v_c(t) = 4|H(j1)|\cos(t + \pi/4 + \angle H(j1))$$
$$= 2\sqrt{2}\cos(t)$$

which coincides with the solution found using phasors.   ■

■ Example 4.2

An ideal communication system provides as output the input signal with only a possible delay in the transmission. Such an ideal system does not cause any distortion to the input signal beyond the delay. Find the frequency response of the ideal communication system, and use it to determine the steady-state response when the delay caused by the system is $\tau = 3$ sec, and the input is $x(t) = 2\cos(4t - \pi/4)$.

Solution

The impulse response of the ideal system is $h(t) = \delta(t - \tau)$ where τ is the delay of the transmission. In fact, the output according to the convolution integral gives

$$y(t) = \int_0^\infty \underbrace{\delta(\rho - \tau)}_{h(\rho)} x(t - \rho)d\rho = x(t - \tau)$$

as expected. Let us then find the frequency response of the ideal communication system. According to the eigenvalue property, if the input is $x(t) = e^{j\Omega_0 t}$, then the output is

$$y(t) = e^{j\Omega_0 t} H(j\Omega_0)$$

but also

$$y(t) = x(t - \tau) = e^{j\Omega_0(t-\tau)}$$

so that comparing these equations we have that

$$H(j\Omega_0) = e^{-j\tau\Omega_0}$$

For a generic frequency $0 \leq \Omega < \infty$, we would get

$$H(j\Omega) = e^{-j\tau\Omega}$$

which is a complex function of Ω, with a unity magnitude $|H(j\Omega)| = 1$, and a linear phase $\angle H(j\Omega) = -\tau\Omega$. This system is called an *all-pass system*, since it allows all frequency components of the input to go through with a phase change only.

Consider the case when $\tau = 3$, and that we input into this system $x(t) = 2\cos(4t - \pi/4)$, then $H(j\Omega) = 1e^{-j3\Omega}$, so that the output in the steady state is

$$y(t) = 2|H(j4)|\cos(4t - \pi/4 + \angle H(j4))$$
$$= 2\cos(4(t - 3) - \pi/4)$$
$$= x(t - 3)$$

where we used $H(j4) = 1e^{-j12}$ (i.e., $|H(j4)| = 1$ and $\angle H(j4) = 12$). ∎

■ Example 4.3

Although there are better methods to compute the frequency response of a system represented by a differential equation, the eigenfunction property can be easily used for that. Consider the RC circuit shown in Figure 4.1 where the input is

$$v_s(t) = 1 + \cos(10{,}000t)$$

with components of low frequency, $\Omega = 0$, and of large frequency, $\Omega = 10{,}000$ rad/sec. The output $v_c(t)$ is the voltage across the capacitor in steady state. We wish to find the frequency response of this circuit to verify that it is a *low-pass filter* (it allows low-frequency components to go through, but filters out high-frequency components).

Solution

Using Kirchhoff's voltage law, this circuit is represented by a first-order differential equation,

$$v_s(t) = v_c(t) + \frac{dv_c(t)}{dt}$$

Now, if the input is $v_s(t) = e^{j\Omega t}$, for a generic frequency Ω, then the output is $v_c(t) = e^{j\Omega t}H(j\Omega)$. Replacing these in the differential equation, we have

$$e^{j\Omega t} = e^{j\Omega t}H(j\Omega) + \frac{de^{j\Omega t}H(j\Omega)}{dt}$$

$$= e^{j\Omega t}H(j\Omega) + j\Omega e^{j\Omega t}H(j\Omega)$$

so that

$$H(j\Omega) = \frac{1}{1 + j\Omega}$$

or the frequency response of the filter for any frequency Ω. The magnitude of $H(j\Omega)$ is

$$|H(j\Omega)| = \frac{1}{\sqrt{1 + \Omega^2}}$$

which is close to one for small values of the frequency, and tends to zero when the frequency values are large—the characteristics of a low-pass filter.

For the input

$$v_s(t) = 1 + \cos(10{,}000t) = \cos(0t) + \cos(10{,}000t)$$

(i.e., it has a zero frequency component and a 10,000-rad/sec frequency component) using Euler's identity, we have that

$$v_s(t) = 1 + 0.5 \left(e^{j10{,}000t} + e^{-j10{,}000t} \right)$$

and the steady-state output of the circuit is

$$v_c(t) = 1H(j0) + 0.5H(j10{,}000)e^{j10{,}000t} + 0.5H(-j10{,}000)e^{-j10{,}000t}$$

$$\approx 1 + \frac{1}{10{,}000} \cos(10{,}000t - \pi/2) \approx 1$$

since

$$H(j0) = 1$$

$$H(j10,000) \approx \frac{1}{j\,10^4} = \frac{-j}{10,000}$$

$$H(-j10,000) \approx \frac{1}{-j\,10^4} = \frac{j}{10,000}$$

Thus, this circuit acts like a low-pass filter by keeping the DC component (with the low frequency $\Omega = 0$) and essentially getting rid of the high-frequency ($\Omega = 10,000$) component of the signal.

Notice that the frequency response can also be obtained by considering the phasor ratio for a generic frequency Ω, which by voltage division is

$$\frac{V_c}{V_s} = \frac{1/j\Omega}{1 + 1/j\Omega} = \frac{1}{1 + j\Omega}$$

which for $\Omega = 0$ is 1 and for $\Omega = 10,000$ is approximately $-j/10,000$ (i.e., corresponding to $H(j0)$ and $H(j10,000) = H^*(j10,000)$). ∎

Fourier and Laplace

French mathematician Jean-Baptiste-Joseph Fourier (1768–1830) was a contemporary of Laplace with whom he shared many scientific and political experiences [2, 7]. Like Laplace, Fourier was from very humble origins but he was not as politically astute. Laplace and Fourier were affected by the political turmoil of the French Revolution and both came in close contact with Napoleon Bonaparte, French general and emperor. Named chair of the mathematics department of the Ecole Normale, Fourier led the most brilliant period of mathematics and science education in France. His main work was "The Mathematical Theory of Heat Conduction" where he proposed the harmonic analysis of periodic signals. In 1807 he received the grand prize from the French Academy of Sciences for this work. This was despite the objections of Laplace, Lagrange, and Legendre, who were the referees and who indicated that the mathematical treatment lacked rigor. Following Galton's advice of "Never resent criticism, and never answer it," Fourier disregarded these criticisms and made no change to his 1822 treatise in heat conduction. Although Fourier was an enthusiast for the Revolution and followed Napoleon on some of his campaigns, in the Second Restoration he had to pawn his belongings to survive. Thanks to his friends, he became secretary of the French Academy, the final position he held.

4.3 COMPLEX EXPONENTIAL FOURIER SERIES

The Fourier series is a representation of a periodic signal $x(t)$ in terms of complex exponentials or sinusoids of frequency multiples of the fundamental frequency of $x(t)$. The advantage of using the Fourier series to represent periodic signals is not only the spectral characterization obtained, but in finding the response for these signals when applied to LTI systems by means of the eigenfunction property.

Mathematically, the Fourier series is an expansion of periodic signals in terms of normalized orthogonal complex exponentials. The concept of orthogonality of functions is similar to the concept of perpendicularity of vectors: Perpendicular vectors cannot be represented in terms of each other, as orthogonal functions provide mutually exclusive information. The perpendicularity of two vectors can be established using the *dot or scalar* product of the vectors, and the orthogonality of functions is established by the *inner product*, or the integration of the product of the function and its conjugate. Consider a set of complex functions $\{\psi_k(t)\}$ defined in an interval $[a, b]$, and such that for any pair of these functions, let's say $\psi_\ell(t)$ and $\psi_m(t)$, $\ell \neq m$, their inner product is

$$\int_a^b \psi_\ell(t)\psi_m^*(t)dt = \begin{cases} 0 & \ell \neq m \\ 1 & \ell = m \end{cases} \tag{4.8}$$

Such a set of functions is called *orthonormal* (i.e., orthogonal and normalized).

A finite-energy signal $x(t)$ defined in $[a, b]$ can be approximated by a series

$$\hat{x}(t) = \sum_k a_k \psi_k(t) \tag{4.9}$$

according to some error criterion. For instance, we could minimize the energy of the error function $\varepsilon(t) = x(t) - \hat{x}(t)$ or

$$\int_a^b |\varepsilon(t)|^2 dt = \int_a^b \left| x(t) - \sum_k a_k \psi_k(t) \right|^2 dt \tag{4.10}$$

The expansion can be finite or infinite, and may not approximate the signal point by point.

Fourier proposed sinusoids as the functions $\{\psi_k(t)\}$ to represent periodic signals, and solved the quadratic minimization posed in Equation (4.10) to obtain the coefficients of the representation. For most signals, the resulting Fourier series has an infinite number of terms and coincides with the signal pointwise. We will start with a more general expansion that uses complex exponentials and from it obtain the sinusoidal form. In Chapter 5 we extend the Fourier series to represent aperiodic signals—leading to the Fourier transform that is in turn connected with the Laplace transform.

Recall that a periodic signal $x(t)$ is such that

■   It is defined for $-\infty < t < \infty$ (i.e., it has an infinite support).
■   For any integer k, $x(t + kT_0) = x(t)$, where T_0 is the fundamental period of the signal or the smallest positive real number that makes this possible.

The *Fourier series representation* of a periodic signal $x(t)$, of period T_0, is given by an infinite sum of weighted complex exponentials (cosines and sines) with frequencies multiples of the signal's fundamental frequency $\Omega_0 = 2\pi/T_0$ rad/sec, or

$$x(t) = \sum_{k=-\infty}^{\infty} X_k e^{jk\Omega_0 t} \qquad \Omega_0 = \frac{2\pi}{T_0} \tag{4.11}$$

where the Fourier coefficients X_k are found according to

$$X_k = \frac{1}{T_0} \int_{t_0}^{t_0+T_0} x(t)e^{-jk\Omega_0 t}dt \tag{4.12}$$

for $k = 0, \pm 1, \pm 2, \ldots$, and any t_0. The form of Equation (4.12) indicates that the information needed for the Fourier series can be obtained from any period of $x(t)$.

Remarks

- *The Fourier series uses the Fourier basis $\{e^{jk\Omega_0 t}, \ k \text{ integer}\}$ to represent the periodic signal $x(t)$ of period T_0. The Fourier basis functions are also periodic of period T_0 (i.e., for an integer m,*

$$e^{jk\Omega_0(t+mT_0)} = e^{jk\Omega_0 t}e^{jkm2\pi} = e^{jk\Omega_0 t}$$

as $e^{jkm2\pi} = 1$).
- *The Fourier basis functions are* orthonormal *over a period—that is,*

$$\frac{1}{T_0} \int_{t_0}^{t_0+T_0} e^{jk\Omega_0 t}[e^{j\ell\Omega_0 t}]^* dt = \begin{cases} 1 & k = \ell \\ 0 & k \neq \ell \end{cases} \tag{4.13}$$

That is, $e^{jk\Omega_0 t}$ and $e^{j\ell\Omega_0 t}$ are said to be orthogonal *when for $k \neq \ell$ the above integral is zero, and they are* normal *(or normalized) when for $k = \ell$ the above integral is unity. The functions $e^{jk\Omega_0 t}$ and $e^{j\ell\Omega_0 t}$ are orthogonal since*

$$\frac{1}{T_0} \int_{t_0}^{t_0+T_0} e^{jk\Omega_0 t}[e^{j\ell\Omega_0 t}]^* \, dt = \frac{1}{T_0} \int_{t_0}^{t_0+T_0} e^{j(k-\ell)\Omega_0 t}dt$$

$$= \frac{1}{T_0} \int_{t_0}^{t_0+T_0} [\cos((k-\ell)\Omega_0 t) + j\sin((k-\ell)\Omega_0 t)] \, dt$$

$$= 0 \quad k \neq \ell$$

The above integrals are zero given that the integrands are sinusoids and the limits of the integrals cover one or more periods of the integrands. The normality of the Fourier functions is easily shown when for $k = \ell$ the above integral is

$$\frac{1}{T_0} \int_{t_0}^{t_0+T_0} e^{j0t}dt = 1$$

■ *The Fourier coefficients $\{X_k\}$ are easily obtained using the orthonormality of the Fourier functions: First, we multiply the expression for $x(t)$ in Equation (4.11) by $e^{-j\ell\Omega_0 t}$ and then integrate over a period to get*

$$\int_{T_0} x(t)e^{-j\ell\Omega_0 t}\,dt = \sum_k X_k \int_{T_0} e^{j(k-\ell)\Omega_0 t}\,dt$$

$$= \sum_k X_k T_0 \delta(k-\ell)$$

$$= X_\ell T_0$$

given that when $k = \ell$, then $\int_{T_0} e^{j(k-\ell)\Omega_0 t}\,dt = T_0$; otherwise it is zero according to the orthogonality of the Fourier exponentials. This then gives us the expression for the Fourier coefficients $\{X_\ell\}$ in Equation (4.12). You need to recognize that the k and ℓ are dummy variables in the Fourier series, and as such the expression for the coefficients is the same regardless of whether we use ℓ or k.

■ *It is important to realize from the given Fourier series equations that for a periodic signal $x(t)$, of period T_0, any period*

$$x(t), \quad t_0 \le t \le t_0 + T_0$$

provides all the necessary information in the time-domain characterizing $x(t)$. In an equivalent way the coefficients and their corresponding frequencies $\{X_k, k\Omega_0\}$ provide all the necessary information about $x(t)$ in the frequency domain.

4.4 LINE SPECTRA

The Fourier series provides a way to determine the frequency components of a periodic signal and the significance of these frequency components. Such information is provided by the power spectrum of the signal. For periodic signals, the power spectrum provides information as to how the power of the signal is distributed over the different frequencies present in the signal. We thus learn not only what frequency components are present in the signal but also the strength of these frequency components. In practice, the power spectrum can be computed and displayed using a spectrum analyzer, which will be described in Chapter 5.

4.4.1 Parseval's Theorem—Power Distribution over Frequency

Although periodic signals are infinite-energy signals, they have finite power. The Fourier series provides a way to find how much of the signal power is in a certain band of frequencies.

The power P_x of a periodic signal $x(t)$, of period T_0, can be equivalently calculated in either the time or the frequency domain:

$$P_x = \frac{1}{T_0} \int_{T_0} |x(t)|^2\,dt = \sum_k |X_k|^2 \tag{4.14}$$

The power of a periodic signal $x(t)$ of period T_0 is given by

$$P_x = \frac{1}{T_0} \int_{T_0} |x(t)|^2 dt$$

Replacing the Fourier series of $x(t)$ in the power equation we have that

$$\frac{1}{T_0} \int_{T_0} |x(t)|^2 dt = \frac{1}{T_0} \int_{T_0} \sum_k \sum_m X_k X_m^* e^{j\Omega_0 kt} e^{-j\Omega_0 mt} dt$$

$$= \sum_k \sum_m X_k X_m^* \frac{1}{T_0} \int_{T_0} e^{j\Omega_0 kt} e^{-j\Omega_0 mt} dt$$

$$= \sum_k |X_k|^2$$

after we apply the orthonormality of the Fourier exponentials. Even though $x(t)$ is real, we let $|x(t)|^2 = x(t)x^*(t)$ in the above equations, permitting us to express them in terms of X_k and its conjugate. The above indicates that the power of $x(t)$ can be computed in either the time or the frequency domain giving exactly the same result.

Moreover, considering the signal to be a sum of harmonically related components or

$$x(t) = \sum_k X_k e^{jk\Omega_0 t} = \sum_k x_k(t)$$

the power of each of these components is given by

$$\frac{1}{T_0} \int_{T_0} |x_k(t)|^2 dt = \frac{1}{T_0} \int_{T_0} |X_k e^{jk\Omega_0 t}|^2 dt = \frac{1}{T_0} \int_{T_0} |X_k|^2 dt = |X_k|^2$$

and the power of $x(t)$ is the sum of the powers of the Fourier series components. This indicates that the power of the signal is distributed over the harmonic frequencies $\{k\Omega_0\}$. A plot of $|X_k|^2$ versus the harmonic frequencies $k\Omega_0$, $k = 0, \pm1, \pm2, \ldots$, displays how the power of the signal is distributed over the harmonic frequencies. Given the discrete nature of the harmonic frequencies $\{k\Omega_0\}$ this plot consists of a line at each frequency and as such it is called the *power line spectrum* (that is, a periodic signal has no power in nonharmonic frequencies). Since $\{X_k\}$ are complex, we define two additional spectra, one that displays the magnitude $|X_k|$ versus $k\Omega_0$, called the *magnitude line spectrum*, and the *phase line spectrum* or $\angle X_k$ versus $k\Omega_0$ showing the phase of the coefficients $\{X_k\}$ for $k\Omega_0$. The power line spectrum is simply the magnitude spectrum squared.

A periodic signal $x(t)$, of period T_0, is represented in the frequency by its

$$\text{Magnitude line spectrum:} \qquad |X_k| \ vs \ k\Omega_0 \qquad\qquad (4.15)$$

$$\text{Phase line spectrum:} \qquad \angle X_k \ vs \ k\Omega_0 \qquad\qquad (4.16)$$

The power line spectrum $|X_k|^2$ versus $k\Omega_0$ of $x(t)$ displays the distribution of the power of the signal over frequency.

4.4.2 Symmetry of Line Spectra

For a real-valued periodic signal $x(t)$, of period T_0, represented in the frequency domain by the Fourier coefficients $\{X_k = |X_k|e^{j\angle X_k}\}$ at harmonic frequencies $\{k\Omega_0 = 2\pi k/T_0\}$, we have that

$$X_k = X_{-k}^*$$ (4.17)

or equivalently that

1. $|X_k| = |X_{-k}|$ (i.e., magnitude $|X_k|$ is even function of $k\Omega_0$)
2. $\angle X_k = -\angle X_{-k}$ (i.e., phase $\angle X_k$ is odd function of $k\Omega_0$) (4.18)

Thus, for real-valued signals we only need to display for $k \geq 0$ the
    Magnitude line spectrum: Plot of $|X_k|$ versus $k\Omega_0$
    Phase line spectrum: Plot of $\angle X_k$ versus $k\Omega_0$

For a real signal $x(t)$, the Fourier series of its complex conjugate $x^*(t)$ is

$$x^*(t) = \left[\sum_\ell X_\ell e^{j\ell\Omega_0 t} \right]^*$$

$$= \sum_\ell X_\ell^* e^{-j\ell\Omega_0 t} = \sum_k X_{-k}^* e^{jk\Omega_0 t}$$

Since $x(t) = x^*(t)$, the above equation is equal to

$$x(t) = \sum_k X_k e^{jk\Omega_0 t}$$

Comparing the Fourier series coefficients in the expressions, we have that $X_{-k}^* = X_k$, which means that if $X_k = |X_k|e^{j\angle X_k}$, then

$$|X_k| = |X_{-k}|$$

$$\angle X_k = -\angle X_{-k}$$

or that the magnitude is an even function of k, while the phase is an odd function of k. Thus, the line spectra corresponding to real-valued signals is given for only positive harmonic frequencies, with the understanding that for negative values of the harmonic frequencies the magnitude line spectrum is even and the phase line spectrum is odd.

4.5 TRIGONOMETRIC FOURIER SERIES

The *trigonometric Fourier series* of a real-valued, periodic signal $x(t)$, of period T_0, is an equivalent representation that uses sinusoids rather than complex exponentials as the basis functions. It is given by

$$x(t) = X_0 + 2 \sum_{k=1}^{\infty} |X_k| \cos(k\Omega_0 t + \theta_k)$$

$$= c_0 + 2 \sum_{k=1}^{\infty} [c_k \cos(k\Omega_0 t) + d_k \sin(k\Omega_0 t)] \qquad \Omega_0 = \frac{2\pi}{T_0} \qquad (4.19)$$

where $X_0 = c_0$ is called the *DC component*, and $\{2|X_k| \cos(k\Omega_0 t + \theta_k)\}$ are the kth *harmonics* for $k = 1, 2 \ldots$. The frequencies $\{k\Omega_0\}$ are said to be harmonically related. The coefficients $\{c_k, d_k\}$ are obtained from $x(t)$ as follows:

$$c_k = \frac{1}{T_0} \int_{t_0}^{t_0+T_0} x(t) \cos(k\Omega_0 t)\, dt \qquad k = 0, 1, \ldots$$

$$d_k = \frac{1}{T_0} \int_{t_0}^{t_0+T_0} x(t) \sin(k\Omega_0 t)\, dt \qquad k = 1, 2, \ldots \qquad (4.20)$$

The coefficients $X_k = |X_k| e^{j\theta_k}$ are connected with the coefficients c_k and d_k by

$$|X_k| = \sqrt{c_k^2 + d_k^2}$$

$$\theta_k = -\tan^{-1}\left[\frac{d_k}{c_k}\right]$$

The functions $\{\cos(k\Omega_0 t),\ \sin(k\Omega_0 t)\}$ are orthonormal.

Using the relation $X_k = X_{-k}^*$, obtained in the previous section, we express the exponential Fourier series of a real-valued periodic signal $x(t)$ as

$$x(t) = X_0 + \sum_{k=1}^{\infty} [X_k e^{jk\Omega_0 t} + X_{-k} e^{-jk\Omega_0 t}]$$

$$= X_0 + \sum_{k=1}^{\infty} \left[|X_k| e^{j(k\Omega_0 t + \theta_k)} + |X_k| e^{-j(k\Omega_0 t + \theta_k)} \right]$$

$$= X_0 + 2 \sum_{k=1}^{\infty} |X_k| \cos(k\Omega_0 t + \theta_k)$$

which is the top equation in Equation (4.19).

Let us then show how the coefficients c_k and d_k can be obtained directly from the signal. Using the relation $X_k = X_{-k}^*$ and the fact that for a complex number $z = a + jb$, then $z + z^* = (a + jb) + (a - jb) = 2a = 2Re(z)$, we have that

$$x(t) = X_0 + \sum_{k=1}^{\infty} [X_k e^{jk\Omega_0 t} + X_{-k} e^{-jk\Omega_0 t}]$$

$$= X_0 + \sum_{k=1}^{\infty} [X_k e^{jk\Omega_0 t} + X_k^* e^{-jk\Omega_0 t}]$$

$$= X_0 + \sum_{k=1}^{\infty} 2Re[X_k e^{jk\Omega_0 t}]$$

Since X_k is complex (verify this!),

$$2Re[X_k e^{jk\Omega_0 t}] = 2Re[X_k] \cos(k\Omega_0 t) - 2Im[X_k] \sin(k\Omega_0 t)$$

Now, if we let

$$c_k = Re[X_k] = \frac{1}{T_0} \int_{t_0}^{t_0+T_0} x(t) \cos(k\Omega_0 t) \, dt \qquad k = 1, 2, \ldots$$

$$d_k = -Im[X_k] = \frac{1}{T_0} \int_{t_0}^{t_0+T_0} x(t) \sin(k\Omega_0 t) \, dt \qquad k = 1, 2, \ldots$$

we then have

$$x(t) = X_0 + \sum_{k=1}^{\infty} \left(2Re[X_k] \cos(k\Omega_0 t) - 2Im[X_k] \sin(k\Omega_0 t) \right)$$

$$= X_0 + 2 \sum_{k=1}^{\infty} \left(c_k \cos(k\Omega_0 t) + d_k \sin(k\Omega_0 t) \right)$$

and since the average $X_0 = c_0$ we obtain the second form of the trigonometric Fourier series shown in Equation (4.19). Notice that $d_0 = 0$ and so it is not necessary to define it.

The coefficients $X_k = |X_k| e^{j\theta_k}$ are connected with the coefficients c_k and d_k by

$$|X_k| = \sqrt{c_k^2 + d_k^2}$$

$$\theta_k = -\tan^{-1} \left[\frac{d_k}{c_k} \right]$$

This can be shown by adding the phasors corresponding to $c_k \cos(k\Omega_0 t)$ and $d_k \sin(k\Omega_0 t)$ and finding the magnitude and phase of the resulting phasor.

Finally, since the exponential basis $\{e^{jk\Omega_0 t}\} = \{\cos(k\Omega_0 t) + j\sin(k\Omega_0 t)\}$, the sinusoidal bases $\cos(k\Omega_0 t)$ and $\sin(k\Omega_0 t)$ just like the exponential basis are periodic, of period T_0, and orthonormal.

■ Example 4.4

Find the Fourier series of a raised-cosine signal $(B \geq A)$,

$$x(t) = B + A\cos(\Omega_0 t + \theta)$$

which is periodic of period T_0 and fundamental frequency $\Omega_0 = 2\pi/T_0$. Call $y(t) = B + A\cos(\Omega_0 t - \pi/2)$. Find its Fourier series coefficients and compare them to those for $x(t)$. Use symbolic MATLAB to compute the Fourier series of $y(t) = 1 + \sin(100t)$. Find and plot its magnitude and phase line spectra.

Solution

In this case we do not need to compute the Fourier coefficients since $x(t)$ is already in the trigonometric form. From Equation (4.19) its dc value is B, and A is the coefficient of the first harmonic in the trigonometric Fourier series, so that $X_0 = B$, $|X_1| = A/2$, and $\angle X_1 = \theta$. Likewise, using Euler's identity we obtain that

$$x(t) = B + \frac{A}{2}\left[e^{j(\Omega_0 t + \theta)} + e^{-j(\Omega_0 t + \theta)}\right]$$

$$= B + \frac{Ae^{j\theta}}{2}e^{j\Omega_0 t} + \frac{Ae^{-j\theta}}{2}e^{-j\Omega_0 t}$$

which gives

$$X_0 = B$$

$$X_1 = \frac{Ae^{j\theta}}{2}$$

$$X_{-1} = X_1^*$$

If we let $\theta = -\pi/2$ in $x(t)$, we get

$$y(t) = B + A\sin(\Omega_0 t)$$

Its Fourier series coefficients are $Y_0 = B$ and $Y_1 = Ae^{-j\pi/2}/2$ so that $|Y_1| = |Y_{-1}| = A/2$ and $\angle Y_1 = -\angle Y_{-1} = -\pi/2$. The magnitude and phase line spectra of the raised cosine $(\theta = 0)$ and of the raised sine $(\theta = -\pi/2)$ are shown in Figure 4.2. For both $x(t)$ and $y(t)$ there are only two frequencies—the dc frequency and Ω_0—and as such the power of the signal is concentrated at those two frequencies as shown in Figure 4.2. The difference between the line spectra of $x(t)$ and $y(t)$ is in the phase.

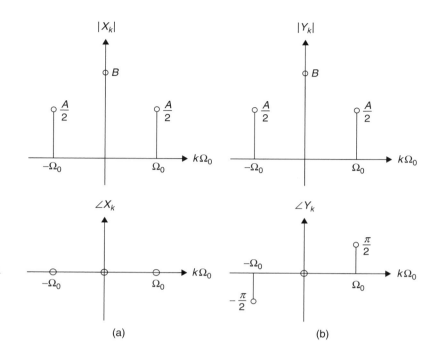

FIGURE 4.2
(a) Magnitude (top left) and phase (bottom left) line spectra of raised cosine and (b) magnitude (top right) and phase (bottom right) line spectra of raised sine.

(a)   (b)

Using symbolic MATLAB integration we can easily find the Fourier series coefficients, and the corresponding magnitude and phase are then plotted using stem to obtain the line spectra. Using our MATLAB function fourierseries the magnitude and phase of the line spectrum corresponding to the periodic raised sine $y(t) = 1 + \sin(100t)$ is shown in Figure 4.3.

```
function [X, w] = fourierseries(x, T0, N)
%%%%%
% symbolic Fourier Series computation
% x: periodic signal
% T0: period
% N: number of harmonics
% X,w: Fourier series coefficients at harmonic frequencies
%%%%%
syms t
% computation of N Fourier series coefficients
for k = 1:N,
    X1(k) = int(x * exp(-j * 2 * pi * (k - 1) * t/T0), t, 0, T0)/T0;
    X(k) = subs(X1(k));
    w(k) = (k-1) * 2 * pi/T0; % harmonic frequencies
end
```

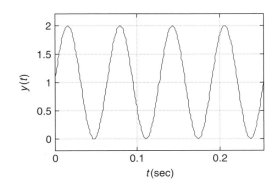

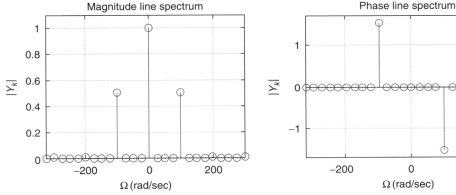

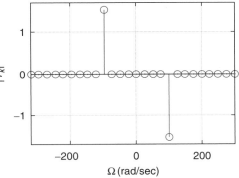

FIGURE 4.3

Line spectra of Fourier series of $y(t) = 1 + \sin(100t)$ (top figure). Notice the even and the odd symmetries of the magnitude and the phase spectra. The phase is $-\pi/2$ at $\Omega = 100$ rad/sec. ■

Remarks *Just because a signal is a sum of sinusoids, which are always periodic, is not enough for it to have a Fourier series. The signal should be periodic. The signal $x(t) = \cos(t) - \sin(\pi t)$ has components with periods $T_1 = 2\pi$ and $T_2 = 2$ so that the ratio $T_1/T_2 = \pi$ is not a rational number. Thus, $x(t)$ is not periodic and no Fourier series for it is possible.*

4.6 FOURIER COEFFICIENTS FROM LAPLACE

The computation of the X_k coefficients (see Eq. 4.12) requires integration that for some signals can be rather complicated. The integration can be avoided whenever we know the Laplace transform of a period of the signal as we will show. In general, the Laplace transform of a period of the signal exists over the whole s-plane, given that it is a finite-support signal. In some cases, the dc coefficient cannot be computed with the Laplace transform, but the dc term is easy to compute directly.

For a periodic signal $x(t)$, of period T_0, if we know or can easily compute the Laplace transform of a period of $x(t)$,

$$x_1(t) = x(t)[u(t_0) - u(t - t_0 - T_0)] \qquad \text{for any } t_0$$

Then the Fourier coefficients of $x(t)$ are given by

$$X_k = \frac{1}{T_0} \mathcal{L}[x_1(t)]_{s=jk\Omega_0} \qquad \Omega_0 = \frac{2\pi}{T_0} \text{ fundamental frequency} \qquad (4.21)$$

This can be seen by comparing the equation for the X_k coefficients with the Laplace transform of a period $x_1(t) = x(t)[u(t_0) - u(t - t_0 - T_0)]$ of $x(t)$. Indeed, we have that

$$X_k = \frac{1}{T_0} \int_{t_0}^{t_0+T_0} x(t)e^{-jk\Omega_0 t}dt$$

$$= \frac{1}{T_0} \int_{t_0}^{t_0+T_0} x(t)e^{-st}dt \,\big|_{s=jk\Omega_0}$$

$$= \frac{1}{T_0} \mathcal{L}[x_1(t)]_{s=jk\Omega_0}$$

subbed s for jkΩ₀

$$\int x(t)e^{-st}dt$$

$$\mathcal{L}[x(t)]_{s=jkΩ_0}$$

■ Example 4.5

Consider the periodic pulse train $x(t)$, of period $T_0 = 1$, shown in Figure 4.4. Find its Fourier series.

Solution

Before finding the Fourier coefficients, we see that this signal has a dc component of 1, and that $x(t) - 1$ (zero-average signal) is well represented by cosines, given its even symmetry, and as such

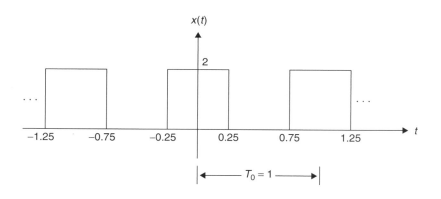

FIGURE 4.4

Train of rectangular pulses.

the Fourier coefficients will be real. Doing this analysis before the computations is important so we know what to expect.

The Fourier coefficients are obtained directly using their integral formulas or from the Laplace transform of a period. Since $T_0 = 1$, the fundamental frequency of $x(t)$ is $\Omega_0 = 2\pi$ rad/sec. Using the integral expression for the Fourier coefficients we have

remember

$$x_h = \frac{1}{T_0}\int_{t_0}^{t_0+T_0} x(t)e^{-j\Omega_0 kt}\,dt$$

$$X_k = \frac{1}{T_0}\int_{-1/4}^{3/4} x(t)e^{-j\Omega_0 kt}\,dt = \int_{-1/4}^{1/4} 2e^{-j2\pi kt}\,dt \quad \leftarrow \text{integrates to}$$

this don't forget this prop

thus $T_0 = \frac{1}{2}$ plugген in gives

$$= \frac{2}{\pi k}\left[\frac{e^{j\pi k/2} - e^{-j\pi k/2}}{2j}\right] = \frac{\sin(\pi k/2)}{(\pi k/2)}$$

$\sin x = \dfrac{e^{x} - e^{-x}}{2j}$

and dont forget to distribute the "$-$"

2 ...

which are real as we predicted. The Fourier series is then

$$x(t) = \sum_{k=-\infty}^{\infty} \frac{\sin(\pi k/2)}{(\pi k/2)} e^{jk2\pi t}$$

To find the Fourier coefficients with the Laplace transform, let the period be $x_1(t) = x(t)$ for $-0.5 \le t \le 0.5$. Delaying it by 0.25 we get $x_1(t - 0.25) = 2[u(t) - u(t - 0.5)]$ with a Laplace transform

$$e^{-0.25s}X_1(s) = \frac{2}{s}(1 - e^{-0.5s})$$

so that $X_1(s) = (2/s)[e^{0.25s} - e^{-0.25s}]$, and therefore

$$X_k = \frac{1}{T_0}\mathcal{L}[x_1(t)]\Big|_{s=jk\Omega_0}$$

$$= \frac{2}{jk\Omega_0 T_0}2j\sin(k\Omega_0/4)$$

and for $\Omega_0 = 2\pi$, $T_0 = 1$, we get

$$X_k = \frac{\sin(\pi k/2)}{\pi k/2} \qquad k \ne 0$$

Since the above equation gives zero over zero when $k = 0$ (i.e., it is undefined), the dc value is found from the integral formula as

$$X_0 = \int_{-1/4}^{1/4} 2\,dt = 1$$

These Fourier coefficients coincide with the ones found before.

The following script is used to find the Fourier coefficients with our function fourierseries and to plot the magnitude and phase line spectra.

```
%%%%%%%%%%%%%%%%%
% Example 4.5---Fourier series of train of pulses
%%%%%%%%%%%%%%%%%%
clear all;clf
syms t
T0 = 1; m = heaviside(t) − heaviside(t − T0/4) + heaviside(t − 3 * T0/4);x = 2 * m
[X,w] = fourierseries(x,T0,20);
subplot(221); ezplot(x,[0 T0]); grid
subplot(223); stem(w,abs(X))
subplot(224); stem(w,angle(X))
```

Notice that in this case:

1. The X_k Fourier coefficients of the train of pulses are given in terms of the $\sin(x)/x$ or the sinc function. This function was presented in Chapter 1. Recall that the sinc is
   - Even—that is, $\sin(x)/x = \sin(-x)/(-x)$.
   - The value at $x = 0$ is found by means of L'Hôpital's rule because the numerator and the denominator of sinc are zero for $x = 0$, so

$$\lim_{x \to 0} \frac{\sin(x)}{x} = \lim_{x \to 0} \frac{d\sin(x)/dx}{dx/dx} = 1$$

   - It is bounded, indeed

$$\frac{-1}{x} \leq \frac{\sin(x)}{x} \leq \frac{1}{x}$$

2. Since the dc component of $x(t)$ is 1, once it is subtracted it is clear that the rest of the series can be represented as a sum of cosines:

$$x(t) = 1 + \sum_{k=-\infty, k \neq 0}^{\infty} \frac{\sin(\pi k/2)}{(\pi k/2)} e^{jk2\pi t}$$

$$= 1 + 2 \sum_{k=1}^{\infty} \frac{\sin(\pi k/2)}{(\pi k/2)} \cos(2\pi kt)$$

   This can also be seen by considering the trigonometric Fourier series of $x(t)$. Since $x(t)\sin(k\Omega_0 t)$ is odd, as $x(t)$ is even and $\sin(k\Omega_0 t)$ is odd, then the coefficients corresponding to the sines in the expansion will be zero. On the other hand, $x(t)\cos(k\Omega_0 t)$ is even and gives nonzero Fourier coefficients. See Equations (4.20).

3. In general, the Fourier coefficients are complex and as such need to be represented by their magnitudes and phases. In this case, the X_k coefficients are real-valued, and in particular zero when $k\pi/2 = \pm m\pi$, m an integer, or when $k = \pm 2, \pm 4, \ldots$. Since the X_k values are real, the corresponding phase would be zero when $X_k \geq 0$, and $\pm \pi$ when $X_k < 0$. In Figure 4.5 we show a period of the signal, and the magnitude and phase line spectra displayed only for positive values of frequency (with the understanding that the magnitude spectrum is even and the phase is odd functions of the frequency).

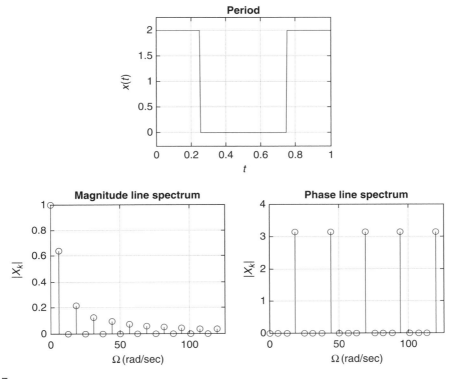

FIGURE 4.5

Period of train of rectangular pulses (top) and its magnitude and phase line spectra (bottom).

4. The X_k coefficients and its squares, related to the power line spectrum, are obtained using the fourierseries function (see Figure 4.5):

k	$X_k = X_{-k}$	X_k^2
0	1	1
1	0.64	0.41
2	0	0
3	−0.21	0.041
4	0	0
5	0.13	0.016
6	0	0
7	−0.09	0.008

Notice that about 11 of them (including the zero values), or the dc value and 5 harmonics, provide a very good approximation of the pulse train, and would occupy a bandwidth of approximately 10π rad/sec. The power contribution, as indicated by X_k^2 after $k = \pm 6$, is relatively small.   ∎

∎ Example 4.6

Find the Fourier series of the full-wave rectified signal $x(t) = |\cos(\pi t)|$ shown in Figure 4.6. This signal is used in the design of dc sources. The rectification of an ac signal is the first step in this design.

Solution

The integral to find the Fourier coefficients is

$$X_k = \int_{-0.5}^{0.5} \cos(\pi t)e^{-j2\pi kt}\,dt$$

which can be computed by using Euler's identity or any other method. We want to show that this can be avoided by using the Laplace transform.

A period $x_1(t)$ of $x(t)$ can be expressed as

$$x_1(t - 0.5) = \sin(\pi t)u(t) + \sin(\pi(t-1))u(t-1)$$

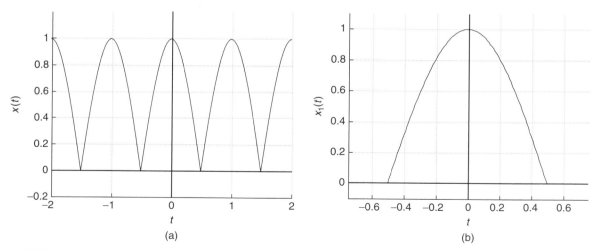

FIGURE 4.6
(a) Full-wave rectified signal $x(t)$ and (b) one of its periods $x_1(t)$.

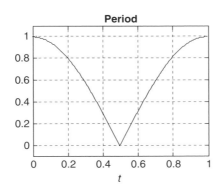

Period

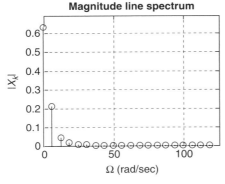

Magnitude line spectrum

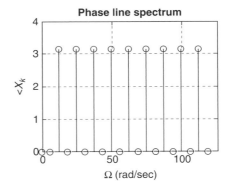

Phase line spectrum

FIGURE 4.7

Period of full-wave rectified signal $x(t)$ and its magnitude and phase line spectra.

and using the Laplace transform we have

$$X_1(s)e^{-0.5s} = \frac{\pi}{s^2 + \pi^2}[1 + e^{-s}]$$

so that

$$X_1(s) = \frac{\pi}{s^2 + \pi^2}[e^{0.5s} + e^{-0.5s}]$$

The Fourier coefficients are then

$$X_k = \frac{1}{T_0}X_1(s)|_{s=j\Omega_0 k}$$

where $T_0 = 1$ and $\Omega_0 = 2\pi$, giving

$$X_k = \frac{\pi}{(j2\pi k)^2 + \pi^2} 2\cos(2\pi k/2)$$

$$= \frac{2(-1)^k}{\pi(1 - 4k^2)}$$

Handwritten annotations:

$\cos(\theta) = \dfrac{e^{\theta} + e^{-\theta}}{2}$

$2\cos(\theta) = e^{\theta} + e^{-\theta}$

In this problem because of ratio...

$\overbrace{[e^{0.5s} + e^{-0.5s}]}$ Interval

$\underbrace{}_{2[t]}$

All multiplied by 2 do at end.

$X_K = \dfrac{\pi}{j^2 \cdot 4\pi^2 K^2 + \pi^2} = \dfrac{\pi(-1)^K}{\pi(4\pi K^2 + \pi)} = \dfrac{(-1)^K}{\pi(4K^2 + 1)} \cdot 2 = \dfrac{2(-1)^K}{\pi(1 - 4K^2)}$

since $\cos(\pi k) = (-1)^k$. The DC value of the full-wave rectified signal is $X_0 = 2/\pi$. Notice that the Fourier coefficients are real given that the signal is even.

The MATLAB script used in the previous example can be used again with the following modification for the generation of a period of $x(t)$. The results are shown in Figure 4.7.

```
%%%%%%%%%%%%%%%%%
% Example 4.6---Fourier series of full-wave rectified signal
%%%%%%%%%%%%%%%%%
% period generation
T0 = 1;
m = heaviside(t) − heaviside(t − T0);x = abs(cos(pi * t)) * m
```

■ Example 4.7

Computing the derivative of a signal enhances higher harmonics. To illustrate this consider the train of triangular pulses $y(t)$ (Figure 4.8) with fundamental period $T_0 = 2$. Let $x(t) = dy(t)/dt$. Find its Fourier series and compare $|X_k|$ with $|Y_k|$ to determine which of these signals is smoother—that is, which one has lower frequency components.

Solution

A period of $y(t)$, $-1 \le t \le 1$, is given by

$$y_1(t) = r(t+1) - 2r(t) + r(t-1)$$

with a Laplace transform

Func of time / of s

r(t) $\dfrac{1}{s^2}$

t+a $e^{-\alpha s}$

$$Y_1(s) = \frac{1}{s^2}\left[e^s - 2 + e^{-s}\right]$$

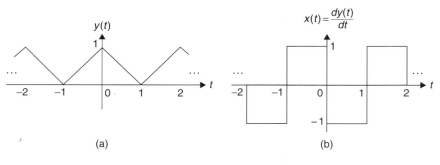

FIGURE 4.8
(a) Train of triangular pulses $y(t)$ and (b) its derivative $x(t)$. Notice that $y(t)$ is a continuous function while $x(t)$ is discontinuous.

so that the Fourier coefficients are given by ($T_0 = 2, \Omega_0 = \pi$):

$$Y_k = \frac{1}{T_0}Y_1(s)|_{s=j\Omega_0 k} = \frac{1}{2(j\pi k)^2}[2\cos(\pi k) - 2]$$

$$= \frac{1 - \cos(\pi k)}{\pi^2 k^2} = \frac{1 - (-1)^k}{\pi^2 k^2} \qquad k \neq 0$$

Handwritten margin notes:
$$\cos(\pi k) = (-1)^k$$
$$\frac{1}{2}(j\pi k)^2 \cdot 2[\cos(\pi k) - 1]$$
two cancel and denom j^2
turns to -1 thus
$$-\left(\frac{\cos(\pi k) - 1}{\pi^2 k^2}\right) = \frac{1 - \cos(\pi k)}{\pi^2 k^2}$$
$$= \frac{1 - (-1)^k}{\pi^2 k^2}$$

This is also equal to

$$Y_k = 0.5\left[\frac{\sin(\pi k/2)}{(\pi k/2)}\right]^2 \tag{4.22}$$

using the identity $1 - \cos(\pi k) = 2\sin^2(\pi k/2)$. By observing $y(t)$ we deduce that its DC value is $Y_0 = 0.5$.

Let us then consider the periodic signal $x(t) = dy(t)/dt$ (shown in Fig. 4.8(b)) with a dc value $X_0 = 0$. For $-1 \leq t \leq 1$, its period is $x_1(t) = u(t + 1) - 2u(t) + u(t - 1)$ and

$$X_1(s) = \frac{1}{s} \cdot [e^s - 2 + e^{-s}]$$

which gives the Fourier series coefficients ($T_0 = 2, \Omega_0 = \pi$, i.e., the period and the fundamental frequency are equal to the ones for $y(t)$)

$$X_k = \frac{\sin^2(k\pi/2)}{k\pi/2}j \tag{4.23}$$

since $X_k = \frac{1}{2}X_1(s)|_{s=j\pi k}$.

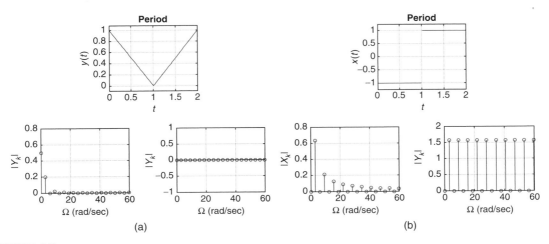

FIGURE 4.9

Magnitude and phase line spectra of (a) triangular signal $y(t)$ (top left) and (b) its derivative $x(t)$ (top right). Ignoring the dc values, the $\{|Y_k|\}$ decay faster to zero than the $\{|X_k|\}$, thus $y(t)$ is smoother than $x(t)$.

For $k \neq 0$ we have $|Y_k| = |X_k|/(\pi k)$, so that as k increases the frequency components of $y(t)$ decrease in magnitude faster than the corresponding ones of $x(t)$. Thus, $y(t)$ is smoother than $x(t)$. The magnitude line spectrum $|Y_k|$, ignoring its average, goes faster to zero than the magnitude line spectrum $|X_k|$, as seen in Figure 4.9.

Notice that in this case $y(t)$ is even and its Fourier coefficients Y_k are real, while $x(t)$ is odd and its Fourier coefficients X_k are purely imaginary. If we subtract the average of $y(t)$, the signal $y(t)$ can be clearly approximated as a series of cosines, thus the need for real coefficients in its complex exponential Fourier series. The signal $x(t)$ is zero-average and as such it can be clearly approximated by a series of sines requiring its Fourier coefficients X_k to be imaginary. ■

■ Example 4.8

Integration of a periodic signal, provided it has zero mean, gives a smoother signal. To see this, find and compare the magnitude line spectra of a sawtooth signal $x(t)$, of period $T_0 = 2$, and its integral

$$y(t) = \int x(t)dt$$

shown Figure 4.10.

Solution

Before doing any calculations it is important to realize that the integral would not exist if the dc is not zero. Using the following script we can compute the Fourier series coefficients of $x(t)$ and $y(t)$.

A period of $x(t)$ is

$$x_1(t) = tw(t) + (t - 2)w(t - 1) \quad 0 \leq t \leq 2$$

where $w(t) = u(t) - u(t - 1)$ is a rectangular window.

FIGURE 4.10
(a) Sawtooth signal $x(t)$
and (b) its integral $y(t)$.
Notice that $x(t)$ is a
discontinuous function
while $y(t)$ is continuous.

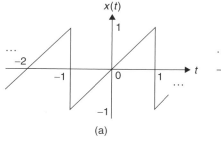

(a)

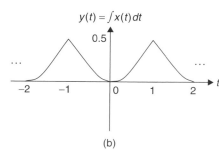

(b)

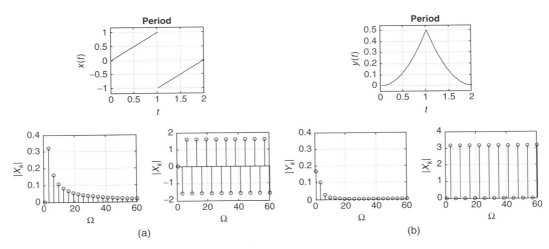

FIGURE 4.11

(a) Periods of the sawtooth signal $x(t)$ and (b) its integral $y(t)$ and their magnitude and phase line spectra.

```
%%%%%%%%%%%%%%%%%%
% Example 4.8---Saw-tooth signal and its integral
%%%%%%%%%%%%%%%%%%
syms t
T0 = 2;
m = heaviside(t) − heaviside(t − T0/2);
m1 = heaviside(t − T0/2) - heaviside(t − T0);
x = t * m + (t − 2) * m1;
y = int(x);
[X, w] = fourierseries(x, T0, 20);
[Y, w] = fourierseries(y, T0, 20);
```

The signal $y(t)$ is smoother than $x(t)$; $y(t)$ is a continuous function of time, while $x(t)$ is discontinuous. This is indicated as well by the magnitude line spectra of the two signals. Ignoring the dc components, the $\{|Y_k|\}$ of $y(t)$ decay a lot faster to zero than the $\{|X_k|\}$ of $x(t)$ (See Figure 4.11). As we will see in Section 4.10, computing the derivative of a periodic signal is equivalent to multiplying its Fourier series coefficients by $j\Omega_0 k$, which emphasizes the higher harmonics. If the periodic signal is zero-mean so that its integral exists, the Fourier coefficients of the integral can be found by dividing them by $j\Omega_0 k$ so that now the low harmonics are emphasized. ∎

4.7 CONVERGENCE OF THE FOURIER SERIES

It can be said, without overstating it, that any periodic signal of practical interest has a Fourier series. Only very strange signals would not have a converging Fourier series. Establishing convergence is necessary because the Fourier series has an infinite number of terms. To establish some general conditions under which the series converges, we need to classify signals with respect to their smoothness.

A signal $x(t)$ is said to be *piecewise smooth* if it has a finite number of discontinuities, while a *smooth* signal has a derivative that changes continuously. Thus, smooth signals can be considered special cases of piecewise smooth signals.

The Fourier series of a piecewise smooth (continuous or discontinuous) periodic signal $x(t)$ converges for all values of t. The mathematician Dirichlet showed that for the Fourier series to converge to the periodic signal $x(t)$, the signal should satisfy the following sufficient (not necessary) conditions over a period:

- Be absolutely integrable.
- Have a finite number of maxima, minima, and discontinuities.

The infinite series equals $x(t)$ at every continuity point and equals the average

$$0.5[x(t + 0+) + x(t + 0-)]$$

of the right limit $x(t + 0+)$ and the left limit $x(t + 0-)$ at every discontinuity point. If $x(t)$ is continuous everywhere, then the series converges absolutely and uniformly.

Although the Fourier series converges to the arithmetic average at discontinuities, it can be observed that there is some ringing before and after the discontinuity points. This is called the *Gibb's phenomenon*. To understand this phenomenon it is necessary to explain how the Fourier series can be seen as an approximation to the actual signal, and how when a signal has discontinuities the convergence is not uniform around them. It will become clear that the smoother the signal $x(t)$ is, the easier it is to approximate it with a Fourier series with a finite number of terms.

When the signal is continuous everywhere, the convergence is such that at each point t the series approximates the actual value $x(t)$ as we increase the number of terms in the approximation. However, that is not the case when discontinuities occur in the signal. This is despite the fact that a minimum mean-square approximation seems to indicate that the approximation could give a zero error. Let

$$x_N(t) = \sum_{k=-N}^{N} X_k e^{jk\Omega_0 t} \tag{4.24}$$

be the Nth-order approximation of a periodic signal $x(t)$, of fundamental frequency Ω_0, that minimizes the average quadratic error over a period

$$E_N = \frac{1}{T_0} \int_{T_0} |x(t) - x_N(t)|^2 dt \tag{4.25}$$

with respect to the Fourier coefficients X_k. To minimize E_N with respect to the coefficients X_k we set its derivative with respect to X_k to zero. Let $\varepsilon(t) = x(t) - x_N(t)$, so that

$$\frac{dE_N}{dX_k} = \frac{1}{T_0} \int_{T_0} 2\varepsilon(t) \frac{d\varepsilon^*(t)}{dX_k} dt$$

$$= -\frac{1}{T_0} \int_{T_0} 2[x(t) - x_N(t)] e^{-jk\Omega_0 t} dt$$

$$= 0$$

which after replacing $x_N(t)$ and using the orthogonality of the Fourier exponentials gives

$$X_k = \frac{1}{T_0} \int_{T_0} x(t) e^{-j\Omega_0 k t} dt \tag{4.26}$$

corresponding to the Fourier coefficients of $x(t)$ for $-N \leq k \leq N$. As $N \to \infty$ the average error $E_N \to 0$.

The only issue left is how $x_N(t)$ converges to $x(t)$. As indicated before, if $x(t)$ is smooth $x_N(t)$ approximates $x(t)$ at every point, but if there are discontinuities the approximation is in an average fashion. The Gibb's phenomenon indicates that around discontinuities there will be ringing, regardless of the order N of the approximation, even though the average quadratic error E_N goes to zero as N increases. This phenomenon will be explained in Chapter 5 as the effect of using a rectangular window to obtain a finite-frequency representation of a periodic signal.

■ Example 4.9

To illustrate the Gibb's phenomenon consider the approximation of a train of pulses $x(t)$ with zero mean and period $T_0 = 1$ (see the dashed signal in Figure 4.12) with a Fourier series $x_N(t)$ with $N = 1, \ldots, 20$.

Solution

We compute analytically the Fourier coefficients of $x(t)$ and use them to obtain an approximation $x_N(t)$ of $x(t)$ having a zero DC component and up to 20 harmonics. The dashed-line plot in Figure 4.12 is $x(t)$ and the solid–line plot is $x_N(t)$ when $N = 20$. The discontinuities of the pulse train cause the Gibb's phenomenon. Even if we increase the number of harmonics there is an overshoot in the approximation around the discontinuities.

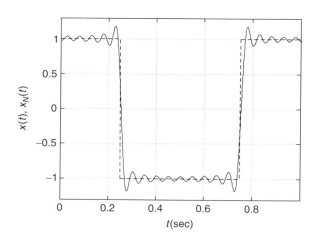

FIGURE 4.12

Approximate Fourier series $x_N(t)$ of the pulse train $x(t)$ (discontinuous) using the DC component and 20 harmonics. The approximate $x_N(t)$ displays the Gibb's phenomenon around the discontinuities.

```
%%%%%%%%%%%%%%%%%%
% Example 4.9---Simulation of Gibb's phenomenon
%%%%%%%%%%%%%%%%%%
clf; clear all
w0 = 2 * pi; DC = 0; N = 20;  % parameters of periodic signal
% computation of Fourier series coefficients
for k = 1:N,
  X(k) = sin(k * pi/2)/(k * pi/2);
end
X = [DC X]; % Fourier series coefficients
% computation of periodic signal
Ts = 0.001; t = 0:Ts:1 − Ts;
L = length(t); x = [ones(1, L/4) zeros(1, L/2) ones(1, L/4)]; x = x − 0.5;
% computation of approximate
xN = X(1)*ones(1,length(t));
for k = 2:N,
  xN = xN + 2 * X(k) * cos(2 * pi * (k − 1). * t); % approximate signal
  plot(t, xN); axis([0  max(t) 1.1 * min(xN) 1.1 * max(xN)])
  hold on; plot(t, x, 'r')
  ylabel('x(t), x_N(t)'); xlabel('t (sec)');grid
  hold off
  pause(0.1)
end
```

When you execute the above script, it pauses to display the approximation for an increasing number of terms in the approximation. At each of these values ringing around the discontinuities the Gibb's phenomenon is displayed. ∎

■ Example 4.10

Consider the mean-square error optimization to obtain an approximation of the periodic signal $x(t)$ shown in Figure 4.4 from Example 4.5. We wish to obtain an approximate $x_2(t) = \alpha + 2\beta \cos(\Omega_0 t)$, given that it is clear that $x(t)$ has an average, and that once we subtract it from the signal the resulting signal is approximated by a cosine function. Minimize the mean-square error

$$E_2 = \frac{1}{T_0} \int_{T_0} |x(t) - x_2(t)|^2 dt$$

with respect to α and β to find these values.

Solution

To minimize E_2 we set to zero its derivatives with respect to α and β to get

$$\frac{dE_2}{d\alpha} = -\frac{1}{T_0} \int_{T_0} 2[x(t) - \alpha - 2\beta \cos(\Omega_0 t)] dt = -\frac{1}{T_0} \int_{T_0} 2[x(t) - \alpha] dt = 0$$

$$\frac{dE_2}{d\beta} = -\frac{1}{T_0} \int_{T_0} 2[x(t) - \alpha - 2\beta \cos(\Omega_0 t)] \cos(\Omega_0 t) dt = 0$$

which, after getting rid of $\frac{2}{T_0}$ of both sides of the above equations and applying the orthogonality of the Fourier basis, gives

$$\alpha = \frac{1}{T_0} \int_{T_0} x(t) dt$$

$$\beta = \frac{1}{T_0} \int_{T_0} x(t) \cos(\Omega_0 t) dt$$

For the signal in Figure 4.4 we obtain

$$\alpha = 1$$

$$\beta = \frac{2}{\pi}$$

giving as approximation the signal

$$x_2(t) = 1 + \frac{4}{\pi} \cos(2\pi t)$$

which at $t = 0$ gives $x_2(0) = 2.27$ instead of the expected 2; $x_2(0.25) = 1$ (because of the discontinuity at this point, this value is the average of 2 and 0, the values, respectively, before and after the discontinuity) instead of 2 and $x_2(0.5) = -0.27$ instead of the expected 0.   ■

■ **Example 4.11**

Consider the train of pulses in Example 4.5. Determine how many Fourier coefficients are necessary to get a representation containing 97% of the power of the periodic signal.

Solution

The desired 97% of the power of $x(t)$ is

$$0.97 \frac{1}{T_0} \int_{T_0} x^2(t)dt = 0.97 \int_{-0.25}^{0.25} 4dt = 1.94$$

and so we need to find an integer N such that

$$\sum_{k=-N}^{N} |X_k|^2 = \sum_{k=-N}^{N} \left| \frac{\sin(\pi k/2)}{(\pi k/2)} \right|^2 = 1.94$$

The value of N is found by trial and error, adding consecutive values of the magnitude squared of Fourier coefficients. Using MATLAB, it is found that for $N = 5$ (dc and 5 harmonics) the Fourier series approximation has a power of 1.93. Thus, 11 Fourier coefficients give a very good approximation to the periodic train of pulses, with about 97% of the signal power.   ■

4.8 TIME AND FREQUENCY SHIFTING

Time shifting and frequency shifting are duals of each other.

■   Time-shifting: A periodic signal $x(t)$, of period T_0, remains periodic of the same period when shifted in time. If X_k are the Fourier coefficients of $x(t)$, the Fourier coefficients for $x(t - t_0)$ are

$$\left\{ X_k e^{-jk\Omega_0 t_0} = |X_k| e^{j(\angle X_k - k\Omega_0 t_0)} \right\} \tag{4.27}$$

That is, only a change in phase is caused by the time shift. The magnitude spectrum remains the same.

■   Frequency-shifting: When a periodic signal $x(t)$, of period T_0, modulates a complex exponential $e^{j\Omega_1 t}$:
- The modulated signal $x(t)e^{j\Omega_1 t}$ is periodic of period T_0 if $\Omega_1 = M\Omega_0$ for an integer $M \geq 1$.
- The Fourier coefficients X_k are shifted to frequencies $k\Omega_0 + \Omega_1$.
- The modulated signal is real-valued by multiplying $x(t)$ by $\cos(\Omega_1 t)$.

If we delay or advance in time a periodic signal, the resulting signal is periodic of the same period. Only a change in the phase of the coefficients occurs to accommodate for the shift. Indeed, if

$$x(t) = \sum_k X_k e^{jk\Omega_0 t}$$

we then have that

$$x(t - t_0) = \sum_k X_k e^{jk\Omega_0(t-t_0)} = \sum_k \left[X_k e^{-jk\Omega_0 t_0} \right] e^{jk\Omega_0 t}$$

$$x(t + t_0) = \sum_k X_k e^{jk\Omega_0(t+t_0)} = \sum_k \left[X_k e^{jk\Omega_0 t_0} \right] e^{jk\Omega_0 t}$$

so that the Fourier coefficients $\{X_k\}$ corresponding to $x(t)$ are changed to $\{X_k e^{\mp jk\Omega_0 t_0}\}$ for $x(t \mp t_0)$. In both cases, they have the same magnitude $|X_k|$ but different phases.

In a dual way, if we multiply the above periodic signal $x(t)$ by a complex exponential of frequency Ω_1, $e^{j\Omega_1 t}$, we obtain a so-called *modulated signal* $y(t)$ and its spectrum is shifted in frequency by Ω_1 with respect to the spectrum of the periodic signal $x(t)$. In fact,

$$y(t) = x(t)e^{j\Omega_1 t}$$
$$= \sum_k X_k e^{j(\Omega_0 k + \Omega_1)t}$$

indicating that the harmonic frequencies are shifted by Ω_1. The signal $y(t)$ is not necessarily periodic. Since T_0 is the period of $x(t)$, then

$$y(t + T_0) = x(t + T_0)e^{j\Omega_1(t+T_0)}$$

and for it to be equal to $y(t)$, then $\Omega_1 T_0 = 2\pi M$, for an integer $M \neq 0$ or

$$\Omega_1 = M\Omega_0 \qquad M >> 1$$

which goes along with the condition that the modulating frequency Ω_1 is chosen much larger than Ω_0. The modulated signal is then given by

$$y(t) = \sum_k X_k e^{j(\Omega_0 k + \Omega_1)t} = \sum_k X_k e^{j\Omega_0(k+M)t} = \sum_\ell X_{\ell-M} e^{j\Omega_0 \ell t}$$

so that the Fourier coefficients are shifted to new frequencies $\Omega_0(k + M)$.

To keep the modulated signal real-valued, one multiplies the periodic signal $x(t)$ by a cosine of frequency $\Omega_1 = M\Omega_0$ for $M >> 1$ to obtain a modulated signal

$$y_1(t) = x(t)\cos(\Omega_1 t)$$
$$= \sum_k 0.5 X_k [e^{j(k\Omega_0 + \Omega_1)t} + e^{j(k\Omega_0 - \Omega_1)t}]$$

so that the harmonic components are now centered around $\pm\Omega_1$.

■ Example 4.12

To illustrate the modulation property using MATLAB consider modulating a sinusoid $\cos(20\pi t)$ with a train of square pulses

$$x_1(t) = 0.5[1 + \text{sign}(\sin(\pi t))]$$

and with a sinusoid

$$x_2(t) = \cos(\pi t)$$

Use our function fourierseries to find the Fourier series of the modulated signals and plot their magnitude line spectra.

Solution

The function sign is defined as

$$\text{sign}(x(t)) = \begin{cases} -1 & x(t) < 0 \\ 1 & x(t) \geq 0 \end{cases} \tag{4.28}$$

That is, it determines the sign of the signal. Thus, $0.5[1 + \text{sign}(\sin(\pi t))] = u(t) - u(t-1)$ equals 1 for $0 \leq t \leq 1$, and 0 for $1 < t \leq 2$, which corresponds to a period of a train of square pulses.

The following script allows us to compute the Fourier coefficients of the two modulated signals.

```
%%%%%%%%%%%%%%%%%%
% Example 4.12---Modulation
%%%%%%%%%%%%%%%%%%%
syms t
T0 = 2;
m = heaviside(t) − heaviside(t − T0/2);
m1 = heaviside(t) − heaviside(t − T0);
x = m * cos(20 * pi * t);
x1 = m1 * cos(pi * t) * cos(20 * pi * t);
[X, w] = fourierseries(x, T0, 60);
[X1, w1] = fourierseries(x1, T0, 60);
```

The modulated signals and their corresponding magnitude line spectra are shown in Figure 4.13. The Fourier coefficients of the modulated signals are now clustered around the frequency 20π.

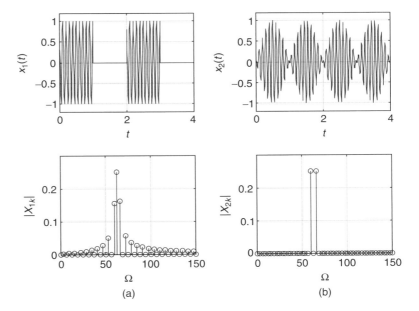

FIGURE 4.13
(a) Modulated square-wave $x_1(t) \cos(20\pi t)$ and (b) cosine $x_2(t) \cos(20\pi t)$ and their respective magnitude line spectra.

4.9 RESPONSE OF LTI SYSTEMS TO PERIODIC SIGNALS

The most important property of LTI systems is the eigenfunction property.

Eigenfunction property: In steady state, the response to a complex exponential (or a sinusoid) of a certain frequency is the same complex exponential (or sinusoid), but its amplitude and phase are affected by the frequency response of the system at that frequency.

Suppose that the impulse response of an LTI system is $h(t)$ and that $H(s) = \mathcal{L}[h(t)]$ is the corresponding transfer function. If the input to this system is a periodic signal $x(t)$, of period T_0, with Fourier series

$$x(t) = \sum_{k=-\infty}^{\infty} X_k e^{jk\Omega_0 t} \qquad \Omega_0 = \frac{2\pi}{T_0} \tag{4.29}$$

then according to the eigenfunction property the output in the steady state is

$$y_{ss}(t) = \sum_{k=-\infty}^{\infty} [X_k H(jk\Omega_0)] e^{jk\Omega_0 t} \tag{4.30}$$

If we call $Y_k = X_k H(jk\Omega_0)$ we have a Fourier series representation of $y_{ss}(t)$ with Y_k as its Fourier coefficients.

4.9.1 Sinusoidal Steady State

If the input of a stable and causal LTI system, with impulse response $h(t)$, is $x(t) = Ae^{j\Omega_0 t}$, the output is

$$y(t) = \int_0^\infty h(\tau)x(t-\tau)d\tau = Ae^{j\Omega_0 t} \int_0^\infty h(\tau)e^{-j\Omega_0 \tau}d\tau$$

$$= Ae^{j\Omega_0 t}H(j\Omega_0) = A|H(j\Omega_0)|e^{j\Omega_0 t + \angle H(j\Omega_0)} \tag{4.31}$$

The limits of the first integral indicate that the system is causal (the $h(\tau) = 0$ for $\tau < 0$) and that the input $x(t-\tau)$ is applied from $-\infty$ (when $\tau = \infty$) to t (when $\tau = 0$); thus $y(t)$ is the steady-state response of the system. If the input is a sinusoid—for example,

$$x_1(t) = \mathcal{Re}[x(t) = Ae^{j\Omega_0 t}] = A\cos(\Omega_0 t) \tag{4.32}$$

then the corresponding steady-state response is

$$y_1(t) = \mathcal{Re}[A|H(j\Omega_0)|e^{j\Omega_0 t + \angle H(j\Omega_0)}]$$

$$= A|H(j\Omega_0)|\cos(\Omega_0 t + \angle H(j\Omega_0)). \tag{4.33}$$

As in the eigenfunction property, the frequency of the output coincides with the frequency of the input, however, the magnitude and the phase of the input signal is changed by the response of the system at the input frequency.

The following script simulates the convolution of a sinusoid $x(t)$ of frequency $\Omega = 20\pi$, amplitude 10, and random phase with the impulse response $h(t)$ (a modulated decaying exponential) of an LTI system. The convolution integral is approximated using the MATLAB function conv.

```
%%%%%%%%%%%%%%%%%%%
% Simulation of Convolution
%%%%%%%%%%%%%%%%%%%
clear all; clf
Ts = 0.01; Tend = 2; t = 0:Ts:Tend;
x = 10 * cos(20 * pi * t + pi * (rand(1, 1) − 0.5));    % input signal
h = 20 * exp(−10.*t). * cos(40 * pi * t);       % impulse response
% approximate convolution integral
y = Ts * conv(x, h);
```

```
M = length(x);
figure(1)
x1 = [zeros(1, 5) x(1:M)];
z = y(1); y1 = [zeros(1, 5) z zeros(1, M − 1)];
t0 = −5 * Ts:Ts:Tend;
for k = 0:M − 6,
   pause(0.05)
   h0 = fliplr(h);
   h1 = [h0(M - k - 5:M) zeros(1, M - k - 1)];
   subplot(211)
   plot(t0, h1, 'r')
   hold on
   plot(t0, x1, 'k')
   title('Convolution of x(t) and h(t)')
   ylabel('x(τ), h(t-τ)'); grid; axis([min(t0) max(t0) 1.1*min(x) 1.1*max(x)])
    hold off
   subplot(212)
   plot(t0, y1, 'b')
   ylabel('y(t) = (x * h)(t)'); grid; axis([min(t0) max(t0) 0.1 * min(x) 0.1 * max(x)])
   z = [z y(k + 2)];
   y1 = [zeros(1, 5) z zeros(1, M - length(z))];
end
```

Figure 4.14 displays the last step of the convolution integral simulation. Notice that the steady state is attained in a very short time (around $t = 0.5$ sec). The transient changes every time that the script is executed due to the random phase.

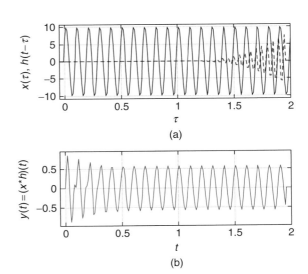

FIGURE 4.14

Convolution simulation: (a) input $x(\tau)$ (solid line) and $h(t − \tau)$ (dashed line), and (b) output $y(t)$: transient and steady-state response.

If the input $x(t)$ of a causal and stable LTI system, with impulse response $h(t)$, is periodic of period T_0 and has the Fourier series

$$x(t) = X_0 + 2 \sum_{k=1}^{\infty} |X_k| \cos(k\Omega_0 t + \angle X_k) \qquad \Omega_0 = \frac{2\pi}{T_0} \qquad (4.34)$$

the steady-state response of the system is

$$y(t) = X_0 |H(j0)| \cos(\angle H(j0)) + 2 \sum_{k=1}^{\infty} |X_k| |H(jk\Omega_0)| \cos(k\Omega_0 t + \angle X_k + \angle H(jk\Omega_0)) \qquad (4.35)$$

where

$$H(jk\Omega_0) = \int_0^{\infty} h(\tau) e^{-jk\Omega_0 \tau} d\tau \qquad (4.36)$$

is the *frequency response* of the system at $k\Omega_0$.

Remarks

- *If the input signal $x(t)$ is a combination of sinusoids of frequencies that are not harmonically related, the signal is not periodic, but the eigenfunction property still holds. For instance, if*

$$x(t) = \sum_k A_k \cos(\Omega_k t + \theta_k)$$

and the frequency response of the LTI system is $H(j\Omega)$, the steady-state response is

$$y(t) = \sum_k A_k |H(j\Omega_k)| \cos(\Omega_k t + \theta_k + \angle H(j\Omega_k))$$

- *It is important to realize that if the LTI system is represented by a differential equation and the input is a sinusoid, or combination of sinusoids, it is not necessary to use the Laplace transform to obtain the complete response and then let $t \to \infty$ to find the sinusoidal steady-state response. The Laplace transform is only needed to find the transfer function of the system, which can then be used in Equation (4.35) to find the sinusoidal steady state.*

4.9.2 Filtering of Periodic Signals

According to Equation (4.35) if we know the frequency response of the system (Eq. 4.36), at the harmonic frequencies of the periodic input, $H(jk\Omega_0)$, we have that in the steady state the output of the system $y(t)$ is as follows:

- Periodic of the same period as the input.
- Its Fourier coefficients are those of the input X_k multiplied by the frequency response at the harmonic frequencies, $H(jk\Omega_0)$.

■ Example 4.13

To illustrate the filtering of a periodic signal, consider a zero-mean pulse train

$$x(t) = \sum_{k=-\infty, \neq 0}^{\infty} \frac{\sin(k\pi/2)}{k\pi/2} e^{j2k\pi t}$$

as the driving source of an RC circuit that realizes a low-pass filter (i.e., a system that tries to keep the low-frequency harmonics and get rid of the high-frequency harmonics of the input). The transfer function of the RC low-pass filter is

$$H(s) = \frac{1}{1 + s/100}$$

Solution

The following script computes the frequency response of the filter at the harmonic frequencies $H(jk\Omega_0)$ (see Figure 4.15).

```
%%%%%%%%%%%%%%%%%
% Example 4.13
%%%%%%%%%%%%%%%%%
% Freq response of H(s)=1/(s/scale+1) -- low-pass filter
w0 = 2 * pi;    % fundamental frequency of input
M = 20; k = 0:M - 1; w1 = k. * w0;   % harmonic frequencies
H = 1./(1 + j * w1/100);  Hm = abs(H); Ha = angle(H); % frequency response
subplot(211)
stem(w1, Hm, 'filled'); grid; ylabel('—H(jω)—')
axis([0 max(w1) 0 1.3])
subplot(212)
stem(w1, Ha, 'filled'); grid
axis([0 max(w1) -1 0])
ylabel('¡H(j ω)'); xlabel('w (rad/sec)')
```

The response due to the pulse train can be found by finding the response to each of its Fourier series components and adding them. Approximating $x(t)$ using $N = 20$ harmonics by

$$x_N(t) = \sum_{k=-20, \neq 0}^{20} \frac{\sin(k\pi/2)}{k\pi/2} e^{j2k\pi t}$$

Then the output voltage across the capacitor is given in the steady state,

$$y_{ss}(t) = \sum_{k=-20, \neq 0}^{20} H(j2k\pi) \frac{\sin(k\pi/2)}{k\pi/2} e^{j2k\pi t}$$

Because the magnitude response of the low-pass filter changes very little in the range of frequencies of the input, the output signal is very much like the input (see Figure 4.15). The following script is used to find the response.

```
% low-pass filtering
% FS coefficients of input
X(1) = 0; % mean value
for k = 2:M - 1,
    X(k) = sin((k − 1) * pi/2)/((k − 1) * pi/2);
end
% periodic signal
Ts = 0.001; t1 = 0:Ts:1 - Ts;L = length(t1);
x1 = [ones(1, L /4) zeros(1, L /2) ones(1, L /4)]; x1 = x1 − 0.5; x = [x1 x1];
% output of filter
t = 0:Ts:2 − Ts;
y = X(1) * ones(1, length(t)) * Ha(1);
plot(t, y); axis([0 max(t)  − .6 .6])
for k = 2:M - 1,
    y = y + X(k) * Hm(k) * cos(w0 * (k − 1). * t + Ha(k));
    plot(t, y); axis([0 max(t)  − .6 .6]); hold on
    plot(t, x, 'r'); axis([0 max(t)  − 0.6 0.6]); grid
    ylabel('x(t), y(t)'); xlabel('t (sec)') ; hold off
    pause(0.1)
end
```

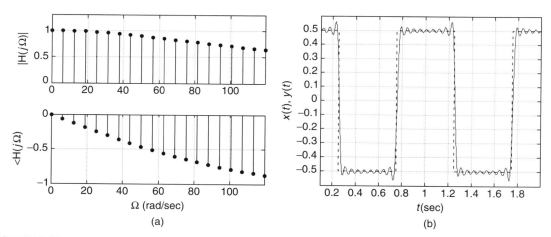

(a)

(b)

FIGURE 4.15

(a) Magnitude and phase response of the low-pass RC filter $H(s)$ at harmonic frequencies, and (b) response due to a train of pulses.

4.10 OTHER PROPERTIES OF THE FOURIER SERIES

In this section we present additional properties of the Fourier series that will help us with its computation and with our understanding of the relation between time and frequency. We are in particular interested in showing that even and odd signals have special representations, and that it is possible to find the Fourier series of the sum, product, derivative, and integral of periodic signals without the integration required by the definition of the series.

4.10.1 Reflection and Even and Odd Periodic Signals

If the Fourier series of $x(t)$, periodic with fundamental frequency Ω_0, is

$$x(t) = \sum_k X_k e^{jk\Omega_0 t}$$

then the one for its reflected version $x(-t)$ is

$$x(-t) = \sum_m X_m e^{-jm\Omega_0 t} = \sum_k X_{-k} e^{jk\Omega_0 t} \tag{4.37}$$

so that the Fourier coefficients of $x(-t)$ are X_{-k} (remember that m and k are just dummy variables). This can be used to simplify the computation of Fourier series of even and odd signals.

For an even signal $x(t)$, we have that $x(t) = x(-t)$, and as such $X_k = X_{-k}$ and therefore $x(t)$ is naturally represented in terms of cosines and a dc term. Indeed, its Fourier series is

$$x(t) = X_0 + \sum_{k=-\infty}^{-1} X_k e^{jk\Omega_0 t} + \sum_{k=1}^{\infty} X_k e^{jk\Omega_0 t}$$

$$= X_0 + \sum_{k=1}^{\infty} X_k [e^{jk\Omega_0 t} + e^{-jk\Omega_0 t}]$$

$$= X_0 + 2\sum_{k=1}^{\infty} X_k \cos(k\Omega_0 t) \tag{4.38}$$

indicating that X_k are real-valued. This is also seen from

$$X_k = \frac{1}{T_0} \int_{T_0} x(t) e^{-jk\Omega_0 t} dt = \frac{1}{T_0} \int_{T_0} x(t) [\cos(k\Omega_0 t) - j\sin(k\Omega_0)] dt$$

$$= \frac{1}{T_0} \int_{T_0} x(t) \cos(k\Omega_0 t) dt$$

because $x(t) \sin(k\Omega_0 t)$ is odd and their integral is zero. It will be similar for an odd function for which $x(t) = -x(-t)$, or $X_k = -X_{-k}$, in which case the Fourier series has a zero dc value and sine harmonics.

The X_k are purely imaginary. Indeed, for an odd $x(t)$,

$$X_k = \frac{1}{T_0} \int_{T_0} x(t)e^{-jk\Omega_0 t}dt = \frac{1}{T_0} \int_{T_0} x(t)[\cos(k\Omega_0 t) - j\sin(k\Omega_0)]dt$$

$$= \frac{-j}{T_0} \int_{T_0} x(t)\sin(k\Omega_0 t)dt$$

since $x(t)\cos(k\Omega_0 t)$ is odd. The Fourier series of an odd function can thus be written as

$$x(t) = 2\sum_{k=1}^{\infty}(jX_k)\sin(k\Omega_0 t) \tag{4.39}$$

According to the even and odd decomposition, any periodic signal $x(t)$ can be expressed as

$$x(t) = x_e(t) + x_o(t)$$

where $x_e(t)$ is the even and $x_o(t)$ is the odd component of $x(t)$. Finding the Fourier coefficients of $x_e(t)$, which will be real, and those of $x_o(t)$, which will be purely imaginary, we would then have $X_k = X_{ek} + X_{ok}$ since

$$x_e(t) = 0.5[x(t) + x(-t)] \quad \Rightarrow \quad X_{ek} = 0.5[X_k + X_{-k}]$$
$$x_o(t) = 0.5[x(t) - x(-t)] \quad \Rightarrow \quad X_{ok} = 0.5[X_k - X_{-k}] \tag{4.40}$$

- *Reflection*: If the Fourier coefficients of a periodic signal $x(t)$ are $\{X_k\}$ then those of $x(-t)$, the time-reversed signal with the same period as $x(t)$, are $\{X_{-k}\}$.
- *Even periodic signal* $x(t)$: Its Fourier coefficients X_k are real, and its trigonometric Fourier series is

$$x(t) = X_0 + 2\sum_{k=1}^{\infty}X_k\cos(k\Omega_0 t) \tag{4.41}$$

- *Odd periodic signal* $x(t)$: Its Fourier coefficients X_k are imaginary, and its trigonometric Fourier series is

$$x(t) = 2\sum_{k=1}^{\infty}jX_k\sin(k\Omega_0 t) \tag{4.42}$$

For any periodic signal $x(t) = x_e(t) + x_o(t)$ where $x_e(t)$ and $x_o(t)$ are the even and odd component of $x(t)$, then

$$X_k = X_{ek} + X_{ok} \tag{4.43}$$

where $\{X_{ek}\}$ are the Fourier coefficients of $x_e(t)$ and $\{X_{ok}\}$ are the Fourier coefficients of $x_o(t)$.

■ Example 4.14

Consider the periodic signals $x(t)$ and $y(t)$ shown in Figure 4.16. Determine their Fourier coefficients by using the symmetry conditions and the even–odd decomposition.

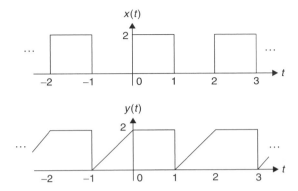

FIGURE 4.16

Nonsymmetric periodic signals.

Solution

The given signal $x(t)$ is neither even nor odd, but the advance signal $x(t + 0.5)$ is even with a period of $T_0 = 2$, $\Omega_0 = \pi$. Then between -1 and 1 the shifted period is

$$x_1(t + 0.5) = 2[u(t + 0.5) - u(t - 0.5)]$$

so that its Laplace transform is

$$X_1(s)e^{0.5s} = \frac{2}{s}\left[e^{0.5s} - e^{-0.5s}\right]$$

which gives the Fourier coefficients

$$X_k = \frac{1}{2}\frac{2}{jk\pi}\left[e^{jk\pi/2} - e^{-jk\pi/2}\right]e^{-jk\pi/2}$$

$$= \frac{1}{0.5\pi k}\sin(0.5\pi k)e^{-jk\pi/2}$$

after replacing s by $jk\Omega_o = jk\pi$ and dividing by the period $T_0 = 2$. These coefficients are complex as corresponding to a signal that is neither even nor odd. The dc coefficient is $X_0 = 1$.

The given signal $y(t)$ is neither even nor odd, and cannot be made even or odd by shifting. The even and odd components of a period of $y(t)$ are shown in Figure 4.17. The even and odd components of a period $y_1(t)$ between -1 and 1 are

$$y_{1e}(t) = \underbrace{[u(t + 1) - u(t - 1)]}_{\text{rectangular pulse}} + \underbrace{[r(t + 1) - 2r(t) + r(t - 1)]}_{\text{triangle}}$$

$$y_{1o}(t) = t[u(t + 1) - u(t - 1)] = [(t + 1)u(t + 1) - u(t + 1)] - [(t - 1)u(t - 1) + u(t - 1)]$$

$$= r(t + 1) - r(t - 1) - u(t + 1) - u(t - 1)$$

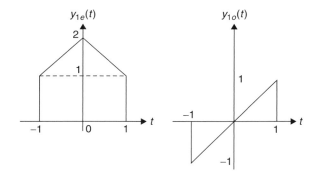

FIGURE 4.17
Even and odd components of the period of $y(t)$,
$-1 \leq t \leq 1$.

Thus, the mean value of $y_e(t)$ is the area under $y_{1e}(t)$ divided by 2 or 1.5, and for $k \neq 0$,

$$Y_{ek} = \frac{1}{T_0} Y_{1e}(s) \Big|_{s=jk\Omega_0} = \frac{1}{2}\left[\frac{1}{s}(e^s - e^{-s}) + \frac{1}{s^2}(e^s - 2 + e^{-s})\right]_{s=jk\pi}$$

$$= \frac{\sin(k\pi)}{\pi k} + \frac{1 - \cos(k\pi)}{(k\pi)^2} = 0 + \frac{1 - \cos(k\pi)}{(k\pi)^2} = \frac{1 - (-1)^k}{(k\pi)^2}$$

The mean value of $y_o(t)$ is zero, and for $k \neq 0$,

$$Y_{ok} = \frac{1}{T_0} Y_{1o}(s) \Big|_{s=jk\Omega_0} = \frac{1}{2}\left[\frac{e^s - e^{-s}}{s^2} - \frac{e^s + e^{-s}}{s}\right]_{s=jk\pi}$$

$$= -j\frac{\sin(k\pi)}{(k\pi)^2} + j\frac{\cos(k\pi)}{k\pi} = 0 + j\frac{\cos(k\pi)}{k\pi} = j\frac{(-1)^k}{k\pi}$$

Finally, the Fourier series coefficients of $y(t)$ are

$$Y_k = \begin{cases} Y_{e0} + Y_{o0} = 1.5 + 0 = 1.5 & k = 0 \\ Y_{ek} + Y_{ok} = (1 - (-1)^k)/(k\pi)^2 + j(-1)^k/(k\pi) & k \neq 0 \end{cases}$$

∎

4.10.2 Linearity of Fourier Series—Addition of Periodic Signals

■ *Same fundamental frequency:* If $x(t)$ and $y(t)$ are periodic signals with the same fundamental frequency Ω_0, then the Fourier series coefficients of $z(t) = \alpha x(t) + \beta y(t)$ for constants α and β are

$$Z_k = \alpha X_k + \beta Y_k \tag{4.44}$$

where X_k and Y_k are the Fourier coefficients of $x(t)$ and $y(t)$.
■ *Different fundamental frequencies:* If $x(t)$ is periodic of period T_1, and $y(t)$ is periodic of period T_2 such that $T_2/T_1 = N/M$, for nondivisible integers N and M, then $z(t) = \alpha x(t) + \beta y(t)$ is periodic of period

$T_0 = MT_2 = NT_1$, and its Fourier coefficients are

$$Z_k = \alpha X_{k/N} + \beta Y_{k/M} \qquad \text{for integers } k \text{ such that } k/N, \text{ and } k/M \text{ are integers} \qquad (4.45)$$

where X_k and Y_k are the Fourier coefficients of $x(t)$ and $y(t)$.

If $x(t)$ and $y(t)$ are periodic signals of the same period T_0, the Fourier coefficients of $z(t) = \alpha x(t) + \beta y(t)$ (also periodic of period T_0) are then $Z_k = \alpha X_k + \beta Y_k$ where X_k and Y_k are the Fourier coefficients of $x(t)$ and $y(t)$, respectively.

In general, if $x(t)$ is periodic of period T_1, and $y(t)$ is periodic of period T_2, their sum $z(t) = \alpha x(t) + \beta y(t)$ is periodic if the ratio T_2/T_1 is a rational number (i.e., $T_2/T_1 = N/M$ for some nondivisible integers N and M). If so, the period of $z(t)$ is $T_0 = MT_2 = NT_1$. The fundamental frequency of $z(t)$ would be $\Omega_0 = \Omega_1/N = \Omega_2/M$ for Ω_1 the fundamental frequency of $x(t)$ and Ω_2 the fundamental frequency of $y(t)$. The Fourier series of $z(t)$ is then

$$z(t) = \alpha x(t) + \beta y(t) = \alpha \sum_k X_k e^{j\Omega_1 kt} + \beta \sum_m Y_m e^{j\Omega_2 mt}$$

$$= \alpha \sum_k X_k e^{jN\Omega_0 kt} + \beta \sum_m Y_m e^{jM\Omega_0 mt}$$

$$= \alpha \sum_{n=0,\pm N,\pm 2N,\ldots} X_{n/N} e^{j\Omega_0 nt} + \beta \sum_{\ell=0,\pm M,\pm 2M,\ldots} Y_{\ell/M} e^{j\Omega_0 \ell t}$$

Thus, the coefficients are

$$Z_k = \alpha X_{k/N} + \beta Y_{k/M}$$

for integers k such that k/N and k/M are integers.

■ Example 4.15

Consider the sum $z(t)$ of a periodic signal $x(t)$ of period $T_1 = 2$, with a periodic signal $y(t)$ with period $T_2 = 0.2$. Find the Fourier coefficients Z_k of $z(t)$ in terms of the Fourier coefficients X_k and Y_k of $x(t)$ and $y(t)$.

Solution

The ratio $T_2/T_1 = 1/10 = N/M$ is rational, so $z(t)$ is periodic of period $T_0 = T_1 = 10T_2 = 2$. The fundamental frequency of $z(t)$ is $\Omega_0 = \Omega_1 = \pi$, and $\Omega_2 = 10\Omega_0 = 10\pi$ is the fundamental frequency of $y(t)$. Thus, the Fourier coefficients of $z(t)$ are

$$Z_k = \begin{cases} X_k + Y_{k/10} & \text{when } k = 0, \pm 10, \pm 20, \ldots \\ X_k & \text{otherwise} \end{cases}$$

■

4.10.3 Multiplication of Periodic Signals

If $x(t)$ and $y(t)$ are periodic signals of same period T_0, then their product

$$z(t) = x(t)y(t) \tag{4.46}$$

is also periodic of period T_0, and with Fourier coefficients that are the *convolution sum* of the Fourier coefficients of $x(t)$ and $y(t)$:

$$Z_k = \sum_{\ell} X_\ell Y_{k-\ell} \tag{4.47}$$

If $x(t)$ and $y(t)$ are periodic with the same period T_0, then $z(t) = x(t)y(t)$ is also periodic of period T_0, since $z(t + kT_0) = x(t + kT_0)y(t + kT_0) = x(t)y(t) = z(t)$. Furthermore,

$$x(t)y(t) = \sum_k X_k e^{jk\Omega_0 t} \sum_\ell Y_\ell e^{j\ell\Omega_0 t} = \sum_k \sum_\ell X_k Y_\ell e^{j(k+\ell)\Omega_0 t}$$

$$= \sum_m \left[\sum_k X_k Y_{m-k} \right] e^{jm\Omega_0 t} = z(t)$$

where we let $m = k + \ell$. The coefficients of the Fourier series of $z(t)$ are then

$$Z_m = \sum_k X_k Y_{m-k}$$

or the convolution sum of the sequences X_k and Y_k, to be formally defined in Chapter 8.

■ Example 4.16

Consider the train of rectangular pulses $x(t)$ shown in Figure 4.4. Let $z(t) = 0.25x^2(t)$. Use the Fourier series of $z(t)$ to show that

$$X_k = \alpha \sum_m X_m X_{k-m}$$

for some constant α. Determine α.

Solution

The signal $0.5x(t)$ is a train of pulses of unit amplitude, so that $z(t) = (0.5x(t))^2 = 0.5x(t)$. Thus, $Z_k = 0.5X_k$, but also as a product of $0.5x(t)$ with itself we have that

$$Z_k = \sum_m [0.5X_m][0.5X_{k-m}]$$

and thus

$$\underbrace{0.5X_k}_{Z_k} = 0.25 \sum_m X_m X_{k-m} \quad \Rightarrow \quad X_k = \frac{1}{2} \sum_m X_m X_{k-m} \qquad (4.48)$$

so that $\alpha = 0.5$.

The Fourier series of $z(t) = 0.5x(t)$ according to the results in Example 4.5 is

$$z(t) = 0.5x(t) = \sum_{k=-\infty}^{\infty} \frac{\sin(\pi k/2)}{\pi k} e^{jk2\pi t}$$

If we define

$$S(k) = 0.5X_k = \frac{\sin(k\pi/2)}{k\pi} \quad \Rightarrow \quad X_k = 2S(k)$$

we have from Equation (4.48) the interesting result

$$S(k) = \sum_{m=-\infty}^{\infty} S(m)S(k-m)$$

or the convolution sum of the discrete sinc function $S(k)$ with itself is $S(k)$. ∎

4.10.4 Derivatives and Integrals of Periodic Signals

■ *Derivative*: The derivative $dx(t)/dt$ of a periodic signal $x(t)$, of period T_0, is periodic of the same period T_0. If $\{X_k\}$ are the coefficients of the Fourier series of $x(t)$, the Fourier coefficients of $dx(t)/dt$ are

$$jk\Omega_0 X_k \qquad (4.49)$$

where Ω_0 is the fundamental frequency of $x(t)$.

■ *Integral*: For a zero-mean, periodic signal $y(t)$, of period T_0, the integral

$$z(t) = \int_{-\infty}^{t} y(\tau)d\tau$$

is periodic of the same period as $y(t)$, with Fourier coefficients

$$Z_k = \frac{Y_k}{jk\Omega_0} \qquad k \neq 0$$

$$Z_0 = -\sum_{m \neq 0} Y_m \frac{1}{jm\Omega_0} \qquad \Omega_0 = \frac{2\pi}{T_0} \qquad (4.50)$$

These properties come naturally from the Fourier series representation of the periodic signal. Once we find the Fourier series of a periodic signal, we can differentiate it or integrate it (only when the dc value is zero). The derivative of a periodic signal is obtained by computing the derivative of each of the terms of its Fourier series—that is, if

$$x(t) = \sum_k X_k e^{jk\Omega_0 t}$$

then

$$\frac{dx(t)}{dt} = \sum_k X_k \frac{de^{jk\Omega_0 t}}{dt} = \sum_k \left[jk\Omega_0 X_k \right] e^{jk\Omega_0 t}$$

indicating that if the Fourier coefficients of $x(t)$ are X_k, the Fourier coefficients of $dx(t)/dt$ are $jk\Omega_0 X_k$.

To obtain the integral property we assume $y(t)$ is a zero-mean signal so that its integral $z(t)$ is finite. If for some integer M, $MT_0 \le t < (M+1)T_0$, then

$$z(t) = \int_{-\infty}^{t} y(\tau)d\tau = \int_{-\infty}^{MT_0} y(\tau)d\tau + \int_{MT_0}^{t} y(\tau)d\tau$$

$$= 0 + \int_{MT_0}^{t} y(\tau)d\tau$$

Replacing $y(t)$ by its Fourier series gives

$$z(t) = \int_{MT_0}^{t} y(\tau)d\tau = \int_{MT_0}^{t} \sum_{k \ne 0} Y_k e^{jk\Omega_0 \tau} d\tau$$

$$= \sum_{k \ne 0} Y_k \int_{MT_0}^{t} e^{jk\Omega_0 \tau} d\tau = \sum_{k \ne 0} Y_k \frac{1}{jk\Omega_0} \left[e^{jk\Omega_0 t} - 1 \right]$$

$$= -\sum_{k \ne 0} Y_k \frac{1}{jk\Omega_0} + \sum_{k \ne 0} Y_k \frac{1}{jk\Omega_0} e^{jk\Omega_0 t}$$

where the first term corresponds to the average Z_0 and $Z_k = Y_k/(jk\Omega_0)$, $k \ne 0$, are the rest of the Fourier coefficients of $z(t)$.

Remarks *It should be now clear why the derivative of a periodic signal $x(t)$ enhances its higher harmonics. Indeed, the Fourier coefficients of the derivative $dx(t)/dt$ are those of $x(t)$, X_k, multiplied by $j\Omega_0 k$, which increases with k. Likewise, the integration of a zero-mean periodic signal $x(t)$ does the opposite—that is, it makes the signal smoother, as we multiply X_k by decreasing terms $1/(jk\Omega_0)$ as k increases.*

■ Example 4.17

Let $g(t)$ be the derivative of a triangular train of pulses $f(t)$, of period $T_0 = 1$. The period of $f(t)$, $0 \le t \le 1$, is

$$f_1(t) = 2r(t) - 4r(t - 0.5) + 2r(t - 1)$$

Use the Fourier series of $g(t)$ to find the Fourier series of $f(t)$.

Solution

According to the derivative property we have that

$$F_k = \frac{G_k}{jk\Omega_0} \quad k \ne 0$$

are the Fourier coefficients of $f(t)$. The signal $g(t) = df(t)/dt$ has a corresponding period $g_1(t) = df_1(t)/dt = 2u(t) - 4u(t - 0.5) + 2u(t - 1)$. The Fourier series coefficients of $g(t)$ are

$$G_k = \frac{2e^{-0.5s}}{s}\left(e^{0.5s} - 2 + e^{-0.5s}\right)|_{s=j2\pi k} = 2(-1)^k \frac{\cos(\pi k) - 1}{j\pi k} \quad k \ne 0$$

which are used to obtain the coefficients F_k for $k \ne 0$. The dc component of $f(t)$ is found to be 0.5 from its plot as $g(t)$ does not provide it. ■

■ Example 4.18

Consider the reverse of Example 4.17. That is, given the periodic signal $g(t)$ of period $T_0 = 1$ and Fourier coefficients

$$G_k = 2(-1)^k \frac{\cos(\pi k) - 1}{j\pi k} \quad k \ne 0$$

and $G_0 = 0$. Find the integral

$$z(t) = \int_{-\infty}^{t} g(\tau)d\tau$$

Solution

As shown above, $z(t)$ is also periodic of the same period as $g(t)$ (i.e., $T_0 = 1$). The Fourier coefficients of $z(t)$ are

$$Z_k = \frac{G_k}{j\Omega_0 k} = (-1)^k \frac{4(\cos(\pi k) - 1)}{(j2\pi k)^2} = (-1)^{(k+1)} \frac{\cos(\pi k) - 1}{\pi^2 k^2} \quad k \ne 0$$

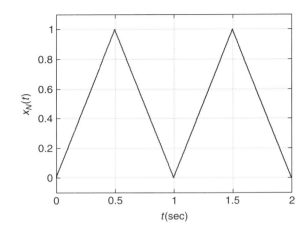

FIGURE 4.18

Two periods of the approximate triangular signal $x_N(t)$ using 100 harmonics.

and the average term is

$$Z_0 = -\sum_{m\neq 0} G_m \frac{1}{j2m\pi} = \sum_{m\neq 0} (-1)^m \frac{\cos(\pi m) - 1}{(\pi m)^2}$$

$$= 0.5 \sum_{m=-\infty,m\neq 0}^{\infty} (-1)^{m+1} \left[\frac{\sin(\pi m/2)}{(\pi m/2)} \right]^2$$

where we used $1 - \cos(\pi m) = 2\sin^2(\pi m/2)$. We used the following script to obtain the average, and to approximate the triangular signal using 100 harmonics (see Figure 4.18). The mean is obtained as 0.498.

```
%%%%%%%%%%%%%%%%%
% Example 4.18
%%%%%%%%%%%%%%%%%
clf; clear all
w0 = 2 * pi; N = 100;  % parameters of periodic signal
% computation of mean value
DC = 0;
for m = 1:N,
    DC = DC + 2 * (-1)^(m) * (cos(pi * m) -1)/(pi * m)^2;
end
% computation of Fourier series coefficients
Ts = 0.001; t = 0:Ts:2 - Ts;
for k = 1:N,
  X(k) = (-1)^(k + 1)*(cos(pi * k) - 1)/((pi * k)^2);
end
X = [DC X]; % Fourier series coefficients
xa = X(1)*ones(1,length(t));
figure(1)
```

```
for k = 2:N,
  xa = xa + 2 * abs(X(k)) * cos(w0 * (k - 1). * t + angle(X(k))); % approximate signal
end
```

4.11 WHAT HAVE WE ACCOMPLISHED? WHERE DO WE GO FROM HERE?

Periodic signals are not to be found in practice, so where did Fourier get the intuition to come up with a representation for them? As you will see, the fact that periodic signals are not found in practice does not mean that they are not useful. The Fourier representation of periodic signals will be fundamental in finding a representation for nonperiodic signals.

A very important concept you have learned in this chapter is that the inverse relation between time and frequency provides complementary information for the signal. The frequency domain constitutes the other side of the coin in representing signals. As mentioned before, it is the eigenfunction property of linear time-invariant systems that holds the theory together. It will provide the fundamental principle for filtering. You should have started to experience *déjà vu* in terms of the properties of the Fourier series; some look like a version of the ones in the Laplace transform. This is due to the connection existing between these transforms. You should have also noticed the usefulness of the Laplace transform in finding the Fourier coefficients, avoiding integration whenever possible. Table 4.1 provides the basic properties of the Fourier series for continuous–time periodic signals.

Chapter 5 will extend some of the results obtained in this chapter, thus unifying the treatment of periodic and nonperiodic signals and the concept of spectrum. Also the frequency representation of

Table 4.1 Basic Properties of Fourier Series

	Time Domain	Frequency Domain
Signals and constants	$x(t), y(t)$ periodic with period T_0, α, β	X_k, Y_k
Linearity	$\alpha x(t) + \beta y(t)$	$\alpha X_k + \beta Y_k$
Parseval's power relation	$P_x = \frac{1}{T_0} \int_{T_0} |x(t)|^2 dt$	$P_x = \sum_k |X_k|^2$
Differentiation	$\dfrac{dx(t)}{dt}$	$jk\Omega_0 X_k$
Integration	$\int_{-\infty}^{t} x(t')dt'$ only if $X_0 = 0$	$\dfrac{X_k}{jk\Omega_0} k \neq 0, -\sum_{m\neq 0} \dfrac{X_m}{jm\Omega_0},$ dc
Time shifting	$x(t - \alpha)$	$e^{-j\alpha\Omega_0} X_k$
Frequency shifting	$e^{jM\Omega_0 t} x(t)$	X_{k-M}
Symmetry	$x(t)$ real	$|X_k| = |X_{-k}|$ even function of k $\angle X_k = -\angle X_{-k}$ odd function of k

systems will be introduced and exemplified by its application in filtering. Modulation is the basic tool in communications and can be easily explained in the frequency domain.

PROBLEMS

4.1. Eigenfunctions and LTI systems
The eigenfunction property is only valid for LTI systems. Consider the cases of nonlinear and of time-varying systems.
(a) A system represented by the following input–output equation is nonlinear:

$$y(t) = x^2(t)$$

Let $x(t) = e^{j\pi t/4}$. Find the corresponding system output $y(t)$. Does the eigenfunction property hold? Explain.
(b) Consider a time-varying system

$$y(t) = x(t)[u(t) - u(t-1)]$$

Let $x(t) = e^{j\pi t/4}$. Find the corresponding system output $y(t)$. Does the eigenfunction property hold? Explain.

4.2. Eigenfunctions and LTI systems
The output of an LTI system is

$$y(t) = \int_0^t h(\tau)x(t-\tau)d\tau$$

where the input $x(t)$ and the impulse response $h(t)$ of the system are assumed to be causal. Let $x(t) = 2\cos(2\pi t)u(t)$. Compute the output $y(t)$ in the steady state and determine if the eigenfunction property holds.

4.3. Eigenfunctions and frequency response of LTI systems
The input–output equation for an analog averager is

$$y(t) = \frac{1}{T} \int_{t-T}^t x(\tau)d\tau$$

Let $x(t) = e^{j\Omega_0 t}$. Since the system is LTI, then the output should be

$$y(t) = e^{j\Omega_0 t}H(j\Omega_0)$$
(a) Find $y(t)$ for the given input and then compare it with the above equation to find $H(j\Omega_0)$, the response of the averager at frequency Ω_0.
(b) Find $H(s)$ and verify the frequency response value $H(j\Omega_0)$ obtained above.

4.4. Generality of eigenfunctions
The eigenfunction property holds for any input signal, periodic or not, that can be expressed in sinusoidal form.
(a) Consider the input $x(t) = \cos(t) + \cos(2\pi t)$, $-\infty < t < \infty$, into an LTI system. Is $x(t)$ periodic? If so, indicate its period.

(b) Suppose that the system is represented by a first-order differential equation,

$$y'(t) + 5y(t) = x(t)$$

where $y(t)$ is the output of the system and the given $x(t)$ is the input of the system. Find the steady-state response $y(t)$ due to $x(t)$ using the eigenfunction property.

4.5. Steady state of LTI systems

The transfer function of an LTI system is

$$H(s) = \frac{Y(s)}{X(s)} = \frac{s+1}{s^2 + 3s + 2}$$

If the input to this system is $x(t) = 1 + \cos(t + \pi/4)$, $-\infty < t < \infty$, what is the output $y(t)$ in the steady state?

4.6. Eigenfunction property of LTI systems and Laplace

The transfer function of an LTI system is given by

$$H(s) = \frac{Y(s)}{X(s)} = \frac{1}{s^2 + 3s + 2}$$

and its input is

$$x(t) = 4u(t)$$

(a) Use the eigenfunction property of LTI systems to find the steady-state response $y(t)$ of this system.
(b) Verify your result in (a) by means of the Laplace transform.

4.7. Different ways to compute the Fourier coefficients—MATLAB

We would like to find the Fourier series of a sawtooth periodic signal $x(t)$ of period $T_0 = 1$. The period of $x(t)$ is

$$x_1(t) = r(t)[u(t) - u(t-1)]$$

(a) Carefully plot $x(t)$ and compute the Fourier coefficients X_k using the integral definition.
(b) An easier way to do this is to use the Laplace transform of $x_1(t)$. Find X_k this way.
(c) Use MATLAB to plot the signal $x(t)$ and its magnitude and phase line spectra.
(d) Obtain a trigonometric Fourier series $\hat{x}(t)$ consisting of the DC term and 40 harmonics to approximate $x(t)$. Use MATLAB to find the values of $\hat{x}(t)$ for $t = 0$ to 10 in steps of 0.001. How does it compare with $x(t)$?

4.8. Addition of periodic signals—MATLAB

Consider a sawtooth signal $x(t)$ with period $T_0 = 2$ and period

$$x_1(t) = \begin{cases} t & 0 \leq t < 1 \\ 0 & \text{otherwise} \end{cases}$$

(a) Find the Fourier coefficients X_k using the Laplace transform. Consider the cases when k is odd and even ($k \neq 0$). You need to compute X_0 directly from the signal.
(b) Let $y(t) = x(-t)$. Find the Fourier coefficients Y_k.
(c) The sum $z(t) = x(t) + y(t)$ is a triangular function. Find the Fourier coefficients Z_k and compare them to $X_k + Y_k$.
(d) Use MATLAB to plot $x(t)$, $y(t)$, and $z(t)$ and their corresponding magnitude line spectra. Find an approximate of z(t) using the dc value and10 harmonics and plot it.

4.9. Fourier series coefficients via Laplace—MATLAB

The computation of the Fourier series coefficients is simplified by the relation between the formula for these coefficients and the Laplace transform of a period of the periodic signal.

(a) A periodic signal $x(t)$, of period $T_0 = 2$ sec, has as period the signal $x_1(t) = u(t) - u(t-1)$, so that $x(t)$ can be represented as

$$x(t) = \sum_{m=-\infty}^{\infty} x_1(t - mT_0)$$

Expand this sum, and use the information for $x_1(t)$ and T_0 to carefully plot the periodic signal $x(t)$.

(b) Find the Laplace transform of $x_1(t)$, and let $s = jk\Omega_0$, where $\Omega_0 = 2\pi/T_0$ is the fundamental frequency, to obtain the Fourier coefficients of $x(t)$.

(c) Use MATLAB to plot the magnitude line spectrum of $x(t)$. Find an approximate of $x(t)$ using the dc and 40 harmonics. Plot it.

4.10. Half- and full-wave rectifying and Fourier—MATLAB

Rectifying a sinusoid provides a way to create a dc source. In this problem we consider the Fourier series of the full- and half-wave rectified signals. The full-wave rectified signal $x_f(t)$ has a period $T_0 = 1$ and its period from 0 to 1 is

$$x_1(t) = \sin(\pi t) \qquad 0 \leq t \leq 1$$

while the period for the half-wave rectifier signal $x_h(t)$ is

$$x_2(t) = \begin{cases} \sin(\pi t) & 0 \leq t \leq 1 \\ 0 & 1 < t \leq 2 \end{cases}$$

with period $T_1 = 2$.

(a) Obtain the Fourier coefficients for both of these periodic signals.

(b) Use the even and odd decomposition of $x_h(t)$ to obtain its Fourier coefficients. This computation of the Fourier coefficients of $x_h(t)$ avoids some difficulties when you attempt to plot its magnitude line spectrum. Use MATLAB and your analytic results here to plot the magnitude line spectrum of the half-wave signal and use the dc and 40 harmonics to obtain an approximation of the half-wave signal.

4.11. Smoothness and Fourier series—MATLAB

The smoothness of a period determines the way the magnitude line spectrum decays. Consider the following periodic signals $x(t)$ and $y(t)$, both of period $T_0 = 2$ sec, and with a period from $0 \leq t \leq T_0$ equal to

$$x_1(t) = u(t) - u(t-1)$$
$$y_1(t) = r(t) - 2r(t-1) + r(t-2)$$

Find the Fourier series coefficients of $x(t)$ and $y(t)$ and use MATLAB to plot their magnitude line spectrum for $k = 0, \pm1, \pm2, \ldots, \pm20$. Determine which of these spectra decays faster and how it relates to the smoothness of the period. (To see this relate $|X_k|$ to the corresponding $|Y_k|$.)

4.12. Time support and frequency content—MATLAB

The support of a period of a periodic signal relates inversely to the support of the line spectrum. Consider two periodic signals: $x(t)$ of period $T_0 = 2$ and $y(t)$ of period $T_1 = 1$, and with periods

$$x_1(t) = u(t) - u(t-1) \qquad 0 \leq t \leq 2$$
$$y_1(t) = u(t) - u(t-0.5) \qquad 0 \leq t \leq 1$$

(a) Find the Fourier series coefficients for $x(t)$ and $y(t)$.
(b) Use MATLAB to plot the magnitude line spectra of the two signals from 0 to 40π rad/sec. Plot them on the same figure so you can determine which has a broader support. Indicate which signal is smoother and explain how it relates to its line spectrum.

4.13. Derivatives and Fourier Series

Given the Fourier series representation for a periodic signal,

$$x(t) = \sum_{k=-\infty}^{\infty} X_k e^{jk\Omega_0 t}$$

we can compute derivatives of it, just like for any other signal.

(a) Consider the periodic train of pulses shown in Figure 4.19. Compute its derivative

$$y(t) = \frac{dx(t)}{dt}$$

and carefully plot it. Find the Fourier series of $y(t)$.

(b) Use the Fourier series representation of $x(t)$ and find its derivative to obtain the Fourier series of $y(t)$. How does it compare to the Fourier series obtained above?

4.14. Fourier series of sampling delta

The periodic signal

$$\delta_{T_s}(t) = \sum_{m=-\infty}^{\infty} \delta(t - mT_s)$$

will be very useful in the sampling of continuous-time signals.

(a) Find the Fourier series of this signal—that is,

$$\delta_{T_s}(t) = \sum_{k=-\infty}^{\infty} \Delta_k e^{jk\Omega_s t}$$

find the Fourier coefficients Δ_k.

(b) Plot the magnitude line spectrum of this signal.

(c) Plot $\delta_{T_s}(t)$ and its corresponding line spectrum Δ_k as functions of time and frequency. Are they both periodic? How are their periods related? Explain.

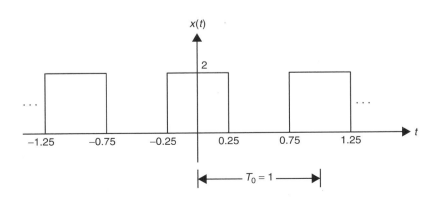

FIGURE 4.19

Problem 4.13: train of rectangular pulses.

4.15. Figuring out Fourier's idea

Fourier proposed to represent a periodic signal as a sum of sinusoids, perhaps an infinite number of them. For instance, consider the representation of a periodic signal $x(t)$ as a sum of cosines of different frequencies

$$x(t) = \sum_{k=0}^{\infty} A_k \cos(\Omega_k t + \theta_k)$$

(a) If $x(t)$ is periodic of period T_0, what should the frequencies Ω_k be?

(b) Consider $x(t) = 2 + \cos(2\pi t) - 3\cos(6\pi t + \pi/4)$. Is this signal periodic? If so, what is its period T_0? Determine its trigonometric Fourier series as given above by specifying the values of A_k and θ_k for all values of $k = 0, 1, \ldots$.

(c) Let the signal $x_1(t) = 2 + \cos(2\pi t) - 3\cos(20t + \pi/4)$ (this signal is almost like $x(t)$ given above, except that the frequency 6π rad/sec of the second cosine has been approximated by 20 rad/sec). Is this signal periodic? Can you determine its Fourier series as given above by specifying the values of A_k and θ_k for all values of $k = 0, 1, \ldots$? Explain.

4.16. DC output from a full-wave rectified signal—MATLAB

Consider a full-wave rectifier that has as output a periodic signal $x(t)$ of period $T_0 = 1$ and a period of it is given as

$$x_1(t) = \begin{cases} \cos(\pi t) & -0.5 \le t \le 0.5 \\ 0 & \text{otherwise} \end{cases}$$

(a) Obtain the Fourier coefficients X_k.

(b) Suppose we pass $x(t)$ through an ideal filter of transfer function $H(s)$. Determine the values of this filter at harmonic frequencies $2\pi k, = 0, \pm 1, \pm 2, \ldots$, so that its output is a constant (i.e., we have a dc source).

(c) Use MATLAB to plot the signal $x(t)$ and its magnitude line spectrum.

4.17. Fourier series of sum of periodic signals

Suppose you have the Fourier series of two periodic signals $x(t)$ and $y(t)$ of periods T_1 and T_2, respectively. Let X_k and Y_k be the Fourier series coefficients corresponding to $x(t)$ and $y(t)$.

(a) If $T_1 = T_2$, what would be the Fourier series coefficients of $z(t) = x(t) + y(t)$ in terms of X_k and Y_k?

(b) If $T_1 = 2T_2$, determine the Fourier series coefficients of $w(t) = x(t) + y(t)$ in terms of X_k and Y_k?

4.18. Manipulation of periodic signals

Let the following be the Fourier series of a periodic signal $x(t)$ of period T_0 (fundamental frequency $\Omega_0 = 2\pi/T_0$):

$$x(t) = \sum_{k=-\infty}^{\infty} X_k e^{j\Omega_0 k t}$$

Consider the following functions of $x(t)$, and determine if they are periodic and what are their periods if so:

- $y(t) = 2x(t) - 3$
- $z(t) = x(t-2) + x(t)$
- $w(t) = x(2t)$

Express the Fourier series coefficients Y_k, Z_k, and W_k in terms of X_k.

4.19. Using properties to find the Fourier series

Use the Fourier series of a square train of pulses (done in this chapter) to compute the Fourier series of the triangular signal $x(t)$ with a period,

$$x_1(t) = r(t) - 2r(t-1) + r(t-2)$$

(a) Find the derivative of $x(t)$ or $y(t) = dx(t)/dt$ and carefully plot it. Plot also $z(t) = y(t) + 1$. Use the Fourier series of the square train of pulses to compute the Fourier series coefficients of $y(t)$ and $z(t)$.
(a) Obtain the trigonometric Fourier series of $y(t)$ and $z(t)$ and explain why they are represented by sines and why $z(t)$ has a nonzero mean.
(c) Obtain the Fourier series coefficients of $x(t)$ from those of $y(t)$.
(d) Obtain the sinusoidal form of $x(t)$ and explain why the cosine representation is more appropriate for this signal than a sine representation.

4.20. Applying Parseval's result—MATLAB

We wish to approximate the triangular signal $x(t)$ in Problem 4.19 by its Fourier series with a finite number of terms, let's say $2N + 1$. This approximation should have 95% of the average power of the triangular signal. Use MATLAB to find the value of N.

4.21. Fourier series of multiplication of periodic signals

Consider the Fourier series of two periodic signals,

$$x(t) = \sum_{k=-\infty}^{\infty} X_k e^{j\Omega_0 kt}$$

$$y(t) = \sum_{k=-\infty}^{\infty} Y_k e^{j\Omega_1 kt}$$

(a) Let $\Omega_1 = \Omega_0$. Is $z(t) = x(t)y(t)$ periodic? If so, what is its period and its Fourier series coefficients?
(b) If $\Omega_1 = 2\Omega_0$. Is $w(t) = x(t)y(t)$ periodic? If so, what is its period and its Fourier series coefficients?

4.22. Integration of periodic signals

Consider now the integral of the Fourier series of the pulse signal $p(t) = x(t) - 1$ of period $T_0 = 1$, where $x(t)$ is given in Figure 4.20.

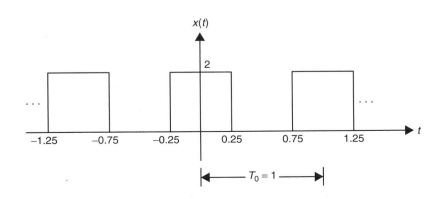

FIGURE 4.20

Problem 4.22: train of rectangular pulses.

(a) Given that an integral of $p(t)$ is the area under the curve, find and plot the function

$$s(t) = \int_{-\infty}^{t} p(t)dt \qquad t \le 1$$

Indicate the values of $s(t)$ for $t = 0, 0.25, 0.5, 0.75$, and 1.

(b) Find the Fourier series of $p(t)$ and $s(t)$ and relate their Fourier series coefficients.

(c) Suppose you want to compute the integral

$$\int_{-T_0/2}^{T_0/2} p(t)dt$$

using the Fourier series of $p(t)$. What is the integral equal to?

(d) You can also compute the integral from the plot of $p(t)$:

$$\int_{-T_0/2}^{T_0/2} p(t)dt$$

What is it? Does it coincide with the result obtained using the Fourier series? Explain.

4.23. Full-wave rectifying and DC sources

Let $x(t) = \sin^2(2\pi t)$, a periodic signal of period $T_0 = 0.5$, and $y(t) = |\sin(2\pi t)|$, which is also periodic of period $T_1 = 0.5$.

(a) A trigonometric identity gives that

$$x(t) = \frac{1}{2}[1 - \cos(4\pi t)]$$

Use this result to find its complex exponential Fourier series.

(b) Use the Laplace transform to find the Fourier series of $y(t)$.

(c) Are $x(t)$ and $y(t)$ identical? Explain.

(d) Indicate how you would use an ideal low-pass filter to get a DC source of unit value from $x(t)$ and $y(t)$. Indicate the bandwidth and the magnitude of the filters. Compare these two signals in terms of advantages or disadvantages in generating the desired DC source.

4.24. Windowing and music sounds—MATLAB

In the computer generation of musical sounds, pure tones need to be windowed to make them more interesting. Windowing mimics the way a musician would approach the generation of a certain sound. Increasing the richness of the harmonic frequencies is the result of the windowing, as we will see in this problem. Consider the generation of a musical note with frequencies around $f_A = 880$ Hz. Assume our "musician" while playing this note uses three strokes corresponding to a window $w_1(t) = r(t) - r(t - T_1) - r(t - T_2) + r(t - T_0)$, so that the resulting sound would be the multiplication, or windowing, of a pure sinusoid $\cos(2\pi f_A t)$ by a periodic signal $w(t)$, with $w_1(t)$ a period that repeats every $T_0 = 5T$ where T is the period of the sinusoid. Let $T_1 = T_0/4$ and $T_2 = 3T_0/4$.

(a) Analytically determine the Fourier series of the window $w(t)$ and plot its line spectrum using MATLAB. Indicate how you would choose the number of harmonics needed to obtain a good approximation to $w(t)$.

(b) Use the modulation or the convolution properties of the Fourier series to obtain the coefficients of the product $s(t) = \cos(2\pi f_A t)w(t)$. Use MATLAB to plot the line spectrum of this periodic signal and again determine how many harmonic frequencies you would need to obtain a good approximation to $s(t)$.

(c) The line spectrum of the pure tone $p(t) = \cos(2\pi f_A t)$ only displays one harmonic, the one corresponding to the $f_A = 880$ Hz frequency. How many more harmonics does $s(t)$ have? To listen to the richness in harmonics use the MATLAB function sound to play the sinusoid $p(t)$ and $s(t)$ (use $F_s = 2 \times 880$ Hz to play both).

(d) Consider a combination of notes in a certain scale; for instance, let

$$p(t) = \sin(2\pi \times 440t) + \sin(2\pi \times 550t) + \sin(2\pi \times 660t)$$

Use the same windowing $w(t)$, and let $s(t) = p(t)w(t)$. Use MATLAB to plot $p(t)$ and $s(t)$ and to compute and plot their corresponding line spectra. Use sound to play $p(nT_s)$ and $s(nT_s)$ using $F_s = 1000$.

4.25. Computation of π—MATLAB

As you know, π is an irrational number that can only be approximated by a number with a finite number of decimals. How to compute this value recursively is a problem of theoretical interest. In this problem we show that the Fourier series can provide that formulation.

(a) Consider a train of rectangular pulses $x(t)$, with a period

$$x_1(t) = 2[u(t + 0.25) - u(t - 0.25)] - 1 \qquad -0.5 \leq t \leq 0.5$$

and period $T_0 = 1$. Plot the periodic signal and find its trigonometric Fourier series.

(b) Use the above Fourier series to find an infinite sum for π.

(c) If π_N is an approximation of the infinite sum with N coefficients, and π is the value given by MATLAB, find the value of N so that π_N is 95% of the value of π given by MATLAB.

4.26. Square error approximation of periodic signals—MATLAB

To understand the Fourier series consider a more general problem, where a periodic signal $x(t)$, of period T_0, is approximated as a finite sum of terms,

$$\hat{x}(t) = \sum_{k=-N}^{N} \hat{X}_k \phi_k(t)$$

where $\{\phi_k(t)\}$ are orthonormal functions. To pose the problem as an optimization problem, consider the square error

$$\varepsilon = \int_{T_0} |x(t) - \hat{x}(t)|^2 dt$$

and look for the coefficients $\{\hat{X}(k)\}$ that minimize ε.

(a) Assume that $x(t)$ as well as $\hat{x}(t)$ are real valued, and that $x(t)$ is even so that the Fourier series coefficients X_k are real. Show that the error can be expressed as

$$\varepsilon = \int_{T_0} x^2(t)dt - 2\sum_{k=-N}^{N} \hat{X}_k \int_{T_0} x(t)\phi_k(t)dt + \sum_{\ell=-N}^{N} |\hat{X}_\ell|^2 T_0$$

(b) Compute the derivative of ε with respect to $\hat{X}_n$ and set it to zero to minimize the error. Find $\hat{X}_n$.

(c) In the Fourier series the $\{\phi_k(t)\}$ are the complex exponentials and the $\{\hat{X}_n\}$ coincide with the Fourier series coefficients. To illustrate the above procedure consider the case of the pulse signal $x(t)$, of period $T_0 = 1$, and a period

$$x_1(t) = 2[u(t + 0.25) - u(t - 0.25)]$$

Use MATLAB to compute and plot the approximation $\hat{x}(t)$ and the errror ε for increasing values of N from 1 to 100.

(d) Concentrate your plot of $\hat{x}(t)$ around the one of the discontinuities, and observe the Gibb's phenomenon. Does it disappear when N is very large. Plot $\hat{x}(t)$ around the discontinuity for $N = 1000$.

4.27. Walsh functions—MATLAB

As seen in Problem 4.26, the Fourier series is one of a possible class of representations in terms of orthonormal functions. Consider the case of the *Walsh functions*, which are a set of rectangular pulse signals that are orthonormal in a finite time interval $[0, 1]$. These functions are such that: (1) they take only 1 and -1 values, (2) $\phi_k(0) = 1$ for all k, and (3) they are ordered according to the number of sign changes.

(a) Consider obtaining the functions $\{\phi_k\}_{k=0,\ldots,5}$. The Walsh functions are clearly normal since when squared they are unity for $t \in [0, 1]$. Let $\phi_0(t) = 1$ for $t \in [0, 1]$ and zero elsewhere. Obtain $\phi_1(t)$ with one change of sign and that is orthogonal to $\phi_0(t)$. Find then $\phi_2(t)$, which has two changes of sign and is orthogonal to both $\phi_0(t)$ and $\phi_1(t)$. Continue this process. Carefully plot the $\{\phi_i(t)\}$, $i = 0, \ldots, 5$. Use the MATLAB function stairs to plot these Walsh functions.

(b) Consider the Walsh functions obtained above as sequences of 1s and -1s of length 8, and carefully write these six sequences. Observe the symmetry of the sequences corresponding to $\{\phi_i(t), i = 0, 1, 3, 5\}$, and determine the circular shift needed to find the sequence corresponding to $\phi_2(t)$ from the sequence from $\phi_1(t)$, and $\phi_4(t)$ from $\phi_3(t)$. Write a MATLAB script that generates a matrix Φ with entries as the sequences. Find the product $(1/8)\Phi\Phi^T$, and explain how this result connects with the orthonormality of the Walsh functions.

(c) We wish to approximate a ramp function $x(t) = r(t)$, $0 \le t \le 1$, using $\{\phi_k\}_{k=0,\ldots,5}$. This could be written as

$$\mathbf{r} = \Phi\mathbf{a}$$

where $\mathbf{r}$ is a vector of $x(nT) = r(nT)$ where $T = 1/8$, $\mathbf{a}$ are the coefficients of the expansion, and Φ is the Walsh matrix found above. Determine the vector $\mathbf{a}$ and use it to obtain an approximation of $x(t)$. Plot $x(t)$ and the approximation $\hat{x}(t)$ (use stairs for this signal).

Frequency Analysis: The Fourier Transform

Imagination is the beginning of creation.
You imagine what you desire, you will what you imagine,
and at last you create what you will.
George Bernard Shaw (1856–1950)
Irish dramatist

5.1 INTRODUCTION

In this chapter we continue the frequency analysis of signals. In particular, we will concentrate in the following issues:

- *Generalization of the Fourier series*—The frequency representation of signals as well as the frequency response of systems are tools of great significance in signal processing, communications, and control theory. In this chapter we will complete the Fourier representation of signals by extending it to aperiodic signals. By a limiting process the harmonic representation of periodic signals is extended to the Fourier transform, a frequency-dense representation for nonperiodic signals. The concept of spectrum introduced for periodic signals is generalized for both finite-power and finite-energy signals. Thus, the Fourier transform measures the frequency content of a signal, and unifies the representation of periodic and aperiodic signals.

- *Laplace and Fourier transform*—In this chapter the connection between the Laplace and the Fourier transforms will be highlighted for computational and analytical reasons. The Fourier transform turns out to be a very important case of the Laplace transform for signals of which the region of convergence includes the $j\Omega$ axis. There are, however, signals where the Fourier transform cannot be obtained from the Laplace transform; for those cases, properties of the Fourier transform will be used. The duality of the direct and inverse transforms is of special interest in computing the Fourier transform.

- *Basics of filtering*—Filtering is an important application of the Fourier transform. The Fourier representation of signals and the eigenfunction property of LTI systems provide the tools to change

Signals and Systems Using MATLAB®. DOI: 10.1016/B978-0-12-374716-7.00008-9

the frequency content of a signal by processing it with an LTI system with a desired frequency response.

- *Modulation and communications*—The idea of changing the frequency content of a signal via modulation is basic in analog communications. Modulation allows us to send signals over the airwaves using antennas of reasonable sizes. Voice and music are relative low-frequency signals that cannot be easily radiated without the help of modulation. Continuous-wave modulation changes the amplitude, the frequency, or the phase of a sinusoidal carrier of frequency much greater than the frequencies present in the message we wish to transmit.

5.2 FROM THE FOURIER SERIES TO THE FOURIER TRANSFORM

In practice there are no periodic signals—such signals would have infinite supports and exact periods, which are not possible. Since only finite-support signals can be processed numerically, signals in practice are treated as aperiodic. To obtain the Fourier representation of aperiodic signals, we use the Fourier series representation in a limiting process.

An aperiodic, or nonperiodic, signal $x(t)$ can be thought of as a periodic signal $\tilde{x}(t)$ with an infinite period. Using the Fourier series representation of this signal and a limiting process we obtain a pair

$$x(t) \quad \Leftrightarrow \quad X(\Omega)$$

where the signal $x(t)$ is transformed into a function $X(\Omega)$ in the frequency domain by the

$$\text{Fourier transform:} \qquad X(\Omega) = \int_{-\infty}^{\infty} x(t)e^{-j\Omega t}dt \qquad (5.1)$$

while $X(\Omega)$ is transformed into a signal $x(t)$ in the time domain by the

$$\text{Inverse Fourier transform:} \qquad x(t) = \frac{1}{2\pi} \int_{-\infty}^{\infty} X(\Omega)e^{j\Omega t}d\Omega \qquad (5.2)$$

Any aperiodic signal can be assumed to be periodic with an infinite period. That is, an aperiodic signal $x(t)$ can be expressed as

$$x(t) = \lim_{T_0 \to \infty} \tilde{x}(t)$$

where $\tilde{x}(t)$ is a periodic signal of period T_0. The Fourier series representation of $\tilde{x}(t)$ is

$$\tilde{x}(t) = \sum_{n=-\infty}^{\infty} X_n e^{jn\Omega_0 t} \qquad \Omega_0 = \frac{2\pi}{T_0}$$

$$X_n = \frac{1}{T_0} \int_{-T_0/2}^{T_0/2} \tilde{x}(t)e^{-jn\Omega_0 t}dt$$

As $T_0 \to \infty$, X_n will tend to zero. To avoid this we define $X(\Omega_n) = T_0 X_n$ where $\{\Omega_n = n\Omega_0\}$ are the harmonic frequencies.

Letting $\Delta\Omega = 2\pi/T_0 = \Omega_0$ be the frequency interval between harmonics, we can then write the above equations as

$$\tilde{x}(t) = \sum_{n=-\infty}^{\infty} \frac{X(\Omega_n)}{T_0} e^{j\Omega_n t} = \sum_{n} X(\Omega_n) e^{j\Omega_n t} \frac{\Delta\Omega}{2\pi}$$

$$X(\Omega_n) = \int_{-T_0/2}^{T_0/2} \tilde{x}(t) e^{-j\Omega_n t} dt$$

As $T_0 \to \infty$, then $\Delta\Omega \to d\Omega$, the line spectrum becomes denser—that is, the lines in the line spectrum get closer, the sum becomes an integral, and $\Omega_n = n\Omega_0 = n\Delta\Omega \to \Omega$, so that in the limit we obtain

$$x(t) = \frac{1}{2\pi} \int_{-\infty}^{\infty} X(\Omega) e^{j\Omega t} d\Omega$$

$$X(\Omega) = \int_{-\infty}^{\infty} x(t) e^{-j\Omega t} dt$$

which are the inverse and the direct Fourier transforms, respectively. The first equation transforms a function in the frequency domain $X(\Omega)$ into a signal in the time domain $x(t)$, while the other equation does the opposite.

The Fourier transform measures the frequency content of a signal. As we will see, time and frequency are complementary, thus the characterization in one domain provides information that is not clearly available in the other.

Remarks

- *Although we have obtained the Fourier transform from the Fourier series, the Fourier transform of a periodic signal cannot be obtained from the above integral. Consider $x(t) = \cos(\Omega_0 t)$, $-\infty < t < \infty$, which is periodic of period $2\pi/\Omega_0$. If you attempt to compute its Fourier transform using the integral you do not have a well-defined problem (try to obtain the integral to convince yourself). But it is known from the line spectrum that the power of this signal is concentrated at the frequencies $\pm\Omega_0$, so somehow we should be able to find its Fourier transform. Sinusoids are basic functions.*
- *On the other hand, if you consider a decaying exponential $x(t) = e^{-|a|t}$ signal, which has finite energy and is absolutely integrable and has a Laplace transform that is valid on the $j\Omega$ axis (i.e., the region of convergence $X(s)$ includes this axis), then its Fourier transform is simply the Laplace transform $X(s)$ computed at $s = j\Omega$, as we will see. There is no need for the integral formula in this case, although if you apply it your result coincides with the one from the Laplace transform.*
- *Finally, consider finding the Fourier transform of a sinc function (which is the impulse response of a low-pass filter as we see later). Neither the integral nor the Laplace transform can be used to find it. For this signal, we need to exploit the duality that exists between the direct and the inverse Fourier transforms (Notice the duality in Equations (5.1) and (5.2)).*

5.3 EXISTENCE OF THE FOURIER TRANSFORM

For the Fourier transform to exist, $x(t)$ must be *absolutely integrable*—that is,

$$|X(\Omega)| \leq \int_{-\infty}^{\infty} |x(t)e^{-j\Omega t}| dt = \int_{-\infty}^{\infty} |x(t)| dt < \infty$$

Moreover, $x(t)$ must have only a finite number of discontinuities and a finite number of minima and maxima in any finite interval. (Given the limiting connection between the Fourier transform and the Fourier series, it is not surprising that the above conditions coincide with the existence conditions for the Fourier series.)

The Fourier transform

$$X(\Omega) = \int_{-\infty}^{\infty} x(t)e^{-j\Omega t} dt$$

of a signal $x(t)$ exists (i.e., we can calculate its Fourier transform via this integral) provided
- $x(t)$ is absolutely integrable or the area under $|x(t)|$ is finite.
- $x(t)$ has only a finite number of discontinuites as well as maxima and minima.

From the definitions of the direct and the inverse Fourier transforms—both being infinite integrals— one wonders whether they exist in general, and if so how to most efficiently compute them. Commenting on the existence conditions, Professor E. Craig [17] wrote:

> It appears that almost nothing has a Fourier transform—nothing except practical communi- cation signals. No signal amplitude goes to infinity and no signal lasts forever; therefore, no practical signal can have infinite area under it, and hence all have Fourier transforms.

Indeed, signals of practical interest have Fourier transforms and their spectra can be displayed using a *spectrum analyzer* (or better yet, any signal for which we can display its spectrum will have a Fourier transform). A spectrum analyzer is a device that displays the energy or the power of a signal distributed over frequencies.

5.4 FOURIER TRANSFORMS FROM THE LAPLACE TRANSFORM

The region of convergence of the Laplace transform $X(s)$ indicates the region in the s-plane where $X(s)$ is defined. The following applies to signals whether they are causal, anti-causal, or noncausal.

If the region of convergence (ROC) of $X(s) = \mathcal{L}[x(t)]$ contains the $j\Omega$ axis, so that $X(s)$ can be defined for $s = j\Omega$, then

$$\mathcal{F}[x(t)] = \mathcal{L}[x(t)]|_{s=j\Omega} = \int_{-\infty}^{\infty} x(t)e^{-j\Omega t} dt$$

$$= X(s)\,|_{s=j\Omega} \tag{5.3}$$

The following rules of thumb will help you get a better understanding of the time-frequency relationship of a signal and its Fourier transform, and the best way to compute it. On a first reading the use of these rules might not be obvious, but they will be helpful in understanding the discussions that follow and you might want to come back to these rules.

Rules of Thumb for Computing the Fourier Transform of a Signal $x(t)$

- If $x(t)$ has a finite time support and in that support $x(t)$ is finite, its Fourier transform exists. To find it use the integral definition or the Laplace transform of $x(t)$.
- If $x(t)$ has a Laplace transform $X(s)$ with a region of convergence including the $j\Omega$ axis, its Fourier transform is $X(s)|_{s=j\Omega}$.
- If $x(t)$ is periodic of infinite energy but finite power, its Fourier transform is obtained from its Fourier series using delta functions.
- If $x(t)$ is none of the above, if it has discontinuities (e.g., $x(t) = u(t)$) or it has discontinuities and it is not finite energy (e.g., $x(t) = \cos(\Omega_0 t)u(t)$), or it has possible discontinuities in the frequency domain even though it has finite energy (e.g., $x(t) = \text{sinc}(t)$), use properties of the Fourier transform.

Keep in mind to

- Consider the Laplace transform if the interest is in transients and steady state, and the Fourier transform if steady-state behavior is of interest.
- Represent periodic signals by their Fourier series before considering their Fourier transforms.
- Attempt other methods before performing integration to find the Fourier transform.

■ Example 5.1

Discuss whether it is possible to obtain the Fourier transform of the following signals using their Laplace transforms:

$$\text{(a)} \quad x_1(t) = u(t)$$
$$\text{(b)} \quad x_2(t) = e^{-2t}u(t)$$
$$\text{(c)} \quad x_3(t) = e^{-|t|}$$

Solution

(a) The Laplace transform of $x_1(t)$ is $X_1(s) = 1/s$ with a region of convergence corresponding to the open right s-plane, or ROC $= \{s = \sigma + j\Omega : \sigma > 0, -\infty < \Omega < \infty\}$, which does not include the $j\Omega$ axis, so the Laplace transform cannot be used to find the Fourier transform of $x_1(t)$.

(b) The signal $x_2(t)$ has as Laplace transform $X_2(s) = 1/(s+2)$ with a region of convergence ROC $= \{s = \sigma + j\Omega : \sigma > -2, -\infty < \Omega < \infty\}$ containing the $j\Omega$ axis. Then the Fourier transform of $x_2(t)$ is

$$X_2(\Omega) = \frac{1}{s+2}\Big|_{s=j\Omega} = \frac{1}{j\Omega + 2}$$

(c) The Laplace transform of $x_3(t)$ is

$$X_3(s) = \frac{1}{s+1} + \frac{1}{-s+1} = \frac{2}{1-s^2}$$

with a region of convergence ROC $= \{s = \sigma + j\Omega : -1 < \sigma < 1, -\infty < \Omega < \infty\}$ that contains the $j\Omega$ axis. Then the Fourier transform of $x_3(t)$ is

$$X_3(\Omega) = X_3(s)|_{s=j\Omega} = \frac{2}{1-(j\Omega)^2} = \frac{2}{1+\Omega^2}$$

■

5.5 LINEARITY, INVERSE PROPORTIONALITY, AND DUALITY

Many of the properties of the Fourier transform are very similar to those of the Fourier series or of the Laplace transform, which is to be expected given the strong connection among these transformations. The linearity and the duality between time and frequency of the Fourier transform will help us determine the transform of signals that do not satisfy the existence conditions given before.

5.5.1 Linearity

Just like the Laplace transform, the Fourier transform is linear.

If $\mathcal{F}[x(t)] = X(\Omega)$ and $\mathcal{F}[y(t)] = Y(\Omega)$, for constants α and β, we have that

$$\mathcal{F}[\alpha x(t) + \beta y(t)] = \alpha \mathcal{F}[x(t)] + \beta \mathcal{F}[y(t)]$$

$$= \alpha X(\Omega) + \beta Y(\Omega) \tag{5.4}$$

■ Example 5.2

Suppose you create a periodic sine

$$x(t) = \sin(\Omega_0 t) \quad -\infty < t < \infty$$

by adding a causal sine $v(t) = \sin(\Omega_0 t)u(t)$ and an anti-causal sine $y(t) = \sin(\Omega_0 t)u(-t)$, for each of which you can find Laplace transforms $V(s)$ and $Y(s)$. Discuss what would be wrong with this approach to find the Fourier transform of $x(t)$ by letting $s = j\Omega$.

Solution

The Laplace transforms of $v(t)$ and $y(t)$ are

$$V(s) = \frac{\Omega_0}{s^2 + \Omega_0^2} \qquad \text{ROC: } \mathcal{Re}\,[s] > 0$$

$$Y(s) = \frac{-\Omega_0}{(-s)^2 + \Omega_0^2} \qquad \text{ROC: } \mathcal{Re}\,[s] < 0$$

giving $X(s) = V(s) + Y(s) = 0$. Moreover, the region of convergence of $X(s)$ is the intersection of the two given ROCs, which is null, so it is not possible to obtain the Fourier transform of $x(t)$ this way. This is so even though the time signals add correctly to $x(t)$. The Fourier transform of the sine signal will be found using the periodicity of $x(t)$ or the duality property. ∎

5.5.2 Inverse Proportionality of Time and Frequency

It is very important to realize that frequency is inversely proportional to time, and that as such, time and frequency signal characterizations are complementary. Consider the following examples to illustrate this.

- The impulse signal $x_1(t) = \delta(t)$, although not a regular signal, has finite support (its support is only at $t = 0$ as the signal is zero everywhere else). It is also absolutely integrable, so it has a Fourier transform

$$X_1(\Omega) = \mathcal{F}[\delta(t)] = \int_{-\infty}^{\infty} \delta(t)e^{-j\Omega t}dt = e^{-j0}\int_{-\infty}^{\infty} \delta(t)dt = 1 \quad -\infty < \Omega < \infty$$

displaying infinite support. (The Fourier transform could have also been obtained from the Laplace transform $\mathcal{L}[\delta(t)] = 1$ for all values of s. For $s = j\Omega$, we have that $\mathcal{F}[\delta(t)] = 1$.) This result means that since $\delta(t)$ changes so fast in such a short time, its Fourier transform has all possible frequency components.

- Consider then the opposite case: A signal that is constant for all times, that does not change, or a dc signal $x_2(t) = A$, $-\infty < t < \infty$. We know that the frequency of $\Omega = 0$ is assigned to it since the signal does not vary at all. The Fourier transform cannot be found by means of the integral because $x_2(t)$ is not absolutely integrable, but we can verify that it is given by $X_2(\Omega) = 2\pi A\delta(\Omega)$ (we will formally show this using the duality property). In fact, the inverse Fourier transform is

$$\frac{1}{2\pi}\int_{-\infty}^{\infty} X_2(\Omega)e^{j\Omega t}d\Omega = \frac{1}{2\pi}\int_{-\infty}^{\infty} 2\pi A\delta(\Omega)e^{j\Omega t}d\Omega = A$$

Notice the complementary nature of $x_1(t)$ and $x_2(t)$: $x_1(t) = \delta(t)$ has a one-point support, while $x_2(t) = A$ has infinite support. Their corresponding Fourier transforms $X_1(\Omega) = 1$ and $X_2(\Omega) = 2\pi A\delta(\Omega)$ have infinite and one-point support in the frequency domain, respectively.

- To appreciate the transition from the dc signal to the impulse signal, consider a pulse signal $x_3(t) = A[u(t + \tau/2) - u(t - \tau/2)]$. This signal has finite energy, and its Fourier transform can be found using its Laplace transform. We have

$$X_3(s) = \frac{A}{s}\left[e^{s\tau/2} - e^{-s\tau/2}\right]$$

with the whole s-plane as its region of convergence, so that

$$X_3(\Omega) = X(s)|_{s=j\Omega}$$

$$= A\frac{(e^{j\Omega\tau/2} - e^{-j\Omega\tau/2})}{j\Omega}$$

$$= A\tau\frac{\sin(\Omega\tau/2)}{\Omega\tau/2} \tag{5.5}$$

or a sinc function where $A\tau$ corresponds to the area under $x_3(t)$. The Fourier transform $X_3(\Omega)$ is an even function of Ω. At $\Omega = 0$ using L'Hôpital's rule we find that $X_3(0) = A\tau$. Finally, the Fourier transform of the pulse becomes zero when $\Omega = 2k\pi/\tau$, $k = \pm 1, \pm 2, \ldots$.

If we let $A = 1/\tau$ (so that the area of the pulse is unity), and let $\tau \to 0$, the pulse $x_3(t)$ becomes a delta function $\delta(t)$ in the limit and the sinc function expands (for $\tau \to 0$, $X_3(\Omega)$ is not zero for any finite value) to become unity. On the other hand, if we let $\tau \to \infty$, the pulse becomes a constant signal A extending from $-\infty$ to ∞, and the Fourier transform gets closer and closer to $\delta(\Omega)$ (the sinc function becomes zero at values very close to zero and the amplitude at $\Omega = 0$ becomes larger and larger, although the area under the curve remains constant). As shown above, $X_3(\Omega) = 2\pi A\delta(\Omega)$ is the transform of $x_3(t) = A$, $-\infty < t < \infty$.

■ To illustrate the transition in the Fourier transform as the time support increases, we used the following MATLAB script to compute the Fourier transform of pulses of the same amplitude $A = 1$ but different time supports 1 and 4. The script below shows the case when the support is 1, but it can be easily changed to get the support of 4. The symbolic MATLAB function fourier computes the Fourier transform. The results are shown in Figure 5.1.

```
%%%%%%%%%%%%%%%%%%%%%%
% Time-frequency relation
%%%%%%%%%%%%%%%%%%%%%%
syms t w
x = heaviside(t + 0.5) − heaviside(t − 0.5);
subplot(211)
ezplot(x, [− 3]);axis([−3 3 − 0.1 1.1]);grid
X = fourier(x)  % Fourier transform
subplot(212)
ezplot(X, [−50 50]); axis([−50 50 −1 5]);grid
```

In summary, the support of $X(\Omega)$ is inversely proportional to the support of $x(t)$. If $x(t)$ has a Fourier transform $X(\Omega)$ and $\alpha \neq 0$ is a real number, then $x(\alpha t)$ is an

■ Contracted ($\alpha > 1$),
■ Contracted and reflected ($\alpha < -1$),
■ Expanded ($0 < \alpha < 1$),
■ Expanded and reflected ($-1 < \alpha < 0$), or
■ Simply reflected ($\alpha = -1$)

signal, and we have the pair

$$x(\alpha t) \quad \Leftrightarrow \quad \frac{1}{|\alpha|} X\left(\frac{\Omega}{\alpha}\right) \tag{5.6}$$

First let us mention that the symbol $\Leftrightarrow$ means that to a signal $x(t)$ in the time domain (on the left) there corresponds a Fourier transform $X(\Omega)$ in the frequency domain (on the right). This is *not* an equality—far from it!

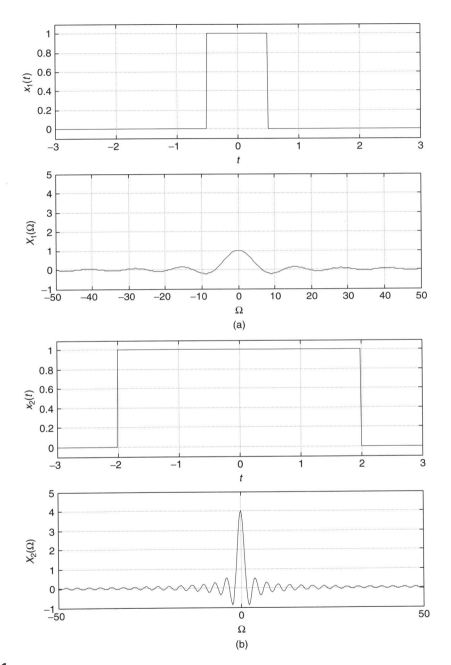

FIGURE 5.1

Fourier transform of pulses (a) $x_1(t)$ with $A = 1$ and $\tau = 1$, and (b) $x_2(t)$ with $A = 1$ and $\tau = 4$. Notice the wider the pulse the more concentrated in frequency its Fourier transform, and that $X_i(0) = A\tau$, $i = 1, 2$, is the area under the pulses.

This property is shown by a change of variable in the integration,

$$\mathcal{F}[x(\alpha t)] = \int_{-\infty}^{\infty} x(\alpha t)e^{-j\Omega t}dt = \begin{cases} \frac{1}{\alpha} \int_{-\infty}^{\infty} x(\rho)e^{-j\Omega\rho/\alpha}d\rho & \alpha > 0 \\ \\ -\frac{1}{\alpha} \int_{-\infty}^{\infty} x(\rho)e^{j\Omega\rho/\alpha}d\rho & \alpha < 0 \end{cases}$$

$$= \frac{1}{|\alpha|}X\left(\frac{\Omega}{\alpha}\right)$$

by change of variable $\rho = \alpha t$. If $|\alpha| > 1$, when compared with $x(t)$ the signal $x(\alpha t)$ contracts while its corresponding Fourier transform expands. Likewise, when $0 < |\alpha| < 1$, the signal $x(\alpha t)$ expands, as compared with $x(t)$, and its Fourier transform contracts. If $\alpha < 0$, the corresponding contraction or expansion is accompanied by a reflection in time. In particular, if $\alpha = -1$, the reflected signal $x(-t)$ has $X(-\Omega)$ as its Fourier transform.

■ Example 5.3

Consider a pulse $x(t) = u(t) - u(t - 1)$. Find the Fourier transform of $x_1(t) = x(2t)$.

Solution

The Laplace transform of $x(t)$ is

$$X(s) = \frac{1 - e^{-s}}{s}$$

with the whole s-plane as its region of convergence. Thus, its Fourier transform is

$$X(\Omega) = \frac{1 - e^{-j\Omega}}{j\Omega} = \frac{e^{-j\Omega/2}(e^{j\Omega/2} - e^{-j\Omega/2})}{2j\Omega/2}$$

$$= \frac{\sin(\Omega/2)}{\Omega/2}e^{-j\Omega/2}$$

To the finite-support signal $x(t)$ corresponds $X(\Omega)$ of infinite support. Then,

$$x_1(t) = x(2t) = u(2t) - u(2t - 1) = u(t) - u(t - 0.5)$$

and its Fourier transform is found, again using its Laplace transform, to be

$$X_1(\Omega) = \frac{1 - e^{-j\Omega/2}}{j\Omega} = \frac{e^{-j\Omega/4}(e^{j\Omega/4} - e^{-j\Omega/4})}{j\Omega}$$

$$= \frac{1}{2}\frac{\sin(\Omega/4)}{\Omega/4}e^{-j\Omega/4} = \frac{1}{2}X(\Omega/2)$$

which is an expanded version of $X(\Omega)$ in the frequency domain and coincides with the result from the property. See Figure 5.2.

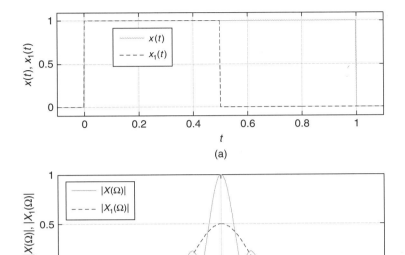

FIGURE 5.2

(a) Pulse $x(t)$ and its compressed version $x_1(t) = x(2t)$, and (b) the magnitude of their Fourier transforms. Notice that when the signal contracts in time it expands in frequency.

The Fourier transforms can be found from the integral definitions. Thus, for $x(t)$,

$$X(\Omega) = \int_0^1 1e^{j\Omega t}dt = \frac{e^{-j\Omega t}}{-j\Omega}\Big|_0^1 = \frac{\sin(\Omega/2)}{\Omega/2}e^{-j\Omega/2}$$

Likewise, for $x_1(t)$,

$$X_1(\Omega) = \int_0^{0.5} 1e^{j\Omega t}dt = 0.5\frac{\sin(\Omega/4)}{\Omega/4}e^{-j\Omega/4}$$

∎

■ Example 5.4

Apply the reflection property to find the Fourier transform of $x(t) = e^{-a|t|}$, $a > 0$. For $a = 1$, plot using MATLAB the signal and its magnitude and phase spectra.

Solution

The signal $x(t)$ can be expressed as $x(t) = e^{-at}u(t) + e^{at}u(-t) = x_1(t) + x_1(-t)$. The Fourier transform of $x_1(t)$ is

$$X_1(\Omega) = \frac{1}{s+a}\Big|_{s=j\Omega} = \frac{1}{j\Omega + a}$$

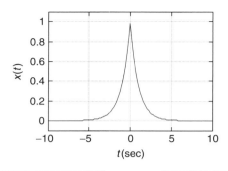

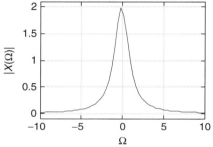
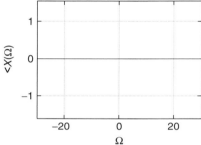

FIGURE 5.3
Magnitude and phase spectrum of two-sided signal $x(t) = e^{-|t|}$. The magnitude spectrum indicates $x(t)$ is low pass. Notice the phase is zero.

and according to the given result $x_1(-t)$ ($\alpha = -1$), we have that

$$\mathcal{F}[x_1(-t)] = \frac{1}{-j\Omega + a}$$

so that

$$X(\Omega) = \frac{1}{j\Omega + a} + \frac{1}{-j\Omega + a} = \frac{2a}{a^2 + \Omega^2}$$

If $a = 1$, using MATLAB the signal $x(t) = e^{-|t|}$ and its magnitude and phase spectra are computed and plotted as shown in Figure 5.3. Since $X(\Omega)$ is real and positive, the corresponding phase spectrum is zero. This signal is called *low-pass* since its energy is concentrated in the low frequencies. ∎

5.5.3 Duality

Besides the inverse relationship of frequency and time, by interchanging the frequency and the time variables in the definitions of the direct and the inverse Fourier transform (see Eqs. 5.1 and 5.2) similar equations are obtained. Thus, the direct and the inverse Fourier transforms are dual.

To the Fourier transform pair

$$x(t) \quad \Leftrightarrow \quad X(\Omega) \tag{5.7}$$

corresponds the following dual–Fourier transform pair

$$X(t) \quad \Leftrightarrow \quad 2\pi x(-\Omega) \tag{5.8}$$

This can be shown by considering the inverse Fourier transform

$$x(t) = \frac{1}{2\pi} \int_{-\infty}^{\infty} X(\rho)e^{j\rho t}d\rho$$

and replacing t by $-\Omega$ and multiplying by 2π to get

$$2\pi x(-\Omega) = \int_{-\infty}^{\infty} X(\rho)e^{-j\rho\Omega}d\rho$$

$$= \int_{-\infty}^{\infty} X(t)e^{-j\Omega t}dt$$

$$= \mathcal{F}[X(t)]$$

To understand the above equations you need to realize that ρ and t are dummy variables inside the integral, and as such they are not reflected outside the integral.

Remarks

- *This duality property allows us to obtain the Fourier transform of signals for which we already have a Fourier pair and that would be difficult to obtain directly. It is thus one more method to obtain the Fourier transform, besides the Laplace transform and the integral definition of the Fourier transform.*
- *When computing the Fourier transform of a constant signal, $x(t) = A$, we indicated that it would be $X(\Omega) = 2\pi A\delta(\Omega)$. Indeed, we have the dual pairs*

$$A\delta(t) \quad \Leftrightarrow \quad A$$
$$A \quad \Leftrightarrow \quad 2\pi A\delta(-\Omega) = 2\pi A\delta(\Omega) \tag{5.9}$$

where in the second equation we use the fact that $\delta(\Omega)$ is even.

■ Example 5.5

Use the duality property to find the Fourier transform of the sinc signal

$$x(t) = A\frac{\sin(0.5t)}{0.5t} = A\,\text{sinc}(0.5t) \qquad -\infty < t < \infty$$

Solution

The Fourier transform of the sinc signal cannot be found using the Laplace transform or the integral definition of the Fourier transform. The duality property provides a way to obtain it. We found before, for $\tau = 1$, the following pair of Fourier transforms:

$$A[u(t+0.5) - u(t-0.5)] \quad \Leftrightarrow \quad A\frac{\sin(0.5\Omega)}{0.5\Omega} = A\,\text{sinc}(0.5\Omega)$$

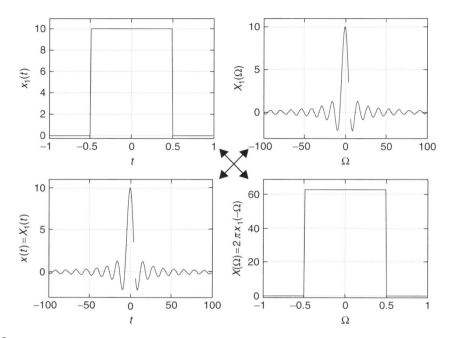

FIGURE 5.4

Application of duality to find the Fourier transform of $x(t) = 10 \operatorname{sinc}(0.5t)$. Notice that
$X(\Omega) = 2\pi x_1(\Omega) \approx 6.28 x_1(\Omega) = 62.8[u(\Omega + 0.5) - u(\Omega - 0.5)]$.

Then according to the duality property, the Fourier transform of $x(t)$ is

$$X(\Omega) = 2\pi A[u(-\Omega + 0.5) - u(-\Omega - 0.5)] = 2\pi A[u(\Omega + 0.5) - u(\Omega - 0.5)]$$

given that the function is even with respect to Ω. So the Fourier transform of the sinc is a rectangular pulse in frequency, in the same way that the Fourier transform of a pulse in time is a sinc function in frequency. Figure 5.4 shows the dual pairs for $A = 10$.   ■

■ Example 5.6

Find the Fourier transform of $x(t) = A \cos(\Omega_0 t)$ using duality.

Solution

The Fourier transform of $x(t)$ cannot be computed using the integral definition since this signal is not absolutely integrable, or using the Laplace transform since $x(t)$ does not have a Laplace transform. As a periodic signal, $x(t)$ has a Fourier series representation and we will use it later to find its Fourier transform. For now, let us consider the Fourier pair

$$\delta(t - \rho_0) + \delta(t + \rho_0) \iff e^{-j\rho_0\Omega} + e^{j\rho_0\Omega} = 2\cos(\rho_0\Omega)$$

where we used the Laplace transform of $\delta(t - \rho_0) + \delta(t + \rho_0)$, which is $e^{-s\rho_0} + e^{s\rho_0}$ defined over the whole s-plane. At $s = j\Omega$, we get $2\cos(\rho_0\Omega)$. According to the duality property, we thus have the following Fourier pair:

$$2\cos(\rho_0 t) \quad \Leftrightarrow \quad 2\pi[\delta(-\Omega - \rho_0) + \delta(-\Omega + \rho_0)] = 2\pi[\delta(\Omega + \rho_0) + \delta(\Omega - \rho_0)]$$

Replacing ρ_0 by Ω_0 and canceling the 2 in both sides, and multiplying by A both sides, we have

$$x(t) = A\cos(\Omega_0 t) \quad \Leftrightarrow \quad X(\Omega) = \pi A[\delta(\Omega + \Omega_0) + \delta(\Omega - \Omega_0)] \tag{5.10}$$

indicating that it only exists at $\pm\Omega_0$. ■

5.6 SPECTRAL REPRESENTATION

In this section, we consider first how to find the Fourier transform of periodic signals using the modulation property, and then consider Parseval's result for finite-energy signals. With these results, we will unify the spectral representation of both periodic and aperiodic signals.

5.6.1 Signal Modulation

One of the most significant properties of the Fourier transform is modulation. Its application to signal transmission is fundamental in communications.

■ Frequency shift: If $X(\Omega)$ is the Fourier transform of $x(t)$, then we have the pair

$$x(t)e^{j\Omega_0 t} \quad \Leftrightarrow \quad X(\Omega - \Omega_0) \tag{5.11}$$

■ Modulation: The Fourier transform of the modulated signal

$$x(t)\cos(\Omega_0 t) \tag{5.12}$$

is given by

$$0.5\left[X(\Omega - \Omega_0) + X(\Omega + \Omega_0)\right] \tag{5.13}$$

That is, $X(\Omega)$ is shifted to frequencies Ω_0 and $-\Omega_0$, and multiplied by 0.5.

The frequency shifting property is easily shown:

$$\mathcal{F}[x(t)e^{j\Omega_0 t}] = \int_{-\infty}^{\infty} [x(t)e^{j\Omega_0 t}]e^{-j\Omega t}dt$$

$$= \int_{-\infty}^{\infty} x(t)e^{-j(\Omega - \Omega_0)t}dt$$

$$= X(\Omega - \Omega_0)$$

Applying the frequency shifting to

$$x(t)\cos(\Omega_0 t) = 0.5x(t)e^{j\Omega_0 t} + 0.5x(t)e^{-j\Omega_0 t}$$

we obtain the Fourier transform of the modulated signal (Eq. 5.13). In communications, the *message* $x(t)$ (typically of lower frequency content than the frequency of the cosine) modulates the *carrier* $\cos(\Omega_0 t)$ to obtain the *modulated signal* $x(t)\cos(\Omega_0 t)$. Modulation is an important application of the Fourier transform, as it allows us to change the original frequencies of a message to much higher frequencies, making it possible to transmit the signal over the airwaves.

Remarks

- *As indicated before, amplitude modulation consists in multiplying an incoming signal $x(t)$, or message, by a sinusoid of frequency higher than the maximum frequency of the incoming signal. The modulated signal is*

$$x(t)\cos(\Omega_0 t) = 0.5[x(t)e^{j\Omega_0 t} + x(t)e^{-j\Omega_0 t}]$$

  with a Fourier transform, according to the frequency shifting property, of

$$\mathcal{F}[x(t)\cos(\Omega_0 t)] = 0.5[X(\Omega - \Omega_0) + X(\Omega + \Omega_0)]$$

  Thus, modulation shifts the frequencies of $x(t)$ to frequencies around $\pm\Omega_0$.
- *Modulation using a sine, instead of a cosine, changes the phase of the Fourier transform of the incoming signal besides performing the frequency shift. Indeed,*

$$\mathcal{F}[x(t)\sin(\Omega_0 t)] = \mathcal{F}\left[\frac{x(t)e^{j\Omega_0 t} - x(t)e^{-j\Omega_0 t}}{2j}\right]$$

$$= \frac{1}{2j}X(\Omega - \Omega_0) - \frac{1}{2j}X(\Omega + \Omega_0)$$

$$= \frac{-j}{2}X(\Omega - \Omega_0) + \frac{j}{2}X(\Omega + \Omega_0)$$

  where the $-j$ and j terms add $-\pi/2$ and $\pi/2$, respectively, radians to the signal phase.
- *According to the eigenfunction property of LTI systems, modulation systems are not LTI. Modulation shifts the frequencies at the input to new frequencies at the output. Nonlinear or time-varying systems are typically used as amplitude modulation transmitters.*

■ Example 5.7

Consider modulating a carrier $\cos(10t)$ with the following signals:

1. $x_1(t) = e^{-|t|}$, $-\infty < t < \infty$. Use MATLAB to find the Fourier transform of $y_1(t) = x_1(t)\cos(10t)$ and plot $y_1(t)$ and its magnitude and phase spectra.
2. $x_2(t) = 0.2[r(t + 5) - 2r(t) + r(t + 5)]$, where $r(t)$ is the ramp signal. Use MATLAB to plot $x_2(t)$ and $y_2(t) = x_2(t)\cos(10t)$ and compute and plot the magnitude of their Fourier transforms.

Solution

The modulated signals are

- $y_1(t) = x_1(t)\cos(10t) = e^{-|t|}\cos(10t)$ $-\infty < t < \infty$
- $y_2(t) = x_2(t)\cos(10t) = 0.2[r(t+5) - 2r(t) + r(t+5)]\cos(10t)$

The signal $x_1(t)$ is very smooth, although of infinite support, and thus most of its frequency components are of low frequency. The signal $x_2(t)$ is not as smooth and has a finite support, so that its frequency components are mostly low pass but its spectrum also displays higher frequencies.

The MATLAB scripts used to compute the Fourier transform of the modulated signals and to plot the signals, their modulated versions, and the magnitude and phase of the Fourier transforms are very similar. The following script indicates how to generate $y_1(t)$ and how to find the magnitude and phase of its Fourier transform $Y_1(\Omega)$. Notice the way the phase is computed.

```
%%%%%%%%%%%%%%%%%%%%%%
% Example 5.7---Modulation
%%%%%%%%%%%%%%%%%%%%%%
y1 = exp(−abs(t)). * cos(10 * t);
% magnitude and phase of Y1(Omega)
Y1 = fourier(y1); Ym = abs(Y1); Ya = atan(imag(Y1)/real(Y1));
```

The signal $x_2(t)$ is a triangular signal. The following script shows how to generate the signal $x_2(t)$. Instead of multiplying $x_2(t)$ by the cosine, we multiply it by the cosine-equivalent representation in complex exponentials, which will give better plots of the Fourier transforms when using ezplot.

```
m = heaviside(t + 5) − heaviside(t);
m1 = heaviside(t) − heaviside(t−5);
x2 = (t + 5) * m + m1 * (−t + 5); x2 = x2/5;
x = x2 * exp(−j * 10 * t)/2; y = x2 * exp(+j * 10 * t)/2;
X = fourier(x); Y = fourier(y);
Y2m = abs(X) + abs(Y); % magnitude of Y_2(Omega)
X2 = fourier(x2); X2m = abs(X2); % magnitude of X_2(Omega)
```

The results are shown in Figure 5.5.   ∎

Why Modulation?

The use of modulation to change the frequency content of a message from its baseband frequencies to higher frequencies makes its transmission over the airwaves possible. Let us explore why it is necessary to use modulation to transmit a music or a speech signal. Typically, acoustic signals such as music are audible up to frequencies of about 22 KHz, while speech signals typically display frequencies from about 100 Hz to about 5 KHz. Thus, music and speech signals are relatively low-frequency

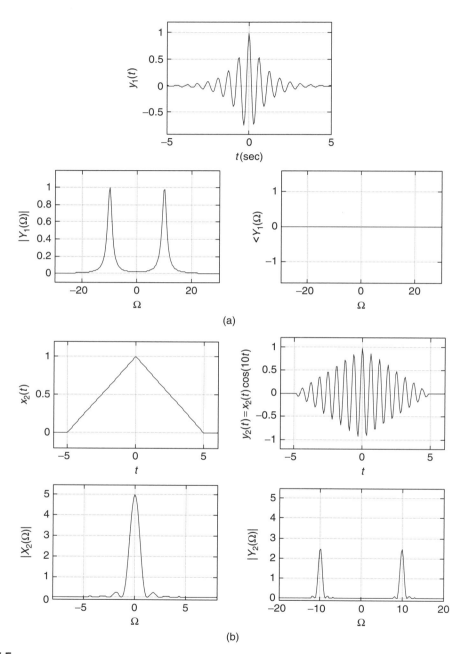

FIGURE 5.5

(a) The modulated signal $y_1(t) = e^{-|t|}\cos(10t)$ and its magnitude and phase spectra. Notice that the phase is zero. (b) The triangular signal, its modulated version, and their magnitude spectra.

signals. When radiating a signal with an antenna, the length of the antenna is about a quarter of the wavelength,

$$\lambda = \frac{3 \times 10^8}{f} \text{ meters}$$

where f is the frequency in hertz of the signal being radiated. Thus, if we assume that frequencies up to $f = 30$ KHz are present in the signal (this would allow us to include music and speech in the signal) the wavelength is 10 kilometers and the size of the antenna is 2.5 kilometers—a 1.5-mile long antenna! Thus, for a music or a speech signal to be transmitted with a reasonable-size antenna requires increasing the frequencies present in the signal. Modulation provides an efficient way to shift an acoustic or speech signal to a desirable frequency.

5.6.2 Fourier Transform of Periodic Signals

By applying the frequency-shifting property to compute the Fourier transform of periodic signals, we are able to unify the Fourier representation of aperiodic as well as periodic signals.

For a periodic signal $x(t)$ of period T_0, we have the Fourier pair

$$x(t) = \sum_k X_k e^{jk\Omega_0 t} \quad \Leftrightarrow \quad X(\Omega) = \sum_k 2\pi X_k \delta(\Omega - k\Omega_0) \tag{5.14}$$

obtained by representing $x(t)$ by its Fourier series.

Since a periodic signal $x(t)$ is not absolutely integrable, its Fourier transform cannot be computed using the integral formula. But we can use its Fourier series

$$x(t) = \sum_k X_k e^{jk\Omega_0 t}$$

where the $\{X_k\}$ are the Fourier coefficients, and $\Omega_0 = 2\pi/T_0$ is the fundamental frequency of the periodic signal $x(t)$ of period T_0. As such, according to the linearity and the frequency-shifting properties of the Fourier transform, we obtain

$$X(\Omega) = \sum_k \mathcal{F}[X_k e^{jk\Omega_0 t}]$$

$$= \sum_k 2\pi X_k \delta(\Omega - k\Omega_0)$$

where we used that X_k as a constant has a Fourier transform $2\pi X_k \delta(\Omega)$. Notice that for a periodic signal the Fourier coefficients $\{X_k\}$ still characterize its frequency representation: The Fourier transform of a periodic signal is a sequence of impulses in frequency at the harmonic frequencies, $\{\delta(\Omega - k\Omega_0)\}$, with amplitudes $\{2\pi X_k\}$.

Remarks

- *When plotting $|X(\Omega)|$ versus Ω, which we call the Fourier magnitude spectrum, for a periodic signal $x(t)$, we notice it is analogous to its line spectrum discussed before. Both indicate that the signal power is concentrated in multiples of the fundamental frequency, the only difference being in how the information is provided at each of the frequencies. The line spectrum displays the Fourier series coefficients at their corresponding frequencies, while the spectrum from the Fourier transform displays the concentration of the power at the harmonic frequencies by means of delta functions with amplitudes of 2π times the Fourier series coefficients. Thus, there is a clear relation between these two spectra, showing exactly the same information in slightly different form.*
- *The Fourier transform of a cosine signal can now be computed directly as*

$$\mathcal{F}[\cos(\Omega_0 t)] = \mathcal{F}[0.5e^{j\Omega_0 t} + 0.5e^{-j\Omega_0 t}]$$

$$= \pi\delta(\Omega - \Omega_0) + \pi\delta(\Omega + \Omega_0)$$

and for a sine (compare this result with the one obtained before),

$$\mathcal{F}[\sin(\Omega_0 t)] = \mathcal{F}\left[\frac{0.5}{j}e^{j\Omega_0 t} - \frac{0.5}{j}e^{-j\Omega_0 t}\right]$$

$$= \frac{\pi}{j}\delta(\Omega - \Omega_0) - \frac{\pi}{j}\delta(\Omega + \Omega_0)$$

$$= \pi e^{-j\pi/2}\delta(\Omega - \Omega_0) + \pi e^{j\pi/2}\delta(\Omega + \Omega_0)$$

The magnitude spectra of the two signals coincide, but the cosine has a zero-phase spectrum, while the phase spectrum for the sine displays a phase of $\mp\pi/2$ at frequencies $\pm\Omega_0$.

■ Example 5.8

Consider a periodic signal $x(t)$ with a period

$$x_1(t) = r(t) - 2r(t - 0.5) + r(t - 1)$$

If the fundamental frequency is $\Omega_0 = 2\pi$, determine the Fourier transform $X(\Omega)$ analytically and using MATLAB. Plot several periods of the signal and its Fourier transform.

Solution

The given period $x_1(t)$ corresponds to a triangular signal. Its Laplace transform is

$$X_1(s) = \frac{1}{s^2}\left(1 - 2e^{-0.5s} + e^{-s}\right) = \frac{e^{-0.5s}}{s^2}\left(e^{0.5s} - 2 + e^{-0.5s}\right)$$

so that the Fourier coefficients of $x(t)$ are ($T_0 = 1$):

$$X_k = \frac{1}{T_0} X_1(s)|_{s=j2\pi k} = \frac{1}{(j2\pi k)^2} 2(\cos(\pi k) - 1)e^{-j\pi k}$$

$$= (-1)^{(k+1)} \frac{\cos(\pi k) - 1}{2\pi^2 k^2} = (-1)^k \frac{\sin^2(\pi k/2)}{\pi^2 k^2}$$

after using the identity $\cos(2\theta) - 1 = -2\sin^2(\theta)$. The DC term is $X_0 = 0.5$. The Fourier transform of $x(t)$ is then

$$X(\Omega) = 2\pi X_0 \delta(\Omega) + \sum_{k=-\infty, \neq 0}^{\infty} 2\pi X_k \delta(\Omega - 2k\pi)$$

To compute the Fourier transform using symbolic MATLAB, we approximate $x(t)$ by its Fourier series by means of its average and $N = 10$ harmonics (the Fourier coefficients are computed using the fourierseries function from Chapter 4). We then create a sequence $\{2\pi X_k\}$ and the corresponding harmonic frequencies $\{\Omega_k = k\Omega_0\}$ and plot them as the spectrum $X(\Omega)$ (see Figure 5.6). The following script gives some of the necessary steps to generate the periodic signal and to find its Fourier transform. The MATLAB function fliplr is used to reflect the Fourier coefficients.

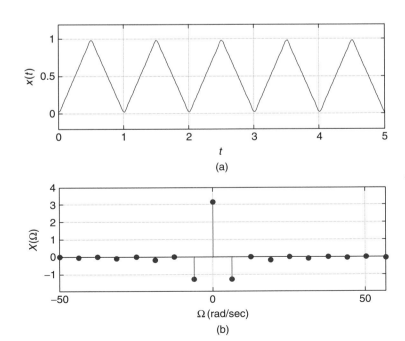

FIGURE 5.6

(a) Triangular periodic signal $x(t)$, and (b) its Fourier transform $X(\Omega)$, which is zero except at harmonic frequencies where it is an impulse of magnitude $2\pi X_k$ where X_k is a Fourier coefficient of $x(t)$.

```
%%%%%%%%%%%%%%%%%%%%%
% Example 5.8---Fourier series
%%%%%%%%%%%%%%%%%%%%%%
   T0 = 1; N = 10; w0 = 2 * pi/T0;
   m = heaviside(t) − heaviside(t − T0/2);
   m1 = heaviside(t − T0/2) − heaviside(t − T0);
   x = t * m + m1 * (−t + T0); x = 2 * x; % periodic signal
   [Xk, w] = fourierseries(x, T0, N); % Fourier coefficients, harmonic frequencies
   % Fourier series approximation
   for k = 1:N,
       if k == 1;
          x1 = abs(Xk(k));
       else
          x1 = x1 + 2 * abs(Xk(k)) * cos(w0 * (k−1) * t + angle(Xk(k)));
       end
   end
   % sequence of Fourier coefficients and harmonic frequencies
   k = 0:N−1; Xk1 = 2 * pi * Xk; wk = [−fliplr(k(2:N−1)) k] * w0; Xk = [fliplr(Xk1(2:N − 1)) Xk1];
```

In this case, the Laplace transform simplifies the computation of the X_k values. Indeed, the Fourier series coefficients are given by

$$X_k = \int_0^{0.5} t e^{-j2\pi kt} dt + \int_{0.5}^{1} (1 - t) e^{-j2\pi kt} dt$$

which need to be found using integration by parts. ∎

5.6.3 Parseval's Energy Conservation

We saw in Chapter 4 that for periodic signals having finite power but infinite energy, Parseval's theorem indicates how the power of the signal is distributed among the harmonic components. Likewise, for aperiodic signals of finite energy, an energy version of Parseval's result indicates how the signal energy is distributed over frequencies.

For a finite-energy signal $x(t)$ with Fourier transform $X(\Omega)$, its energy is conserved when going from the time to the frequency domain, or

$$E_x = \int_{-\infty}^{\infty} |x(t)|^2 dt = \frac{1}{2\pi} \int_{-\infty}^{\infty} |X(\Omega)|^2 d\Omega \qquad (5.15)$$

Thus, $|X(\Omega)|^2$ is an energy density indicating the amount of energy at each of the frequencies Ω.

The plot $|X(\Omega)|^2$ versus Ω is called the energy spectrum of $x(t)$, and it displays how the energy of the signal is distributed over frequency.

This energy conservation property is shown using the inverse Fourier transform. The finite-energy signal of $x(t)$ can be computed in the frequency domain by

$$\int_{-\infty}^{\infty} x(t)x^*(t)dt = \int_{-\infty}^{\infty} x(t) \left[\frac{1}{2\pi} \int_{-\infty}^{\infty} X^*(\Omega)e^{-j\Omega t}d\Omega \right] dt$$

$$= \frac{1}{2\pi} \int_{-\infty}^{\infty} X^*(\Omega) \left[\int_{-\infty}^{\infty} x(t)e^{-j\Omega t}dt \right] d\Omega$$

$$= \frac{1}{2\pi} \int_{-\infty}^{\infty} |X(\Omega)|^2 d\Omega$$

■ Example 5.9

Parseval's result helps us to understand better the nature of an impulse $\delta(t)$. It is clear from its definition that the area under an impulse is unity, which means $\delta(t)$ is absolutely integrable, but does it have finite energy? Show how Parseval's result can help resolve this issue.

Solution

Let's consider this from the frequency point of view, using Parseval's result. The Fourier transform of $\delta(t)$ is unity for all values of frequency and as such its energy is infinite. Such a result seems puzzling, because $\delta(t)$ was defined as the limit of a pulse of finite duration and unity area. This is what happens if

$$p_\Delta(t) = \frac{1}{\Delta}[u(t + \Delta/2) - u(t - \Delta/2)]$$

is a pulse of the unity area from which we obtain the impulse by letting $\Delta \to 0$. The signal

$$p_\Delta^2(t) = \frac{1}{\Delta^2}[u(t + \Delta/2) - u(t - \Delta/2)]$$

is a pulse of area $1/\Delta$. If we then let $\Delta \to 0$, the squared pulse $p_\Delta^2(t)$ will tend to infinity with an infinite area under it. Thus, $\delta(t)$ is not finite energy. ■

■ Example 5.10

Consider a pulse $p(t) = u(t + 1) - u(t - 1)$. Use its Fourier transform $P(\Omega)$ and Parseval's result to show that

$$\int_{-\infty}^{\infty} \left(\frac{\sin(\Omega)}{\Omega} \right)^2 d\Omega = \pi$$

Solution

The energy of the pulse is $E = 2$ (the area under the pulse). But according to Parseval's result the energy computed in the frequency domain is given by

$$\frac{1}{2\pi} \int_{-\infty}^{\infty} \left(\frac{2\sin(\Omega)}{\Omega} \right)^2 d\Omega = E_x$$

since $2\sin(\Omega)/\Omega = P(\Omega) = \mathcal{F}(p(t))$. Replacing E_x, we obtain the interesting and not obvious result

$$\int_{-\infty}^{\infty} \left(\frac{\sin(\Omega)}{\Omega} \right)^2 d\Omega = \pi$$

This is one more way to compute π! ∎

5.6.4 Symmetry of Spectral Representations

Now that the Fourier representation of aperiodic and periodic signals is unified, we can think of just one spectrum that accommodates both finite-energy as well as infinite-energy signals. The word *spectrum* is loosely used to mean different aspects of the frequency representation. In the following we provide definitions and the symmetry characteristic of the spectrum of real-valued signals.

If $X(\Omega)$ is the Fourier transform of a real-valued signal $x(t)$, periodic or aperiodic, the magnitude $|X(\Omega)|$ is an even function of Ω:

$$|X(\Omega)| = |X(-\Omega)| \tag{5.16}$$

and the phase $\angle X(\Omega)$ is an odd function of Ω:

$$\angle X(\Omega) = -\angle X(-\Omega) \tag{5.17}$$

We then have:

Magnitude spectrum: $|X(\Omega)|$ versus Ω

Phase spectrum: $\angle X(\Omega)$ versus Ω

Energy/power spectrum: $|X(\Omega)|^2$ versus Ω

To show this, consider the inverse Fourier transform of a real-valued signal $x(t)$,

$$x(t) = \frac{1}{2\pi} \int_{-\infty}^{\infty} X(\Omega) e^{j\Omega t} d\Omega$$

which, because of being real, is identical to

$$x^*(t) = \frac{1}{2\pi} \int_{-\infty}^{\infty} X^*(\Omega) e^{-j\Omega t} d\Omega = \frac{1}{2\pi} \int_{-\infty}^{\infty} X^*(-\Omega) e^{j\Omega t} d\Omega$$

since the integral can be thought of as an infinite sum of complex values. Comparing the two integrals, we have that

$$X(\Omega) = X^*(-\Omega)$$

or

$$|X(\Omega)|e^{j\theta(\Omega)} = |X(-\Omega)|e^{-j\theta(-\Omega)}$$

where $\theta(\Omega) = \angle(X(\Omega))$ is the phase of $X(\Omega)$. We can then see that

$$|X(\Omega)| = |X(-\Omega)|$$
$$\theta(\Omega) = -\theta(-\Omega)$$

or that the magnitude is an even function of Ω and the phase is an odd function of Ω. It can also be seen that

$$\mathcal{Re}[X(\Omega)] = \mathcal{Re}[X(-\Omega)]$$
$$\mathcal{Im}[X(\Omega)] = -\mathcal{Im}[X(-\Omega)]$$

or that the real part of the Fourier transform is an even function and that the imaginary part of the Fourier transform is an odd function of Ω.

Remarks

- *Clearly, if the signal is complex, the above symmetry will not hold. For instance, if $x(t) = e^{j\Omega_0 t} = \cos(\Omega_0 t) + j\sin(\Omega_0 t)$, using the frequency-shift property its Fourier transform is*

$$X(\Omega) = 2\pi\delta(\Omega - \Omega_0)$$

  which occurs at $\Omega = \Omega_0$ only, so the symmetry in the magnitude and phase does not exist.
- *It is important to recognize the meaning of "negative" frequencies. In reality, only positive frequencies exist and can be measured, but as shown the spectrum, magnitude or phase, of a real-valued signal requires negative frequencies. It is only under this context that negative frequencies should be understood as necessary to generate "real-valued" signals.*

■ Example 5.11

Use MATLAB to compute the Fourier transform of the following signals:

$$\text{(a) } x_1(t) = u(t) - u(t-1)$$
$$\text{(b) } x_2(t) = e^{-t}u(t)$$

Plot their magnitude and phase spectra.

Solution

Three possible ways to compute the Fourier transforms of these signals using MATLAB are: (1) find their Laplace transforms as in Chapter 3 using laplace and compute the magnitude and phase function by letting $s = j\Omega$, (2) by using the symbolic function fourier, and (3) sample $x(t)$ and find its Fourier transform (this requires sampling theory—see Chapter 7).

The following script is used to compute and plot the signal $x_2(t) = e^{-t}u(t)$ and the magnitude and phase of its Fourier transform using symbolic MATLAB. A similar script is used for $x_1(t)$.

```
%%%%%%%%%%%%%%%%%%%%%%%
% Example 5.11
%%%%%%%%%%%%%%%%%%%%%%%
symm t
x2 = exp(-t) * heaviside(t);
X2 = fourier(x2)
X2m = sqrt((real(X2))^2 + (imag(X2))^2; % magnitude
X2p = imag(log(X2); % phase
```

Notice the way that the magnitude and the phase are computed. To compute the magnitude we use

$$|X_2(\Omega)| = \sqrt{\mathcal{R}e[X_2(\Omega)]^2 + \mathcal{I}m[X_2(\Omega)]^2}.$$

The computation of the phase is complicated by the lack of the function atan2 in symbolic MAT-LAB, which extends the principal values of the inverse tangent to $(-\pi, \pi]$ by considering the sign of the real part. The phase computation can be done by using the log function; indeed

$$\log(X_2(\Omega)) = \log\left(|X_2(\Omega)|e^{j\angle X_2(\Omega)}\right) = \log(|X_2(\Omega)|) + j\angle X_2(\Omega)$$

so that

$$\angle X_2(\Omega) = \mathcal{I}m[\log(X_2(\Omega))]$$

Analytically, the phase of the Fourier transform of $x_1(t) = u(t) - u(t-1)$ can be found by considering the advanced signal $z(t) = x_1(t + 0.5) = u(t + 0.5) - u(t - 0.5)$ with Fourier transform

$$Z(\Omega) = \frac{\sin(\Omega/2)}{\Omega/2}$$

Given that $Z(\Omega)$ is real, its phase is either zero when $Z(\Omega) \geq 0$ and $\pm\pi$ when $Z(\Omega) < 0$ (using these values so that the phase is an odd function of Ω). Since $z(t) = x_1(t + 0.5)$, then $Z(\Omega) = X_1(\Omega)e^{j0.5\Omega}$, so that

$$X_1(\Omega) = e^{-j0.5\Omega}Z(\Omega)$$

and as such

$$\angle X_1(\Omega) = \angle Z(\Omega) - 0.5\Omega = \begin{cases} -0.5\Omega & Z(\Omega) \geq 0 \\ \pm\pi - 0.5\Omega & Z(\Omega) < 0 \end{cases}$$

The Fourier transform of $x_2(t) = e^{-t}u(t)$ is

$$X_2(\Omega) = \frac{1}{1 + j\Omega}$$

The magnitude and phase are given by

$$|X_2(\Omega)| = \frac{1}{\sqrt{1 + \Omega^2}}$$

$$\theta(\Omega) = -\tan^{-1}\Omega$$

When we compute these in terms of Ω, we have

Ω	$|X_2(\Omega)|$	$\theta(\Omega)$
0	1	0
1	$\dfrac{1}{\sqrt{2}}$	$-\pi/4$
∞	0	$-\pi/2$

That is, the magnitude spectrum decays as Ω increases. The signal $x_2(t)$ is called *low-pass* given that the magnitude of its Fourier transform is concentrated in the low frequencies. This also implies that the signal $x_2(t)$ is rather smooth. See Figure 5.7 for results. ∎

■ Example 5.12

It is not always the case that the Fourier transform is a complex-valued function. Consider the signals

$$\text{(a)} \quad x(t) = 0.5e^{-|t|}$$

$$\text{(b)} \quad y(t) = e^{-|t|}\cos(\Omega_0 t)$$

Find their Fourier transforms. Discuss the smoothness of these signals.

Solution

(a) The Fourier transform of $x(t)$ is

$$X(\Omega) = \frac{1}{\Omega^2 + 1}$$

which is a real-valued function of Ω. Indeed,

Ω	$|X(\Omega)| = X(\Omega)$	$\theta(\Omega)$
0	1	0
1	0.5	0
∞	0	0

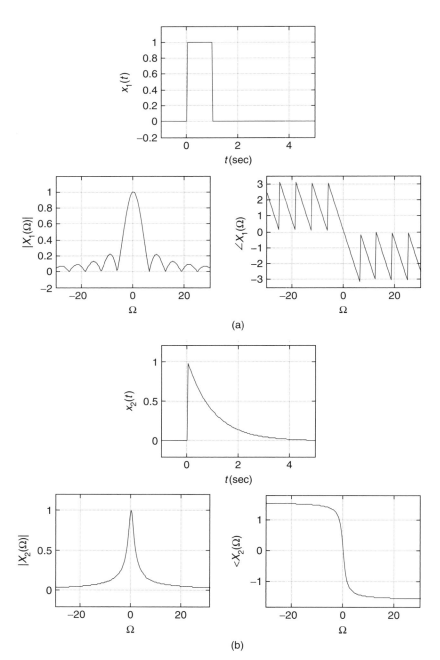

FIGURE 5.7

Fourier transforms of (a) pulse $x_1(t)$ and of (b) decaying exponential $x_2(t)$.

This is also a low-pass signal like $x_2(t)$ in Example 5.11, but this is "smoother" than that one because the magnitude response is more concentrated in the low frequencies. Compare the values of the magnitude responses to verify this. Also this signal has zero phase, because its Fourier transform is real and positive for all values of Ω.

(b) The signal $y(t) = 2x(t)\cos(\Omega_0 t)$ is a *band-pass* signal. It is not as smooth as the signals in the above example given that the concentration of

$$Y(\Omega) = X(\Omega - \Omega_0) + X(\Omega + \Omega_0)$$

is around the frequency Ω_0, a frequency typically higher than the frequencies in $x(t)$. The higher this frequency, the more variation is displayed by the signal. In communications, low-pass signals are called *base-band* signals. ∎

The bandwidth of a signal $x(t)$ is the support—on the positive frequencies—of its Fourier transform $X(\Omega)$. There are different definitions of the bandwidth of a signal depending on how the support of its Fourier transform is measured. We will discuss some of the bandwidth measures used in filtering and in communications in Chapter 6.

The bandwidth together with the information about the signal being low-pass or band-pass provides a good characterization of the signal. The concept of the bandwidth of a filter that was discussed in circuit theory is one of its possible definitions; other possible definitions will be introduced in Chapter 6. The spectrum analyzer, a device used to measure the spectral characteristics of a signal, will be presented in section 5.7.4 after considering filtering.

5.7 CONVOLUTION AND FILTERING

The modulation and the convolution integral properties are the most important properties of the Fourier transform. Modulation is essential in communications, and the convolution property is basic in the analysis and design of filters.

If the input $x(t)$ (periodic or aperiodic) to a stable LTI system has a Fourier transform $X(\Omega)$, and the system has a frequency response $H(j\Omega) = \mathcal{F}[h(t)]$ where $h(t)$ is the impulse response of the system, the output of the LTI system is the convolution integral $y(t) = (x * h)(t)$, with Fourier transform

$$Y(\Omega) = X(\Omega)\, H(j\Omega) \qquad (5.18)$$

In particular, if the input signal $x(t)$ is periodic, the output is also periodic with Fourier transform

$$Y(\Omega) = \sum_{k=-\infty}^{\infty} 2\pi\, X_k\, H(jk\,\Omega_0)\delta(\Omega - k\Omega_0) \qquad (5.19)$$

where X_k are the Fourier series coefficients of $x(t)$ and Ω_0 its fundamental frequency.

This can be shown by considering the eigenfunction property of LTI systems. The Fourier representation of $x(t)$, if aperiodic, is an infinite summation of complex exponentials $e^{j\Omega t}$ multiplied by complex constants $X(\Omega)$, or

$$x(t) = \frac{1}{2\pi} \int_{-\infty}^{\infty} X(\Omega)e^{j\Omega t}d\Omega$$

According to the eigenfunction property, the response of an LTI system to each term $X(\Omega)e^{j\Omega t}$ is $X(\Omega)e^{j\Omega t}H(j\Omega)$ where $H(j\Omega)$ is the frequency response of the system, and thus by superposition the response $y(t)$ is

$$y(t) = \frac{1}{2\pi} \int_{-\infty}^{\infty} [X(\Omega)H(j\Omega)]\, e^{j\Omega t}d\Omega$$

$$= \frac{1}{2\pi} \int_{-\infty}^{\infty} Y(\Omega)e^{j\Omega t}d\Omega$$

so that $Y(\Omega) = X(\Omega)H(j\Omega)$.

If $x(t)$ is periodic of period T_0 (or fundamental frequency $\Omega_0 = 2\pi/T_0$), then

$$X(\Omega) = \sum_{k=-\infty}^{\infty} 2\pi X_k \delta(\Omega - k\Omega_0)$$

so that the output $y(t)$ has as its Fourier transform

$$Y(\Omega) = X(\Omega)H(j\Omega)$$

$$= \sum_{k=-\infty}^{\infty} 2\pi X_k H(j\Omega)\delta(\Omega - k\Omega_0)$$

$$= \sum_{k=-\infty}^{\infty} 2\pi X_k H(jk\Omega_0)\delta(\Omega - k\Omega_0)$$

Therefore, the output is periodic—that is,

$$y(t) = \sum_{k=-\infty}^{\infty} Y_k e^{jk\Omega_0 t}$$

where $Y_k = X_k H(jk\Omega_0)$.

An important consequence of the convolution property, just like in the Laplace transform, is that the ratio of the Fourier transforms of the input and the output gives the frequency response of the system, or

$$H(j\Omega) = \frac{Y(\Omega)}{X(\Omega)} \tag{5.20}$$

The magnitude and the phase of $H(j\Omega)$ are the magnitude and phase frequency responses of the system, or how the system responds to each particular frequency.

Remarks

- *It is important to keep in mind the following connection between the impulse response h(t), the transfer function H(s), and the frequency response H(jΩ) that characterize an LTI system:*

$$H(j\Omega) = \mathcal{L}[h(t)]|_{s=j\Omega}$$
$$= H(s)|_{s=j\Omega}$$
$$= \frac{Y(s)}{X(s}|_{s=j\Omega}$$

- *As the Fourier transform of a real-valued function, the impulse response h(t), the function H(jΩ) has a magnitude |H(jΩ)| and a phase ∠H(jΩ), which are even and odd functions of the frequency Ω.*
- *The convolution property relates to the processing of an input signal by an LTI system. But it is possible, in general, to consider the case of convolving two signals x(t) and y(t) to get z(t) = [x ∗ y](t), in which case we have that Z(Ω) = X(Ω)Y(Ω) where X(Ω) and Y(Ω) are the Fourier transforms of x(t) and y(t).*

5.7.1 Basics of Filtering

The most important application of LTI systems is filtering. Filtering consists in getting rid of undesirable components of a signal. A typical example is when noise $\eta(t)$ is added to a desired signal $x(t)$ (i.e., $y(t) = x(t) + \eta(t)$), and the spectral characteristics of $x(t)$ and the noise $\eta(t)$ are known. The problem then is to design a filter, or an LTI system, that will get rid of the noise as much as possible. The filter design consists in finding a transfer function $H(s) = B(s)/A(s)$ that satisfies certain specifications that will allow getting rid of the noise. Such specifications are typically given in the frequency domain. This is a *rational approximation* problem, as we look for the coefficients of the numerator and denominator of $H(s)$ that make $H(j\Omega)$ in magnitude and phase approximate the filter specifications. The designed filter should be implementable and stable. In this section we discuss the basics of filtering and in Chapter 6 we introduce the filter design.

Frequency-discriminating filters keep the frequency components of a signal in a certain frequency band and attenuate the rest. Filtering an aperiodic signal $x(t)$ represented by its Fourier transform $X(\Omega)$ with magnitude $|X(\Omega)|$ and phase $\angle X(\Omega)$, using a filter with frequency response $H(j\Omega)$, gives an output $y(t)$ with a Fourier transform of

$$Y(\Omega) = H(j\Omega)X(\Omega)$$

Thus, the output $y(t)$ is composed of only those frequency components of the input that are not filtered out by the filter. When designing the filter, we assign appropriate values to the magnitude in the desirable frequency band or bands, and let it be close to zero in those frequencies we would not want in the output signal.

If the input signal $x(t)$ is periodic of period T_0, or fundamental frequency $\Omega_0 = 2\pi/T_0$, the Fourier transform of the output is

$$Y(\Omega) = X(\Omega)H(j\Omega)$$

$$= 2\pi \sum_k X_k H(jk\Omega_0)\delta(\Omega - k\Omega_0) \qquad (5.21)$$

where the magnitude and the phase of each of the Fourier series coefficients are changed by the frequency response of the filter at the harmonic frequencies. Indeed, X_k corresponding to the frequency $k\Omega_0$ is changed into

$$X_k H(jk\Omega_0) = |X_k||H(jk\Omega_0)|e^{j(\angle X_k + \angle H(jk\Omega_0))}$$

The filter output $y(t)$ is also periodic of period T_0 but is missing the harmonics of the input that have been filtered out.

The above shows that independent of whether the input signal $x(t)$ is periodic or aperiodic, the output signal $y(t)$ has the frequency components allowed through by the filter.

■ Example 5.13

Consider how to obtain a dc source using a full-wave rectifier and a low-pass filter (it keeps only the low-frequency components). Let the full-wave rectified signal $x(t)$ be the input of the filter and let the output of the filter be $y(t)$. We want $y(t) = 1$ volt. The rectifier and the low-pass filter constitute a system that converts alternating into direct voltage.

Solution

We found in Chapter 4 that the Fourier series coefficients of $x(t) = |\cos(\pi t)|$, $-\infty < t < \infty$, are given by

$$X_k = \frac{2(-1)^k}{\pi(1 - 4k^2)}$$

so that the average of $x(t)$ is $X_0 = 2/\pi$. To filter out all the harmonics and leave only the average component, we need an ideal low-pass filter with a magnitude A and a cut-off frequency $0 < \Omega_c < \Omega_0$ where $\Omega_0 = 2\pi/T_0 = 2\pi$ is the fundamental frequency of $x(t)$. Thus, the filter is given by $H(j\Omega) = A$ for $-\Omega_0 < \Omega_c < \Omega_0$ and zero otherwise. According to the convolution property, then

$$Y(\Omega) = H(j\Omega)X(\Omega) = H(j\Omega)\left[2\pi X_0\delta(\Omega) + \sum_{k\neq 0} 2\pi X_k\delta(\Omega - k\Omega_0)\right]$$

$$= 2\pi A X_0\delta(\Omega)$$

so that $AX_0 = 1$, or $A = 1/X_0 = \pi/2$, to get the output to have a unit amplitude. Although the proposed filter is not realizable, the above provides what needs to be done to obtain a dc source from a full-wave rectified signal. ∎

■ Example 5.14

Windowing is a time-domain process by which we select part of a signal. This is done by multiplying the signal by a "window" signal $w(t)$. Consider the rectangular window

$$w(t) = u(t + \Delta) - u(t - \Delta) \qquad \Delta > 0$$

For a given signal $x(t)$, the windowed signal is given by

$$y(t) = x(t)w(t)$$

Discuss how windowing relates to the convolution property.

Solution

Windowing is the dual of filtering. In this case, the signal $y(t)$ has the support determined by the window, or $-\Delta \leq t \leq \Delta$, and as such it is zero outside this interval. The window gets rid of parts of the signal outside its support. The signal $y(t)$ can be written as

$$y(t) = w(t)x(t) = w(t)\frac{1}{2\pi}\int_{-\infty}^{\infty} X(\rho)e^{j\rho t}d\rho$$

$$= \frac{1}{2\pi}\int_{-\infty}^{\infty} X(\rho)w(t)e^{j\rho t}d\rho$$

Considering the integral an infinite summation, the Fourier transform of $y(t)$ is

$$Y(\Omega) = \frac{1}{2\pi}\int_{-\infty}^{\infty} X(\rho)\mathcal{F}[w(t)e^{j\rho t}]d\rho$$

$$= \frac{1}{2\pi}\int_{-\infty}^{\infty} X(\rho)W(\Omega - \rho)d\rho$$

using the frequency-shifting property. Thus, we have that the windowing, or multiplication in the time domain, $y(t) = w(t)x(t)$ gives $Y(\Omega)$ as the convolution of $X(\Omega) = \mathcal{F}[x(t)]$ and

$$W(\Omega) = \mathcal{F}[w(t)] = \frac{1}{s}\left[e^{\Delta s} - e^{-\Delta s}\right]_{s=j\Omega} = \frac{2\sin(\Omega\Delta)}{\Omega}$$

multiplied by a constant. This is one more example of the inverse relationship between time and frequency. In this case, the support of the result of the windowing is finite, while the convolution in the frequency domain gives an infinite support for $Y(\Omega)$ given that $W(\Omega)$ has an infinite support. ∎

5.7.2 Ideal Filters

Frequency-discriminating filters that keep low-, middle-, and high-frequency components, or a combination of these, are called *low-pass, band-pass, high-pass*, and *multiband* filters, respectively. A *band-eliminating* or *notch* filter gets rid of middle-frequency components. It is also possible to have an *all-pass* filter that although it does not filter out any of the input frequency components, it changes the phase of the input signal.

The magnitude frequency response of an ideal low-pass filter is given by

$$|H_{lp}(j\Omega)| = \begin{cases} 1 & -\Omega_1 \le \Omega \le \Omega_1 \\ 0 & \text{otherwise} \end{cases}$$

and the phase frequency response of this filter is

$$\angle H_{lp}(j\Omega) = -\alpha\Omega$$

which as a function of Ω is a straight line with slope $-\alpha$, thus its term *linear phase*. The frequency Ω_1 is called the *cut-off frequency* of the low-pass filter. The above magnitude and phase responses only need to be given for positive frequencies, given that the magnitude and the phase responses are the even and odd function of Ω. The rest of the frequency response is obtained by symmetry.

An ideal band-pass filter has a magnitude response

$$|H_{bp}(j\Omega)| = \begin{cases} 1 & \Omega_1 \le \Omega \le \Omega_2 \quad \text{and} \quad -\Omega_2 \le \Omega \le -\Omega_1 \\ 0 & \text{otherwise} \end{cases}$$

with cut-off frequencies Ω_1 and Ω_2. The magnitude response of an ideal high-pass filter is given by

$$|H_{hp}(j\Omega)| = \begin{cases} 1 & \Omega \ge \Omega_2 \quad \text{and} \quad \Omega \le -\Omega_2 \\ 0 & \text{otherwise} \end{cases}$$

with a cut-off frequency of Ω_2. For both of these filters, i is assumed the phase is linear in the pass-band (band of frequencies where the magnitude is unity).

From these definitions, we have that the ideal band-stop filter has as magnitude response of

$$|H_{bs}(j\Omega)| = 1 - |H_{bp}(j\Omega)|$$

The sum of the magnitude responses of the given low-, band-, and high-pass filters gives the magnitude response of an ideal all-pass filter

$$|H_{ap}(j\Omega)| = |H_{lp}(j\Omega)| + |H_{bp}(j\Omega)| + |H_{hp}(j\Omega)| = 1$$

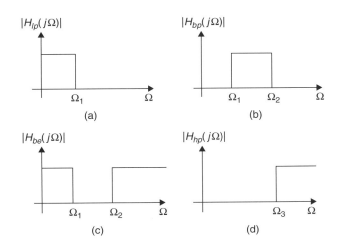

FIGURE 5.8
Ideal filters: (a) low pass, (b) band pass, (c) band eliminating, and (d) high pass.

for all frequencies, since in this case we chose the frequencies Ω_1 and Ω_2 so that the response of these filters add to unity. An ideal multi-band filter can be obtained as a combination of the low-, band-, and high-pass filters. Figure 5.8 displays the frequency responses of the ideal filters discussed here.

Remarks

- *If $h_{lp}(t)$ is the impulse response of a low-pass filter, applying the modulation property we get that $2h_{lp}(t)\cos(\Omega_0 t)$ (where $\Omega_0 >> \Omega_1$ and Ω_1 is the cut-off frequency of the low-pass filter) corresponds to the impulse response of a band-pass filter centered around Ω_0. Indeed, its Fourier transform is given by*

$$\mathcal{F}[2h_{lp}(t)\cos(\Omega_0 t)] = H_{lp}(j(\Omega - \Omega_0)) + H_{lp}(j(\Omega + \Omega_0))$$

which is the frequency response of the low-pass filter shifted to new center frequencies Ω_0 and $-\Omega_0$, making it a band-pass filter.
- *A zero-phase ideal low-pass filter $H_{lp}(j\Omega) = u(\Omega + \Omega_1) - u(\Omega - \Omega_1)$ has as impulse response a sinc function with a support from $-\infty$ to ∞. This ideal low-pass filter is clearly noncausal as its impulse response is not zero for negative values of time t. To make it causal we could approximate its impulse response by a function $h_1(t) = h_{lp}(t)w(t)$ where $w(t) = u(t + \tau) - u(t - \tau)$ is a rectangular window where the value of τ is chosen so that outside the window the values of the impulse response $h_{lp}(t)$ are very close to zero. Although the Fourier transform of $h_1(t)$ is a very good approximation of the desired frequency response, the frequency response of $h_1(t)$ displays ringing around the cut-off frequency Ω_1 because of the rectangular window. Finally, we delay $h_1(t)$ by τ to get a causal filter with linear phase. That is, $h_1(t - \tau)$ has as its magnitude response $|H_1(j\Omega)| \approx |H_{lp}(j\Omega)|$ and its phase response is $\angle H_1(j\Omega) = -\tau\Omega$. Although the above procedure is a valid way to obtain approximate low-pass filters with linear phase, they are not guaranteed to be rational and would be difficult to implement. Thus, other methods are used to design filters.*
- *Since ideal filters are not causal they cannot be used in real-time applications—that is when the input signal needs to be processed as it comes to the filter. Imposing causality on the filter restricts the frequency response of the filter in significant ways. According to the Paley-Wiener integral condition, a causal*

and stable filter with frequency response $H(j\Omega)$ should satisfy the condition

$$\int_{-\infty}^{\infty} \frac{|\log(H(j\Omega))|}{1 + \Omega^2} d\Omega < \infty \tag{5.22}$$

To satisfy this condition, $H(j\Omega)$ cannot be zero in any band of frequencies, because in such cases the numerator of the integrand would be infinite. The Paley-Wiener integral condition is clearly not satisfied by ideal filters. So they cannot be implemented or used in actual situations, but they can be used as models for designing filters.

■ *That ideal filters are not realizable can be understood also by considering what it means to make the magnitude response of a filter zero in some frequency bands. A measure of attenuation is given by the* loss function *in decibels, defined as*

$$\alpha(\Omega) = -10 \log_{10} |H(j\Omega)|^2$$

$$= -20 \log_{10} |H(j\Omega)| \text{ dB}$$

Thus, when $|H(j\Omega)| = 1$ and there is no attenuation the loss is 0 dB, and when $|H(j\Omega)| = 10^{-5}$ for a large attenuation the loss is 100 dB. You quickly convince yourself that if a filter achieves a magnitude response of 0 at any frequency this would mean a loss or attenuation at that frequency of ∞ dBs! Values of 60 to 100 dB attenuation are considered extremely good, and to obtain that the signal needs to be attenuated by a factor of 10^{-3} to 10^{-5}. A curious term JND *or "just noticeable difference" is used by experts in human hearing to characterize the smallest sound intensity that can be judged by a human as different. Such a value varies from 0.25 to 1 dB. To illustrate what is loud in the dB scale, consider the following cases: A sound pressure level higher than 130 dB causes pain; 110 dB is generated by an amplified rock band performance [65].*

■ Example 5.15

The Gibb's phenomenon, which we mentioned when discussing the Fourier series of periodic signals with discontinuities, consists in ringing around these discontinuities. To see this, consider a periodic train of square pulses $x(t)$ of period T_0 displaying discontinuities at $kT_0/2$, for $k = \pm 1, \pm 2, \ldots$. Show how the Gibb's phenomenon is due to ideal low-pass filtering.

Solution

Choosing $2N + 1$ of the Fourier series coefficients to approximate the signal $x(t)$ is equivalent to passing $x(t)$ through an ideal low-pass filter,

$$H(j\Omega) = \begin{cases} 1 & -\Omega_c \le \Omega \le \Omega_c \\ 0 & \text{otherwise} \end{cases}$$

having as impulse response a sinc function $h(t)$. If the Fourier transform of the periodic signal $x(t)$ of fundamental frequency $\Omega_0 = 2\pi/T_0$ is

$$X(\Omega) = \sum_{k=-\infty}^{\infty} 2\pi X_k \delta(\Omega - k\Omega_0)$$

the output of the filter is the signal

$$x_N(t) = \mathcal{F}^{-1}[X(\Omega)H(j\Omega)]$$

$$= \mathcal{F}^{-1}\left[\sum_{k=-N}^{N} 2\pi X_k \delta(\Omega - k\Omega_0)\right]$$

or the inverse Fourier transform of $X(\Omega)$ multiplied by a low-pass filter with an ideal magnitude response of 1 for $-\Omega_c < \Omega < \Omega_c$ where the cut-off frequency Ω_c is chosen so that $N\Omega_0 < \Omega_c < (N+1)\Omega_0$. As such, $x_N(t)$ is the convolution

$$x_N(t) = [x * h](t)$$

where $h(t)$ is the inverse Fourier transform of $H(j\Omega)$, or a sinc signal of infinite support. The convolution around the discontinuities of $x(t)$ causes ringing before and after them, and this ringing appears independent of the value of N. ■

■ Example 5.16

Obtain different filters from an RLC circuit (Figure 5.9) by choosing different outputs. Let the input be a voltage source with Laplace transform $V_i(s)$. For simplicity, let $R = 1\ \Omega$, $L = 1\ H$, and $C = 1\ F$, and assume the initial conditions to be zero.

Solution

■ *Low-pass filter:* Let the output be the voltage across the capacitor; by voltage division we have that

$$V_C(s) = \frac{V_i(s)/s}{1 + s + 1/s} = \frac{V_i(s)}{s^2 + s + 1}$$

so that the transfer function is

$$H_{lp}(s) = \frac{V_C(s)}{V_i(s)} = \frac{1}{s^2 + s + 1}$$

This is the transfer function of a second-order low-pass filter. If the input is a dc source, so that its frequency is $\Omega = 0$, the inductor is a short circuit (its impedance would be 0) and

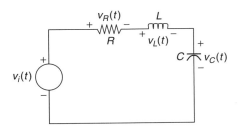

FIGURE 5.9

RLC circuit for implementing different filters.

the capacitor is an open circuit (its impedance would be infinite), so that the voltage in the capacitor is equal to the voltage in the source. On the other hand, if the frequency of the input source is very high, then the inductor is an open circuit and the capacitor a short circuit (its impedance is zero) so that the capacitor voltage is zero. This is a low-pass filter. Notice that this filter has no finite zeros, and complex conjugate poles.

■ *High-pass filter:* Suppose then that we let the output be the voltage across the inductor. Then again by voltage division the transfer function

$$H_{hp}(s) = \frac{V_L(s)}{V_i(s)} = \frac{s^2}{s^2 + s + 1}$$

is that of a high-pass filter. Indeed, for a dc input (frequency zero) the impedance in the inductor is zero, so that the inductor voltage is zero, and for very high frequency the impedance of the inductor is very large so that it can be considered open circuit and the voltage in the inductor equals that of the source. This filter has the same poles of the low-pass filter (this is determined by the overall impedance of the circuit, which has not changed) and double zeros at zero. It is these zeros that make the frequency response for low frequencies be close to zero.

■ *Band-pass filter:* Letting the output be the voltage across the resistor, its transfer function is

$$H_{bp}(s) = \frac{V_R(s)}{V_i(s)} = \frac{s}{s^2 + s + 1}$$

or the transfer function of a band-pass filter. For zero frequency, the capacitor is an open circuit so the current is zero and the voltage across the resistor is zero. Similarly, for very high frequency the impedance of the inductor is very large, or an open circuit, making the voltage across the resistor zero because again the current is zero. For some middle frequency the serial combination of the inductor and the capacitor resonates and will have zero impedance. At the resonance frequency, the current achieves its largest value and the voltage across the resistor does too. This behavior is that of a band-pass filter. This filter again has the same poles as the other two, but only one zero at zero.

■ *Band-stop filter:* Finally, suppose we consider as output the voltage across the connection of the inductor and the capacitor. At low and high frequencies, the impedance of the LC connection is very high, or open circuit, and so the output voltage is the input voltage. At the resonance frequency $\Omega_r = 1$ the impedance of the LC connection is zero, so the output voltage is zero. The resulting filter is a band-stop filter with the transfer function

$$H_{bs}(s) = \frac{s^2 + 1}{s^2 + s + 1}$$

Second-order filters can then be easily identified by the numerator of their transfer functions. Second-order low-pass filters have no zeros, and the numerator is $N(s) = 1$; band-pass filters have a zero at $s = 0$ so $N(s) = s$, and so on. We will see next that such a behavior can be easily seen from a geometric approach.

5.7.3 Frequency Response from Poles and Zeros

Given a rational transfer function $H(s) = B(s)/A(s)$, to calculate its frequency response we let $s = j\Omega$ and find the magnitude and phase for a discrete set of frequencies. This can be done using MATLAB. A geometric way to obtain an approximate magnitude and phase frequency responses is using the effects of zeros and poles on the frequency response of a system.

Consider a function

$$G(s) = \frac{s - z}{s - p}$$

with a zero z and a pole p, as shown in Figure 5.10. The frequency response corresponding to $G(s)$ at some frequency Ω_0 is found by letting $s = j\Omega_0$, or

$$G(s)|_{s=j\Omega_0} = \frac{j\Omega_0 - z}{j\Omega_0 - p}$$

Representing $j\Omega_0$, z, and p, which are complex numbers, as vectors coming from the origin, then the vector $\vec{Z}(\Omega_0) = j\Omega_0 - z$ (adding to $\vec{Z}(\Omega_0)$ the vector corresponding to z gives a vector corresponding to $j\Omega_0$) goes from the zero z to $j\Omega_0$, and likewise the vector $\vec{P}(\Omega_0) = j\Omega_0 - p$ goes from the pole p to $j\Omega_0$. The argument Ω_0 in the vectors indicates that the magnitude and phase of these vectors depend on the frequency at which we are finding the frequency response. As we change the frequency at which we are finding the frequency response, the lengths and the phases of these vectors change. Therefore,

$$G(j\Omega_0) = \frac{\vec{Z}(\Omega_0)}{\vec{P}(\Omega_0)} = \frac{|\vec{Z}(\Omega_0)|}{|\vec{P}(\Omega_0)|} e^{j(\angle\vec{Z}(\Omega_0) - \angle\vec{P}(\Omega_0))}$$

and the magnitude response is

$$|G(j\Omega_0)| = \frac{|\vec{Z}(\Omega_0)|}{|\vec{P}(\Omega_0)|} \tag{5.23}$$

and the phase response is

$$\angle G(j\Omega_0) = \angle\vec{Z}(\Omega_0) - \angle\vec{P}(\Omega_0) \tag{5.24}$$

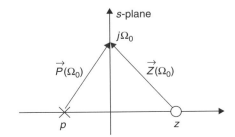

FIGURE 5.10

Geometric interpretation of poles and zeros.

So that for $0 \leq \Omega_0 < \infty$, if we compute the length and the angle of $\vec{Z}(\Omega_0)$ and $\vec{P}(\Omega_0)$, the ratio of these lengths gives the magnitude response and the difference of their angles gives the phase response.

For a filter with a transfer function

$$H(s) = \frac{\prod_i (s - z_i)}{\prod_k (s - p_k)}$$

where z_i, p_k are zeros and poles of H(s) with vectors $\vec{Z}_i(\Omega) = j\Omega - z_i$ and $\vec{P}_k(\Omega) = j\Omega - p_k$, going from each of the zeros and poles to the frequency at which we are computing the magnitude and phase response in the $j\Omega$ axis, gives

$$H(j\Omega) = H(s)|_{s=j\Omega} = \frac{\prod_i \vec{Z}_i(\Omega)}{\prod_k \vec{P}_k(\Omega)}$$

$$= \underbrace{\frac{\prod_i |\vec{Z}_i(\Omega)|}{\prod_k |\vec{P}_k(\Omega)|}}_{|H(j\Omega)|} \underbrace{e^{j\left[\sum_i \angle(\vec{Z}_i(\Omega)) - \sum_k \angle(\vec{P}_k(\Omega))\right]}}_{e^{j\angle H(j\Omega)}} \quad (5.25)$$

■ Example 5.17

Consider series RC circuit with a voltage source $v_i(t)$. Choose the output to obtain low-pass and high-pass filters and use the poles and zeros of the transfer functions to determine their frequency responses. Let $R = 1\ \Omega$, $C = 1$ F, and the initial conditions be zero.

Solution

■ *Low-pass filter:* Let the output be the voltage across the capacitor. By voltage division, we obtain that the transfer function of the filter is

$$H(s) = \frac{V_C(s)}{V_i(s)} = \frac{1/Cs}{R + 1/Cs}$$

For dc frequency, the capacitor behaves as an open circuit so that the output voltage equals the input voltage, and for very high frequencies the impedance of the capacitor tends to zero so that the voltage across the capacitor also goes to zero. This is a low-pass filter.
Let $R = 1\ \Omega$ and $C = 1$ F, so

$$H(j\Omega) = \frac{1}{1 + j\Omega} = \frac{1}{\vec{P}(\Omega)}$$

Drawing a vector from the pole $s = -1$ to any point on the $j\Omega$ axis gives $\vec{P}(\Omega)$, and for different frequencies we get

$$\Omega = 0 \qquad \vec{P}(0) = 1e^{j0}$$
$$\Omega = 1 \qquad \vec{P}(1) = \sqrt{2}e^{j\pi/4}$$
$$\Omega = \infty \qquad \vec{P}(\infty) = \infty\, e^{j\pi/2}$$

Since there are no zeros, the frequency response of this filter depends inversely on the behavior of the pole vector $\vec{P}(\Omega)$. The frequency responses for these three frequencies are:

$$H(j0) = 1e^{j0}$$
$$H(j1) = 0.707e^{-j\pi/4}$$
$$H(j\infty) = 0e^{-j\pi/2}$$

Thus, the magnitude response is unity at $\Omega = 0$ and it decays as the frequency increases. The phase is zero at $\Omega = 0$, $-\pi/4$ at $\Omega = 1$, and $-\pi/2$ at $\Omega \to \infty$. The magnitude response is even and the phase response is odd.

■ *High-pass filter:* Consider then the output being the voltage across the resistor. Again by voltage division we obtain the transfer function of this circuit as

$$H(s) = \frac{V_r(s)}{V_s(s)} = \frac{CRs}{CRs + 1}$$

Again, let $C = R = 1$, so the frequency response is

$$H(j\Omega) = \frac{j\Omega}{1 + j\Omega} = \frac{\vec{Z}(\Omega)}{\vec{P}(\Omega)}$$

The vector $\vec{Z}(\Omega)$ goes from zero at the origin $s = 0$ to $j\Omega$ in the $j\Omega$ axis, and the vector $\vec{P}(\Omega)$ goes from the pole $s = -1$ to $j\Omega$ in the $j\Omega$ axis. The vectors and the frequency response, at three different frequencies, are given by

$$\Omega = 0 \quad \vec{P}(0) = 1e^{j0} \quad \vec{Z}(0) = 0e^{j\pi/2} \quad H(j0) = \frac{\vec{Z}(0)}{\vec{P}(0)} = 0e^{j\pi/2}$$

$$\Omega = 1 \quad \vec{P}(1) = \sqrt{2}e^{j\pi/4} \quad \vec{Z}(1) = 1e^{j\pi/2} \quad H(j1) = \frac{\vec{Z}(1)}{\vec{P}(1)} = 0.707e^{j\pi/4}$$

$$\Omega = \infty \quad \vec{P}(\infty) = \infty\, e^{j\pi/2} \quad \vec{Z}(\infty) = \infty\, e^{j\pi/2} \quad H(j\infty) = \frac{\vec{Z}(\infty)}{\vec{P}(\infty)} = 1e^{j0}$$

Thus, the magnitude response is zero at $\Omega = 0$ (this is due to the zero at $s = 0$, making $\vec{Z}(0) = 0$ as it is right on top of the zero), and it grows to unity as the frequency increases (at very high frequency, the lengths of the pole and the zero vectors are alike and so the magnitude response is unity and the phase response is zero). ■

Remarks

■ *Poles create "hills" at frequencies in the $j\Omega$ axis in front of the poles imaginary parts. The closer the pole is to the $j\Omega$ axis, the narrower and higher the hill. If, for instance, the poles are on the $j\Omega$ axis (this would correspond to an unstable and useless filter) the frequency response at the frequency of the poles will be infinity.*

■ *Zeros create "valleys" at the frequencies in the $j\Omega$ axis in front of the zeros imaginary parts. The closer the zero is to the $j\Omega$ axis (from its left or its right, as the zeros are not restricted by stability to be in the open left-hand s-plane) the closer the frequency response is to zero. If the zeros are on the $j\Omega$ axis, the frequency response at the frequency of the zeros is zero. Thus, poles produce frequency responses that look like hills (or like the main pole in a circus) around the frequencies of the poles, and zeros make the frequency response go to zero in the form of valleys around the frequencies of the zeros.*

■ Example 5.18

Use MATLAB to find and plot the poles and zeros and the corresponding magnitude and phase frequency responses of:

(a) A second-order band-pass filter and a high-pass filter realized using a series connection of a resistor, an inductor, and a capacitor, each with unit resistance, inductance, and capacitance. Let the input be a voltage source $v_s(t)$ and initial conditions be zero.

(b) An all-pass filter with a transfer function

$$H(s) = \frac{s^2 - 2.5s + 1}{s^2 + 2.5s + 1}$$

Solution

Our function freq resp_s computes and plots the poles and the zeros of the filter transfer function and the corresponding frequency response (the function requests the coefficients of its numerator and denominator in decreasing order of powers of s).

(a) As from a Example 5.16, the transfer functions of the band-pass and high-pass second-order filters are

$$H_{bp}(s) = \frac{s}{s^2 + s + 1}$$

$$H_{hp}(s) = \frac{s^2}{s^2 + s + 1}$$

The denominator in the two cases is exactly the same since the values of $R, L,$ and C remain the same for the two filters—the only difference is in the numerator.

To compute the frequency response of these filters and to plot their poles and zeros, we used the following script, which uses two functions: freqresp_s, which we give below, and splane, which plots the poles and zeros. The coefficients of the numerator and the denominator correspond to the coefficients, from the highest to the lowest order of s, of the transfer function.

```
%%%%%%%%%%%%%%%%%%%%%%
% Example 5.18---Frequency response
%%%%%%%%%%%%%%%%%%%%%%
n = [0 1 0]; % numerator coefficients -- bandpass
% n = [1 0  0]; % numerator coefficients -- highpass
d = [1 1 1]; % denominator coefficients
```

```
wmax = 10; % maximum frequency
[w, Hm, Ha] = freqresp_s(n, d, wmax); % frequency response
splane(n, d) % plotting of poles and zeros
```

The following is the function freqresp_s used to compute the magnitude and phase response of the filter with the given numerator and denominator coefficients.

```
function [w, Hm, Ha] = freqresp_s(b, a, wmax)
w = 0:0.01:wmax;
H = freqs(b, a, w);
Hm = abs(H); % magnitude
Ha = angle(H) * 180/pi; % phase in degrees
```

- *Band-pass filter*: Letting the output of the filter be the voltage across resistor, we find that the transfer function has a zero at zero, so that the frequency response is zero at $\Omega = 0$. When Ω goes to infinity, one of the two poles cancels the zero effect so that the other pole makes the frequency response tend to zero.

- *High-pass filter*: When the output of the filter is the voltage across the inductor the filter is high pass. In this case there is a double zero at $s = 0$, and the poles are located as before. Thus, when $\Omega = 0$ the magnitude response is zero due to the double zeros at zero, and when Ω goes to infinity the effect of two poles and the two zeros cancel out giving a constant magnitude response, which corresponds to a high-pass filter.

The results for the band-pass and the high-pass filters are shown in Figure 5.11. Notice that the frequency response of the band-pass and the high-pass filter is determined by the 'number' of zeros at the origin. The 'location' of zeros, like in the all-pass filter we consider next, also determines the frequency response.

(b) *All-pass filter*: The poles and the zeros of an all-pass filter have the same imaginary parts, but the negative of its real part. At any frequency in the $j\Omega$-axis the lengths of the vectors from the poles equal the length of the vectors from the zeros to the frequency in the $j\Omega$ axis. Thus the magnitude response of the filter is unity. The following changes to the above script are needed for the all-pass filter:

```
clear all
clf
n = [1  −2.5 1];
d = [1 2.5 1];
wmax = 10;
freq_resp_s(n, d, wmax)
```

The results are shown in Figure 5.12. ∎

5.7.4 Spectrum Analyzer

A *spectrum analyzer* is a device that measures the spectral characteristics of a signal. It can be implemented as a bank of narrow band band-pass filters with fixed bandwidths covering the desired frequencies (see Figure 5.13). The power at the output of each filter is computed and displayed at the corresponding center frequency. Another possible implementation is using a band-pass filter with an adjustable center frequency, with the power in its bandwidth being computed and displayed [16].

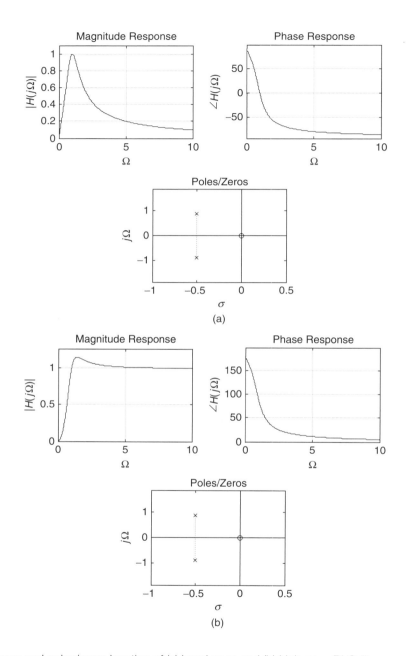

FIGURE 5.11

Frequency response and poles/zeros location of (a) band-pass and (b) high-pass RLC filters.

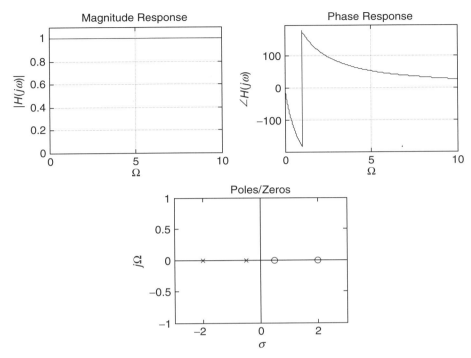

FIGURE 5.12
Frequency response and poles/zeros location of the all-pass filter.

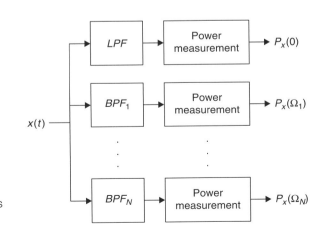

FIGURE 5.13
Bank-of-filter spectrum analyzer. LPF stands for low-pass filter, and BPF_i corresponds to band-pass filters, $i = 1, \ldots, N$.

If the input of the spectrum analyzer is $x(t)$, the output of either the fixed- or the adjustable-bandpass filters in the implementations—assumed to have a very narrow bandwidth $\Delta\Omega$—would be

$$y(t) = \frac{1}{2\pi} \int_{\Omega_0 - 0.5\Delta\Omega}^{\Omega_0 + 0.5\Delta\Omega} X(\Omega)e^{j\Omega t}d\Omega$$

$$\approx \frac{1}{2\pi}\Delta\Omega \; X(\Omega_0)e^{j\Omega_0 t}$$

Computing the mean square of this signal we get

$$\frac{1}{T}\int_T |y(t)|^2 dt = \left(\frac{\Delta\Omega}{2\pi}\right)^2 |X(\Omega_0)|^2$$

which is proportional to the power or the energy of the signal in $\Omega_0 \pm 0.5\Delta\Omega$. A similar computation can be done at each of the frequencies of the input signal.

Remarks

- *The bank-of-filter spectrum analyzer is used for the audio range of the spectrum.*
- *Radio frequency spectrum analyzers resemble an AM demodulator. It usually consists of a single narrow-band intermediate frequency (IF) bandpass filter fed by a mixer. The local oscillator sweeps across the desired band, and the power at the output of the filter is computed and displayed on a monitor.*

5.8 ADDITIONAL PROPERTIES

We consider now some additional properties of the Fourier transform, some of which look like those of the Laplace transform when $s = j\Omega$ and some are different.

5.8.1 Time Shifting

If $x(t)$ has a Fourier transform $X(\Omega)$, then

$$x(t - t_0) \Leftrightarrow X(\Omega)e^{-j\Omega t_0}$$
$$x(t + t_0) \Leftrightarrow X(\Omega)e^{j\Omega t_0} \tag{5.26}$$

The Fourier transform of $x(t - t_0)$ is

$$\mathcal{F}[x(t - t_0)] = \int_{-\infty}^{\infty} x(t - t_0)e^{-j\Omega t}dt$$

$$= \int_{-\infty}^{\infty} x(\tau)e^{-j\Omega(\tau + t_0)}d\tau = e^{-j\Omega t_0}X(\Omega)$$

where we changed the variable to $\tau = t - t_0$. Likewise for $x(t + t_0)$.

It is important to realize that shifting in time does not change the frequency content of the signal—that is, the signal does not change when delayed or advanced. This is clear when we see that the magnitude of the two transforms, corresponding to the original and the shifted signals, is the same,

$$|X(\Omega)| = |X(\Omega)e^{\pm j\Omega t_0}|$$

and the effect of the time shift is only in the phase spectrum.

■ Example 5.19

Consider the signal

$$x(t) = A[\delta(t - \tau) + \delta(t + \tau)]$$

Find its Fourier transform $X(\Omega)$. Use this Fourier pair and the duality property to verify the Fourier transform of a $\cos(\Omega_0 t)$ obtained before.

Solution

Applying the time-shift property, we have

$$X(\Omega) = A[1e^{-j\Omega\tau} + 1e^{j\Omega\tau}]$$
$$= 2A\cos(\Omega\tau)$$

giving the Fourier transform pair

$$x(t) = A[\delta(t - \tau) + \delta(t + \tau)] \quad \Leftrightarrow \quad X(\Omega) = 2A\cos(\Omega\tau)$$

Using the duality property, we then have

$$X(t) = 2A\cos(t\tau) \quad \Leftrightarrow \quad 2\pi x(-\Omega) = 2\pi A[\delta(-\Omega - \tau) + \delta(-\Omega + \tau)]$$

Let $\tau = \Omega_0$, use the evenness of $\delta(\Omega)$ to get

$$A\cos(\Omega_0 t) \quad \Leftrightarrow \quad \pi A[\delta(\Omega + \Omega_0) + \delta(\Omega - \Omega_0)] \qquad ■$$

■ Example 5.20

Consider computing the Fourier transform of $y(t) = \sin(\Omega_0 t)$ using the Fourier transform of the cosine signal $x(t) = \cos(\Omega_0 t)$ we just found.

Solution

Since $y(t) = \cos(\Omega_0 t - \pi/2) = x(t - \pi/(2\Omega_0))$, applying the time-shifting property, we then get

$$\mathcal{F}[\sin(\Omega_0 t)] = \mathcal{F}[x(t - \pi/2\Omega_0)]$$
$$= \pi[\delta(\Omega - \Omega_0) + \delta(\Omega + \Omega_0)]e^{-j\Omega\pi/(2\Omega_0)}$$
$$= \pi\delta(\Omega - \Omega_0)e^{-j\pi/2} + \pi\delta(\Omega + \Omega_0)e^{j\pi/2}$$
$$= -j\pi\delta(\Omega - \Omega_0) + j\pi\delta(\Omega + \Omega_0)$$

after applying the sifting property of $\delta(\Omega)$. The above shows that this Fourier transform is different from the one for the cosine in the phase only. ∎

5.8.2 Differentiation and Integration

If $x(t)$, $-\infty < t < \infty$, has a Fourier tranform $X(\Omega)$, then

$$\frac{d^N x(t)}{dt^N} \quad \Leftrightarrow \quad (j\Omega)^N X(\Omega) \tag{5.27}$$

$$\int_{-\infty}^{t} x(\sigma)d\sigma \quad \Leftrightarrow \quad \frac{X(\Omega)}{j\Omega} + \pi X(0)\delta(\Omega) \tag{5.28}$$

where

$$X(0) = \int_{-\infty}^{\infty} x(t)dt$$

From the inverse Fourier transform given by

$$x(t) = \frac{1}{2\pi} \int_{-\infty}^{\infty} X(\Omega)e^{j\Omega t}d\Omega$$

we then have that

$$\frac{dx(t)}{dt} = \frac{1}{2\pi} \int_{-\infty}^{\infty} X(\Omega)\frac{d\,e^{j\Omega t}}{dt}d\Omega$$

$$= \frac{1}{2\pi} \int_{-\infty}^{\infty} [X(\Omega)j\Omega]\,e^{j\Omega t}d\Omega$$

indicating that

$$\frac{dx(t)}{dt} \Leftrightarrow j\Omega X(\Omega)$$

and similarly for higher derivatives.

The proof of the integration property can be done in two parts:

1. The convolution of $u(t)$ and $x(t)$ gives the integral—that is

$$\int_{-\infty}^{t} x(\tau)d\tau = \int_{-\infty}^{\infty} x(\tau)u(t-\tau)d\tau = [x * u](t)$$

since $u(t - \tau)$ as a function of τ equals

$$u(t - \tau) = \begin{cases} 1 & \tau < t \\ 0 & \tau > t \end{cases}$$

We thus have that

$$\mathcal{F}\left[\int_{-\infty}^{t} x(\tau)d\tau\right] = X(\Omega)\mathcal{F}[u(t)] \tag{5.29}$$

2. Since the unit-step signal is not absolutely integrable its Fourier transform cannot be found from the integral definition, and we cannot use its Laplace transform either because its ROC does not include the $j\Omega$ axis. Let's transform it into an absolutely integrable signal by subtracting 1/2 and multiplying the result by 2. This gives the *sign* signal:

$$\text{sgn}(t) = 2[u(t) - 0.5] = \begin{cases} 1 & t > 0 \\ -1 & t < 0 \end{cases}$$

The derivative of $\text{sgn}(t)$ is

$$\frac{d\text{sgn}(t)}{dt} = 2\delta(t)$$

and thus if $S(\Omega) = \mathcal{F}[\text{sgn}(t)]$,

$$S(\Omega) = \frac{2}{j\Omega}$$

using the derivative property. The linearity of the Fourier transform applied to the definition of $\text{sgn}(t)$ gives

$$\mathcal{F}[\text{sgn}(t)] = 2\mathcal{F}[u(t)] - 2\pi\delta(\Omega) \quad \Rightarrow \quad \mathcal{F}[u(t)] = \frac{1}{j\Omega} + \pi\delta(\Omega) \tag{5.30}$$

Replacing the Fourier transform of $u(t)$ in Equation (5.29), we get

$$\mathcal{F}\left[\int_{-\infty}^{t} x(\tau)d\tau\right] = X(\Omega)\left[\frac{1}{j\Omega} + \pi\delta(\Omega)\right]$$

$$= \frac{X(\Omega)}{j\Omega} + \pi X(0)\delta(\Omega) \tag{5.31}$$

Remarks

- *Just like in the Laplace transform where the operator s corresponds to the derivative operation in time of the signal, in the Fourier transform $j\Omega$ becomes the corresponding operator for the derivative operation in time of the signal.*

■ If $X(0)$ (i.e., the dc value of $X(\Omega)$) is zero, then the operator $1/(j\Omega)$ corresponds to integration in time of $x(t)$, just like $1/s$ in the Laplace domain. For $X(0)$ to be zero the integral of the signal from $-\infty$ to ∞ must be zero.

■ Example 5.21

Suppose a system is represented by a second-order differential equation with constant coefficients:

$$2y(t) + 3\frac{dy(t)}{dt} + \frac{d^2y(t)}{dt^2} = x(t)$$

and that the initial conditions are zero. Let $x(t) = \delta(t)$. Find $y(t)$.

Solution

Computing the Fourier transform of this equation, we get

$$[2 + 3j\Omega + (j\Omega)^2]Y(\Omega) = X(\Omega)$$

Replacing $X(\Omega) = 1$ and solving for $Y(\Omega)$, we have

$$Y(\Omega) = \frac{1}{2 + 3j\Omega + (j\Omega)^2}$$
$$= \frac{1}{(j\Omega + 1)(j\Omega + 2)}$$
$$= \frac{1}{(j\Omega + 1)} + \frac{-1}{(j\Omega + 2)}$$

and the inverse Fourier transform of these terms gives

$$y(t) = [e^{-t} - e^{-2t}]u(t)$$

■

■ Example 5.22

Find the Fourier transform of the triangular pulse

$$x(t) = r(t) - 2r(t-1) + r(t-2)$$

which is piecewise linear, using the derivative property.

Solution

A first derivative gives

$$\frac{dx(t)}{dt} = u(t) - 2u(t-1) + u(t-2)$$

and a second derivative gives

$$\frac{d^2x(t)}{dt^2} = \delta(t) - 2\delta(t-1) + \delta(t-2)$$

Using the time-shift and the derivative properties, we get from the expression for the second derivative and letting $X(\Omega)$ be the Fourier transform of $x(t)$:

$$(j\Omega)^2 X(\Omega) = 1 - 2e^{-j\Omega} + e^{-j2\Omega}$$
$$= e^{-j\Omega}[e^{j\Omega} - 2 + e^{-j\Omega}]$$

so that

$$X(\Omega) = \frac{2e^{-j\Omega}}{\Omega^2}[1 - \cos(\Omega)]$$

■

■ Example 5.23

Consider the integral

$$y(t) = \int_{-\infty}^{t} x(\tau)d\tau \quad -\infty < t < \infty$$

where $x(t) = u(t+1) - u(t-1)$. Find the Fourier transform $Y(\Omega)$ directly and from the integration property.

Solution

The integral is

$$y(t) = \begin{cases} 0 & t < -1 \\ t+1 & -1 \le t < 1 \\ 2 & t \ge 1 \end{cases}$$

or

$$y(t) = \underbrace{[r(t+1) - r(t-1) - 2u(t-1)]}_{y_1(t)} + 2u(t-1)$$

The Fourier transform of $y_1(t)$ is

$$Y_1(\Omega) = \left[\frac{e^s - e^{-s}}{s^2} - \frac{2e^{-s}}{s}\right]_{s=j\Omega} = \frac{-2j\sin(\Omega)}{\Omega^2} + j\frac{2e^{-j\Omega}}{\Omega}$$

The Fourier transform of $2u(t-1)$ is $-2je^{-j\Omega}/\Omega + 2\pi\delta(\Omega)$ so that

$$Y(\Omega) = \frac{-2j\sin(\Omega)}{\Omega^2} + j\frac{2e^{-j\Omega}}{\Omega} - j\frac{2e^{-j\Omega}}{\Omega} + 2\pi\delta(\Omega)$$

$$= \frac{-2j\sin(\Omega)}{\Omega^2} + 2\pi\delta(\Omega)$$

To use the integration property we first need $X(\Omega)$, which is

$$X(\Omega) = \frac{2\sin(\Omega)}{\Omega}$$

and according to the property,

$$Y(\Omega) = \frac{X(\Omega)}{j\Omega} + \pi X(0)\delta(\Omega)$$

$$= \frac{-2j\sin(\Omega)}{\Omega^2} + 2\pi\delta(\Omega)$$

since $X(0) = 2$ (using L'Hôpital's rule). As expected, the two results coincide.   ■

5.9 WHAT HAVE WE ACCOMPLISHED? WHAT IS NEXT?

You should by now have a very good understanding of the frequency representation of signals and systems. In this chapter, we have unified the treatment of periodic and nonperiodic signals and their spectra, and consolidated the concept of frequency response of a linear time-invariant system.

Basic properties of the Fourier transform and important Fourier pairs are given in Tables 5.1 and 5.2. Two significant applications are in filtering and communications. We introduced the basics of filtering in this chapter and will expand on them in Chapter 6. The fundamentals of modulation provided in this chapter will be illustrated in Chapter 6 where we will consider their application in communications.

Certainly the next step is to find out where the Laplace and the Fourier analyses apply, which will be done in Chapter 6. After that, we will go into discrete-time signals and systems. We will show that sampling, quantization, and coding bridge the continuous-time and the digital signal processing, and that transformations similar to the Laplace and the Fourier transforms will permit us to do processing of discrete–time signals and systems.

PROBLEMS

5.1. Fourier series versus Fourier transform—MATLAB

The connection between the Fourier series and the Fourier transform can be seen by considering what happens when the period of a periodic signal increases to a point at which the periodicity is not clear as only one period is seen. Consider a train of pulses $x(t)$ with a period $T_0 = 2$, and a period of $x(t)$ is $x_1(t) = u(t+0.5) - u(t-0.5)$. Let T_0 be increased to 4, 8, and 16.

Table 5.1 Basic Properties of the Fourier Transform

	Time Domain	Frequency Domain
Signals and constants	$x(t), y(t), z(t), \alpha, \beta$	$X(\Omega), Y(\Omega), Z(\Omega)$
Linearity	$\alpha x(t) + \beta y(t)$	$\alpha X(\Omega) + \beta Y(\Omega)$
Expansion/contraction in time	$x(\alpha t), \alpha \neq 0$	$\frac{1}{|\alpha|} X\left(\frac{\Omega}{\alpha}\right)$
Reflection	$x(-t)$	$X(-\Omega)$
Parseval's energy relation	$E_x = \int_{-\infty}^{\infty} |x(t)|^2 dt$	$E_x = \frac{1}{2\pi} \int_{-\infty}^{\infty} |X(\Omega)|^2 d\Omega$
Duality	$X(t)$	$2\pi x(-\Omega)$
Time differentiation	$\frac{d^n x(t)}{dt^n}, \ n \geq 1, \ \text{integer}$	$(j\Omega)^n X(\Omega)$
Frequency differentiation	$-jtx(t)$	$\frac{dX(\Omega)}{d\Omega}$
Integration	$\int_{-\infty}^{t} x(t')dt'$	$\frac{X(\Omega)}{j\Omega} + \pi X(0)\delta(\Omega)$
Time shifting	$x(t - \alpha)$	$e^{-j\alpha\Omega} X(\Omega)$
Frequency shifting	$e^{j\Omega_0 t} x(t)$	$X(\Omega - \Omega_0)$
Modulation	$x(t)\cos(\Omega_c t)$	$0.5[X(\Omega - \Omega_c) + X(\Omega + \Omega_c)]$
Periodic signals	$x(t) = \sum_k X_k e^{jk\Omega_0 t}$	$X(\Omega) = \sum_k 2\pi X_k \delta(\Omega - k\Omega_0)$
Symmetry	$x(t)$ real	$|X(\Omega)| = |X(-\Omega)|$
		$\angle X(\Omega) = -\angle X(-\Omega)$
Convolution in time	$z(t) = [x * y](t)$	$Z(\Omega) = X(\Omega) Y(\Omega)$
Windowing/multiplication	$x(t)y(t)$	$\frac{1}{2\pi}[X * Y](\Omega)$
Cosine transform	$x(t)$ even	$X(\Omega) = \int_{-\infty}^{\infty} x(t)\cos(\Omega t)dt, \ \text{real}$
Sine transform	$x(t)$ odd	$X(\Omega) = -j\int_{-\infty}^{\infty} x(t)\sin(\Omega t)dt, \ \text{imaginary}$

Table 5.2 Fourier Transform Pairs

	Function of Time	Function of Ω
1	$\delta(t)$	1
2	$\delta(t - \tau)$	$e^{-j\Omega\tau}$
3	$u(t)$	$\frac{1}{j\Omega} + \pi\delta(\Omega)$
4	$u(-t)$	$\frac{-1}{j\Omega} + \pi\delta(\Omega)$
5	$\text{sgn}(t) = 2[u(t) - 0.5]$	$\frac{2}{j\Omega}$
6	$A, \ -\infty < t < \infty$	$2\pi A\delta(\Omega)$
7	$Ae^{-at}u(t), \ a > 0$	$\frac{A}{j\Omega + a}$
8	$Ate^{-at}u(t), \ a > 0$	$\frac{A}{(j\Omega + a)^2}$
9	$e^{-a|t|}, \ a > 0$	$\frac{2a}{a^2 + \Omega^2}$
10	$\cos(\Omega_0 t), \ -\infty < t < \infty$	$\pi[\delta(\Omega - \Omega_0) + \delta(\Omega + \Omega_0)]$
11	$\sin(\Omega_0 t), \ -\infty < t < \infty$	$-j\pi[\delta(\Omega - \Omega_0) - \delta(\Omega + \Omega_0)]$
12	$A[u(t + \tau) - u(t - \tau)], \ \tau > 0$	$2A\tau \frac{\sin(\Omega\tau)}{\Omega\tau}$
13	$\frac{\sin(\Omega_0 t)}{\pi t}$	$u(\Omega + \Omega_0) - u(\Omega - \Omega_0)$
14	$x(t)\cos(\Omega_0 t)$	$0.5[X(\Omega - \Omega_0) + X(\Omega + \Omega_0)]$

(a) Find the Fourier series coefficient X_0 for each of the values of T_0 and indicate how it changes for the different values of T_0.

(b) Find the Fourier series coefficients for $x(t)$ and carefully plot the magnitude line spectrum for each of the values of T_0. Explain what is happening in these spectra.

(c) If you were to let T_0 be very large, what would you expect to happen to the Fourier coefficients? Explain.

(d) Write a MATLAB script that simulates the conversion from the Fourier series to the Fourier transform of a sequence of rectangular pulses as the period is increased. The Fourier coefficients need to be multiplied by the period so that they do not become insignificant. Plot using `stem` the magnitude line spectrum for pulse sequences with periods T_0 from 4 to 62.

5.2. Fourier transform from Laplace transform—MATLAB

The Fourier transform of finite-support signals, which are absolutely integrable or finite energy, can be obtained from their Laplace transform rather than doing the integral. Consider the following signals:

$$x_1(t) = u(t + 0.5) - u(t - 0.5)$$

$$x_2(t) = \sin(2\pi t)[u(t) - u(t - 0.5)]$$

$$x_3(t) = r(t + 1) - 2r(t) + r(t - 1)$$

(a) Plot each of the signals.

(b) Find the Fourier transforms $\{X_i(\Omega)\}$ for $i = 1, 2$, and 3 using the Laplace transform.

(c) Use MATLAB's symbolic function `fourier` to compute the Fourier transform of the given signals. Plot the magnitude spectrum corresponding to each of the signals.

5.3. Fourier transform from Laplace transform of infinite-support signals—MATLAB

For signals with infinite support, their Fourier transforms cannot be derived from the Laplace transform unless they are absolutely integrable or the region of convergence of the Laplace transform contains the $j\Omega$ axis. Consider the signal $x(t) = 2e^{-2|t|}$.

(a) Plot the signal $x(t)$ for $-\infty < t < \infty$.

(b) Use the evenness of the signal to find the integral

$$\int_{-\infty}^{\infty} |x(t)|dt$$

and determine whether this signal is absolutely integrable or not.

(c) Use the integral definition of the Fourier transform to find $X(\Omega)$.

(d) Use the Laplace transform of $x(t)$ to verify the above found Fourier transform.

(e) Use MATLAB's symbolic function `fourier` to compute the Fourier transform of $x(t)$. Plot the magnitude spectrum corresponding to $x(t)$.

5.4. Fourier and Laplace transforms—MATLAB

Consider the signal $x(t) = 2e^{-2t}\cos(2\pi t)u(t)$.

(a) Use the fact this signal is bounded by the exponential $\pm 2e^{-2t}u(t)$ to show that the integral

$$\int_{-\infty}^{\infty} |x(t)|dt$$

is finite, indicating the signal is absolutely integrable and also finite energy.

(b) Use the Laplace transform to find the Fourier transform $X(\Omega)$ of $x(t)$.

(c) Use the MATLAB function fourier to compute the magnitude and phase spectrum of $X(\Omega)$.

5.5. Fourier transform of causal signals

Any causal signal $x(t)$ having a Laplace transform with poles in the open-left s-plane (i.e., not including the $j\Omega$ axis) has, as we saw before, a region of convergence that includes the $j\Omega$ axis, and as such its Fourier transform can be found from its Laplace transform. Consider the following signals:

$$x_1(t) = e^{-2t}u(t)$$

$$x_2(t) = r(t)$$

$$x_3(t) = x_1(t)x_2(t)$$

(a) Determine the Laplace transform of the above signals (use properties of the Laplace transform) indicating the corresponding region of convergence.

(b) Determine for which of these signals you can find its Fourier transform from its Laplace transform. Explain.

(c) Give the Fourier transform of the signals that can be obtained from their Laplace transform.

5.6. Duality of Fourier transform

There are some signals for which the Fourier transforms cannot be found directly by either the integral definition or the Laplace transform, and for those we need to use the properties of the Fourier transform, in particular the duality property. Consider, for instance,

$$x(t) = \frac{\sin(t)}{t}$$

or the sinc signal. Its importance is that it is the impulse response of an ideal low-pass filter.

(a) Let $X(\Omega) = A[u(\Omega + \Omega_0) - u(\Omega - \Omega_0]$ be a possible Fourier transform of $x(t)$. Find the inverse Fourier transform of $X(\Omega)$ using the integral equation to determine the values of A and Ω_0.

(b) How could you use the duality property of the Fourier transform to obtain $X(\Omega)$? Explain.

5.7. Cosine and sine transforms

The Fourier transforms of even and odd functions are very important. The reason is that they are computationally simpler than the Fourier transform. Let $x(t) = e^{-|t|}$ and $y(t) = e^{-t}u(t) - e^{t}u(-t)$.

(a) Plot $x(t)$ and $y(t)$, and determine whether they are even or odd.

(b) Show that the Fourier transform of $x(t)$ is found from

$$X(\Omega) = \int_{-\infty}^{\infty} x(t) \cos(\Omega t)dt$$

which is a real function of Ω, thus its computational importance. Show that $X(\Omega)$ is also even as a function of Ω.

(c) Find $X(\Omega)$ from the above equation (called the cosine transform).

(d) Show that the Fourier transform of $y(t)$ is found from

$$Y(\Omega) = -j \int_{-\infty}^{\infty} y(t) \sin(\Omega t)dt$$

which is imaginary function of Ω, thus its computational importance. Show that $Y(\Omega)$ is also odd as a function of Ω.

(e) Find $Y(\Omega)$ from the above equation (called the sine transform). Verify that your results are correct by finding the Fourier transform of $z(t) = x(t) + y(t)$ directly and using the above results.

(f) What advantages do you see to the cosine and sine transforms? How would you use the cosine and the sine transforms to compute the Fourier transform of any signal, not necessarily even or odd? Explain.

5.8. Time versus frequency—MATLAB

The supports in time and in frequency of a signal $x(t)$ and its Fourier transform $X(\Omega)$ are inversely proportional. Consider a pulse

$$x(t) = \frac{1}{T_0}[u(t) - u(t - T_0)]$$

(a) Let $T_0 = 1$ and $T_0 = 10$ and find and compare the corresponding $|X(\Omega)|$.

(b) Use MATLAB to simulate the changes in the magnitude spectrum when $T_0 = 10^k$ for $k = 0, \ldots, 4$ for $x(t)$. Compute $X(\Omega)$ and plot its magnitude spectra for the increasing values of T_0 on the same plot. Explain the results.

5.9. Smoothness and frequency content—MATLAB

The smoothness of the signal determines the frequency content of its spectrum. Consider the signals

$$x(t) = u(t + 0.5) - u(t - 0.5)$$

$$y(t) = (1 + \cos(\pi t))[u(t + 0.5) - u(t - 0.5)]$$

(a) Plot these signals. Can you tell which one is smoother?

(b) Find $X(\Omega)$ and carefully plot its magnitude $|X(\Omega)|$ versus frequency Ω.

(c) Find $Y(\Omega)$ (use the Fourier transform properties) and carefully plot its magnitude $|Y(\Omega)|$ versus frequency Ω.

(d) Which one of these two signals has higher frequencies? Can you now tell which of the signals is smoother? Use MATLAB to decide. Make $x(t)$ and $y(t)$ have unit energy. Plot $20 \log_{10} |Y(\Omega)|$ and $20 \log_{10} |X(\Omega)|$ using MATLAB and see which of the spectra shows lower frequencies.

5.10. Smoothness and frequency—MATLAB

Let the signals $x(t) = r(t + 1) - 2r(t) + r(t - 1)$ and $y(t) = dx(t)/dt$.

(a) Plot $x(t)$ and $y(t)$.

(b) Find $X(\Omega)$ and carefully plot its magnitude spectrum. Is $X(\Omega)$ real? Explain.

(c) Find $Y(\Omega)$ (use properties of Fourier transform) and carefully plot its magnitude spectrum. Is $Y(\Omega)$ real? Explain.

(d) Determine from the above spectra which of these two signals is smoother. Use MATLAB to plot $20 \log_{10} |Y(\Omega)|$ and $20 \log_{10} |X(\Omega)|$ and decide. Would you say in general that computing the derivative of a signal generates high frequencies or possible discontinuities?

5.11. Integration and smoothing—MATLAB

Consider the signal

$$x(t) = u(t + 1) - 2u(t) + u(t - 1)$$

and let

$$y(t) = \int\limits_{-\infty}^{t} x(\tau)d\tau$$

(a) Plot $x(t)$ and $y(t)$.

(b) Find $X(\Omega)$ and carefully plot its magnitude spectrum. Is $X(\Omega)$ real? Explain. (Use MATLAB to do the plotting.)

(c) Find $Y(\Omega)$ and carefully plot its magnitude spectrum. Is $Y(\Omega)$ real? Explain. (Use MATLAB to do the plotting.)

(d) Determine from the above spectra which of these two signals is smoother. Use MATLAB to decide. Would you say that in general by integrating a signal you get rid of higher frequencies, or smooth out a signal?

5.12. Duality of Fourier transforms

As indicated by the derivative property, if we multiply a Fourier transform by $(j\Omega)^N$, it corresponds to computing an Nth derivative of its time signal. Consider the dual of this property—that is, if we compute the derivative of $X(\Omega)$, what would happen to the signal in the time?

(a) Let $x(t) = \delta(t-1) + \delta(t+1)$. Find its Fourier transform (using properties) $X(\Omega)$.

(b) Compute $dX(\Omega)/d\Omega$ and determine its inverse Fourier transform.

5.13. Periodic functions in frequency

The duality property provides interesting results. Consider the signal

$$x(t) = \delta(t + T_1) + \delta(t - T_1)$$

(a) Find $X(\Omega) = \mathcal{F}[x(t)]$ and plot both $x(t)$ and $X(\Omega)$.

(b) Suppose you then generate a signal

$$y(t) = \delta(t) + \sum_{k=1}^{\infty} [\delta(t + kT_0) + \delta(t - kT_0)]$$

Find its Fourier transform $Y(\Omega)$ and plot both $y(t)$ and $Y(\Omega)$.

(c) Are $y(t)$ and the corresponding Fourier transform $Y(\Omega)$ periodic in time and in frequency? If so, determine their periods.

5.14. Sampling signal

The sampling signal

$$\delta_{T_s}(t) = \sum_{n=-\infty}^{\infty} \delta(t - nT_s)$$

will be important in the sampling theory later on.

(a) As a periodic signal of period T_s, express $\delta_{T_s}(t)$ by its Fourier series.

(b) Determine then the Fourier transform $\Delta(\Omega) = \mathcal{F}[\delta_{T_s}(t)]$.

(c) Plot $\delta_{T_s}(t)$ and $\Delta(\Omega)$ and comment on the periodicity of these two functions.

5.15. Piecewise linear signals

The derivative property can be used to simplify the computation of some Fourier transforms. Let

$$x(t) = r(t) - 2r(t-1) + r(t-2)$$

(a) Find and plot the second derivative with respect to t of $x(t)$, or $y(t) = d^2 x(t)/dt^2$.

(b) Find $X(\Omega)$ from $Y(\Omega)$ using the derivative property.

(c) Verify the above result by computing the Fourier transform $X(\Omega)$ directly from $x(t)$ using the Laplace transform.

5.16. Periodic signal-equivalent representations

Applying the time and frequency shifts it is possible to get different but equivalent Fourier transforms of periodic signals. Assume a period of a periodic signal $x(t)$ of period T_0 is $x_1(t)$, so that

$$x(t) = \sum_k x_1(t - kT_0)$$

and as seen in Chapter 4 the Fourier series coefficients of $x(t)$ are found as $X_k = X_1(jk\Omega_0)/T_0$, so that $x(t)$ can also be represented as

$$x(t) = \frac{1}{T_0} \sum_k X_1(jk\Omega_0)e^{jk\Omega_0 t}$$

(a) Find the Fourier transform of the first expression given above for $x(t)$ using the time-shift property.
(b) Find the Fourier transform of the second expression for $x(t)$ using the frequency-shift property.
(c) Compare the two expressions and comment on your results.

5.17. Modulation property

Consider the raised-cosine pulse

$$x(t) = [1 + \cos(\pi t)](u(t + 1) - u(t - 1))$$

(a) Carefully plot $x(t)$.
(b) Find the Fourier transform of the pulse $p(t) = u(t + 1) - u(t - 1)$.
(c) Use the definition of the pulse $p(t)$ and the modulation property to find the Fourier transform of $x(t)$ in terms of $P(\Omega) = \mathcal{F}[p(t)]$.

5.18. Solution of differential equations

An analog averager is characterized by the relationship

$$\frac{dy(t)}{dt} = 0.5[x(t) - x(t - 2)]$$

where $x(t)$ is the input and $y(t)$ the output. If $x(t) = u(t) - 2u(t - 1) + u(t - 2)$:
(a) Find the Fourier transform of the output $Y(\Omega)$.
(b) Find $y(t)$ from $Y(\Omega)$.

5.19. Generalized AM

Consider the following generalization of amplitude modulation where instead of multiplying by a cosine we multiply by a periodic signal with harmonic frequencies much higher than those of the message. Suppose the carrier $c(t)$ is a periodic signal with fundamental frequency Ω_0, let's say

$$c(t) = \sum_{k=4}^{6} 2\cos(k\Omega_0 t)$$

and that the message is a sinusoid of frequency $\Omega_0 = 2\pi$, or $x(t) = \cos(\Omega_0 t)$.
(a) Find the AM signal $s(t) = x(t)c(t)$.
(b) Determine the Fourier transform $S(\Omega)$.
(c) What would be a possible advantage of this generalized AM system? Explain.

5.20. Filter for half-wave rectifier

Suppose you want to design a dc source using a half-wave rectified signal $x(t)$ and an ideal filter. Let $x(t)$ be periodic, $T_0 = 2$, and with a period

$$x_1(t) = \begin{cases} \sin(\pi t) & 0 \leq t \leq 1 \\ 0 & 1 < t \leq 2, \end{cases}$$

(a) Find the Fourier transform $X(\Omega)$ of $x(t)$, and plot the magnitude spectrum including the dc and the first three harmonics.

(b) Determine the magnitude and cut-off frequency of an ideal low-pass filter $H(j\Omega)$ such that when we have $x(t)$ as its input, the output is $y(t) = 1$. Plot the magnitude response of the ideal low-pass filter. (For simplicity assume the phase is zero.)

5.21. Passive RLC filters—MATLAB

Consider an RLC series circuit with a voltage source $v_s(t)$. Let the values of the resistor, capacitor, and inductor be unity. Plot the poles and zeros and the corresponding frequency responses of the filters with the output the voltage across the

(a) Capacitor

(b) Inductor

(c) Resistor

Indicate the type of filter obtained in each case. Use MATLAB to plot the poles and zeros, the magnitude, and the phase response of each of the filters obtained above.

5.22. AM modulation and demodulation

A pure tone $x(t) = 4\cos(1000t)$ is transmitted using an AM communication system with a carrier $\cos(10,000t)$. The output of the AM system is

$$y(t) = x(t)\cos(10,000t)$$

At the receiver, to recover $x(t)$ the sent signal $y(t)$ needs first to be separated from the thousands of other signals. This is done with a band-pass filter with a center frequency equal to the carrier frequency, and the output of this filter then needs to be demodulated.

(a) Consider an ideal band-pass filter $H(j\Omega)$. Let its phase be zero. Determine its bandwidth, center frequency, and amplitude so we get as its output $10y(t)$. Plot the spectrum of $x(t)$, $10y(t)$, and the magnitude frequency response of $H(j\Omega)$.

(b) To demodulate $10y(t)$, we multiply it by $\cos(10,000t)$. You need then to pass the resulting signal through an ideal low-pass filter to recover the original signal $x(t)$. Plot the spectrum of

$$z(t) = 10y(t)\cos(10,000t)$$

and from it determine the frequency response of the low-pass filter $G(j\Omega)$ needed to recover $x(t)$. Plot the magnitude response of $G(j\Omega)$.

5.23. Ideal low-pass filter—MATLAB

Consider an ideal low-pass filter $H(s)$ with zero phase and magnitude response

$$|H(j\Omega)| = \begin{cases} 1 & -\pi \leq \Omega \leq \pi \\ 0 & \text{otherwise} \end{cases}$$

(a) Find the impulse response $h(t)$ of the low-pass filter. Plot it and indicate whether this filter is a causal system or not.

(b) Suppose you wish to obtain a band-pass filter $G(j\Omega)$ from $H(j\Omega)$. If the desired center frequency of $|G(j\Omega)|$ is 5π, and its desired magnitude is 1 at the center frequency, how would you process $h(t)$ to get the desired filter? Explain your procedure.

(c) Use symbolic MATLAB to find $h(t), g(t)$, and $G(j\Omega)$. Plot $|H(j\Omega)|, h(t), g(t)$, and $|G(j\Omega)|$.

5.24. Magnitude response from poles and zeros—MATLAB

Consider the following filters with the given poles and zeros and dc constant:

$$H_1(s): \quad K = 1 \text{ poles} \quad p_1 = -1, p_{2,3} = -1 \pm j\pi; \quad \text{zeros} \quad z_1 = 1, z_{2,3} = 1 \pm j\pi$$

$$H_2(s): \quad K = 1 \text{ poles} \quad p_1 = -1, p_{2,3} = -1 \pm j\pi; \quad \text{zeros} \quad z_{1,3} = \pm j\pi$$

$$H_3(s): \quad K = 1 \text{ poles} \quad p_1 = -1, p_{2,3} = -1 \pm j\pi; \quad \text{zero} \quad z_1 = 1$$

Use MATLAB to plot the magnitude responses of these filters and indicate the type of filters they are.

5.25. Different types of AM modulations—MATLAB

Let the signal

$$m(t) = \sin(2\pi t)[u(t) - u(t-1)]$$

be the message or input to different types of AM systems with the output the following signals. Carefully plot $m(t)$ and the following outputs in $0 \le t \le 1$ and their corresponding spectra using MATLAB. Let the sampling period be $T_s = 0.001$.

(a) $y_1(t) = m(t)\cos(20\pi t)$

(b) $y_2(t) = [1 + m(t)]\cos(20\pi t)$

5.26. Windows—MATLAB

The signal $x(t)$ in Problem 5.17 is called a raised-cosine window. Notice that it is a very smooth signal and that it decreases at both ends. The rectangular window is the signal $y(t) = u(t+1) - u(t-1)$.

(a) Use MATLAB to compute the magnitude spectrum of $x(t)$ and $y(t)$ and indicate which is the smoother of the two by considering the presence of high frequencies as an indication of roughness.

(b) When computing the Fourier transform of a very long signal it makes sense to break it up into smaller sections and compute the Fourier transform of each. In such a case, windows are used to smooth out the transition from one section to the other. Consider a sinusoid $z(t) = \cos(2\pi t)$ for $0 \le t \le 1000$ sec. Divide the signal into two sections of duration 500 sec. Multiply the corresponding signal in each of the sections by a raised-cosine $x(t)$ and rectangular $y(t)$ windows of length 500 and compute using MATLAB the corresponding Fourier transforms. Compare them to the Fourier transform of the whole signal and comment on your results. Sample all the signals using $T_s = 1/(4\pi)$ as the sampling period.

(c) Consider the computation of the Fourier transform of the acoustic signal corresponding to a train whistle, which MATLAB provides as a sampled signal in "train.mat" using the discrete approximation of the Fourier transform. The frequency content of the whole signal (hard to find) would not be as meaningful as the frequency content of a smaller section of it as they change with time. Compute the Fourier transform of sections of 1000 samples by windowing the signal with the raised-cosine window (sampled with the same sampling period as the "train.mat" signal or $T_s = 1/F_s$ where F_s is the sampling frequency given for "train.mat"). Plot the spectra of a few of these segments and comment on the change in the frequency content as time changes.

Application to Control and Communications

6.1 INTRODUCTION

Control and communications are areas in electrical engineering where the Laplace and the Fourier analyses apply. In this chapter, we illustrate how these transform methods and the concepts of transfer function, frequency response, and spectrum connect with the classical theories of control and communications.

In classical control, the objective is to change the dynamics of a given system to be able to achieve a desired response by frequency-domain methods. This is typically done by means of a feedback connection of a controller to a plant. The plant is a system such as a motor, a chemical plant, or an automobile we would like to control so that it responds in a certain way. The controller is a system we design to make the plant follow a prescribed input or reference signal. By feeding back the response of the system to the input, it can be determined how the plant responds to the controller. The commonly used negative feedback generates an error signal that permits us to judge the performance of the controller. The concepts of transfer function, stability of systems, and different types of responses obtained through the Laplace transform are very useful in the analysis and design of classical control systems.

A communication system consists of three components: a transmitter, a channel, and a receiver. The objective of communications is to transmit a message over a channel to a receiver. The message is a signal, for instance, a voice or a music signal, typically containing low frequencies. Transmission of the message can be done over the airwaves or through a line connecting the transmitter to the receiver, or a combination of the two—constituting channels with different characteristics. Telephone communication can be done with or without wires, and radio and television are wireless. The concepts of

Signals and Systems Using MATLAB®. DOI: 10.1016/B978-0-12-374716-7.00009-0

frequency, bandwidth, spectrum, and modulation developed by means of the Fourier transform are fundamental in the analysis and design of communication systems.

The aim of this chapter is to serve as an introduction to problems in classical control and communications and to link them with the Laplace and Fourier analyses. More in-depth discussion of these topics can be found in many excellent texts in control and communications.

The other topic covered in this chapter is an introduction to analog filter design. Filtering is a very important application of LTI systems in communications, control, and digital signal processing. The material in this chapter will be complemented by the design of discrete filters in Chapter 11. Important issues related to signals and system are illustrated in the design and implementation of filters.

6.2 SYSTEM CONNECTIONS AND BLOCK DIAGRAMS

Control and communication systems consist of interconnection of several subsystems. As we indicated in Chapter 2, there are three important connections of LTI systems:

- Cascade
- Parallel
- Feedback

Cascade and parallel result from properties of the convolution integral, while the feedback connection relates the output of the overall system to its input. With the background of the Laplace transform we present now a transform characterization of these connections that can be related to the time-domain characterizations given in Chapter 2.

The connection of two LTI continuous-time systems with transfer functions $H_1(s)$ and $H_2(s)$ (and corresponding impulse responses $h_1(t)$ and $h_2(t)$) can be done in:

- Cascade (Figure 6.1(a)): Provided that the two systems are isolated, the transfer function of the overall system is

$$H(s) = H_1(s)H_2(s) \qquad (6.1)$$

- Parallel (Figure 6.1(b)): The transfer function of the overall system is

$$H(s) = H_1(s) + H_2(s) \qquad (6.2)$$

- Negative feedback (Figure 6.4): The transfer function of the overall system is

$$H(s) = \frac{H_1(s)}{1 + H_2(s)H_1(s)} \qquad (6.3)$$

  - Open-loop transfer function: $H_{o\ell}(s) = H_1(s)$.
  - Closed-loop transfer function: $H_{c\ell}(s) = H(s)$.

Cascading of LTI Systems

Given two LTI systems with transfer functions $H_1(s) = \mathcal{L}[h_1(t)]$ and $H_2(s) = \mathcal{L}[h_2(t)]$ where $h_1(t)$ and $h_2(t)$ are the corresponding impulse responses of the systems, the *cascading* of these systems gives a new system with transfer function

$$H(s) = H_1(s)H_2(s) = H_2(s)H_1(s)$$

provided that these systems are isolated from each other (i.e., they do not load each other). A graphical representation of the cascading of two systems is obtained by representing each of the systems with blocks with their corresponding transfer function (see Figure 6.1(a)). Although cascading of systems is a simple procedure, it has some disadvantages:

- It requires isolation of the systems.
- It causes delay as it processes the input signal, possibly compounding any errors in the processing.

Remarks

- *Loading, or lack of system isolation, needs to be considered when cascading two systems. Loading does not allow the overall transfer function to be the product of the transfer functions of the connected systems. Consider the cascade connection of two resistive voltage dividers (Figure 6.2), each with a simple transfer function $H_i(s) = 1/2$, $i = 1, 2$. The cascade in Figure 6.2(b) clearly will not have as transfer function $H(s) = H_1(s)H_2(s) = (1/2)(1/2)$ unless we include a buffer (such as an operational amplifier voltage*

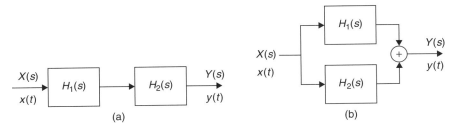

(a)   (b)

FIGURE 6.1

(a) Cascade and (b) parallel connections of systems with transfer function $H_1(s)$ and $H_2(s)$. The input and output are given in the time or in the frequency domains.

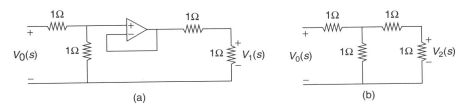

(a)   (b)

FIGURE 6.2

Cascading of two voltage dividers: (a) using a voltage follower gives $V_1(s)/V_0(s) = (1/2)(1/2)$ with no loading effect, and (b) using no voltage follower $V_2(s)/V_0(s) = 1/5 \neq V_1(s)/V_0(s)$ due to loading.

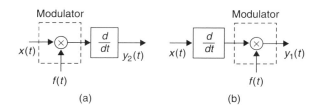

FIGURE 6.3

Cascading of (a) an LTV and (b) an LTI system.
The outputs are different, $y_1(t) \neq y_2(t)$.

follower) in between (see Figure 6.2(a)). The cascading of the two voltage dividers without the voltage follower gives a transfer function $H_1(s) = 1/5$, as can be easily shown by doing mesh analysis on the circuit.

■ The block diagrams of the cascade of two or more LTI systems can be interchanged with no effect on the overall transfer function, provided the connection is done with no loading. That is not true if the systems are not LTI. For instance, consider cascading a modulator (LTV system) and a differentiator (LTI) as shown in Figure 6.3. If the modulator is first, Figure 6.3(a), the output of the overall system is

$$y_2(t) = \frac{dx(t)f(t)}{dt} = f(t)\frac{dx(t)}{dt} + x(t)\frac{df(t)}{dt}$$

while if we put the differentiator first, Figure 6.3(b), the output is

$$y_1(t) = f(t)\frac{dx(t)}{dt}$$

It is obvious that if $f(t)$ is not a constant, the two responses are very different.

Parallel Connection of LTI Systems

According to the distributive property of the convolution integral, the *parallel* connection of two or more LTI systems has the same input and its output is the sum of the outputs of the systems being connected (see Figure 6.1(b)). The parallel connection is better than the cascade, as it does not require isolation between the systems, and reduces the delay in processing an input signal. The transfer function of the parallel system is

$$H(s) = H_1(s) + H_2(s)$$

Remarks

■ Although a communication system can be visualized as the cascading of three subsystems—the transmitter, the channel, and the receiver—typically none of these subsystems is LTI. As we discussed in Chapter 5, the low-frequency nature of the message signals requires us to use as the transmitter a system that can generate a signal with much higher frequencies, and that is not possible with LTI systems (recall the eigenfunction property). Transmitters are thus typically nonlinear or linear time varying. The receiver is also not LTI. A wireless channel is typically time varying.

■ Some communication systems use parallel connections (see quadrature amplitude modulation (QAM) later in this chapter). To make it possible for several users to communicate over the same channel, a combination of parallel and cascade connections are used (see frequency division multiplexing (FDM) systems later in this chapter). But again, it should be emphasized that these subsystems are not LTI.

FIGURE 6.4

Negative-feedback connection of systems with transfer function $H_1(s)$ and $H_2(s)$. The input and the output are $x(t)$ and $y(t)$, respectively, and $e(t)$ is the error signal.

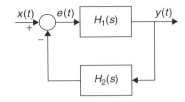

Feedback Connection of LTI Systems

In control, *feedback* connections are more appropriate than cascade or parallel connections. In the feedback connection, the output of the first system is fed back through the second system into the input (see Figure 6.4). In this case, like in the parallel connection, beside the blocks representing the systems we use *adders* to add/subtract two signals.

It is possible to have *positive-* or *negative-feedback* systems depending on whether we add or subtract the signal being fed back to the input. Typically, negative feedback is used, as positive feedback can greatly increase the gain of the system. (Think of the screeching sound created by an open microphone near a loud-speaker: the microphone continuously picks up the amplified sound from the loud-speaker, increasing the volume of the produced signal. This is caused by positive feedback.) For negative feedback, the connection of two systems is done by putting one in the feedforward loop, $H_1(s)$, and the other in the feedback loop, $H_2(s)$ (there are other possible connections). To find the overall transfer function we consider the Laplace transforms of the error signal $e(t)$, $E(s)$, and of the output $y(t)$, $Y(s)$, in terms of the Laplace transform of the input $x(t)$, $X(s)$, and the transfer functions $H_1(s)$ and $H_2(s)$ of the systems:

$$E(s) = X(s) - H_2(s)Y(s)$$

$$Y(s) = H_1(s)E(s)$$

Replacing $E(s)$ in the second equation gives

$$Y(s)[1 + H_1(s)H_2(s)] = H_1(s)X(s)$$

and the transfer function of the feedback system is then

$$H(s) = \frac{Y(s)}{X(s)} = \frac{H_1(s)}{1 + H_1(s)H_2(s)} \tag{6.4}$$

As you recall, in Chapter 2 we were not able to find an explicit expression for the impulse response of the overall system and now you can understand why.

6.3 APPLICATION TO CLASSIC CONTROL

Because of different approaches, the theory of control systems can be divided into classic and modern control. Classic control uses frequency-domain methods, while modern control uses time-domain methods. In classic linear control, the transfer function of the plant we wish to control is available;

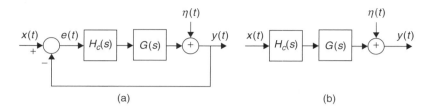

FIGURE 6.5

(a) Closed- and (b) open-loop control systems. The transfer function of the plant is $G(s)$ and the transfer function of the controller is $H_c(s)$.

let us call it $G(s)$. The controller, with a transfer function $H_c(s)$, is designed to make the output of the overall system perform in a specified way. For instance, in a cruise control the plant is the car, and the desired performance is to automatically set the speed of the car to a desired value. There are two possible ways the controller and the plant are connected: in open-loop or in closed-loop (see Figure 6.5).

Open-Loop Control

In the *open-loop* approach the controller is cascaded with the plant (Figure 6.5(b)). To make the output $y(t)$ follow the reference signal at the input $x(t)$, we minimize an error signal

$$e(t) = y(t) - x(t)$$

Typically, the output is affected by a disturbance $\eta(t)$, due to modeling or measurement errors. If we assume initially no disturbance, $\eta(t) = 0$, we find that the Laplace transform of the output of the overall system is

$$Y(s) = \mathcal{L}[y(t)] = H_c(s)G(s)X(s)$$

and that of the error is

$$E(s) = Y(s) - X(s) = [H_c(s)G(s) - 1]X(s)$$

To make the error zero, so that $y(t) = x(t)$, it would require that $H_c(s) = 1/G(s)$ or the inverse of the plant, making the overall transfer function of the system $H_c(s)G(s)$ unity.

Remarks

Although open-loop systems are simple to implement, they have several disadvantages:

- *The controller $H_c(s)$ must cancel the poles and the zeros of $G(s)$ exactly, which is not very practical. In actual systems, the exact location of poles and zeros is not known due to measurement errors.*
- *If the plant $G(s)$ has zeros on the right-hand s-plane, then the controller $H_c(s)$ will be unstable, as its poles are the zeros of the plant.*
- *Due to ambiguity in the modeling of the plant, measurement errors, or simply the presence of noise, the output $y(t)$ is typically affected by a disturbance signal $\eta(t)$ mentioned above ($\eta(t)$ is typically random— we are going to assume for simplicity that it is deterministic so we can compute its Laplace transform).*

The Laplace transform of the overall system output is

$$Y(s) = H_c(s)G(s)X(s) + \eta(s)$$

where $\eta(s) = \mathcal{L}[\eta(t)]$. In this case, $E(s)$ is given by

$$E(s) = [H_c(s)G(s) - 1]X(s) + \eta(s)$$

Although we can minimize this error by choosing $H_c(s) = 1/G(s)$ as above, in this case $e(t)$ cannot be made zero—it remains equal to the disturbance $\eta(t)$ and we have no control over this.

Closed-Loop Control

Assuming $y(t)$ and $x(t)$ in the open-loop control are the same type of signals, (e.g., both are voltages, or temperatures), if we feed back $y(t)$ and compare it with the input $x(t)$ we obtain a closed-loop control. Considering the case of negative-feedback system (see Figure 6.5(a)), and assuming no disturbance ($\eta(t) = 0$), we have that

$$E(s) = X(s) - Y(s)$$

$$Y(s) = H_c(s)G(s)E(s)$$

and replacing $Y(s)$ gives

$$E(s) = \frac{X(s)}{1 + G(s)H_c(s)}$$

If we wish the error to go to zero in the steady state, so that $y(t)$ tracks the input, the poles of $E(s)$ should be in the open left-hand s-plane.

If a disturbance signal $\eta(t)$ (consider it for simplicity deterministic and with Laplace transform $\eta(s)$) is present (See Figure 6.5(a)), the above analysis becomes

$$E(s) = X(s) - Y(s)$$

$$Y(s) = H_c(s)G(s)E(s) + \eta(s)$$

so that

$$E(s) = X(s) - H_c(s)G(s)E(s) - \eta(s)$$

or solving for $E(s)$,

$$E(s) = \frac{X(s)}{1 + G(s)H_c(s)} - \frac{\eta(s)}{1 + G(s)H_c(s)}$$

$$= E_1(s) + E_2(s)$$

If we wish $e(t)$ to go to zero in the steady state, then poles of $E_1(s)$ and $E_2(s)$ should be in the open left-hand s-plane. Different from the open-loop control, the closed-loop control offers more flexibility in achieving this by minimizing the effects of the disturbance.

Remarks

A control system includes two very important components:

- Transducer: *Since it is possible that the output signal y(t) and the reference signal x(t) might not be of the same type, a transducer is used to change the output so as to be compatible with the reference input. Simple examples of a transducer are: lightbulbs, which convert voltage into light; a thermocouple, which converts temperature into voltage.*
- Actuator: *A device that makes possible the execution of the control action on the plant, so that the output of the plant follows the reference input.*

■ Example 6.1: Controlling an unstable plant

Consider a dc motor modeled as an LTI system with a transfer function

$$G(s) = \frac{1}{s(s+1)}$$

The motor is not BIBO stable given that its impulse response $g(t) = (1 - e^{-t})u(t)$ is not absolutely integrable. We wish the output of the motor $y(t)$ to track a given reference input $x(t)$, and propose using a so-called *proportional controller* with transfer $H_c(s) = K > 0$ to control the motor (see Figure 6.6). The transfer function of the overall negative-feedback system is

$$H(s) = \frac{Y(s)}{X(s)} = \frac{KG(s)}{1 + KG(s)}$$

Suppose that $X(s) = 1/s$, or the reference signal is $x(t) = u(t)$. The question is: What should be the value of K so that in the steady state the output of the system $y(t)$ coincides with $x(t)$? Or, equivalently, is the error signal in the steady state zero? We have that the Laplace transform of the error signal $e(t) = x(t) - y(t)$ is

$$E(s) = X(s)[1 - H(s)] = \frac{1}{s(1 + G(s)K)} = \frac{s+1}{s(s+1) + K}$$

The poles of $E(s)$ are the roots of the polynomial $s(s+1) + K = s^2 + s + K$, or

$$s_{1,2} = -0.5 \pm 0.5\sqrt{1 - 4K}$$

For $0 < K \leq 0.25$ the roots are real, and complex for $K > 0.25$, and in either case in the left-hand s-plane. The partial fraction expansion corresponding to $E(s)$ would be

$$E(s) = \frac{B_1}{s - s_1} + \frac{B_2}{s - s_2}$$

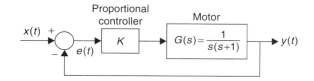

FIGURE 6.6

Proportional control of a motor.

for some values B_1 and B_2. Given that the real parts of s_1 and s_2 are negative, their corresponding inverse Laplace terms will have a zero steady-state response. Thus,

$$\lim_{t \to \infty} e(t) \to 0$$

This can be found also with the final value theorem, $e(0)$ is

$$sE(s)|_{s=0} = 0$$

So for any $K > 0$, $y(t) \to x(t)$ in steady state.

Suppose then that $X(s) = 1/s^2$ or that $x(t) = tu(t)$, a ramp signal. Intuitively, this is a much harder situation to control, as the output needs to be continuously growing to try to follow the input. In this case, the Laplace transform of the error signal is

$$E(s) = \frac{1}{s^2(1 + G(s)K)} = \frac{s+1}{s(s(s+1) + K)}$$

In this case, even if we choose K to make the roots of $s(s+1) + K$ be in the left-hand s-plane, we have a pole at $s = 0$. Thus, in the steady state, the partial fraction expansion terms corresponding to poles s_1 and s_2 will give a zero steady-state response, but the pole $s = 0$ will give a constant steady-state response A where

$$A = E(s)s|_{s=0} = 1/K$$

In the case of a ramp as input, it is not possible to make the output follow exactly the input command, although by choosing a very large gain K we can get them to be very close. ∎

Choosing the values of the gain K of the open-loop transfer function

$$G(s)H_c(s) = \frac{KN(s)}{D(s)}$$

to be such that the roots of

$$1 + G(s)H_c(s) = 0$$

are in the open left-hand s-plane, is the *root-locus* method, which is of great interest in control theory.

■ Example 6.2: A cruise control

Suppose we are interested in controlling the speed of a car or in obtaining a cruise control. How to choose the appropriate controller is not clear. We consider initially a proportional plus integral (PI) controller $H_c(s) = 1 + 1/s$ and will ask you to consider the proportional controller as an exercise. See Figure 6.7.

Suppose we want to keep the speed of the car at V_0 miles/hour for $t \geq 0$ (i.e., $x(t) = V_0 u(t)$), and that the model for a car in motion is a system with transfer function

$$H_p(s) = \beta/(s + \alpha)$$

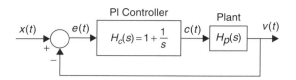

FIGURE 6.7
Cruise control system: reference speed
$x(t) = V_0 u(t)$ and output speed of car $v(t)$.

with both β and α positive values related to the mass of the car and the friction coefficient. For simplicity, let $\alpha = \beta = 1$. The question is: Can this be achieved with the PI controller? The Laplace transform of the output speed $v(t)$ of the car is

$$V(s) = \frac{H_c(s)H_p(s)}{1 + H_c(s)H_p(s)} X(s)$$

$$= \frac{V_0(s+1)}{s(s^2 + 2s + 1)} = \frac{V_0}{s(s+1)}$$

The poles of $V(s)$ are $s = 0$ and $s = -1$ on the left-hand s-plane. We can then write $V(s)$ as

$$V(s) = \frac{B}{s+1} + \frac{A}{s}$$

where $A = V_0$. The steady-state response is

$$\lim_{t \to \infty} v(t) = V_0$$

since the inverse Laplace transform of the first term goes to zero due to its poles being in the left-hand s-plane. The error signal $e(t) = x(t) - v(t)$ in the steady state is zero. The controlling signal $c(t)$ (see Figure 6.7) that changes the speed of the car is

$$c(t) = e(t) + \int_0^t e(\tau)d\tau$$

so that even if the error signal becomes zero at some point—indicating the desired speed had been reached—the value of $c(t)$ is not necessarily zero. The values of $e(t)$ at $t = 0$ and at steady–state can be obtained using the initial- and the final-value theorems of the Laplace transform applied to

$$E(s) = X(s) - V(s) = \frac{V_0}{s}\left[1 - \frac{1}{s+1}\right]$$

The final-value theorem gives that the steady-state error is

$$\lim_{t \to \infty} e(t) = \lim_{s \to 0} sE(s) = 0$$

coinciding with our previous result. The initial value is found as

$$e(0) = \lim_{s \to \infty} sE(s)$$

$$= \lim_{s \to \infty} V_0 \left[1 - \frac{1/s}{1 + 1/s}\right] = V_0$$

The PI controller used here is one of various possible controllers. Consider a simpler and cheaper controller such as a proportional controller with $H_c(s) = K$. Would you be able to obtain the same results? Try it. ∎

6.3.1 Stability and Stabilization

A very important question related to the performance of systems is: How do we know that a given causal system has finite zero-input, zero-state, or steady-state responses? This is the stability problem of great interest in control. Thus, if the system is represented by a linear differential equation with constant coefficients the stability of the system determines that the zero-input, the zero-state, as well as the steady-state responses may exist. The stability of the system is also required when considering the frequency response in the Fourier analysis. It is important to understand that only the Laplace transform allows us to characterize stable as well as unstable systems; the Fourier transform does not.

Two possible ways to look at the stability of a causal LTI system are:

■ When there is no input so that the response of the system depends on initial energy in the system. This is related to the zero-input response of the system.
■ When there is a bounded input and no initial condition. This is related to the zero-state response of the system.

Relating the zero-input response of a causal LTI system to stability leads to *asymptotic* stability. An LTI system is said to be asymptotically stable if the zero-input response (due only to initial conditions in the system) goes to zero as t increases—that is,

$$y_{zi}(t) \to 0 \qquad t \to \infty \tag{6.5}$$

for all possible initial conditions.

The second interpretation leads to the *bounded-input bounded-output* (BIBO) stability, which we defined in Chapter 2. A causal LTI system is BIBO stable if its response to a bounded input is also bounded. The condition we found in Chapter 2 for a causal LTI system to be BIBO stable was that the impulse response of the system be absolutely integrable—that is

$$\int_0^\infty |h(t)|dt < \infty \tag{6.6}$$

Such a condition is difficult to test, and we will see in this section that it is equivalent to the poles of the transfer function being in the open left-hand s-plane, a condition that can be more easily visualized and for which algebraic tests exist.

Consider a system being represented by the differential equation

$$y(t) + \sum_{k=1}^{N} a_k \frac{d^k y(t)}{dt^k} = b_0 x(t) + \sum_{\ell=1}^{M} b_\ell \frac{d^\ell x(t)}{dt^\ell} \qquad M < N$$

For some initial conditions and input $x(t)$, with Laplace transform $X(s)$, we have that the Laplace transform of the output is

$$Y(s) = Y_{zi}(s) + Y_{zs}(s) = \mathcal{L}[y(t)] = \frac{I(s)}{A(s)} + \frac{X(s)B(s)}{A(s)}$$

$$A(s) = 1 + \sum_{k=1}^{N} a_k s^k, \quad B(s) = b_0 + \sum_{m=1}^{M} b_m s^m$$

where $I(s)$ is due to the initial conditions. To find the poles of $H_1(s) = 1/A(s)$, we set $A(s) = 0$, which corresponds to the characteristic equation of the system and its roots (real, complex conjugate, simple, and multiple) are the natural modes or eigenvalues of the system.

A causal LTI system with transfer function $H(s) = B(s)/A(s)$ exhibiting no pole-zero cancellation is said to be:

■ Asymptotically stable if the all-pole transfer function $H_1(s) = 1/A(s)$, used to determine the zero-input response, has all its poles in the open left-hand s-plane (the $j\Omega$ axis excluded), or equivalently

$$A(s) \neq 0 \quad \text{for} \quad \mathcal{R}e[s] \geq 0 \tag{6.7}$$

■ BIBO stable if all the poles of $H(s)$ are in the open left-hand s-plane (the $j\Omega$ axis excluded), or equivalently

$$A(s) \neq 0 \quad \text{for} \quad \mathcal{R}e[s] \geq 0 \tag{6.8}$$

■ If $H(s)$ exhibits pole-zero cancellations, the system can be BIBO stable but not necessarily asymptotically stable.

Testing the stability of a causal LTI system thus requires finding the location of the roots of $A(s)$, or the poles of the system. This can be done for low-order polynomials $A(s)$ for which there are formulas to find the roots of a polynomial exactly. But as shown by Abel,[1] there are no equations to find the roots of higher than fourth-order polynomials. Numerical methods to find roots of these polynomials only provide approximate results that might not be good enough for cases where the poles are close to the $j\Omega$ axis. The Routh stability criterion [53] is an algebraic test capable of determining whether the roots of $A(s)$ are on the left-hand s-plane or not, thus determining the stability of the system.

■ Example 6.3: Stabilization of a plant

Consider a plant with a transfer function $G(s) = 1/(s - 2)$, which has a pole in the right-hand s-plane and therefore is unstable. Let us consider stabilizing it by cascading it with an all-pass filter (Figure 6.8(a)) so that the overall system is not only stable but also keeps its magnitude response.

---

[1] Niels H. Abel (1802–1829) was a Norwegian mathematician who accomplished brilliant work in his short lifetime. At *age 19*, he showed there is no general algebraic solution for the roots of equations of degree greater than four, in terms of explicit algebraic operations.

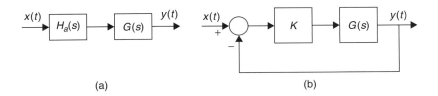

FIGURE 6.8

Stabilization of an unstable plant $G(s)$ using (a) an all-pass filter and (b) a proportional controller of gain K.

To get rid of the pole at $s = 2$ and to replace it with a new pole at $s = -2$, we let the all-pass filter be

$$H_a(s) = \frac{s - 2}{s + 2}$$

To see that this filter has a constant magnitude response consider

$$H_a(s)H_a(-s) = \frac{(s - 2)(-s - 2)}{(s + 2)(-s + 2)}$$

$$= \frac{(s - 2)(s + 2)}{(s + 2)(s - 2)} = 1$$

If we let $s = j\Omega$, the above gives the magnitude-squared function

$$H_a(j\Omega)H(-j\Omega) = H_a(j\Omega)H^*(j\Omega)$$

$$= |H(j\Omega)|^2$$

which is unity for all values of frequency. The cascade of the unstable system with the all-pass system gives a stable system

$$H(s) = G(s)H_a(s) = \frac{1}{s + 2}$$

with the same magnitude response as $G(s)$. This is an open-loop stabilization and it depends on the all-pass system having a zero exactly at 2 so that it cancels the pole causing the instability. Any small change on the zero and the overall system would not be stabilized. Another problem with the cascading of an all-pass filter to stabilize a filter is that it does not work when the pole causing the unstability is at the origin, as we cannot obtain an all-pass filter able to cancel that pole.

Consider then a negative-feedback system (Figure 6.8(b)). Suppose we use a proportional controller with a gain K, then the overall system transfer function is

$$H(s) = \frac{KG(s)}{1 + KG(s)} = \frac{K}{s + (K - 2)}$$

and if the gain K is chosen so that $K - 2 > 0$ or $K > 2$, the feedback system will be stable. ■

6.3.2 Transient Analysis of First- and Second-Order Control Systems

Although the input to a control system is not known a-priori, there are many applications where the system is frequently subjected to a certain type of input and thus one can select a test signal. For instance, if a system is subjected to intense and sudden inputs, then an impulse signal might be the appropriate test input for the system; if the input applied to a system is constant or continuously increasing, then a unit step or a ramp signal would be appropriate. Using test signals such as an impulse, a unit-step, a ramp, or a sinusoid, mathematical and experimental analyses of systems can be done.

When designing a control system its stability becomes its most important attribute, but there are other system characteristics that need to be considered. The transient behavior of the system, for instance, needs to be stressed in the design. Typically, as we drive the system to reach a desired response, the system's response goes through a transient before reaching the desired response. Thus, how fast the system responds and what steady-state error it reaches need to be part of the design considerations.

First-Order Systems

As an example of a first-order system consider an RC serial circuit with a voltage source $v_i(t) = u(t)$ as input (Figure 6.9), and as the output the voltage across the capacitor, $v_c(t)$. By voltage division, the transfer function of the circuit is

$$H(s) = \frac{V_c(s)}{V_i(s)} = \frac{1}{1 + RCs}$$

Considering the RC circuit, a feedback system with input $v_i(t)$ and output $v_c(t)$, the feedforward transfer function $G(s)$ in Figure 6.9 is $1/RCs$. Indeed, from the feedback system we have

$$E(s) = V_i(s) - V_c(s)$$

$$V_c(s) = E(s)G(s)$$

Replacing $E(s)$ in the second of the above equations, we have that

$$\frac{V_c(s)}{V_i(s)} = \frac{G(s)}{1 + G(s)} = \frac{1}{1 + 1/G(s)}$$

so that the open-loop transfer function, when we compare the above equation to $H(s)$, is

$$G(s) = \frac{1}{RCs}$$

The RC circuit can be seen as a feedback system: the voltage across the capacitor is constantly compared with the input voltage, and if found smaller, the capacitor continues charging until its voltage coincides with it. How fast depends on the RC value.

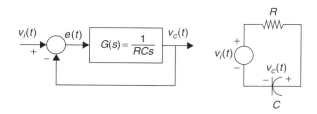

FIGURE 6.9
Feedback modeling of an RC circuit in series.

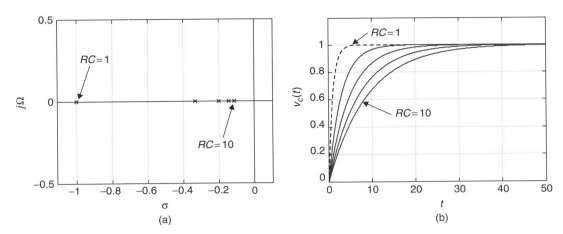

FIGURE 6.10

(a) Clustering of poles and (b) time responses of a first-order feedback system for $1 \leq RC \leq 10$.

For $v_i(t) = u(t)$, so that $V_i(s) = 1/s$, then the Laplace transform of the output is

$$V_c(s) = \frac{1}{s(sRC + 1)} = \frac{1/RC}{s(s + 1/RC)} = \frac{1}{s} - \frac{1}{s + 1/RC}$$

so that

$$v_c(t) = (1 - e^{-t/RC})u(t)$$

The following MATLAB script plots the poles $V_c(s)/V_i(s)$ and simulates the transients of $v_c(t)$ for $1 \leq RC \leq 10$, shown in Figure 6.10. Thus, if we wish the system to respond fast to the unit-step input we locate the system pole far from the origin.

```
%%%%%%%%%%%%%%%%%%%%%%
% Transient analysis
%%%%%%%%%%%%%%%%%%%%%%%
clf; clear all
syms s t
num = [0 1];
for RC = 1:2:10,
  den = [ RC 1];
  figure(1)
  splane(num, den) % plotting of poles and zeros
  hold on
  vc = ilaplace(1/(RC * s^2 + s)) %  inverse Laplace
  figure(2)
  ezplot(vc, [0, 50]); axis([0 50 0 1.2]); grid
  hold on
end
hold off
```

Second-Order System

A series RLC circuit with the input a voltage source, $v_s(t)$, and the output the voltage across the capacitor, $v_c(t)$, has a transfer function

$$\frac{V_c(s)}{V_s(s)} = \frac{1/Cs}{R + Ls + 1/Cs} = \frac{1/LC}{s^2 + (R/L)s + 1/LC}$$

If we define

$$\text{Natural frequency}: \Omega_n = \frac{1}{\sqrt{CL}} \tag{6.9}$$

$$\text{Damping ratio}: \psi = 0.5R\sqrt{\frac{C}{L}} \tag{6.10}$$

we can write

$$\frac{V_c(s)}{V_s(s)} = \frac{\Omega_n^2}{s^2 + 2\psi\Omega_n s + \Omega_n^2} \tag{6.11}$$

A feedback system with this transfer function is given in Figure 6.11 where the feedforward transfer function is

$$G(s) = \frac{\Omega_n^2}{s(s + 2\psi\Omega_n)}$$

Indeed, the transfer function of the feedback system is given by

$$H(s) = \frac{V_c(s)}{V_s(s)} = \frac{G(s)}{1 + G(s)}$$

$$= \frac{\Omega_n^2}{s^2 + 2\psi\Omega_n s + \Omega_n^2}$$

The dynamics of a second-order system can be described in terms of the parameters Ω_n and ψ, as these two parameters determine the location of the poles of the system and thus its response. We adapted the previously given script to plot the cluster of poles and the time response of the second-order system.

Assume $\Omega_n = 1$ rad/sec and let $0 \leq \psi \leq 1$ (so that the poles of $H(s)$ are complex conjugate for $0 \leq \psi < 1$ and double real for $\psi = 1$). Let the input be a unit-step signal so that $V_s(s) = 1/s$. We then have:

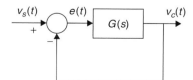

FIGURE 6.11
Second-order feedback system.

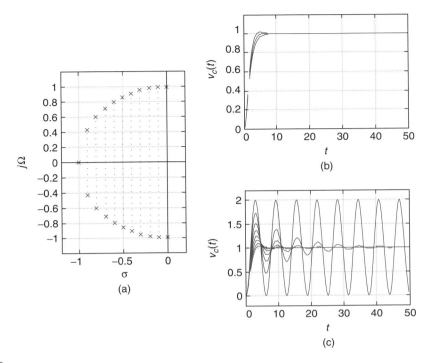

FIGURE 6.12
(a) Clustering of poles and time responses $v_c(t)$ of second-order feedback system for (b) $\sqrt{2}/2 \leq \psi \leq 1$ and
(c) $0 \leq \psi \leq \sqrt{2}/2$.

(a) If we plot the poles of $H(s)$ as ψ changes from 0 (poles in $j\Omega$ axis) to 1 (double real poles) the
     response $y(t)$ in the steady state changes from a sinusoid shifted up by 1 to a damped signal. The
     locus of the poles is a semicircle of radius $\Omega_n = 1$. Figure 6.12 shows this behavior of the poles
     and the responses.
(b) As in the first-order system, the location of the poles determines the response of the system. The
     system is useless if the poles are on the $j\Omega$ axis, as the response is completely oscillatory and
     the input will never be followed. On the other extreme, the response of the system is slow when
     the poles become real. The designer would have to choose a value in between these two for ψ.
(c) For values of ψ between $\sqrt{2}/2$ to 1 the oscillation is minimum and the response is relatively
     fast (see Figure 6.12(b)). For values of ψ from 0 to $\sqrt{2}/2$ the response oscillates more and more,
     giving a large steady-state error (see Figure 6.12(c)).

■ **Example 6.4**

In this example we find the response of an LTI system to different inputs by using functions in the
control toolbox of MATLAB. You can learn more about the capabilities of this toolbox, or set of
specialized functions for control, by running the demo respdemo and then using help to learn more
about the functions tf, impulse, step, and pzmap, which we will use here.

We want to create a MATLAB function that has as inputs the coefficients of the numerator $N(s)$ and of the denominator $D(s)$ of the system's transfer function $H(s) = N(s)/D(s)$ (the coefficients are ordered from the highest order to the lowest order or constant term). The other input of the function is the type of response t where $t = 1$ corresponds to the impulse response, $t = 2$ to the unit-step response, and $t = 3$ to the response to a ramp. The output of the function is the desired response. The function should show the transfer function, the poles, and zeros, and plot the corresponding response. We need to figure out how to compute the ramp response using the step function.

Consider the following transfer functions:

$$\text{(a)} \quad H_1(s) = \frac{s+1}{s^2+s+1}$$

$$\text{(b)} \quad H_2(s) = \frac{s}{s^3+s^2+s+1}$$

Determine the stability of these systems.

Solution

The following script is used to look at the desired responses of the two systems and the location of their poles and zeros. We consider the second system; you can run the script for the first system by putting % at the numerator and the denominator after $H_2(s)$ and getting rid of % after $H_1(s)$ in the script. The function response computes the desired responses (in this case the impulse, step, and ramp responses).

```
%%%%%%%%%%%%%%%%%%%%%
% Example 6.4 -- Control toolbox
%%%%%%%%%%%%%%%%%%%%%
clear all; clf
% % H_1(s)
% nu = [1 1]; de = [1 1 1];
%% H_2(s)
nu = [1 0]; de = [1 1 1 1]; %  unstable
h = response(nu, de, 1);
s = response(nu, de, 2);
r = response(nu, de, 3);

  function y = response(N, D, t)
sys = tf(N, D)
poles = roots(D)
zeros = roots(N)
figure(1)
pzmap(sys);grid
if t == 3,
    D1 = [D 0]; % for ramp response
end
```

```
figure(2)
if t == 1,
    subplot(311)
    y = impulse(sys,20);
    plot(y);title(' Impulse response');ylabel('h(t)');xlabel('t'); grid
elseif t == 2,
    subplot(312)
    y = step(sys, 20);
    plot(y);title(' Unit-step response');ylabel('s(t)');xlabe('t');grid
else
    subplot(313)
    sys = tf(N, D1); % ramp response
    y = step(sys, 40);
    plot(y); title(' Ramp response'); ylabel('q(t)'); xlabel('t');grid
end
```

The results for $H_2(s)$ are as follows.

```
Transfer function:
       s
-----------------------
s^3 + s^2 + s + 1
poles =
  -1.0000
  -0.0000 + 1.0000i
  -0.0000 - 1.0000i
zeros =
      0
```

As you can see, two of the poles are on the $j\Omega$ axis, and so the system corresponding to $H_2(s)$ is unstable. The other system is stable. Results for both systems are shown in Figure 6.13. ∎

6.4 APPLICATION TO COMMUNICATIONS

The application of the Fourier transform in communications is clear. The representation of signals in the frequency domain and the concept of modulation are basic in communications. In this section we show examples of linear (amplitude modulation or AM) as well as nonlinear (frequency modulation or FM, and phase modulation or PM) modulation methods. We also consider important extensions such as frequency-division multiplexing (FDM) and quadrature amplitude modulation (QAM).

Given the low-pass nature of most message signals, it is necessary to shift in frequency the spectrum of the message to avoid using a very large antenna. This can be attained by means of modulation, which is done by changing either the magnitude or the phase of a carrier:

$$A(t) \cos(2\pi f_c + \theta(t)) \qquad (6.12)$$

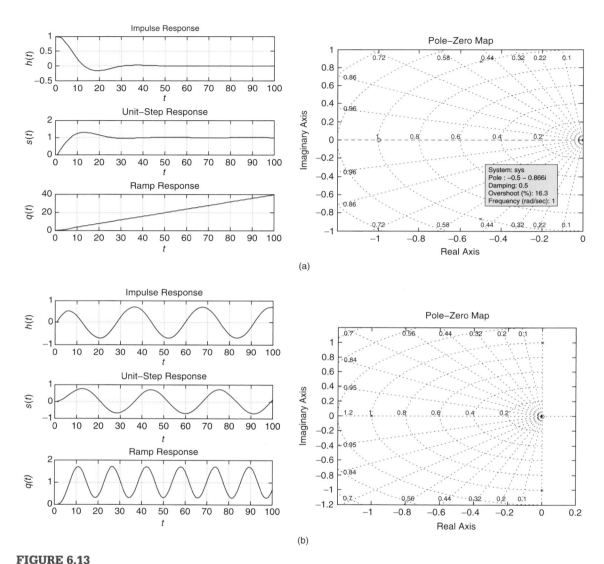

FIGURE 6.13

Impulse, unit-step, and ramp responses and poles and zeros for system with transfer function (a) $H_1(s)$ and (b) $H_2(s)$.

When $A(t)$ is proportional to the message, for constant phase, we have amplitude modulation (AM). On the other hand, if we let $\theta(t)$ change with the message, keeping the amplitude constant, we then have frequency modulation (FM) or phase modulation (PM), which are called *angle modulations*.

6.4.1 AM with Suppressed Carrier

Consider a message signal $m(t)$ (e.g., voice or music, or a combination of the two) modulating a cosine carrier $\cos(\Omega_c t)$ to give an amplitude modulated signal

$$s(t) = m(t)\cos(\Omega_c t) \tag{6.13}$$

The carrier frequency $\Omega_c >> 2\pi f_0$ where f_0 (Hz) is the maximum frequency in the message (for music f_0 is about 22 KHz). The signal $s(t)$ is called an *amplitude modulated with suppressed carrier* (AM-SC) signal (the last part of this denomination will become clear later). According to the modulation property of the Fourier transform, the transform of $s(t)$ is

$$S(\Omega) = \frac{1}{2}\left[M(\Omega - \Omega_c) + M(\Omega + \Omega_c)\right] \tag{6.14}$$

where $M(\Omega)$ is the spectrum of the message. The frequency content of the message is now shifted to a much larger frequency Ω_c (rad/sec) than that of the baseband signal $m(t)$. Accordingly, the antenna needed to transmit the amplitude modulated signal is of reasonable length. An AM-SC system is shown in Figure 6.14.

At the receiver, we need to first detect the desired signal among the many signals transmitted by several sources. This is possible with a tunable band-pass filter that selects the desired signal and rejects the others. Suppose that the signal obtained by the receiver, after the band-pass filtering, is exactly $s(t)$—we then need to demodulate this signal to get the original message signal $m(t)$. This is done by multiplying $s(t)$ by a cosine of exactly the same frequency of the carrier in the transmitter (i.e., Ω_c), which will give $r(t) = 2s(t)\cos(\Omega_c t)$, which again according to the modulation property has a Fourier transform

$$R(\Omega) = S(\Omega - \Omega_c) + S(\Omega + \Omega_c) = M(\Omega) + \frac{1}{2}\left[M(\Omega - 2\Omega_c) + M(\Omega + 2\Omega_c)\right] \tag{6.15}$$

The spectrum of the message, $M(\Omega)$, is obtained by passing the received signal $r(t)$ through a low-pass filter that rejects the other terms $M(\Omega \pm 2\Omega_c)$. The obtained signal is the desired message $m(t)$.

The above is a simplification of the actual processing of the received signal. Besides the many other transmitted signals that the receiver encounters, there is channel noise caused by interferences from

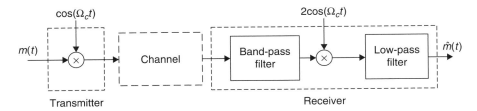

FIGURE 6.14
AM-SC transmitter, channel, and receiver.

equipment in the transmission path and interference from other signals being transmitted around the carrier frequency. This noise will also be picked up by the band-pass filter and a perfect recovery of $m(t)$ will not be possible. Furthermore, the sent signal has no indication of the carrier frequency Ω_c, which is suppressed in the sent signal, and so the receiver needs to guess it and any deviation would give errors.

Remarks

- *The transmitter is linear but time varying. AM-SC is thus called a linear modulation. The fact that the modulated signal displays frequencies much higher than those in the message indicates the transmitter is not LTI—otherwise it would satisfy the eigenfunction property.*

- *A more general characterization than $\Omega_c \gg 2\pi f_0$ where f_0 is the largest frequency in the message is given by $\Omega_c \gg BW$ where BW (rad/sec) is the bandwidth of the message. You probably recall the definition of bandwidth of filters used in circuit theory. In communications there are several possible definitions for bandwidth. The bandwidth of a signal is the width of the range of positive frequencies for which some measure of the spectral content is satisfied. For instance, two possible definitions are:*

    - *The* half-power *or* 3-dB bandwidth *is the width of the range of positive frequencies where a peak value at zero or infinite frequency (low-pass and high-pass signals) or at a center frequency (band-pass signals) is attenuated to 0.707, the value at the peak. This corresponds to the frequencies for which the power at dc, infinity, or center frequency reduces to half.*

    - *The* null-to-null bandwidth *determines the width of the range of positive frequencies of the spectrum of a signal that has a main lobe containing a significant part of the energy of the signal. If a low-pass signal has a clearly defined maximum frequency, then the bandwidth are frequencies from zero to the maximum frequency, and if the signal is a band-pass signal and has a minimum and a maximum frequency, its bandwidth is the maximum minus the minimum frequency.*

- *In AM-SC demodulation it is important to know exactly the carrier frequency. Any small deviation would cause errors when recovering the message. Suppose, for instance, that there is a small error in the carrier frequency—that is, instead of Ω_c the demodulator uses $\Omega_c + \Delta$—so that the received signal in that case has the Fourier transform*

$$\tilde{R}(\Omega) = S(\Omega - \Omega_c - \Delta) + S(\Omega + \Omega_c + \Delta)$$

$$= \frac{1}{2}[M(\Omega + \Delta) + M(\Omega - \Delta)]$$

$$+ \frac{1}{2}[M(\Omega - 2(\Omega_c + \Delta/2)) + M(\Omega + 2(\Omega_c + \Delta/2))]$$

The low-pass filtered signal will not be the message.

6.4.2 Commercial AM

In commercial broadcasting, the carrier is added to the AM signal so that information of the carrier is available at the receiver helping in the identification of the radio station. For demodulation, such information is not important, as commercial AM uses *envelope detectors* to obtain the message. By making the envelope of the modulated signal look like the message, detecting this envelope is all that is needed. Thus, the commercial AM signal is of the form

$$s(t) = [K + m(t)] \cos(\Omega_c t)$$

where the AM modulation index K is chosen so that $K + m(t) > 0$ for all values of t so that the envelope of $s(t)$ is proportional to the message $m(t)$. The Fourier transform is given by

$$S(\Omega) = K\pi \left[\delta(\Omega - \Omega_c) + \delta(\Omega + \Omega_c)\right] + \frac{1}{2}\left[M(\Omega - \Omega_c) + M(\Omega + \Omega_c)\right]$$

The receiver for this type of AM is an *envelope receiver*, which basically detects the message by finding the envelope of the received signal.

Remarks

- *The advantage of adding the carrier to the message, which allows the use of a simple envelope detector, comes at the expense of increasing the power in the transmitted signal.*
- *The demodulation in commercial AM is called noncoherent.* Coherent demodulation *consists in multiplying—or mixing—the received signal with a sinusoid of the same frequency and phase of the carrier. A local oscillator generates this sinusoid.*
- *A disadvantage of commercial as well as suppressed-carrier AM is the doubling of the bandwidth of the transmitted signal compared to the bandwidth of the message. Given the symmetry of the spectrum, in magnitude as well as in phase, it becomes clear that it is not necessary to send the upper and the lower sidebands of the spectrum to get back the signal in the demodulation. It is thus possible to have upper- and lower-sideband AM modulations, which are more efficient in spectrum utilization.*
- *Most AM receivers use the* superheterodyne receiver technique *developed by Armstrong and Fessenden.*[2]

■ Example 6.5: Simulation of AM modulation with MATLAB

For simulations, MATLAB provides different data files, such as "train.mat" (the extension mat indicates it is a data file) used here. Suppose the analog signal $y(t)$ is a recording of a "choo-choo" train, and we wish to use it to modulate a cosine $\cos(\Omega_c t)$ to create an amplitude modulated signal $z(t)$. Because the *train* $y(t)$ signal is given in a sampled form, the simulation requires discrete-time processing, and so we will comment on the results here and leave the discussion of the issues related to the code for the next chapters.

The carrier frequency is chosen to be $f_c = 20.48$ KHz. For the envelope detector to work at the transmitter we add a constant K to the message to ensure this sum is positive. The envelope of the AM-modulated signal should resemble the message. Thus, the AM signal is

$$z(t) = [K + y(t)] \cos(\Omega_c t) \qquad \Omega_c = 2\pi f_c$$

In Figure 6.15 we show the train signal, a segment of the signal, and the corresponding modulated signal displaying the envelope, as well as the Fourier transform of the segment and of its modulated

---

[2] Reginald Fessenden was the first to suggest the heterodyne principle: mixing the radio-frequency signal using a local oscillator of different frequency, resulting in a signal that could drive the diaphragm of an earpiece at an audio frequency. Fessenden could not make a practical success of the heterodyne receiver, which was accomplished by Edwin H. Armstrong in the 1920s using electron tubes.

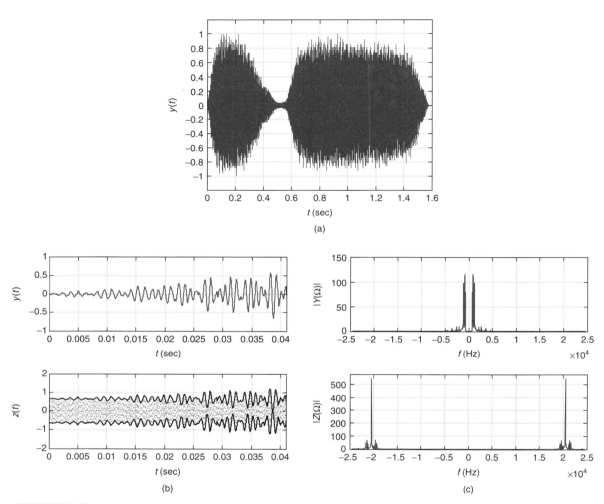

FIGURE 6.15

Commercial AM modulation: (a) original signal, (b) part of original signal and corresponding AM-modulated signal, and (c) spectrum of the original signal, and of the modulated signal.

version. Notice that the envelope resembles the original signal. Also from the spectrum of the segment of the train signal its bandwidth is about 5 Khz, while the spectrum of the modulated segment displays the frequency-shifted spectrum plus the large spectrum at f_c corresponding to the carrier.   ∎

6.4.3 AM Single Sideband

The message $m(t)$ is typically a real-valued signal that, as indicated before, has a symmetric spectrum—that is, the magnitude and the phase of the Fourier transform $M(\Omega)$ are even and odd

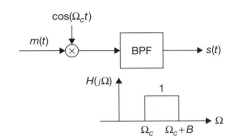

FIGURE 6.16
Upper sideband AM transmitter. Ω_c is the carrier frequency and B is the bandwidth in rad/sec of the message.

functions of frequency. When using AM modulation the resulting spectrum has redundant information by providing the upper and the lower sidebands. To reduce the bandwidth of the transmitted signal, we could get rid of either the upper or the lower sideband of the AM signal. The resulting modulation is called *AM single sideband* (AM-SSB) (upper or lower sideband depending on which of the two sidebands is kept). This type of modulation is used whenever the quality of the received signal is not as important as the advantages of a narrowband and having less noise in the frequency band of the received signal. AM-SSB is used by amateur radio operators.

As shown in Figure 6.16, the upper sideband modulated signal is obtained by band-pass filtering the upper sideband in the modulated signal. At the receiver, band-pass filtering the received signal the output is then demodulated like in an AM-SC system, and the result is then low-pass filtered using the bandwidth of the message.

6.4.4 Quadrature AM and Frequency-Division Multiplexing

Quadrature amplitude modulation (QAM) and frequency division multiplexing (FDM) are the precursors of many of the new communication systems. QAM and FDM are of great interest for their efficient use of the radio spectrum.

Quadrature Amplitude Modulation

QAM enables two AM-SC signals to be transmitted over the same frequency band, conserving bandwidth. The messages can be separated at the receiver. This is accomplished by using two orthogonal carriers, such as a cosine and a sine (see Figure 6.17). The QAM-modulated signal is given by

$$s(t) = m_1(t)\cos(\Omega_c t) + m_2(t)\sin(\Omega_c t) \tag{6.16}$$

where $m_1(t)$ and $m_2(t)$ are the messages. You can think of $s(t)$ as having a phasor representation that is the sum of two phasors perpendicular to each other (the cosine leading the sine by $\pi/2$); indeed,

$$s(t) = \mathcal{R}e[(m_1(t)e^{j0} + m_2(t)e^{-j\pi/2})e^{j\Omega_c t}].$$

Since

$$m_1(t)e^{j0} + m_2(t)e^{-j\pi/2} = m_1(t) - jm_2(t)$$

we could interpret the QAM signal as the result of amplitude modulating the real and the imaginary parts of a complex message $m(t) = m_1(t) - jm_2(t)$.

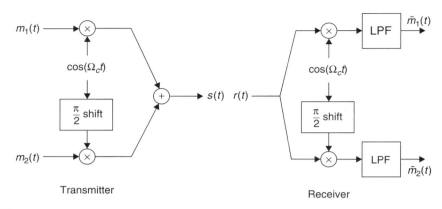

FIGURE 6.17
QAM transmitter and receiver: $s(t)$ is the transmitted signal and $r(t)$ is the received signal.

To simplify the computation of the spectrum of $s(t)$, let us consider the message $m(t) = m_1(t) - jm_2(t)$ (i.e., a complex message) with spectrum $M(\Omega) = M_1(\Omega) - jM_2(\Omega)$ so that

$$s(t) = \mathcal{R}e[m(t)e^{j\Omega_c t}]$$
$$= 0.5[m(t)e^{j\Omega_c t} + m^*(t)e^{-j\Omega_c t}]$$

where $*$ stands for complex conjugate. The spectrum of $s(t)$ is then given by

$$S(\Omega) = 0.5[M(\Omega - \Omega_c) + M^*(\Omega + \Omega_c)]$$
$$= 0.5[M_1(\Omega - \Omega_c) - jM_2(\Omega - \Omega_c) + M_1^*(\Omega + \Omega_c) + jM_2^*(\Omega + \Omega_c)]$$

where the superposition of the spectra of the two messages is clearly seen. At the receiver, if we multiply the received signal (for simplicity assume it to be $s(t)$) by $\cos(\Omega_c t)$, we get

$$r_1(t) = s(t)\cos(\Omega_c t)$$
$$= 0.25[m(t) + m^*(t)] + 0.25[m(t)e^{j2\Omega_c t} + m^*(t)e^{-j2\Omega_c t}]$$

which when passed through a low-pass filter, with the appropriate bandwidth, gives

$$0.25[m(t) + m^*(t)] = 0.25[m_1(t) - jm_2(t) + m_1(t) + jm_2(t)]$$
$$= 0.5m_1(t)$$

Likewise, to get the second message we multiply $s(t)$ by $\sin(\Omega_c t)$ and pass the resulting signal through a low-pass filter.

Frequency-Division Multiplexing
Frequency-division multiplexing (FDM) implements sharing of the spectrum by several users by allocating a specific frequency band to each. One could, for instance, think of the commercial AM or FM

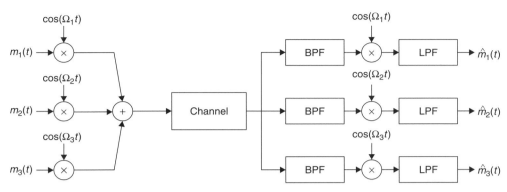

FIGURE 6.18
FDM system: transmitter (left), channel, and receiver (right).

locally as an FDM sytem. In the United States, the Federal Communication Commission (FCC) is in charge of the spectral allocation. In telephony, using a bank of filters it is possible to also get several users in the same system—it is, however, necessary to have a similar system at the receiver to have a two-way communication.

To illustrate an FDM system (Figure 6.18), consider we have a set of messages of known finite bandwidth (we could low-pass filter the messages to satisfy this condition) that we wish to transmit. Each of the messages modulate different carriers so that the modulated signals are in different frequency bands without interfering with each other (if needed a frequency guard could be used to make sure of this). These frequency-multiplexed messages can now be transmitted. At the receiver, using a bank of band-pass filters centered at the carrier frequencies in the transmitter and followed by appropriate demodulators recover the different messages (see FDM receiver in Figure 6.18). Any of the AM modulation techniques could be used in the FDM system.

6.4.5 Angle Modulation

Amplitude modulation is said to be linear modulation, because as a system it behaves like a linear system. Frequency and phase, or angle, modulation systems on the other hand are nonlinear. The interest in angle modulation is due to the decreasing effect of noise or interferences on it, when compared with AM, although at the cost of a much wider bandwidth and greater complexity in implementation. The nonlinear behavior of angle modulation systems makes their analysis more difficult than that for AM. The spectrum of an FM or PM signal is much harder to obtain than that of an AM signal. In the following we consider the case of the so-called *narrowband FM* where we are able to find its spectrum directly.

Professor Edwin H. Armstrong developed the first successful frequency modulation system—narrowband FM.[3] If $m(t)$ is the message signal, and we modulate a carrier signal of frequency

---

[3] Edwind H. Armstrong (1890–1954), professor of electrical engineering at Columbia University, and inventor of some of the basic electronic circuits underlying all modern radio, radar, and television, was born in New York. His inventions and developments form the backbone of radio communications as we know it.

Ω_c (rad/sec) with $m(t)$, the transmitted signal $s(t)$ in angle modulation is of the form

$$s(t) = A\cos(\Omega_c t + \theta(t)) \tag{6.17}$$

where the angle $\theta(t)$ depends on the message $m(t)$. In the case of *phase modulation*, the angle function is proportional to the message $m(t)$—that is,

$$\theta(t) = K_f\, m(t) \tag{6.18}$$

where $K_f > 0$ is called the *modulation index*. If the angle is such that

$$\frac{d\theta(t)}{dt} = \Delta\Omega\, m(t) \tag{6.19}$$

this relation defines *frequency modulation*. The *instantaneous frequency*, as a function of time, is the derivative of the argument of the cosine or

$$IF(t) = \frac{d[\Omega_c t + \theta(t)]}{dt} \tag{6.20}$$

$$= \Omega_c + \frac{d\theta(t)}{dt} \tag{6.21}$$

$$= \Omega_c + \Delta\Omega\, m(t) \tag{6.22}$$

indicating how the frequency is changing with time. For instance, if $\theta(t)$ is a constant—so that the carrier is just a sinusoid of frequency Ω_c and constant phase θ—the instantaneous frequency is simply Ω_c. The term $\Delta\Omega\, m(t)$ relates to the spreading of the frequency about Ω_c. Thus, the *modulation paradox* Professor E. Craig proposed in his book [17]:

> In *amplitude* modulation the bandwidth depends on the *frequency* of the message, while in *frequency* modulation the bandwidth depends on the *amplitude* of the message.

Thus, the modulated signals are

$$PM: \quad s_{PM}(t) = \cos(\Omega_c t + K_f m(t)) \tag{6.23}$$

$$FM: \quad s_{FM}(t) = \cos\left(\Omega_c t + \Delta\Omega \int_{-\infty}^{t} m(\tau)d\tau\right) \tag{6.24}$$

Narrowband FM
In this case the angle $\theta(t)$ is small, so that $\cos(\theta(t)) \approx 1$ and $\sin(\theta(t)) \approx \theta(t)$, simplifying the spectrum of the transmitted signal:

$$S(\Omega) = \mathcal{F}[\cos(\Omega_c t + \theta(t))]$$

$$= \mathcal{F}[\cos(\Omega_c t)\cos(\theta(t)) - \sin(\Omega_c t)\sin(\theta(t))]$$

$$\approx \mathcal{F}[\cos(\Omega_c t) - \sin(\Omega_c t)\theta(t)] \tag{6.25}$$

Using the spectrum of a cosine and the modulation theorem, we get

$$S(\Omega) \approx \pi \left[\delta(\Omega - \Omega_c) + \delta(\Omega + \Omega_c)\right] - \frac{1}{2j} \left[\Theta(\Omega - \Omega_c) - \Theta(\Omega + \Omega_c)\right] \qquad (6.26)$$

where $\Theta(\Omega)$ is the spectrum of the angle, which is found to be (using the derivative property of the Fourier transform)

$$\Theta(\Omega) = \frac{\Delta\Omega}{j\Omega} M(\Omega) \qquad (6.27)$$

If the angle $\theta(t)$ is not small, we have *wideband FM* and its spectrum is more difficult to obtain.

■ Example 6.6: Simulation of FM modulation with MATLAB

In these simulations we will concern ourselves with the results and leave the discussion of issues related to the code for the next chapter since the signals are approximated by discrete-time signals. For the narrowband FM we consider a sinusoidal message

$$m(t) = 80 \sin(20\pi t)u(t),$$

and a sinusoidal carrier of frequency $f_c = 100$ Hz, so that for $\Delta\Omega = 0.1\pi$ the FM signal is

$$x(t) = \cos(2\pi f_c t + 0.1\pi \int_{-\infty}^{t} m(\tau)d\tau)$$

Figure 6.19 shows on the top left the message and the narrowband FM signal $x(t)$ right below it, and on the top right their corresponding magnitude spectra $|M(\Omega)|$ and below $|X(\Omega)|$. The narrowband FM has only shifted the frequency of the message. The instantaneous frequency (the derivative of the argument of the cosine) is

$$IF(t) = 2\pi f_c + 0.1\pi m(t) = 200\pi + 8\pi \sin(20\pi t) \approx 200\pi$$

That is, it remains almost constant for all times. For the narrowband FM, the spectrum of the modulated signal remains the same for all times. To illustrate this we computed the spectrogram of $x(t)$. Simply, the spectrogram can be thought of as the computation of the Fourier transform as the signal evolves with time (see Figure 6.19(c)).

To illustrate the wideband FM, we consider two messages,

$$m_1(t) = 80 \sin(20\pi t)u(t)$$

$$m_2(t) = 2000tu(t)$$

giving FM signals,

$$x_i(t) = \cos(2\pi f_{ci} t + 50\pi \int_{-\infty}^{t} m_i(\tau)d\tau) \qquad i = 1, 2$$

where $f_{c1} = 2500$ Hz and $f_{c2} = 25$ Hz. In this case, the instantaneous frequency is

$$IF_i(t) = 2\pi f_{ci} + 50\pi m_i(t) \qquad i = 1, 2$$

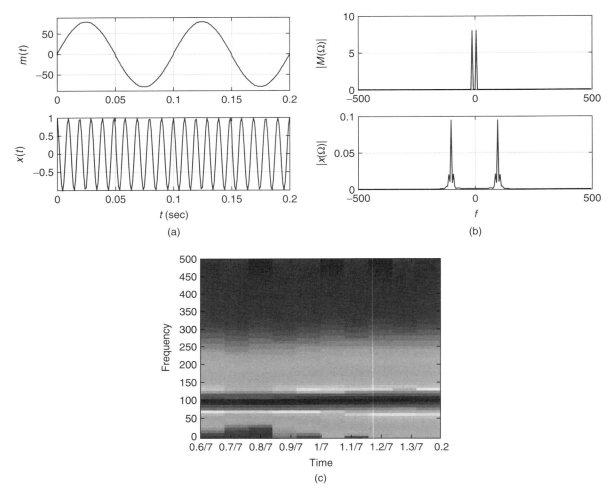

FIGURE 6.19

Narrowband frequency modulation: (a) message $m(t)$ and narrowband FM signal $x(t)$; (b) magnitude spectra of $m(t)$ and $x(t)$; and (c) spectrogram of $x(t)$ displaying evolution of its Fourier transform with respect to time.

These instantaneous frequencies are not almost constant as before. The frequency of the carrier is now continuously changing with time. For instance, for the ramp message the instantaneous frequency is

$$IF_2(t) = 50\pi + 10^5 t\pi$$

so that for a small time interval $[0, 0.1]$ we get a chirp (sinusoid with time-varying frequency), as shown in Figure 6.20(b). Figure 6.20 display the messages, the FM signals, and their corresponding magnitude spectra and their spectrograms. These FM signals are broadband, occupying a band of frequencies much larger than the messages, and their spectrograms show that their spectra change with time. ∎

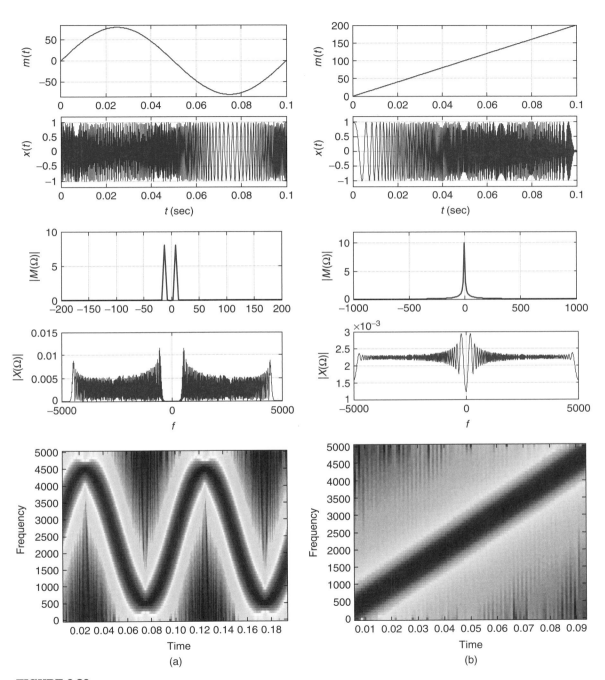

FIGURE 6.20

Wideband frequency modulation, from top to bottom, for (a) the sinusoidal message and for (b) the ramp message: messages, FM-modulated signals, spectra of messages, spectrum of FM signals, and spectrogram of FM signals.

6.5 ANALOG FILTERING

The basic idea of filtering is to get rid of frequency components of a signal that are not desirable. Application of filtering can be found in control, in communications, and in signal processing. In this section we provide a short introduction to the design of analog filters. Chapter 11 is dedicated to the design of discrete filters and to some degree that chapter will be based on the material in this section.

According to the eigenfunction property of LTI systems (Figure 6.21) the steady-state response of an LTI system to a sinusoidal input—with a certain magnitude, frequency, and phase—is a sinusoid of the same frequency as the input, but with magnitude and phase affected by the response of the system at the frequency of the input. Since periodic as well as aperiodic signals have Fourier representations consisting of sinusoids of different frequencies, the frequency components of any signal can be modified by appropriately choosing the frequency response of the LTI system, or filter. Filtering can thus be seen as a way of changing the frequency content of an input signal.

The appropriate filter for a certain application is specified using the spectral characterization of the input and the desired spectral characteristics of the output. Once the specifications of the filter are set, the problem becomes one of approximation as a ratio of polynomials in s. The classical approach in filter design is to consider low-pass prototypes, with normalized frequency and magnitude responses, which may be transformed into other filters with the desired frequency response. Thus, a great deal of effort is put into designing low-pass filters and into developing frequency transformations to map low-pass filters into other types of filters. Using cascade and parallel connections of filters also provide a way to obtain different types of filters.

The resulting filter should be causal, stable, and have real-valued coefficients so that it can be used in real-time applications and realized as a passive or an active filter. Resistors, capacitors, and inductors are used in the realization of passive filters, while resistors, capacitors, and operational amplifiers are used in active filter realizations.

6.5.1 Filtering Basics

A filter $H(s) = B(s)/A(s)$ is an LTI system having a specific frequency response. The convolution property of the Fourier transform gives that

$$Y(\Omega) = X(\Omega)H(j\Omega) \tag{6.28}$$

where

$$H(j\Omega) = H(s)|_{s=j\Omega}$$

Thus, the frequency content of the input, represented by the Fourier transform $X(\Omega)$, is changed by the frequency response $H(j\Omega)$ of the filter so that the output signal with spectrum $Y(\Omega)$ only has desirable frequency components.

FIGURE 6.21
Eigenfunction property of continuous LTI systems.

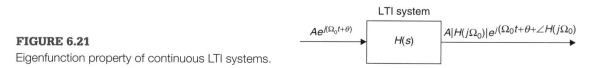

Magnitude Squared Function

The magnitude-squared function of an analog low-pass filter has the general form

$$|H(j\Omega)|^2 = \frac{1}{1 + f(\Omega^2)} \tag{6.29}$$

where for low frequencies $f(\Omega^2) \approx 0$ so that $|H(j\Omega)|^2 \approx 1$, and for high frequencies $f(\Omega^2) \to \infty$ so that $|H(j\Omega)|^2 \to 0$. Accordingly, there are two important issues to consider:

■ Selection of the appropriate function $f(.)$.
■ The factorization needed to get $H(s)$ from the magnitude-squared function.

As an example of the above steps, consider the *Butterworth low-pass analog filter*. The Butterworth magnitude-squared response of order N is

$$|H_N(j\Omega)|^2 = \frac{1}{1 + \left[\frac{\Omega}{\Omega_{hp}}\right]^{2N}} \tag{6.30}$$

where Ω_{hp} is the half-power frequency of the filter. We then have that for $\Omega << \Omega_{hp}$, $|H_N(j\Omega)| \approx 1$, and for $\Omega >> \Omega_{hp}$, then $|H_N(j\Omega)| \to 0$. To find $H(s)$ we need to factorize Equation (6.30). Letting S be a normalized variable $S = s/\Omega_{hp}$, the magnitude-squared function (Eq. 6.30) can be expressed in terms of the S variable by letting $S/j = \Omega/\Omega_{hp}$ to obtain

$$H(S)H(-S) = \frac{1}{1 + (-S^2)^N}$$

since $|H(j\Omega')|^2 = H(j\Omega')H^*(j\Omega') = H(j\Omega')H(-j\Omega')$. As we will see, the poles of $H(S)H(-S)$ are symmetrically clustered in the s-plane with none on the $j\Omega$ axis. The factorization then consists of assigning poles in the open left-hand s-plane to $H(S)$, and the rest to $H(-S)$. We thus obtain

$$H(S)H(-S) = \frac{1}{D(S)} \frac{1}{D(-S)}$$

so that the final form of the filter is

$$H(S) = \frac{1}{D(S)}$$

where $D(S)$ has roots on the left-hand s-plane. A final step is the replacement of S by the unnormalized variable s, to obtain the final form of the filter transfer function:

$$Butterworth\ low\text{-}pass\ filter:\quad H(s) = H(S)|_{S=s/\Omega_{hp}} \tag{6.31}$$

Filter Specifications

Although an ideal low-pass filter is not realizable (recall the Paley-Wiener condition in Chapter 5) its magnitude response can be used as prototype for specifying low-pass filters. Thus, the desired magnitude is specified as

$$1 - \delta_2 \leq |H(j\Omega)| \leq 1 \qquad 0 \leq \Omega \leq \Omega_p \quad \text{(passband)}$$
$$0 \leq |H(j\Omega)| \leq \delta_1 \qquad \Omega \geq \Omega_s \quad \text{(stopband)} \tag{6.32}$$

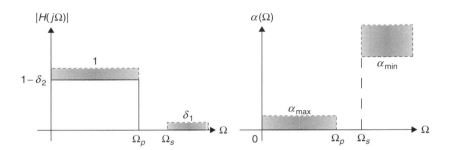

FIGURE 6.22

Magnitude specifications for a low-pass filter.

for some small values δ_1 and δ_2. There is no specification in the transition region $\Omega_p < \Omega < \Omega_s$. Also the phase is not specified, although we wish it to be linear at least in the passband. See Figure 6.22.

To simplify the computation of the filter parameters, and to provide a scale that has more resolution and physiological significance than the specifications given above, the magnitude specifications are typically expressed in a logarithmic scale. Defining the loss function (in decibels, or dBs) as

$$\alpha(\Omega) = -10 \log_{10} |H(j\Omega)|^2$$
$$= -20 \log_{10} |H(j\Omega)| \qquad \text{dBs} \qquad (6.33)$$

an equivalent set of specifications to those in Equation (6.32) is

$$0 \leq \alpha(\Omega) \leq \alpha_{max} \qquad 0 \leq \Omega \leq \Omega_p \quad \text{(passband)}$$
$$\alpha(\Omega) \geq \alpha_{min} \qquad \Omega \geq \Omega_s \quad \text{(stopband)} \qquad (6.34)$$

where $\alpha_{max} = -20 \log_{10}(1 - \delta_2)$ and $\alpha_{min} = -20 \log_{10}(\delta_1)$.

In the above specifications, the dc loss was 0 dB corresponding to a normalized dc gain of 1. In more general cases, $\alpha(0) \neq 0$ and the loss specifications are given as $\alpha(0) = \alpha_1$, α_2 in the passband and α_3 in the stopband. To normalize these specifications we need to subtract α_1, so that the loss specifications are

$$\alpha(0) = \alpha_1 \quad \text{(dc loss)}$$
$$\alpha_{max} = \alpha_2 - \alpha_1 \quad \text{(maximum attenuation in passband)}$$
$$\alpha_{min} = \alpha_3 - \alpha_1 \quad \text{(minimum attenuation in stopband)}$$

Using $\{\alpha_{max}, \Omega_p, \alpha_{min}, \Omega_s\}$ we proceed to design a magnitude-normalized filter, and then use α_1 to achieve the desired dc gain.

The design problem is then: Given the magnitude specifications in the passband ($\alpha(0)$, α_{max}, and Ω_p) and in the stopband (α_{min} and Ω_s) we then

1. Choose the rational approximation method (e.g., Butterworth).
2. Solve for the parameters of the filter to obtain a magnitude-squared function that satisfies the given specifications.

3. Factorize the magnitude-squared function and choose the poles on the left-hand s-plane, guaranteeing the filter stability, to obtain the transfer function $H_N(s)$ of the filter.

6.5.2 Butterworth Low-Pass Filter Design

The magnitude-squared approximation of a low-pass Nth-order Butterworth filter is given by

$$|H_N(j\Omega')|^2 = \frac{1}{1 + (\Omega/\Omega_{hp})^{2N}} \qquad \Omega' = \frac{\Omega}{\Omega_{hp}} \tag{6.35}$$

where Ω_{hp} is the half-power or -3-dB frequency. This frequency response is normalized with respect to the half-power frequency (i.e., the normalized frequency is $\Omega' = \Omega/\Omega_{hp}$) and normalized in magnitude as the dc gain is $|H(j0)| = 1$. The frequency $\Omega' = \Omega/\Omega_{hp} = 1$ is the normalized half-power frequency since $|H_N(j1)|^2 = 1/2$. The given magnitude-squared function is thus normalized with respect to frequency (giving a unity half-power frequency) and in magnitude (giving a unity DC gain for the low-pass filter). The approximation improves (i.e., gets closer to the ideal filter) as the order N increases.

Remarks

■ *The half-power frequency is called the -3-dB frequency because in the case of the low-pass filter with a dc gain of 1, at the half-power frequency Ω_{hp} the magnitude-squared function is*

$$|H(j\Omega_{hp})|^2 = \frac{|H(j0)|^2}{2} = \frac{1}{2}. \tag{6.36}$$

In the logarithmic scale we have

$$10\log_{10}(|H(j\Omega_{hp})|^2) = -10\log_{10}(2) \approx -3 \quad (dB) \tag{6.37}$$

This corresponds to a loss of 3 dB.

■ *It is important to understand the significance of the frequency and magnitude normalizations typical in filter design. Having a low-pass filter with normalized magnitude, its dc gain is 1, if one desires a filter with a DC gain $K \neq 1$ it can be obtained by multiplying the magnitude-normalized filter by the constant K. Likewise, a filter $H(S)$ designed with a normalized frequency, say $\Omega' = \Omega/\Omega_{hp}$ so that the normalized half-power frequency is 1, is converted into a denormalized filter $H(s)$ with a desired Ω_{hp} by replacing $S = s/\Omega_{hp}$ in $H(S)$.*

Factorization

To obtain a filter that satisfies the specifications and that is stable we need to factorize the magnitude-squared function. By letting $S = s/\Omega_{hp}$ be a normalized Laplace variable, then $S/j = \Omega' = \Omega/\Omega_{hp}$ and

$$H(S)H(-S) = \frac{1}{1 + (-S^2)^N}$$

If the denominator can be factorized as

$$D(S)D(-S) = 1 + (-S^2)^N \tag{6.38}$$

we let $H(S) = 1/D(S)$—that is, we assign to $H(S)$ the poles in the left-hand s-plane so that the resulting filter is stable. The roots of $D(S)$ in Equation (6.38) are

$$S_k^{2N} = \frac{e^{j(2k-1)\pi}}{e^{-j\pi N}} = e^{j(2k-1+N)\pi} \quad \text{for integers } k = 1, \ldots, 2N$$

after replacing $-1 = e^{j(2k-1)\pi}$ and $(-1)^N = e^{-j\pi N}$. The $2N$ roots are then

$$S_k = e^{j(2k-1+N)\pi/(2N)} \qquad k = 1, \ldots, 2N \tag{6.39}$$

Remarks

- *Since $|S_k| = 1$, the poles of the Butterworth filter are on a circle of unit radius. De Moivre's theorem guarantees that the poles are also symmetrically distributed around the circle, and because of the condition that complex poles should be complex conjugate pairs, the poles are symmetrically distributed with respect to the σ axis. Letting $S = s/\Omega_{hp}$ be the normalized Laplace variable, then $s = S\Omega_{hp}$, so that the denormalized filter $H(s)$ has its poles in a circle of radius Ω_{hp}.*
- *No poles are on the $j\Omega'$ axis, as can be seen by showing that the angle of the poles are not equal to $\pi/2$ or $3\pi/2$. In fact, for $1 \leq k \leq N$, the angle of the poles are bounded below and above by letting $1 \leq k$ and then $k \leq N$ to get*

$$\frac{\pi}{2}\left(1 + \frac{1}{N}\right) \leq \frac{(2k-1+N)\pi}{2N} \leq \frac{\pi}{2}\left(3 - \frac{1}{N}\right)$$

and for integers $N \geq 1$ the above indicates that the angle will not be equal to either $\pi/2$ or $3\pi/2$, or on the $j\Omega'$ axis.
- *Consecutive poles are separated by π/N radians from each other. In fact, subtracting the angles of two consecutive poles can be shown to give $\pm\pi/N$.*

Using the above remarks and the fact that the poles must be in conjugate pairs, since the coefficients of the filter are real-valued, it is easy to determine the location of the poles geometrically.

■ Example 6.7

A second-order low-pass Butterworth filter, normalized in magnitude and in frequency, has a transfer function of

$$H(S) = \frac{1}{S^2 + \sqrt{2}S + 1}$$

We would like to obtain a new filter $H(s)$ with a dc gain of 10 and a half-power frequency $\Omega_{hp} = 100$ rad/sec.

The DC gain of $H(S)$ is unity—in fact, when $\Omega = 0$, $S = j0$ gives $H(j0) = 1$. The half-power frequency of $H(S)$ is unity, indeed letting $\Omega' = 1$, then $S = j1$ and

$$H(j1) = \left[\frac{1}{j^2 + j\sqrt{2} + 1}\right] = \frac{1}{j\sqrt{2}}$$

so that $|H(j1)|^2 = |H(j0)|^2/2 = 1/2$, or $\Omega' = 1$ is the half-power frequency.

Thus, the desired filter with a dc gain of 10 is obtained by multiplying $H(S)$ by 10. Furthermore, if we let $S = s/100$ be the normalized Laplace variable when $S = j\Omega'_{hp} = j1$, we get that $s = j\Omega_{hp} = j100$, or $\Omega_{hp} = 100$, the desired half-power frequency. Thus, the denormalized filter in frequency $H(s)$ is obtained by replacing $S = s/100$. The denormalized filter in magnitude and frequency is then

$$H(s) = \frac{10}{(s/100)^2 + \sqrt{2}(s/100) + 1} = \frac{10^5}{s^2 + 100\sqrt{2}s + 10^4}$$

■

Design

For the Butterworth low-pass filter, the design consists in finding the parameters N, the minimum order, and Ω_{hp}, the half-power frequency, of the filter from the constrains in the passband and in the stopband.

The loss function for the low-pass Butterworth is

$$\alpha(\Omega) = -10\log_{10}|H_N(\Omega/\Omega_{hp})|^2 = 10\log_{10}(1 + (\Omega/\Omega_{hp})^{2N})$$

The loss specifications are

$$0 \le \alpha(\Omega) \le \alpha_{max} \qquad 0 \le \Omega \le \Omega_p$$
$$\alpha_{min} \le \alpha(\Omega) < \infty \qquad \Omega \ge \Omega_s$$

At $\Omega = \Omega_p$, we have that

$$10\log_{10}(1 + (\Omega_p/\Omega_{hp})^{2N}) \le \alpha_{max}$$

so that

$$\left(\frac{\Omega_p}{\Omega_{hp}}\right)^{2N} \le 10^{0.1\alpha_{max}} - 1 \tag{6.40}$$

and similarly for $\Omega = \Omega_s$, we have that

$$10^{0.1\alpha_{min}} - 1 \le \left(\frac{\Omega_s}{\Omega_{hp}}\right)^{2N} \tag{6.41}$$

We then have that from Equation (6.40) and (6.41), the half-power frequency is in the range

$$\frac{\Omega_p}{(10^{0.1\alpha_{max}} - 1)^{1/2N}} \le \Omega_{hp} \le \frac{\Omega_s}{(10^{0.1\alpha_{min}} - 1)^{1/2N}} \tag{6.42}$$

and from the log of the two extremes of Equation (6.42), we have that

$$N \ge \frac{\log_{10}[(10^{0.1\alpha_{min}} - 1)/(10^{0.1\alpha_{max}} - 1)]}{2\log_{10}(\Omega_s/\Omega_p)} \tag{6.43}$$

Remarks

- *According to Equation (6.43) when either*
  - *The transition band is narrowed (i.e., $\Omega_p \to \Omega_s$), or*
  - *The loss $\alpha_{\min}$ is increased, or*
  - *The loss $\alpha_{\max}$ is decreased*

  the quality of the filter is improved at the cost of having to implement a filter with a high order N.
- *The minimum order N is an integer larger or equal to the right side of Equation (6.43). Any integer larger than the minimum N also satisfies the specifications but increases the complexity of the filter.*
- *Although there is a range of possible values for the half-power frequency, it is typical to make the frequency response coincide with either the passband or the stopband specifications giving a value for the half-power frequency in the range. Thus, we can have either*

$$\Omega_{hp} = \frac{\Omega_p}{(10^{0.1\alpha_{\max}} - 1)^{1/2N}} \tag{6.44}$$

or

$$\Omega_{hp} = \frac{\Omega_s}{(10^{0.1\alpha_{\min}} - 1)^{1/2N}} \tag{6.45}$$

as possible values for the half-power frequency.
- *The design aspect is clearly seen in the flexibility given by the equations. We can select out of an infinite possible set of values of N and of half-power frequencies. The optimal order is the smallest value of N and the half-power frequency can be taken as one of the extreme values.*
- *After the factorization, or the formation of D(S) from the poles, we need to denormalize the obtained transfer function $H_N(S) = 1/D(S)$ by letting $S = s/\Omega_{hp}$ to get $H_N(s) = 1/D(s/\Omega_{hp})$, the filter that satisfies the specifications. If the desired DC gain is not unit, the filter needs to be denormalized in magnitude by multiplying it by an appropriate gain K.*

6.5.3 Chebyshev Low-Pass Filter Design

The normalized magnitude-squared function for the Chebyshev low-pass filter is given by

$$|H_N(\Omega')|^2 = \frac{1}{1 + \varepsilon^2 C_N^2(\Omega/\Omega_p)} \qquad \Omega' = \frac{\Omega}{\Omega_p} \tag{6.46}$$

where the frequency is normalized with respect to the passband frequency Ω_p so that $\Omega' = \Omega/\Omega_p$, N stands for the order of the filter, ε is a ripple factor, and $C_N(.)$ are the Chebyshev orthogonal[4] polynomials of the first kind defined as

$$C_N(\Omega') = \begin{cases} \cos(N\cos^{-1}(\Omega')) & |\Omega'| \leq 1 \\ \cosh(N\cosh^{-1}(\Omega')) & |\Omega'| > 1 \end{cases} \tag{6.47}$$

The definition of the Chebyshev polynomials depends on the value of Ω'. Indeed, whenever $|\Omega'| > 1$, the definition based in the cosine is not possible since the inverse would not exist; thus the cosh(.)

---

[4]Pafnuty Chebyshev (1821–1894), a brilliant Russian mathematician, was probably the first one to recognize the general concept of orthogonal polynomials.

definition is used. Likewise, whenever $|\Omega'| \leq 1$, the definition based in the hyperbolic cosine would not be possible since the inverse of this function only exists for values of Ω' bigger or equal to 1 and so the cos(.) definition is used. From the definition it is not clear that $C_N(\Omega')$ is an Nth-order polynomial in Ω'. However, if we let $\theta = \cos^{-1}(\Omega')$ or $\Omega' = \cos(\theta)$ when $|\Omega'| \leq 1$, we have that $C_N(\Omega') = \cos(N\theta)$ and

$$C_{N+1}(\Omega') = \cos((N+1)\theta) = \cos(N\theta)\cos(\theta) - \sin(N\theta)\sin(\theta)$$

$$C_{N-1}(\Omega') = \cos((N-1)\theta) = \cos(N\theta)\cos(\theta) + \sin(N\theta)\sin(\theta)$$

so that adding them we get

$$C_{N+1}(\Omega') + C_{N-1}(\Omega') = 2\cos(\theta)\cos(N\theta) = 2\Omega' C_N(\Omega')$$

This gives a three-term expression for computing $C_N(\Omega')$, or a difference equation

$$C_{N+1}(\Omega') + C_{N-1}(\Omega') = 2\Omega' C_N(\Omega') \qquad N \geq 0 \tag{6.48}$$

with initial conditions

$$C_0(\Omega') = \cos(0) = 1$$

$$C_1(\Omega') = \cos(\cos^{-1}(\Omega')) = \Omega'$$

We can then see that

$$C_0(\Omega') = 1$$

$$C_1(\Omega') = \Omega'$$

$$C_2(\Omega') = -1 + 2\Omega'^2$$

$$C_3(\Omega') = -3\Omega' + 4\Omega'^3$$

$$\vdots$$

which are polynomials in Ω' of order $N = 0, 1, 2, 3, \ldots$. In Chapter 0 we gave a script to compute and plot these polynomials using symbolic MATLAB.

Remarks

- *Two fundamental characteristics of the $C_N(\Omega')$ polynomials are: (1) they vary between 0 and 1 in the range $\Omega' \in [-1, 1]$, and (2) they grow outside this range (according to their definition, the Chebyshev polynomials outside this range become cosh(.) functions, which are functions always bigger than 1). The first characteristic generates ripples in the passband, while the second makes these filters have a magnitude response that goes to zero faster than Butterworth's.*
- *There are other characteristics of interest for the Chebyshev polynomials. The Chebyshev polynomials are unity at $\Omega' = 1$ (i.e., $C_N(1) = 1$ for all N). In fact, $C_0(1) = 1$, $C_1(1) = 1$, and if we assume that $C_{N-1}(1) = C_N(1) = 1$, we then have that $C_{N+1}(1) = 1$ according to the three-term recursion. This indicates that the magnitude-square function is $|H_N(j1)|^2 = 1/(1 + \varepsilon^2)$ for any N.*

- Different from the Butterworth filter that has a unit dc gain, the dc gain of the Chebyshev filter depends on the order of the filter. This is due to the property of the Chebyshev polynomial of being $|C_N(0)| = 0$ if N is odd and 1 if N is even. Thus, the dc gain is 1 when N is odd, but $1/\sqrt{1 + \varepsilon^2}$ when N is even. This is due to the fact that the Chebyshev polynomials of odd order do not have a constant term, and those of even order have 1 or -1 as the constant term.
- Finally, the polynomials $C_N(\Omega')$ have N real roots between -1 and 1. Thus, the Chebyshev filter displays $N/2$ ripples between 1 and $\sqrt{1 + \varepsilon^2}$ for normalized frequencies between 0 and 1.

Design

The loss function for the Chebyshev filter is

$$\alpha(\Omega') = 10 \log_{10} \left[1 + \varepsilon^2 C_N^2(\Omega') \right] \qquad \Omega' = \frac{\Omega}{\Omega_p} \tag{6.49}$$

The design equations for the Chebyshev filter are obtained as follows:

- Ripple factor ε and ripple width (RW): From $C_N(1) = 1$, and letting the loss equal α_{max} at that normalized frequency, we have that

$$\varepsilon = \sqrt{10^{0.1\alpha_{max}} - 1}$$

$$RW = 1 - \frac{1}{\sqrt{1 + \varepsilon^2}} \tag{6.50}$$

- Minimum order: The loss function at Ω'_s is bigger or equal to α_{min}, so that solving for the Chebyshev polynomial we get after replacing ε,

$$C_N(\Omega'_s) = \cosh(N \cosh^{-1}(\Omega'_s))$$

$$\geq \left(\frac{10^{0.1\alpha_{min}} - 1}{10^{0.1\alpha_{max}} - 1} \right)^{0.5}$$

where we used the cosh(.) definition of the Chebyshev polynomials since $\Omega'_s > 1$. Solving for N we get

$$N \geq \frac{\cosh^{-1}\left(\left[\frac{10^{0.1\alpha_{min}} - 1}{10^{0.1\alpha_{max}} - 1} \right]^{0.5} \right)}{\cosh^{-1}\left(\frac{\Omega_s}{\Omega_p} \right)} \tag{6.51}$$

- Half-power frequency: Letting the loss at the half-power frequency equal 3 dB and using that $10^{0.3} \approx 2$, we obtain from Equation 6.49 the Chebyshev polynomial at that normalized frequency to be

$$C_N(\Omega'_{hp}) = \frac{1}{\varepsilon}$$

$$= \cosh\left(N \cosh^{-1}(\Omega'_{hp}) \right)$$

where the last term is the definition of the Chebyshev polynomial for $\Omega'_{hp} > 1$. Thus, we get

$$\Omega_{hp} = \Omega_p \cosh\left[\frac{1}{N}\cosh^{-1}\left(\frac{1}{\varepsilon}\right)\right] \tag{6.52}$$

Factorization

The factorization of the magnitude-squared function is a lot more complicated for the Chebyshev filter than for the Butterworth filter. If we let the normalized variable $S = s/\Omega_p$ equal $j\Omega'$, the magnitude-squared function can be written as

$$H(S)H(-S) = \frac{1}{1 + \varepsilon^2 C_N^2(S/j)} = \frac{1}{D(S)D(-S)}$$

As before in the Butterworth case, the poles in the left-hand s-plane gives $H(S) = 1/D(S)$, a stable filter.

The poles of the $H(S)$ can be found to be in an ellipse. They can be connected with the poles of the corresponding order Butterworth filter by an algorithm due to Professor Ernst Guillemin. The poles of $H(S)$ are given by the following equations for $k = 1, \ldots, N$, with N the minimal order of the filter:

$$a = \frac{1}{N}\sinh^{-1}\left(\frac{1}{\varepsilon}\right)$$

$$\sigma_k = -\sinh(a)\cos(\psi_k) \qquad \text{(real part)}$$

$$\Omega'_k = \pm\cosh(a)\sin(\psi_k) \qquad \text{(imaginary part)} \tag{6.53}$$

where $0 \leq \psi_k < \pi/2$ (refer to Equation 6.39) are the angles corresponding to the Butterworth filters (measured with respect to the negative real axis of the s-plane).

Remarks

- *The dc gain of the Chebyshev filter is not easy to determine as in the Butterworth filter, as it depends on the order N. We can, however, set the desired dc value by choosing the appropriate value of a gain K so that $\hat{H}(S) = K/D(S)$ satisfies the dc gain specification.*
- *The poles of the Chebyshev filter depend now on the ripple factor ε and so there is no simple way to find them as it was in the case of the Butterworth.*
- *The final step is to replace the normalized variable $S = s/\Omega_p$ in $H(S)$ to get the desired filter $H(s)$.*

■ Example 6.8

Consider the low-pass filtering of an analog signal $x(t) = [-2\cos(5t) + \cos(10t) + 4\sin(20t)]u(t)$ with MATLAB. The filter is a third-order low-pass Butterworth filter with a half-power frequency $\Omega_{hp} = 5$ rad/sec—that is, we wish to attenuate the frequency components of the frequencies 10 and 20 rad/sec. Design the desired filter and show how to do the filtering.

The design of the filter is done using the MATLAB function butter where besides the specification of the desired order, $N = 3$, and half-power frequency, $\Omega_{hp} = 5$ rad/sec, we also need to indicate that the filter is analog by including an 's' as one of the arguments. Once the coefficients of the filter are obtained, we could then either solve the differential equation from these coefficients or use the Fourier transform, which we choose to do. Symbolic MATLAB is thus used to compute the Fourier transform of the input $X(\Omega)$, and after generating the frequency response function $H(j\Omega)$ from the filter coefficients, we multiply these two to get $Y(\Omega)$, which is inversely transformed to obtain $y(t)$. To obtain $H(j\Omega)$ symbolically we multiply the coefficients of the numerator and denominator obtained from butter by variables $(j\Omega)^n$ where n corresponds to the order of the coefficient in the numerator or the denominator, and then add them. The poles of the designed filter and its magnitude response are shown in Figure 6.23, as well as the input $x(t)$ and the output $y(t)$. The following script was used for the filter design and the filtering of the given signal.

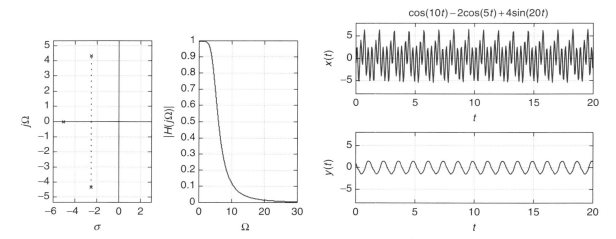

FIGURE 6.23

Filtering of an analog signal $x(t)$ using a low-pass Butterworth filter. Notice that the output of the filter is approximately the sinusoid of 5 rad/sec in $x(t)$, as the other two components have been attenuated.

```
%%%%%%%%%%%%%%%%%%%%%%
% Example 6.8 -- Filtering with Butterworth filter
%%%%%%%%%%%%%%%%%%%%%%
clear all; clf
syms t w
x = cos(10 * t) - 2 * cos(5 * t) + 4 * sin(20 * t);  % input signal
X = fourier(x);
N = 3; Whp = 5;   % filter parameters
[b, a] = butter(N, Whp, 's'); % filter design
W = 0:0.01:30; Hm = abs(freqs(b, a, W));  % magnitude response in W
% filter output
n = N:-1:0;  U = (j * w).^n
num = b - conj(U'); den = a - conj(U');
```

```
H = num/den; % frequency response
Y = X * H; % convolution property
y = ifourier(Y, t); % inverse Fourier
```
■

■ Example 6.9

In this example we will compare the performance of Butterworth and Chebyshev low-pass filters in the filtering of an analog signal $x(t) = [-2\cos(5t) + \cos(10t) + 4\sin(20t)]u(t)$ using MATLAB. We would like the two filters to have the same half-power frequency.

The magnitude specifications for the low-pass Butterworth filter are

$$\alpha_{max} = 0.1\text{dB}, \ \Omega_p = 5\text{rad/sec} \tag{6.54}$$

$$\alpha_{min} = 15\text{dB}, \ \Omega_s = 10 \ \text{rad/sec} \tag{6.55}$$

and a dc loss of 0 dB. Once this filter is designed, we would like the Chebyshev filter to have the same half-power frequency. In order to obtain this, we need to change the Ω_p specification for the Chebyshev filter. To do that we use the formulas for the half-power frequency of this type of filter to find the new value for Ω_p.

The Butterworth filter is designed by first determining the minimum order N and the half-power frequency Ω_{hp} using the function buttord, and then finding the filter coefficients by means of the function butter. Likewise, for the design of the Chebyshev filter we use the function cheb1ord to find the minimum order and the cut-off frequency (the new Ω_p is obtained from the half-power frequency). The filtering is implemented using the Fourier transform as before.

There are two significant differences between the designed Butterworth and Chebyshev filters. Although both of them have the same half-power frequency, the transition band of the Chebyshev filter is narrower, [6.88 10], than that of the Butterworth filter, [5 10], indicating that the Chebyshev is a better filter. The narrower transition band is compensated by a lower minimum order of five for the Chebyshev compared to the six-order Butterworth. Figure 6.24 displays the poles of the Butterworth and the Chebyshev filters, their magnitude responses, as well as the input signal $x(t)$ and the output $y(t)$ for the two filters (the two perform very similarly).

```
%%%%%%%%%%%%%%%%%%%
% Example 6.9 -- Filtering with Butterworth and Chebyshev filters
%%%%%%%%%%%%%%%%%%%%
clear all;clf
syms t w
x = cos(10 * t) − 2 * cos(5 * t) + 4 * sin(20 * t); X = fourier(x);
wp = 5;ws = 10;alphamax = 0.1;alphamin = 15; % filter parameters
% butterworth filter
[N, whp] = buttord(wp, ws, alphamax, alphamin, 's')
[b, a] = butter(N, whp, 's')
% cheby1 filter
epsi = sqrt(10^(alphamax/10) −1)
```

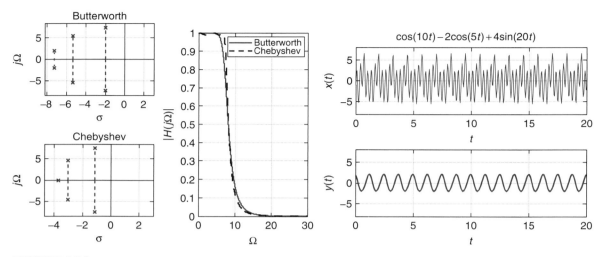

FIGURE 6.24
Comparison of filtering of an analog signal $x(t)$ using a low-pass Butterworth and Chebyshev filter with the same half-power frequency.

```
wp = whp/cosh(acosh(1/epsi)/N) % recomputing wp to get same whp
[N1, wn] = cheb1ord(wp, ws, alphamax, alphamin, 's');
[b1, a1] = cheby1(N1, alphamax, wn, 's');
% frequency responses
W = 0:0.01:30;
Hm = abs(freqs(b, a, W));
Hm1 = abs(freqs(b1, a1, W));
% generation of frequency response from coefficients
n = N:−1:0;  n1 = N1:−1:0;
U = (j * w).^n;  U1 = (j * w).^n1
num = b * conj(U'); den = a * conj(U');
num1 = b1 * conj(U1'); den1 = a1 * conj(U1')
H = num/den; % Butterworth LPF
H1 = num1/den1; % Chebyshev LPF
% output of filter
Y = X * H;
Y1 = X * H1;
y = ifourier(Y, t)
y1 = ifourier(Y1, t)
```

6.5.4 Frequency Transformations

As indicated before, the design of an analog filter is typically done by transforming the frequency of a normalized prototype low-pass filter. The frequency transformations were developed by Professor Ronald Foster [72] using the properties of reactance functions. The frequency transformations for the basic filters are given by:

$$Low\ pass\text{-}low\ pass : \qquad S = \frac{s}{\Omega_0}$$

$$Low\ pass\text{-}high\ pass : \qquad S = \frac{\Omega_0}{s}$$

$$Low\ pass\text{-}band\ pass : \qquad S = \frac{s^2 + \Omega_0^2}{s\ BW}$$

$$Low\ pass\text{-}band\ eliminating : \qquad S = \frac{s\ BW}{s^2 + \Omega_0^2} \qquad (6.56)$$

where S is the normalized and s the final variables, while Ω_0 is a desired cut-off frequency and BW is a desired bandwidth.

Remarks

- *The low-pass to low-pass (LP-LP) and low-pass to high-pass (LP-HP) transformations are linear in the numerator and denominator; thus the number of poles and zeros of the prototype low-pass filter is preserved. On the other hand, the low-pass to band-pass (LP-BP) and low-pass to band-eliminating (LP-BE) transformations are quadratic in either the numerator or the denominator, so that the number of poles/zeros is doubled. Thus, to obtain a 2Nth-order band-pass or band-eliminating filter the prototype low-pass filter should be of order N. This is an important observation useful in the design of these filters with MATLAB.*
- *It is important to realize that only frequencies are transformed, and the magnitude of the prototype filter is preserved. Frequency transformations will be useful also in the design of discrete filters, where these transformations are obtained in a completely different way, as no reactance functions would be available in that domain.*

■ Example 6.10

To illustrate how the above transformations can be used to convert a prototype low-pass filter we use the following script. First a low-pass prototype filter is designed using butter, and then to this filter we apply the lowpass to highpass transformation with $\Omega_0 = 40$ (rad/sec) to obtain a high-pass filter. Let then $\Omega_0 = 6.32$ (rad/sec) and $BW = 10$ (rad/sec) to obtain a band-pass and a band-eliminating filters using the appropriate transformations. The following is the script used. The magnitude responses are plotted with ezplot. Figure 6.25 shows the results.

```
clear all; clf
syms w
N = 5; [b, a] = butter(N, 1, 's') % low-pass prototype
omega0 = 40;BW = 10; omega1=sqrt(omega0); % transformation parameters
% low-pass prototype
n = N:−1:0;
U = (j * w).^n;  num = b * conj(U'); den = a * conj(U');
H = num/den;
% low-pass to high-pass
```

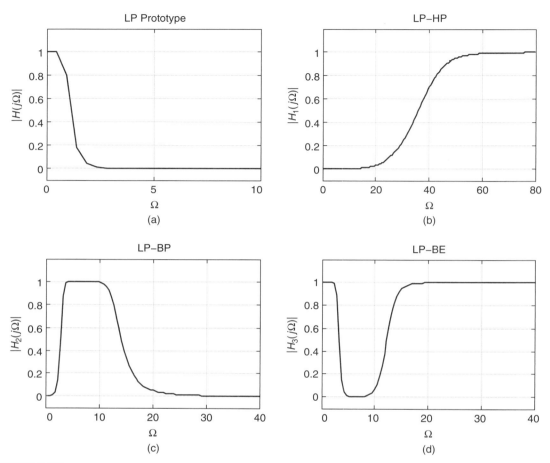

FIGURE 6.25

Frequency transformations: (a) prototype low-pass filter, (b) low-pass to high-pass transformation, (c) low-pass to band-pass transformation, and (d) low-pass to band-eliminating transformation.

```
U1 = (omega0/(j * w)).^n;
num1 = b * conj(U1'); den1 = a * conj(U1');
H1 = num1/den1;
% low-pass to band-pass
U2 = ((-w^2 + omega1^2)/(BW * j * w)).^n
num2 = b * conj(U2'); den2 = a * conj(U2');
H2 = num2/den2;
% low-pass to band-eliminating
U3 = ((BW * j * w)/(-w^2 + omega1^2)).^n
num3 = b * conj(U3'); den3 = a * conj(U3');
H3 = num3/den3
```

6.5.5 Filter Design with MATLAB

The design of filters, analog and discrete, is simplified by the functions that MATLAB provides. Functions to find the filter parameters from magnitude specifications, as well as functions to find the filter poles/zeros and to plot the designed filter magnitude and phase responses, are available.

Low-Pass Filter Design

The design procedure is similar for all of the approximation methods (Butterworth, Chebyshev, elliptic) and consists of both

- Finding the filter parameters from loss specifications.
- Obtaining the filter coefficients from these parameters.

Thus, to design an analog low-pass filter using the Butterworth approximation, the loss specifications α_{max} and α_{min}, and the frequency specifications, Ω_p and Ω_s are first used by the function buttord to determine the minimum order N and the half-power frequency Ω_{hp} of the filter that satisfies the specifications. Then the function butter uses these two values to determine the coefficients of the numerator and the denominator of the designed filter. We can then use the function freqs to plot the designed filter magnitude and phase. Similarly, this applies for the design of low-pass filters using the Chebyshev or the elliptic design methods. To include the design of low-pass filters using the Butterworth, Chebyshev (two versions), and the elliptic methods we wrote the function analogfil.

```
function [b, a] = analogfil(Wp, Ws, alphamax, alphamin, Wmax, ind)
%%
%     Analog filter design
%     Parameters
%       Input: loss specifications (alphamax, alphamin), corresponding
%              frequencies (Wp,Ws), frequency range [0,Wmax] and indicator ind (1 for
%              Butterworth, 2 for Chebyshev1, 3 for Chebyshev2 and 4 for elliptic).
%     Output: coefficients of designed filter.
%     Function plots magnitude, phase responses, poles and zeros of filter, and
%       loss specifications
%%%
if ind == 1,% Butterworth low-pass
    [N, Wn] = buttord(Wp, Ws, alphamax, alphamin, 's')
    [b, a] = butter(N, Wn, 's')
elseif ind == 2, % Chebyshev low-pass
    [N, Wn] = cheb1ord(Wp, Ws, alphamax, alphamin, 's')
    [b, a] = cheby1(N, alphamax, Wn, 's')
elseif ind == 3, % Chebyshev2 low-pass
    [N, Wn] = cheb2ord(Wp, Ws, alphamax, alphamin, 's')
    [b, a] = cheby2(N, alphamin, Wn, 's')
else % Elliptic low-pass
    [N, Wn] = ellipord(Wp, Ws, alphamax, alphamin, 's')
    [b, a] = ellip(N, alphamax, alphamin, Wn, 's')
end
```

```
W = 0:0.001:Wmax; % frequency range for plotting
H = freqs(b, a, W); Hm = abs(H); Ha = unwrap(angle(H)) % magnitude (Hm) and phase (Ha)
N = length(W); alpha1 = alphamax * ones(1, N); alpha2 = alphamin * ones(1, N); % loss specs
subplot(221)
plot(W, Hm); grid; axis([0 Wmax 0 1.1 * max(Hm)])
subplot(222)
plot(W, Ha); grid; axis([0 Wmax 1.1 * min(Ha) 1.1 * max(Ha)])
subplot(223)
splane(b, a)
subplot(224)
plot(W, -20 * log10(abs(H))); hold on
plot(W, alpha1, 'r', W, alpha2, 'r'); grid; axis([0 max(W) -0.1 100])
hold off
```

■ Example 6.11

To illustrate the use of analogfil consider the design of low-pass filters using the Chebyshev2 and the Elliptic design methods. The specifications for the designs are

$$\alpha(0) = 0, \quad \alpha_{max} = 0.1, \quad \alpha_{min} = 60 \quad \text{dB}$$

$$\Omega_p = 10, \quad \Omega_s = 15 \quad \text{rad/sec}$$

We wish to find the coefficients of the designed filters, plot their magnitude and phase, and plot the loss function for each of the filters and verify that the specifications have been met. The results are shown in Figure 6.26.

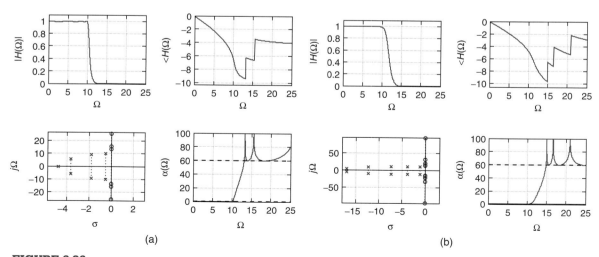

FIGURE 6.26

(a) Elliptic and (b) Chebyshev2 low-pass filter designs using analogfil function. Clockwise: magnitude, phase, loss function, and poles and zeros are shown for each design.

```
%%%%%%%%%%%%%%%%%%%%
% Example 6.11 -- Filter design using analogfil
%%%%%%%%%%%%%%%%%%%%
clear all; clf
alphamax = 0.1;
alphamin = 60;
Wp =10; Ws = 15;
Wmax = 25;
ind = 4 % elliptic design
% ind = 3 %  chebyshev2 design
[b, a] = analogfil(Wp, Ws, alphamax, alphamin, Wmax, ind)
```

The elliptic design is illustrated above. To obtain the Chebyshev2 design get rid of the comment symbol % in front of the corresponding indicator and put it in front of the one for the elliptic design. ∎

General comments on the design of low-pass filters using Butterworth, Chebyshev (1 and 2), and Elliptic methods are:

- The Butterworth and the Chebyshev2 designs are flat in the passband, while the others display ripples in that band.
- For identical specifications, the obtained order of the Butterworth filter is much greater than the order of the other filters.
- The phase of all of these filters is approximately linear in the passband, but not outside it. Because of the rational transfer functions for these filters, it is not possible to have linear phase over all frequencies. However, the phase response is less significant in the stopband where the magnitude response is very small.
- The filter design functions provided by MATLAB can be used for analog or discrete filters. When designing an analog filter there is no constrain in the values of the frequency specifications and an 's' indicates that the filter being designed is analog.

General Filter Design

The filter design programs butter, cheby1, cheby2, and ellip allow the design of other filters besides low-pass filters. Conceptually, a prototype low-pass filter is designed and then transformed into the desired filter by means of the frequency transformations given before. The filter is specified by the order and cut-off frequencies. In the case of low-pass and high-pass filters the specified cut-off frequencies are scalar, while for band-pass and stopband filters the specified cut-off frequencies are given as a vector. Also recall that the frequency transformations double the order of the low-pass prototype for the band-pass and band-eliminating filters, so when designing these filters half of the desired order should be given.

∎ Example 6.12

To illustrate the general design consider:

(a) Using the cheby2 method, design a band-pass filter with the following specifications:
- order $N = 10$
- $\alpha(\Omega) = 60$ dB in the stopband

- passband frequencies [10, 20] rad/sec
- unit gain in the passband

(b) Using the ellip method, design a band-stop filter with unit gain in the passbands and the following specifications:

- order $N = 20$
- $\alpha(\Omega) = 0.1$ dB in the passband
- $\alpha(\Omega) = 40$ dB in the stopband
- passband frequencies [10, 11] rad/sec

The following script is used.

```
%%%%%%%%%%%%%%%%%%%%%%%%%%%%%%%%
% Example 6.12 --- general filter design
%%%%%%%%%%%%%%%%%%%%%%%%%%%%%%%%
clear all;clf
N = 10;
[b, a] = ellip(N/2, 0.1, 40, [10 11], 'stop', 's') % elliptic band-stop
%[b, a] = cheby2(N, 60, [10 20], 's') % cheby2 bandpass
W = 0:0.01:30;
H = freqs(b, a, W);
```

Notice that the order given to ellip is 5 and 10 to cheby2 since a quadratic transformation will be used to obtain the notch and the band-pass filters from a prototype low-pass filter. The magnitude and phase responses of the two designed filters are shown in Figure 6.27. ■

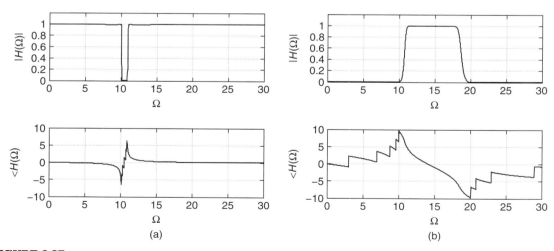

FIGURE 6.27
Design of (a) a notch filter using ellip and of (b) a band-pass filter using cheby2.

6.6 WHAT HAVE WE ACCOMPLISHED? WHAT IS NEXT?

In this chapter we have illustrated the application of the Laplace and the Fourier analysis to the theories of control, communications, and filtering. As you can see, the Laplace transform is very appropriate for control problems where transients as well as steady-state responses are of interest. On the other hand, in communications and filtering there is more interest in steady-state responses and frequency characterizations, which are more appropriately treated using the Fourier transform. It is important to realize that stability can only be characterized in the Laplace domain, and that it is necessary when considering steady-state responses. The control examples show the importance of the transfer function and transient and steady-state computations. Block diagrams help to visualize the interconnection of the different systems. Different types of modulation systems are illustrated in the communication examples. Finally, this chapter provides an introduction to the design of analog filters. In all the examples, the application of MATLAB was illustrated.

Although the material in this chapter does not have sufficient depth, reserved for texts in control, communications, and filtering, it serves to connect the theory of continuous-time signals and systems with applications. In the next part of the book, we will consider how to process signals using computers and how to apply the resulting theory again in control, communications, and signal processing problems.

PROBLEMS

6.1. Cascade implementation and loading

The transfer function of a filter $H(s) = 1/(s + 1)^2$ is to be implemented by cascading two first-order filters $H_i(s) = 1/(s + 1), i = 1, 2$.

(a) Implement $H_i(s)$ as a series RC circuit with input $v_i(t)$ and output $v_{i+1}(t)$, $i = 1, 2$. Cascade two of these circuits and find the overall transfer function $V_3(s)/V_1(s)$. Carefully draw the circuit.

(b) Use a voltage follower to connect the two circuits when cascaded and find the overall transfer function $V_3(s)/V_1(s)$. Carefully draw the circuit.

(c) Use the voltage follower circuit to implement a new transfer function

$$G(s) = \frac{1}{(s + 1000)(s + 1)}$$

Carefully draw your circuit.

6.2. Cascading LTI and LTV systems

The receiver of an AM system consists of a band-pass filter, a demodulator, and a low-pass filter. The received signal is

$$r(t) = m(t) \cos(40000\pi t) + q(t)$$

where $m(t)$ is a desired voice signal with bandwidth $BW = 5$ KHz that modulates the carrier $\cos(40,000\pi t)$ and $q(t)$ is the rest of the signals available at the receiver. The low-pass filter is ideal with magnitude 1 and bandwidth BW. Assume the band-pass filter is also ideal and that the demodulator is $\cos(\Omega_c t)$.

(a) What is the value of Ω_c in the demodulator?

(b) Suppose we input the received signal into the band-pass filter cascaded with the demodulator and the low-pass filter. Determine the magnitude response of the band-pass filter that allows us to recover $m(t)$. Draw the overall system and indicate which of the components are LTI and which are LTV.

(c) By mistake we input the received signal into the demodulator, and the resulting signal into the cascade of the band-pass and the low-pass filters. If you use the band-pass filter obtained above, determine the recovered signal (i.e., the output of the low-pass filter). Would you get the same result regardless of what $m(t)$ is? Explain.

6.3. Op-amps as feedback systems

An ideal operational amplifier circuit can be shown to be equivalent to a negative-feedback system. Consider the amplifier circuit in Figure 6.28 and its two-port network equivalent circuit to obtain a feedback system with input $V_i(s)$ and output $V_0(s)$. What is the effect of $A \to \infty$ on the above circuit?

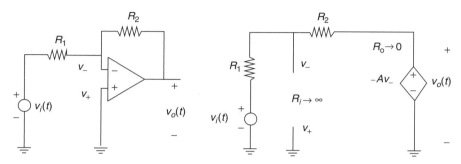

FIGURE 6.28

6.4. RC circuit as feedback system

Consider a series RC circuit with input a voltage source $v_i(t)$ and output the voltage across the capacitor $v_0(t)$.

(a) Draw a negative-feedback system for the circuit using an integrator, a constant multiplier, and an adder.

(b) Let the input be a battery (i.e., $v_i(t) = Au(t)$). Find the steady-state error $e(t) = v_i(t) - v_0(t)$.

6.5. RLC circuit as feedback system

A resistor R, a capacitor C, and an inductor L are connected in series with a source $v_i(t)$. Consider the output of the voltage across the capacitor $v_0(t)$. Let $R = 1\Omega$, $C = 1$ F and $L = 1$ H.

(a) Use integrators and adders to implement the differential equation that relates the input $v_i(t)$ and the output $v_0(t)$ of the circuit.

(b) Obtain a negative-feedback system block diagram with input $V_i(s)$ and output $V_0(s)$. Determine the feedforward transfer function $G(s)$ and the feedback transfer function $H(s)$ of the feedback system.

(c) Find an equation for the error $E(s) = V_i(s) - V_0(s)H(s)$ and determine its steady-state response when the input is a unit-step signal (i.e., $V_i(s) = 1/s$).

6.6. Ideal and lossy integrators

An ideal integrator has a transfer function $1/s$, while a lossy integrator has a transfer function $1/(s + K)$.

(a) Determine the feedforward transfer function $G(s)$ and the feedback transfer function $H(s)$ of a negative-feedback system that implements the overall transfer function

$$\frac{Y(s)}{X(s)} = \frac{K}{K + s}$$

where $X(s)$ and $Y(s)$ are the Laplace transforms of the input $x(t)$ and the output $y(t)$ of the feedback system. Sketch the magnitude response of this system and determine the type of filter it is.

(b) If we let $G(s) = s$ in the previous feedback system, determine the overall transfer function $Y(s)/X(s)$ where $X(s)$ and $Y(s)$ are the Laplace transforms of the input $x(t)$ and the output $y(t)$ of this new feedback system. Sketch the magnitude response of the overall system and determine the type of filter it is.

6.7. Feedback implementation of an all-pass system

Suppose you would like to obtain a feedback implementation of an all-pass filter

$$T(s) = \frac{s^2 - \sqrt{2}s + 1}{s^2 + \sqrt{2}s + 1}$$

(a) Determine if the $T(s)$ is the transfer function corresponding to an all-pass filter by means of the poles and zeros of $T(s)$.
(b) Determine the feedforward transfer function $G(s)$ and the feedback transfer function $H(s)$ of a negative-feedback system that has $T(s)$ as its overall transfer function.
(c) Would it be possible to implement $T(s)$ using a positive-feedback system? If so, indicate its feedforward transfer function $G(s)$ and the feedback transfer function $H(s)$.

6.8. Filter stabilization

The transfer function of a designed filter is

$$G(s) = \frac{1}{(s+1)(s-1)}$$

which is unstable given that one of its poles is in the right-hand s-plane.
(a) Consider stabilizing $G(s)$ by means of negative feedback with a gain $K > 0$ in the feedback. Determine the range of values of K that would make the stabilization possible.
(b) Use the cascading of an all-pass filter $H_a(s)$ with the given $G(s)$ to stabilize it. Give $H_a(s)$. Would it be possible for the resulting filter to have the same magnitude response as $G(s)$?

6.9. Error and feedforward transfer function

Suppose the feedforward transfer function of a negative-feedback system is $G(s) = N(s)/D(s)$, and the feedback transfer function is unity.
(a) Given that the Laplace transform of the error is

$$E(s) = X(s)[1 - H(s)]$$

where $H(s) = G(s)/(1 + G(s))$ is the overall transfer function of the feedback system, find an expression for the error in terms of $X(s)$, $N(s)$, and $D(s)$. Use this equation to determine the conditions under which the steady-state error is zero for $x(t) = u(t)$.
(b) If the input is $x(t) = u(t)$, the denominator $D(s) = (s+1)(s+2)$, and the numerator $N(s) = 1$, find an expression for $E(s)$ and from it determine the initial value $e(0)$ and the final value $\lim_{t \to \infty} e(t)$ of the error.

6.10. Product of polynomials in s—MATLAB

Given a transfer function

$$\frac{Y(s)}{X(s)} = \frac{N(s)}{D(s)}$$

where $Y(s)$ and $X(s)$ are the Laplace transforms of the output $y(t)$ and of the input $x(t)$ of an LTI system, and $N(s)$ and $D(s)$ are polynomials in s, to find the output

$$Y(s) = X(s)\frac{N(s)}{D(s)}$$

we need to multiply polynomials to get $Y(s)$ before we perform partial fraction expansion to get $y(t)$.

(a) Find out about the MATLAB function conv and how it relates to the multiplication of polynomials. Let $P(s) = 1 + s + s^2$ and $Q(s) = 2 + 3s + s^2 + s^3$. Obtain analytically the product $Z(s) = P(s)Q(s)$ and then use conv to compute the coefficients of $Z(s)$.

(b) Suppose that $X(s) = 1/s^2$, and we have $N(s) = s + 1$, $D(s) = (s + 1)((s + 4)^2 + 9)$. Use conv to find the numerator and the denominator polynomials of $Y(s) = N_1(s)/D_1(s)$. Use MATLAB to find $y(t)$, and to plot it.

(c) Create a function that takes as input the values of the coefficients of the numerators and denominators of $X(s)$ and of the transfer function $H(s)$ of the system and provides the response of the system. Show your function, and demonstrate its use with the $X(s)$ and $H(s)$ given above.

6.11. Feedback error—MATLAB

Control systems attempt to follow the reference signal at the input, but in many cases they cannot follow particular types of inputs. Let the system we are trying to control have a transfer function $G(s)$, and the feedback transfer function be $H(s)$. If $X(s)$ is the Laplace transform of the reference input signal, and $Y(s)$ the Laplace transform of the output, then the close-loop transfer function is

$$\frac{Y(s)}{X(s)} = \frac{G(s)}{1 + G(s)H(s)}$$

The Laplace transform of the error signal is $E(s) = X(s) - Y(s)H(s)$,

$$G(s) = \frac{1}{s(s + 1)(s + 2)} \quad \text{and} \quad H(s) = 1$$

(a) Find an expression for $E(s)$ in terms of $X(s)$, $G(s)$, and $H(s)$.

(b) Let $x(t) = u(t)$ and the Laplace transform of the corresponding error be $E_1(s)$. Use the final value property of the Laplace transform to obtain the steady-state error e_{1ss}.

(c) Let $x(t) = tu(t)$ (i.e., a ramp signal) and $E_2(s)$ be the Laplace transform of the corresponding error signal. Use the final value property of the Laplace transform to obtain the steady-state error e_{2ss}. Is this error value larger than the one above? Which of the two inputs $u(t)$ and $r(t)$ is easier to follow?

(d) Use MATLAB to find the partial fraction expansions of $E_1(s)$ and $E_2(s)$ and use them to find $e_1(t)$ and $e_2(t)$ and then plot them.

6.12. Wireless transmission—MATLAB

Consider the transmission of a sinusoid $x(t) = \cos(2\pi f_0 t)$ through a channel affected by multipath and Doppler. Let there be two paths, and assume the sinusoid is being sent from a moving transmitter so that a Doppler frequency shift occurs. Let the received signal be

$$r(t) = \alpha_0 \cos(2\pi (f_0 - \nu)(t - L_0/c)) + \alpha_1 \cos(2\pi (f_0 - \nu)(t - L_1/c))$$

where $0 \leq \alpha_i \leq 1$ are attenuations, L_i are the distances from the transmitter to the receiver that the signal travels in the ith path $i = 1, 2, c = 3 \times 10^8$ m/sec, and the frequency shift ν is caused by the Doppler effect.

(a) Let $f_0 = 2$ KHz, $\nu = 50$ Hz, $\alpha_0 = 1$, $\alpha_1 = 0.9$, and $L_0 = 10,000$ meters. What would be L_1 if the two sinusoids have a phase difference of $\pi/2$?

(b) Is the received signal $r(t)$, with the parameters given above but $L_1 = 10,000$, periodic? If so, what would be its period and how much does it differ from the period of the original sinusoid? If $x(t)$ is the input and $r(t)$ the output of the transmission channel, considered a system, is it linear and time invariant? Explain.

(c) Sample the signals $x(t)$ and $r(t)$ using a sampling frequency $F_s = 10$ KHz. Plot the sampled sent $x(nT_s)$ and received $r(nT_s)$ signals for $n = 0$ to 2000.

(d) Consider the situation where $f_0 = 2\,\text{KHz}$, but the parameters of the paths are random, trying to simulate real situations where these parameters are unpredictable, although somewhat related. Let

$$r(t) = \alpha_0 \cos(2\pi(f_0 - v)(t - L_0/c)) + \alpha_1 \cos(2\pi(f_0 - v)(t - L_1/c))$$

where $v = 50\eta\,\text{Hz}$, $L_0 = 1{,}000\eta$, $L_1 = 10{,}000\eta$, $\alpha_0 = 1 - \eta$, $\alpha_1 = \alpha_0/10$, and η is a random number between 0 and 1 with equal probability of being any of these values (this can be realized by using the rand MATLAB function). Generate the received signal for 10 different events, use $F_s = 10{,}000\,\text{Hz}$ as the sampling rate, and plot them together to observe the effects of the multipath and Doppler.

6.13. RLC implementation of low-pass Butterworth filters

Consider the RLC circuit shown in Figure 6.29 where $R = 1\,\Omega$.

(a) Determine the values of the inductor and the capacitor so that the transfer function of the circuit when the output is the voltage across the capacitor is

$$\frac{V_o(s)}{V_i(s)} = \frac{1}{s^2 + \sqrt{2}s + 1}$$

That is, it is a second-order Butterworth filter.

(b) Find the transfer function of the circuit, with the values obtained in (a) for the capacitor and the inductor, when the output is the voltage across the resistor. Carefully sketch the corresponding frequency response and determine the type of filter it is.

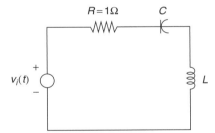

FIGURE 6.29

6.14. Design of low-pass Butterworth/Chebyshev filters

The specifications for a low-pass filter are:

- $\Omega_p = 1500\,\text{rad/sec}$, $\alpha_{\text{max}} = 0.5\,\text{dBs}$
- $\Omega_s = 3500\,\text{rad/sec}$, $\alpha_{\text{min}} = 30\,\text{dBs}$

  (a) Determine the minimum order of the low-pass Butteworth filter and compare it to the minimum order of the Chebyshev filter that satisfy the specifications. Which is the smaller of the two?

  (b) Determine the half-power frequencies of the designed Butterworth and Chebyshev low-pass filters by letting $\alpha(\Omega_p) = \alpha_{\text{max}}$. Use the minimum orders obtained above.

  (c) For the Butterworth and the Chebyshev designed filters, find the loss function values at Ω_p and Ω_s. How are these values related to the α_{max} and α_{min} specifications? Explain.

  (d) If new specifications for the passband and stopband frequencies are $\Omega_p = 750\,\text{rad/sec}$ and $\Omega_s = 1750\,\text{rad/sec}$, respectively, are the minimum orders of the Butterworth and the Chebyshev filters changed? Explain.

6.15. Low-pass Butterworth filters

The loss at a frequency $\Omega = 2000\,\text{rad/sec}$ is $\alpha(2000) = 19.4\,\text{dBs}$ for a fifth-order low-pass Butterworth filter. If we let $\alpha(\Omega_p) = \alpha_{\text{max}} = 0.35\,\text{dBs}$, determine

- The half-power frequency Ω_{hp} of the filter.
- The passband frequency Ω_p of the filter.

6.16. Design of low-pass Butterworth/Chebyshev filters

The specifications for a low-pass filter are:

- $\alpha(0) = 20\,\text{dBs}$
- $\Omega_p = 1500\,\text{rad/sec}, \alpha_1 = 20.5\,\text{dBs}$
- $\Omega_s = 3500\,\text{rad/sec}, \alpha_2 = 50\,\text{dBs}$

(a) Determine the minimum order of the low-pass Butterworth and Chebyshev filters, and determine which is smaller.

(b) Give the transfer function of the designed low-pass Butterworth and Chebyshev filters (make sure the dc loss is as specified).

(c) Determine the half-power frequency of the designed filters by letting $\alpha(\Omega_p) = \alpha_{max}$.

(d) Find the loss function values provided by the designed filters at Ω_p and Ω_s. How are these values related to the α_{max} and α_{min} specifications? Explain. Which of the two filters provides more attenuation in the stopband?

(e) If new specifications for the passband and stopband frequencies are $\Omega_p = 750\,\text{rad/sec}$ and $\Omega_s = 1750\,\text{rad/sec}$, respectively, are the minimum orders of the filter changed? Explain.

6.17. Butterworth, Chebyshev, and Elliptic filters—MATLAB

Design an analog low-pass filter satisfying the following magnitude specifications:

- $\alpha_{max} = 0.5\,\text{dB}; \alpha_{min} = 20\,\text{dB}$
- $\Omega_p = 1000\,\text{rad/sec}; \Omega_s = 2000\,\text{rad/sec}$

(a) Use the Butterworth method. Plot the poles and zeros and the magnitude and phase of the designed filter. Verify that the specifications are satisfied by plotting the loss function.

(b) Use the Chebyshev method cheby1. Plot the poles and zeros and the magnitude and phase of the designed filter. Verify that the specifications are satisfied by plotting the loss function.

(c) Use the elliptic method. Plot the poles and zeros and the magnitude and phase of the designed filter. Verify that the specifications are satisfied by plotting the loss function.

(d) Compare the three filters and comment on their differences.

6.18. Chebyshev filter design—MATLAB

Consider the following low-pass filter specifications:

- $\alpha_{max} = 0.1\,\text{dB}; \alpha_{min} = 60\,\text{dB}$
- $\Omega_p = 1000\,\text{rad/sec}; \Omega_s = 2000\,\text{rad/sec}$

(a) Use MATLAB to design a Chebyshev low-pass filter that satisfies the above specifications. Plot the poles and zeros and the magnitude and phase of the designed filter. Verify that the specifications are satisfied by plotting the loss function.

(b) Compute the half-power frequency of the designed filter.

6.19. Getting rid of 60-Hz hum with different filters—MATLAB

A desirable signal

$$x(t) = \cos(100\pi t) - 2\cos(50\pi t)$$

is recorded as $y(t) = x(t) + \cos(120\pi t)$—that is, as the desired signal but with a 60-Hz hum. We would like to get rid of the hum and recover the desired signal. Use symbolic MATLAB to plot $x(t)$ and $y(t)$. Consider the following three different alternatives (use symbolic MATLAB to implement the filtering and use any method to design the filters):

(a) Design a band-eliminating filter to get rid of the 60-Hz hum in the signal. Plot the output of the band-eliminating filter.

(b) Design a high-pass filter to get the hum signal and then subtract it from $y(t)$. Plot the output of the high-pass filter.

(c) Design a band-pass filter to get rid of the hum. Plot the output of the band-pass filter.

(d) Is any of these alternatives better than the others? Explain.

6.20. Demodulation of AM—MATLAB

The signal at the input of an AM receiver is

$$u(t) = m_1(t) \cos(20t) + m_2(t) \cos(100t)$$

where the messages $m_i(t), i = 1, 2$ are the outputs of a low-pass Butterworth filter with inputs

$$x_1(t) = r(t) - 2r(t - 1) + r(t - 2)$$
$$x_2(t) = u(t) - u(t - 2)$$

respectively. Suppose we are interested in recovering the message $m_1(t)$.

(a) Design a 10th-order low-pass Butterworth filter with half-power 10 rad/sec. Implement this filter using MATLAB and find the two messages $m_i(t)$, $i = 1, 2$ using the indicated inputs $x_i(t)$, $i = 1, 2$, and plot them.

(b) To recover the desired message $m_1(t)$, first use a band-pass filter to obtain the desired signal containing $m_1(t)$ and to suppress the other. Design a band-pass Butterworth filter with a bandwidth of 10 rad/sec, centered at 20 rad/sec and order 10 that will pass the signal $m_1(t) \cos(20t)$ and reject the other signal.

(c) Multiply the output of the band-pass filter by a sinusoid $\cos(20t)$ (exactly the carrier in the transmitter), and low-pass filter the output of the mixer (the system that multiplies by the carrier frequency cosine). Design a low-pass Butterworth filter of bandwidth 10 rad/sec, and order 10 to filter the output of the mixer.

(d) Use MATLAB to display the different spectra. Compute and plot the spectrum of $m_1(t)$, $u(t)$, the output of the band-pass filter, the output of the mixer, and the output of the low-pass filter. Write numeric functions to compute the analog Fourier transform and its inverse.

6.21. Quadrature AM—MATLAB

Suppose we would like to send the two messages $m_i(t)$, $i = 1, 2$, created in Problem 6.20 using the same bandwidth and to recover them separately. To implement this, consider the QAM approach where the transmitted signal is

$$s(t) = m_1(t) \cos(50t) + m_2(t) \sin(50t)$$

Suppose that at the receiver we receive $s(t)$ and that we only need to demodulate it to obtain $m_i(t), i = 1, 2$. Design a low-pass Butterworth filter of order 10 and a half-power frequency 10 rad/sec (the bandwidth of the messages).

(a) Use MATLAB to plot $s(t)$ and its magnitude spectrum $|S(\Omega)|$. Write numeric functions to compute the analog Fourier transform and its inverse.

(b) Multiply $s(t)$ by $\cos(50t)$, and filter the result using the low-pass filter designed before. Use MATLAB to plot the result and to find and plot its magnitude spectrum.

(c) Multiply $s(t)$ by $\sin(50t)$, and filter the result using the low-pass filter designed before. Use MATLAB to plot the result and to find and plot its magnitude spectrum.

(d) Comment on your results.

PART

Theory and Application of Discrete-Time Signals and Systems

CHAPTER 7

Sampling Theory

*The pure and simple truth
is rarely pure and never simple.*
Oscar Wilde (1854–1900)
Irish writer and poet

7.1 INTRODUCTION

Since many of the signals found in applications such as communications and control are analog, if we wish to process these signals with a computer it is necessary to sample, quantize, and code them to obtain digital signals. Once the analog signal is sampled in time, the amplitude of the obtained discrete-time signal is quantized and coded to give a binary sequence that can be either stored or processed with a computer.

The main issues considered in this chapter are:

- *How to sample*—As we will see, it is the inverse relation between time and frequency that provides the solution to the problem of preserving the information of an analog signal when it is sampled. When sampling an analog signal one could choose an extremely small value for the sampling period so that there is no significant difference between the analog and the discrete signals— visually as well as from the information content point of view. Such a representation would, however, give redundant values that could be spared without losing the information provided by the analog signal. If, on the other hand, we choose a large value for the sampling period, we achieve data compression but at the risk of losing some of the information provided by the analog signal. So how do we choose an appropriate value for the sampling period? The answer is not clear in the time domain. It does become clear when considering the effects of sampling in the frequency domain: The sampling period depends on the maximum frequency present in the analog signal. Furthermore, when using the correct sampling period the information in the analog signal will remain in the discrete signal after sampling, thus allowing the reconstruction of the original signal from the samples. These results, introduced by Nyquist and Shannon, constitute

Signals and Systems Using MATLAB®. DOI: 10.1016/B978-0-12-374716-7.00011-9

the bridge between analog and discrete signals and systems and were the starting point for digital signal processing as a technical area.

■ *Practical aspects of sampling*—The device that samples, quantizes, and codes an analog signal is called an *analog-to-digital converter* (ADC), while the device that converts digital signals into analog signals is called a *digital-to-analog converter* (DAC). These devices are far from ideal and thus some practical aspects of sampling and reconstruction need to be considered. Besides the possibility of losing information by choosing too large of a sampling period, the ADC also loses information in the quantization process. The quantization error is, however, made less significant by increasing the number of bits used to represent each sample. The DAC interpolates and smooths out the digital signal, converting it back into an analog signal. These two devices are essential in the processing of continuous-time signals with computers.

7.2 UNIFORM SAMPLING

The first step in converting a continuous-time signal $x(t)$ into a digital signal is to discretize the time variable—that is, to consider samples of $x(t)$ at uniform times $t = nT_s$, or

$$x(nT_s) = x(t)|_{t=nT_s} \qquad n \text{ integer} \tag{7.1}$$

where T_s is the sampling period. The sampling process can be thought of as a modulation process, in particular connected with pulse amplitude modulation (PAM), a basic approach in digital communications. A pulse amplitude modulated signal consists of a sequence of narrow pulses with amplitudes the values of the continuous-time signal within the pulse. Assuming that the width of the pulses is much narrower than the sampling period T_s permits a simpler analysis based on impulse sampling.

7.2.1 Pulse Amplitude Modulation

A PAM system can be visualized as a switch that closes every T_s seconds for Δ seconds, and remains open otherwise. The PAM signal is thus the multiplication of the continuous-time signal $x(t)$ by a periodic signal $p(t)$ consisting of pulses of width Δ, amplitude $1/\Delta$, and period T_s. Thus, $x_{PAM}(t)$ consists of narrow pulses with the amplitudes of the signal within the pulse width. For a small pulse width Δ, the PAM signal is approximately a train of pulses with amplitudes $x(mT_s)$—that is,

$$x_{PAM}(t) = x(t)p(t) \approx \frac{1}{\Delta} \sum_m x(mT_s)[u(t - mT_s) - u(t - mT_s - \Delta)] \tag{7.2}$$

Now, as a periodic signal we represent $p(t)$ by its Fourier series

$$p(t) = \sum_k P_k e^{jk\Omega_0 t} \qquad \Omega_0 = \frac{2\pi}{T_s}$$

where P_k are the Fourier series coefficients. Thus, the PAM signal can be expressed as

$$x_{PAM}(t) = \sum_k P_k \, x(t) \, e^{jk\Omega_0 t}$$

and its Fourier transform is

$$X_{PAM}(\Omega) = \sum_k P_k X(\Omega - k\Omega_0)$$

showing that PAM is a modulation of the train of pulses $p(t)$ by the signal $x(t)$. The spectrum of $x_{PAM}(t)$ is the spectrum of $x(t)$ shifted in frequency by $\{k\Omega_0\}$, weighted by P_k, and superposed.

7.2.2 Ideal Impulse Sampling

Given that the pulse width Δ is much smaller than T_s, $p(t)$ can be replaced by a periodic sequence of impulses of period T_s (see Figure 7.1) or $\delta_{T_s}(t)$. This simplifies considerably the analysis and makes the results easier to grasp. Later in the chapter we consider the effects of having pulses instead of impulses, a more realistic assumption.

The *sampling function* $\delta_{T_s}(t)$, or a periodic sequence of impulses of period T_s, is

$$\delta_{T_s}(t) = \sum_n \delta(t - nT_s) \tag{7.3}$$

where $\delta(t - nT_s)$ is an approximation of the normalized pulse $[u(t - nT_s) - u(t - nT_s - \Delta)]/\Delta$ when $\Delta << T_s$. The sampled signal is then given by

$$x_s(t) = x(t)\delta_{T_s}(t)$$
$$= \sum_n x(nT_s)\delta(t - nT_s) \tag{7.4}$$

as illustrated in Figure 7.1.

There are two equivalent ways to view the sampled signal $x_s(t)$ in the frequency domain:

■ *Modulation*: Since $\delta_{T_s}(t)$ is periodic, of fundamental frequency $\Omega_s = 2\pi/T_s$, its Fourier series is

$$\delta_{T_s}(t) = \sum_{k=-\infty}^{\infty} D_k e^{jk\Omega_s t}$$

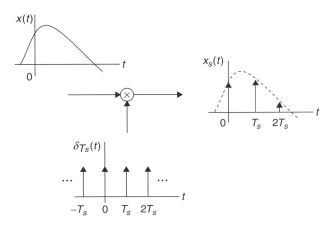

FIGURE 7.1
Ideal impulse sampling.

where the Fourier coefficients $\{D_k\}$ are

$$D_k = \frac{1}{T_s} \int_{-T_s/2}^{T_s/2} \delta_{T_s}(t) e^{-jk\Omega_s t} dt = \frac{1}{T_s} \int_{-T_s/2}^{T_s/2} \delta(t) e^{-jk\Omega_s t} dt$$

$$= \frac{1}{T_s} \int_{-T_s/2}^{T_s/2} \delta(t) e^{-j0} dt = \frac{1}{T_s}$$

The last equation is obtained using the sifting property of the $\delta(t)$ and that the area of the impulse is unity. Thus, the Fourier series of the sampling signal is

$$\delta_{T_s}(t) = \sum_{k=-\infty}^{\infty} \frac{1}{T_s} e^{jk\Omega_s t} \tag{7.5}$$

and the sampled signal $x_s(t) = x(t)\delta_{T_s}(t)$ is then expressed as

$$x_s(t) = \frac{1}{T_s} \sum_{k=-\infty}^{\infty} x(t) e^{jk\Omega_s t}$$

with Fourier transform

$$X_s(\Omega) = \frac{1}{T_s} \sum_{k=-\infty}^{\infty} X(\Omega - k\Omega_s) \tag{7.6}$$

where we used the frequency-shift property of the Fourier transform, and let $X(\Omega)$ and $X_s(\Omega)$ be the Fourier transforms of $x(t)$ and $x_s(t)$, respectively.

■ *Discrete-time Fourier transform*: The Fourier transform of the sum representation of $x_s(t)$ in the second equation in Equation (7.4) is

$$X_s(\Omega) = \sum_{n} x(nT_s) e^{-j\Omega T_s n} \tag{7.7}$$

where we used the Fourier transform of a shifted impulse. This equation is equivalent to Equation (7.6) and will be used later in deriving the Fourier transform of discrete-time signals.

Remarks

■ *The spectrum $X_s(\Omega)$ of the sampled signal, according to Equation (7.6), is a superposition of shifted analog spectra $\{X(\Omega - k\Omega_s)\}$ multiplied by $1/T_s$ (i.e., the modulation process involved in the sampling).*
■ *Considering that the output of the sampler displays frequencies that are not present in the input, according to the eigenfunction property the sampler is not LTI. It is a time-varying system. Indeed, if sampling $x(t)$ gives $x_s(t)$, sampling $x(t - \tau)$ where $\tau \neq kT_s$ for an integer k will not be $x_s(t - \tau)$. The sampler is, however, a linear system.*

■ *Equation (7.7) provides the relation between the continuous frequency Ω (rad/sec) of $x(t)$ and the discrete frequency ω (rad) of the discrete-time signal $x(nT_s)$ or $x[n]$[1]:*

$$\omega = \Omega T_s \qquad [rad/sec] \times [sec] = [rad]$$

Sampling a continuous-time signal $x(t)$ at uniform times $\{nT_s\}$ gives a sampled signal

$$x_s(t) = \sum_n x(nT_s)\delta(t - nT_s) \qquad (7.8)$$

or a sequence of samples $\{x(nT_s)\}$. Sampling is equivalent to modulating the sampling signal

$$\delta_{T_s}(t) = \sum_n \delta(t - nT_s) \qquad (7.9)$$

periodic of period T_s (the sampling period) with $x(t)$.

If $X(\Omega)$ is the Fourier transform of $x(t)$, the Fourier transform of the sampled signal $x_s(t)$ is given by the equivalent expressions

$$X_s(\Omega) = \frac{1}{T_s}\sum_k X(\Omega - k\Omega_s)$$

$$= \sum_n x(nT_s)e^{-j\Omega T_s n} \qquad \Omega_s = \frac{2\pi}{T_s} \qquad (7.10)$$

Depending on the maximum frequency present in the spectrum of $x(t)$ and on the chosen sampling frequency Ω_s (or the sampling period T_s) it is possible to have overlaps when the spectrum of $x(t)$ is shifted and added to obtain the spectrum of the sampled signal. We have three possible situations:

■ *If the signal has a low-pass spectrum of finite support—that is, $X(\Omega) = 0$ for $|\Omega| > \Omega_{max}$ (see Figure 7.2(a)) where Ω_{max} is the maximum frequency present in the signal—such a signal is called band limited. As shown in Figure 7.2(b), for band-limited signals it is possible to choose Ω_s so that the spectrum of the sampled signal consists of shifted nonoverlapping versions of $(1/T_s)X(\Omega)$. Graphically (see Figure 7.2(b)), this can be accomplished by letting $\Omega_s - \Omega_{max} \geq \Omega_{max}$, or*

$$\Omega_s \geq 2\Omega_{max}$$

which is called the Nyquist sampling rate condition. As we will see later, in this case we are able to recover $X(\Omega)$, or $x(t)$, from $X_s(\Omega)$ or from the sampled signal $x_s(t)$. Thus, the information in $x(t)$ is preserved in the sampled signal $x_s(t)$.

■ *On the other hand, if the signal $x(t)$ is band limited but we let $\Omega_s < 2\Omega_{max}$, then when creating $X_s(\Omega)$ the shifted spectra of $x(t)$ overlap (see Figure 7.2(c)). In this case, due to the overlap it will not be*

---

[1]To help the reader visualize the difference between a continuous-time signal, which depends on a continuous variable t, or a real number, and a discrete-time signal, which depends on the integer variable n, we will use square brackets for these. Thus, $\eta(t)$ is a continuous-time signal, while $\rho[n]$ is a discrete-time signal.

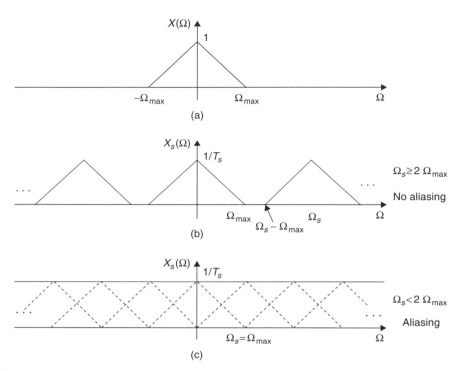

FIGURE 7.2

(a) Spectrum of band-limited signal, (b) spectrum of sampled signal when satisfying the Nyquist sampling rate condition, and (c) spectrum of sampled signal with aliasing (superposition of spectra, shown in dashed lines, gives a constant shown by continuous line).

possible to recover the original continuous-time signal from the sampled signal, and thus the sampled signal does not share the same information with the original continuous-time signal. This phenomenon is called frequency aliasing *since due to the overlapping of the spectra some frequency components of the original continuous-time signal acquire a different frequency value or an "alias."*

■ *When the spectrum of x(t) does not have a finite support (i.e., the signal is not band limited) sampling using any sampling period T_s generates a spectrum of the sampled signal consisting of overlapped shifted spectra of x(t). Thus, when sampling non-band-limited signals frequency aliasing is always present. The only way to sample a non-band-limited signal x(t) without aliasing—at the cost of losing information provided by the high-frequency components of x(t) — is by obtaining an approximate signal $x_a(t)$ that lacks the high-frequency components of x(t), thus permitting us to determine a maximum frequency for it. This is accomplished by* antialiasing filtering *commonly used in samplers.*

A band-limited signal $x(t)$—that is, its low-pass spectrum $X(\Omega)$ is such that

$$|X(\Omega)| = 0 \text{ for } |\Omega| > \Omega_{\max} \tag{7.11}$$

where Ω_{max} is the maximum frequency in $x(t)$—can be sampled uniformly and without frequency aliasing using a sampling frequency

$$\Omega_s = \frac{2\pi}{T_s} \geq 2\Omega_{max} \tag{7.12}$$

This is called the *Nyquist sampling rate condition.*

■ Example 7.1

Consider the signal $x(t) = 2\cos(2\pi t + \pi/4)$, $-\infty < t < \infty$. Determine if it is band limited or not. Use $T_s = 0.4$, 0.5, and 1 sec/sample as sampling periods, and for each of these find out whether the Nyquist sampling rate condition is satisfied and if the sampled signal looks like the original signal or not.

Solution

Since $x(t)$ only has the frequency 2π, it is band limited with $\Omega_{max} = 2\pi$ rad/sec. For any T_s the sampled signal is given as

$$x_s(t) = \sum_{n=-\infty}^{\infty} x(nT_s)\delta(t - nT_s) \qquad T_s \text{ sec/sample} \tag{7.13}$$

with $x(nT_s) = x(t)|_{t=nT_s}$.

Using $T_s = 0.4$ sec/sample the sampling frequency in rad/sec is $\Omega_s = 2\pi/T_s = 5\pi > 2\Omega_{max} = 4\pi$, satisfying the Nyquist sampling rate condition. The samples in Equation (7.13) are then

$$x(nT_s) = 2\cos(2\pi\ 0.4n + \pi/4) = 2\cos\left(\frac{4\pi}{5}n + \frac{\pi}{4}\right) \qquad -\infty < n < \infty$$

The sampled signal $x_s(t)$ repeats periodically every five samples. Indeed, for $T_s = 0.4$,

$$x_s(t + 5T_s) = \sum_{n=-\infty}^{\infty} x(nT_s)\delta(t - (n-5)T_s) \quad \text{letting } m = n - 5$$

$$= \sum_{m=-\infty}^{\infty} x((m+5)T_s)\delta(t - mT_s) = x_s(t)$$

since $x((m+5)T_s) = x(mT_s)$. Looking at Figure 7.3(b), we see that there are three samples in each period of the analog sinusoid, and it is not obvious that the information of the continuous-time signal is preserved. We will show in the next section that it is actually possible to recover $x(t)$ from this sampled signal $x_s(t)$, which allows us to say that $x_s(t)$ has the same information as $x(t)$.

When $T_s = 0.5$ the sampling frequency is $\Omega_s = 2\pi/T_s = 4\pi = 2\Omega_{max}$, barely satisfying the Nyquist sampling rate condition. The samples in Equation (7.13) are now

$$x(nT_s) = 2\cos(2\pi n0.5 + \pi/4) = 2\cos\left(\frac{2\pi}{2}n + \frac{\pi}{4}\right) \qquad -\infty < n < \infty$$

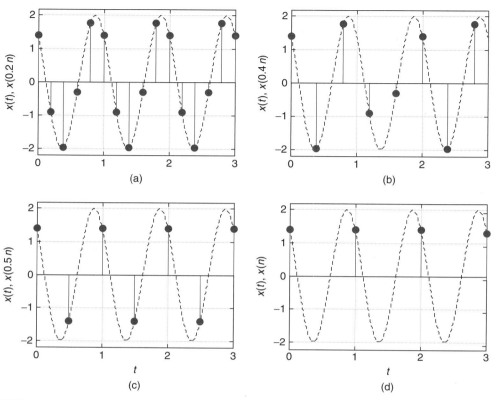

FIGURE 7.3

Sampling of $x(t) = 2\cos(2\pi t + \pi/4)$ with sampling periods (a) $T_s = 0.2$, (b) $T_s = 0.4$, (c) $T_s = 0.5$, and (d) $T_s = 1$ sec/sample.

In this case it can be shown that the sampled signal repeats periodically every two samples, since $x((n+2)T_s) = x(nT_s)$, which can be easily checked. According to the Nyquist sampling rate condition, this is the minimum number of samples per period allowed before we start having aliasing. In fact, if we let $\Omega_s = \Omega_{max} = 2\pi$ corresponding to the sampling period $T_s = 1$, the samples in Equation (7.13) are

$$x(nT_s) = 2\cos(2\pi n + \pi/4) = 2\cos(\pi/4) = \sqrt{2}$$

and the sampled signal is $\sqrt{2}\delta_{T_s}(t)$. With $T_s = 1$, the sampled signal cannot be possibly converted back into an analog sinusoid. Thus, we have lost the information provided by the sinusoid. Undersampling (getting too few samples per unit time) has changed the nature of the original signal.

We use MATLAB to plot the continuous signal and four sampled signals (see Figure 7.3) for different values of T_s. Clearly, when $T_s = 1$ sec/sample there is no similarity between the analog and the discrete signals due to frequency aliasing. ∎

■ Example 7.2

Consider the following signals:

$$\text{(a)} \quad x_1(t) = u(t + 0.5) - u(t - 0.5)$$

$$\text{(b)} \quad x_2(t) = e^{-t}u(t)$$

Determine if they are band limited or not. If not, determine the frequency for which the energy of the non-band-limited signal corresponds to 99% of its total energy and use this result to approximate its maximum frequency.

Solution

(a) The signal $x_1(t) = u(t + 0.5) - u(t - 0.5)$ is a unit pulse signal. Clearly, this signal can be easily sampled by choosing any value of $T_s \ll 1$. For instance, $T_s = 0.01$ sec would be a good value, giving a discrete-time signal $x_1(nT_s) = 1$, for $0 \le nT_s = 0.01n \le 1$ or $0 \le n \le 100$. There seems to be no problem in sampling this signal; however, we have that the Fourier transform of $x_1(t)$,

$$X_1(\Omega) = \frac{e^{j0.5\Omega} - e^{-j0.5\Omega}}{j\Omega} = \frac{\sin(0.5\Omega)}{0.5\Omega}$$

does not have a maximum frequency and so $x_1(t)$ is not band limited. Thus, any chosen value of T_s will cause aliasing. Fortunately, the values of the sinc function go fast to zero, so that one could compute an approximate maximum frequency that covers 99% of the energy of the signal.

Using Parseval's energy relation we have that the energy of $x_1(t)$ (the area under $x_1^2(t)$) is 1 and if we wish to find a value Ω_M, such that 99% of this energy is in the frequency band $[-\Omega_M, \Omega_M]$, we need to look for the limits of the following integral so it equals 0.99:

$$0.99 = \frac{1}{2\pi} \int_{-\Omega_M}^{\Omega_M} \left[\frac{\sin(0.5\Omega)}{0.5\Omega} \right]^2 d\Omega$$

Since this integral is difficult to find analytically, we use the following script in MATLAB to approximate it.

```
%%%%%%%%%%%%%%%%%%%%%%%
% Example 7.2 --- Parseval's relation and sampling
%%%%%%%%%%%%%%%%%%%%%%%
    syms W
    for k = 1:23;
    E(k) = int((sin(0.5*W)/(0.5*W))^2,0,k*pi)/pi
    if E(k)> = 0.9900,
      k
      return
    end
    end
```

We found that for $\Omega_M = 20\pi$ rad/sec 98.9% of the energy of the signal is included, and thus it could be used to determine that $T_s < \pi/\Omega_M = 0.05$ sec/sample.

(b) For the causal exponential

$$x(t) = e^{-t}u(t)$$

its Fourier transform is

$$X(\Omega) = \frac{1}{1 + j\Omega} \quad \text{so that} \quad |X(\Omega)| = \frac{1}{\sqrt{1 + \Omega^2}}$$

which does not go to zero for any finite Ω, then $x(t)$ is not band limited. To find a frequency Ω_M so that 99% of the energy is in $-\Omega_M \leq \Omega \leq \Omega_M$, we let

$$\frac{1}{2\pi} \int_{-\Omega_M}^{\Omega_M} |X(\Omega)|^2 d\Omega = \frac{0.99}{2\pi} \int_{-\infty}^{\infty} |X(\Omega)|^2 d\Omega$$

which gives

$$2 \tan^{-1}(\Omega)|_0^{\Omega_M} = 2 \times 0.99 \tan^{-1}(\Omega)|_0^{\infty} \quad \text{or} \quad \Omega_M = \tan\left(\frac{0.99\pi}{2}\right) = 63.66$$

If we choose $\Omega_s = 2\pi/T_s = 5\Omega_M$ or $T_s = 2\pi/(5 \times 63.66) \approx 0.02$, there will be hardly any aliasing or loss of information. ∎

7.2.3 Reconstruction of the Original Continuous-Time Signal

If the signal $x(t)$ to be sampled is band limited with Fourier transform $X(\Omega)$ and maximum frequency Ω_{max}, by choosing the sampling frequency Ω_s to satisfy the Nyquist sampling rate condition, or $\Omega_s > 2\Omega_{max}$, the spectrum of the sampled signal $x_s(t)$ displays a superposition of shifted versions of the spectrum of $x(t)$, multiplied by $1/T_s$, but with no overlaps. In such a case, it is possible to recover the original analog signal from the sampled signal by filtering. Indeed, if we consider an ideal low-pass analog filter $H_{lp}(j\Omega)$ with magnitude T_s in the pass-band $-\Omega_s/2 < \Omega < \Omega_s/2$, and zero elsewhere—that is,

$$H_{lp}(j\Omega) = \begin{cases} T_s & -\Omega_s/2 < \Omega < \Omega_s/2 \\ 0 & \text{elsewhere} \end{cases} \tag{7.14}$$

the Fourier transform of the output of the filter is $X_r(\Omega) = H_{lp}(j\Omega)X_s(\Omega)$ or

$$X_r(\Omega) = \begin{cases} X(\Omega) & -\Omega_s/2 < \Omega < \Omega_s/2 \\ 0 & \text{elsewhere} \end{cases}$$

which coincides with the Fourier transform of the original signal $x(t)$. So that when sampling a band-limited signal, using a sampling period T_s that satisfies the Nyquist sampling rate, the signal can be recovered exactly from the sampled signal by means of an ideal low-pass filter.

Bandlimited or Not?

The following, taken from David Slepian's paper "On Bandwidth" [66], clearly describes the uncertainty about bandlimited signals:

The Dilemma—Are signals really bandlimited? They seem to be, and yet they seem not to be.

On the one hand, a pair of solid copper wires will not propagate electromagnetic waves at optical frequencies and so the signals I receive over such a pair must be bandlimited. In fact, it makes little physical sense to talk of energy received over wires at frequencies higher than some finite cutoff W, say 10^{20} Hz. It would seem, then, that signals must be bandlimited.

On the other hand, however, signals of limited bandwith W are finite Fourier transforms,

$$s(t) = \int_{-W}^{W} e^{2\pi i f t} S(f) df$$

and irrefutable mathematical arguments show them to be extremely smooth. They possess derivatives of all orders. Indeed, such integrals are entire functions of t, completely predictable from any little piece, and they cannot vanish on any t interval unless they vanish everywhere. Such signals cannot start or stop, but must go on forever. Surely *real signals* start and stop, and they cannot be bandlimited!

Thus we have a dilemma: to assume that real signals must go on forever in time (a consequence of bandlimitedness) seems just as unreasonable as to assume that real signals have energy at arbitrary high frequencies (no bandlimitation). Yet one of these alternatives must hold if we are to avoid mathematical contradiction, for either signals are bandlimited or they are not: there is no other choice. Which do you think they are?

Remarks

- *In practice, the exact recovery of the original signal may not be possible for several reasons. One could be that the continuous-time signal is not exactly band limited, so that it is not possible to obtain a maximum frequency causing frequency aliasing in the sampling. Second, the sampling is not done exactly at uniform times—random variation of the sampling times may occur. Third, the filter required for the exact recovery is an ideal low-pass filter, which in practice cannot be realized; only an approximation is possible. Although this indicates the limitations of sampling, in most cases where: (1) the signal is band limited or approximately band limited, (2) the Nyquist sampling rate condition is satisfied in the sampling, and (3) the reconstruction filter approximates well the ideal low-pass filter, the recovered signal closely approximates the original signal.*

- *For signals that do not satisfy the band-limitedness condition, one can obtain an approximate signal that satisfies that condition. This is done by passing the non-band-limited signal through an ideal low-pass filter. The filter output is guaranteed to have as maximum frequency the cut-off frequency of the filter (see Figure 7.4). Because of the low-pass filtering, the filtered signal is a smoothed version of the original signal—high frequencies of the signal have been removed. The low-pass filter is called an* antialiasing filter, *since it makes the approximate signal band limited, thus avoiding aliasing in the frequency domain.*

- *In applications, the cut-off frequency of the antialiasing filter is set according to prior knowledge. For instance, when sampling speech, it is known that speech has frequencies ranging from about 100 Hz to*

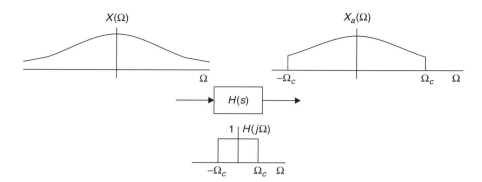

FIGURE 7.4
Anti-aliasing filtering of non-band-limited signal.

about 5 KHz (this range of frequencies provides understandable speech in phone conversations). Thus, when sampling speech an anti-aliasing filter with a cut-off frequency of 5 KHz is chosen and the sampling rate is then set to 10,000 samples/sec. Likewise, it is also known that an acceptable range of frequencies from 0 to 22 KHz provides music with good fidelity, so that when sampling music signals the anti-aliasing filter cut-off frequency is set to 22 KHz and the sampling rate to 44 K samples/sec or higher to provide good-quality music.

Origins of the Sampling Theory—Part 1

The sampling theory has been attributed to many engineers and mathematicians. It seems as if mathematicians and researchers in communications engineering came across these results from different perspectives. In the engineering community, the sampling theory has been attributed traditionally to Harry Nyquist and Claude Shannon, although other famous researchers such as V. A. Kotelnikov, E. T. Whittaker, and D. Gabor came out with similar results. Nyquist's work did not deal directly with sampling and reconstruction of sampled signals but it contributed to advances by Shannon in those areas. Harry Nyquist was born in Sweden in 1889 and died in 1976 in the United States. He attended the University of North Dakota at Grand Forks and received his Ph.D. from Yale University in 1917. He worked for the American Telephone and Telegraph (AT&T) Company and the Bell Telephone Laboratories, Inc. He received 138 patents and published 12 technical articles. Nyquist's contributions range from the fields of thermal noise, stability of feedback amplifiers, telegraphy, and television, to other important communications problems. His theoretical work on determining the bandwidth requirements for transmitting information provided the foundations for Claude Shannon's work on sampling theory [33].
As Hans D. Luke [44] concludes in his paper "The Origins of the Sampling Theorem," regarding the attribution of the sampling theorem to many authors:

> This history also reveals a process which is often apparent in theoretical problem in technology or physics: first the practicians put forward a rule of thumb, then theoreticians develop the general solution, and finally someone discovers that the mathematicians have long since solved the mathematical problem which it contains, but in "splendid isolation."

■ Example 7.3

Consider the two sinusoids

$$x_1(t) = \cos(\Omega_0 t) \qquad -\infty \le t \le \infty$$

$$x_2(t) = \cos((\Omega_0 + \Omega_s)t) \qquad -\infty \le t \le \infty$$

Show that if we sample these signals using $T_s = 2\pi/\Omega_s$, we cannot differentiate the sampled signals (i.e., $x_1(nT_s) = x_2(nT_s)$). Use MATLAB to show the above graphically when $\Omega_0 = 1$ and $\Omega_s = 7$. Explain the significance of this.

Solution

Sampling the two signals using $T_s = 2\pi/\Omega_s$, we have

$$x_1(nT_s) = \cos(\Omega_0 nT_s) \qquad -\infty \le n \le \infty$$

$$x_2(nT_s) = \cos((\Omega_0 + \Omega_s)nT_s) \qquad -\infty \le n \le \infty$$

but since $\Omega_s T_s = 2\pi$, the sinusoid $x_2(nT_s)$ can be written as

$$x_2(nT_s) = \cos((\Omega_0 T_s + 2\pi)n)$$

$$= \cos(\Omega_0 T_s n) = x_1(nT_s)$$

The following script shows the aliasing effect when $\Omega_0 = 1$ and $\Omega_s = 7$ rad/sec. Notice that $x_1(t)$ is sampled satisfying the Nyquist sampling rate condition ($\Omega_s = 7 > 2\Omega_0 = 2$ rad/sec), while $x_2(t)$ is not ($\Omega_s = 7 < 2(\Omega_0 + \Omega_s) = 16$ rad/sec).

```
%%%%%%%%%%%%%%%%%%%%%%%%%%%%%%%%%%%%%%%%
% Example 7.3 ---Two sinusoids of different frequencies being sampled
% with same sampling period -- aliasing for signal with higher frequency
%%%%%%%%%%%%%%%%%%%%%%%%%%%%%%%%%%%%%%%%
clear all; clf
% sinusoids
omega_0 = 1;omega_s = 7;
T = 2 * pi/omega_0; t = 0:0.001:T; % a period of x1
x1 = cos(omega_0 * t); x2 = cos((omega_0 + omega_s) * t);
N = length(t); Ts = 2 * pi/omega_s; % sampling period
M = fix(Ts/0.001); imp = zeros(1,N);
for k = 1:M:N − 1.
   imp(k) = 1; % sequence of impulses
end
xs = imp. * x1; % sampled signal
plot(t,x1,'b',t,x2,'k'); hold on
stem(t,imp. * x1,'r','filled');axis([0 max(t) − 1.1 1.1]); xlabel('t'); grid
```

Figure 7.5 shows the two sinusoids and the sampled signal that coincides for the two signals. The result in the frequency domain is shown in Figure 7.6: The spectra of the two sinusoids are different but the spectra of the sampled signals are identical. ∎

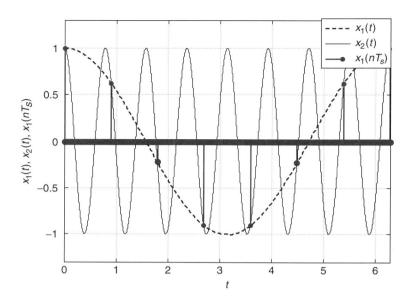

FIGURE 7.5
Sampling of two sinusoids of frequencies $\Omega_0 = 1$ and $\Omega_0 + \Omega_s = 8$ with $T_s = 2\pi/\Omega_s$. The higher-frequency signal is undersampled, causing aliasing, which makes the two sampled signals coincide.

FIGURE 7.6
(a) Spectra of sinusoids $x_1(t)$ and $x_2(t)$.
(b) The spectra of the sampled signals $x_{1s}(t)$ and $x_{2s}(t)$ look exactly the same due to the undersampling of $x_2(t)$.

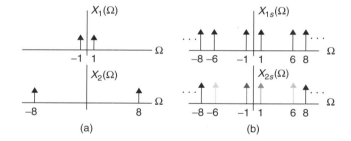

7.2.4 Signal Reconstruction from Sinc Interpolation

The analog signal reconstruction from the samples can be shown to be an interpolation using sinc signals. First, the ideal low-pass filter $H_{lp}(s)$ in Equation (7.14) has as impulse response

$$h_{lp}(t) = \frac{T_s}{2\pi} \int_{-\Omega_s/2}^{\Omega_s/2} e^{j\Omega t} d\Omega = \frac{\sin(\pi t/T_s)}{\pi t/T_s} \qquad (7.15)$$

which is a sinc function that has an infinite time support and decays symmetrically with respect to the origin $t = 0$. The reconstructed signal $x_r(t)$ is the convolution of the sampled signal $x_s(t)$ and $h_{lp}(t)$, which is found to be

$$x_r(t) = [x_s * h_{lp}](t) = \int_{-\infty}^{\infty} x_s(\tau)h_{lp}(t - \tau)d\tau$$

$$= \int_{-\infty}^{\infty} \left[\sum_n x(nT_s)\delta(\tau - nT_s) \right] h_{lp}(t - \tau)d\tau$$

$$= \sum_n x(nT_s)\frac{\sin(\pi(t - nT_s)/T_s)}{\pi(t - nT_s)/T_s} \tag{7.16}$$

after replacing $x_s(\tau)$ and applying the sifting property of the delta function. The recovered signal is thus an interpolation in terms of time-shifted sinc signals with amplitudes the samples $\{x(nT_s)\}$. In fact, if we let $t = kT_s$, we can see that

$$x_r(kT_s) = \sum_n x(nT_s)\frac{\sin(\pi(k - n))}{\pi(k - n)} = x(kT_s)$$

since

$$\frac{\sin(\pi(k - n))}{\pi(k - n)} = \begin{cases} 1 & k - n = 0 \text{ or } k = n \\ 0 & k \neq n \end{cases}$$

This is because the above sinc function by L'Hôpital's rule is shown to be unity when $k = n$, and it is 0 when $k \neq n$ since the sine is zero at multiples of π. Thus, the values at $t = kT_s$ are recovered exactly, and the rest are interpolated by a sum of sinc signals.

7.2.5 Sampling Simulation with MATLAB

The simulation of sampling with MATLAB is complicated by the representation of analog signals and the numerical computation of the analog Fourier transform. Two sampling rates are needed: one being the sampling rate under study, f_s, and the other being the one used to simulate the analog signal, $f_{sim} \gg f_s$. The computation of the analog Fourier transform of $x(t)$ can be done approximately using the fast Fourier transform (FFT) multiplied by the sampling period. For now, think of the FFT as an algorithm to compute the Fourier transform of a discretized signal.

To illustrate the sampling procedure consider sampling a sinusoid $x(t) = \cos(2\pi f_0 t)$ where $f_0 = 1$ KHz. To simulate this as an analog signal we choose a sampling period $T_{sim} = 0.5 \times 10^{-4}$ sec/sample or a sampling frequency $f_{sim} = 20,000$ samples/sec.

No aliasing sampling—If we sample $x(t)$ with a sampling frequency $f_s = 6000 > 2 f_0 = 2000$ Hz, the sampled signal $y(t)$ will not display aliasing in its frequency representation, as we are satisfying the Nyquist sampling rate condition. Figure 7.7(a) displays the signal $x(t)$ and its sampled version $y(t)$, as well as their approximate Fourier transforms. The magnitude spectrum $|X(\Omega)|$ corresponds to the sinusoid $x(t)$, while $|Y(\Omega)|$ is the first period of the spectrum of the sampled signal (recall the spectrum of the sampled signal is periodic of period $\Omega_s = 2\pi f_s$). In this case, when no aliasing occurs, the first period of the spectrum of $y(t)$ coincides with the spectrum of $x(t)$ (notice that as a sinusoid, the magnitude spectrum $|X(\Omega)|$ is zero except at the frequency of the sinusoid or ± 1 KHz; likewise $|Y(\Omega)|$ is zero except at ± 1 KHz and the range of frequencies is $[-f_s/2, f_s/2] = [-3, 3]$ KHz). In Figure 7.7(b) we show the sinc interpolation of three samples of $y(t)$; the solid line is the interpolated values or the sum of sincs centered at the three samples. At the bottom of that figure we show the sinc interpolation, for all the samples, obtained using our function sincinterp. The sampling is implemented using our function sampling.

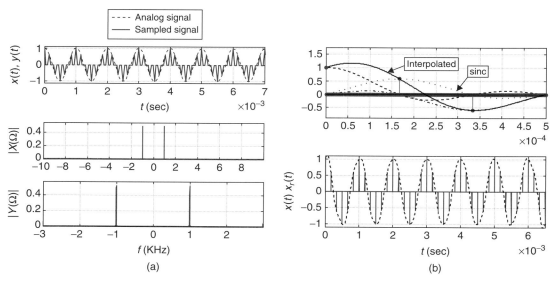

FIGURE 7.7

No aliasing: sampling simulation of $x(t) = \cos(2000\pi t)$ using $f_s = 6000$ samples/sec. (a) Plots are of the signal $x(t)$ and the sampled signal $y(t)$, and their spectra ($|Y(\Omega)|$ is periodic and so a period is shown). (b) The top plot illustrates the sinc interpolation of three samples, and the bottom plot is the sinc-interpolated signal $x_r(t)$ and the sampled signal. In this case $x_r(t)$ is very close to the original signal.

Sampling with aliasing—In Figure 7.8 we show the case when the sampling frequency is $f_s = 800 < 2f_s = 2000$, so that in this case we have aliasing. This can be seen in the sampled signal $y(t)$ in the top plot of Figure 7.8(a), which appears as if we were sampling a sinusoid of lower frequency. It can also be seen in the spectra of $x(t)$ and $y(t)$: $|X(\Omega)|$ is the same as in the previous case, but now $|Y(\Omega)|$, which is a period of the spectrum of the sampled signal $y(t)$, displays a frequency of 200 Hz, lower than that of $x(t)$, within the frequency range $[-400, 400]$ Hz or $[-f_s/2, f_s/2]$. Aliasing has occurred. Finally, the sinc interpolation gives a sinusoid of frequency 0.2 KHz, different from $x(t)$.

Similar situations occur when a more complex signal is sampled. If the signal to be sampled is $x(t) = 2 - \cos(\pi f_0 t) - \sin(2\pi f_0 t)$ where $f_0 = 500$ Hz, if we use a sampling frequency of $f_s = 6000 > 2 f_{max} = 2 f_0 = 1000$ Hz, there will be no aliasing. On the other hand, if the sampling frequency is $f_s = 800 < 2f_{max} = 2f_0 = 1000$ Hz, frequency aliasing will occur. In the no aliasing sampling, the spectrum $|Y(\Omega)|$ (in a frequency range $[-3000, 3000] = [-f_s/2, f_s/2]$) corresponding to a period of the Fourier transform of the sampled signal $y(t)$ shows the same frequencies as $|X(\Omega)|$. The reconstructed signal equals the original signal. See Figure 7.9(a). When we use $f_s = 800$ Hz, the given signal $x(t)$ is undersampled and aliasing occurs. The spectrum $|Y(\Omega)|$ corresponding to a period of the Fourier transform of the undersampled signal $y(t)$ does not show the same frequencies as $|X(\Omega)|$. The reconstructed signal shown in the bottom right plot of Figure 7.9(b) does not resemble the original signal.

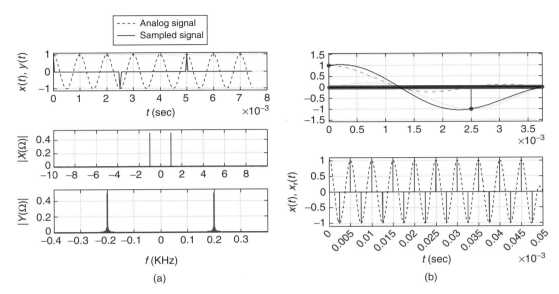

FIGURE 7.8

Aliasing: sampling simulation of $x(t) = \cos(2000\pi t)$ using $f_s = 800$ samples/sec. (a) Plots display the original signal $x(t)$ and the sampled signal $y(t)$ (it looks like a lower-frequency signal being sampled). The sprectra of $x(t)$ and $y(t)$ are shown below ($|Y(\Omega)|$ is periodic and displays a lower frequency than $|X(\Omega)|$). (b) Sinc interpolation for three samples and the whole signal. The reconstructed signal $x_r(t)$ is a sinusoid of period 0.5×10^{-2} or 200-Hz frequency due to aliasing.

The following function implements the sampling and computes the Fourier transform of the analog signal and of the sampled signal using the fast Fourier transform. It gives the range of frequencies for each of the spectra.

```
function [y,y1,X,fx,Y,fy] = sampling(x,L,fs)
%
% Sampling
%   x  analog signal
%   L  length of simulated x
%   fs sampling rate
%   y  sampled signal
%   X,Y magnitude spectra of x,y
%   fx,fy frequency ranges for X,Y
%
fsim = 20000; % analog signal sampling frequency
% sampling with rate fsim/fs
delta = fsim/fs;
y1 = zeros(1,L);
```

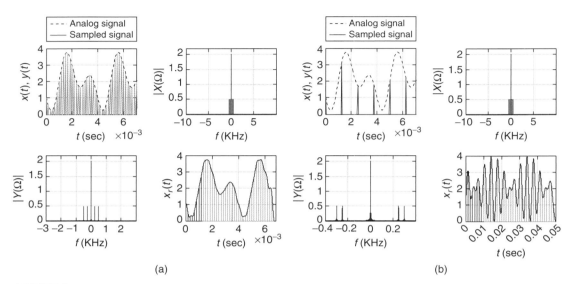

FIGURE 7.9

Sampling of $x(t) = 2 - \cos(500\pi t) - \sin(1000\pi t)$ with (a) no aliasing ($f_s = 6000$ samples/sec) and (b) with aliasing ($f_s = 800$ samples/sec).

```
y1(1:delta:L) = x(1:delta:L);
y = x(1:delta:L);
% analog FT and DTFT of signals
dtx = 1/fsim;
X = fftshift(abs(fft(x))) * dtx;
N = length(X); k = 0:(N − 1); fx = 1/N.*k; fx = fx * fsim/1000 − fsim/2000;
dty = 1/fs;
Y = fftshift(abs(fft(y))) * dty;
N = length(Y); k = 0:(N − 1); fy = 1/N.*k; fy = fy * fs/1000 − fs/2000;
```

The following function computes the sinc interpolation of the samples.

```
function [t,xx,xr] = sincinterp(x,Ts)
%
% Sinc interpolation
% x sampled signal
% Ts sampling period of x
% xx,xr original samples and reconstructed in range t
%
N = length(x)
t = 0:dT:N;
xr = zeros(1,N * 100 + 1);
for k = 1:N,
```

```
   xr = xr +  x(k) * sinc(t − (k − 1));
end
xx(1:100:N * 100) = x(1:N);
xx = [xx zeros(1,99)];
NN = length(xx)
t = 0:NN − 1;t = t * Ts/100;
```

7.3 THE NYQUIST-SHANNON SAMPLING THEOREM

If a low-pass continuous-time signal $x(t)$ is band limited (i.e., it has a spectrum $X(\Omega)$ such that $X(\Omega) = 0$ for $|\Omega| > \Omega_{max}$, where Ω_{max} is the maximum frequency in $x(t)$), we then have:

■ $x(t)$ is uniquely determined by its samples $x(nT_s) = x(t)|_{t=nT_s}$, $n = 0, \pm 1, \pm 2, \cdots$, provided that the sampling frequency Ω_s (rad/sec) is such that

$$\Omega_s \geq 2\Omega_{max} \quad \text{Nyquist sampling rate condition} \tag{7.17}$$

or equivalently if the sampling rate f_s (samples/sec) or the sampling period T_s (sec/sample) are given by

$$f_s = \frac{1}{T_s} \geq \frac{\Omega_{max}}{\pi} \tag{7.18}$$

■   When the Nyquist sampling rate condition is satisfied, the original signal $x(t)$ can be reconstructed by passing the sampled signal $x_s(t)$ through an ideal low-pass filter with the following frequency response:

$$H(\Omega) = \begin{cases} T_s & \frac{-\Omega_s}{2} < \Omega < \frac{\Omega_s}{2} \\ 0 & \text{elsewhere} \end{cases}$$

The reconstructed signal is given by the following sinc interpolation from the samples:

$$x_r(t) = \sum_n x(nT_s) \frac{\sin(\pi(t - nT_s)/T_s)}{\pi(t - nT_s)/T_s} \tag{7.19}$$

Remarks

■   *The value $2\Omega_{max}$ is called the* Nyquist sampling rate. *The value $\Omega_s/2$ is called the* folding rate.

■   *The units of the sampling frequency f_s are samples/sec and as such the units of T_s are sec/sample. Considering the number of samples available, every second or the time at which each sample is available we can get a better understanding of the data storage requirements, the speed limitations imposed by real-time processing, and the need for data compression algorithms. For instance, music being sampled at 44,000 samples/sec, with each sample represented by 8 bits/sample, for every second of music we would need to store $44 \times 8 = 352$ Kbits/sec, and in an hour of sampling we would have $3600 \times 44 \times 8$ Kbits. If you want better quality, let's say 16 bits/sample, then double that quantity, and if you want more fidelity increase the sampling rate but be ready to provide more storage or to come up with some data compression algorithm. Likewise, if you were to process the signal you would have a new sample every $T_s = 0.0227$ msec, so that any real-time processing would have to be done very fast.*

Origins of the Sampling Theory — Part 2

As mentioned in Chapter 0, the theoretical foundations of digital communications theory were given in the paper "A Mathematical Theory of Communication" by Claude E. Shannon in 1948 [51]. His results on sampling theory made possible the new areas of digital communications and digital signal processing.

Shannon was born in 1916 in Petoskey, Michigan. He studied electrical engineering and mathematics at the University of Michigan, pursued graduate studies in electrical engineering and mathematics at MIT, and then joined Bell Telephone Laboratories. In 1956, he returned to MIT to teach.

Besides being a celebrated researcher, Shannon was an avid chess player. He developed a juggling machine, rocket-powered frisbees, motorized Pogo sticks, a mind-reading machine, a mechanical mouse that could navigate a maze, and a device that could solve the Rubik's Cube™ puzzle. At Bell Labs, he was remembered for riding the halls on a unicycle while juggling three balls [23, 52].

7.3.1 Sampling of Modulated Signals

The given Nyquist sampling rate condition applies to low-pass or baseband signals. Sampling of band-pass signals is used for simulation of communication systems and in the implementation of modulation systems in software radio. For modulated signals it can be shown that the sampling rate depends on the bandwidth of the message or modulating signal, not on the absolute frequencies involved. This result provides a significant savings in the sampling, as it is independent of the carrier. A voice message transmitted via a satellite communication system with a carrier of 6 GHz, for instance, would only need to be sampled at about a 10-KHz rate, rather than at 12 GHz as determined by the Nyquist sampling rate condition when we consider the frequencies involved.

Consider a modulated signal $x(t) = m(t) \cos(\Omega_c t)$ where $m(t)$ is the message and $\cos(\Omega_c t)$ is the carrier with carrier frequency

$$\Omega_c >> \Omega_{max}$$

where Ω_{max} is the maximum frequency present in the message. The sampling of $x(t)$ with a sampling period T_s generates in the frequency domain a superposition of the spectrum of $x(t)$ shifted in frequency by Ω_s and multiplied by $1/T_s$. Intuitively, to avoid aliasing the shifting in frequency should be such that there is no overlapping of the shifted spectra, which would require that

$$\Omega_c + \Omega_{max} - \Omega_s < \Omega_c - \Omega_{max} \quad \Rightarrow \Omega_s > 2\Omega_{max} \quad \text{or} \quad T_s < \frac{\pi}{\Omega_{max}}$$

Thus, the sampling period depends on the bandwidth Ω_{max} of the message $m(t)$ rather than on the maximum frequency present in the modulated signal $x(t)$. A formal proof of this result requires the quadrature representation of band-pass signals typically considered in communication theory [16].

If the message $m(t)$ of a modulated signal $x(t) = m(t) \cos(\Omega_c)$ has a bandwidth B Hz, $x(t)$ can be reconstructed from samples taken at a sampling rate

$$f_s \geq 2B$$

independent of the frequency Ω_c of the carrier $\cos(\Omega_c t)$.

■ Example 7.4

Consider the development of an AM transmitter that uses a computer to generate the modulated signal and is capable of transmitting music and speech signals. Indicate how to implement the transmitter.

Solution

Let the message be $m(t) = x(t) + y(t)$ where $x(t)$ is a speech signal and $y(t)$ is a music signal. Since music signals display larger frequencies than speech signals, the maximum frequency of $m(t)$ is that of the music signals, or $f_{max} = 22$ KHz. To transmit $m(t)$ using AM, we modulate it with a sinusoid of frequency $f_c > f_{max}$, say $f_c = 3f_{max} = 66$ KHz.

To satisfy the Nyquist sampling rate condition, the maximum frequency of the modulated signal would be $f_c + f_{max} = (66 + 22)$ KHz $= 88$ KHz, and so we would choose $T_s = 10^{-3}/176$ sec/sample as the sampling period. However, according to the above results we can also choose $T_s = 1/(2B)$ where B is the bandwidth of $m(t)$ in hertz or $B = f_{max} = 22$ KHz, which gives $T_s = 10^{-3}/44$ — four times larger than the previous sampling period, so we choose this as the sampling period.

The analog signal $m(t)$ to be transmitted is inputted into an ADC in the computer, capable of sampling at $44,000$ samples/sec. The output of the converter is then multiplied by a computer-generated sinusoid

$$\cos(2\pi f_c n T_s) = \cos(2\pi \times 66 \times 10^3 \times (10^{-3}/44)n) = \cos(3\pi n) = (-1)^n$$

to obtain the AM signal. The AM digital signal can then be inputted into a DAC and its output sent to an antenna for broadcasting. ■

7.4 PRACTICAL ASPECTS OF SAMPLING

To process analog signals with computers it is necessary to convert analog into digital signals and digital into analog signals. The analog-to-digital and digital-to-analog conversions are done by ADCs and DACs. In practice, these converters differ from the ideal versions we have discussed so far where the sampling is done with impulses, the discrete-time samples are assumed representable with infinite precision, and the reconstruction is performed by an ideal low-pass filter. Pulses rather than impulses are needed, and the discrete-time signals need to be discretized also in amplitude and the reconstruction filter needs to be reconsidered.

7.4.1 Sample-and-Hold Sampling

In an actual ADC the time required to do the sampling, quantization, and coding needs to be considered. Therefore, the width Δ of the sampling pulses cannot be zero as assumed. A *sample-and-hold sampling system* takes the sample and holds it long enough for quantization and coding to be done before the next sample is acquired. The question is then how does this affect the sampling process and how does it differ from the ideal results obtained before? We hinted at the effects when we considered the PAM before, except that now the resulting pulses are flat.

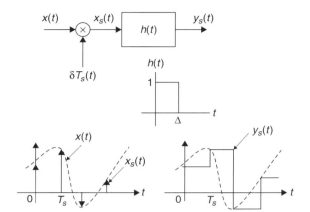

FIGURE 7.10
Sampling using a sample-and-hold system ($\delta = T_s$).

The system shown in Figure 7.10 generates the desired signal. Basically, we are modulating the ideal sampling signal $\delta_{T_s}(t)$ with the analog input $x(t)$, giving an ideally sampled signal $x_s(t)$. This signal is then passed through a *zero-order hold filter*, an LTI system having as impulse response $h(t)$ a pulse of the desired width $\Delta \leq T_s$. The output of the sample-and-hold system is a weighted sequence of shifted versions of the impulse response. In fact, the output of the ideal sampler is $x_s(t) = x(t)\delta_{T_s}(t)$, and using the linearity and time invariance of the zero-order hold system its output is

$$y_s(t) = (x_s * h)(t) \tag{7.20}$$

with a Fourier transform of

$$Y_s(\Omega) = X_s(\Omega)H(j\Omega)$$

$$= \left[\frac{1}{T_s}\sum_k X(\Omega - k\Omega_s)\right] H(j\Omega) \tag{7.21}$$

where the term in the brackets is the spectrum of the ideally sampled signal and

$$H(j\Omega) = \frac{e^{-\Delta s/2}}{s}(e^{\Delta s/2} - e^{-\Delta s/2})|_{s=j\Omega}$$

$$= \frac{\sin(\Delta\Omega/2)}{\Omega/2}e^{-j\Omega\Delta/2} \tag{7.22}$$

is the frequency response of the LTI system.

Remarks

- *Equation (7.20) can be written as*

$$y_s(t) = \sum_n x(nT_s)h(t - nT_s)$$

That is, $y_s(t)$ is a train of pulses $h(t) = u(t) - u(t - \Delta)$ shifted and weighted by the sample values $x(nT_s)$, a more realistic representation of the sampled signal.

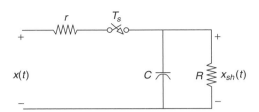

FIGURE 7.11
Sample-and-hold circuit.

- *Two significant changes due to considering the pulses of width $\Delta > 0$ in the sampling are:*
  - *The spectrum of the ideal sampled signal $x_s(t)$ is now weighted by the sinc function of the frequency response $H(j\Omega)$ of the zero-order hold filter. Thus, the spectrum of the sampled signal using the sample-and-hold system will not be periodic and will decay as Ω increases.*
  - *The reconstruction of the original signal $x(t)$ requires a more complex filter than the one used in the ideal sampling. Indeed, the concatenation of the zero-order hold filter with the reconstruction filter should be such that $H(s)H_r(s) = 1$, or that $H_r(s) = 1/H(s)$.*
- *A circuit used for implementing the sample-and-hold system is shown in Figure 7.11. In this circuit the switch closes every T_s seconds and remains closed for a short time Δ. If the time constant $rC << \Delta$, the capacitor charges very fast to the value of the sample attained when the switch closes at some nT_s, and by setting the time constant $RC >> T_s$ when the switch opens Δ seconds later, the capacitor slowly discharges. The cycle repeats providing a signal that approximates the output of the sample-and-hold system explained before.*
- *The DAC also uses a holder to generate an analog signal from the discrete signal coming out of the decoder into the DAC. There are different possible types of holders, providing an interpolation that will make the final smoothing of the signal a lot easier. The so-called zero-order hold basically expands the sample value in between samples, providing a rough approximation of the discrete signal, which is then smoothed out by a low-pass filter to provide the analog signal.*

7.4.2 Quantization and Coding

Amplitude discretization of the sampled signal $x_s(t)$ is accomplished by a quantizer consisting of a number of fixed amplitude levels against which the sample amplitudes $\{x(nT_s)\}$ are compared. The output of the quantizer is one of the fixed amplitude levels that best represents $x(nT_s)$ according to some approximation scheme. The quantizer is a nonlinear system.

Independent of how many levels, or equivalently of how many bits are allocated to represent each level of the quantizer, there is a possible error in the representation of each sample. This is called the *quantization error*. To illustrate this, consider a 2-bit or 2^2-level quantizer shown in Figure 7.12. The input of the quantizer are the samples $x(nT_s)$, which are compared with the values in the bins $[-2\Delta, -\Delta]$, $[-\Delta, 0]$, $[0, \Delta]$, and $[\Delta, 2\Delta]$, and depending on which of these bins the sample falls in it is replaced by the corresponding levels -2Δ, $-\Delta$, 0, or Δ. The value of the quantization step Δ for the four-level quantizer is

$$\Delta = \frac{2 \max|x(t)|}{2^2} \qquad (7.23)$$

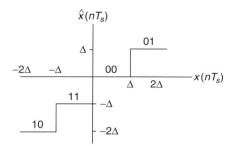

FIGURE 7.12

Four-level quantizer and coder.

That is, Δ is assigned so as to cover the possible peak-to-peak range of values of the signal, or its dynamic range. To each of the levels a binary code is assigned. The code assigned to each of the levels uniquely represents the different levels $[-2\Delta, -\Delta, 0, \Delta]$. As to the way to approximate the given sample to one of these levels, it can be done by *rounding* or by *truncating*. The quantizer shown in Figure 7.12 approximates by truncation—that is, if the sample $k\Delta \leq x(nT_s) < (k+1)\Delta$, for $k = -2, -1, 0, 1$, then it is approximated by the level $k\Delta$.

To see the quantization, coding, and quantization error, let the sampled signal be

$$x(nT_s) = x(t)|_{t=nT_S}$$

The given four-level quantizer is such that

$$k\Delta \leq x(nT_s) < (k+1)\Delta \quad \Rightarrow \quad \hat{x}(nT_s) = k\Delta \quad k = -2, -1, 0, 1 \tag{7.24}$$

where the sampled signal $x(nT_s)$ is the input and the quantized signal $\hat{x}(nT_s)$ is the output. Therefore,

$$-2\Delta \leq x(nT_s) < -\Delta \quad \Rightarrow \quad \hat{x}(nT_s) = -2\Delta$$
$$-\Delta \leq x(nT_s) < 0 \quad \Rightarrow \quad \hat{x}(nT_s) = -\Delta$$
$$0 \leq x(nT_s) < \Delta \quad \Rightarrow \quad \hat{x}(nT_s) = 0$$
$$\Delta \leq x(nT_s) < 2\Delta \quad \Rightarrow \quad \hat{x}(nT_s) = \Delta$$

To transform the quantized values into unique binary 2-bit values, one could use a code such as

$$\begin{aligned}
\hat{x}(nT_s) &= -2\Delta &\Rightarrow& \quad 10 \\
\hat{x}(nT_s) &= -\Delta &\Rightarrow& \quad 11 \\
\hat{x}(nT_s) &= 0\Delta &\Rightarrow& \quad 00 \\
\hat{x}(nT_s) &= \Delta &\Rightarrow& \quad 01
\end{aligned}$$

which assigns a unique 2-bit binary number to each of the four quantization levels.

If we define the quantization error as

$$\varepsilon(nT_s) = x(nT_s) - \hat{x}(nT_s)$$

and use the characterization of the quantizer given in Equation (7.24), we have then that the error $\varepsilon(nT_s)$ is obtained from

$$\hat{x}(nT_s) \leq x(nT_s) \leq \hat{x}(nT_s) + \Delta \text{ by subtracting } \hat{x}(nT_s) \Rightarrow 0 \leq \varepsilon(nT_s) \leq \Delta \qquad (7.25)$$

indicating that one way to decrease the quantization error is to make the quantization step Δ very small. That clearly depends on the quality of the ADC. Increasing the number of bits of the ADC makes Δ smaller (see Equation (7.23) where the denominator is 2 raised to the number of bits), which will make the quantization error smaller.

In practice, the quantization error is considered random, and so it needs to be characterized probabilistically. This characterization becomes meaningful only when the number of bits is large, and the input signal is not a deterministic signal. Otherwise, the error is predictable and thus not random. Comparing the energy of the input signal to the energy of the error, by means of the so-called signal-to-noise ratio (SNR), it is possible to determine the number of bits that are needed in a quantizer to get a reasonable quantization error.

■ Example 7.5

Suppose we are trying to decide between an 8- and a 9-bit ADC for a certain application. The signals in this application are known to have frequencies that do not exceed 5 KHz. The amplitude of the signals is never more than 5 volts (i.e., the dynamic range of the signals is 10 volts, so that the signal is bounded as $-5 \leq x(t) \leq 5$). Determine an appropriate sampling period and compare the percentage of error for the two ADCs of interest.

Solution

The first consideration in choosing the ADC is the sampling period, so we need to get an ADC capable of sampling at $f_s = 1/T_s > 2f_{max}$ samples/sec. Choosing $f_s = 4f_{max} = 20$ K samples/sec, then $T_s = 1/20$ msec/sample. Suppose then we look at an 8-bit ADC, which means that the quantizer would have $2^8 = 256$ levels so that the quantization step is $\Delta = 10/256$ volts. If we use the truncation quantizer given above the quantization error would be

$$0 \leq \varepsilon(nTs) \leq 10/256$$

If we find that objectionable we can then consider a 9-bit ADC, with a quantizer of $2^9 = 512$ levels and the quantization step is $\Delta = 10/512$ or half that of the 8-bit ADC

$$0 \leq \varepsilon(nT_s) \leq 10/512$$

So that by increasing 1 bit we cut the quantization error in half from the previous quantizer (in practice, one of the 8 or 9 bits is used to determine the sign of the sampled value). Inputting a signal of constant amplitude 5 into the 9-bit ADC gives a quantization error of $[(10/512)/5] \times 100\% = (100/256)\% \approx 0.4\%$ in representing the input signal. For the 8-bit ADC it would correspond to a 0.8% error. ■

7.4.3 Sampling, Quantizing, and Coding with MATLAB

The conversion of an analog signal into a digital signal consists of three steps: sampling, quantization, and coding. These are the three operations an ADC does. To illustrate them consider a sinusoid $x(t) = 4\cos(2\pi t)$. Its sampling period, according to the Nyquist sampling rate condition, is

$$T_s \leq \pi / \Omega_{\max} = 0.5 \text{ sec/sample}$$

as the maximum frequency of $x(t)$ is $\Omega_{\max} = 2\pi$. We let $T_s = 0.01$ (sec/sample) to obtain a sampled signal $x_s(nT_s) = 4\cos(2\pi nT_s) = 4\cos(2\pi n/100)$, a discrete sinusoid of period 100. The following script is used to get the sampled $x[n]$ and the quantized $x_q[n]$ signals and the quantization error $\varepsilon[n]$ (see Figure 7.13).

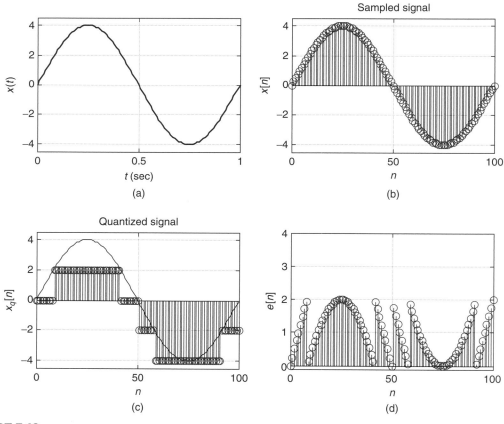

FIGURE 7.13

(a) Sinusoid, (b) sampled sinusoid using $T_s = 0.01$, (c) quantized sinusoid using four levels, and (d) quantization error.

```
%%%%%%%%%%%%%%%%%%%%%%%%%%%%%%%
% Sampling, quantization and coding
%%%%%%%%%%%%%%%%%%%%%%%%%%%%%%%
clear all; clf
% analog signal
t = 0:0.01:1; x = 4 * sin(2 * pi * t);
% sampled signal
Ts = 0.01; N = length(t); n = 0:N − 1;
xs = 4 * sin(2 * pi * n * Ts);
% quantized signal
Q = 2;      % quantization levels is 2Q
[d,y,e] = quantizer(x,Q);
% binary signal
z = coder(y,d)
```

The quantization of the sampled signal is implemented with the function quantizer which compares each of the samples $x_s(nT_s)$ with four levels and assigns to each the corresponding level. Notice the appproximation of the values given by the quantized signal samples to the actual values of the signal. The difference between the original and the quantized signal, or the quantization error, $\varepsilon(nT_s)$, is also computed and shown in Figure 7.13.

```
function [d,y,e] = quantizer(x,Q)
%  Input: x, signal to be quantized at 2Q levels
%  Outputs: y quantized signal
%          e, quantization error
%          d quantum
%  USE [d,y,e] = quantizer(x,Q)
%
N = length(x);
d = max(abs(x))/Q;
for k = 1:N,
  if x(k)> = 0,
    y(k) = floor(x(k)/d)*d;
  else
    if x(k) == min(x),
      y(k) = (x(k)/abs(x(k))) * (floor(abs(x(k))/d) * d);
    else
    y(k) = (x(k)/abs(x(k))) * (floor(abs(x(k))/d) * d + d);
    end
  end
  if y(k) == 2 * d,
    y(k) = d;
  end
end
```

The binary signal corresponding to the quantized signal is computed using the function coder which assigns the binary codes '10','11','00', and '01' to the four possible levels of the quantizer. The result is a sequence of 0s and 1s, each pair of digits sequentially corresponding to each of the samples of the quantized signal. The following is the function used to effect this coding.

```
function z1 = coder(y,delta)
%  Coder for 4-level quantizer
%  input: y quantized signal
%  output: z1 binary sequence
%  USE z1 = coder(y)
%
z1 = '00';  % starting code
N = length(y);
for n = 1:N,
 y(n)
   if y(n) == delta
      z = '01';
   elseif y(n) == 0
      z = '00';
   elseif y(n) == -delta
      z = '11';
   else
      z = '10';
   end
   z1 = [z1 z];
end
M = length(z1);
z1 = z1(3:M) % get rid of starting code
```

7.5 WHAT HAVE WE ACCOMPLISHED? WHERE DO WE GO FROM HERE?

The material in this chapter is the bridge between analog and digital signal processing. The sampling theory provides the necessary information to convert a continuous-time signal into a discrete-time signal and then into a digital signal with minimum error. It is the frequency representation of an analog signal that determines the way in which it can be sampled and reconstructed. Analog-to-digital and digital-to-analog converters are the devices that in practice convert an analog signal into a digital signal and back. Two parameters characterizing these devices are the sampling rate and the number of bits each sample is coded into. The rate of change of a signal determines the sampling rate, while the precision in representing the samples determines the number of levels of the quantizer and the number of bits assigned to each sample.

In the following chapters we will consider the analysis of discrete-time signals, as well as the analysis and synthesis of discrete systems. The effect of quantization in the processing and design of systems is an important problem that is left for texts in digital signal processing. We will, however, develop the theory of discrete-time signals.

PROBLEMS

7.1. Sampling actual signals

Consider the sampling of real signals.

(a) Typically, a speech signal that can be understood over a telephone shows frequencies from about 100 Hz to about 5 KHz. What would be the sampling frequency f_s (samples/sec) that would be used to sample speech without aliasing? How many samples would you need to save when storing an hour of speech? If each sample is represented by 8 bits, how many bits would you have to save for the hour of speech?

(b) A music signal typically displays frequencies from 0 up to 22 KHz. What would be the sampling frequency f_s that would be used in a CD player?

(c) If you have a signal that combines voice and musical instruments, what sampling frequency would you use to sample this signal? How would the signal sound if played at a frequency lower than the Nyquist sampling frequency?

7.2. Sampling of band-limited signals

Consider the sampling of a sinc signal and related signals.

(a) For the signal $x(t) = \sin(t)/t$, find its magnitude spectrum $|X(\Omega)|$ and determine if this signal is band limited or not.

(b) Suppose you want to sample $x(t)$. What would be the sampling period T_s you would use for the sampling without aliasing?

(c) For a signal $y(t) = x^2(t)$, what sampling frequency f_s would you use to sample it without aliasing? How does this frequency relate to the sampling frequency used to sample $x(t)$?

(d) Find the sampling period T_s to sample $x(t)$ so that the sampled signal $x_s(0) = 1$, otherwise $x_s(nT_s) = 0$ for $n \neq 0$.

7.3. Sampling of time-limited signals—MATLAB

Consider the signals $x(t) = u(t) - u(t-1)$ and $y(t) = r(t) - 2r(t-1) + r(t-2)$.

(a) Are either of these signals band limited? Explain.

(b) Use Parseval's theorem to determine a reasonable value for a maximum frequency for these signals (choose a frequency that would give 90% of the energy of the signals). Use MATLAB.

(c) If we use the sampling period corresponding to $y(t)$ to sample $x(t)$, would aliasing occur? Explain.

(d) Determine a sampling period that can be used to sample both $x(t)$ and $y(t)$ without causing aliasing in either signal.

7.4. Uncertainty in time and frequency—MATLAB

Signals of finite time support have infinite support in the frequency domain, and a band-limited signal has infinite time support. A signal cannot have finite support in both domains.

(a) Consider $x(t) = (u(t+0.5) - u(t-0.5))(1 + \cos(2\pi t))$. Find its Fourier transform $X(\Omega)$. Compute the energy of the signal, and determine the maximum frequency of a band-limited approximation signal $\hat{x}(t)$ that would give 95% of the energy of the original signal.

(b) The fact that a signal cannot be of finite support in both domains is expressed well by the *uncertainty principle*, which says that

$$\Delta(t)\Delta(\Omega) \geq \frac{1}{4\pi}$$

where

$$\Delta(t) = \left[\frac{\int_{-\infty}^{\infty} t^2 \, |x(t)|^2 \, dt}{E_x} \right]^{0.5}$$

measures the duration of the signal for which the signal is significant in time, and

$$\Delta(\Omega) = \left[\frac{\int_{-\infty}^{\infty} \Omega^2 \, |X(\Omega)|^2 \, d\Omega}{E_x} \right]^{0.5}$$

measures the frequency support for which the Fourier representation is significant. The energy of the signal is represented by E_x. Compute $\Delta(t)$ and $\Delta(\Omega)$ for the given signal $x(t)$ and verify that the uncertainty principle is satisfied.

7.5. Nyquist sampling rate condition and aliasing

Consider the signal

$$x(t) = \frac{\sin(0.5t)}{0.5t}$$

(a) Find the Fourier transform $X(\Omega)$ of $x(t)$.
(b) Is $x(t)$ band limited? If so, find its maximum frequency Ω_{max}.
(c) Suppose that $T_s = 2\pi$. How does Ω_s relate to the Nyquist frequency $2\Omega_{max}$? Explain.
(d) What is the sampled signal $x(nT_s)$ equal to? Carefully plot it and explain if $x(t)$ can be reconstructed.

7.6. Anti-aliasing

Suppose you want to find a reasonable sampling period T_s for the noncausal exponential

$$x(t) = e^{-|t|}$$

(a) Find the Fourier transform of $x(t)$, and plot $|X(\Omega)|$. Is $x(t)$ band limited?
(b) Find a frequency Ω_0 so that 99% of the energy of the signal is in $-\Omega_0 \le \Omega \le \Omega_0$.
(c) If we let $\Omega_s = 2\pi/T_s = 5\Omega_0$, what would be T_s?
(d) Determine the magnitude and bandwidth of an anti-aliasing filter that would change the original signal into the band-limited signal with 99% of the signal energy.

7.7. Sampling of modulated signals

Assume you wish to sample an amplitude modulated signal

$$x(t) = m(t) \cos(\Omega_c t)$$

where $m(t)$ is the message signal and $\Omega_c = 2\pi \, 10^4$ rad/sec is the carrier frequency.
(a) If the message is an acoustic signal with frequencies in a band of $[0, 22]$ KHz, what would be the maximum frequency present in $x(t)$?
(b) Determine the range of possible values of the sampling period T_s that would allow us to sample $x(t)$ satisfying the Nyquist sampling rate condition.
(c) Given that $x(t)$ is a band-pass signal, compare the above sampling period with the one that can be used to sample band-pass signals.

7.8. Sampling output of nonlinear system

The input–output relation of a nonlinear system is

$$y(t) = x^2(t)$$

where $x(t)$ is the input and $y(t)$ is the output.

(a) The signal $x(t)$ is band limited with a maximum frequency $\Omega_M = 2000\pi$ rad/sec. Determine if $y(t)$ is also band limited, and if so, what is its maximum frequency Ω_{max}?

(b) Suppose that the signal $y(t)$ is low-pass filtered. The magnitude of the low-pass filter is unity and the cut-off frequency is $\Omega_c = 5000\pi$ rad/sec. Determine the value of the sampling period T_s according to the given information.

(c) Is there a different value for T_s that would satisfy the Nyquist sampling rate condition for both $x(t)$ and $y(t)$ and that is larger than the one obtained above? Explain.

7.9. Signal reconstruction

You wish to recover the original analog signal $x(t)$ from its sampled form $x(nT_s)$.

(a) If the sampling period is chosen to be $T_s = 1$ so that the Nyquist sampling rate condition is satisfied, determine the magnitude and cut-off frequency of an ideal low-pass filter $H(j\Omega)$ to recover the original signal and plot them.

(b) What would be a possible maximum frequency of the signal? Consider an ideal and a nonideal low-pass filter to reconstruct $x(t)$. Explain.

7.10. CD player versus record player

Explain why a CD player cannot produce the same fidelity of music signals as a conventional record player. (If you do not know what these are, ignore this problem, or get one to find out what they do or ask your grandparents about LPs and record players!)

7.11. Two-bit analog-to-digital converter—MATLAB

Let $x(t) = 0.8 \; \cos(2\pi t) + 0.15,\ 0 \le t \le 1$, and zero otherwise, be the input to a 2-bit analog-to-digital converter.

(a) For a sampling period $T_s = 0.025$ sec determine and plot using MATLAB the sampled signal,

$$x(nT_s) = x(t)|_{t=nT_S}$$

(b) The four-level quantizer (see Figure 1.2) corresponding to the 2-bit ADC is defined as

$$k\Delta \le x(nT_s) < (k+1)\Delta \quad \rightarrow \quad \hat{x}(nT_s) = k\Delta \qquad k = -2, -1, 0, 1 \qquad (7.26)$$

where $x(nT_s)$, found above, is the input and $\hat{x}(nT_s)$ is the output of the quantizer. Let the quantization step be $\Delta = 0.5$. Plot the input–output characterization of the quantizer, and find the quantized output for each of the sample values of the sampled signal $x(nT_s)$.

(c) To transform the quantized values into unique binary 2-bit values, consider the following code:

$$\hat{x}(nT_s) = -2\Delta \quad \rightarrow \quad 10$$
$$\hat{x}(nT_s) = -\Delta \quad \rightarrow \quad 11$$
$$\hat{x}(nT_s) = 0\Delta \quad \rightarrow \quad 00$$
$$\hat{x}(nT_s) = \Delta \quad \rightarrow \quad 01$$

Obtain the digital signal corresponding to $x(t)$.

CHAPTER 8

Discrete-Time Signals and Systems

It's like déjà-vu,
all over again.
Lawrence "Yogi" Berra (1925)
Yankees baseball player

8.1 INTRODUCTION

As you will see in this chapter, the basic theory of discrete-time signals and systems is very much like that for continuous-time signals and systems. However, there are significant differences that need to be understood. Specifically in this chapter we will consider the following contrasting issues:

■ Discrete-time signals resulting from sampling of continuous-time signals are only available at uniform times determined by the sampling period; they are not defined in-between sampling periods. It is important to emphasize the significance of sampling according to the Nyquist sampling rate condition since the characteristics of discrete-time signals will depend on it. Given the knowledge of the sampling period, discrete-time signals depend on an integer variable n, which unifies the treatment of discrete-time signals obtained from analog signals by sampling and those that are naturally discrete. It will also be seen that the frequency in the discrete domain differs from the analog frequency. The radian discrete frequency cannot be measured, and depends on the sampling period used whenever the discrete-time signals result from sampling.

■ Although the concept of periodicity of discrete-time signals coincides with that for continuous-time signals, there are significant differences. As functions of an integer variable, discrete-time periodic signals must have integer periods. This imposes some restrictions that do not exist in continuous-time periodic signals. For instance, continuous-time sinusoids are always periodic as their period can be a positive real number; however, that will not be the case for discrete-time sinusoids. It is possible to have discrete-time sinusoids that are not periodic, even if they resulted from the uniform sampling of continuous-time sinusoids.

■ Characteristics such as energy, power, and symmetry of continuous-time signals are conceptually the same for discrete-time signals. Integrals are replaced by sums, derivatives by finite differences, and differential equations by difference equations. Likewise, one can define a set of basic signals

Signals and Systems Using MATLAB®. DOI: 10.1016/B978-0-12-374716-7.00012-0

just like those for continuous-time signals. However, some of these basic signals do not display the mathematical complications of their continuous-time counterparts. For instance, the discrete-impulse signal is defined at every integer value in contrast with the continuous-impulse response, which is not defined at zero.

■ The discrete approximation of derivatives and integrals provides an approximation of differential equations, representing dynamic continuous-time systems by difference equations. Extending the concept of linear time invariance to discrete-time systems, we obtain a convolution sum to represent LTI systems. Thus, dynamic discrete-time systems can be represented by difference equations and convolution sums. A computationally significant difference with continuous-time systems is that the solution of difference equations can be recursively obtained, and that the convolution sum provides a class of systems that do not have a counterpart in the analog domain.

8.2 DISCRETE-TIME SIGNALS

A *discrete-time signal x[n]* can be thought of as a real- or complex-valued function of the integer sample index n:

$$x[.] : \mathcal{I} \to \mathcal{R} \ (\mathcal{C})$$

$$n \quad x[n] \qquad\qquad (8.1)$$

The above means that for discrete-time signals the independent variable is an integer n, the sample index, and that the value of the signal at n, $x[n]$, is either a real- or a complex-value function. Thus, the signal is only defined at integer values n—no definition exists for values between the integers.

Remarks

■ *It should be understood that a sampled signal $x(nT_s) = x(t)|_{t=nT_s}$ is a discrete-time signal $x[n]$ that is a function of n only. Once the value of T_s is known, the sampled signal only depends on n, the sample index. However, this should not prevent us in some situations from considering a discrete-time signal obtained through sampling as a function of time t where the signal values only exist at discrete times $\{nT_s\}$.*

■ *Although in many situations discrete-time signals are obtained from continuous-time signals by sampling, that is not always the case. There are many signals that are inherently discrete—think, for instance, of a signal consisting of the final values attained daily by the shares of a company in the stock market. Such a signal would consist of the values reached by the share in the days when the stock market opens. This signal is naturally discrete. A signal generated by a random number generator in a computer would be a sequence of real values and can be considered a discrete-time signal. Telemetry signals, consisting of measurements—for example, voltages, temperatures, pressures—from a certain process, taken at certain times, are also naturally discrete.*

■ Example 8.1

Consider a sinusoidal signal

$$x(t) = 3\cos(2\pi t + \pi/4) \qquad -\infty < t < \infty$$

Determine an appropriate sampling period T_s according to the Nyquist sampling rate condition, and obtain the discrete-time signal $x[n]$ corresponding to the largest allowed sampling period.

Solution

To sample $x(t)$ so that no information is lost, the Nyquist sampling rate condition indicates that the sampling period should be

$$T_s \leq \frac{\pi}{\Omega_{max}} = \frac{\pi}{2\pi} = 0.5$$

For the largest allowed sampling period $T_s = 0.5$, we obtain

$$x[n] = 3\cos(2\pi t + \pi/4)|_{t=0.5n} = 3\cos(\pi n + \pi/4) \qquad -\infty < n < \infty$$

which is a function of the integer n. ∎

■ Example 8.2

To generate the celebrated Fibonacci sequence of numbers, $\{x[n]\}$, we use the recursive equation

$$x[n] = x[n-1] + x[n-2] \qquad n \geq 2$$
$$x[0] = 0$$
$$x[1] = 1$$

which is a difference equation with zero input and two initial conditions. The Fibonacci sequence has been used to model different biological systems.[1] Find the Fibonacci sequence.

Solution

The given equation allows us to compute the Fibonacci sequence recursively. For $n \geq 2$, we find

$$x[2] = 1 + 0 = 1$$
$$x[3] = 1 + 1 = 2$$
$$x[4] = 2 + 1 = 3$$
$$x[5] = 3 + 2 = 5$$
$$\vdots$$

where we are simply adding the previous two numbers in the sequence. The sequence is purely discrete as it is not related to a continuous-time signal. ∎

---

[1] Leonardo of Pisa (also known as Fibonacci) in his book *Liber Abaci* described how his sequence could be used to model the reproduction of rabbits over a number of months assuming bunnies begin breeding when they are a few months old.

8.2.1 Periodic and Aperiodic Signals

A discrete-time signal $x[n]$ is *periodic* if
- It is defined for all possible values of n, $-\infty < n < \infty$.
- There is a positive integer N, the period of $x[n]$, such that

$$x[n + kN] = x[n] \tag{8.2}$$

for any integer k.

Periodic discrete-time sinusoids, of period N, are of the form

$$x[n] = A \cos\left(\frac{2\pi m}{N} n + \theta\right) \qquad -\infty < n < \infty \tag{8.3}$$

where the discrete frequency is $\omega_0 = 2\pi m/N$ rad, for positive integers m and N, which are not divisible by each other, and θ is the phase angle.

The definition of a discrete-time periodic signal is similar to that of continuous-time periodic signals, except for the period being an integer. That discrete-time sinusoids are of the given form can be easily shown: Shifting the sinusoid in Equation (8.3) by a multiple k of the period N, we have

$$x[n + kN] = A \cos\left(\frac{2\pi m}{N}(n + kN) + \theta\right)$$

$$= A \cos\left(\frac{2\pi m}{N} n + 2\pi m k + \theta\right) = x[n]$$

since we add to the original angle a multiple mk (an integer) of 2π, which does not change the angle.

Remarks

- *The units of the discrete frequency ω is radians. Moreover, discrete frequencies repeat every 2π (i.e., $\omega = \omega + 2\pi k$ for any integer k), and as such we only need to consider the range $-\pi \leq \omega < \pi$. This is in contrast with the analog frequency Ω, which has rad/sec as units, and its range is from $-\infty$ to ∞.*
- *If the frequency of a periodic sinusoid is*

$$\omega = \frac{2\pi}{N} m$$

for nondivisible integers m and $N > 0$, the period is N. If the frequency of the sinusoid cannot be written like this, the discrete sinusoid is not periodic.

■ Example 8.3

Consider the sinusoids

$$x_1[n] = 2 \cos(\pi n - \pi/3)$$

$$x_2[n] = 3 \sin(3\pi n + \pi/2) \qquad -\infty < n < \infty$$

From their frequencies determine if these signals are periodic, and if so, determine their corresponding periods.

Solution

The frequency of $x_1[n]$ can be written as

$$\omega_1 = \pi = \frac{2\pi}{2}$$

where $m = 1$ and $N = 2$, so that $x_1[n]$ is periodic of period $N_1 = 2$. Likewise, the frequency of $x_2[n]$ can be written as

$$\omega_2 = 3\pi = \frac{2\pi}{2}3$$

where $m = 3$ and $N = 2$, so that $x_2[n]$ is also periodic of period $N_2 = 2$, which can be verified as follows:

$$x_2[n+2] = 3\sin(3\pi(n+2) + \pi/2) = 3\sin(3\pi n + 6\pi + \pi/2) = x[n]$$

∎

■ Example 8.4

What is true for continuous-time sinusoids—that they are always periodic—is not true for discrete-time sinusoids. These sinusoids can be nonperiodic even if they result from uniformly sampling a continuous-time sinusoid. Consider the discrete signal $x[n] = \cos(n + \pi/4)$, which is obtained by sampling the analog sinusoid $x(t) = \cos(t + \pi/4)$ with a sampling period $T_s = 1$ sec/sample. Is $x[n]$ periodic? If so, indicate its period. Otherwise, determine values of the sampling period, satisfying the Nyquist sampling rate condition, that when used in sampling $x(t)$ result in periodic signals.

Solution

The sampled signal $x[n] = x(t)|_{t=nT_s} = \cos(n + \pi/4)$ has a discrete frequency $\omega = 1$ rad that cannot be expressed as $2\pi m/N$ for any integers m and N because π is an irrational number. So $x[n]$ is not periodic.

Since the frequency of the continuous-time signal $x(t)$ is $\Omega = 1$ (rad/sec), then the sampling period, according to the Nyquist sampling rate condition, should be

$$T_s \le \frac{\pi}{\Omega} = \pi$$

and for the sampled signal $x(t)|_{t=nT_s} = \cos(nT_s + \pi/4)$ to be periodic of period N or

$$\cos((n+N)T_s + \pi/4) = \cos(nT_s + \pi/4) \quad \text{is necessary that} \quad NT_s = 2k\pi$$

for an integer k (i.e., a multiple of 2π). Thus, $T_s = 2k\pi/N \le \pi$ satisfies the Nyquist sampling condition at the same time that it ensures the periodicity of the sampled signal. For instance, if we wish to have a sinusoid with period $N = 10$, then $T_s = 0.2k\pi$ for k chosen so the Nyquist sampling rate condition is satisfied—that is,

$$0 < T_s = k\pi/5 \leq \pi \quad \text{so that} \quad 0 < k \leq 5.$$

From these possible values for k we choose $k = 1$ and 3 so that N and k are not divisible by each other and we get the desired period $N = 10$ (the values $k = 2$ and 4 would give 5 as the period, and $k = 5$ would give a period of 2 instead of 10). Indeed, if we let $k = 1$, then $T_s = 0.2\pi$ satisfies the Nyquist sampling rate condition, and we obtain the sampled signal

$$x[n] = \cos(0.2n\pi + \pi/4) = \cos\left(\frac{2\pi}{10}n + \frac{\pi}{4}\right)$$

which according to its frequency is periodic of period 10. This is the same for $k = 3$.   ■

When sampling an analog sinusoid

$$x(t) = A\cos(\Omega_0 t + \theta) \qquad -\infty < t < \infty \tag{8.4}$$

of period $T_0 = 2\pi/\Omega_0$, $\Omega_0 > 0$, we obtain a *periodic discrete sinusoid*,

$$x[n] = A\cos(\Omega_0 T_s n + \theta) = A\cos\left(\frac{2\pi T_s}{T_0}n + \theta\right) \tag{8.5}$$

provided that

$$\frac{T_s}{T_0} = \frac{m}{N} \tag{8.6}$$

for positive integers N and m, which are not divisible by each other. To avoid frequency aliasing the sampling period should also satisfy

$$T_s \leq \frac{\pi}{\Omega_0} = \frac{T_0}{2} \tag{8.7}$$

Indeed, sampling a continuous-time signal $x(t)$ using as sampling period T_s, we obtain

$$x[n] = A\cos(\Omega_0 T_s n + \theta)$$

$$= A\cos\left(\frac{2\pi T_s}{T_0}n + \theta\right)$$

where the discrete frequency is $\omega_0 = 2\pi T_s/T_0$. For this signal to be periodic we should be able to express this frequency as $2\pi m/N$ for nondivisible positive integers m and N. This requires that

$$\frac{T_0}{T_s} = \frac{N}{m}$$

be a rational number, or that

$$mT_0 = NT_s \tag{8.8}$$

which says that a period ($m = 1$) or several periods ($m > 1$) should be divided into $N > 0$ segments of duration T_s seconds. If the condition in Equation (8.6) is not satisfied, then the discretized sinusoid is not periodic. To avoid frequency aliasing the sampling period should be chosen so that

$$T_s \leq \frac{\pi}{\Omega_0} = \frac{T_0}{2}$$

The sum $z[n] = x[n] + y[n]$ of periodic signals $x[n]$ with period N_1, and $y[n]$ with period N_2 is periodic if the ratio of periods of the summands is rational—that is,

$$\frac{N_2}{N_1} = \frac{p}{q}$$

where p and q are integers not divisible by each other. If so, the period of $z[n]$ is $qN_2 = pN_1$.

If $qN_2 = pN_1$, we then have that

$$z[n + pN_1] = x[n + pN_1] + y[n + pN_1]$$
$$= x[n] + y[n + qN_2]$$
$$= x[n] + y[n] = z[n]$$

since pN_1 and qN_2 are multiples of the periods of $x[n]$ and $y[n]$.

■ Example 8.5

The signal

$$z[n] = v[n] + w[n] + y[n]$$

is the sum of three periodic signals $v[n]$, $w[n]$, and $y[n]$ of periods $N_1 = 2$, $N_2 = 3$, and $N_3 = 4$, respectively. Determine if $z[n]$ is periodic, and if so, determine its period.

Solution

Let $x[n] = v[n] + w[n]$, so that $z[n] = x[n] + y[n]$. The signal $x[n]$ is periodic since $N_2/N_1 = 3/2$ is a rational number and 3 and 2 are non-divisible by each other, and its period is $N_4 = 3N_1 = 2N_2 = 6$. The signal $z[n]$ is also periodic since

$$\frac{N_4}{N_3} = \frac{6}{4} = \frac{3}{2}$$

Its period is $N = 2N_4 = 3N_3 = 12$. Thus, $z[n]$ is periodic of period 12, indeed

$$z[n + 12] = v[n + 6N_1] + w[n + 4N_2] + y[n + 3N_3] = v[n] + w[n] + y[n] = z[n]$$

■

■ Example 8.6

Determine if the signal

$$x[n] = \sum_{m=0}^{\infty} X_m \cos(m\omega_0 n) \qquad \omega_0 = \frac{2\pi}{N_0}$$

is periodic, and if so, determine its period.

Solution

The signal $x[n]$ consists of the sum of a constant X_0 and cosines of frequency

$$m\omega_0 = \frac{2\pi m}{N_0} \qquad m = 1, 2, \ldots$$

The periodicity of $x[n]$ depends on the periodicity of the cosines. According to the frequency of the cosines, they are periodic of period N_0. Thus, $x[n]$ is periodic of period N_0. Indeed

$$x[n + N_0] = \sum_{m=0}^{\infty} X_m \cos(m\omega_0(n + N_0))$$

$$= \sum_{m=0}^{\infty} X_m \cos(m\omega_0 n + 2\pi m) = x[n]$$

■

8.2.2 Finite-Energy and Finite-Power Discrete-Time signals

For discrete-time signals, we obtain definitions for energy and power similar to those for continuous-time signals by replacing integrals by summations.

For a discrete-time signal $x[n]$, we have the following definitions:

$$\text{Energy:} \qquad \varepsilon_x = \sum_{n=-\infty}^{\infty} |x[n]|^2 \tag{8.9}$$

$$\text{Power:} \qquad P_x = \lim_{N \to \infty} \frac{1}{2N + 1} \sum_{n=-N}^{N} |x[n]|^2 \tag{8.10}$$

■ $x[n]$ is said to have *finite energy* or to be *square summable* if $\varepsilon_x < \infty$.
■ $x[n]$ is called *absolutely summable* if

$$\sum_{n=-\infty}^{\infty} |x[n]| < \infty \tag{8.11}$$

■ $x[n]$ is said to have *finite power* if $P_x < \infty$.

■ Example 8.7

A "causal" sinusoid, obtained from a signal generator after it is switched on, is

$$x(t) = \begin{cases} 2\cos(\Omega_0 t - \pi/4) & t \geq 0 \\ 0 & \text{otherwise} \end{cases}$$

The signal $x(t)$ is sampled using a sampling period of $T_s = 0.1$ sec to obtain a discrete-time signal

$$x[n] = x(t)|_{t=0.1n} = 2\cos(0.1\Omega_0 n - \pi/4) \qquad n \geq 0$$

and zero otherwise. Determine if this discrete-time signal has finite energy and finite power and compare these characteristics with those of the continuous-time signal $x(t)$ when $\Omega_0 = \pi$ and when $\Omega_0 = 3.2$ rad/sec (an upper approximation of π).

Solution

The continuous-time signal $x(t)$ has infinite energy, and so does the discrete-time signal $x[n]$, for both values of Ω_0. Indeed, its energy is

$$\varepsilon_x = \sum_{n=-\infty}^{\infty} x[n]^2 = \sum_{n=0}^{\infty} 4\cos^2(0.1\Omega_0 n - \pi/4) \to \infty$$

Although the continuous-time and the discrete-time signals have infinite energy, they have finite power. That the continuous-time signal has finite power can be shown as indicated in Chapter 1. For the discrete-time signal $x[n]$, we have for the two frequencies:

1. For $\Omega_0 = \pi$, $x_1[n] = 2\cos(\pi n/10 - \pi/4) = 2\cos(2\pi n/20 - \pi/4)$ for $n \geq 0$ and zero otherwise. Thus, $x[n]$ repeats every $N_0 = 20$ samples for $n \geq 0$, and its power is

$$P_x = \lim_{N \to \infty} \frac{1}{2N+1} \sum_{n=-N}^{N} |x_1[n]|^2 = \lim_{N \to \infty} \frac{1}{2N+1} \sum_{n=0}^{N} |x_1[n]|^2$$

$$= \lim_{N \to \infty} \frac{1}{2N+1} N \underbrace{\left[\frac{1}{N_0} \sum_{n=0}^{N_0-1} |x_1[n]|^2 \right]}_{\text{power of period, } n \geq 0} = \frac{1}{2N_0} \sum_{n=0}^{N_0-1} |x_1[n]|^2 < \infty$$

where we used the causality of the signal ($x_1[n] = 0$ for $n < 0$), and considered N periods of $x_1[n]$ for $n \geq 0$, and for each computed its power to get the final result. Thus, for $\Omega_0 = \pi$ the discrete-time signal $x_1[n]$ has finite power and can be computed using a period for $n \geq 0$.

To find the power we use the trigonometric identity (or Euler's equation) $\cos^2(\theta) = 0.5(1 + \cos(2\theta))$, and so replacing $x_1[n]$, we have

$$P_x = \frac{4}{2N_0} 0.5 \left[\sum_{n=0}^{N_0-1} 1 + \sum_{n=0}^{N_0-1} \cos(0.2\pi n - \pi/2) \right] \qquad N_0 = 20$$

$$= \frac{2}{40}[20 + 0] = 1$$

where the sum of the cosine is zero, as we are adding the values of the periodic cosine over a period.

2. For $\Omega_0 = 3.2$, $x_2[n] = 2\cos(3.2n/10 - \pi/4)$ for $n \geq 0$ and zero otherwise. The signal now does not repeat periodically after $n = 0$, as the frequency $3.2/10$ (which equals the rational $32/100$) cannot be expressed as $2\pi m/N$ (which due to π is an irrational value) for integers m and N. Thus, in this case we do not have a close form for the power, so we can simply say that the power is

$$P_x = \lim_{N \to \infty} \frac{1}{2N+1} \sum_{n=-N}^{N} |x_2[n]|^2$$

and conjecture that because the corresponding analog signal has finite power, so would $x_2[n]$. Thus, we use MATLAB to compute the power for both cases.

```
%%%%%%%%%%%%%%%%%%%%%%%%
% Example 8.7 --- Power
%%%%%%%%%%%%%%%%%%%%%%%%
clear all;clf
n = 0:100000;
x2 = 2*cos(0.1*n*3.2 – pi/4); % non-periodic for positive n
x1 = 2*cos(0.1*n*pi – pi/4);     % periodic for positive n
N = length(x1)
Px1 = sum(x1.^2)/(2*N+1)  % power of x1
Px2 = sum(x2.^2)/(2*N+1)  % power of x2
P1 = sum(x1(1:20).^2)/(20);  %power in period of x1
```

The signal $x_1[n]$ in the script has unit power and so does the signal $x_2[n]$ when we consider 100,001 samples. The two signals and their squared magnitudes are shown in Figure 8.1.   ∎

■ Example 8.8

Determine if a discrete-time exponential

$$x[n] = 2(0.5)^n \qquad n \geq 0$$

and zero otherwise, has finite energy, finite power, or both.

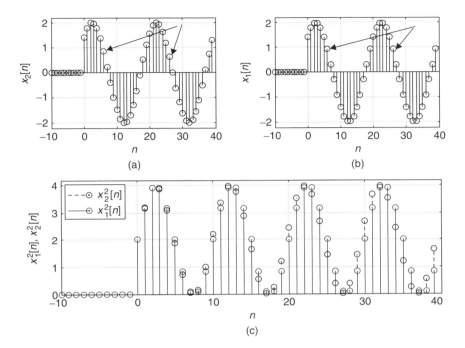

FIGURE 8.1

(a) Signal $x_2[n]$ (nonperiodic for $n \geq 0$) and (b) signal $x_1[n]$ (periodic for $n \geq 0$). The arrows show that the values are not equal for $x_2[n]$ and equal for $x_1[n]$. (c) The square of the signals differ slightly, suggesting that if $x_1[n]$ has finite power so does $x_2[n]$.

Solution

The energy is given by

$$\varepsilon_x = \sum_{n=0}^{\infty} 4(0.5)^{2n} = 4 \sum_{n=0}^{\infty} (0.25)^n = \frac{4}{1 - 0.25} = \frac{16}{3}$$

thus, $x[n]$ is a finite-energy signal. Just as with continuous-time signals, a finite-energy signal is a finite-power (actually zero power) signal. ■

8.2.3 Even and Odd Signals

Time shifting and scaling of discrete-time signals are very similar to the continuous-time cases, the only difference being that the operations are now done using integers.

A discrete-time signal $x[n]$ is said to be
- *Delayed* by N (an integer) samples if $x[n - N]$ is $x[n]$ shifted to the right N samples.
- *Advanced* by M (an integer) samples if $x[n + M]$ is $x[n]$ shifted to the left M samples.
- *Reflected* if the variable n in $x[n]$ is negated (i.e., $x[-n]$).

The shifting to the right or the left can be readily seen by considering where $x[0]$ is attained. For $x[n - N]$, this is when $n = N$ (i.e., N samples to the right of the origin), or $x[n]$ is delayed by N samples.

Likewise, for $x[n+M]$ the $x[0]$ appears advanced by M samples, or shifted to the left (i.e., when $n = -M$). Negating the variable n flips over the signal with respect to the origin.

■ Example 8.9

A triangular discrete pulse is defined as

$$x[n] = \begin{cases} n & 0 \leq n \leq 10 \\ 0 & \text{otherwise} \end{cases}$$

Find an expression for $y[n] = x[n+3] + x[n-3]$ and $z[n] = x[-n] + x[n]$ in terms of n and carefully plot them.

Solution

Replacing n by $n+3$ and $n-3$ in the definition of $x[n]$, we get the advanced and delayed signals

$$x[n+3] = \begin{cases} n+3 & -3 \leq n \leq 7 \\ 0 & \text{otherwise} \end{cases}$$

and

$$x[n-3] = \begin{cases} n-3 & 3 \leq n \leq 13 \\ 0 & \text{otherwise} \end{cases}$$

so that when added, we get

$$y[n] = x[n+3] + x[n-3] = \begin{cases} n+3 & -3 \leq n \leq 2 \\ 2n & 3 \leq n \leq 7 \\ n-3 & 8 \leq n \leq 13 \\ 0 & \text{otherwise} \end{cases}$$

Likewise, we have that

$$z[n] = x[n] + x[-n] = \begin{cases} n & 1 \leq n \leq 10 \\ 0 & n = 0 \\ -n & -10 \leq n \leq -1 \\ 0 & \text{otherwise} \end{cases}$$

The results are shown in Figure 8.2.   ■

■ Example 8.10

We will see that in the convolution sum we need to figure out how a signal $x[n-k]$ behaves as a function of k for different values of n. Consider the signal

$$x[k] = \begin{cases} k & 0 \leq k \leq 3 \\ 0 & \text{otherwise} \end{cases}$$

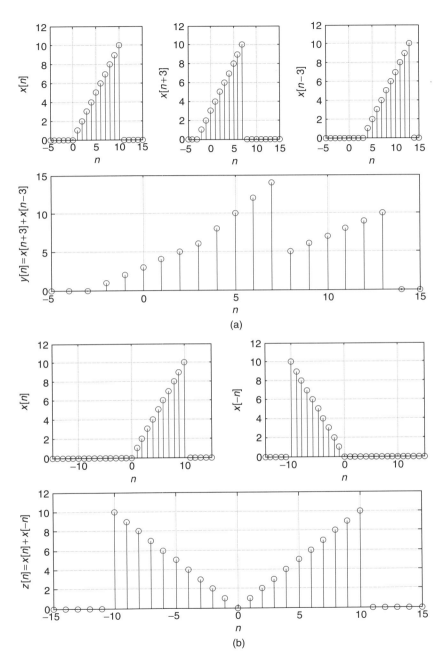

FIGURE 8.2

Generation of (a) $y[n] = x[n + 3] + x[n - 3]$ and (b) $z[n] = x[n] + x[-n]$.

Obtain an expression for $x[n - k]$ for $-2 \leq n \leq 2$ and determine in which direction it shifts as n increases from -2 to 2.

Solution

Although $x[n - k]$, as a function of k, is reflected it is not clear if it is advanced or delayed as n increases from -2 to 2. If $n = 0$,

$$x[-k] = \begin{cases} -k & -3 \leq k \leq 0 \\ 0 & \text{otherwise} \end{cases}$$

For $n \neq 0$, we have that

$$x[n - k] = \begin{cases} n - k & n - 3 \leq k \leq n \\ 0 & \text{otherwise} \end{cases}$$

As n increases from -2 to 2 the supports of $x[n - k]$ move to the right. For $n = -2$ the support of $x[-2 - k]$ is $-5 \leq k \leq -2$, while for $n = 0$ the support of $x[-k]$ is $-3 \leq k \leq 0$, and for $n = 2$ the support of $x[2 - k]$ is $-1 \leq k \leq 2$, each shifted to the right. ∎

We can thus use the above to define even and odd signals and obtain a general decomposition of any signal in terms of even and odd signals.

Even and odd discrete-time signals are defined as

$$x[n] \text{ is } even: \quad \Leftrightarrow \quad x[n] = x[-n] \tag{8.12}$$

$$x[n] \text{ is } odd: \quad \Leftrightarrow \quad x[n] = -x[-n] \tag{8.13}$$

Any discrete-time signal $x[n]$ can be represented as the sum of an even and an odd component,

$$x[n] = \underbrace{\frac{1}{2}\left(x[n] + x[-n]\right)}_{x_e[n]} + \underbrace{\frac{1}{2}\left(x[n] - x[-n]\right)}_{x_o[n]}$$

$$= x_e[n] + x_o[n] \tag{8.14}$$

The even and odd decomposition can be easily seen. The even component $x_e[n] = 0.5(x[n] + x[-n])$ is even since $x_e[-n] = 0.5(x[-n] + x[n])$ equals $x_e[n]$, and the odd component $x_o[n] = 0.5(x[n] - x[-n])$ is odd since $x_o[-n] = 0.5(x[-n] - x[n]) = -x_o[n]$.

■ Example 8.11

Find the even and the odd components of the discrete-time signal

$$x[n] = \begin{cases} 4 - n & 0 \leq n \leq 4 \\ 0 & \text{otherwise} \end{cases}$$

Solution

The even component of $x[n]$ is given by

$$x_e[n] = 0.5(x[n] + x[-n])$$

When $n = 0$ then $x_e[0] = 0.5 \times 2x[0] = 4$, when $n > 0$ then $x_e[n] = 0.5x[n]$, and when $n < 0$ then $x_e[n] = 0.5x[-n]$, giving

$$x_e[n] = \begin{cases} 2 + 0.5n & -4 \le n \le -1 \\ 4 & n = 0 \\ 2 - 0.5n & 1 \le n \le 4 \\ 0 & \text{otherwise} \end{cases}$$

The odd component

$$x_o[n] = 0.5(x[n] - x[-n])$$

gives 0 when $n = 0$, $0.5x[n]$ for $n > 0$, and $-0.5x[-n]$ when $n < 0$, or

$$x_o[n] = \begin{cases} -2 - 0.5n & -4 \le n \le -1 \\ 0 & n = 0 \\ 2 - 0.5n & 1 \le n \le 4 \\ 0 & \text{otherwise} \end{cases}$$

The sum of these two components gives $x[n]$.   ∎

Remarks *Expansion and compression of discrete-time signals is more complicated than in the continuous-time signals. In the discrete domain, expansion and compression can be related to the change of the sampling period in the sampling. Thus, if a continuous-time signal $x(t)$ is sampled using a sampling period T_s, by changing the sampling period to MT_s for an integer $M > 1$, we obtain fewer samples, and by changing the sampling period to T_s/L for an integer $L > 1$, we increase the number of samples. For the corresponding discrete-time signal $x[n]$, increasing the sampling period would give $x[Mn]$, which is called the* downsampling *of $x[n]$ by M. Unfortunately, because the argument of discrete-time signals must be integers, it is not clear what $x[n/L]$ is unless the values for n are multiples of L (i.e., $n = \pm 0, \pm L, \pm 2L, \ldots$) without a clear definition when n takes other values. This leads to the definition of the* upsampled *signal*

$$x_u[n] = \begin{cases} x[n/L] & n = \pm 0, \pm L, \pm 2L, \ldots \\ 0 & \text{otherwise} \end{cases} \tag{8.15}$$

To replace the zero entries with the values obtained by decreasing the sampling period we need to low-pass filter the upsampled signal. MATLAB provides the functions decimate *and* interp *to implement the downsampling and upsampling without losing information due to possible frequency aliasing. In Chapter 10, we will continue the discussion of these operations including their frequency characterization.*

8.2.4 Basic Discrete-Time Signals

The representation of discrete-time signals via basic signals is simpler than in the continuous-time domain. This is due to the lack of ambiguity in the definition of the impulse and the unit-step discrete-time signals. The definitions of impulses and unit-step signals in the continuous-time domain are more abstract.

Discrete-Time Complex Exponential

Given complex numbers $A = |A|e^{j\theta}$ and $\alpha = |\alpha|e^{j\omega_0}$, a *discrete-time complex exponential* is a signal of the form

$$x[n] = A\alpha^n$$
$$= |A||\alpha|^n e^{j(\omega_0 n + \theta)}$$
$$= |A||\alpha|^n [\cos(\omega_0 n + \theta) + j\sin(\omega_0 n + \theta)] \tag{8.16}$$

where ω_0 is a discrete frequency in radians.

Remarks

- *The discrete-time complex exponential looks different from the continuous-time complex exponential. This can be explained by sampling the continuous-time complex exponential*

$$x(t) = Ae^{(-a+j\Omega_0)t}$$

  (for simplicity we let A be real) using as sampling period T_s. The sampled signal is

$$x[n] = x(nT_s) = Ae^{(-anT_s + j\Omega_0 nT_s)} = A(e^{-aT_s})^n e^{j(\Omega_0 T_s)n}$$
$$= A\alpha^n e^{j\omega_0 n}$$

  where we let $\alpha = e^{-aT_s}$ and $\omega_0 = \Omega_0 T_s$.
- *Just as with the continuous-time complex exponential, we obtain different signals depending on the chosen parameters A and α. For instance, the real part of $x[n]$ in Equation (8.16) is a real signal*

$$g[n] = \mathcal{R}e[x[n]] = |A||\alpha|^n \cos(\omega_0 n + \theta)$$

  where when $|\alpha| < 1$ it is a damped sinusoid, and when $|\alpha| > 1$ it is a growing sinusoid (see Figure 8.3). If $\alpha = 1$ then the above signal is a sinusoid.
- *It is important to realize that for $\alpha > 0$ the real exponential*

$$x[n] = (-\alpha)^n = (-1)^n \alpha^n = \alpha^n \cos(\pi n)$$

■ Example 8.12

Given the analog signal

$$x(t) = e^{-at} \cos(\Omega_0 t)u(t)$$

determine the values of $a > 0$, Ω_0, and T_s that permit us to obtain a discrete-time signal

$$y[n] = \alpha^n \cos(\omega_0 n) \qquad n \geq 0$$

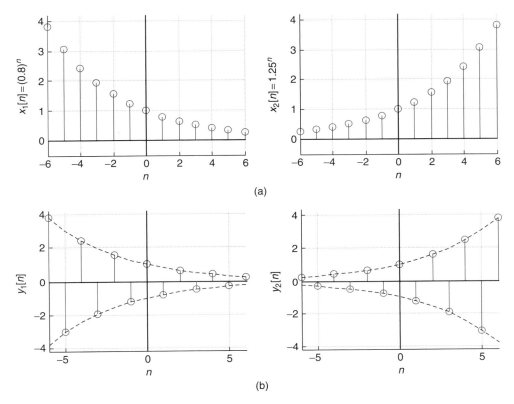

FIGURE 8.3

(a) Real exponential $x_1[n] = 0.8^n$, $x_2[n] = 1.25^n$, and (b) modulated exponential $y_1[n] = x_1[n]\cos(\pi n)$ and $y_2[n] = x_2[n]\cos(\pi n)$.

and zero otherwise. Consider the case when $\alpha = 0.9$ and $\omega_0 = \pi/2$. Find a, Ω_0, and T_s that will permit us to obtain $y[n]$ from $x(t)$ by sampling. Plot $x[n]$ and $y[n]$ using MATLAB.

Solution

Comparing the sampled continuous-time signal $x(nT_s) = (e^{-aT_s})^n \cos((\Omega_0 T_s)n)u[n]$ with $y[n]$ we obtain the following two equations:

$$\alpha = e^{-aT_s}$$
$$\omega_0 = \Omega_0 T_s$$

with three unknowns (a, Ω_0, and T_s), so there is no unique solution. According to the Nyquist sampling rate condition,

$$T_s \le \frac{\pi}{\Omega_{max}}$$

Assuming the maximum frequency is $\Omega_{max} = N\Omega_0$ for $N \geq 2$ (since the signal is not band limited the maximum frequency is not known; to estimate it we could use Parseval's result as indicated in Chapter 7, instead we are assuming that it is a multiple of Ω_0), if we let $T_s = \pi/N\Omega_0$ after replacing it in the above equations, we get

$$\alpha = e^{-a\pi/N\Omega_0}$$
$$\omega_0 = \Omega_0\pi/N\Omega_0 = \pi/N$$

If we want $\alpha = 0.9$ and $\omega_0 = \pi/2$, we have that $N = 2$ and

$$a = -\frac{2\Omega_0}{\pi} \log 0.9$$

for any frequency $\Omega_0 > 0$. For instance, if $\Omega_0 = 2\pi$, then $a = -4\log 0.9$ and $T_s = 0.25$. Figure 8.4 displays the continuous- and the discrete-time signals generated using the above parameters. The following script is used. The continuous-time and the discrete-time signals coincide at the sample times.

```
%%%%%%%%%%%%%%%%%%%%%%%
% Example 8.12
%%%%%%%%%%%%%%%%%%%%%%%
a = −4*log(0.9);Ts = 0.25; % parameters
alpha = exp(−a*Ts);
n = 0:30; y = alpha.^n.*cos(pi*n/2); % discrete-time signal
t = 0:0.001:max(n)*Ts; x = exp(−a*t).*cos(2*pi*t); % analog signal
stem(n, y, 'r'); hold on
plot(t/Ts, x); grid; legend('y[n]', 'x(t)'); hold off
```

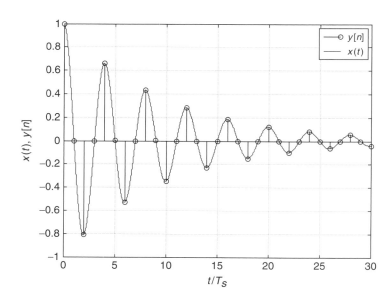

FIGURE 8.4

Determination of parameters for a continuous-time signal $x(t)$ that when sampled gives a desired discrete-time signal $y[n]$.

■ Example 8.13

Show how to obtain the discrete-time exponential $x[n] = (-1)^n$ for $n \geq 0$ and zero otherwise, by sampling a continuous-time signal $x(t)$.

Solution

Because the values of $x[n]$ are 1 and -1, $x[n]$ cannot be generated by sampling a real exponential signal $e^{-at}u(t)$; indeed, $e^{-at} > 0$ for any values of a and t. The discrete signal can be written as

$$x[n] = (-1)^n = \cos(\pi n)$$

for $n \geq 0$. If we sample an analog signal $x(t) = \cos(\Omega_0 t)u(t)$ with a sampling period T_s, we get

$$x[n] = x(nT_s) = \cos(\Omega_0 nT_s) = \cos(\pi n) \qquad n \geq 0$$

and zero otherwise. Thus, $\Omega_0 T_s = \pi$, giving $T_s = \pi / \Omega_0$. For instance, for $\Omega_0 = 2\pi$, then $T_s = 0.5$. ■

Discrete-Time Sinusoids

Discrete-time sinusoids are a special case of the complex exponential. Letting $\alpha = e^{j\omega_0}$ and $A = |A|e^{j\theta}$, we have according to Equation (8.16),

$$x[n] = A\alpha^n = |A|e^{j(\omega_0 n + \theta)} = |A|\cos(\omega_0 n + \theta) + j|A|\sin(\omega_0 n + \theta) \tag{8.17}$$

so the real part of $x[n]$ is a cosine, while the imaginary part is a sine. As indicated before, discrete sinusoids of amplitude A and phase shift θ are periodic if they can be expressed as

$$A\cos(\omega_0 n + \theta) = A\sin(\omega_0 n + \theta + \pi/2) \qquad -\infty < n < \infty \tag{8.18}$$

where $w_0 = 2\pi m/N$ rad is the discrete frequency for integers m and $N > 0$, which are not divisible by each other. Otherwise, discrete-time sinusoids are not periodic.

Because ω is given in radians, it repeats periodically with 2π as the period—that is,

$$\omega = \omega + 2\pi k \qquad k \text{ integer} \tag{8.19}$$

To avoid this ambiguity, we will let $-\pi < \omega \leq \pi$ as the possible range of discrete frequencies. This is possible since

$$\omega = \begin{cases} \omega - 2\pi k & \text{when } \omega > 2\pi, \text{ for some } k > 0 \text{ integer} \\ \omega - 2\pi & 0 \leq \omega \leq 2\pi \end{cases} \tag{8.20}$$

See Figure 8.5. Thus, $\sin(3\pi n)$ equals $\sin(\pi n)$, and $\sin(1.5\pi n)$ equals $\sin(-0.5\pi n) = -\sin(0.5\pi n)$.

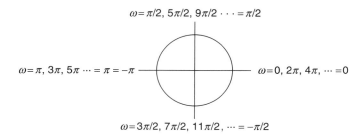

FIGURE 8.5
Discrete frequencies ω.

■ Example 8.14

Consider the following four sinusoids:

$$\text{(a) } x_1[n] = \sin(0.1\pi n)$$
$$\text{(b) } x_2[n] = \sin(0.2\pi n)$$
$$\text{(c) } x_3[n] = \sin(0.6\pi n)$$
$$\text{(d) } x_4[n] = \sin(0.7\pi n)$$

Find if they are periodic, and if so, determine their periods. Are these signals harmonically related? Use MATLAB to plot these signals from $n = 0, \dots, 40$. Comment on which of these signals resemble sampled analog sinusoids.

Solution

To find if they are periodic, rewrite the given signals as follows indicating that the signals are periodic of periods 20, 10, 10, and 20:

$$\text{(a) } x_1[n] = \sin(0.1\pi n) = \sin\left(\frac{2\pi}{20}n\right)$$

$$\text{(b) } x_2[n] = \sin(0.2\pi n) = \sin\left(\frac{2\pi}{10}n\right)$$

$$\text{(c) } x_3[n] = \sin(0.6\pi n) = \sin\left(\frac{2\pi}{10}3n\right)$$

$$\text{(d) } x_4[n] = \sin(0.7\pi n) = \sin\left(\frac{2\pi}{20}7n\right)$$

If we let $\omega_1 = 2\pi/20$, the frequencies of $x_2[n]$, $x_3[n]$, and $x_4[n]$ are $2\omega_1$, $6\omega_1$, and $7\omega_1$, respectively; thus they are harmonically related. Also, one could consider the frequencies of $x_1[n]$ and $x_4[n]$ harmonically related (i.e., the frequency of $x_4[n]$ is seven times that of $x_1[n]$), and likewise the frequencies of $x_2[n]$ and $x_3[n]$ are also harmonically related, with the frequency of $x_3[n]$ being

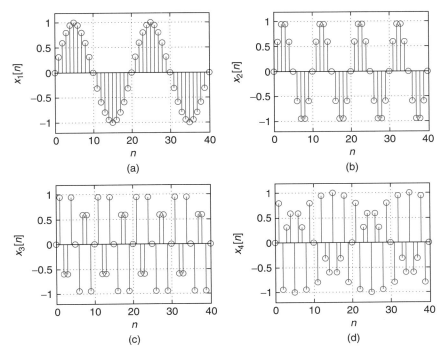

FIGURE 8.6

Periodic signals $x_i[n]$, (a) $i = 1$, (b) $i = 2$, (c) $i = 3$, and (d) $i = 4$, given in Example 8.14.

three times that of $x_2[n]$. When plotting these signals using MATLAB, the first two resemble analog sinusoids but not the other two. See Figure 8.6.   ■

Remarks

- *The discrete-time sine and cosine signals, as in the continuous-time case, are out of phase $\pi/2$ radians.*
- *The discrete frequency ω is given in radians since n, the sample index, does not have units. This can also be seen when we sample a sinusoid using a sampling period T_s so that*

$$\cos(\Omega_0 t)|_{t=nT_s} = \cos(\Omega_0 T_s n) = \cos(\omega_0 n)$$

where we defined $\omega_0 = \Omega_0 T_s$, and since Ω_0 has rad/sec as units and T_s has seconds as units, then ω_0 has radians as units.

- *The frequency Ω of analog sinusoids can vary from 0 (dc frequency) to ∞. Discrete frequencies ω as radian frequencies can only vary from 0 to π. Negative frequencies are needed in the analysis of real-valued signals; thus $-\infty < \Omega < \infty$ and $-\pi < \omega \leq \pi$. A discrete-time cosine of frequency 0 is constant for all n, and a discrete-time cosine of frequency π varies from -1 to 1 from sample to sample, giving the largest variation possible for the discrete-time signal.*

Discrete-Time Unit-Step and Unit-Sample Signals

The unit-step $u[n]$ and the unit-sample $\delta[n]$ discrete-time signals are defined as

$$u[n] = \begin{cases} 1 & n \geq 0 \\ 0 & n < 0 \end{cases} \tag{8.21}$$

$$\delta[n] = \begin{cases} 1 & n = 0 \\ 0 & \text{otherwise} \end{cases} \tag{8.22}$$

These two signals are related as follows:

$$\delta[n] = u[n] - u[n-1] \tag{8.23}$$

$$u[n] = \sum_{k=0}^{\infty} \delta[n-k] = \sum_{m=-\infty}^{n} \delta[m] \tag{8.24}$$

It is easy to see the relation between the two signals $u[n]$ and $\delta[n]$:

$$\delta[n] = u[n] - u[n-1]$$
$$u[n] = \delta[n] + \delta[n-1] + \cdots$$
$$= \sum_{k=0}^{\infty} \delta[n-k] = \sum_{m=-\infty}^{n} \delta[m]$$

where the last expression is obtained by a change of variable, $m = n - k$. These two equations should be contrasted with the ones for $u(t)$ and $\delta(t)$. Instead of the derivative relation $\delta(t) = du(t)/dt$, we have a difference relation, and instead of the integral connection

$$u(t) = \int_{-\infty}^{t} \delta(\tau) d\tau$$

we now have a summation relation between $u[n]$ and $\delta[n]$.

Remarks *Notice that there is no ambiguity in the definition of $u[n]$ or $\delta[n]$ as there is for their continuous-time counterparts $u(t)$ and $\delta(t)$. Moreover, the definitions of these functions do not depend on $u(t)$ or $\delta(t)$, and $u[n]$ and $\delta[n]$ are not sampled versions of $u(t)$ and $\delta(t)$.*

Generic Representation of Discrete-Time Signals

Any discrete-time signal $x[n]$ is represented using unit-sample signals as

$$x[n] = \sum_{k=-\infty}^{\infty} x[k]\delta[n-k] \tag{8.25}$$

The representation of any signal $x[n]$ in terms of $\delta[n]$ results from the *sifting property* of the unit-sample signal:

$$x[n]\delta[n - n_0] = x[n_0]\delta[n - n_0]$$

which is due to

$$\delta[n - n_0] = \begin{cases} 1 & n = n_0 \\ 0 & \text{otherwise} \end{cases}$$

Thus, considering $x[n]$ a sequence of samples

$$\ldots \; x[-1] \; x[0] \; x[1] \; \ldots$$

at times $\ldots \; -1, \; 0, \; 1, \ldots$, we can write $x[n]$ as

$$x[n] = \cdots + x[-1]\delta[n + 1] + x[0]\delta[n] + x[1]\delta[n - 1] + \cdots$$

$$= \sum_{k=-\infty}^{\infty} x[k]\delta[n - k]$$

The generic representation (Eq. 8.25) of any signal $x[n]$ will be useful in finding the output of a discrete-time linear time-invariant system.

■ Example 8.15

Use the generic representations in terms of unit-sample signals to represent the *ramp signal* $r[n]$ defined as

$$r[n] = nu[n]$$

and from it show that

$$r[n] = \sum_{m=0}^{n} (n - m) - \sum_{m=1}^{n} (n - m)$$

Solution

Using the unit-sample signal generic representation, we have

$$r[n] = \sum_{k=-\infty}^{\infty} (ku[k])\delta[n-k] = \sum_{k=0}^{\infty} k\delta[n-k] = 0\delta[n] + 1\delta[n-1] + 2\delta[n-2] + \cdots$$

Letting $m = n - k$, we write the above equation as

$$r[n] = \sum_{m=-\infty}^{n} (n-m)\delta[m] = \sum_{m=-\infty}^{n} (n-m)(u[m] - u[m-1])$$

$$= \sum_{m=0}^{n} (n-m) - \sum_{m=1}^{n} (n-m) = n + \sum_{m=1}^{n} (n-m) - \sum_{m=1}^{n} (n-m) = n \qquad n \geq 0$$

For $n < 0$, $r[n] = 0$. ■

■ Example 8.16

Consider a discrete pulse

$$x[n] = \begin{cases} 1 & 0 \leq n \leq N-1 \\ 0 & \text{otherwise} \end{cases}$$

Obtain representations of $x[n]$ using unit-sample and unit-step signals.

Solution

The signal $x[n]$ can be represented as

$$x[n] = \sum_{k=0}^{N-1} \delta[n-k]$$

and using $\delta[n] = u[n] - u[n-1]$, we obtain a representation of the discrete pulse in terms of unit-step signals,

$$x[n] = \sum_{k=0}^{N-1} (u[n-k] - u[n-k-1]) = (u[n] - u[n-1]) + (u[n-1] - u[n-2])$$

$$+ \cdots - u[n-N]$$

$$= u[n] - u[n-N]$$

because of the cancellation of consecutive terms. ■

■ Example 8.17

Consider how to generate a train of triangular, discrete-time pulses $t[n]$, which is periodic of period $N = 11$. A period of $t[n]$ is

$$\tau[n] = \begin{cases} n & 0 \le n \le 5 \\ -n + 10 & 6 \le n \le 10 \\ 0 & \text{otherwise} \end{cases}$$

Find then an expression for its finite difference $d[n] = t[n] - t[n - 1]$.

Solution

The periodic signal can be generated by adding shifted versions of $\tau[n]$, or

$$t[n] = \cdots + \tau[n + 11] + \tau[n] + \tau[n - 11] + \cdots = \sum_{k=-\infty}^{\infty} \tau[n - 11k]$$

The finite difference $d[n]$ is then

$$d[n] = t[n] - t[n - 1]$$

$$= \sum_{k=-\infty}^{\infty} (\tau[n - 11k] - \tau[n - 1 - 11k])$$

The signal $d[n]$ is also periodic of the same period $N = 11$ as $t[n]$. If we let

$$s[n] = \tau[n] - \tau[n - 1] = \begin{cases} 1 & 0 \le n \le 5 \\ -1 & 6 \le n \le 10 \\ 0 & \text{otherwise} \end{cases}$$

then

$$d[n] = \sum_{k=-\infty}^{\infty} s[n - 11k]$$

When sampled, these two signals look very much like the continuous-time train of triangular pulses, and its derivative. ■

■ Example 8.18

Consider the discrete-time signal

$$y[n] = 3r(t + 3) - 6r(t + 1) + 3r(t) - 3u(t - 3)|_{t=0.15n}$$

obtained by sampling a continuous-time signal formed by ramp and unit-step signals with a sampling period $T_s = 0.15$. Write MATLAB functions to generate the ramp and the unit-step signals and obtain $y[n]$. Then write a MATLAB function that provides the even and the odd decomposition of $y[n]$.

Solution

The real-valued signal is obtained by sequentially adding the different signals as we go from $-\infty$ to ∞:

$$
y(t) = \begin{cases}
0 & t < -3 \\
3r(t+3) = 3t + 9 & -3 \leq t < -1 \\
3t + 9 - 6r(t+1) = -3t + 3 & -1 \leq t < 0 \\
-3t + 3 + 3r(t) = 3 & 0 \leq t < 3 \\
3 - 3 = 0 & t \geq 3
\end{cases}
$$

The three functions ramp, ustep, and evenodd for this example are shown below. The following script shows how they can be used to generate the ramp signals, with the appropriate slopes and time shifts, as well as the unit-step signals with the desired delay, and then how to compute the even and the odd decomposition of $y[n]$.

```
%%%%%%%%%%%%%%%%%%%%%
% Example 8.18
%%%%%%%%%%%%%%%%%%%%%%
Ts = 0.15; % sampling period
t = -5:Ts:5; % time support
y1 = ramp(t, 3, 3); y2 = ramp(t, -6, 1); y3 = ramp(t, 3, 0); % ramp signals
y4 = -3*ustep(t, -3); % unit-step signal
y = y1 + y2 + y3 + y4;
[ze, zo] = evenodd(t, y);
```

We choose as support $-5 \leq t \leq 5$ for the continuous-time signal $y(t)$, which translates into a support $-5 \leq 0.15n \leq 5$ or $-5/0.15 \leq n \leq 5/0.15$ for the discrete-time signal. Since the limits are not integers, to make them integers (as required because n is an integer) we use the MATLAB function ceil to find integers larger than $-5/0.15$ and $5/0.15$ giving a range $[-33, 33]$. This is used when plotting $y[n]$.

The following function generates a ramp signal for a range of time values, for different slopes and time shifts.

```
function y = ramp(t, m, ad)
% ramp generation
% t: time support
% m: slope of ramp
% ad : advance (positive), delay (negative) factor
```

```
N = length(t);
y = zeros(1, N);
for i = 1:N,
    if t(i)> = −ad,
    y(i) = m∗(t(i) + ad);
    end
end
```

Likewise, the following function generates unit-step signals with different time shifts (notice the similarities with the ramp function).

```
function y = ustep(t, ad)
% generation of unit step
% t: time support
% ad : advance (positive), delay (negative)
N = length(t);
y = zeros(1, N);
for i = 1:N,
    if t(i) >= −ad,
    y(i) = 1;
    end
end
```

Finally, the following function can be used to compute the even and the odd decomposition of a discrete-time signal. The MATLAB function fliplr reflects the signal as needed in the generation of the even and the odd components.

```
function [ye, yo] = evenodd(y)
% even/odd decomposition
% NOTE: the support of the signal should be
%           symmetric about the origin
% y: analog signal
% ye, yo: even and odd components
yr = fliplr(y);
ye = 0.5∗(y + yr);
yo = 0.5∗(y − yr);
```

The results are shown in Figure 8.7. The discrete-time signal is given as

$$y[n] = \begin{cases} 0 & n \le -21 \\ 0.45n + 9 & -20 \le n \le -6 \\ -0.45n + 3 & -7 \le n \le 0 \\ 3 & 1 \le n \le 19 \\ 0 & n \ge 20 \end{cases}$$

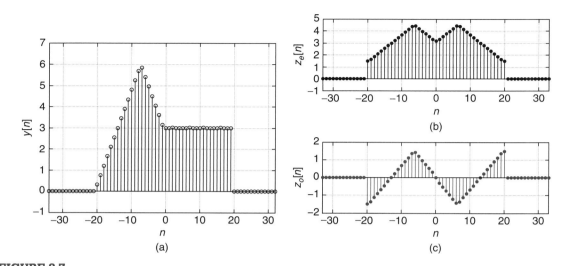

FIGURE 8.7

(a) Discrete-time signal, and (b) even and (c) odd components.

8.3 DISCRETE-TIME SYSTEMS

Just as with continuous-time systems, a discrete-time system is a transformation of a discrete-time input signal $x[n]$ into a discrete-time output signal $y[n]$—that is,

$$y[n] = S\{x[n]\} \tag{8.26}$$

Just as we were when we studied the continuous-time systems, we are interested in dynamic systems $S\{.\}$ having the following properties:

- Linearity
- Time invariance
- Stability
- Causality

A discrete-time system S is said to be

- *Linear:* If for inputs $x[n]$ and $v[n]$ and constants a and b, it satisfies the following
  - *Scaling:* $S\{ax[n]\} = aS\{x[n]\}$
  - *Additivity:* $S\{x[n] + v[n]\} = S\{x[n]\} + S\{v[n]\}$

  or equivalently if *superposition* applies—that is,

$$S\{ax[n] + bv[n]\} = aS\{x[n]\} + bS\{v[n]\} \tag{8.27}$$

- *Time-invariant:* If for an input $x[n]$ with a corresponding output $y[n] = S\{x[n]\}$, the output corresponding to a delayed or advanced version of $x[n]$, $x[n \pm M]$, is $y[n \pm M] = S\{x[n \pm M]\}$ for an integer M.

■ Example 8.19

A square-root computation system. The input–output relation characterizing a discrete-time system is nonlinear if there are nonlinear terms that include the input $x[n]$, the output $y[n]$, or both (e.g., a square root of $x[n]$, products of $x[n]$ and $y[n]$, etc.). Consider the development of an iterative algorithm to compute the square root of a positive real number α. If the result of the algorithm is $y[n]$ as $n \to \infty$, then $y^2[n] = \alpha$ and likewise $y^2[n-1] = \alpha$, thus $y[n] = 0.5(y[n-1] + y[n-1])$. Replacing $y[n-1] = \alpha/y[n-1]$ in this equation, the following difference equation, with some initial condition $y[0]$, can be used to find the square root of α:

$$y[n] = 0.5\left[y[n-1] + \frac{\alpha}{y[n-1]}\right] \qquad n > 0$$

Find recursively the solution of this difference equation. Use the results of finding the square roots of 4 and 2 to show the system is nonlinear. Solve the difference equation and plot the results for $\alpha = 4, 2$ with MATLAB.

Solution

The given difference equation is first order, nonlinear (expanding it you get the product of $y[n]$ with $y[n-1]$ and $y^2[n-1]$, which are nonlinear terms) with constant coefficients. This equation can be solved recursively for $n > 0$ by replacing $y[0]$ to get $y[1]$, and use this to get $y[2]$ and so on—that is,

$$y[1] = 0.5\left[y[0] + \frac{\alpha}{y[0]}\right]$$

$$y[2] = 0.5\left[y[1] + \frac{\alpha}{y[1]}\right]$$

$$y[3] = 0.5\left[y[2] + \frac{\alpha}{y[2]}\right]$$

$$\vdots$$

For instance, let $y[0] = 1$ and $\alpha = 4$ (i.e., we wish to find the square root of 4),

$$y[0] = 1$$

$$y[1] = 0.5\left[1 + \frac{4}{1}\right] = 2.5$$

$$y[2] = 0.5\left[2.5 + \frac{4}{2.5}\right] = 2.05$$

$$\vdots$$

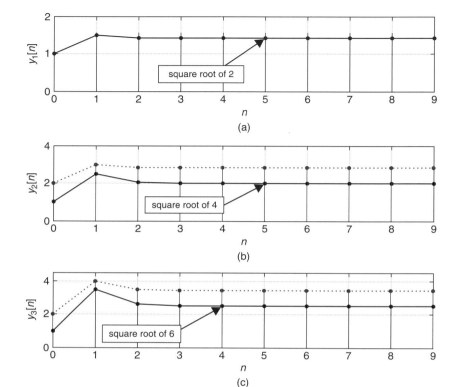

FIGURE 8.8

Nonlinear system:
(a) square root of 2,
(b) square root of 4
compared with twice the
square root of 2, and
(c) sum of previous
responses with response
when computing square
root of $2 + 4$. Figure (b)
shows scaling does not
hold and (c) shows that
additivity does not hold
either. The system is
nonlinear.

which is converging to 2, the square root of 4 (see Figure 8.8). Thus, as indicated before, when $n \to \infty$, then $y[n] = y[n-1] = Y$, for some value Y, which according to the difference equation satisfies the relation $Y = 0.5Y + 0.5(4/Y)$ or $Y = \sqrt{4} = 2$.

Suppose then that the input is $\alpha = 2$, half of what it was before. If the system is linear, we should get half the previous output according to the scaling property. That is not the case, however. For the same initial condition $y[0] = 1$, we obtain recursively for $\alpha[n] = 2u[n-1]$:

$$y[0] = 1$$
$$y[1] = 0.5[1+2] = 1.5$$
$$y[2] = 0.5\left[1.5 + \frac{2}{1.5}\right] = 1.4167$$
$$\vdots$$

This solution is clearly not half of the previous one. Moreover, as $n \to \infty$, we expect $y[n] = y[n-1] = Y$, for Y that satisfies the relation $Y = 0.5Y + 0.5(2/Y)$ or $Y = \sqrt{2} = 1.4142$, so that the solution is tending to $\sqrt{2}$ and not to 2 as it would if the system were linear. Finally, if we add the signals in the above two cases and compare the resulting signal with the one we obtain when finding the square root of the previous two values of α, 2 and 4, they do not coincide. The additive condition is not satisfied either, so the system is not linear. ∎

8.3.1 Recursive and Nonrecursive Discrete-Time Systems

Depending on the relation between the input $x[n]$ and the output $y[n]$ two types of discrete-time systems of interest are:

- *Recursive system*:

$$y[n] = -\sum_{k=1}^{N-1} a_k y[n-k] + \sum_{m=0}^{M-1} b_m x[n-m] \qquad n \geq 0$$

$$\text{initial conditions } y[-k], \ k = 1, \ldots, N-1 \tag{8.28}$$

This system is also called *infinite-impulse response* (IIR).

- *Nonrecursive system*:

$$y[n] = \sum_{m=0}^{M-1} b_m x[n-m] \tag{8.29}$$

This system is also called *finite-impulse response* (FIR).

The recursive system is analogous to a continuous-time system represented by a differential equation. For this type of system the discrete-time input $x[n]$ and the discrete-time output $y[n]$ are related by an $(N-1)$th-order difference equation. If such a difference equation is linear, with constant coefficients, zero initial conditions, and the input is zero for $n < 0$, then it represents a linear and time-invariant system. For these systems the output at a present time n, $y[n]$, depends or recurs on previous values of the output $\{y[n-k], \ k = 1, \ldots, N-1\}$, and thus they are called *recursive*. We will see that these systems are also called *infinite-impulse response* or *IIR* because their impulse responses are typically of infinite length.

On the other hand, if the output $y[n]$ does not depend on previous values of the output, but only on weighted and shifted inputs $\{b_m x[n-m], \ m = 0, \ldots, M-1\}$, the system is called *nonrecursive*. We will see that the impulse response of nonrecursive systems is of finite length; as such, these systems are also called *finite impulse response* or *FIR*.

∎ Example 8.20

Moving-average discrete filter: A third-order moving-average filter (also called a *smoother* since it smooths out the input signal) is an FIR filter for which the input $x[n]$ and the output $y[n]$ are related by

$$y[n] = \frac{1}{3}(x[n] + x[n-1] + x[n-2])$$

Show that this system is linear time invariant.

Solution

This is a nonrecursive system that uses a present sample, $x[n]$, and two past values, $x[n-1]$ and $x[n-2]$, of the input to get an average, $y[n]$, at every n. Thus, its name, *moving-average filter*.

Linearity: If we let the input be $ax_1[n] + bx_2[n]$, and assume that $\{y_i[n], \; i = 1, 2\}$ are the corresponding outputs to $\{x_i[n], \; i = 1, 2\}$, the filter output is

$$\frac{1}{3}[(ax_1[n] + bx_2[n]) + (ax_1[n-1] + bx_2[n-1]) + (ax_1[n-2] + bx_2[n-2])] = ay_1[n] + by_2[n]$$

That is, the system is linear.

Time invariance: If the input is $x_1[n] = x[n-N]$, the corresponding output to it is

$$\frac{1}{3}(x_1[n] + x_1[n-1] + x_1[n-2]) = \frac{1}{3}(x[n-N] + x[n-N-1] + x[n-N-2])$$
$$= y[n-N]$$

That is, the system is time invariant. ∎

■ Example 8.21

Autoregressive discrete filter: The recursive discrete-time system represented by the first-order difference equation (with initial condition $y[-1]$)

$$y[n] = ay[n-1] + bx[n] \qquad n \geq 0, \; y[-1]$$

is also called an *autoregressive* (AR) filter. "Autoregressive" refers to the feedback in the output—that is, the present value of the output $y[n]$ depends on its previous value $y[n-1]$.

Find recursively the solution of the difference equation and determine under what conditions the system represented by this difference equation is linear and time invariant.

Solution

Let's first discuss why the initial condition is $y[-1]$. The initial condition is the value needed to compute $y[0]$, which according to the difference equation

$$y[0] = ay[-1] + bx[0]$$

is $y[-1]$ since $x[0]$ is known.

Assume that the initial condition is $y[-1] = 0$, and that the input is $x[n] = 0$ for $n < 0$ (i.e., the system is not energized for $n < 0$). The solution of the difference equation when the input $x[n]$ is not defined can be found by a repetitive substitution of the input–output relationship. Thus, replacing $y[n-1] = ay[n-2] + bx[n-1]$ in the difference equation, and then replacing

$y[n-2] = ay[n-3] + bx[n-2]$, and so on, we obtain

$$y[n] = a(ay[n-2] + bx[n-1]) + bx[n]$$
$$= a(a(ay[n-3] + bx[n-2])) + abx[n-1] + bx[n]$$
$$= \cdots$$
$$= \cdots a^3 bx[n-3] + a^2 bx[n-2] + abx[n-1] + bx[n]$$

until we reach $x[0]$. The solution can be written as

$$y[n] = \sum_{k=0}^{n} ba^k x[n-k] \tag{8.30}$$

which we will see in the next section is the convolution sum of the impulse response of the system and the input.

To see that Equation (8.30) is actually the solution of the given difference equation, we need to show that when replacing the above expression for $y[n]$ in the right term of the difference equation we obtain the left term $y[n]$. Indeed, we have that

$$ay[n-1] + bx[n] = a \left[\sum_{k=0}^{n-1} ba^k x[n-1-k] \right] + bx[n]$$

$$= \sum_{m=1}^{n} ba^m x[n-m] + bx[n] = \sum_{m=0}^{n} ba^m x[n-m] = y[n]$$

where the dummy variable in the sum was changed to $m = k + 1$, so that the limits of the summation became $m = 1$ when $k = 0$, and $m = n$ when $k = n - 1$. The final equation is identical to $y[n]$.

To establish if the system represented by the difference equation is linear, we use the solution Eq. (8.30) with input $x[n] = \alpha x_1[n] + \beta x_2[n]$, where the outputs $\{y_i[n], \ i = 1, 2\}$ correspond to inputs $\{x_i[n], \ i = 1, 2\}$, and α and β are constants. The output for $x[n]$ is

$$\sum_{k=0}^{n} ba^k x[n-k] = \sum_{k=0}^{n} ba^k \left(\alpha x_1[n-k] + \beta x_2[n-k] \right)$$

$$= \alpha \sum_{k=0}^{n} ba^k x_1[n-k] + \beta \sum_{k=0}^{n} ba^k x_2[n-k] = \alpha y_1[n] + \beta y_2[n]$$

So the system is linear.

The time invariance is shown by letting the input be $v[n] = x[n - N]$, $n \geq N$, and zero otherwise. The corresponding output according to Equation (8.30) is

$$\sum_{k=0}^{n} ba^k v[n - k] = \sum_{k=0}^{n} ba^k x[n - N - k]$$

$$= \sum_{k=0}^{n-N} ba^k x[n - N - k] + \sum_{k=n-N+1}^{n} ba^k x[n - N - k] = y[n - N]$$

since the summation

$$\sum_{k=n-N+1}^{n} ba^k x[n - N - k] = 0$$

given that $x[-N] = \cdots = x[-1] = 0$ is assumed. Thus, the system represented by the above difference equation is linear and time invariant. As in the continuous-time case, however, if the initial condition $y[-1]$ is not zero, or if $x[n] \neq 0$ for $n < 0$, the system characterized by the difference equation is not LTI. ∎

■ Example 8.22

Autoregressive moving average filter: The recursive system represented by the first-order difference equation

$$y[n] = 0.5y[n - 1] + x[n] + x[n - 1] \qquad n \geq 0, \ y[-1]$$

is called the *autoregressive moving average* given that it is the combination of the two systems discussed before. Consider two cases:

■ Let the initial condition be $y[-1] = -2$, and the input be $x[n] = u[n]$ first and then $x[n] = 2u[n]$.

■ Let the initial condition be $y[-1] = 0$, and the input be $x[n] = u[n]$ first and then $x[n] = 2u[n]$.

Determine in each of these cases if the system is linear.

Find the steady-state response—that is,

$$\lim_{n \to \infty} y[n]$$

Solution

For an initial condition $y[-1] = -2$ and $x[n] = u[n]$, we get recursively

$$y[0] = 0.5y[-1] + x[0] + x[-1] = 0$$

$$y[1] = 0.5y[0] + x[1] + x[0] = 2$$

$$y[2] = 0.5y[1] + x[2] + x[1] = 3$$

$$\cdots$$

Let us then double the input (i.e., $x[n] = 2u[n]$) and call the response $y_1[n]$. As the initial condition remains the same (i.e., $y_1[-1] = -2$), we get

$$y_1[0] = 0.5y_1[-1] + x[0] + x[-1] = 1$$

$$y_1[1] = 0.5y_1[0] + x[1] + x[0] = 4.5$$

$$y_1[2] = 0.5y_1[1] + x[2] + x[1] = 6.25$$

$$\cdots$$

It is clear that the $y_1[n]$ is not $2y[n]$. Due to the initial condition not being zero, the system is nonlinear.

If the initial condition is set to zero, and the input $x[n] = u[n]$, the response is

$$y[0] = 0.5y[-1] + x[0] + x[-1] = 1$$

$$y[1] = 0.5y[0] + x[1] + x[0] = 2.5$$

$$y[2] = 0.5y[1] + x[2] + x[1] = 3.25$$

$$\cdots$$

If we double the input (i.e., $x[n] = 2u[n]$) and call the response $y_1[n]$, $y_1[-1] = 0$, we obtain

$$y_1[0] = 0.5y_1[-1] + x[0] + x[-1] = 2$$

$$y_1[1] = 0.5y_1[0] + x[1] + x[0] = 5$$

$$y_1[2] = 0.5y_1[1] + x[2] + x[1] = 6.5$$

$$\cdots$$

For the zero initial condition, it is clear that $y_1[n] = 2y[n]$ when we double the input. One can also show that superposition holds for this system. For instance, if we let the input be the sum of the previous inputs, $x[n] = u[n] + 2u[n] = 3u[n]$, and let $y_{12}[n]$ be the response when the initial condition is zero, $y_{12}[0] = 0$, we get

$$y_{12}[0] = 0.5y_{12}[-1] + x[0] + x[-1] = 3$$

$$y_{12}[1] = 0.5y_{12}[0] + x[1] + x[0] = 7.5$$

$$y_{12}[2] = 0.5y_{12}[1] + x[2] + x[1] = 9.75$$

$$\cdots$$

showing that $y_{12}[n]$ is the sum of the responses when the inputs are $u[n]$ and $2u[n]$. Thus, the system represented by the given difference equation with a zero initial condition is linear.

Although when the initial condition is -2 or 0, and $x[n] = u[n]$ we cannot find a closed form for the response, we can see that the response is going toward a final value or a steady-state response. Assuming that as $n \to \infty$ we have that $Y = y[n] = y[n - 1]$, and since $x[n] = x[n - 1] = 1$, according to the difference equation the steady-state value Y is found from

$$Y = 0.5Y + 2 \quad \text{or} \quad Y = 4$$

independent of the initial condition. Likewise, when $x[n] = 2u[n]$, the steady-state solution Y is obtained from $Y = 0.5Y + 4$ or $Y = 8$, independent of the initial condition. ∎

Remarks

- *Like in the continuous-time system, to show that a discrete-time system is linear and time invariant an explicit expression relating the input and the output is needed.*
- *Although the solution of linear difference equations can be obtained in the time domain, just like with differential equations, we will see in the next chapter that their solution can also be obtained using the Z-transform, just like the Laplace transform being used to solve linear differential equations.*

8.3.2 Discrete-Time Systems Represented by Difference Equations

As we saw before, a recursive discrete-time system is represented by a difference equation

$$y[n] = -\sum_{k=1}^{N-1} a_k y[n-k] + \sum_{m=0}^{M-1} b_m x[n-m] \qquad n \geq 0$$

$$\text{initial conditions } y[-k], \ k = 1, \ldots, N-1 \tag{8.31}$$

If the system is discrete-time, the difference equation naturally characterizes the dynamics of the system. On the other hand, the difference equation could be the approximation of a differential equation representing a continuous-time system that is being processed discretely. For instance, to approximate a second-order differential equation by a difference equation, we could approximate the first derivative of a continuous-time function $v_c(t)$ as

$$\frac{dv_c(t)}{dt} \approx \frac{v_c(t) - v_c(t - T_s)}{T_s}$$

and its second derivative as

$$\frac{d^2 v_c(t)}{dt^2} = \frac{d\frac{dv_c(t)}{dt}}{dt} \approx \frac{d(v_c(t) - v_c(t - T_s))/T_s}{dt}$$

$$\approx \frac{v_c(t) - 2v_c(t - T_s) + v_c(t - 2T_s)}{T_s^2}$$

to obtain a second-order difference equation when $t = nT_s$. Choosing a small value for T_s provides an accurate approximation to the differential equation. Other transformations can be used. In Chapter 0 we indicated that approximating integrals by the trapezoidal rule gives the *bilinear transformation*, which can also be used to change differential into difference equations.

Just as in the continuous-time case, the system being represented by the difference equation is not LTI unless the initial conditions are zero and the input is causal. The Z-transform will, however, allow us to find the complete response of the system even when the initial conditions are not zero. When the initial conditions are not zero, just like in the continuous case, these systems are *incrementally linear*.

The complete response of a system represented by the difference equation can be shown to be composed of a *zero-input* and a *zero-state responses*—that is, if $y[n]$ is the solution of the difference Equation (8.31) with initial conditions not necessarily equal to zero, then

$$y[n] = y_{zi}[n] + y_{zs}[n] \tag{8.32}$$

The component $y_{zi}[n]$ is the response when the input $x[n]$ is set to zero, thus it is completely due to the initial conditions. The response $y_{zs}[n]$ is due to the input, as we set the initial conditions equal to zero. The complete response $y[n]$ is thus seen as the superposition of these two responses. The Z-transform provides an algebraic way to obtain the complete response, whether the initial conditions are zero or not.

It is important, as in the continuous-time system, to differentiate the zero-input and the zero-state responses from the *transient* and the *steady-state* responses.

8.3.3 The Convolution Sum

Let $h[n]$ be the impulse response of an LTI discrete-time system, or the output of the system corresponding to an impulse $\delta[n]$ as input and initial conditions (if needed) equal to zero. Using the generic representation of the input $x[n]$ of the LTI system,

$$x[n] = \sum_{k=-\infty}^{\infty} x[k]\delta[n - k] \tag{8.33}$$

the output of the system is given by either of the following two equivalent forms of the *convolution sum*:

$$y[n] = \sum_{k=-\infty}^{\infty} x[k]h[n-k]$$

$$= \sum_{m=-\infty}^{\infty} x[n-m]h[m] \qquad (8.34)$$

The impulse response $h[n]$ of a discrete-time system is due exclusively to an input $\delta[n]$; as such, the initial conditions are set to zero. In some cases there are no initial conditions, as in the case of nonrecursive systems.

Now, if $h[n]$ is the response due to $\delta[n]$, by time invariance the response to $\delta[n-k]$ is $h[n-k]$. By superposition, the response due to $x[n]$ with the generic representation

$$x[n] = \sum_{k} x[k]\delta[n-k]$$

is the sum of responses due to $x[k]\delta[n-k]$, which is $x[k]h[n-k]$ ($x[k]$ is not a function of n), or

$$y[n] = \sum_{k} x[k]h[n-k]$$

i.e., the convolution sum of the input $x[n]$ with the impulse response $h[n]$ of the system. The second expression of the convolution sum in Equation (8.34) is obtained by a change of variable $m = n - k$.

Remarks

■ *The output of nonrecursive or FIR systems is the convolution sum of the input and the impulse response of the system. The input–output expression of an FIR system is*

$$y[n] = \sum_{k=0}^{N-1} b_k x[n-k] \qquad (8.35)$$

and its impulse response is found by letting $x[n] = \delta[n]$, which gives

$$h[n] = \sum_{k=0}^{N-1} b_k \delta[n-k] = b_0\delta[n] + b_1\delta[n-1] + \cdots + b_{N-1}\delta[n-(N-1)]$$

so that $h[n] = b_n$ for $n = 0, \ldots, N-1$, and zero otherwise. Replacing the b_k coefficients in Equaiton (8.35) by $h[k]$ we find that the output can be written as

$$y[n] = \sum_{k=0}^{N-1} h[k]x[n-k]$$

or the convolution sum of the input and the impulse response. This is a very important result, indicating that the output of FIR systems is obtained by means of the convolution sum rather than difference equations, which gives great significance to the efficient computation of the convolution sum.

- *Considering the convolution sum as an operator—that is,*

$$y[n] = [h * x][n] = \sum_{k=-\infty}^{\infty} x[k]h[n-k]$$

it is easily shown to be linear. Indeed, whenever the input is $ax_1[n] + bx_2[n]$, and $\{y_i[n]\}$ are the outputs corresponding to $\{x_i[n]\}$ for $i = 1, 2$, then we have that

$$[h * (ax_1 + bx_2)][n] = \sum_k (ax_1[k] + bx_2[k])h[n-k] = a\sum_k x_1[k]h[n-k] + b\sum_k x_2[k]h[n-k]$$

$$= a[h * x_1][n] + b[h * x_2][n] = ay_1[n] + by_2[n]$$

as expected, since the system was assumed to be linear when the expression for the convolution sum was obtained. We will then have that if the output corresponding to $x[n]$ is $y[n]$, given by the convolution sum, then the output corresponding to a shifted version of the input, $x[n-N]$, should be $y[n-N]$. In fact, if we let $x_1[n] = x[n-N]$, the corresponding output is

$$[h * x_1][n] = \sum_k x_1[n-k]h[k] = \sum_k x[n-N-k]h[k]$$

$$= [h * x][n-N] = y[n-N]$$

Again, this result is expected given that the system was considered time invariant when the convolution sum was obtained.

- *From the equivalent representations for the convolution sum we have that*

$$[h * x][n] = \sum_k x[k]h[n-k] = \sum_k x[n-k]h[k]$$

$$= [x * h][n]$$

indicating that the convolution commutes with respect to the input $x[n]$ and the impulse response $h[n]$.

- *Just as with continuous-time systems, when conecting two LTI discrete-time systems (with impulse responses $h_1[n]$ and $h_2[n]$) in cascade or in parallel, their respective impulse responses are given by $[h_1 * h_2][n]$ and $h_1[n] + h_2[n]$. See Figure 8.9 for block diagrams.*
- *There are situations when instead of giving the input and the impulse response to compute the output, the information that it is available is, for instance, the input and the output and we wish to find the impulse response of the system, or we have the output and the impulse response and wish to find the input. This type of problem is called* deconvolution. *We consider this problem later in this chapter after considering causality, and in Chapter 9 where we show that it can be easily solved using the Z-transform.*
- *The computation of the convolution sum is typically difficult. It is made easier when the Z-transform is used, as we will see. MATLAB provides the function* conv *which greatly simplifies the computation.*

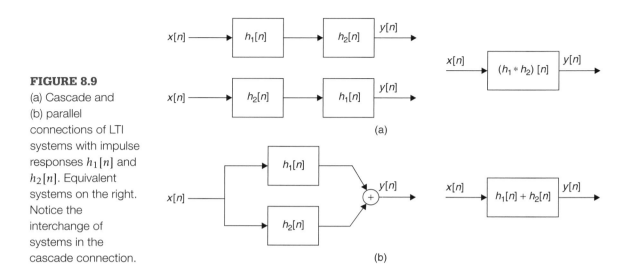

FIGURE 8.9
(a) Cascade and
(b) parallel
connections of LTI
systems with impulse
responses $h_1[n]$ and
$h_2[n]$. Equivalent
systems on the right.
Notice the
interchange of
systems in the
cascade connection.

■ **Example 8.23**

Consider a moving-averaging filter where the input is $x[n]$ and the output is $y[n]$:

$$y[n] = \frac{1}{3}(x[n] + x[n-1] + x[n-2])$$

Find the impulse response $h[n]$ of this filter. Then,

(a) Let $x[n] = u[n]$. Find the output of the filter $y[n]$ using the input–output relation and the convolution sum.

(b) If the input of the filter is $x[n] = A\cos(2\pi n/N)u[n]$, determine the values of A and N, so that the steady-state response of the filter is zero. Explain. Use MATLAB to verify your results.

Solution

(a) If the input is $x[n] = \delta[n]$, the output of the filter is $y[n] = h[n]$, or the impulse response of the system. No initial conditions are needed. We thus have that

$$h[n] = \frac{1}{3}(\delta[n] + \delta[n-1] + \delta[n-2])$$

so that $h[0] = 1/3$ as $\delta[0] = 1$ but $\delta[-1] = \delta[-2] = 0$; likewise, $h[1] = h[2] = 1/3$ so that the coefficients of the filter equal the impulse response of the filter at $n = 0$, 1, and 2.

Now if $x[n]$ is the input to the filter according to the convolution sum, its output is

$$y[n] = \sum_{k=0}^{n} x[n-k]h[k] = h[0]x[n] + h[1]x[n-1] + h[2]x[n-2]$$

$$= \frac{1}{3}(x[n] + x[n-1] + x[n-2])$$

Notice that the lower bound of the sum is set by the impulse response being zero for $n < 0$, while the upper bound is set by the input being zero for $n < 0$ (i.e., if $k > n$, then $n - k < 0$ and $x[n-k] = 0$). The convolution sum coincides with the input–output equation. This holds for any FIR filter.

For any input $x[n]$, let us then find a few values of the convolution sum to see what happens as n grows. If $n < 0$, the arguments of $x[n]$, $x[n-1]$, and $x[n-2]$ are negative giving zero values, and so the output is also zero (i.e., $y[n] = 0$, $n < 0$). For $n \geq 0$, we have

$$y[0] = \frac{1}{3}(x[0] + x[-1] + x[-2]) = \frac{1}{3}x[0]$$

$$y[1] = \frac{1}{3}(x[1] + x[0] + x[-1]) = \frac{1}{3}(x[0] + x[1])$$

$$y[2] = \frac{1}{3}(x[2] + x[1] + x[0]) = \frac{1}{3}(x[0] + x[1] + x[2])$$

$$y[3] = \frac{1}{3}(x[3] + x[2] + x[1]) = \frac{1}{3}(x[1] + x[2] + x[3])$$

$$\cdots$$

Thus, if $x[n] = u[n]$, then we have that $y[0] = 1/3$, $y[1] = 2/3$, and $y[n] = 1$ for $n \geq 2$.

(b) Notice that for $n \geq 2$, the output is the average of the present and past two values of the input. Thus, when the input is $x[n] = A\cos(2\pi n/N)$, if we let $N = 3$ and A be any real value, the input repeats every three samples and the local average of three of its values is zero, giving $y[n] = 0$ for $n \geq 2$; thus the steady-state response will be zero.

The following MATLAB script uses the function conv to compute the convolution sum when the input is either $x[n] = u[n]$ or $x[n] = \cos(2\pi n/3)u[n]$.

```
%%%%%%%%%%%%%%%%%%%
% Example 8.23 -- Convolution sum
%%%%%%%%%%%%%%%%%%%
x1 = [0 0 ones(1, 20)] % unit-step input
n = -2:19; n1 = 0:19;
x2 = [0 0 cos(2*pi*n1/3)]; % cosine input

h = (1/3)*ones(1, 3); % impulse response
y = conv(x1, h); y1 = y(1:length(n)); % convolution sums
y = conv(x2, h); y2 = y(1:length(n));
```

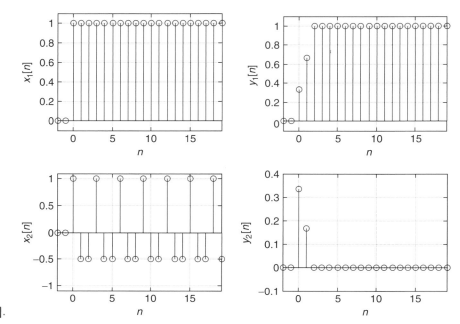

FIGURE 8.10

Convolution sums for a moving-averaging filter with input $x_1[n] = u[n]$ and $x_2[n] = \cos(2\pi n/3)u[n]$.

Notice that each of the input sequences has two zeros at the beginning so that the response can be found at $n \geq -2$. Also, when the input is of infinite support, like when $x[n] = u[n]$, we can only approximate it as a finite sequence in MATLAB, and as such the final values of the convolution obtained from conv are not correct and should not be considered. In this case, the final two values of the convolution results are not correct and are not considered. The results are shown in Figure 8.10. ∎

■ Example 8.24

Consider an autoregressive system represented by a first-order difference equation

$$y[n] = 0.5y[n-1] + x[n] \qquad n \geq 0$$

Find the impulse response $h[n]$ of the system and then compute the response of the system to $x[n] = u[n] - u[n-3]$ using the convolution sum. Verify results with MATLAB.

Solution

The impulse response $h[n]$ can be found recursively. Letting $x[n] = \delta[n]$, $y[n] = h[n]$, and initial condition $y[-1] = h[-1] = 0$, we have

$$h[0] = 0.5h[-1] + \delta[0] = 1$$
$$h[1] = 0.5h[0] + \delta[1] = 0.5$$
$$h[2] = 0.5h[1] + \delta[2] = 0.5^2$$
$$h[3] = 0.5h[2] + \delta[3] = 0.5^3$$

$$\cdots$$

from which a general expression for the impulse response is obtained as $h[n] = 0.5^n u[n]$.

The response to $x[n] = u[n] - u[n - 3]$ using the convolution sum is then given by

$$y[n] = \sum_{k=-\infty}^{\infty} x[k]h[n - k] = \sum_{k=-\infty}^{\infty} (u[k] - u[k - 3])0.5^{n-k}u[n - k]$$

Since as functions of k, $u[k]u[n - k] = 1$ for $0 \leq k \leq n$, zero otherwise, and $u[k - 3]u[n - k] = 1$ for $3 \leq k \leq n$, zero otherwise (in the two cases, draw the two signals as functions of k and verify this is true), $y[n]$ can be expressed as

$$y[n] = 0.5^n \left[\sum_{k=0}^{n} 0.5^{-k} - \sum_{k=3}^{n} 0.5^{-k} \right] u[n]$$

$$= \begin{cases} 0 & n < 0 \\ 0.5^n \sum_{k=0}^{n} 0.5^{-k} & n = 0, 1, 2 \\ 0.5^n \sum_{k=0}^{2} 0.5^{-k} & n \geq 3. \end{cases}$$

Another way to solve this problem is to notice that the input can be rewritten as

$$x[n] = \delta[n] + \delta[n - 1] + \delta[n - 2]$$

and since the system is LTI, the output can be written as

$$y[n] = h[n] + h[n - 1] + h[n - 2] = 0.5^n u[n] + 0.5^{n-1} u[n - 1] + 0.5^{n-2} u[n - 2]$$

which gives

$$y[0] = 0.5^0 = 1$$

$$y[1] = 0.5^1 + 0.5^0 = \frac{3}{2}$$

$$y[2] = 0.5^2 + 0.5^1 + 0.5^0 = \frac{7}{4}$$

$$y[3] = 0..5^3 + 0.5^2 + 0.5 = \frac{7}{8}$$

$$\cdots$$

which coincides with the above more general solution. It should be noticed that even in a simple example like this the computation required by the convolution sum is quite high. We will see that the Z-transform simplifies these types of problems, just like the Laplace transform does in the computation of the convolution integral.

The following MATLAB script is used to verify the above results. The MATLAB function filter is used to compute the impulse response and the response of the filter to the pulse. The output obtained then with filter coincided with the output computed using conv, as it should. Figure 8.11 displays the results.

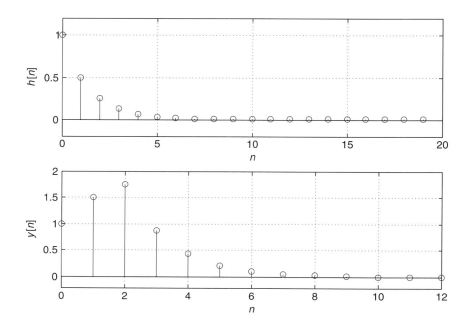

FIGURE 8.11

First-order autoregressive filter impulse response $h[n]$ and response $y[n]$ due to $x[n] = u[n] - u[n-3]$.

```
%%%%%%%%%%%%%%%%%%%%%%
% Example 8.24
%%%%%%%%%%%%%%%%%%%%%%%%%
a = [1 −0.5]; b = 1;  % coefficients of the difference equation
d = [1 zeros(1, 99)];  % approximate delta function
h = filter(b, a, d);  % impulse response
x = [ones(1, 3) zeros(1, 10)];  % input
y = filter(b, a, x);  % output from filter function
y1 = conv(h, x); y1 = y1(1:length(y))  % output from conv
```

■

8.3.4 Linear and Nonlinear Filtering with MATLAB

One is not always able to get rid of undesirable components of a signal by means of linear filtering. In this section we will illustrate the possible advantages of using nonlinear filters.

Linear Filtering

To illustrate the way a linear filter works, consider getting rid of a random disturbance $\eta[n]$, which we model as Gaussian noise (this is one of the possible noise signals MATLAB provides) that has been added to a sinusoid $x[n] = \cos(\pi n/16)$. Let $y[n] = x[n] + \eta[n]$. We will use an averaging filter having an input–output equation

$$z[n] = \frac{1}{M} \sum_{k=0}^{M-1} y[n-k]$$

This M-order filter averages M past input values $\{y[n-k], k = 0, \ldots, M-1\}$ and assigns this average to the output $z[n]$. The effect is to smooth out the input signal by attenuating the high-frequency components of the signal due to the noise. The larger the value of M the better the results, but at the expense of more complexity and a larger delay in the output signal (this is due to the linear-phase frequency response of the filter, as we will see later).

We use a third-order and a fifteenth-order filter, implemented by our function averager given below. The denoising is done by means of the following script.

```
%%%%%%%%%%%%%%%%%%%
% Linear filtering
%%%%%%%%%%%%%%%%%%%
N = 200; n = 0:N - 1;
x = cos(pi*n/16);  % input signal
noise = 0.2*randn(1, N);  % noise
y = x + noise;  % noisy signal
z = averager(3, y);  % averaging linear filter with M = 3
z1 = averager(15, y);  % averaging linear filter with M = 15
```

Our function averager defines the coefficients of the averaging filter and then uses the MATLAB function filter to compute the filter response. The inputs of filter are the vector $b = (1/M)[1 \cdots 1]$, the coefficients connected with the input, the unit coefficient connected with the output, and x is a vector with the entries the signal samples we wish to filter. The results of filtering using these two filters are shown in Figure 8.12. As expected, the performance of the filter with $M = 15$ is a lot better, but a delay of 8 samples (or the integer larger than $M/2$) is shown in the filter output.

```
function y = averager(M,x)
% Moving average of signal x
%     M: order of averager
%     x: input signal
%
b = (1/M)*ones(1, M);
y = filter(b, 1, x);
```

Nonlinear Filtering

Is linear filtering always capable of getting rid of noise? The answer is: It depends on the type of noise. In the previous example we showed that a high-order averaging filter, which is linear, performs well for Gaussian noise. Let us now consider an *impulsive* noise that is either zero or a certain value at random. This is the type of noise occurring in communications whenever cracking sounds are heard in the transmission, or the "salt-and-pepper" noise that appears in images.

It will be shown that even the 15th-order averager—that did well before—is not capable of denoising the signal with impulsive noise. A *median filter* considers a certain number of samples (the example shows the case of a 5th-order median filter), orders them according to their amplitudes, and chooses the one in the middle value (i.e., the median) as the output of the filter. Such a filter is nonlinear as it does not satisfy superposition. The following script is used to filter the noisy signal using a linear and

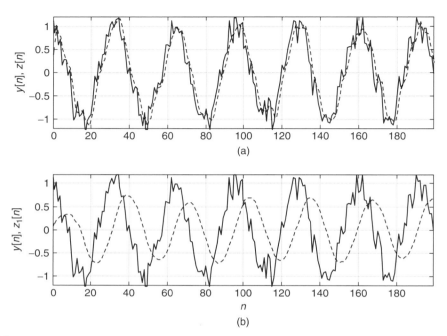

FIGURE 8.12

Averaging filtering with filters of order (a) $M = 3$ and of order (b) $M = 15$ result used to get rid of Gaussian noise added to a sinusoid $x[n] = cos(\pi n/16)$. Solid line corresponds to the noisy signal, while the dashed line is for the filtered signal. The filtered signal is very much like the noisy signal (a) when $M = 3$ is the order of the filter, while the filtered signal looks like the sinusoid, but shifted, (b) when $M = 15$. The plotting in this figure is done using `plot` instead of `stem` to allow a better visualization of the filtering results.

a nonlinear filter, and a comparison of the results is shown in Figure 8.13. In this case the nonlinear filter is able to denoise the signal much better than the linear filter.

```
%%%%%%%%%%%%%%%%%%%%
%  Nonlinear filtering
%%%%%%%%%%%%%%%%%%%%%
clear all; clf
N = 200;n = 0:N − 1;
% impulsive noise
for m = 1:N,
   d = rand(1, 1);
   if d >= 0.95,
      noise(m) = −1.5;
   else
      noise(m) = 0;
   end
end
```

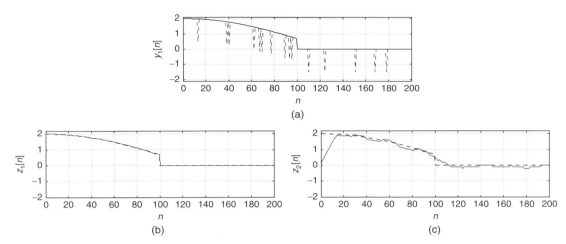

FIGURE 8.13

Top figure (a): noisy signal (dashed blue line) and clean signal (solid line). The clean signal (dashed line) is superposed on the denoised signal (solid blue line) in the bottom plots. The solid line in plot (b) is the result of median filtering, and the solid line in plot (c) is the result of the averager.

```
x = [2*cos(pi*n(1:100)/256) zeros(1, 100)];
y1 = x + noise;
% linear filtering
z2 = averager(15, y1);
% nonlinear filtering -- median filtering
z1(1) = median([0 0 y1(1) y1(2) y1(3)]);
z1(2) = median([0 y1(1) y1(2) y1(3) y1(4)]);
z1(N − 1) = median([y1(N − 3) y1(N − 2) y1(N − 1) y1(N) 0]);
z1(N) = median([y1(N − 2) y1(N − 1) y1(N) 0 0]);
for k = 3:N − 2,
    z1(k) = median([y1(k − 2) y1(k − 1) y1(k) y1(k + 1) y1(k + 2)]);
end
```

Although the theory of nonlinear filtering is beyond the scope of this book, it is good to remember that in cases like this when linear filters do not seem to do well, there are other methods to use.

8.3.5 Causality and Stability of Discrete-Time Systems

As with continuous-time systems, two additional independent properties of discrete-time systems are causality and stability. Causality relates to the conditions under which computation can be performed in real time, while stability relates to the usefulness of the system.

Causality

In many situations signals need to be processed in real time—that is, the processing must be done as the signal comes into the system. In those situations, the system must be causal. In many other situations, real-time processing is not required as the data can be stored and processed without the requirements of real time. Under such circumstances causality is not necessary.

A discrete-time system S is *causal* if:
- Whenever the input $x[n] = 0$, and there are no initial conditions, the output is $y[n] = 0$.
- The output $y[n]$ does not depend on future inputs.

Causality is independent of the linearity and time-invariance properties of a system. For instance, the system represented by the input–output equation

$$y[n] = x^2[n]$$

where $x[n]$ is the input and $y[n]$ is the output is nonlinear but time invariant. According to the above definition it is a causal system: The output is zero whenever the input is zero, and the output depends on the present value of the input. Likewise, an LTI system can be noncausal, as can be seen in the following discrete-time system that computes the moving average of the input:

$$y[n] = \frac{1}{3}(x[n+1] + x[n] + x[n-1]).$$

The input–output equation indicates that at the present time n to compute $y[n]$ we need a present value $x[n]$, a past value $x[n-1]$, and a future value $x[n+1]$. Thus, the system is LTI but noncausal since it requires future values of the input.

- An LTI discrete-time system is *causal* if the impulse response of the system is such that

$$h[n] = 0 \qquad n < 0 \tag{8.36}$$

- A signal $x[n]$ is said to be *causal* if

$$x[n] = 0 \qquad n < 0 \tag{8.37}$$

- For a causal LTI discrete-time system with a causal input $x[n]$ its output $y[n]$ is given by

$$y[n] = \sum_{k=0}^{n} x[k]h[n-k] \qquad n \geq 0 \tag{8.38}$$

where the lower limit of the sum depends on the input causality, $x[k] = 0$ for $k < 0$, and the upper limit depends on the causality of the system, $h[n-k] = 0$ for $n - k < 0$ or $k > n$.

That $h[n] = 0$ for $n < 0$ is the condition for an LTI discrete-time system to be causal is understood by considering that when computing the impulse response, the input $\delta[n]$ only occurs at $n = 0$ and there are no initial conditions, so the response for $n < 0$ should be zero. Extending the notion of causality to signals we can then see that the output of a causal LTI discrete-time system can be written in terms of the convolution sum as

$$y[n] = \sum_{k=-\infty}^{\infty} x[k]h[n-k] = \sum_{k=0}^{\infty} x[k]h[n-k] = \sum_{k=0}^{n} x[k]h[n-k]$$

where we first used the causality of the input ($x[k] = 0$ for $k < 0$) and then that of the system (i.e., $h[n-k] = 0$ whenever $n - k < 0$ or $k > n$). According to this equation the output depends on inputs $\{x[0], \ldots, x[n]\}$, which are past and present values of the input.

■ Example 8.25

So far we have considered the convolution sum as a way of computing the output $y[n]$ of an LTI system with impulse response $h[n]$ for a given input $x[n]$. But it actually can be used to find either of these three variables given the other two. The problem is then called *deconvolution*. Assume the input $x[n]$ and the output $y[n]$ of a causal LTI system are given. Find equations to compute recursively the impulse response $h[n]$ of the system. Consider finding the impulse response $h[n]$ of a causal LTI system with input $x[n] = u[n]$ and output $y[n] = \delta[n]$. Use the MATLAB function deconv to find $h[n]$.

Solution

If the system is causal and LTI, the input $x[n]$ and the output $y[n]$ are connected by the convolution sum

$$y[n] = \sum_{m=0}^{n} h[n-m]x[m] = h[n]x[0] + \sum_{m=1}^{n} h[n-m]x[m]$$

To find $h[n]$ from given $x[n]$ and $y[n]$, under the condition that $x[0] \neq 0$, the above equation can be rewritten as

$$h[n] = \frac{1}{x[0]} \left[y[n] - \sum_{m=1}^{n} h[n-m]x[m] \right]$$

so that the impulse response of the causal LTI can be found recursively as follows:

$$h[0] = \frac{1}{x[0]} y[0]$$

$$h[1] = \frac{1}{x[0]} (y[1] - h[0]x[1])$$

$$h[2] = \frac{1}{x[0]} (y[2] - h[0]x[2] - h[1]x[1])$$

$$\vdots$$

For the given case where $y[n] = \delta[n]$ and $x[n] = u[n]$, we get, according to the above,

$$h[0] = \frac{1}{x[0]} y[0] = 1$$

$$h[1] = \frac{1}{x[0]} \left(y[1] - h[0]x[1] \right) = 0 - 1 = -1$$

$$h[2] = \frac{1}{x[0]} \left(y[2] - h[0]x[2] - h[1]x[1] \right) = 0 - 1 + 1 = 0$$

$$h[3] = \frac{1}{x[0]} \left(y[3] - h[0]x[3] - h[1]x[2] - h[2]x[3] \right) = 0 - 1 + 1 - 0 = 0$$

$$\vdots$$

and, in general, $h[n] = \delta[n] - \delta[n-1]$.

The length of the convolution $y[n]$ is the sum of the lengths of the input $x[n]$ and of the impulse response $h[n]$ minus one. Thus,

$$\text{length of } h[n] = \text{length of } y[n] - \text{length of } x[n] + 1$$

When using deconv we need to make sure that the length of $y[n]$ is always larger than that of $x[n]$. If $x[n]$ is of infinite length, like when $x[n] = u[n]$, this would require an even longer $y[n]$, which is not possible. However, MATLAB can only provide a finite-support input, so we make the support of $y[n]$ larger. In this example we have found analytically that the impulse response $h[n]$ is of length 2, so if the length of $y[n]$ is chosen so that length $y[n]$ is larger than the length of $x[n]$ by one, we get the correct answer (case (a) in the script below); otherwise we do not (case (b)). Run the two cases to verify this (get rid of % symbol to run case (b)).

```
%%%%%%%%%%%%%%%%%%%%%%%%
% Example 8.25 --- Deconvolution
%%%%%%%%%%%%%%%%%%%%%%%%
clear all
x = ones(1, 100);
y = [1 zeros(1, 100)]; % (a) correct h
% y = [1 zeros(1, 99)]; % (b) incorrect h
[h, r] = deconv(y, x)
```

■

Bounded-Input Bounded-Output Stability

Stability characterizes useful systems. A stable system provides well-behaved outputs for well-behaved inputs. Bounded-input bounded-output (BIBO) stability establishes that for a bounded (that is what is meant by well-behaved) input $x[n]$ the output of a BIBO-stable system $y[n]$ is also bounded. This means that if there is a finite bound $M < \infty$ such that $|x[n]| < M$ for all n (you can think of it as an envelope $[-M, M]$ inside which the input is in for all time), the output is also bounded (i.e., $|y[n]| < L$ for $L < \infty$ and all n).

An LTI discrete-time system is said to be BIBO stable if its impulse response $h[n]$ is absolutely summable,

$$\sum_k |h[k]| < \infty \tag{8.39}$$

Assuming that the input $x[n]$ of the system is bounded, or that there is a value $M < \infty$ such that $|x[n]| < M$ for all n, the output $y[n]$ of the system represented by a convolution sum is also bounded, or

$$|y[n]| \leq \left| \sum_{k=-\infty}^{\infty} x[n-k]h[k] \right| \leq \sum_{k=-\infty}^{\infty} |x[n-k]||h[k]|$$

$$\leq M \sum_{k=-\infty}^{\infty} |h[k]| \leq MN < \infty$$

provided that $\sum_{k=-\infty}^{\infty} |h[k]| < N < \infty$, or that the impulse response be absolutely summable.

Remarks

- *Nonrecursive or FIR systems are BIBO stable. Indeed, the impulse response of such a system is of finite length and thus absolutely summable.*
- *For a recursive or IIR system represented by a difference equation, to establish stability we need to find the system impulse response $h[n]$ and determine whether it is absolutely summable or not.*
- *A much simpler way to test the stability of an IIR system will be based on the location of the poles of the Z-transform of $h[n]$, as we will see in Chapter 9.*

■ Example 8.26

Consider an autoregressive system

$$y[n] = 0.5y[n-1] + x[n]$$

Determine if the system is BIBO stable.

Solution

As shown in Example 8.24, the impulse response of the system is $h[n] = 0.5^n u[n]$. Checking the BIBO stability condition, we have

$$\sum_{n=-\infty}^{\infty} |h[n]| = \sum_{n=0}^{\infty} 0.5^n = \frac{1}{1 - 0.5} = 2$$

Thus, the system is BIBO stable. ■

8.4 WHAT HAVE WE ACCOMPLISHED? WHERE DO WE GO FROM HERE?

As you saw in this chapter the theory of discrete-time signals and systems is very similar to the theory of continuous-time signals and systems. Many of the results in the continuous-time theory are changed by swapping integrals for sums and differential equations for difference equations. However, there are significant differences imposed by the way the discrete-time signals and systems are generated. For instance, the discrete frequency can be considered finite but circular, and it depends on the sampling time. Discrete sinusoids, as another example, are not necessarily periodic. Thus, despite the similarities there are also significant differences between the continuous-time and the discrete-time signals and systems.

Now that we have a basic structure for discrete-time signals and systems, we will continue developing the theory of linear time-invariant discrete-time systems using transforms. Again, you will find a great deal of similarity but also some very significant differences. In the next chapters, carefully notice the relation that exists between the Z-transform and the Fourier representations of discrete-time signals and systems, not only with each other but with the Laplace and Fourier transforms. There is a great deal of connection among all of these transforms, and a clear understanding of this would help you with the analysis and synthesis of discrete-time signals and systems.

PROBLEMS

8.1. Discrete sequence—MATLAB
Consider the following formula

$$x[n] = x[n-1] + x[n-3] \qquad n \geq 3$$
$$x[0] = 0$$
$$x[1] = 1$$
$$x[2] = 2$$

Find the rest of the sequence for $0 \leq n \leq 50$ and plot it using the MATLAB function stem.

8.2. Finite-energy signals—MATLAB
Given the discrete signal $x[n] = 0.5^n u[n]$:
(a) Use MATLAB to plot the signal $x[n]$ for $n = -5$ to 200.
(b) Is this a finite-energy discrete-time signal? That is compute the infinite sum

$$\sum_{n=-\infty}^{\infty} |x[n]|^2$$

Hint: Show that

$$\sum_{n=0}^{\infty} \alpha^n = \frac{1}{1-\alpha}$$

or equivalently that

$$(1 - \alpha) \sum_{n=0}^{\infty} \alpha^n = 1$$

provided $|\alpha| < 1$.

(c) Verify your results by using symbolic MATLAB to find an expression for the above sum.

8.3. Periodicity of sampled signals—MATLAB

Consider an analog periodic sinusoid $x(t) = \cos(3\pi t + \pi/4)$ being sampled using a sampling period T_s to obtain the discrete-time signal $x[n] = x(t)|_{t=nT_s} = \cos(3\pi T_s n + \pi/4)$.

(a) Determine the discrete frequency of $x[n]$.

(b) Choose a value of T_s for which the discrete-time signal $x[n]$ is periodic. Use MATLAB to plot a few periods of $x[n]$, and verify its periodicity.

(c) Choose a value of T_s for which the discrete-time signal $x[n]$ is not periodic. Use MATLAB to plot $x[n]$ and choose an appropriate length to show the signal is not periodic.

(d) Determine under what condition the value of T_s makes $x[n]$ periodic.

8.4. Even and odd decomposition and energy—MATLAB

Suppose you sample the analog signal

$$x(t) = \begin{cases} 1 - t & 0 \le t \le 1 \\ 0 & \text{otherwise} \end{cases}$$

with a sampling period $T_s = 0.25$ to generate $x[n] = x(t)|_{t=nT_s}$.

(a) Use MATLAB to plot $x[-n]$ for an appropriate interval.

(b) Find $x_e[n] = 0.5[x[n] + x[-n]]$ and plot it carefully using MATLAB.

(c) Find $x_o[n] = 0.5[x[n] - x[-n]]$ and plot it carefully using MATLAB.

(d) Verify that $x_e[n] + x_o[n] = x[n]$ graphically.

(e) Compute the energy of $x[n]$ and compare it to the sum of the energies of $x_e[n]$ and $x_o[n]$.

8.5. Signal representation in terms of $u[n]$—MATLAB

We have shown how any discrete-time signal can be represented as a sum of weighted and shifted versions of $\delta[n]$. Given that

$$\delta[n] = u[n] - u[n - 1]$$

it should be possible to represent any signal as a combination of unit-step functions. Consider a discrete-time ramp $r[n] = nu[n]$, which in terms of $\delta[n]$ is written as

$$r[n] = \sum_{k=-\infty}^{\infty} r[k]\delta[n - k]$$

Replace $r[k] = ku[k]$ and use $\delta[n] = u[n] - u[n - 1]$ to show that $r[n]$ can be expressed in terms of $u[n]$ as

$$r[n] = \sum_{k=1}^{\infty} u[n - k]$$

Does this equation make sense? Use MATLAB to plot the obtained $r[n]$ to help you answer this.

8.6. Generation of periodic discrete-time signals—MATLAB

Periodic signals can be generated by obtaining a period and adding shifted versions of this period. Suppose we wish to generate a train of triangular pulses. A period of the signal is

$$x[n] = r[n] - 2r[n-1] + r[n-2]$$

where $r[n] = nu[n]$ is the discrete-time ramp signal.
(a) Carefully plot $x[n]$.
(b) Let

$$y[n] = \sum_{k=-\infty}^{\infty} x[n-2k]$$

and carefully plot it. Indicate the period N of $y[n]$.
(c) Write a MATLAB script to generate and plot the periodic signal $y[n]$.

8.7. Expansion and compression of discrete-time signals—MATLAB

Consider the discrete-time signal $x[n] = \cos(2\pi n/7)$.
(a) The discrete-time signal can be compressed by getting rid of some of its samples (*downsampling*). Consider the downsampling by 2. Write a MATLAB script to obtain and plot $z[n] = x[2n]$. Plot also $x[n]$ and compare it with $z[n]$. What happended? Explain.
(b) The expansion for discrete-time signals requires interpolation, and we will see it later. However, a first step of this process is the so-called *upsampling*. Upsampling by 2 consists in defining a new signal $y[n]$ such that $y[n] = x[n/2]$ for n even, and $y[n] = 0$ otherwise. Write a MATLAB script to perform upsampling on $x[n]$. Plot the resulting signal $y[n]$ and explain its relation with $x[n]$.
(c) If $x[n]$ resulted from sampling a continuous-time signal $x(t) = \cos(2\pi t)$ using a sampling period T_s and with no frequency aliasing, determine T_s. How would you sample the analog signal $x(t)$ to get the downsampled signals $z[n]$? That is, choose a value for the sampling period T_s to get $z[n]$ directly from $x(t)$. Can you choose T_s to get $y[n]$ from $x(t)$ directly? Explain.

8.8. Absolutely summable and finite-energy discrete-time signals—MATLAB

Suppose we sample the analog signal $x(t) = e^{-2t}u(t)$ using a sample period $T_s = 1$.
(a) Expressing the sampled signal as $x(nT_s) = x[n] = \alpha^n u[n]$, what is the corresponding value of α? Use MATLAB to plot $x[n]$.
(b) Show that $x[n]$ is absolutely summable—that is, show the following sum is finite:

$$\sum_{n=-\infty}^{\infty} |x[n]|$$

(c) If you know that $x[n]$ is absolutely summable, could you say that $x[n]$ is a finite-energy signal? Use MATLAB to plot $|x[n]|$ and $x^2[n]$ in the same plot to help you decide.
(d) In general, for what values of α are signals $y[n] = \alpha^n u[n]$ finite energy? Explain.

8.9. Discrete-time periodic signals

Determine whether the following discrete-time sinusoids are periodic or not. If periodic, determine its period N_0.

$$x[n] = 2\cos(\pi n - \pi/2)$$
$$y[n] = \sin(n - \pi/2)$$
$$z[n] = x[n] + y[n]$$
$$v[n] = \sin(3\pi n/2)$$

8.10. Periodicity of discrete-time signals

Consider periodic signals $x[n]$, of period $N_1 = 4$, and $y[n]$, of period $N_2 = 6$. What would be the period of

$$z[n] = x[n] + y[n]$$
$$v[n] = x[n]y[n]$$
$$w[n] = x[2n]$$

8.11. Periodicity of sum and product of periodic signals—MATLAB

If $x[n]$ is periodic of period $N_1 > 0$ and $y[n]$ is periodic of period $N_2 > 0$:

(a) What should be the condition for the sum of $x[n]$ and $y[n]$ to be periodic?

(b) What would be the period of the product $x[n]y[n]$?

(c) Would the formula

$$\frac{N_1 N_2}{\gcd(N1, N2)}$$

($\gcd(N_1, N_2)$ stands for the greatest common divisor of N_1 and N_2) give the period of the sum and the product of the two signals $x[n]$ and $y[n]$?

(d) Use MATLAB to plot the signals $x[n] = \cos(2\pi n/3)u[n]$, and $y[n] = (1 + \sin(6\pi n/7))u[n]$, their sum and product, and to find their periods and to verify your analytic results.

8.12. Echoing of music—MATLAB

An effect similar to multipath in acoustics is echoing or reverberation. To see the effects of an echo in an acoustic signal consider the simulation of echoes on the "handel.mat" signal $y[n]$. Pretend that this piece is being played in a round theater where the orchestra is in the middle of two concentric circles and the walls on one half side are at a radial distances of 17.15 meters (corresponding to the inner circle) and 34.3 meters (corresponding to the outer circle) on the other side (yes, an usual theater!) from the orchestra. The speed of sound is 343 meters/sec. Assume that the recorded signal is the sum of the original signal $y[n]$ and attenuated echoes from the two walls so that the recorded signal is given by

$$r[n] = y[n] + 0.8y[n - N_1] + 0.6y[n - N_2]$$

where N_1 is the delay caused by the closest wall and N_2 is the delay caused by the farther wall. The recorder is at the center of the auditorium where the orchestra is and we record for 1.5 seconds.

(a) Find the values of the two delays N_1 and N_2. Give the expression for the discrete-time recorded signal $r[n]$. The sampling frequency F_s of "handel.mat" is given when you load it in MATLAB.

(b) Simulate the echo signal. Plot $r[n]$. Use sound to listen to the original and the echoed signals.

8.13. Envelope modulation—MATLAB

In the generation of music by computer, the process of modulation is extremely important. When playing an instrument, the player typically does it in three stages: (1) rise time or attack, (2) sustained time, and (3) decay time. Suppose we model these three stages as an envelope continuous-time signal given by

$$e(t) = \frac{1}{3}[r(t) - r(t - 3)] - \frac{1}{0.1}[r(t - 20) + r(t - 30)]$$

where $r(t)$ is the ramp signal.

(a) For a simple tone $x(t) = \cos(2\pi/T_0 t)$, the modulated signal is $y(t) = x(t)e(t)$. Find the period T_0 so that 100 cycles of the sinusoid occur for the duration of the envelope signal.

(b) Simulate in MATLAB the modulated signal using the value of $T_0 = 1$ and a simulation sampling time of $0.1T_0$. Plot $y(t)$ and $e(t)$ (discretized with the sampling period $0.1T_0$) and listen to the modulated signal using sound.

8.14. LTI of ADCs

An ADC can be thought of as composed of three subsystems: a sampler, a quantizer, and a coder.

(a) The sampler, as a system, has as input an analog signal $x(t)$ and as output a discrete-time signal $x(nT_s) = x(t)|_{t=nT_s}$ where T_s is the sampling period. Determine whether the sampler is a linear system or not.

(b) Sample $x(t) = \cos(0.5\pi t)u(t)$ and $x(t - 0.5)$ using $T_s = 1$ to get $y(nT_s)$ and $z(nT_s)$, respectively. Plot $x(t)$, $x(t - 0.5)$, and $y(nT_s)$ and $z(nT_s)$. Is $z(nT_s)$ a shifted version of $y(nT_s)$ so that you can say the sampler is time invariant? Explain.

8.15. LTI of ADCs (part 2)

A two-bit quantizer of an ADC has as input $x(nT_s)$ and as output $\hat{x}(nT_s)$, such that if

$$ k\Delta \le x(nT_s) < (k+1)\Delta \quad \rightarrow \quad \hat{x}(nT_s) = k\Delta \qquad k = -2, -1, 0, 1 $$

(a) Is this system time invariant? Explain.

(b) Suppose that the value of Δ in the quantizer is 0.25, and the sampled signal is $x(nT_s) = nT_s$, $T_s = 0.1$ and $-5 \le n \le 5$. Use the sampled signal to determine whether the quantizer is a linear system or not. Explain.

(c) From the results in this and the previous problem, would you say that the ADC is an LTI system? Explain.

8.16. Rectangular windowing system—MATLAB

A window is a signal $w[n]$ that is used to highlight part of another signal. The windowing process consists in multiplying an input signal $x[n]$ by the window signal $w[n]$, so that the output is

$$ y[n] = x[n]w[n] $$

There are different types of windows used in signal processing. One of them is the so-called *rectangular window*, which is given by

$$ w[n] = u[n] - u[n - N] $$

(a) Determine whether the rectangular windowing system is linear. Explain.

(b) Suppose $x[n] = nu[n]$. Plot the output $y[n]$ of the windowing system (with $N = 6$).

(c) Let the input be $x[n - 6]$. Plot the corresponding output of the rectangular windowing system, and indicate whether the rectangular windowing system is time invariant.

8.17. Impulse response of an IIR system—MATLAB

A discrete-time IIR system is represented by the difference equation

$$ y[n] = 0.15y[n - 2] + x[n] \quad n \ge 0 $$

where $x[n]$ is the input and $y[n]$ is the output.

(a) To find the impulse response $h[n]$ of the system, let $x[n] = \delta[n]$, $y[n] = h[n]$, and the initial conditions be zero, $y[n] = h[n] = 0$, $n < 0$. Find recursively the values of $h[n]$ for values of $n \ge 0$.

(b) As a second way to do it, replace the relation between the input and the output given by the difference equation to obtain a convolution sum representation that will give the impulse response $h[n]$. What is $h[n]$?

(c) Use the MATLAB function filter to get the impulse response $h[n]$ (use help to learn about the function filter).

8.18. FIR filter—MATLAB

An FIR filter has a nonrecursive input–output relation

$$y[n] = \sum_{k=0}^{5} kx[n-k]$$

(a) Find and plot using MATLAB the impulse response $h[n]$ of this filter.

(b) Is this a causal and stable filter? Explain.

(c) Find and plot the unit-step response $s[n]$ for this filter.

(d) If the input $x[n]$ for this filter is bounded, i.e., $|x[n]| < 3$, what would be a minimum bound M for the output (i.e., $|y[n]| \le M$)?

(e) Use the MATLAB function filter to compute the impulse response $h[n]$ and the unit-step response $s[n]$ for the given filter and plot them.

8.19. LTI and convolution sum—MATLAB

The impulse response of a discrete-time system is $h[n] = (-0.5)^n u[n]$.

(a) If the input of the system is $x[n] = \delta[n] + \delta[n-1] + \delta[n-2]$, use the linearity and time invariance of the system to find the corresponding output $y[n]$.

(b) Find the convolution sum corresponding to the above input, and show that your solution coincides with the output $y[n]$ obtained above.

(c) Use the MATLAB function conv to find the output $y[n]$ due to the given input $x[n]$. Plot $x[n]$, $h[n]$, and $y[n]$ using MATLAB.

8.20. Steady state of IIR systems—MATLAB

Suppose an IIR system is represented by a difference equation

$$y[n] = ay[n-1] + x[n]$$

where $x[n]$ is the input and $y[n]$ is the output.

(a) If the input $x[n] = u[n]$ and it is known that the steady-state response is $y[n] = 2$, what would be a for that to be possible (in steady state $x[n] = 1$ and $y[n] = y[n-1] = 2$ since $n \to \infty$).

(b) Writing the system input as $x[n] = u[n] = \delta[n] + \delta[n-1] + \delta[n-2] + \cdots$ then according to the linearity and time invariance, the output should be

$$y[n] = h[n] + h[n-1] + h[n-2] + \cdots$$

Use the value for a found above, that the initial condition is zero (i.e., $y[-1] = 0$) and that the input is $x[n] = u[n]$, to find the values of the impulse response $h[n]$ for $n \ge 0$ using the above equation. The system is causal.

(c) Use the MATLAB function filter to compute the impulse response $h[n]$ and compare it with the one obtained above.

8.21. Causal systems and real-time processing

Systems that operate under real-time conditions need to be causal—that is, they can only process present and past inputs. When no real-time processing is needed the system can be noncausal.

(a) Consider the case of averaging an input signal $x[n]$ under real-time conditions. Suppose you are given two different filters,

$$\bullet \quad y[n] = \frac{1}{N} \sum_{k=0}^{N-1} x[n-k]$$

$$\bullet \quad y[n] = \frac{1}{N} \sum_{k=-N+1}^{N-1} x[n-k]$$

Which one of these would you use and why?

(b) If you are given a tape with the data, which of the two filters would you use? Why? Would you use either? Explain.

8.22. IIR versus FIR systems

A significant difference between IIR and FIR discrete-time systems is stability. Consider an IIR filter with the difference equation

$$y_1[n] = x[n] - 0.5y_1[n-1]$$

where $x[n]$ is the input and $y_1[n]$ is the output. Then consider an FIR filter

$$y_2[n] = x[n] + 0.5x[n-1] + 3x[n-2] + x[n-5]$$

where $x[n]$ is the input and $y_2[n]$ is the output.

(a) Since to check the stability of these filters we need their impulse responses, find the impulse responses $h_1[n]$ corresponding to the IIR filter by recursion, and $h_2[n]$ corresponding to the FIR filter.

(b) Use the impulse response $h_1[n]$ to check the stability of the IIR filter.

(c) Use the impulse response $h_2[n]$ to check the stability of the FIR filter.

(d) Since the impulse response of a FIR filter has a finite number of nonzero terms, would it be correct to say that FIR filters are always stable? Explain.

8.23. Unit-step versus impulse response—MATLAB

The unit-step response of a discrete-time LTI system is

$$s[n] = 2[(-0.5)^n - 1]u[n]$$

Use this information to find

(a) The impulse response $h[n]$ of the discrete-time LTI system.

(b) The response of the LTI system to a ramp signal $x[n] = nu[n]$. Use the MATLAB function filter and superposition to find it.

8.24. Convolution sum—MATLAB

A discrete-time system has a unit-impulse response $h[n]$.

(a) Let the input to the discrete-time system be a pulse $x[n] = u[n] - u[n-4]$. Compute the output of the system in terms of the impulse response.

(b) Let $h[n] = 0.5^n u[n]$. What would be the response of the system $y[n]$ to $x[n] = u[n] - u[n-4]$? Plot the output $y[n]$.

(c) Use the convolution sum to verify your response $y[n]$.

(d) Use the MATLAB function conv to compute the response $y[n]$ to $x[n] = u[n] - u[n-4]$. Plot both the input and output.

8.25. Discrete envelope detector—MATLAB

Consider an *envelope detector* that would be used to detect the message sent in an AM system. Consider the envelope detector as a system composed of the cascading of two systems: one which computes the absolute value of the input, and a second one that low-pass filters its input. A circuit that is used as an envelope detector consists of a diode circuit that does the absolute value operation, and an RC circuit that does the low-pass filtering. The following is an implementation of these operations in the discrete-time system.

Let the input to the envelope detector be a sampled signal,

$$x(nT_s) = p(nT_s) \cos(2000\pi nT_s)$$

where

$$p(nT_s) = u(nT_s) - u(nT_s - 20T_s) + u(nT_s - 40T_s) - u(nT_s - 60T_s)$$

where two pulses of duration $20T_s$ and amplitude equal to one.

(a) Choose $T_s = 0.01$, and generate 100 samples of the input signal $x(nT_s)$ and plot it.

(b) Consider then the subsystem that computes the absolute value of the input $x(nT_s)$ and compute and plot 100 samples of $y(nT_s) = |x(nT_s)|$.

(c) Let the low-pass filtering be done by a moving averager of order 15—that is, if $y(nT_s)$ is the input, then the output of the filter is

$$z(nT_s) = \frac{1}{15} \sum_{k=0}^{14} y(nT_s - kT_s)$$

Implement this filter using the MATLAB function filter, and plot the result. Explain your results.

(d) Is this a linear system? Come up with an example using the script developed above to show that the system is linear or not.

CHAPTER 9

The Z-Transform

*I was born not knowing and have had
only a little time to change that here and there.*
Richard P. Feynman, (1918–1988)
Professor and Nobel Prize physicist

9.1 INTRODUCTION

Just as with the Laplace transform for continuous-time signals and systems, the Z-transform provides
a way to represent discrete-time signals and systems, and to process discrete-time signals.

Although the Z-transform can be related to the Laplace transform, the relation is operationally not
very useful. However, it can be used to show that the complex z-plane is in a polar form where the
radius is a damping factor and the angle corresponds to the discrete frequency ω in radians. Thus,
the unit circle in the z-plane is analogous to the $j\Omega$ axis in the Laplace plane, and the inside of the
unit circle is analogous to the left-hand s-plane. We will see that once the connection between the
Laplace plane and the z-plane is established, the significance of poles and zeros in the z-plane can be
obtained like in the Laplace plane.

The representation of discrete-time signals by the Z-transform is very intuitive—it converts a sequence
of samples into a polynomial. The inverse Z-transform can be achieved by many more methods than
the inverse Laplace transform, but the partial fraction expansion is still the most commonly used
method. Using the one-sided Z-transform, for solving difference equations that could result from
the discretization of differential equations, but not exclusively, is an important application of the
Z-transform.

As it was the case with the Laplace transform and the convolution integral, the most important
property of the Z-transform is the implementation of the convolution sum as a multiplication of
polynomials. This is not only important as a computational tool but also as a way to represent a
discrete system by its transfer function. Filtering is again an important application, and as before, the

Signals and Systems Using MATLAB®. DOI: 10.1016/B978-0-12-374716-7.00013-2

localization of poles and zeros determines the type of filter. However, in the discrete domain there is a greater variety of filters than in the analog domain.

9.2 LAPLACE TRANSFORM OF SAMPLED SIGNALS

The Laplace transform of a sampled signal

$$x(t) = \sum_n x(nT_s)\delta(t - nT_s)$$

(9.1)

is given by

$$X(s) = \sum_n x(nT_s)\mathcal{L}[\delta(t - nT_s)]$$

$$= \sum_n x(nT_s)e^{-nsT_s}$$

(9.2)

By letting $z = e^{sT_s}$, we can rewrite Equation (9.2) as

$$\mathcal{Z}[x(nT_s)] = \mathcal{L}[x_s(t)]\big|_{z=e^{sT_s}}$$

$$= \sum_n x(nT_s)z^{-n}$$

(9.3)

which is called the Z-transform of the sampled signal.

Remarks *The function $X(s)$ in Equation (9.2) is different from the Laplace transforms we considered before:*

■ *Letting $s = j\Omega$, $X(\Omega)$ is periodic of period $2\pi/T_s$ (i.e., $X(\Omega + 2\pi/T_s) = X(\Omega)$ for an integer k). Indeed,*

$$X(\Omega + 2\pi/T_s) = \sum_n x(nT_s)e^{-jn(\Omega+2\pi/T_s)T_s} = \sum_n x(nT_s)e^{-jn(\Omega T_s+2\pi)} = X(\Omega)$$

■ *$X(s)$ may have an infinite number of poles or zeros—complicating the partial fraction expansion when finding its inverse. Fortunately, the presence of the $\{e^{-nsT_s}\}$ terms suggests that the inverse should be done using the time-shift property of the Laplace transform instead, giving Equation (9.1).*

■ Example 9.1

To see the possibility of an infinite number of poles and zeros in the Laplace transform of a sampled signal, consider a pulse $x(t) = u(t) - u(t - T_0)$ sampled with a sampling period $T_s = T_0/N$. Find the Laplace transform of the sampled signal and determine its poles and zeros.

Solution

The sampled signal is

$$x(nT_s) = \begin{cases} 1 & 0 \le nT_s \le T_0 \text{ or } 0 \le n \le N \\ 0 & \text{otherwise} \end{cases}$$

with Laplace transform

$$X(s) = \sum_{n=0}^{N} e^{-nsT_s} = \frac{1 - e^{-(N+1)sT_s}}{1 - e^{-sT_s}}$$

The poles are the s_k values that make the denominator zero—that is,

$$e^{-s_k T_s} = 1$$
$$= e^{j2\pi k} \quad k \text{ integer,} \quad -\infty < k < \infty$$

or $s_k = -j2\pi k/T_s$ for any integer k, an infinite number of poles. Similarly, one can show that $X(s)$ has an infinite number of zeros by finding the values s_m that make the numerator zero, or

$$e^{-(N+1)s_m T_s} = 1$$
$$= e^{j2\pi m} \quad m \text{ integer,} \quad -\infty < m < \infty$$

or $s_m = -j2\pi m/((N+1)T_s)$ for any integer m. Such a behavior can be better understood when we consider the connection between the s-plane and the z-plane. ∎

The History of the Z-Transform

The history of the Z-transform goes back to the work of the French mathematician De Moivre, who in 1730 introduced the characteristic function to represent the probability mass function of a discrete random variable. The characteristic function is identical to the Z-transform. Also, the Z-transform is a special case of the Laurent's series, used to represent complex functions.

In the 1950s the Russian engineer and mathematician Yakov Tsypkin (1919–1997) proposed the discrete Laplace transform, which he applied to the study of pulsed systems. Then Professor John Ragazzini and his students Eliahu Jury and Lofti Zadeh at Columbia University developed the Z-transform. Ragazzini (1912–1988) was chairman of the Department of Electrical Engineering at Columbia University. Three of his students are well recognized in electrical engineering for their accomplishments: Jury for the Z-transform, nonlinear systems, and the inners stability theory; Zadeh for the Z-transform and fuzzy set theory; and Rudolf Kalman for the Kalman filtering.

Jury was born in Iraq, and received his doctor of engineering science degree from Columbia University in 1953. He was professor of electrical engineering at the University of California, Berkeley, and at the University of Miami. Among his publications, Professor Jury's "Theory and Application of the Z-transform," is a seminal work on the theory and application of the Z-transform.

Remarks

- *The relation $z = e^{sT_s}$ provides the connection between the s-plane and the z-plane:*

$$z = e^{sT_s} = e^{(\sigma + j\Omega)T_s} = e^{\sigma T_s} e^{j\Omega T_s}$$

Letting $r = e^{\sigma T_s}$ and $\omega = \Omega T_s$, we have that

$$z = re^{j\omega}$$

which is a complex variable in polar form, with radius $0 \leq r < \infty$ and angle ω in radians. The variable r is a damping factor and ω is the discrete frequency in radians, so the z-plane corresponds to circles of radius r and angles $-\pi \leq \omega < \pi$.

■ *Let us see how the relation $z = e^{sT_s}$ maps the s-plane into the z-plane. Consider the strip of width $2\pi/T_s$ across the s-plane shown in Figure 9.1. The width of this strip is related to the Nyquist condition establishing that the maximum frequency of the analog signals we are considering is $\Omega_M = \Omega_s/2 = \pi/T_s$ where Ω_s is the sampling frequency and T_s is the sampling period. If $T_s \to 0$, we would be considering the class of signals with maximum frequency approaching ∞—that is, all signals.*

The relation $z = e^{sT_s}$ maps the real part of $s = \sigma + j\Omega$, $\mathcal{R}e(s) = \sigma$, into the radius $r = e^{\sigma T_s} \geq 0$, and the analog frequencies $-\pi/T_s \leq \Omega \leq \pi/T_s$ into $-\pi \leq \omega < \pi$, according to the frequency connection $\omega = \Omega T_s$. Thus, the mapping of the $j\Omega$ axis in the s-plane, corresponding to $\sigma = 0$, gives a circle of radius $r = 1$ or the unit circle.

The right-hand s-plane, $\sigma > 0$, maps into circles with radius $r > 1$, and the left-hand s-plane, $\sigma < 0$, maps into circles of radius $r < 1$. Points A, B, and C in the strip are mapped into corresponding points in the z-plane as shown in Figure 9.1. So the given strip in the s-plane maps into the whole z-plane— similarly for other strips of the same width. Thus, the s-plane, as a union of these strips, is mapped onto the same z-plane.

■ *The mapping $z = e^{sT_s}$ can be used to illustrate the sampling process. Consider a band-limited signal $x(t)$ with maximum frequency π/T_s with a spectrum in the band $[-\pi/T_s \ \pi/T_s]$. According to the relation $z = e^{sT_s}$ the spectrum of $x(t)$ in $[-\pi/T_s \ \pi/Ts]$ is mapped into the unit circle of the z-plane from $[-\pi,\pi)$ on the unit circle. Going around the unit circle in the z-plane, the mapped frequency response repeats periodically just like the spectrum of the sampled signal.*

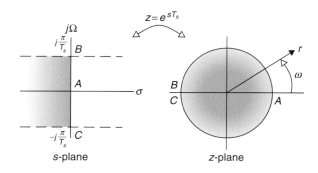

FIGURE 9.1

Mapping of the Laplace plane into the z-plane. Slabs of width $2\pi/T_s$ in the left-hand s-plane are mapped into the inside of a unit circle in the z-plane. The right-hand side of the slab is mapped outside the unit circle. The $j\Omega$-axis in the s-plane is mapped into the unit-circle in the z-plane. The whole s-plane as a union of these slabs is mapped onto the same z-plane.

9.3 TWO-SIDED Z-TRANSFORM

Given a discrete-time signal $x[n]$, $-\infty < n < \infty$, its two-sided Z-transform is

$$X(z) = \sum_{n=-\infty}^{\infty} x[n]z^{-n} \qquad (9.4)$$

defined in a region of convergence (ROC) in the z-plane.

Considering the sampled signal $x(nT_s)$ a function of n in Equation (9.3), we obtain the two-sided Z-transform.

Remarks

■ *The Z-transform can be thought of as the transformation of the sequence $\{x[n]\}$ into a polynomial $X(z)$ (possibly of infinite degree in positive and negative powers of z) where to each $x[n_0]$ we attach a monomial z^{-n_0}. Thus, given a sequence of samples $\{x[n]\}$ its Z-transform simply consists in creating a polynomial with coefficients $x[n]$ corresponding to z^{-n}. Given a Z-transform as in Equation (9.4), its inverse is easily obtained by looking at the coefficients attached to the z^{-n} monomials for positive as well as negative values of the sample value n. Clearly, this inverse is not in a closed form. We will see ways to compute these later in this chapter.*

■ *The two-sided Z-transform is not useful in solving difference equations with initial conditions, just as the two-sided Laplace transform was not useful either in solving differential equations with initial conditions. To include initial conditions in the transformation it is necessary to define the one-sided Z-transform.*

The one-sided Z-transform is defined for a causal signal, $x[n] = 0$ for $n < 0$, or for signals that are made causal by multiplying them with the unit-step signal $u[n]$:

$$X_1(z) = \mathcal{Z}(x[n]u[n]) = \sum_{n=0}^{\infty} x[n]u[n]z^{-n} \qquad (9.5)$$

in a region of convergence $\mathcal{R}_1$.

The two-sided Z-transform can be expressed in terms of the one-sided Z-transform as follows:

$$X(z) = \mathcal{Z}\left(x[n]u[n]\right) + \mathcal{Z}\left(x[-n]u[n]\right)|_z - x[0] \qquad (9.6)$$

The region of convergence of $X(z)$ is

$$\mathcal{R} = \mathcal{R}_1 \cap \mathcal{R}_2$$

where $\mathcal{R}_1$ is the region of convergence of $\mathcal{Z}\left(x[n]u[n]\right)$ and $\mathcal{R}_2$ is the region of convergence of $\mathcal{Z}\left(x[-n]u[n]\right)|_z$.

The one-sided Z-transform coincides with the two-sided Z-transform whenever the discrete-time signal $x[n]$ is causal (i.e., $x[n] = 0$ for $n < 0$). If the signal is noncausal, multiplying it by $u[n]$ makes it causal. To express the two-sided Z-transform in terms of the one-sided Z-transform we separate the sum into two and make each into a causal sum:

$$X(z) = \sum_{n=-\infty}^{\infty} x[n]z^{-n} = \sum_{n=0}^{\infty} x[n]u[n]z^{-n} + \sum_{n=-\infty}^{0} x[n]u[-n]z^{-n} - x[0]$$

$$= \mathcal{Z}\left(x[n]u[n]\right) + \sum_{m=0}^{\infty} x[-m]u[m]z^{m} - x[0]$$

$$= \mathcal{Z}\left(x[n]u[n]\right) + \mathcal{Z}\left(x[-n]u[n]\right)|_z - x[0]$$

where the inclusion of the additional term $x[0]$ in the sum from $-\infty$ to 0 is compensated by subtracting it, and in the same sum a change of variable ($m = -n$) gives a one-sided Z-transform in terms of positive powers of z, as indicated by the notation $\mathcal{Z}\left(x[-n]u[n]\right)|_z$.

9.3.1 Region of Convergence

The infinite summation of the two-sided Z-transform must converge for some values of z. For $X(z)$ to converge it is necessary that

$$|X(z)| = \left|\sum_{n} x[n]z^{-n}\right| \leq \sum_{n} |x[n]||r^{-n}e^{j\omega n}| = \sum_{n} |x[n]||r^{-n}| < \infty$$

Thus, the convergence of $X(z)$ depends on r. The region in the z-plane where $X(z)$ converges, its ROC, connects the signal and its Z-transform uniquely. As with the Laplace transform, the poles of $X(z)$ are connected with its region of convergence.

The poles of a Z-transform $X(z)$ are complex values $\{p_k\}$ such that

$$X(p_k) \to \infty$$

while the zeros of $X(z)$ are the complex values $\{z_k\}$ that make

$$X(z_k) = 0$$

■ Example 9.2

Find the poles and the zeros of the following Z-transforms:

$$\textbf{(a) } X_1(z) = 1 + 2z^{-1} + 3z^{-2} + 4z^{-3}$$

$$\textbf{(b) } X_2(z) = \frac{(z^{-1} - 1)(z^{-1} + 2)^2}{z^{-1}(z^{-2} + \sqrt{2}z^{-1} + 1)}$$

Solution

To see the poles and the zeros more clearly let us express $X_1(z)$ as a function of positive powers of z:

$$X_1(z) = \frac{z^3(1 + 2z^{-1} + 3z^{-2} + 4z^{-3})}{z^3}$$

$$= \frac{z^3 + 2z^2 + 3z + 4}{z^3} = \frac{N_1(z)}{D_1(z)}$$

There are three poles at $z = 0$, the roots of $D_1(z) = 0$, and the zeros are the roots of $N_1(z) = z^3 + 2z^2 + 3z + 4 = 0$.

Likewise, expressing $X_2(z)$ as a function of positive powers of z,

$$X_2(z) = \frac{z^3(z^{-1} - 1)(z^{-1} + 2)^2}{z^3(z^{-1}(z^{-2} + \sqrt{2}z^{-1} + 1))}$$

$$= \frac{(1 - z)(1 + 2z)^2}{1 + \sqrt{2}z + z^2} = \frac{N_2(z)}{D_2(z)}$$

The poles of $X_2(z)$ are the roots of $D_2(z) = 1 + \sqrt{2}z + z^2 = 0$, while the zeros of $X_2(z)$ are the roots of $N_2(z) = (1 - z)(1 + 2z)^2 = 0$. ∎

The region of convergence depends on the support of the signal. If it is finite, the ROC is very much the whole z-plane; if it is infinite, the ROC depends on whether the signal is causal, anti-causal, or noncausal. Something to remember is that in no case does the ROC include any poles of the Z-transform.

ROC of Finite-Support Signals

The ROC of the Z-transform of a signal $x[n]$ of finite support $[N_0, N_1]$ where $-\infty < N_0 \leq n \leq N_1 < \infty$,

$$X(z) = \sum_{n=N_0}^{N_1} x[n]z^{-n} \tag{9.7}$$

is the whole z-plane, excluding the origin $z = 0$ and/or $z = \pm\infty$ depending on N_0 and N_1.

Given the finite support of $x[n]$ its Z-transform has no convergence problem. Indeed, for any $z \neq 0$ (or $z \neq \pm\infty$ if positive powers of z occur in Equation (9.7)), we have

$$|X(z)| \leq \sum_{n=N_0}^{N_1} |x[n]||z^{-n}| \leq (N_1 - N_0 + 1) \max |x[n]| \max |z^{-n}| < \infty$$

The poles of $X(z)$ are either at the origin of the z-plane (e.g., when $N_0 \geq 0$) or there are no poles (e.g., when $N_1 \leq 0$). Thus, only when $z = 0$ or $z = \pm\infty$ would $X(z)$ go to infinity. The ROC is the whole z-plane excluding these values.

■ Example 9.3

Find the Z-transform of a discrete-time pulse

$$x[n] = \begin{cases} 1 & 0 \leq n \leq 9 \\ 0 & \text{otherwise} \end{cases}$$

Determine the region of convergence of $X(z)$.

Solution

The Z-transform of $x[n]$ is

$$X(z) = \sum_{n=0}^{9} 1 \, z^{-n} = \frac{1 - z^{-10}}{1 - z^{-1}} = \frac{z^{10} - 1}{z^{9}(z - 1)} \tag{9.8}$$

That this sum equals the term on the right can be shown by multiplying the left term by the denominator $1 - z^{-1}$ and verifying the result is the same as the numerator in negative powers of z. In fact,

$$(1 - z^{-1}) \sum_{n=0}^{9} 1 \, z^{-n} = \sum_{n=0}^{9} 1 \, z^{-n} - \sum_{n=0}^{9} 1 \, z^{-n-1}$$

$$= (1 + z^{-1} + \cdots + z^{-9}) - (z^{-1} + \cdots + z^{-9} + z^{-10}) = 1 - z^{-10}$$

Since $x[n]$ is a finite sequence there is no problem with the convergence of the sum, although $X(z)$ in Equation (9.8) seems to indicate the need for $z \neq 1$ ($z = 1$ makes the numerator and denominator zero). From the sum, if we let $z = 1$, then $X(1) = 10$, so there is no need to restrict z to be different from 1. This is caused by the pole at $z = 1$ being canceled by a zero. Indeed, the zeros z_k of $X(z)$ (see Eq. 9.8) are the roots of $z^{10} - 1 = 0$, which are $z_k = e^{j2\pi k/10}$ for $k = 0, \ldots, 9$. Therefore, the zero when $k = 0$, or $z_0 = 1$, cancels the pole at 1 so that

$$X(z) = \frac{\prod_{k=1}^{9}(z - e^{j\pi k/5})}{z^{9}}$$

That is, $X(z)$ has nine poles at the origin and nine zeros around the unit circle except at $z = 1$. Thus, the whole z-plane excluding the origin is the region of convergence of $X(z)$. ■

ROC of Infinite-Support Signals

Signals of infinite support are either causal, anti-causal, or a combination of these or noncausal. Now for the Z-transform of a causal signal $x_c[n]$ (i.e., $x_c[n] = 0, n < 0$)

$$X_c(z) = \sum_{n=0}^{\infty} x_c[n] z^{-n} = \sum_{n=0}^{\infty} x_c[n] r^{-n} e^{-jn\omega}$$

to converge we need to determine appropriate values of r, the damping factor. The frequency ω has no effect on the convergence. If R_1 is the radius of the farthest-out pole of $X_c(z)$, then there is an exponential $R_1^n u[n]$ such that $|x_c[n]| < MR_1^n$ for $n \geq 0$ for some value $M > 0$. Then, for $X(z)$ to converge we need that

$$|X_c(z)| \leq \sum_{n=0}^{\infty} |x_c[n]||r^{-n}| < M \sum_{n=0}^{\infty} \left| \frac{R_1}{r} \right|^n < \infty$$

or that $R_1/r < 1$, which is equivalent to $|z| = r > R_1$. As indicated, this ROC does not include any poles of $X_c(z)$—it is the outside of a circle containing all the poles of $X_c(z)$.

Likewise, for an anti-causal signal $x_a[n]$, if we choose a radius R_2 that is smaller than the radius of all the poles of $X_a(z)$, the region of convergence is $|z| = r < R_2$. This ROC does not include any poles of $X_a(z)$—it is the inside of a circle that does not contain any of the poles of $X_a(z)$.

If the signal $x[n]$ is noncausal, it can be expressed as

$$x[n] = x_c[n] + x_a[n]$$

where the supports of $x_a[n]$ and $x_c[n]$ can be finite or infinite or any possible combination of these two. The corresponding ROC of $X(z) = \mathcal{Z}\{x[n]\}$ would then be

$$0 \leq R_1 < |z| < R_2 < \infty$$

This ROC is a torus surrounded on the inside by the poles of the causal component, and in the outside by the poles of the anti-causal component. If the signal has finite support, then $R_1 = 0$ and $R_2 = \infty$, coinciding with the result for finite-support signals.

For the Z-transform $X(z)$ of an infinite-support signal:
- A causal signal $x[n]$ has a region of convergence $|z| > R_1$ where R_1 is the largest radius of the poles of $X(z)$—that is, the region of convergence is the outside of a circle of radius R_1.
- An anti-causal signal $x[n]$ has as region of convergence the inside of the circle defined by the smallest radius R_2 of the poles of $X(z)$, or $|z| < R_2$.
- A noncausal signal $x[n]$ has as region of convergence $R_1 < |z| < R_2$, or the inside of a torus of inside radius R_1 and outside radius R_2 corresponding to the maximum and minimum radii of the poles of $X_c(z)$ and $X_a(z)$, which are the Z-transforms of the causal and anti-causal components of $x[n]$.

■ Example 9.4

The poles of $X(z)$ are $z = 0.5$ and $z = 2$. Find all the possible signals that can be associated with it according to different regions of convergence.

Solution

Possible regions of convergence are:

- $\{\mathcal{R}_1 : |z| > 2\}$—the outside of a circle of radius 2, we associate $X(z)$ with a causal signal $x_1[n]$.

■ $\{\mathcal{R}_2 : |z| < 0.5\}$—the inside of a circle of radius 0.5, an anti-causal signal $x_2[n]$ can be associated with $X(z)$.

■ $\{\mathcal{R}_3 : 0.5 < |z| < 2\}$—a torus of radii 0.5 and 2, a noncausal signal $x_3[n]$ can be associated with $X(z)$.

Three different signals can be connected with $X(z)$ by considering three different regions of convergence. ■

■ Example 9.5

Find the regions of convergence of the Z-transforms of the following signals:

$$\textbf{(a) } x_1[n] = \left(\frac{1}{2}\right)^n u[n]$$

$$\textbf{(b) } x_2[n] = -\left(\frac{1}{2}\right)^n u[-n-1]$$

Determine then the Z-transform of $x_1[n] + x_2[n]$.

Solution

The signal $x_1[n]$ is causal, while $x_2[n]$ is anti-causal. The Z-transform of $x_1[n]$ is

$$X_1(z) = \sum_{n=0}^{\infty} \left(\frac{1}{2}\right)^n z^{-n} = \frac{1}{1 - 0.5z^{-1}} = \frac{z}{z - 0.5}$$

provided that $|0.5z^{-1}| < 1$ or that its region of convergence is $\mathcal{R}_1 : |z| > 0.5$. The region $\mathcal{R}_1$ is the outside of a circle of radius 0.5.

The signal $x_2[n]$ grows as n decreases from -1 to $-\infty$, and the rest of its values are zero. Its Z-transform is found as

$$X_2(z) = -\sum_{n=-\infty}^{-1} \left(\frac{1}{2}\right)^n z^{-n} = -\sum_{m=0}^{\infty} \left(\frac{1}{2}\right)^{-m} z^m + 1$$

$$= -\sum_{m=0}^{\infty} 2^m z^m + 1 = \frac{-1}{1 - 2z} + 1 = \frac{z}{z - 0.5}$$

with a region of convergence of $\mathcal{R}_2 : |z| < 0.5$.

Although the signals are clearly different, their Z-transforms are identical. It is the corresponding regions of convergence that differentiate them. The Z-transform of $x_1[n] + x_2[n]$ does not exist given that the intersection of $\mathcal{R}_1$ and $\mathcal{R}_2$ is empty. ■

Remarks *The uniqueness of the Z-transform requires that the Z-transform of a signal be accompanied by a region of convergence. It is possible to have identical Z-transforms with different regions of convergence, corresponding to different signals.*

■ Example 9.6

Let $c[n] = \alpha^{|n|}$, $0 < \alpha < 1$, be a discrete-time signal (it is actually an autocorrelation function related to the power spectrum of a random signal). Determine its Z-transform.

Solution

To find its two-sided Z-transform $C(z)$ we consider its causal and anti-causal components. First,

$$\mathcal{Z}(c[n]u[n]) = \sum_{n=0}^{\infty} \alpha^n z^{-n} = \frac{1}{1 - \alpha z^{-1}}$$

with the region of convergence of $|\alpha z^{-1}| < 1$ or $|z| > \alpha$. For the anti-causal component,

$$\mathcal{Z}(c[-n]u[n])_z = \sum_{n=0}^{\infty} \alpha^n z^n = \frac{1}{1 - \alpha z}$$

with a region of convergence of $|\alpha z| < 1$ or $|z| < |1/\alpha|$.

Thus, the two-sided Z-transform of $c[n]$ is (notice that the term for $n = 0$ was used twice in the above calculations, so we need to subtract it)

$$C(z) = \frac{1}{1 - \alpha z^{-1}} + \frac{1}{1 - \alpha z} - 1 = \frac{z}{z - \alpha} - \frac{z}{(z - 1/\alpha)}$$

$$= \frac{(\alpha - 1/\alpha)z}{(z - \alpha)(z - 1/\alpha)}$$

with a region of convergence of

$$|\alpha| < |z| < \left| \frac{1}{\alpha} \right|$$

For instance, for $\alpha = 0.5$, we get

$$C(z) = \frac{-1.5z}{(z - 0.5)(z - 2)} \qquad 0.5 < |z| < 2 \qquad ∎$$

9.4 ONE-SIDED Z-TRANSFORM

In most situations where the Z-transform is used the system is causal (its impulse response is $h[n] = 0$ for $n < 0$) and the input signal is also causal ($x[n] = 0$ for $n < 0$). In such cases the one-sided Z-transform is very appropriate. Moreover, as we saw before, the two-sided Z-transform can be expressed in terms of one-sided Z-transforms. Another valid reason to study the one-sided Z-transform in more detail is its use in solving difference equations with initial conditions.

Recall that the one-sided Z-transform is defined as

$$X_1(z) = \mathcal{Z}(x[n]u[n]) = \sum_{n=0}^{\infty} x[n]u[n]z^{-n} \tag{9.9}$$

in a region of convergence $\mathcal{R}_1$. Also recall that the computation of the two-sided Z-transform using the one-sided Z-transform is given in Equation (9.6).

9.4.1 Computing the Z-Transform with Symbolic MATLAB

Similar to the computation of the Laplace transform, the computation of the Z-transform can be done using the symbolic toolbox of MATLAB. The following is the necessary code for computing the Z-transform of

$$h_1[n] = 0.9u[n]$$
$$h_2[n] = u[n] - u[n - 10]$$
$$h_3[n] = \cos(\omega_0 n)u[n]$$
$$h_4[n] = h_1[n]h_3[n]$$

The results are shown at the bottom. (As in the continuous case, in MATLAB the heaviside function is the same as the unit-step function.)

```
%%%%%%%%%%%%%%%%%%%%%
% Z-transform computation
%%%%%%%%%%%%%%%%%%%%%%%
syms n w0
h1 = 0.9.^n; H1 = ztrans(h1)
h2 = heaviside(n) - heaviside(n-10); H2 = ztrans(h2)
h3 = cos(w0*n)*heaviside(n); H3 = ztrans(h3)
H4 = ztrans(h1*h3)

    H1 = 10/9/(10/9*z - 1)*z
    H2 = 1 + 1/z + 1/z^2 + 1/z^3 + 1/z^4 + 1/z^5 + 1/z^6 + 1/z^7 + 1/z^8 + 1/z^9
    H3 = (z - cos(w0))*z/(z^2 - 2*z*cos(w0) + 1)
    H4 = 10/9*(10/9*z - cos(w0))*z/(100/81*z^2 - 20/9*z*cos(w0) + 1)
```

The function iztrans computes the inverse Z-transform. We will illustrate its use later on.

9.4.2 Signal Behavior and Poles

In this section we will use the linearity property of the Z-transform to connect the behavior of the signal with the poles of its Z-transform.

The Z-transform is a linear transformation, meaning that

$$\mathcal{Z}(ax[n] + by[n]) = a\mathcal{Z}(x[n]) + b\mathcal{Z}(y[n]) \qquad (9.10)$$

for signals $x[n]$ and $y[n]$ and constants a and b.

To illustrate the linearity property as well as the connection between the signal and the poles of its Z-transform, consider the signal $x[n] = \alpha^n u[n]$ for real or complex values α. Its Z-transform will be used to compute the Z-transform of the following signals:

- $x[n] = \cos(\omega_0 n + \theta)u[n]$ for frequency $0 \le \omega_0 \le \pi$ and phase θ.
- $x[n] = \alpha^n \cos(\omega_0 n + \theta)u[n]$ for frequency $0 \le \omega_0 \le \pi$ and phase θ.

Show how the poles of the corresponding Z-transform connect with the signals.

The Z-transform of the causal signal $x[n] = \alpha^n u[n]$ is

$$X(z) = \sum_{n=0}^{\infty} \alpha^n z^{-n} = \sum_{n=0}^{\infty} (\alpha z^{-1})^n = \frac{1}{1 - \alpha z^{-1}} = \frac{z}{z - \alpha} \quad \text{ROC: } |z| > |\alpha| \qquad (9.11)$$

Using the last expression in Equation (9.11) the zero of $X(z)$ is $z = 0$ and its pole is $z = \alpha$, since the first value makes $X(0) = 0$ and the second makes $X(\alpha) \to \infty$. For α real, be it positive or negative, the region of convergence is the same, but the poles are located in different places. See Figure 9.2 for $\alpha < 0$.

If $\alpha = 1$ the signal $x[n] = u[n]$ is constant for $n \ge 0$ and the pole of $X(z)$ is at $z = 1e^{j0}$ (the radius is $r = 1$ and the lowest discrete frequency $\omega = 0$ rad). On the other hand, when $\alpha = -1$ the signal is $x[n] = (-1)^n u[n]$, which varies from sample to sample for $n \ge 0$; its Z-transform has a pole at $z = -1 = 1e^{j\pi}$ (a radius $r = 1$ and the highest discrete frequency $\omega = \pi$ rad). As we move the pole toward the center of the z-plane (i.e., $|\alpha| \to 0$), the corresponding signal decays exponentially for $0 < \alpha < 1$, and is a modulated exponential of $|\alpha|^n (-1)^n u[n] = |\alpha|^n \cos(\pi n)u[n]$ for $-1 < \alpha < 0$. When $|\alpha| > 1$ the signal becomes either a growing exponential ($\alpha > 1$) or a growing modulated exponential ($\alpha < -1$).

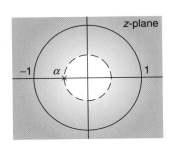

FIGURE 9.2
Region of convergence (shaded area) of $X(z)$ with a pole at $z = \alpha$, $\alpha < 0$ (same ROC if pole is at $z = -\alpha$).

For a real value $\alpha = |\alpha| e^{j\omega_0}$ for $\omega_0 = 0$ or π,

$$x[n] = \alpha^n u[n] \quad \Leftrightarrow \quad X(z) = \frac{1}{1 - \alpha z^{-1}} = \frac{z}{z - \alpha} \quad \text{ROC: } |z| > |\alpha|$$

and the location of the pole of $X(z)$ determines the behavior of the signal:

- When $\alpha > 0$, then $\omega_0 = 0$ and the signal is less and less damped as $\alpha \to \infty$.
- When $\alpha < 0$, then $\omega_0 = \pi$ and the signal is a modulated exponential that grows as $\alpha \to -\infty$.

To compute the Z-transform of $x[n] = \cos(\omega_0 n + \theta) u[n]$, we use Euler's identity to write $x[n]$ as

$$x[n] = \left[\frac{e^{j(\omega_0 n + \theta)}}{2} + \frac{e^{-j(\omega_0 n + \theta)}}{2} \right] u[n]$$

Applying the linearity property and using the above Z-transform when $\alpha = e^{j\omega_0}$ and its conjugate $\alpha^* = e^{-j\omega_0}$, we get

$$X(z) = \frac{1}{2} \left[\frac{e^{j\theta}}{1 - e^{j\omega_0} z^{-1}} + \frac{e^{-j\theta}}{1 - e^{-j\omega_0} z^{-1}} \right]$$

$$= \frac{1}{2} \left[\frac{2 \cos(\theta) - 2 \cos(\omega_0 - \theta) z^{-1}}{1 - 2 \cos(\omega_0) z^{-1} + z^{-2}} \right]$$

$$= \frac{\cos(\theta) - \cos(\omega_0 - \theta) z^{-1}}{1 - 2 \cos(\omega_0) z^{-1} + z^{-2}} \tag{9.12}$$

Expressing $X(z)$ in terms of positive powers of z, we get

$$X(z) = \frac{z(z \cos(\theta) - \cos(\omega_0 - \theta))}{z^2 - 2 \cos(\omega_0) z + 1} = \frac{z(z \cos(\theta) - \cos(\omega_0 - \theta))}{(z - e^{j\omega_0})(z - e^{-j\omega_0})} \tag{9.13}$$

which is valid for any value of θ. If $x[n] = \cos(\omega_0 n) u[n]$, then $\theta = 0$ and the poles of $X(z)$ are a complex conjugate pair on the unit circle at frequency ω_0 radians. The zeros are at $z = 0$ and $z = \cos(\omega_0)$. When $x[n] = \sin(\omega_0 n) u[n] = \cos(\omega_0 n - \pi/2) u[n]$, then $\theta = -\pi/2$ and the poles are at the same location as those for the cosine, but the zeros are at $z = 0$ and $z = \cos(\omega_0 + \pi/2)/\cos(\pi/2) \to \infty$, so there is only one finite zero at zero. For any other value of θ, the poles are located in the same place but there is a zero at $z = 0$ and another at $z = \cos(\omega_0 - \theta)/\cos(\theta)$.

For simplicity, we let $\theta = 0$. If $\omega_0 = 0$, one of the double poles at $z = 1$ is canceled by one of the zeros at $z = 1$, resulting in the poles and the zeros of $\mathcal{Z}([u[n])$. Indeed, the signal when $\omega_0 = 0$ and $\theta = 0$ is $x[n] = \cos(0n) u[n] = u[n]$. When the frequency $\omega_0 > 0$ the poles move along the unit circle from the lowest ($\omega_0 = 0$ rad) to the highest ($\omega_0 = \pi$ rad) frequency.

The Z-transform pairs of a cosine and a sine are, respectively,

$$\cos(\omega_0 n)u[n] \quad \Leftrightarrow \quad \frac{z(z - \cos(\omega_0))}{(z - e^{j\omega_0})(z - e^{-j\omega_0})} \qquad \text{ROC} : |z| > 1 \qquad (9.14)$$

$$\sin(\omega_0 n)u[n] \quad \Leftrightarrow \quad \frac{z\sin(\omega_0)}{(z - e^{j\omega_0})(z - e^{-j\omega_0})} \qquad \text{ROC} : |z| > 1 \qquad (9.15)$$

The Z-transforms for these sinusoids have identical poles $1e^{\pm j\omega_0}$, but different zeros. The frequency of the sinusoid increases from the lowest ($\omega_0 = 0$ rad) to the highest ($\omega_0 = \pi$ rad)) as the poles move along the unit circle from 1 to -1 in its lower and upper parts.

Consider then the signal $x[n] = r^n \cos(\omega_0 n + \theta)u[n]$, which is a combination of the above cases. As before, the signal is expressed as a linear combination

$$x[n] = \left[\frac{e^{j\theta}(re^{j\omega_0})^n}{2} + \frac{e^{-j\theta}(re^{-j\omega_0})^n}{2} \right] u[n]$$

and it can be shown that its Z-transform is

$$X(z) = \frac{z(z\cos(\theta) - r\cos(\omega_0 - \theta))}{(z - re^{j\omega_0})(z - re^{-j\omega_0})} \qquad (9.16)$$

The Z-transform of a sinusoid is a special case of the above (i.e., when $r = 1$). It also becomes clear that as the value of r decreases toward zero, the exponential in the signal decays faster, and that whenever $r > 1$, the exponential in the signal grows making the signal unbound.

The Z-transform pair

$$r^n \cos(\omega_0 n + \theta)u[n] \quad \Leftrightarrow \quad \frac{z(z\cos(\theta) - r\cos(\omega_0 - \theta))}{(z - re^{j\omega_0})(z - re^{-j\omega_0})} \qquad (9.17)$$

shows how complex conjugate pairs of poles inside the unit circle represent the damping indicated by the radius r and the frequency given by ω_0 in radians.

Double poles are related to the derivative of $X(z)$ or to the multiplication of the signal by n. If

$$X(z) = \sum_{n=0}^{\infty} x[n]z^{-n}$$

its derivative with respect to z is

$$\frac{dX(z)}{dz} = \sum_{n=0}^{\infty} x[n]\frac{dz^{-n}}{dz}$$

$$= -z^{-1} \sum_{n=0}^{\infty} nx[n]z^{-n}$$

Or the pair

$$nx[n]u[n] \quad \Leftrightarrow \quad -z\frac{dX(z)}{dz} \tag{9.18}$$

For instance, if $X(z) = 1/(1 - \alpha z^{-1}) = z/(z - \alpha)$, we find that

$$\frac{dX(z)}{dz} = -\frac{\alpha}{(z - \alpha)^2}$$

That is, the pair

$$n\alpha^n u[n] \quad \Leftrightarrow \quad \frac{\alpha z}{(z - \alpha)^2}$$

indicates that double poles correspond to multiplication of $x[n]$ by n.

The above shows that the location of the poles of $X(z)$ provides basic information about the signal $x[n]$. This is illustrated in Figure 9.3, where we display the signal and its corresponding poles.

9.4.3 Convolution Sum and Transfer Function

The most important property of the Z-transform, as it was for the Laplace transform, is the convolution property.

The output $y[n]$ of a causal LTI system is computed using the convolution sum

$$y[n] = [x * h][n] = \sum_{k=0}^{n} x[k]h[n-k] = \sum_{k=0}^{n} h[k]x[n-k] \tag{9.19}$$

where $x[n]$ is a causal input and $h[n]$ is the impulse response of the system. The Z-transform of $y[n]$ is the product

$$Y(z) = \mathcal{Z}\{[x * h][n]\} = \mathcal{Z}\{x[n]\}\mathcal{Z}\{h[n]\} = X(z)H(z) \tag{9.20}$$

and the transfer function of the system is thus defined as

$$H(z) = \frac{Y(z)}{X(z)} = \frac{\mathcal{Z}[\text{output } y[n]]}{\mathcal{Z}[\text{input } x[n]]} \tag{9.21}$$

That is, $H(z)$ transfers the input $X(z)$ into the output $Y(z)$.

Remarks

- *The convolution sum property can be seen as a way to obtain the coefficients of the product of two polynomials. Whenever we multiply two polynomials $X_1(z)$ and $X_2(z)$, of finite or infinite order, the coefficients of the resulting polynomial can be obtained by means of the convolution sum. For instance, consider*

$$X_1(z) = 1 + a_1 z^{-1} + a_2 z^{-2}$$

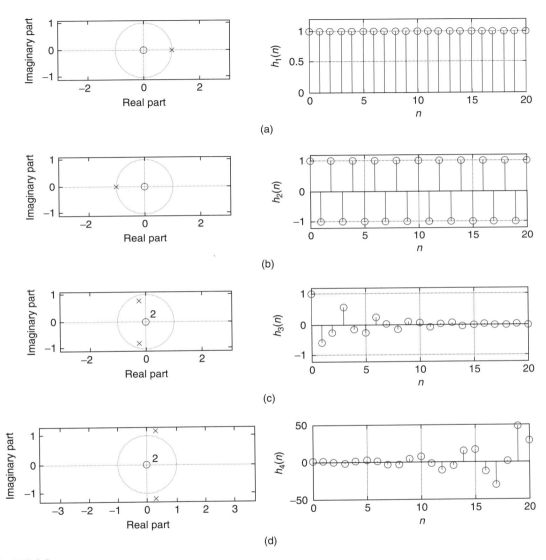

FIGURE 9.3

Effect of pole location on the inverse Z-transform: (a) if the pole is at $z = 1$ the signal is $u[n]$, constant for $n \geq 0$; (b) if the pole is at $z = -1$ the signal is a cosine of frequency π continuously changing, constant amplitude; (c, d) when poles are complex, if inside the unit circle the signal is a decaying modulated exponential, and if outside the unit circle the signal is a growing modulated exponential.

and

$$X_2(z) = 1 + b_1 z^{-1}$$

Their product is

$$X_1(z)X_2(z) = 1 + b_1 z^{-1} + a_1 z^{-1} + a_1 b_1 z^{-2} + a_2 z^{-2} + a_2 b_1 z^{-3}$$

$$= 1 + (b_1 + a_1)z^{-1} + (a_1 b_1 + a_2)z^{-2} + a_2 b_1 z^{-3}$$

The convolution sum of the two sequences $[1 \ a_1 \ a_2]$ and $[1 \ b_1]$, formed by the coefficients of $X_1(z)$ and $X_2(z)$, is $[1 \ (a_1 + b_1) \ (a_2 + b_1 a_1) \ a_2]$, which corresponds to the coefficients of the product of the polynomials $X_1(z)X_2(z)$. Also notice that the sequence of length 3 (corresponding to the first-order polynomial $X_1(z)$) and the sequence of length 2 (corresponding to the second-order polynomial $X_2(z)$) when convolved give a sequence of length $3 + 2 - 1 = 4$ (corresponding to the third-order polynomial $X_1(z)X_2(z)$).

■ *A finite-impulse response or FIR filter is implemented by means of the convolution sum. Consider an FIR with an input–output equation*

$$y[n] = \sum_{k=0}^{N-1} b_k x[n-k] \tag{9.22}$$

where $x[n]$ is the input and $y[n]$ is the output. The impulse response of this filter is (let $x[n] = \delta[n]$ and set initial conditions to zero, so that $y[n] = h[n]$)

$$h[n] = \sum_{k=0}^{N-1} b_k \delta[n-k]$$

giving $h[n] = b_n$ for $n = 0, \ldots, N-1$, and accordingly, we can write Equation (9.22) as

$$y[n] = \sum_{k=0}^{N-1} h[k]x[n-k]$$

which is the convolution of the input $x[n]$ and the impulse response $h[n]$ of the FIR filter. Thus, if $X(z) = \mathcal{Z}(x[n])$ and $H(z) = \mathcal{Z}(h[n])$, then

$$Y(z) = H(z)X(z) \quad \text{and} \quad y[n] = \mathcal{Z}^{-1}[Y(z)]$$

■ The length of the convolution of two sequences of lengths M and N is $M + N - 1$.
■ If one of the sequences is of infinite length, the length of the convolution is infinite. Thus, for an infinite-impulse response (IIR) or recursive filters the output is always of infinite length for any input signal, given that the impulse response of these filters is of infinite length.

■ Example 9.7

Consider computing the output of an FIR filter,

$$y[n] = \frac{1}{2}\left(x[n] + x[n-1] + x[n-2]\right)$$

for an input $x[n] = u[n] - u[n-4]$ using the convolution sum, analytically and graphically, and the Z-transform.

Solution

The impulse response is $h[n] = 0.5(\delta[n] + \delta[n-1] + \delta[n-2])$, so that $h[0]$, $h[1]$, $h[2]$ are, respectively, 0.5, 0.5, and 0.5, and $h[n]$ is zero otherwise.

Convolution sum formula: The equation

$$y[n] = \sum_{k=0}^{n} h[k]x[n-k]$$
$$= x[0]h[n] + x[1]h[n-1] + \cdots + x[n]h[0] \quad n \geq 0$$

with the condition that in each entry the arguments of $x[.]$ and $h[.]$ add to $n \geq 0$, gives

$y[0] = x[0]h[0] = 0.5$

$y[1] = x[0]h[1] + x[1]h[0] = 1$

$y[2] = x[0]h[2] + x[1]h[1] + x[2]h[0] = 1.5$

$y[3] = x[0]h[3] + x[1]h[2] + x[2]h[1] + x[3]h[0] = x[1]h[2] + x[2]h[1] + x[3]h[0] = 1.5$

$y[4] = x[0]h[4] + x[1]h[3] + x[2]h[2] + x[3]h[1] + x[4]h[0] = x[2]h[2] + x[3]h[1] = 1$

$y[5] = x[0]h[5] + x[1]h[4] + x[2]h[3] + x[3]h[2] + x[4]h[1] + x[5]h[0] = x[3]h[2] = 0.5$

and the rest are zero. In the above computations, we notice that the length of $y[n]$ is $4 + 3 - 1 = 6$ since the length of $x[n]$ is 4 and that of $h[n]$ is 3.

Graphical approach: The convolution sum is given by either

$$y[n] = \sum_{k=0}^{n} x[k]h[n-k]$$
$$= \sum_{k=0}^{n} h[k]x[n-k]$$

Choosing one of these equations, let's say the first one, we need $x[k]$ and $h[n-k]$, as functions of k, for different values of n. Multiply them and then add the nonzero values. For instance, for $n = 0$ the sequence $h[-k]$ is the reflection of $h[k]$; multiplying $x[k]$ by $h[-k]$ gives only one value different from zero at $k = 0$, or $y[0] = 1/2$. For $n = 1$, the sequence $h[1-k]$, as a function of k,

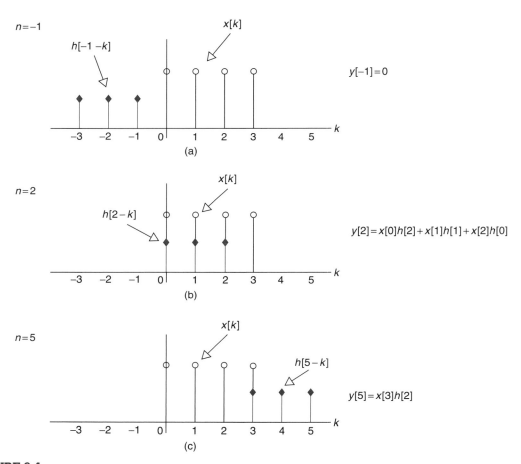

FIGURE 9.4

Graphical approach: convolution sum for (a) $n = -1$, (b) $n = 2$, and (c) $n = 5$ with corresponding outputs $y[-1]$, $y[2]$, and $y[5]$. Both $x[k]$ and $h[n - k]$ are plotted as functions of k for a given value of n. The signal $x[k]$ remains stationary, while $h[n - k]$ moves linearly from left to right. Thus, the convolution sum is also called a linear convolution.

is $h[-k]$ shifted to the right one sample. Multiplying $x[k]$ by $h[1 - k]$ gives two values different from zero, which when added gives $y[1] = 1$, and so on. For increasing values of n we shift to the right one sample to get $h[n - k]$, multiply it by $x[k]$, and then add the nonzero values to obtain the output $y[n]$. Figure 9.4 displays the graphical computation of the convolution sum for $n = -1$, $n = 2$ and $n = 5$.

Convolution sum property: We have

$$X(z) = 1 + z^{-1} + z^{-2} + z^{-3}$$

$$H(z) = \frac{1}{2}\left[1 + z^{-1} + z^{-2}\right]$$

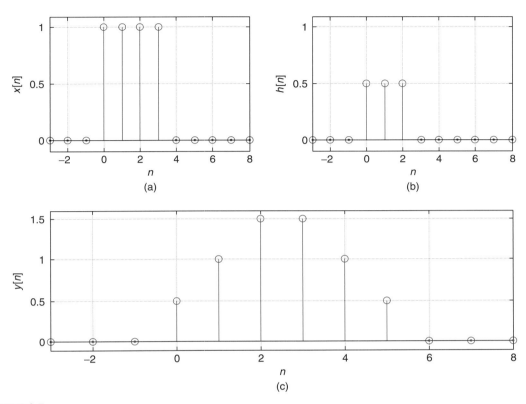

FIGURE 9.5

Convolution sum for an averager FIR: (a) $x[n]$, (b) $h[n]$, and (c) $y[n]$. The output $y[n]$ is of length 6 given that $x[n]$ is of length 4 and $h[n]$ is the impulse response of a second-order FIR filter of length 3.

and according to the convolution sum property,

$$Y(z) = X(z)H(z) = \frac{1}{2}(1 + 2z^{-1} + 3z^{-2} + 3z^{-3} + 2z^{-4} + z^{-5})$$

Thus, $y[0] = 0.5$, $y[1] = 1$, $y[2] = 1.5$, $y[3] = 1.5$, $y[4] = 1$, and $y[5] = 0.5$, just as before. In MATLAB the function conv is used to compute the convolution sum giving the results shown in Figure 9.5, which coincide with the ones obtained in the other approaches. ■

■ Example 9.8

Consider an FIR filter with impulse response

$$h[n] = \delta[n] + \delta[n-1] + \delta[n-2]$$

Find the filter output for an input $x[n] = \cos(2\pi n/3)(u[n] - u[n-14])$. Use the convolution sum to find the output, and verify your results with MATLAB.

Solution

Graphical approach: Let us use the formula

$$y[n] = \sum_{k=0}^{n} x[k]h[n-k]$$

which keeps the more complicated signal $x[k]$ as the unchanged signal. The term $h[n-k]$ is $h[k]$ reversed for $n = 0$, and then shifted to the right for $n \geq 1$. The output is zero for negative values of n, and for $n \geq 0$ we have

$$y[0] = 1$$
$$y[1] = 0.5$$
$$y[n] = 0 \quad 2 \leq n \leq 13$$
$$y[14] = 0.5$$
$$y[15] = -0.5$$

The first value is obtained by reflecting the impulse response to get $h[-k]$, and when multiplied by $x[k]$ we only have the value at $k = 0$ different from zero, therefore $y[0] = x[0]h[0] = 1$. As we shift the impulse response to the right to get $h[1-k]$ for $n = 1$ and multiply it by $x[k]$, we get two values different from zero; when added they equal 0.5. The result for $2 \leq n \leq 13$ is zero because we add three values of -0.5, 1 and -0.5 from the cosine. These results are verified by MATLAB as shown in Figure 9.6. (the cosine does not look like a sampled cosine given that only three values are used per period).

Convolution property approach: By the convolution property, the Z-transform of the output $y[n]$ is

$$Y(z) = X(z)H(z) = X(z)(1 + z^{-1} + z^{-2}) = X(z) + X(z)z^{-1} + X(z)z^{-2}$$

The coefficients of $Y(z)$ can be obtained by adding the coefficients of $X(z)$, $X(z)z^{-1}$, and $X(z)z^{-2}$:

z^0	z^{-1}	z^{-2}	z^{-3}	z^{-4}	z^{-5}	z^{-6}	z^{-7}	z^{-8}	z^{-9}	z^{-10}	z^{-11}	z^{-12}	z^{-13}	z^{-14}	z^{-15}
1	−0.5	−0.5	1	−0.5	−0.5	1	−0.5	−0.5	1	−0.5	−0.5	1	−0.5		
	1	−0.5	−0.5	1	−0.5	−0.5	1	−0.5	−0.5	1	−0.5	−0.5	1	−0.5	
		1	−0.5	−0.5	1	−0.5	−0.5	1	−0.5	−0.5	1	−0.5	−0.5	1	−0.5

Adding these coefficients vertically, we obtain

$$Y(z) = 1 + 0.5z^{-1} + 0z^{-2} + \cdots + 0z^{-13} + 0.5z^{-14} - 0.5z^{-15}$$
$$= 1 + 0.5z^{-1} + 0.5z^{-14} - 0.5z^{-15}$$

Notice from this example that

- The convolution sum is simply calculating the coefficients of the polynomial product $X(z)H(z)$.

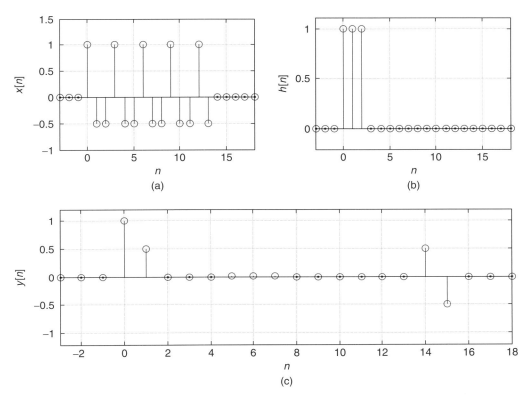

FIGURE 9.6

Convolution sum for FIR filter: (a) $x[n]$, (b) $h[n]$, and (c) $y[n]$.

- The length of the convolution sum = length of $x[n]$ + length of $h[n]$ $-1 = 14 + 3 - 1 =$ 16—that is, $Y(z)$ is a polynomial of order 15. ■

■ Example 9.9

The convolution sum of noncausal signals is more complicated graphically than that of the causal signals we showed in the previous examples. Let

$$h_1[n] = \frac{1}{3} \left(\delta[n+1] + \delta[n] + \delta[n-1] \right)$$

be the impulse response of a noncausal averager FIR filter, and $x[n] = u[n] - u[n-4]$ be the input. Compute the filter output using the convolution sum.

Solution

Graphically, it is a bit confusing to plot $h_1[n-k]$, as a function of k, to do the convolution sum. Using the convolution and the time-shifting properties of the Z-transform we can view the computation more clearly.

According to the convolution property the Z-transform of the output of the noncausal filter is

$$Y_1(z) = X(z)H_1(z)$$
$$= X(z)[zH(z)] \tag{9.23}$$

where we let

$$H_1(z) = \mathcal{Z}[h_1[n]] = \frac{1}{3}(z + 1 + z^{-1})$$
$$= z\left[\frac{1}{3}(1 + z^{-1} + z^{-2})\right] = zH(z)$$

where $H(z) = (1/3)\mathcal{Z}[\delta[n] + \delta[n-1] + \delta[n-2]]$ is the transfer function of a causal filter. Let $Y(z) = X(z)H(z)$ be the Z-transform of the convolution sum $y[n] = [x * h][n]$ of $x[n]$ and $h[n]$, both of which are causal and can be computed as before.

According to Equation (9.23), we then have $Y_1(z) = zY(z)$ or $y_1[n] = [x * h_1][n] = y[n+1]$.

∎

Let $x_1[n]$ be the input to a noncausal LTI system, with an impulse response $h_1[n]$ such that $h_1[n] = 0$ for $n < N_1 < 0$. Assume $x_1[n]$ is also noncausal (i.e., $x_1[n] = 0$ for $n < N_0 < 0$). The output $y_1[n] = [x_1 * h_1][n]$ has a Z-transform of

$$Y_1(z) = X_1(z)H_1(z) = [z^{N_0}X(z)][z^{N_1}H(z)]$$

where $X(z)$ and $H(z)$ are the Z-transforms of a causal signal $x[n]$ and of a causal impulse response $h[n]$. If we let

$$y[n] = [x * h][n] = \mathcal{Z}^{-1}[X(z)H(z)]$$

then

$$y_1[n] = [x_1 * h_1][n] = y[n + N_0 + N_1]$$

Remarks

■ *The impulse response of an IIR system, represented by a difference equation, is found by setting the initial conditions to zero, therefore, the transfer function $H(z)$ also requires a similar condition. If the initial conditions are not zero, the Z-transform of the total response $Y(z)$ is the sum of the Z-transforms of the zero-state and the zero-input responses—that is, its Z-transform is of the form*

$$Y(z) = \frac{X(z)B(z)}{A(z)} + \frac{I_0(z)}{A(z)} \tag{9.24}$$

and it does not permit us to compute the ratio $Y(z)/X(z)$ unless the component due to the initial conditions is $I_0(z) = 0$.

■ It is important to remember the relations

$$H(z) = \mathcal{Z}[h[n]] = \frac{Y(z)}{X(z)} = \frac{\mathcal{Z}[y[n]]}{\mathcal{Z}[x[n]]}$$

where $H(z)$ is the transfer function and $h[n]$ is the impulse response of the system, with $x[n]$ as the input and $y[n]$ as the output.

■ Example 9.10

Consider a discrete-time IIR system represented by the difference equation

$$y[n] = 0.5y[n-1] + x[n] \tag{9.25}$$

with $x[n]$ as the input and $y[n]$ as the output. Determine the transfer function of the system and from it find the impulse and the unit-step responses. Determine under what conditions the system is BIBO stable. If stable, determine the transient and steady-state responses of the system.

Solution

The system transfer function is given by

$$H(z) = \frac{Y(z)}{X(z)} = \frac{1}{1 - 0.5z^{-1}}$$

and its impulse response is

$$h[n] = \mathcal{Z}^{-1}[H(z)] = 0.5^n u[n]$$

The response of the system to any input can be easily obtained by the transfer function. If the input is $x[n] = u[n]$, we have

$$Y(z) = H(z)X(z) = \frac{1}{(1 - 0.5z^{-1})(1 - z^{-1})}$$

$$= \frac{-1}{1 - 0.5z^{-1}} + \frac{2}{1 - z^{-1}}$$

so that the total solution is

$$y[n] = -0.5^n u[n] + 2u[n]$$

From the transfer function $H(z)$ of the LTI system, we can test the stability of the system by finding the location of its poles—very much like in the analog case. An LTI system is BIBO stable if and only if the impulse response of the system is absolutely summable—that is,

$$\sum_n |h[n]| < \infty$$

An equivalent condition is that the poles of $H(z)$ are inside the unit circle. In this case, $h[n]$ is absolutely summable, indeed

$$\sum_{n=0}^{\infty} 0.5^n = \frac{1}{1 - 0.5} = 2$$

On the other hand,

$$H(z) = \frac{1}{1 - 0.5z^{-1}} = \frac{z}{z - 0.5}$$

has a pole at $z = 0.5$, inside the unit circle. Thus, the system is BIBO stable. As such, its transient and steady-state responses exist. As $n \to \infty$, $y[n] = 2$ is the steady-state response, and $-0.5^n u[n]$ is the transient solution. ∎

■ Example 9.11

An FIR system has the input–output equation

$$y[n] = \frac{1}{3}[x[n] + x[n - 1] + x[n - 2]]$$

where $x[n]$ is the input and $y[n]$ is the output. Determine the transfer function and the impulse response of the system, and from them indicate whether the system is BIBO stable or not.

Solution

The transfer function is

$$H(z) = \frac{1}{3}[1 + z^{-1} + z^{-2}]$$

$$= \frac{z^2 + z + 1}{3z^2}$$

and the corresponding impulse response is

$$h[n] = \frac{1}{3}[\delta[n] + \delta[n - 1] + \delta[n - 2]]$$

The impulse response of this system only has three nonzero values, $h[0] = h[1] = h[2] = 1/3$, and the rest of the values are zero. As such, $h[n]$ is absolutely summable and the filter is BIBO stable. FIR filters are always BIBO stable given their impulse responses will be absolutely summable, due to their final support, and equivalently because the poles of the transfer function of these system are at the origin of the z-plane, very much inside the unit circle. ∎

Nonrecursive or FIR systems: The impulse response $h[n]$ of an FIR or nonrecursive system

$$y[n] = b_0 x[n] + b_1 x[n - 1] + \cdots + b_M x[n - M]$$

has finite length and is given by

$$h[n] = b_0\delta[n] + b_1\delta[n-1] + \cdots + b_M\delta[n-M]$$

Its transfer function is

$$H(z) = \frac{Y(z)}{X(z)}$$

$$= b_0 + b_1 z^{-1} + \cdots + b_M z^{-M}$$

$$= \frac{b_0 z^M + b_1 z^{M-1} + \cdots + b_M}{z^M}$$

with all its poles at the origin $z = 0$ (multiplicity M), and as such the system is BIBO stable.

Recursive or IIR systems: The impulse response $h[n]$ of an IIR or recursive system

$$y[n] = -\sum_{k=1}^{N} a_k y[n-k] + \sum_{m=0}^{M} b_m x[n-m]$$

has (possible) infinite length and is given by

$$h[n] = \mathcal{Z}^{-1}[H(z)]$$

$$= \mathcal{Z}^{-1}\left[\frac{\sum_{m=0}^{M} b_m z^{-m}}{1 + \sum_{k=1}^{N} a_k z^{-k}}\right]$$

$$= \mathcal{Z}^{-1}\left[\frac{B(z)}{A(z)}\right]$$

$$= \sum_{\ell=0}^{\infty} h[\ell]\delta[n-\ell]$$

where $H(z)$ is the transfer function of the system. If the poles of $H(z)$ are inside the unit circle, or $A(z) \neq 0$ for $|z| \geq 1$, the system is BIBO stable.

9.4.4 Interconnection of Discrete-Time Systems

Just like with analog systems, two discrete-time LTI systems with transfer functions $H_1(z)$ and $H_2(z)$ (or with impulse responses $h_1[n]$ and $h_2[n]$) can be connected in cascade, parallel, or feedback. The first two forms result from properties of the convolution sum.

The transfer function of the cascading of the two LTI systems is

$$H(z) = H_1(z)H_2(z) = H_2(z)H_1(z) \tag{9.26}$$

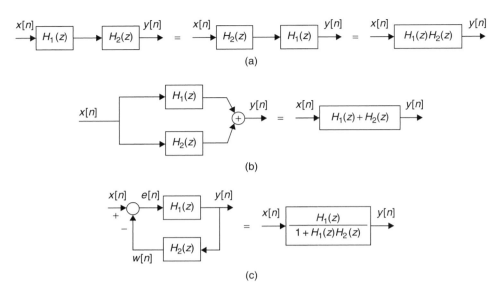

FIGURE 9.7
Connections of LTI systems: (a) cascade, (b) parallel, and (c) negative feedback.

showing that there is no effect on the overall system if we interchange the two systems (see Figure 9.7(a)). Recall that such a property is only valid for LTI systems. In the parallel system, as in Figure 9.7(b), both systems have the same input and the output is the sum of the output of the subsystems. The overall transfer function is

$$H(z) = H_1(z) + H_2(z) \tag{9.27}$$

Finally, the negative feedback connection of the two systems shown in Figure 9.7(c) gives in the feedforward path

$$Y(z) = H_1(z)E(z) \tag{9.28}$$

where $Y(z) = \mathcal{Z}[y[n]]$ is the Z-transform of the output $y[n]$ and $E(z) = X(z) - W(z)$ is the Z-transform of the error function $e[n] = x[n] - w[n]$. The feedback path gives that

$$W(z) = \mathcal{Z}[w[n]] = H_2(z)Y(z)$$

Replacing $W(z)$ in $E(z)$, and then replacing $E(z)$ in Equation (9.28), we obtain the overall transfer function

$$H(z) = \frac{Y(z)}{X(z)} = \frac{H_1(z)}{1 + H_1(z)H_2(z)} \tag{9.29}$$

9.4.5 Initial and Final Value Properties

In some control applications and to check a partial fraction expansion, it is useful to find the initial or the final value of a discrete-time signal $x[n]$ from its Z-transform. These values can be found as shown in the following box.

If $X(z)$ is the Z-transform of a causal signal $x[n]$, then

$$\text{Initial value:} \quad x[0] = \lim_{z \to \infty} X(z)$$

$$\text{Final value:} \quad \lim_{n \to \infty} x[n] = \lim_{z \to 1} (z-1)X(z) \tag{9.30}$$

The initial value results from the definition of the one-sided Z-transform—that is,

$$\lim_{z \to \infty} X(z) = \lim_{z \to \infty} \left(x[0] + \sum_{n \geq 1} \frac{x[n]}{z^n} \right) = x[0]$$

To show the final value, we have that

$$(z-1)X(z) = \sum_{n=0}^{\infty} x[n]z^{-n+1} - \sum_{n=0}^{\infty} x[n]z^{-n}$$

$$= x[0]z + \sum_{n=0}^{\infty} [x[n+1] - x[n]]z^{-n}$$

and thus the limit

$$\lim_{z \to 1}(z-1)X(z) = x[0] + \sum_{n=0}^{\infty}(x[n+1] - x[n])$$

$$= x[0] + (x[1] - x[0]) + (x[2] - x[1]) + (x[3] - x[2]) \cdots$$

$$= \lim_{n \to \infty} x[n]$$

given that the entries in the sum cancel out as n increases, leaving $x[\infty]$.

■ Example 9.12

Consider a negative-feedback connection of a plant with a transfer function $G(z) = 1/(1 - 0.5z^{-1})$ and a constant feedback gain K (see Figure 9.8). If the reference signal is a unit step, $x[n] = u[n]$, determine the behavior of the error signal $e[n]$. What is the effect of the feedback, from the error point of view, on an unstable plant $G(z) = 1/(1 - z^{-1})$?

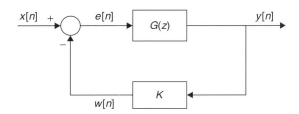

FIGURE 9.8

Negative-feedback system with plant $G(z)$.

Solution

For $G(z) = 1/(1 - 0.5z^{-1})$, the Z-transform of the error signal is

$$E(z) = X(z) - W(z) = X(z) - KG(z)E(z)$$

and for $X(z) = 1/(1 - z^{-1})$,

$$E(z) = \frac{X(z)}{1 + KG(z)} = \frac{1}{(1 - z^{-1})(1 + KG(z))}$$

The initial value of the error signal is then

$$e[0] = \lim_{z \to \infty} E(z) = \frac{1}{1 + K}$$

since $G(\infty) = 1$.

The steady-state or final value of the error is

$$\lim_{n \to \infty} e[n] = \lim_{z \to 1}(z - 1)E(z) = \lim_{z \to 1} \frac{(z - 1)X(z)}{1 + KG(z)}$$

$$= \lim_{z \to 1} \frac{z(z - 1)}{(z - 1)(1 + KG(z))} = \frac{1}{1 + 2K}$$

since $G(1) = 2$. If we want the steady-state error to go to zero, then K must be large. In that case, the initial error is also zero.

If $G(z) = 1/(1 - z^{-1})$ (i.e., the plant is unstable), the initial value of the error function remains the same, $e[0] = 1/(1 + K)$, but the steady state error goes to zero since $G(1) \to \infty$. ■

Tables 9.1 and 9.2 provide a list of one-side Z-transforms and the basic properties of the one-sided Z-transform.

Table 9.1 One-Sided Z-Transforms

	Function of Time	**Function of z, ROC**
1.	$\delta[n]$	1, whole z-plane
2.	$u[n]$	$\dfrac{1}{1 - z^{-1}}$, $\quad |z| > 1$
3.	$nu[n]$	$\dfrac{z^{-1}}{(1 - z^{-1})^2}$, $\quad |z| > 1$
4.	$n^2 u[n]$	$\dfrac{z^{-1}(1 + z^{-1})}{(1 - z^{-1})^3}$, $\quad |z| > 1$
5.	$\alpha^n u[n]$, $|\alpha| < 1$	$\dfrac{1}{1 - \alpha z^{-1}}$, $\quad |z| > |\alpha|$
6.	$n\alpha^n u[n]$, $|\alpha| < 1$	$\dfrac{\alpha z^{-1}}{(1 - \alpha z^{-1})^2}$, $\quad |z| > |\alpha|$
7.	$\cos(\omega_0 n)u[n]$	$\dfrac{1 - \cos(\omega_0)z^{-1}}{1 - 2\cos(\omega_0)z^{-1} + z^{-2}}$, $\quad |z| > 1$
8.	$\sin(\omega_0 n)u[n]$	$\dfrac{\sin(\omega_0)z^{-1}}{1 - 2\cos(\omega_0)z^{-1} + z^{-2}}$, $\quad |z| > 1$
9.	$\alpha^n \cos(\omega_0 n)u[n]$, $|\alpha| < 1$	$\dfrac{1 - \alpha\cos(\omega_0)z^{-1}}{1 - 2\alpha\cos(\omega_0)z^{-1} + z^{-2}}$, $\quad |z| > 1$
10.	$\alpha^n \sin(\omega_0 n)u[n]$, $|\alpha| < 1$	$\dfrac{\alpha\sin(\omega_0)z^{-1}}{1 - 2\alpha\cos(\omega_0)z^{-1} + z^{-2}}$, $\quad |z| > |\alpha|$

Table 9.2 Basic Properties of One-Sided Z-Transform

Causal signals and constants	$\alpha x[n], \beta y[n]$	$\alpha X(z), \beta Y(z)$
Linearity	$\alpha x[n] + \beta y[n]$	$\alpha X(z) + \beta Y(z)$
Convolution sum	$(x * y)[n] = \sum_k x[n]y[n - k]$	$X(z)Y(z)$
Time shifting—causal	$x[n - N]\ N$ integer	$z^{-N}X(z)$
Time shifting—noncausal	$x[n - N]$	$z^{-N}X(z) + x[-1]z^{-N+1}$
	$x[n]$ noncausal, N integer	$\quad + x[-2]z^{-N+2} + \cdots + x[-N]$
Time reversal	$x[-n]$	$X(z^{-1})$
Multiplication by n	$n\, x[n]$	$-z\dfrac{dX(z)}{dz}$
Multiplication by n^2	$n^2\, x[n]$	$z^2\dfrac{d^2 X(z)}{dz^2} + z\dfrac{dX(z)}{dz}$
Finite difference	$x[n] - x[n - 1]$	$(1 - z^{-1})X(z) - x[-1]$
Accumulation	$\sum_{k=0}^{n} x[k]$	$\dfrac{X(z)}{1 - z^{-1}}$
Initial value	$x[0]$	$\lim\limits_{z \to \infty} X(z)$
Final value	$\lim\limits_{n \to \infty} x[n]$	$\lim\limits_{z \to 1}(z - 1)X(z)$

9.5 ONE-SIDED Z-TRANSFORM INVERSE

Different from the inverse Laplace transform, which was done mostly by the partial fraction expansion, the inverse Z-transform can be done in different ways. For instance, if the Z-transform is given as a finite-order polynomial, the inverse can be found by inspection. Indeed, if the given Z-transform is

$$X(z) = \sum_{n=0}^{N} x[n] z^{-n}$$

$$= x[0] + x[1] z^{-1} + x[2] z^{-2} + \cdots + x[N] z^{-N} \tag{9.31}$$

by the definition of the Z-transform, $x[k]$ is the coefficient of the monomial z^{-k} for $k = 0, 1, \ldots, N$; thus the inverse Z-transform is given by the sequence $\{x[0], x[1], \ldots, x[n]\}$. For instance, if we have a Z-transform

$$X(z) = 1 + 2z^{-10} + 3z^{-20}$$

the inverse is a sequence

$$x[n] = \delta[n] + 2\delta[n - 10] + 3\delta[n - 20]$$

so that $x[0] = 1$, $x[10] = 2$, $x[20] = 3$, and $x[n] = 0$ for $n \neq 0,\ 10,\ 20$, respectively. In this case it makes sense to do this because N is finite, but if $N \to \infty$, this way of finding the inverse Z-transform might not be very practical. In that case, the *long-division* method and the *partial fraction expansion* method, which we consider next, are more appropriate. In this section we will consider the inverse of one-sided Z-transforms, and in the next section we consider the inverse of two-sided transforms.

9.5.1 Long-Division Method

When a rational function $X(z) = B(z)/A(z)$, having as ROC the outside of a circle of radius R (i.e., $x[n]$ is causal), is expressed as

$$X(z) = x[0] + x[1] z^{-1} + x[2] z^{-2} + \cdots$$

then the inverse is the sequence $\{x[0], x[1], x[2], \ldots\}$, or

$$x[n] = x[0]\delta[n] + x[1]\delta[n - 1] + x[2]\delta[n - 2] + \cdots$$

To find the inverse we simply divide the polynomial $B(z)$ by $A(z)$ to obtain a possible infinite-order polynomial in negative powers of z^{-1}. The coefficients of this polynomial are the inverse values. The disadvantage of this method is that it does not provide a closed-form solution, unless there is a clear connection between the terms of the sequence. But this method is useful when we are interested in finding some of the initial values of the sequence $x[n]$.

■ Example 9.13

Find the inverse Z-transform of

$$X(z) = \frac{1}{1 + 2z^{-2}} \qquad |z| > \sqrt{2}$$

Solution

We can perform the long division to find the $x[n]$ values, or equivalently let

$$X(z) = x[0] + x[1]z^{-1} + x[2]z^{-2} + \cdots$$

and find the $\{x[n]\}$ samples so that the product $X(z)(1 + 2z^{-2}) = 1$. Thus,

$$
\begin{aligned}
1 &= (1 + 2z^{-2})(x[0] + x[1]z^{-1} + x[2]z^{-2} + \cdots) \\
&= x[0] + x[1]z^{-1} + x[2]z^{-2} + x[3]z^{-3} + \cdots \\
&\quad + 2x[0]z^{-2} + 2x[1]z^{-3} + \cdots
\end{aligned}
$$

and comparing the terms on the two sides of the equality gives

$$
\begin{aligned}
x[0] &= 1 \\
x[1] &= 0 \\
x[2] + 2x[0] &= 0 \;\Rightarrow\; x[2] = -2 \\
x[3] + 2x[1] &= 0 \;\Rightarrow\; x[3] = 0 \\
x[4] + 2x[2] &= 0 \;\Rightarrow\; x[4] = (-2)^2
\end{aligned}
$$

$$\vdots$$

So the inverse Z-transform is $x[0] = 1$ and $x[n] = (-2)^{\log_2(n)}$ for $n > 0$ and even, and zero otherwise. Notice that this sequence grows as $n \to \infty$.

Another possible way to find the inverse is to use the geometric series equation

$$\sum_{n=0}^{\infty} \alpha^n = \frac{1}{1 - \alpha} \qquad |\alpha| < 1$$

with $-\alpha = 2z^{-2}$ (notice that $|\alpha| = 2/|z|^2 < 1$ or $|z| > \sqrt{2}$, the given ROC). Therefore,

$$X(z) = \frac{1}{1 + 2z^{-2}} = 1 + (-2z^{-2})^1 + (-2z^{-2})^2 + (-2z^{-2})^3 + \cdots$$

but this method is not as general as the long division. ■

9.5.2 Partial Fraction Expansion

The basics of partial fraction expansion remain the same for the Z-transform as for the Laplace transform. A rational function is a ratio of polynomials $N(z)$ and $D(z)$ in z or z^{-1}:

$$X(z) = \frac{N(z)}{D(z)}$$

The poles of $X(z)$ are the roots of $D(z) = 0$ and the zeros of $X(z)$ are the roots of the equation $N(z) = 0$.

Remarks

- *The basic characteristic of the partial fraction expansion is that $X(z)$ must be a proper rational function, or that the degree of the numerator polynomial $N(z)$ must be smaller than the degree of the denominator polynomial $D(z)$ (assuming both $N(z)$ and $D(z)$ are polynomials in either z^{-1} or z). If this condition is not satisfied, we perform long division until the residue polynomial is of a degree less than that of the denominator.*
- *It is more common in the Z-transform than in the Laplace transform to find that the numerator and the denominator are of the same degree—this is because $\delta[n]$ is not as unusual as the analog impulse function $\delta(t)$.*
- *The partial fraction expansion is generated, from the poles of the proper rational function, as a sum of terms of which the inverse Z-transforms are easily found in a Z-transform table. By plotting the poles and the zeros of a proper $X(z)$, the location of the poles provides a general form of the inverse within some constants that are found from the poles and the zeros.*
- *Given that the numerator and the denominator polynomials of a proper rational function $X(z)$ can be expressed in terms of positive or negative powers of z, it is possible to do partial fraction expansions in either z or z^{-1}. We will see that the partial fraction expansion in negative powers is more like the partial fraction expansion in the Laplace transform, and as such we will prefer it. Partial fraction expansion in positive powers of z requires more care.*

■ Example 9.14

Consider the nonproper rational function

$$X(z) = \frac{2 + z^{-2}}{1 + 2z^{-1} + z^{-2}}$$

(the numerator and the denominator are of the same degree in powers of z^{-1}). Determine how to obtain an expansion of $X(z)$ containing a proper rational term to find $x[n]$.

Solution

By division we obtain

$$X(z) = 1 + \frac{1 - 2z^{-1}}{1 + 2z^{-1} + z^{-2}}$$

where the second term is proper rational as the denominator is of a higher degree in powers of z^{-1} than the numerator. The inverse Z-transform of $X(z)$ will then be

$$x[n] = \delta[n] + \mathcal{Z}^{-1}\left[\frac{1 - 2z^{-1}}{1 + 2z^{-1} + z^{-2}}\right]$$

The inverse of the proper rational term is done as indicated in this section. ∎

■ Example 9.15

Find the inverse Z-transform of

$$X(z) = \frac{1 + z^{-1}}{(1 + 0.5z^{-1})(1 - 0.5z^{-1})}$$

$$= \frac{z(z + 1)}{(z + 0.5)(z - 0.5)} \qquad |z| > 0.5$$

by using the negative and the positive powers of z expressions.

Solution

Clearly, $X(z)$ is proper if it is considered a function of negative powers z^{-1} (in z^{-1}, the numerator is of degree 1 and the denominator of degree 2), but it is not proper if it is considered a function of positive powers z (the numerator and the denominator are both of degree 2). It is, however, unnecessary to perform long division to make $X(z)$ proper when it is considered as a function of z. One simple approach is to consider $X(z)/z$ as the function. We wish to find its partial fraction expansion—that is,

$$\frac{X(z)}{z} = \frac{z + 1}{(z + 0.5)(z - 0.5)} \qquad (9.32)$$

which is proper. Thus, whenever $X(z)$, as a function of z terms, is not proper it is always possible to divide it by some power in z to make it proper. After obtaining the partial fraction expansion then the z term is put back.

Consider then the partial fraction expansion in z^{-1} terms,

$$X(z) = \frac{1 + z^{-1}}{(1 + 0.5z^{-1})(1 - 0.5z^{-1})}$$

$$= \frac{A}{1 + 0.5z^{-1}} + \frac{B}{1 - 0.5z^{-1}}$$

Given that the poles are real—one at $z = -0.5$ and the other at $z = 0.5$—from the Z-transform table, we get that a general form of the inverse is

$$x[n] = [A\,(-0.5)^n + B\,0.5^n]u[n]$$

The A and B coefficients can be found (by analogy with the Laplace transform partial fraction expansion) as

$$A = X(z)(1 + 0.5z^{-1})|_{z^{-1}=-2} = -0.5$$

$$B = X(z)(1 - 0.5z^{-1})|_{z^{-1}=2} = 1.5$$

so that

$$x[n] = [-0.5(-0.5)^n + 1.5(0.5)^n]u[n]$$

Consider then the partial fraction expansion in positive powers of z. From Equation (9.32) the proper rational function $X(z)/z$ can be expanded as

$$\frac{X(z)}{z} = \frac{z + 1}{(z + 0.5)(z - 0.5)}$$

$$= \frac{C}{z + 0.5} + \frac{D}{z - 0.5}$$

The values of C and D are obtained as follows:

$$C = \frac{X(z)}{z}(z + 0.5)|_{z=-0.5} = -0.5$$

$$D = \frac{X(z)}{z}(z - 0.5)|_{z=0.5} = 1.5$$

We then have that

$$X(z) = \frac{-0.5z}{z + 0.5} + \frac{1.5z}{z - 0.5}$$

which according to the table (if entries are in negative powers of z, convert them into positive powers of z) we get

$$x[n] = [-0.5(-0.5)^n + 1.5(0.5)^n]u[n]$$

which coincides with the above result.

Two simple checks on our result are given by the initial and the final value results. For the initial value,

$$x[0] = 1 = \lim_{z \to \infty} X(z)$$

and

$$\lim_{n \to \infty} x[n] = \lim_{z \to 1}(z - 1)X(z) = 0$$

Both of these check. It is important to recognize that these two checks do not guarantee that we did not make mistakes in computing the inverse, but if the initial or the final values were not to coincide with our results, our inverse would be wrong.

9.5.3 Inverse Z-Transform with MATLAB

Symbolic MATLAB can be used to compute the inverse one-sided Z-transform. The function iztrans provides the sequence that corresponds to its argument. The following script illustrates the use of this function.

```
%%%%%%%%%%%%%%%%%%%%
% Inverse Z-transform
%%%%%%%%%%%%%%%%%%%%
syms n z
        x1 = iztrans((z * (z + 1))/((z + 0.5) * (z - 0.5)))
        x2 = iztrans((2 - z)/(2 * (z - 0.5)))
        x3 = iztrans((8 - 4 * z ^ (-1))/(z ^ (-2) + 6 * z ^ (-1) + 8))
```

The above gives the following results:

```
x1 = 3/2 * (1/2) ^ n - 1/2 * (-1/2) ^ n
x2 = -2 * charfcn[0](n) + 3/2 * (1/2) ^ n
x3 = -3 * (-1/4) ^ n + 4 * (-1/2) ^ n
```

Notice that the Z-transform can be given in positive or negative powers of z, and that when it is nonproper the function charfcn[0] corresponds to $\delta[n]$.

Partial Fraction Expansion with MATLAB

Several numerical functions are available in MATLAB to perform partial fraction expansion of a Z-transform and to obtain the corresponding inverse. In the following we consider the cases of single and multiple poles.

(1) Simple Poles

Consider finding the inverse Z-transform of

$$X(z) = \frac{z(z+1)}{(z-0.5)(z+0.5)} = \frac{(1+z^{-1})}{(1-0.5z^{-1})(1+0.5z^{-1})} \qquad |z| > 0.5$$

The MATLAB function residuez provides the partial fraction expansion coefficients or residues $r[k]$, the poles $p[k]$, and the gain k corresponding to $X(z)$ when the coefficients of its denominator and of its numerator are inputted. If the numerator or the denominator is given in a factored form (as is the case of the denominator above) we need to multiply the terms to obtain the denominator polynomial. Recall that multiplication of polynomials corresponds to convolution of the polynomial coefficients. Thus, to perform the multiplication of the terms in the denominator, we use the MATLAB function conv to obtain the coefficients of the product. The convolution of the coefficients $[1 - 0.5]$ of $p_1(z) = 1 - 0.5z^{-1}$ and $[1 \ \ 0.5]$ of $p_2(z) = 1 + 0.5z^{-1}$ gives the denominator coefficients. By means of the MATLAB function poly we can obtain the polynomials in the numerator and denominator from the zeros and poles. These polynomials are then multiplied as indicated before to obtain the numerator with coefficients $\{b[k]\}$, and the denominator with coefficients $\{a[k]\}$.

To find the poles and the zeros of $X(z)$, given the coefficients $\{b[k]\}$ and $\{a[k]\}$ of the numerator and the denominator, we use the MATLAB function roots. To get a plot of the poles and the zeros of $X(z)$, the MATLAB function zplane, with inputs the coefficients of the numerator and the denominator of $X(z)$, is used (conventionally, an 'x' is used to denote poles and an 'o' for zeros).

Two possible approaches can now be used to compute the inverse Z-transform $x[n]$. We can compute the inverse (below we call it $x_1[n]$ to differentiate it from the other possible solution, which we call $x[n]$) by using the information on the partial fraction expansion (the residues $r[k]$) and the corresponding poles. An alternative is to use the MATLAB function filter, which considers $X(z)$ as a transfer function, with the numerator and the denominator defined by the b and a vectors of coefficients. If we assume the input is a delta function of Z-transform unity, the function filter computes as output the inverse Z-transform $x[n]$ (i.e., we have tricked filter to give us the desired result).

The following script is used to implement the generation of the terms in the numerator and the denominator to obtain the corresponding coefficients, plot them, and find the inverse in the two different ways indicated above. For additional help on the functions used here use help.

```
%%%%%%%%%%%%%%%%%%%%%%%%%%%%%%
% Two methods for inverse Z-transform
%%%%%%%%%%%%%%%%%%%%%%%%%%%%%%
p1 = poly(0.5); p2 = poly(-0.5); % generation of terms in denominator
a = conv(p1,p2)  % denominator coefficients
z1 = poly(0); z2 = poly(-1); % generation of terms in numerator
b = conv(z1,z2)  % numerator coefficients
z = roots(b)  % zeros of X(z)
[r,p,k] = residuez(b,a)  % partial fraction expansion, poles and gain
zplane(b,a)  % plot of poles and zeros
d = [1 zeros(1,99)]; % impulse delta[n]
x = filter(b,a,d);  % x[n] computation from filter
n = 0:99;
x1 = r(1) * p(1).^n + r(2) * p(2).^n;  % x[n] computation from residues
```

```
        a =  1.0000       0  -0.2500
        b =  1    1    0
        z =  0
            -1
        r =  1.5000
            -0.5000
        p =  0.5000
            -0.5000
```

Figure 9.9 displays the plot of the zeros and the poles and the comparison between the inverses $x_1[n]$ and $x[n]$ for $0 \leq n \leq 99$, which coincide sample by sample.

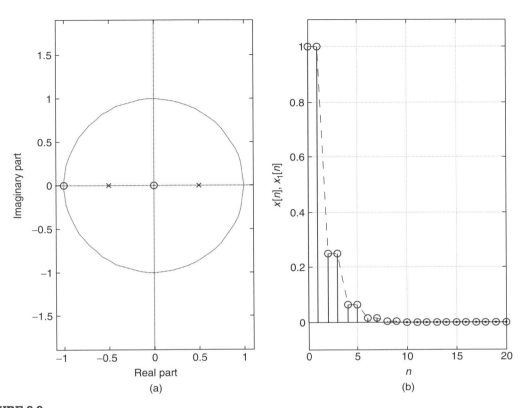

FIGURE 9.9

(a) Poles and zeros of $X(z)$ and (b) inverse Z-transforms $x[n]$ and $x_1[n]$ found using `filter` and the residues.

(2) Multiple Poles

Whenever multiple poles are present one has to be careful in interpreting the MATLAB results. First, use help to get more information on residuez and how the partial fraction expansion for multiple poles is done. Notice from the help file that the residues are ordered the same way the poles are. Furthermore, the residues corresponding to the multiple poles are ordered from the lowest to the highest order. Also notice the difference between the partial fraction expansion of MATLAB and ours. For instance, consider the Z-transform

$$X(z) = \frac{az^{-1}}{(1 - az^{-1})^2} \qquad |z| > a$$

with inverse $x[n] = na^n u[n]$. Writing the partial fraction expansion as MATLAB does gives

$$X(z) = \frac{r_1}{1 - az^{-1}} + \frac{r_2}{(1 - az^{-1})^2} \qquad r_1 = -1, r_2 = 1 \qquad (9.33)$$

where the second term is not found in the Z-transforms table. To write it so that each of the terms in the expansion are in the Z-transforms table, we need to obtain values for A and B in the expansion

$$X(z) = \frac{A}{1 - az^{-1}} + \frac{Bz^{-1}}{(1 - az^{-1})^2} \tag{9.34}$$

so that Equations (9.33) and (9.34) are equal. We find that $A = r_1 + r_2$, while $B - Aa = -r_1 a$ or $B = ar_2$. With these values we find the inverse to be

$$x[n] = [(r_1 + r_2)a^n + nr_2 a^n]u[n] = na^n u[n]$$

as expected.

To illustrate the computation of the inverse Z-transform from the residues in the case of multiple poles, consider the transfer function

$$X(z) = \frac{0.5z^{-1}}{1 - 0.5z^{-1} - 0.25z^{-2} + 0.125z^{-3}}.$$

The following script is used.

```
%%%%%%%%%%%%%%%%%%%%%%%%%%%%%%%
%  Inverse Z-transform --- multiple poles
%%%%%%%%%%%%%%%%%%%%%%%%%%%%%%%
        b = [0 0.5 0 0 ];
        a = [1 -0.5 -0.25 0.125]
        [r,p,k] = residuez(b,a)     % partial fraction expansion, poles and gain
        zplane(b,a) % plot of poles and zeros
        n = 0:99; xx = p(1). ^n; yy = xx. *n;
        x1 = (r(1) + r(2)). *xx + r(2). *yy + r(3) *p(3). ^n; % inverse computation
```

The poles and the zeros and the inverse Z-transform are shown in Figure 9.10—there is a double pole at 0.5. The residues and the corresponding poles are

```
r = -0.2500
     0.5000
    -0.2500
p =  0.5000
     0.5000
    -0.5000
```

Computationally, our method and MATLAB's are comparable but the inverse transform in our method is found directly from the table, while in the case of MATLAB's you need to change the expansion to get it into the forms found in the tables.

9.5.4 Solution of Difference Equations

In this section we will use the shifting in time property of the Z-transform in the solution of difference equations with initial conditions. You will see that the partial fraction expansion used to find the inverse Z-transform is like the one used in the inverse Laplace transform.

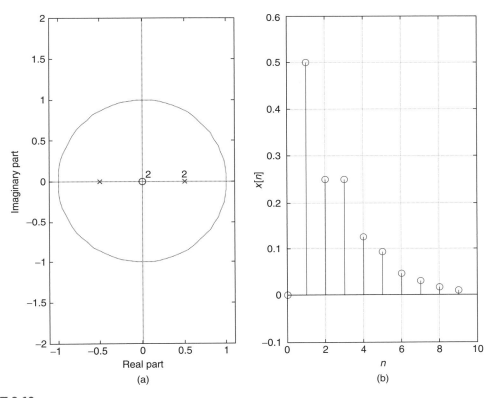

FIGURE 9.10

(a) Poles and zeros of $X(z)$ and (b) inverse Z-transform $x[n]$.

If $x[n]$ has a one-sided Z-transform $X(z)$, then $x[n - N]$ has the following one-sided Z-transform:

$$\mathcal{Z}[x[n - N]] = z^{-N}X(z) + x[-1]z^{-N+1} + x[-2]z^{-N+2} + \cdots + x[-N] \qquad (9.35)$$

Indeed, we have that

$$\mathcal{Z}(x[n - N]) = \sum_{n=0}^{\infty} x[n - N]z^{-n} = \sum_{m=-N}^{\infty} x[m]z^{-(m+N)}$$

$$= z^{-N} \sum_{m=0}^{\infty} x[m]z^{-m} + \sum_{m=-N}^{-1} x[m]z^{-(m+N)}$$

$$= z^{-N}X(z) + x[-1]z^{-N+1} + x[-2]z^{-N+2} + \cdots + x[-N]$$

where we first let $m = n - N$ and then separated the sum into two, one corresponding to the Z-transform of $x[n]$ multiplied by z^{-N} (the delay on the signal) and a second sum that corresponds to initial values $\{x[i], -N \le i \le -1\}$.

Remarks

- *If the signal is causal, so that $\{x[i], -N \le i \le -1\}$ are all zero, we then have that $\mathcal{Z}(x[n - N]) = z^{-N}X(z)$, indicating that the operator z^{-1} is a delay operator. Thus, $x[n - N]$ has been delayed N samples and its Z-transform is then simply $X(z)$ multiplied by z^{-N}.*
- *The shifting in time property is useful in the solution of difference equations, especially when it has nonzero initial conditions as we will see next. On the other hand, if the initial conditions are zero, either the one-sided or the two-sided Z-transforms could be used.*

The analog of differential equations are difference equations, which result directly from the modeling of a discrete system or from discretizing differential equations. The numerical solution of differential equations requires that these equations be converted into difference equations since computers cannot perform integration. Many methods are used to solve differential equations with different degrees of accuracy and sophistication. This is a topic of numerical analysis, outside the scope of this text, and thus only simple methods are illustrated here.

■ Example 9.16

A discrete-time IIR system is represented by a first-order difference equation

$$y[n] = ay[n - 1] + x[n] \qquad n \ge 0 \tag{9.36}$$

where $x[n]$ is the input of the system and $y[n]$ is the output. Discuss how to solve it using recursive methods and the Z-transform. Obtain a general form for the complete solution $y[n]$ in terms of the impulse response $h[n]$ of the system.

For input $x[n] = u[n] - u[n - 11]$, zero initial conditions, and $a = 0.8$, use the MATLAB function filter to find $y[n]$. Plot the input and the output.

Solution

In the time domain a unique solution is obtained by using the recursion given by the difference equation. We would need an initial condition to compute $y[0]$, indeed

$$y[0] = ay[-1] + x[0]$$

and as $x[0]$ is given, we need $y[-1]$ as the initial condition. Given $y[0]$, recursively we find the rest of the solution:

$$y[1] = ay[0] + x[1]$$

$$y[2] = ay[1] + x[2]$$

$$y[3] = ay[2] + x[3]$$

$$\cdots$$

where at each step the needed output values are given by the previous step of the recursion. However, this solution is not in closed form.

To obtain a closed-form solution, we use the Z-transform. Taking the one-sided Z-transform of the two sides of the equation, we get

$$\mathcal{Z}(y[n]) = \mathcal{Z}(ay[n-1]) + \mathcal{Z}[x[n]])$$

$$Y(z) = a(z^{-1}Y(z) + y[-1]) + X(z)$$

Solving for $Y(z)$ in the above equation, we obtain

$$Y(z) = \frac{X(z)}{1 - az^{-1}} + \frac{ay[-1]}{1 - az^{-1}} \tag{9.37}$$

where the first term depends exclusively on the input and the second depends exclusively on the initial condition. If the input $x[n]$ and the initial condition $y[-1]$ are given, we can then find the inverse Z-transform to obtain the complete solution $y[n]$ of the form

$$y[n] = y_{zs}[n] + y_{zi}[n]$$

where the zero-state response $y_{zs}[n]$ is due exclusively to the input $x[n]$ and zero initial conditions, and the zero-input response $y_{zi}[n]$ is the response due to the initial condition $y[-1]$ with zero input.

In this simple case we can obtain the complete solution for any input $x[n]$ and any initial condition $y[-1]$. Indeed, by expressing $1/(1 - az^{-1})$ as its Z-transform sum—that is,

$$\frac{1}{1 - az^{-1}} = \sum_{k=0}^{\infty} a^k z^{-k}$$

Equation (9.37) becomes

$$Y(z) = \sum_{k=0}^{\infty} X(z)a^k z^{-k} + ay[-1] \sum_{k=0}^{\infty} a^k z^{-k}$$

$$= X(z) + aX(z)z^{-1} + a^2 X(z)z^{-2} + \cdots + ay[-1](1 + az^{-1} + a^2 z^{-2} + \cdots)$$

Using the time-shift property we then get the complete solution,

$$y[n] = x[n] + ax[n-1] + a^2 x[n-2] + \cdots + y[-1]a(1 + a\delta[n-1] + a^2\delta[n-2] + \cdots)$$

$$= \sum_{k=0}^{\infty} a^k x[n-k] + ay[-1] \sum_{k=0}^{\infty} a^k \delta[n-k] \tag{9.38}$$

for any input $x[n]$, initial condition $y[-1]$, and a.

To solve the difference equation with $a = 0.8$, $x[n] = u[n] - u[n - 11]$, and zero initial condition $y[-1] = 0$, using MATLAB, we use the following script.

```
%%%%%%%%%%%%%%%%
% Example 9.16
%%%%%%%%%%%%%%%%%
N = 100; n = 0:N - 1; x = [ones(1,10) zeros(1,N - 10)];
den=[1 -0.8]; num = [1 0];
y = filter(num, den,x)
```

The function filter requires that the initial conditions are zero. The results are shown in Figure 9.11.

Let us now find the impulse response $h[n]$ of the system. For that, let $x[n] = \delta[n]$ and $y[-1] = 0$, then we have $y[n] = h[n]$ or $Y(z) = H(z)$. Thus, we have

$$H(z) = \frac{1}{1 - az^{-1}} \quad \text{so that} \quad h[n] = a^n u[n]$$

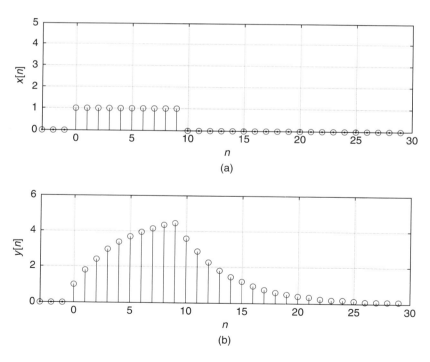

FIGURE 9.11
(a) Solution of the first-order difference equation with (b) input.

You can then see that the first term in Equation (9.38) is a convolution sum, and the second is the impulse response multiplied by $ay[-1]$—that is,

$$y[n] = y_{zs}[n] + y_{zi}[n]$$

$$= \sum_{k=0}^{\infty} h[k]x[n-k] + ay[-1]h[n]$$

∎

■ Example 9.17

Consider a discrete-time system represented by a second-order difference equation with constant coefficients

$$y[n] - a_1 y[n-1] - a_2 y[n-2] = x[n] + b_1 x[n-1] + b_2 x[n-2] \qquad n \geq 0$$

where $x[n]$ is the input, $y[n]$ is the output, and the initial conditions are $y[-1]$ and $y[-2]$. Use the Z-transform to obtain the complete solution.

Solution

Applying the one-sided Z-transform to the two sides of the difference equation, we have

$$\mathcal{Z}(y[n] - a_1 y[n-1] - a_2 y[n-2]) = \mathcal{Z}(x[n] + b_1 x[n-1] + b_2 x[n-2])$$

$$Y(z) - a_1(z^{-1}Y(z) + y[-1]) - a_2(z^{-2}Y(z) + y[-1]z^{-1} + y[-2]) = X(z)(1 + b_1 z^{-1} + b_2 z^{-2})$$

where we used the linearity and the time-shift properties of the Z-transform. It was also assumed that the input is causal, $x[n] = 0$ for $n < 0$, so that the Z-transforms of $x[n-1]$ and $x[n-2]$ are simply $z^{-1}X(z)$ and $z^{-2}X(z)$. Rearranging the above equation, we have

$$Y(z)(1 - a_1 z^{-1} - a_2 z^{-2}) = (y[-1](a_1 + a_2 z^{-1}) + a_2 y[-2]) + X(z)(1 + b_1 z^{-1} + b_2 z^{-2})$$

and solving for $Y(z)$, we have

$$Y(z) = \frac{X(z)(1 + b_1 z^{-1} + b_2 z^{-2})}{1 - a_1 z^{-1} - a_2 z^{-2}} + \frac{y[-1](a_1 + a_2 z^{-1}) + a_2 y[-2])}{1 - a_1 z^{-1} - a_2 z^{-2}}$$

where again the first term is the Z-transform of the zero-state response, due to the input only, and the second term is the Z-transform of the zero-input response, which is due to the initial conditions alone. The inverse Z-transform of $Y(z)$ will give us the complete response. ∎

Remarks *As we saw in Chapter 8, if either the initial conditions are not zero or the input is not causal, the system is not linear time invariant (LTI). However, the time-shift property allows us to find the complete response in that case. We can think of two inputs applied to the system: one due to the initial conditions and the other due to the regular input. By using superposition, we obtain the zero-state and the zero-input responses, which add to the total response.*

Just as with the Laplace transform, the steady-state response of a difference equation

$$y[n] + \sum_{k=1}^{N} a_k y[n-k] = \sum_{m=0}^{M} b_m x[n-m]$$

is due to simple poles of $Y(z)$ on the unit circle. Simple or multiple poles inside the unit circle give a transient, while multiple poles on the unit circle or poles outside the unit circle create an increasing response.

■ Example 9.18

Solve the difference equation

$$y[n] = y[n-1] - 0.25y[n-2] + x[n] \qquad n \geq 0$$

with zero initial conditions and $x[n] = u[n]$.

Solution

The Z-transform of the terms of the difference equation gives

$$Y(z) = \frac{X(z)}{1 - z^{-1} + 0.25z^{-2}}$$

$$= \frac{1}{(1 - z^{-1})(1 - z^{-1} + 0.25z^{-2})} = \frac{z^3}{(z-1)(z^2 - z + 0.25)} \qquad |z| > 1$$

$Y(z)$ has three zeros at $z = 0$, a pole at $z = 1$, and a double pole at $z = 0.5$. The partial fraction expansion of $Y(z)$ is of the form

$$Y(z) = \frac{A}{1 - z^{-1}} + \frac{B(1 - 0.5z^{-1}) + Cz^{-1}}{(1 - 0.5z^{-1})^2} \qquad (9.39)$$

where the terms of the expansion can be found in the Z-transforms table. Within some constants, the complete response is

$$y[n] = Au[n] + [B(0.5)^n + Cn(0.5)^n]u[n]$$

The steady state is then $y_{ss}[n] = A$ (corresponding to the pole on the unit circle $z = 1$) since the other two terms, corresponding to the double pole $z = 0.5$ inside the unit circle, make up the transient response. The value of A is obtained as

$$A = Y(z)(1 - z^{-1})\big|_{z^{-1}=1} = 4$$

To find the complete response $y[n]$ we find the constants in Equation (9.39). Notice in Equation (9.39) that the expansion term corresponding to the double pole $z = 0.5$ has as numerator a first-order polynomial, with constants B and C to be determined, to ensure that the term is proper rational. That term equals

$$\frac{B(1 - 0.5z^{-1}) + Cz^{-1}}{(1 - 0.5z^{-1})^2} = \frac{B}{1 - 0.5z^{-1}} + \frac{Cz^{-1}}{(1 - 0.5z^{-1})^2}$$

which is very similar to the expansion for multiple poles in the inverse Laplace transform. Once we find the values of B and C, the inverse Z-transforms are obtained from the Z-transforms table. A simple method to obtain the coefficients B and C is to first obtain C by multiplying the two sides of Equation (9.39) by $(1 - 0.5z^{-1})^2$ to get

$$Y(z)(1 - 0.5z^{-1})^2 = B(1 - 0.5z^{-1}) + Cz^{-1}$$

and then letting $z^{-1} = 2$ on both sides to find that

$$C = \frac{Y(z)(1 - 0.5z^{-1})^2}{z^{-1}} \Big|_{z^{-1}=2} = -0.5$$

The B value is then obtained by choosing a value for z^{-1} that is different from 1 or 0.5 to compute $Y(z)$. For instance, assume you choose $z^{-1} = 0$ and that you have found A and C, then

$$Y(z)|_{z^{-1}=0} = A + B = 1$$

from which $B = -3$. The complete response is then

$$y[n] = 4u[n] - 3(0.5)^n - 0.5n(0.5)^n u[n]$$

■

■ Example 9.19

Find the complete response of the difference equation

$$y[n] + y[n - 1] - 4y[n - 2] - 4y[n - 3] = 3x[n] \qquad n \geq 0$$
$$y[-1] = 1$$
$$y[-2] = y[-3] = 0$$
$$x[n] = u[n]$$

Determine if the discrete-time system corresponding to this difference equation is BIBO stable or not, and the effect this has in the steady-state response.

Solution

Using the time-shifting and linearity properties of the Z-transform, and replacing the initial conditions, we get

$$Y(z)[1 + z^{-1} - 4z^{-2} - 4z^{-3}] = 3X(z) + [-1 + 4z^{-1} + 4z^{-2}]$$

Letting

$$A(z) = 1 + z^{-1} - 4z^{-2} - 4z^{-3} = (1 + z^{-1})(1 + 2z^{-1})(1 - 2z^{-1})$$

we can write

$$Y(z) = 3\frac{X(z)}{A(z)} + \frac{-1 + 4z^{-1} + 4z^{-2}}{A(z)} \qquad |z| > 2 \tag{9.40}$$

To determine whether the steady-state response exists or not let us first consider the stability of the system associated with the given difference equation. The transfer function $H(z)$ of the system is computed by letting the initial conditions be zero (this makes the second term on the right of the above equation zero) so that we can get the ratio of the Z-transform of the output to the Z-transform of the input. If we do that then

$$H(z) = \frac{Y(z)}{X(z)} = \frac{3}{A(z)}$$

Since the poles of $H(z)$ are the zeros of $A(z)$, which are $z = -1$, $z = -2$, and $z = 2$, then the impulse response $h[n] = \mathcal{Z}^{-1}[H(z)]$ will not be absolutely summable, as required by the BIBO stability, because the poles of $H(z)$ are on and outside the unit circle. Indeed, a general form of the impulse response is

$$h[n] = [C + D\,(2)^n + E\,(-2)^n]u[n]$$

where C, D, and E are constants that can be found by doing a partial fraction expansion of $H(z)$. Thus, $h[n]$ will grow as n increases and it would not be absolutely summable—that is, the system is not BIBO stable.

Since the system is unstable, we expect the total response to grow as n increases. Let us see how we can justify this. The partial fraction expansion of $Y(z)$, after replacing $X(z)$ in Equation (9.40), is given by

$$Y(z) = \frac{2 + 5z^{-1} - 4z^{-3}}{(1 - z^{-1})(1 + z^{-1})(1 + 2z^{-1})(1 - 2z^{-1})}$$

$$= \frac{B_1}{1 - z^{-1}} + \frac{B_2}{1 + z^{-1}} + \frac{B_3}{1 + 2z^{-1}} + \frac{B_4}{1 - 2z^{-1}}$$

$$B_1 = Y(z)(1 - z^{-1})|_{z^{-1}=1} = -\frac{1}{2}$$

$$B_2 = Y(z)(1 + z^{-1})|_{z^{-1}=-1} = -\frac{1}{6}$$

$$B_3 = Y(z)(1 + 2z^{-1})|_{z^{-1}=-1/2} = 0$$

$$B_4 = Y(z)(1 - 2z^{-1})|_{z^{-1}=1/2} = \frac{8}{3}$$

so that

$$y[n] = \left(-0.5 - \frac{1}{6}(-1)^n + \frac{8}{3}2^n\right)u[n]$$

which as expected will grow as n increases — there is no steady-state response.

In a problem like this the chance of making computational errors is large, so it is important to figure out a way to partially check your answer. In this case we can check the value of $y[0]$ using the difference equation, which is $y[0] = -y[-1] + 4y(-2) + 4y(-3) + 3 = -1 + 3 = 2$, and compare it with the one obtained using our solution, which gives $y[0] = -3/6 - 1/6 + 16/6 = 2$. They coincide. Another way to partially check your answer is to use the initial and the final values theorems. ∎

Solution of Differential Equations

The solution of differential equations requires converting them into difference equations, which can then be solved in a closed form by means of the Z-transform.

■ **Example 9.20**

Consider an RLC circuit represented by the second-order differential equation

$$\frac{d^2v_c(t)}{dt^2} + \frac{dv_c(t)}{dt} + v_c(t) = v_s(t)$$

where the voltage across the capacitor $v_c(t)$ is the output and the source $v_s(t) = u(t)$ is the input. Let the initial conditions be zero. Find the voltage across the capacitor $v_c(t)$.

Solution

The Laplace transform of the output is found from the differential equation as

$$V_c(s) = \frac{V_s(s)}{1 + s + s^2}$$

$$= \frac{1}{s(s^2 + s + 1)} = \frac{1}{s((s + 0.5)^2 + 3/4)}$$

where the final equation is obtained after replacing $V_s(s) = 1/s$. The solution of the differential equation is of the general form

$$v_c(t) = [A + Be^{-0.5t}\cos(\sqrt{3}/2t + \theta)]u(t)$$

for constants A, B, and θ. To convert the differential equation into a difference equation we approximate the first derivative as

$$\frac{dv_c(t)}{dt} \approx \frac{v_c(t) - v_c(t - T_s)}{T_s}$$

and the second derivative as

$$\frac{d^2 v_c(t)}{dt^2} = \frac{d \frac{dv_c(t)}{dt}}{dt} \approx \frac{d(v_c(t) - v_c(t - T_s))/T_s)}{dt}$$

$$\approx \frac{v_c(t) - 2v_c(t - T_s) + v_c(t - 2T_s)}{T_s^2}$$

which when replaced in the differential equation, and computing the resulting equation for $t = nT_s$, gives

$$\left(\frac{1}{T_s^2} + \frac{1}{T_s} + 1\right) v_c(nT_s) - \left(\frac{2}{T_s^2} + \frac{1}{T_s}\right) v_c((n-1)T_s) + \left(\frac{1}{T_s^2}\right) v_c((n-2)T_s) = v_s(nT_s)$$

Although we know that we need to choose a very small value for T_s to get a good approximation to the exact result, for simplicity let us first set $T_s = 1$, so that the difference equation is

$$3v_c[n] - 3v_c[n-1] + v_c[n-2] = v_s[n] \qquad n \geq 0$$

For zero initial conditions and unit-step input, we can recursively compute this equation to get

$$v_c[0] = 1/3$$
$$v_c[n] = 1 \qquad n \to \infty$$

A closed-form solution can be obtained using the Z-transform, giving (assuming zero initial conditions)

$$[3 - 3z^{-1} + z^{-2}]V_c(z) = \frac{1}{1 - z^{-1}}$$

so that

$$V_c(z) = \frac{z^3}{(z-1)(3z^2 - 3z + 1)}$$

from which we obtain that there is a triple zero at $z = 0$, and poles at 1 and $-0.5 \pm j\sqrt{3}/6$. The partial fraction expansion will be of the form

$$V_c(z) = \frac{A}{1 - z^{-1}} + \frac{B}{1 + (0.5 + j\sqrt{3}/6)z^{-1}} + \frac{B^*}{1 + (0.5 - j\sqrt{3}/6)z^{-1}}$$

Since the complex poles are inside the unit circle, the steady-state response is due to the input that has a single pole at 1 (i.e., the steady state is $\lim_{n\to\infty} v_c[n] = A = 1$).

We first use the symbolic MATLAB functions ilaplace and ezplot to find the exact solution of the differential equation. We then sample the input signal using a sampling period $T_s = 0.1$ sec and use the approximations of the first and second derivatives to obtain the difference equation, which is computed using filter. The results are shown in Figure 9.12. The exact solution of the

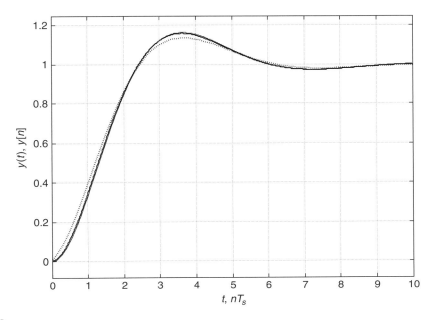

FIGURE 9.12
Solution of the differential equation $d^2 v_c(t)/dt^2 + dv_c(t)/dt + v_c(t) = v_s(t)$ (solid line). Solution of the difference equation approximating the differential equation for $T_s = 0.1$ (dotted line). Exact and approximate solutions are very close.

differential equation is well approximated by the solution of the difference equation obtained by approximating the first and second derivatives.

```
%%%%%%%%%%%%%%%%%
% Example 9.20
%%%%%%%%%%%%%%%%%
   syms s
   vc = ilaplace(1/(s^3 + s^2 + s));  % exact solution
   ezplot(vc,[0,10]);grid; hold on % plotting of exact solution
   Ts = 0.1; % sampling period
   a1 = 1/Ts^2 + 1/Ts + 1;a2 = -2/Ts^2 - 1/Ts;a3 = 1/Ts^2; % coefficients
   a = [1 a2/a1 a3/a1];b = 1;
   t = 0:Ts:10; N = length(t);
   vs=ones(1,N); % input
   vca = filter(b,a,vs);vca = vca/vca(N); % solution
```

9.5.5 Inverse of Two-Sided Z-Transforms

When finding the inverse of a two-sided Z-transform, or a noncausal discrete-time signal, it is important to relate the poles to the causal and the anti-causal components. The region of convergence plays a very important role in making this determination. Once this is done, the inverse is found by looking for the causal and the anti-causal partial fraction expansion components in a Z-transforms table. The coefficients of the partial fraction expansion are calculated like those in the case of causal signals.

■ Example 9.21

Consider finding the inverse Z-transform of

$$X(z) = \frac{2z^{-1}}{(1 - z^{-1})(1 - 2z^{-1})^2} \qquad 1 < |z| < 2$$

which corresponds to a noncausal signal.

Solution

The function $X(z)$ has two zeros at $z = 0$, a pole at $z = 1$, and a double pole at $z = 2$. For the region of convergence to be a torus of internal radius 1 and outer radius 2, we need to associate with the pole at $z = 1$ the region of convergence

$$\mathcal{R}_1 : |z| > 1$$

corresponding to a causal signal, and with the pole at $z = 2$, we associate a region of convergence

$$\mathcal{R}_2 : |z| < 2$$

associated with an anti-causal signal. Thus, we have

$$1 < |z| < 2 = \mathcal{R}_1 \cap \mathcal{R}_2$$

The partial fraction expansion is then done so that

$$X(z) = \underbrace{\frac{A}{1 - z^{-1}}}_{\mathcal{R}_1:|z|>1} + \underbrace{\left[\frac{B}{1 - 2z^{-1}} + \frac{Cz^{-1}}{(1 - 2z^{-1})^2} \right]}_{\mathcal{R}_2:|z|<2}$$

That is, the first term has $\mathcal{R}_1$ as the region of convergence and the terms in the square brackets have $\mathcal{R}_2$ as their region of convergence. The inverse of the first term will be a causal signal, and the inverse of the other two terms will be an anti-causal signal.

The coefficients are found like in the case of causal signals. In this case, we have that

$$A = X(z)(1 - z^{-1})|_{z^{-1}=1} = 2$$

$$C = X(z)\frac{(1 - 2z^{-1})^2}{z^{-1}}|_{z^{-1}=0.5} = 4$$

To calculate B we compute $X(z)$ and its expansion for a value of $z^{-1} \neq 1$ or 0.5. For instance, $z^{-1} = 0$ gives

$$X(0) = A + B = 0$$

so that $B = -A = -2$. The inverse is then found to be

$$x[n] = \underbrace{2u[n]}_{\text{causal}} + \underbrace{\left[-2^{(n+1)}u[-n-1] + 2^{(n+2)}nu[-n-1]\right]}_{\text{anti-causal}}$$

■

■ Example 9.22

Find all the possible impulse responses connected with the following transfer function of a discrete filter having poles at $z = 1$ and $z = 0.5$:

$$H(z) = \frac{1 + 2z^{-1} + z^{-2}}{(1 - 0.5z^{-1})(1 - z^{-1})}$$

determine the cases when the filter is BIBO stable.

Solution

As a function of z^{-1} this function is not proper since both its numerator and denominator are of degree 2. After division we have the following partial fraction expansion:

$$H(z) = B_0 + \frac{B_1}{1 - 0.5z^{-1}} + \frac{B_2}{1 - z^{-1}}$$

There are three possible regions of convergence that can be attached to $H(z)$:

- $\mathcal{R}_1 : |z| > 1$ so that the corresponding impulse response $h_1[n] = \mathcal{Z}^{-1}[H(z)]$ is causal with a general form

$$h_1[n] = B_0\delta[n] + [B_1(0.5)^n + B_2]u[n]$$

  The pole at $z = 1$ makes this filter unstable, as its impulse response is not absolutely summable.
- $\mathcal{R}_2 : |z| < 0.5$ for which the corresponding impulse response $h_2[n] = \mathcal{Z}^{-1}[H(z)]$ is anti-causal with a general form

$$h_2[n] = B_0\delta[n] - (B_1(0.5)^n + B_2)u[-n-1]$$

  The region of convergence $\mathcal{R}_2$ does not include the unit circle and so the impulse response is not absolutely summable ($H(z)$ cannot be defined at $z = 1$ because it is not in the region of convergence $\mathcal{R}_2$). The impulse response $h_2[n]$ grows as n becomes smaller and negative.
- $\mathcal{R}_3 : 0.5 < |z| < 1$, which gives a two-sided impulse response $h_2[n] = \mathcal{Z}^{-1}[H(z)]$ of general form

$$h_3[n] = B_0\delta[n] + \underbrace{B_1((0.5)^n u[n]}_{\text{causal}} \underbrace{-B_2 u[-n-1]}_{\text{anti-causal}}$$

Again, this filter is not stable. ■

9.6 WHAT HAVE WE ACCOMPLISHED? WHERE DO WE GO FROM HERE?

Although the history of the Z-transform is originally connected with probability theory, for discrete-time signals and systems it can be connected with the Laplace transform. The periodicity in the frequency domain and the possibility of an infinite number of poles and zeros makes this connection not very useful. Defining a new complex variable in polar form provides the definition of the Z-transform and the z-plane. As with the Laplace transform, poles of the Z-transform characterize discrete-time signals by means of frequency and attenuation. One- and two-sided Z-transforms are possible, although the one-sided version can be used to obtain the two-sided one. The region of convergence makes the Z-transform have a unique relationship with the signal, and it will be useful in obtaining the discrete Fourier representations in Chapter 10.

Dynamic systems represented by difference equations use the Z-transform for representation by means of the transfer function. The one-sided Z-transform is useful in the solution of difference equations with nonzero initial conditions. As in the continuous-time case, filters can be represented by difference equations. However, discrete filters represented by polynomials are also possible. These nonrecursive filters give significance to the convolution sum, and will motivate us to develop methods that efficiently compute it.

You will see that the reason to present the Z-transform before the Fourier representation of discrete-time signals and systems is to use their connection, thereby simplifying calculations.

PROBLEMS

9.1. Mapping of s-plane into the z-plane

The poles of the Laplace transform $X(s)$ of an analog signal $x(t)$ are

$$p_{1,2} = -1 \pm j1$$
$$p_3 = 0$$
$$p_{4,5} = \pm j1$$

There are no zeros. If we use the transformation $z = e^{sT_s}$ with $T_s = 1$:
(a) Determine where the given poles are mapped into the z-plane.
(b) How would you determine if these poles are mapped inside, on, or outside the unit circle in the z-plane? Explain.
(c) Carefully plot the poles and the zeros of the analog and the discrete-time signals in the Laplace and the z-planes.

9.2. Mapping of z-plane into the s-plane

Consider the inverse relation given by $z = e^{sT_s}$—that is, how to map the z-plane into the s-plane.
(a) Find an expression for s in terms of z from the relation $z = e^{sT_s}$.
(b) Consider the mapping of the unit circle (i.e., $z = 1e^{j\omega}$, $-\pi \le \omega < \pi$). Obtain the segment in the s-plane resulting from the mapping.
(c) Consider the mapping of the inside and the outside of the unit circle. Determine the regions in the s-plane resulting from the mappings.

(d) From the above results, indicate the region in the s-plane to which the whole z-plane is mapped into. Since $\omega = \omega + 2\pi$, is this mapping unique? Explain.

9.3. Z-transform and ROCs

Consider the noncausal sequence

$$s[n] = s_1[n] + s_2[n]$$

where $s_1[n] = u[n]$ is causal and $s_2[n] = -u[-n]$ is anti-causal. This signal is the signum, or sign function, that extracts the sign of a real-valued signal—that is,

$$s[n] = \text{sgn}(x[n]) = \begin{cases} -1 & x[n] < 0 \\ 0 & x[n] = 0 \\ 1 & x[n] > 0 \end{cases}$$

(a) Find the Z-transforms of $s_1[n]$ and $s_2[n]$, indicating the corresponding ROC.

(b) Determine the Z-transform $S(z)$.

9.4. Z-transform and ROC

Given the anti-causal signal

$$x[n] = -\alpha^n u[-n]$$

(a) Determine the Z-transform $X(z)$, and carefully plot the ROC when $\alpha = 0.5$ and $\alpha = 2$. For which of the two values of α does $X(e^{j\omega})$ exist?

(b) Find the signal that corresponds to the derivative $dX(z)/dz$. Express it in terms of α.

9.5. Significance of ROC

Consider a causal signal $x_1[n] = u[n]$ and an anti-causal signal $x_2[n] = -u[-n-1]$.

(a) Find the Z-transforms $X_1(z)$ and $X_2(z)$ and carefully plot their ROCs. If the ROCs are not included with the Z-transforms, would you be able to tell which is the correct inverse? Explain.

(b) Determine if it is possible to find the Z-transform of $x_1[n] + x_2[n]$.

9.6. Fibonacci sequence generation—MATLAB

Consider the Fibonacci sequence generated by the difference equation

$$f[n] = f[n-1] + f[n-2] \qquad n \geq 0$$

with initial conditions $f[-1] = 1, f[-2] = -1$.

(a) Find the Z-transform of $f[n]$, or $F(z)$.

(b) Find the poles ϕ_1 and ϕ_2 and the zeros of $F(z)$ and plot them. How are the poles connected? How are they related to the "golden ratio"?

(c) The Fibonacci difference equation has zero input, but its response is a sequence of ever-increasing integers. Obtain a partial fraction expansion of $F(z)$ and find $f[n]$ in terms of the poles ϕ_1 and ϕ_2, and show that the result is always integer. Use MATLAB to implement the inverse in term of the poles.

9.7. Laplace and Z-transforms of sampled signals

An analog pulse $x(t) = u(t) - u(t-1)$ is sampled using a sampling period $T_s = 0.1$.

(a) Obtain the discrete-time signal $x(nT_s) = x(t)|_{t=nT_s}$ and plot it as a function of nT_s.

(b) If the sampled signal is represented as an analog signal as

$$x_s(t) = \sum_{n=0}^{N-1} x(nT_s)\delta(t - nT_s)$$

determine the value of N in the above equation.

(c) Compute the Laplace transform of the sampled signal (i.e., $X_s(s) = \mathcal{L}[x_s(t)]$).

(d) Determine the Z-transform of $x(nT_s)$, or $X(z)$.

(e) Indicate how to transform $X_s(s)$ into $X(z)$

9.8. Computation of Z-transform—MATLAB

Consider a discrete-time pulse $x[n] = u[n] - u[n - 10]$.

(a) Plot $x[n]$ as a function of n and use the definition of the Z-transform to find $X(z)$.

(b) Use the Z-transform of $u[n]$ and properties of the Z-transform to find $X(z)$. Verify that the expressions obtained above for $X(z)$ are identical.

(c) Find the poles and the zeros of $X(z)$ and plot them in the z-plane. Use MATLAB to plot the poles and zeros.

9.9. Computation of Z-transform

A causal exponential $x(t) = 2e^{-2t}u(t)$ is sampled using a sampling period $T_s = 1$. The corresponding discrete-time signal is $x[n] = 2e^{-2n}u[n]$.

(a) Express the discrete-time signal as $x[n] = 2\alpha^n u[n]$ and give the value of α.

(b) Find the Z-transform $X(z)$ of $x[n]$ and plot its poles and zeros in the z-plane.

9.10. Computation of Z-transform

Consider the signal $x[n] = 0.5(1 + [-1]^n)u[n]$.

(a) Plot $x[n]$ and use the definition of the Z-transform to obtain its Z-transform, $X(z)$.

(b) Use the linearity property and the Z-transforms of $u[n]$ and $[-1]^n u[n]$ to find the Z-transform $X(z) = \mathcal{Z}[x[n]]$.

(c) Determine and plot the poles and the zeros of $X(z)$.

9.11. Solution of difference equations with Z-transform

Consider a system represented by the first-order difference equation

$$y[n] = x[n] - 0.5y[n - 1]$$

where $y[n]$ is the output and $x[n]$ is the input.

(a) Find the Z-transform $Y(z)$ in terms of $X(z)$ and the initial condition $y[-1]$.

(b) Find an input $x[n] \neq 0$ and an initial condition $y[-1] \neq 0$ so that the output is $y[n] = 0$ for $n \geq 0$. Verify you get this result by solving the difference equation recursively.

(c) For zero initial conditions, find the input $x[n]$ so that $y[n] = \delta[n] + 0.5\delta[n - 1]$.

9.12. Transfer function, stability, and impulse response—MATLAB

Consider a second-order discrete-time system represented by the difference equation

$$y[n] - 2r\cos(\omega_0)y[n - 1] + r^2 y[n - 2] = x[n] \qquad n \geq 0$$

where $r > 0$ and $0 \leq \omega_0 \leq 2\pi$, $y[n]$ is the output, and $x[n]$ is the input.

(a) Find the transfer function $H(z)$ of this system.

(b) Find the value of ω_0 and determine the values of r that would make the system stable. Use the MATLAB function zplane to plot the poles and the zeros for $r = 0.5$ and $\omega_0 = \pi/2$ radians.

(c) Let $\omega_0 = \pi/2$. Find the corresponding impulse response $h[n]$ of the system. What other value of ω_0 would get the same impulse response?

9.13. Generation of discrete-time sinusoid—MATLAB

Given that the Z-transform of a discrete-time cosine $A\cos(\omega_0 n)u[n]$ is

$$\frac{A(1 - \cos(\omega_0)z^{-1})}{1 - 2\cos(\omega_0)z^{-1} + z^{-2}}$$

(a) Use the given Z-transform to find a difference equation for which the output $y[n]$ is a discrete-time cosine $A \cos(\omega_0 n)$ and the input is $x[n] = \delta[n]$. What should you use as initial conditions?

(b) Verify your algorithm by generating a signal $y[n] = 2 \cos(\pi n/2)u[n]$ by implementing your algorithm in MATLAB. Plot the input and the output signals $x[n]$ and $y[n]$.

(c) Indicate how to change your previous algorithm to generate a sine function $y[n] = 2 \sin(\pi n/2)u[n]$. Use MATLAB to find $y[n]$, and to plot it.

9.14. Inverse Z-transform and poles and zeros

When finding the inverse Z-transform of functions with z^{-1} terms in the numerator, the fact that z^{-1} can be thought of as a delay operator can be used to simplify the computation. Consider

$$X(z) = \frac{1 - z^{-10}}{1 - z^{-1}}$$

(a) Use the Z-transform of $u[n]$ and the properties of the Z-transform to find $x[n]$.

(b) If we consider $X(z)$ a polynomial in negative powers of z, what would be its degree and the values of its coefficients?

(c) Find the poles and the zeros of $X(z)$ and plot them on the z-plane. Is there a pole or zero at $z = 1$? Explain.

9.15. Initial conditions and steady state

Consider a second-order system represented by the difference equation

$$y[n] = 0.25y[n - 2] + x[n]$$

where $x[n]$ is the input and $y[n]$ is the output.

(a) For the zero-input case (i.e., when $x[n] = 0$), find the initial conditions $y[-1]$ and $y[-2]$ so that $y[n] = 0.5^n u[n]$.

(b) Suppose the input is $x[n] = u[n]$. Without solving the difference equation can you find the corresponding steady state $y_{ss}[n]$? Explain how and give the steady-state output. Verify by inverse Z-transform that the steady-state response $y_{ss}[n]$ is the one obtained.

9.16. Initial conditions and impulse response

A second-order system has the difference equation

$$y[n] = 0.25y[n - 2] + x[n]$$

where $x[n]$ is the input and $y[n]$ is the output.

(a) Find the input $x[n]$ so that for zero initial conditions, the output is given as $y[n] = 0.5^n u[n]$.

(b) If $x[n] = \delta[n] + 0.5\delta[n - 1]$ is the input to the above difference equation, find the impulse response of the system.

9.17. Convolution sum and product of polynomials

The convolution sum is a fast way to find the coefficients of the polynomial resulting from the multiplication of two polynomials.

(a) Suppose $x[n] = u[n] - u[n - 3]$. Find its Z-transform $X(z)$, a second-order polynomial in z^{-1}.

(b) Multiply $X(z)$ by itself to get a new polynomial $Y(z) = X(z)X(z) = X^2(z)$. Find $Y(z)$.

(c) Graphically show the convolution of $x[n]$ with itself and verify that the result coincides with the coefficients of $Y(z)$.

9.18. Inverse Z-transform

Find the inverse Z-transform of

$$X(z) = \frac{8 - 4z^{-1}}{z^{-2} + 6z^{-1} + 8}$$

and determine $x[n]$ as $n \to \infty$. Assume $x[n]$ is causal.

9.19. Z-transform properties and inverse transform

Sometimes the partial fraction expansion is not needed in finding the inverse Z-transform—instead the properties of the transform can be used. Consider the function

$$F(z) = \frac{z + 1}{z^2(z - 1)}$$

(a) Determine whether $F(z)$ is a proper rational function as a function of z and of z^{-1}.
(b) Verify that $F(z)$ can be written as

$$F(z) = \frac{z^{-2}}{1 - z^{-1}} + \frac{z^{-3}}{1 - z^{-1}}$$

Find the inverse Z-transform $f[n]$ using the above expression.

9.20. Inverse Z-transform—MATLAB

We are interested in the unit-step solution of a system represented by the difference equation

$$y[n] = y[n - 1] - 0.5y[n - 2] + x[n] + x[n - 1]$$

(a) Find an expression for $Y(z)$.
(b) Do a partial expansion of $Y(z)$.
(c) Find the inverse Z-transform $y[n]$ and verify your results using MATLAB.

9.21. Padé approximation

Suppose we are given a finite-length sequence $h[n]$ (it could be part of an infinite-length impulse response from a discrete system that has been windowed) and would like to obtain a rational approximation for it. This means that if $H(z) = \mathcal{Z}[h[n]]$, a rational approximation of it would be $H(z) = B(z)/A(z)$, from which we get

$$H(z)A(z) = B(z)$$

Letting

$$B(z) = \sum_{k=0}^{M-1} b_k z^{-k}$$

$$A(z) = 1 + \sum_{k=1}^{N-1} a_k z^{-k}$$

for some choice of M and N, equations from $H(z)A(z) = B(z)$ should allow us to find the $M + N - 1$ coefficients $\{a_k \ b_k\}$.

(a) Find a matrix equation that would allow us to find the coefficients of $B(z)$ and $A(z)$.

(b) Let $h[n] = 0.5^n(u[n] - u[n - 101])$ be the sequence we wish to obtain a rational approximation and let $B(z) = b_0$ while $A(z) = a_0 + a_1 z^{-1}$. Find the equations to solve for the coefficients $\{b_0, a_0, a_1\}$.

9.22. Prony's rational approximation—MATLAB

The Padé approximants provide an exact matching of $M + N - 1$ values of $h[n]$ where M and N are, respectively, the orders of the numerator and the denominator of the rational approximation. But there is no method for choosing the numerator and the denominator orders, M and N. Also, there is no guarantee on how well the rest of the signal is matched. Prony's rational approximation considers how well the rest of the signal is approximated when finding the approximation. Let $h[n] = 0.9^n u[n]$ be the exact impulse response for which we wish to find a rational approximation. Take the first 100 values of this signal as the impulse response.[1]

(a) Assume the order of the numerator and the denominator are equal, $M = N = 1$. Use the MATLAB function prony to obtain the rational approximation, and then use filter to verify that the impulse response of the rational approximation is close to the given 100 values. Plot the error between $h[n]$ and the impulse response of the rational approximation for the first 200 samples.

(b) Plot the poles and the zeros of the rational approximation and compare them to the poles and the zeros of $H(z) = \mathcal{Z}(h[n])$.

(c) Suppose that $h[n] = (h_1 * h_2)[n]$—that is, the convolution of $h_1[n] = 0.9^n u[n]$ and $h_2[n] = 0.8^n u[n]$. Use again prony to find the rational approximation when the first 100 values of $h[n]$ are available. Use conv from MATLAB to compute $h[n]$. Compare the impulse response of the rational approximation to $h[n]$. Plot the poles and the zeros of $H(z) = \mathcal{Z}(h[n])$ and of the rational approximation.

(d) Consider the $h[n]$ given above, and perform the Prony approximation using orders $M = N = 3$. Explain your results. Plot the poles and the zeros.

9.23. MATLAB partial fraction expansion

Consider the partial fraction expansion that MATLAB uses.

(a) Find the inverse Z-transform of $a/(1 - az^{-1})^2$.

(b) Suppose that the partial fraction expansion given by MATLAB is

$$X(z) = \frac{-1}{1 - 0.5z^{-1}} + \frac{1}{(1 - 0.5z^{-1})^2}$$

Determine the inverse $x[n]$.

9.24. MATLAB partial fraction expansion

Consider finding the inverse Z-transform of

$$X(z) = \frac{2z^{-1}}{(1 - z^{-1})(1 - 2z^{-1})^2} \qquad |z| > 2$$

MATLAB does the partial fraction expansion as

$$X(z) = \frac{A}{1 - z^{-1}} + \frac{B}{1 - 2z^{-1}} + \frac{C}{(1 - 2z^{-1})^2}$$

while we do it in the following form:

$$X(z) = \frac{D}{1 - z^{-1}} + \frac{E}{1 - 2z^{-1}} + \frac{Fz^{-1}}{(1 - 2z^{-1})^2}$$

Show that the two partial fraction expansions give the same result.

---

[1] Gaspar de Prony (1765–1839) was a French mathematician and engineer, while Henri Padé (1863–1953) was a French mathematician interested in rational approximations.

9.25. Prony method and Z-Transform—MATLAB

Consider finding the Z-transform of a noncausal signal $h[n] = 0.5^n u[n + 1]$ using the Prony approximation.

(a) Use the prony function to find a rational approximation for $h[n]$ (i.e., the Z-transform $H(z) = B(z)/A(z)$). Use a first order for the numerator and the denominator.

(b) Separate the signal into its causal and anti-causal components, and use prony to find the rational approximation of the causal and then add the anti-causal component to correct the above result.

CHAPTER 10

Fourier Analysis of Discrete-Time Signals and Systems

Diligence is the mother of good luck.
Benjamin Franklin (1706–1790)
Printer, inventor, scientist, and diplomat

I am a great believer in luck,
and I find the harder I work,
the more I have of it.
President Thomas Jefferson (1743–1826)
Main author of the U.S. Declaration of Independence

10.1 INTRODUCTION

In this chapter we will consider the Fourier representation of discrete-time signals and systems. Similar to the connection between the Laplace and the Fourier transforms of continuous-time signals and systems, if the region of convergence of the Z-transform of a signal or of the transfer function of a discrete system includes the unit circle, then the discrete-time Fourier transform (DTFT) of the signal or the frequency response of the system is easily found. Duality in time and frequency is used whenever signals and systems do not satisfy this condition. We can thus obtain the Fourier representation of most discrete-time signals and systems.

Two computational disadvantages of the DTFT are that the direct DTFT is a function of a continuously varying frequency, and the inverse DTFT requires integration. These disadvantages can be removed by sampling in frequency the DTFT, resulting in the so-called discrete Fourier transform (DFT) (notice the difference in the naming of these two related frequency representations). An interesting connection determines their computational feasibility: a *discrete–time* signal has a *periodic continuous-frequency* transform—the DTFT—while a *periodic discrete-time* signal has a *periodic and discrete-frequency* transform—the DFT. As we will discuss in this chapter, any periodic or aperiodic signal can be represented by the DFT, a computationally feasible transformation where both time and frequency are discrete and no integration is required, and that can be implemented very efficiently by the Fast Fourier Transform (FFT) algorithm.

Signals and Systems Using MATLAB®. DOI: 10.1016/B978-0-12-374716-7.00014-4

In this chapter, we will see that a great deal of the Fourier representation of discrete-time signals and characterization of discrete systems can be obtained from our knowledge of the Z-transform. To obtain the DFT, which is of great significance in digital signal processing, we will proceed in an opposite direction as in the continuous-time analysis. First, we consider the Fourier representation of aperiodic signals and then that of periodic discrete-time signals, and finally use this representation to obtain the DFT.

10.2 DISCRETE-TIME FOURIER TRANSFORM

The discrete-time Fourier transform (DTFT) of a discrete-time signal $x[n]$,

$$X(e^{j\omega}) = \sum_n x[n]e^{-j\omega n} \qquad -\pi \le \omega < \pi \tag{10.1}$$

converts $x[n]$ into a function $X(e^{j\omega})$ of the discrete frequency ω (rad), while the inverse transform gives back $x[n]$ from $X(e^{j\omega})$ according to

$$x[n] = \frac{1}{2\pi} \int_{-\pi}^{\pi} X(e^{j\omega})e^{j\omega n} d\omega \tag{10.2}$$

Remarks

■ *The DTFT measures the frequency content of a discrete-time signal. When using the DTFT, it is important to remember some of the differences between the continuous and the discrete domains. Discrete-time signals are only defined for uniform sample times nT_s or integers n, and the discrete frequency is such that it repeats every 2π radians (i.e., $\omega = \omega + 2\pi k$ for any integer k), so that $X(e^{j\omega})$ is periodic and only the frequencies $[-\pi, \pi)$ need to be considered.*

■ *The DTFT $X(e^{j\omega})$ is periodic of period 2π. Indeed, for an integer k,*

$$X(e^{j(\omega+2\pi k)}) = \sum_n x[n]e^{-j(\omega+2\pi k)n} = X(e^{j\omega})$$

since $e^{-j(\omega+2\pi k)n} = e^{-j\omega n}e^{-j2\pi kn} = e^{-j\omega n}$. Thus, one can think of Equation (10.1) as the Fourier series of $X(e^{j\omega})$: If $\varphi = 2\pi$ is the period, the Fourier series coefficients are given by

$$x[n] = \frac{1}{\varphi} \int_{\varphi} X(e^{j\omega})e^{j2\pi n\omega/\varphi} d\omega = \frac{1}{2\pi} \int_{-\pi}^{\pi} X(e^{j\omega})e^{jn\omega} d\omega$$

■ *For the DTFT to converge, as an infinite sum, it is necessary that*

$$|X(e^{j\omega})| \le \sum_n |x[n]||e^{j\omega n}| = \sum_n |x[n]| < \infty$$

or that $x[n]$ be absolutely summable. This means that only for those signals we can use the above direct (Eq. 10.1) and inverse (Eq. 10.2) DTFT definitions. We will see in the next section how to obtain the DTFT of signals that do not satisfy the absolutely summable condition.

10.2.1 Sampling, Z-Transform, Eigenfunctions, and the DTFT

The connection of the DTFT with sampling, eigenfunctions, and the Z-transform can be shown as follows:

- *Sampling and the DTFT.* When sampling an analog signal $x(t)$, the sampled signal $x_s(t)$ can be written as

$$x_s(t) = \sum_n x(nT_s)\delta(t - nT_s)$$

Its Fourier transform is then

$$\mathcal{F}[x_s(t)] = \sum_n x(nT_s)\mathcal{F}[\delta(t - nT_s)]$$

$$= \sum_n x(nT_s)e^{-jn\Omega T_s}$$

Letting $\omega = \Omega T_s$, the discrete frequency in radians, the above equation can be written as

$$X_s(e^{j\omega}) = \mathcal{F}[x_s(t)] = \sum_n x(nT_s)e^{-jn\omega} \tag{10.3}$$

coinciding with the DTFT of the discrete-time signal $x(nT_s) = x(t)|_{t=nT_s}$ or $x[n]$.
At the same time, the spectrum of the sampled signal can be equally represented as

$$X_s(e^{j\Omega T_s}) = X_s(e^{j\omega}) = \sum_k \frac{1}{T_s}X\left(\frac{\omega}{T_s} - \frac{2\pi k}{T_s}\right) \tag{10.4}$$

which is a periodic repetition, with period $2\pi/T_s$, of the spectrum of the analog signal being sampled. Thus, sampling converts a continuous-time signal into a discrete-time signal with a periodic spectrum varying continuously in frequency.

- *Z-transform and the DTFT.* If in the above we ignore T_s and consider $x(nT_s)$ a function of n, we can see that

$$X_s(e^{j\omega}) = X(z)|_{z=e^{j\omega}} \tag{10.5}$$

That is, it is the Z-transform computed on the unit circle. For the above to happen, $X(z)$ must have a region of convergence (ROC) that includes the unit circle. There are discrete-time signals for which we cannot find their DTFTs from the Z-transform because they are not absolutely summable—that is, their ROCs do not include the unit circle. However, any discrete-time signal $x[n]$, of finite support in time, has a Z-transform $X(z)$ with a region of convergence the whole z-plane, excluding either the origin or infinity, and as such its DTFT $X(e^{j\omega})$ is computed from $X(z)$ by letting $z = e^{j\omega}$.

■ *Eigenfunctions and the DTFT.* The frequency representation of a discrete-time linear time-invariant (LTI) system is shown to be the DTFT of the impulse response of the system. Indeed, according to the eigenfunction property of LTI systems, if the input of such a system is a complex exponential, $x[n] = e^{j\omega_0 n}$, the steady-state output, calculated with the convolution sum, is given by

$$y[n] = \sum_k h[k]x[n-k] = \sum_k h[k]e^{j\omega_0(n-k)} = e^{j\omega_0 n}H(e^{j\omega_0}) \tag{10.6}$$

where

$$H(e^{j\omega_0}) = \sum_k h[k]e^{-j\omega_0 k} \tag{10.7}$$

or the DTFT of the impulse response $h[n]$ of the system computed at $\omega = \omega_0$. As with continuous-time systems, the system needs to be bounded-input bounded-output (BIBO) stable. Without the stability of the system, there is no guarantee that there will be a steady-state response.

■ Example 10.1

Consider the noncausal signal $x[n] = \alpha^{|n|}$ with $|\alpha| < 1$. Determine its DTFT. Use the obtained DTFT to find

$$\sum_{n=-\infty}^{\infty} \alpha^{|n|}$$

Solution

The Z-transform of $x[n]$ is

$$X(z) = \sum_{n=0}^{\infty} \alpha^n z^{-n} + \sum_{m=0}^{\infty} \alpha^m z^m - 1$$

$$= \frac{1}{1 - \alpha z^{-1}} + \frac{1}{1 - \alpha z} - 1 = \frac{1 - \alpha^2}{1 - \alpha(z + z^{-1}) + \alpha^2}$$

where the first term has an ROC of $|z| > |\alpha|$, and the ROC of the second term is $|z| < 1/|\alpha|$. Thus, the region of convergence of $X(z)$ is

$$\text{ROC: } |\alpha| < |z| < \frac{1}{|\alpha|}$$

and that includes the unit circle. Thus, the DTFT is

$$X(e^{j\omega}) = \frac{1 - \alpha^2}{(1 + \alpha^2) - 2\alpha \cos(\omega)} \tag{10.8}$$

According to the formula for the DTFT at $\omega = 0$, we have that

$$X(e^{j0}) = \sum_{n=-\infty}^{\infty} x[n]e^{j0n} = \sum_{n=-\infty}^{\infty} \alpha^{|n|} = \frac{2}{1-\alpha} - 1 = \frac{1+\alpha}{1-\alpha}$$

and according to Equation (10.8), equivalently we have

$$X(e^{j0}) = \frac{1-\alpha^2}{1-2\alpha+\alpha^2} = \frac{1-\alpha^2}{(1-\alpha)^2} = \frac{1+\alpha}{1-\alpha}$$ ∎

10.2.2 Duality in Time and Frequency

In practice, there are many signals of interest that do not satisfy the absolute summability condition, and so we cannot find their DTFTs with the definition given in the previous section. Duality in the time and frequency representation of signals permits us to obtain the DTFT of those signals.

Consider the DTFT of the signal $\delta[n-k]$ for some integer k. Since $\mathcal{Z}[\delta[n-k]] = z^{-k}$ with ROC the whole z-plane except for the origin, the DTFT of $\delta[n-k]$ is $e^{-j\omega k}$. By duality, as in the continuous-time domain, we would expect that the signal $e^{-j\omega_0 n}$, $-\pi \leq \omega_0 < \pi$, would have $2\pi\delta(\omega+\omega_0)$ (where $\delta(\omega)$ is the analog delta function) as its DTFT. Indeed, the inverse DTFT of $2\pi\delta(\omega+\omega_0)$ gives

$$\frac{1}{2\pi}\int_{-\pi}^{\pi} 2\pi\delta(\omega+\omega_0)e^{j\omega n}d\omega = e^{-j\omega_0 n}\int_{-\pi}^{\pi}\delta(\omega+\omega_0)d\omega = e^{-j\omega_0 n}$$

Using these results, we have the following dual pairs:

$$\sum_{k=-\infty}^{\infty} x[k]\delta[n-k] \Leftrightarrow \sum_{k=-\infty}^{\infty} x[k]e^{-j\omega k}$$

$$\sum_{k=-\infty}^{\infty} X[k]e^{-j\omega_k n} \Leftrightarrow \sum_{k=-\infty}^{\infty} 2\pi X[k]\delta(\omega+\omega_k) \qquad (10.9)$$

The top left equation is the generic representation of a discrete-time signal $x[n]$ and the corresponding term on the right is its DTFT $X(e^{j\omega})$, one more verification of Equation (10.1). The bottom pair is a dual of the above.[1] Using Equation (10.9), we then have the following dual pairs as special cases:

$$x[n] = A\delta[n] \Leftrightarrow X(e^{j\omega}) = A$$

$$y[n] = A, -\infty < n < \infty \Leftrightarrow Y(e^{j\omega}) = 2\pi A\delta(\omega) \qquad -\pi \leq \omega < \pi$$

The signal $y[n]$ is not absolutely summable, and as a constant it does not change from $-\infty$ to ∞, so that its frequency is $\omega = 0$, thus its DTFT $Y(e^{j\omega})$ is concentrated in that frequency. Consider then

---

[1] Calling this a "dual" is not completely correct given that the ω_k are discrete values of frequency instead of continuous as expressed by ω, and that the delta functions are not the same in the continuous and the discrete domains, but a duality of some sort exists in these two pairs, which we would like to take advantage of.

a sinusoid $x[n] = \cos(\omega_0 n + \theta)$, which is not absolutely summable. According to the above pairs, we get

$$x[n] = \frac{1}{2}\left[e^{j(\omega_0 n + \theta)} + e^{-j(\omega_0 n + \theta)}\right] \Leftrightarrow X(e^{j\omega}) = \pi\left[e^{j\theta}\delta(\omega - \omega_0) + e^{-j\theta}\delta(\omega + \omega_0)\right]$$

The DTFT of the cosine indicates that its power is concentrated at the frequency ω_0.

The "dual" pairs

$$\delta[n - k], \quad \text{integer } k \Leftrightarrow e^{-j\omega k} \tag{10.10}$$

$$e^{-j\omega_0 n}, \quad -\pi \leq \omega_0 < \pi \Leftrightarrow 2\pi\delta(\omega + \omega_0) \tag{10.11}$$

allow us to obtain the DTFT of signals that do not satisfy the absolutely summable condition. Thus, in general, we have

$$\sum_k X[k]e^{-j\omega_k n} \Leftrightarrow \sum_k 2\pi X[k]\delta(\omega + \omega_k) \tag{10.12}$$

The linearity of the DTFT and the above result give that for a signal that is not absolutely summable

$$x[n] = \sum_\ell A_\ell \cos(\omega_\ell n + \theta_\ell)$$

its DTFT is

$$X(e^{j\omega}) = \sum_\ell \pi A_\ell \left[e^{j\theta_\ell}\delta(\omega - \omega_\ell) + e^{-j\theta_\ell}\delta(\omega + \omega_\ell)\right] \quad -\pi \leq \omega < \pi$$

If $x[n]$ is periodic, the discrete frequencies are harmonically related, i.e., $\omega_\ell = \ell\omega_0$ where ω_0 is the fundamental frequency of $x[n]$.

■ Example 10.2

The DTFT of a signal $x[n]$ is

$$X(e^{j\omega}) = 1 + \delta(\omega - 4) + \delta(\omega + 4) + 0.5\delta(\omega - 2) + 0.5\delta(\omega + 2)$$

The signal $x[n] = A + B\cos(\omega_0 n)\cos(\omega_1 n)$ is given as a possible signal that gives $X(e^{j\omega})$. Determine whether you can find A, B, and ω_0 and ω_1 to obtain the desired DTFT. If not, provide a better $x[n]$.

Solution

Using trigonometric identities or Euler's identity, we have that

$$\cos(\omega_0 n)\cos(\omega_1 n) = 0.5\cos((\omega_0 + \omega_1)n) + 0.5\cos((\omega_1 - \omega_0)n)$$

so that $x[n] = A + 0.5B \cos((\omega_0 + \omega_1)n) + 0.5B \cos((\omega_1 - \omega_0)n)$. Letting $\omega_2 = \omega_0 + \omega_1$ and $\omega_3 = \omega_1 - \omega_0$, the DTFT of $x[n]$ is

$$X(e^{j\omega}) = 2\pi A + \pi B(\delta(\omega - \omega_2) + \delta(\omega + \omega_2)) + \pi B(\delta(\omega - \omega_3) + \delta(\omega + \omega_3))$$

Comparing this DTFT with the given one, we find that

$$2\pi A = 1 \Rightarrow A = 1/(2\pi)$$

$$\omega_2 = \omega_0 + \omega_1 = 4$$

$$\omega_3 = \omega_1 - \omega_0 = 2$$

$$\pi B = 1, \; 0.5 \Rightarrow \quad \text{no unique value for } B$$

Although we find that $A = 1/(2\pi)$ and $\omega_0 = 1$ and $\omega_1 = 3$, there is no unique value for B, so the given $x[n]$ is not the correct answer. The correct signal should be $x[n] = (1/\pi)(0.5 + \cos(4n)) + (1/2\pi) \cos(2n)$, which has the desired DTFT (verify it!). ∎

10.2.3 Computation of the DTFT Using MATLAB

According to the definitions of the direct and the inverse DTFT, their computation needs to be done for a continuous frequency $\omega \in [-\pi, \pi)$ and requires integration. In MATLAB the DTFT is approximated in a discrete set of frequency values, and summation instead of integration is used. As we will see later, this can be done by sampling in frequency the DTFT to obtain the discrete Fourier transform (DFT), which in turn is efficiently implemented by an algorithm called the Fast Fourier Transform (FFT). We will introduce the FFT in Chapter 12, and so for now consider the FFT as a black box capable of giving a discrete approximation of the DTFT.

To understand the use of the MATLAB function fft in the script below consider the following issues:

■ The command X = fft(x), where x is a vector with the entries the sample values $x[n]$, $n = 0, \ldots, L - 1$, computes the FFT values $X[k]$, $k = 0, \ldots, L - 1$, or the DTFT $X(e^{j\omega})$ at discrete frequencies $\{2\pi k/L\}$.
■ The $\{k\}$ values correspond to the discretized frequencies $\{\omega_k = 2\pi k/L\}$, which go from 0 to $2\pi(L - 1)/L$ (close to 2π for large L). This is a discretization of the frequency $\omega \in [0, \; 2\pi)$.
■ To find an equivalent representation of the frequency $\omega \in [-\pi, \; \pi)$, we simply subtract π from $\omega_k = 2\pi k/L$ to get a band of frequencies,

$$\tilde{\omega}_k = \omega_k - \pi = \pi \frac{2k - L}{L}, \; k = 0, \ldots, L - 1 \; \text{ or } \; -\pi \leq \tilde{\omega}_k < \pi$$

The frequency $\tilde{\omega}$ can be normalized to $[-1, 1)$ with no units by dividing by π. This change in the frequency scale requires a corresponding shift of the magnitude and the phase spectra. This is done by means of the MATLAB function fftshift.
■ When plotting the signal, which is discrete in time, the function stem is more appropriate than plot. However, the plot function is more appropriate for plotting the magnitude and the phase frequency response functions, which are supposed to be continuously varying with respect to frequency.

- The function abs computes the magnitude and the function angle computes the phase of the frequency response. The magnitude and the phase are even and odd symmetric when plotted in $\omega \in [-\pi, \pi)$ or in the normalized frequency $\omega/\pi \in [-1, 1)$.

The three signals in the following script are a rectangular pulse, a windowed sinusoid, and a chirp. In the script, to process one of the signals you delete the corresponding comment % and keep it for the other two. The length of the FFT is set to $L = 256$, which is larger or equal to the length of either of the three signals.

```
%%%%%%%%%%%%%%%%%%%%%%%
% DTFT of aperiodic signals
%%%%%%%%%%%%%%%%%%%%%%%%
% signals
L = 256; % length of FFT with added zeros
% N = 21; x = [ones(1,N) zeros(1,L - N)];                    % pulse
% N = 200; n = 0:N - 1; x = [cos(4 * pi * n/N) zeros(1,L - N)];   % windowed sinusoid
n = 0:L - 1; x = cos(pi * n. ^ 2/(4 * L));                       % chirp
X = fft(x);
w = 0:2 * pi/L:2 * pi - 2 * pi/L;w1 = (w - pi)/pi; % normalized frequency
n = 0:length(x) - 1;
subplot(311)
stem(n,x); axis([0 length(n) - 1 1.1 * min(x) 1.1 * max(x)]); grid;
xlabel('n'); ylabel('x(n)')
subplot(312)
plot(w1,fftshift(abs(X))); axis([min(w1) max(w1) 0 1.1 * max(abs(X))]);
ylabel('|X|'); grid
subplot(313)
plot(w1,fftshift(angle(X))); ylabel('<X'); xlabel('ω/π'); grid
axis([min(w1) max(w1) 1.1 * min(angle(X)) 1.1 * max(angle(X))])
```

As expected, the magnitude spectrum for the rectangular pulse is like a sinc. The windowed sinusoid has a spectrum that resembles that of the sinusoid but the rectangular window makes it broader. Finally, a chirp is a sinusoid with time-varying frequency; thus its magnitude spectrum displays components over a range of frequencies. We will comment on the phase spectra later. The results are shown in Figure 10.1.

Sampled Signals

When computing the DTFT of a sampled signal, it is important to display the frequency in radians/second or in hertz rather than the discrete frequency in radians. The discrete frequency ω (rad) is converted into the analog signal Ω (rad/sec) according to the relation $\omega = \Omega T_s$ where T_s is the sampling period used. Thus,

$$\Omega = \omega/T_s \ \text{rad/sec} \tag{10.13}$$

If the signal is sampled according to the Nyquist sampling rate condition, the discrete-frequency range $\omega \in [-\pi, \pi)$ (rad) corresponds to $\Omega \in [-\pi/T_s, \pi/T_s)$ or $[-\Omega_s/2, \Omega_s/2)$, where $\Omega_s/2 \geq \Omega_{max}$

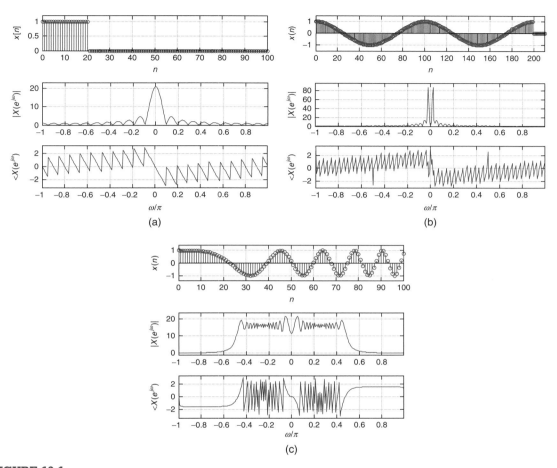

FIGURE 10.1

MATLAB computation of the DTFT of (a) a pulse, (b) a windowed sinusoid, and (c) a chirp: magnitude and phase spectra are shown for each.

for Ω_s the sampling frequency in radians/second and Ω_{max} the maximum frequency in the signal being sampled.

To illustrate this, we sampled a signal $x(t) = 5^{-2t}u(t)$ with $T_s = 0.01$ sec/sample, created a vector of 256 values from the signal, and computed its FFT as before. The above script is modified to consider the change of scale. The changes are as follows:

```
%%%%%%%%%%%%%%%%%%%%%%%%%
% DTFT of sampled signal
%%%%%%%%%%%%%%%%%%%%%%%%%
L = 256;Ts = 0.01; t = 0:Ts:(L - 1) * Ts; x = 5.^( - 2 * t);  % sampling of signal
X = fft(x);
w = 0:2 * pi/L:2 * pi - 2 * pi/L;W = (w - pi)/Ts;   % W is analog frequency
```

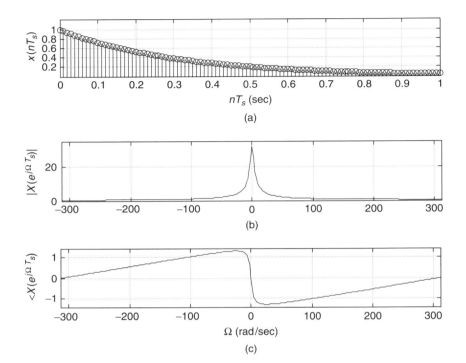

FIGURE 10.2
DTFT of (a) a sampled
signal, and (b) the
magnitude of DTFT and
(c) the phase of DTFT as
functions of Ω (rad/sec).

The results are shown in Figure 10.2. Given that the signal is very smooth, most of the frequency
components have low frequencies.

10.2.4 Time and Frequency Supports

The Fourier representation of a discrete-time signal gives a complementary characterization to its
time representation, just like in the continuous-time case. The following examples illustrate the
complementary nature of the DTFT of discrete-time signals.

> Just like with analog signals, the frequency support of the DTFT of a discrete-time signal is inversely
> proportional to the time support of the signal.

■ Example 10.3

Consider a discrete pulse

$$p[n] = u[n] - u[n - N]$$

Find its DTFT $P(e^{j\omega})$ and discuss the relation between its frequency support and the time support
of $p[n]$ when $N = 1$ and when $N \to \infty$.

Solution

Since $p[n]$ has a finite support, its Z-transform has as region of convergence the whole z-plane, except for $z = 0$, and we can find its DTFT from it. We have

$$P(z) = \sum_{n=0}^{N-1} z^{-n} = 1 + z^{-1} + \cdots + z^{-(N-1)} = \frac{1 - z^{-N}}{1 - z^{-1}}$$

The DTFT of $p[n]$ is then given by

$$P(e^{j\omega}) = 1 + e^{-j\omega} + \cdots + e^{-j\omega(N-1)}$$

or equivalently

$$P(e^{j\omega}) = \frac{1 - e^{-j\omega N}}{1 - e^{-j\omega}} = \frac{e^{-j\omega N/2}[e^{j\omega N/2} - e^{-j\omega N/2}]}{e^{-j\omega/2}[e^{j\omega/2} - e^{-j\omega/2}]}$$

$$= e^{-j\omega((N-1)/2)} \frac{\sin(\omega N/2)}{\sin(\omega/2)}$$

The function $\sin(\omega N/2)/\sin(\omega/2)$ is the discrete counterpart of the sinc function in frequency. It can be shown that like the sinc function, this function is:

- Even function of ω, as both the numerator and the denominator are odd functions of ω.
- $0/0$ at $\omega = 0$, so that using L'Hôpital's rule its value at $\omega = 0$ is

$$\lim_{\omega \to 0} \frac{\sin(\omega N/2)}{\sin(\omega/2)} = N$$

- Zero at $\omega = 2\pi k/N$ for integer values $k \neq 0$, as $\sin(\omega N/2)|_{\omega=2\pi k/N} = \sin(\pi k) = 0$.
- Periodic of period 2π when N is odd, which can be seen from its equivalent representation,

$$\frac{\sin(\omega N/2)}{\sin(\omega/2)} = e^{j\omega((N-1)/2)} P(e^{j\omega})$$

since $P(e^{j\omega})$ is periodic of period 2π and

$$e^{j(\omega+2\pi)((N-1)/2)} = e^{j\omega((N-1)/2)} e^{j\pi(N-1)} = e^{j\omega((N-1)/2)}$$

when N is odd (e.g., $N = 2M + 1$ for some integer M). When N is even, $e^{j\pi(N-1)} = -1$, so $\sin(\omega N/2)/\sin(\omega/2)$ is not periodic of period 2π.

Consider the discrete pulse when $N = 1$, then $p[n] = u[n] - u[n-1] = \delta[n]$, or the discrete impulse. The DTFT is then $P(e^{j\omega}) = 1$. In this case, the support of $p[n]$ is one point, while the support of $P(e^{j\omega})$ is all discrete frequencies, or $[-\pi, \pi)$.

As we let $N \to \infty$, the pulse tends to a constant, 1, making $p[n]$ not absolutely summable. In the limit, we find that $P(e^{j\omega}) = 2\pi\delta(\omega)$, $-\pi \leq \omega < \pi$. Indeed, the inverse DTFT is found to be

$$\frac{1}{2\pi} \int_{-\pi}^{\pi} 2\pi\delta(\omega)e^{j\omega n} d\omega = \int_{-\pi}^{\pi} \delta(\omega)d\omega = 1$$

The time support of $p[n] = 1$ is infinite, while $P(e^{j\omega}) = 2\pi\delta(\omega)$ exists at only one frequency.   ∎

Downsampling and Upsampling

Although the expanding and contracting of discrete-time signals is not as obvious as in the continuous time, the dual effects of contracting and expanding in time and frequency also occur in the discrete case. Contracting and expanding of discrete-time signals relates to downsampling and upsampling.

Downsampling a signal $x[n]$ means getting rid of samples (i.e., contracting the signal). The signal downsampled by an integer factor $M > 1$ is given by

$$x_d[n] = x[Mn] \tag{10.14}$$

If $x[n]$ has a DTFT $X(e^{j\omega})$, $-\pi/M \le \omega \le \pi/M$, and zero otherwise in $[-\pi, \pi)$ (analogous to bandlimited signals in continuous time), by replacing n by Mn in the inverse DTFT, $x[n]$ gives

$$x[Mn] = \frac{1}{2\pi} \int_{-\pi/M}^{\pi/M} X(e^{j\omega})e^{jMn\omega}d\omega = \frac{1}{2\pi} \int_{-\pi}^{\pi} \frac{1}{M}X(e^{j\rho/M})e^{jn\rho}d\rho$$

where we let $\rho = M\omega$. Thus, the DTFT of $x_d[n]$ is $\frac{1}{M}X(e^{j\omega/M})$—that is, an expansion by a factor of M of the DTFT of $x[n]$.

Upsampling a signal $x[n]$, on the other hand, consists in adding $L - 1$ zeros for some integer $L > 1$ in between its samples—that is, the upsampled signal is

$$x_u[n] = \begin{cases} x[n/L] & n = 0, \pm L, \pm 2L, \ldots \\ 0 & \text{otherwise} \end{cases} \tag{10.15}$$

thus expanding the original signal. The DTFT of the upsampled signal, $x_u[n]$, is found to be

$$X_u(e^{j\omega}) = X(e^{jL\omega}) \qquad -\pi \le \omega < \pi$$

Indeed, the DTFT of $x_u[n]$ is

$$X_u(e^{j\omega}) = \sum_{n=0,\pm L,\ldots} x[n/L]e^{-j\omega n} = \sum_{m=-\infty}^{\infty} x[m]e^{-j\omega Lm} = X(e^{jL\omega}) \tag{10.16}$$

indicating that it is a contraction of the DTFT of $x[n]$.

- A signal $x[n]$, bandlimited to π/M in $[-\pi, \pi)$ or $|X(e^{j\omega})| = 0$, $|\omega| > \pi/M$ for an integer $M > 1$, can be downsampled to generate a discrete-time signal

$$x_d[n] = x[Mn] \quad \text{with} \quad X_d(e^{j\omega}) = \frac{1}{M}X(e^{j\omega/M}) \tag{10.17}$$

  which is an expanded version of $X(e^{j\omega})$.
- A signal $x[n]$ is upsampled to generate a signal $x_u[n] = x[n/L]$ for $n = \pm kL$, $k = 0, 1, 2, \ldots$, and zero otherwise. The DTFT of $x_u[n]$ is $X(e^{jL\omega})$, or a compressed version of $X(e^{j\omega})$.

■ Example 10.4

Consider the frequency response of an ideal low-pass filter,

$$H(e^{j\omega}) = \begin{cases} 1 & -\pi/2 \le \omega \le \pi/2 \\ 0 & -\pi \le \omega < -\pi/2 \quad \text{and} \quad \pi/2 < \omega \le \pi \end{cases}$$

which is the DTFT of an impulse response $h[n]$. Determine $h[n]$. Suppose that we downsample $h[n]$ with a factor of $M = 2$. Find the downsampled impulse response $h_d[n] = h[2n]$ and its corresponding frequency response $H_d(e^{j\omega})$.

Solution

The impulse response $h[n]$ corresponding to the ideal low-pass filter is found to be

$$h[n] = \frac{1}{2\pi} \int_{-\pi/2}^{\pi/2} e^{j\omega n} d\omega = \begin{cases} 0.5 & n = 0 \\ \sin(\pi n/2)/(\pi n) & n \ne 0 \end{cases}$$

The downsampled impulse response is given by

$$h_d[n] = h[2n] = \begin{cases} 0.5 & n = 0 \\ \sin(\pi n)/(2\pi n) = 0 & n \ne 0 \end{cases}$$

or $h_d[n] = 0.5\delta[n]$, with a DTFT of $H_d(e^{j\omega}) = 0.5$ for $-\pi < \omega \le \pi$ (i.e., an all-pass filter). This agrees with the downsampling theory, which gives that

$$H_d(e^{j\omega}) = \frac{1}{2}H(e^{j\omega/2}) = \frac{1}{2}, \qquad -\pi \le \omega < \pi$$

That is, $H(e^{j\omega})$ multiplied by $1/M = 1/2$ and expanded by $M = 2$. ■

■ Example 10.5

A discrete pulse is given by $x[n] = u[n] - u[n - 4]$. Suppose we downsample $x[n]$ by a factor of $M = 2$, so that the length 4 of the original signal is reduced to 2, giving

$$x_d[n] = x[2n] = u[2n] - u[2n - 4] = u[n] - u[n - 2]$$

Find the corresponding DTFTs for $x[n]$ and $x_d[n]$, and determine how they are related.

Solution

The Z-transform of $x[n]$ is

$$X(z) = 1 + z^{-1} + z^{-2} + z^{-3}$$

with the whole z-plane (except for the origin) as its region of convergence. Thus, the DTFT of $x[n]$ is

$$X(e^{j\omega}) = e^{-j(\frac{3}{2}\omega)}\left[e^{j(\frac{3}{2}\omega)} + e^{j(\frac{1}{2}\omega)} + e^{-j(\frac{1}{2}\omega)} + e^{-j(\frac{3}{2}\omega)}\right]$$

$$= 2e^{-j(\frac{3}{2}\omega)}\left[\cos\left(\frac{\omega}{2}\right) + \cos\left(\frac{3\omega}{2}\right)\right]$$

The Z-transform of the downsampled signal ($M = 2$) is

$$X_d(z) = 1 + z^{-1}$$

and the DTFT of $x_d[n]$ is

$$X_d(e^{j\omega}) = e^{-j(\frac{1}{2}\omega)}\left[e^{j(\frac{1}{2}\omega)} + e^{-j(\frac{1}{2}\omega)}\right]$$

$$= 2e^{-j(\frac{1}{2}\omega)}\cos\left(\frac{\omega}{2}\right)$$

Clearly, this is not equal to $0.5X(e^{j\omega/2})$. This is caused by aliasing: The maximum frequency of $x[n]$ is not $\pi/M = \pi/2$ and so $X_d(e^{j\omega})$ is the sum of superposed and shifted $X(e^{j\omega})$.

Suppose we pass $x[n]$ through an ideal low-pass filter $H(e^{j\omega})$ with cut-off frequency $\pi/2$. Its output would be a signal $x_1[n]$ with a maximum frequency of $\pi/2$, and downsampling it with $M = 2$ would give a signal with a DTFT of $0.5X_1(e^{j\omega/2})$. ∎

■ Example 10.6

Discuss the effects of downsampling a discrete signal that is not band limited versus the case of one that is. Consider a unit rectangular pulse of length $N = 10$. Downsample it by a factor of $M = 2$, and compute and compare the DTFTs of the pulse and its downsampled version. Do a similar procedure to a sinusoid of discrete frequency $\pi/4$ and comment on the results. Explain the difference between the above two cases. Use the MATLAB function decimate (low-pass filtering is used to avoid aliasing followed by downsampling) to perform similar operations and comment on the differences with downsampling. Use the MATLAB function interp to interpolate (upsampling with smoothing by a low-pass filter) the downsampled signals. See Figure 10.3 for illustrations of downsampler and upsampler and decimator and interpolator.

Solution

As indicated, when we downsample a discrete-time signal $x[n]$ by a factor of M, in order not to have aliasing in frequency the signal must be band limited to π/M. If the signal satisfies this condition, the spectrum of the downsampled signal is an expanded version of the spectrum of $x[n]$. To illustrate this in the following script we downsample by a factor of $M = 2$ first a signal that is not band limited to $\pi/2$, and then another that is.

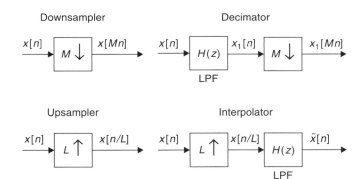

FIGURE 10.3

Top: Downsampler and decimator. Bottom: upsampler and interpolator.

```
%%%%%%%%%%%%%%%%%%%%%%%%%%%%
% Example 10.6---down--sampling and decimation
%%%%%%%%%%%%%%%%%%%%%%%%%%%%%
x = [ones(1,10) zeros(1,100)];Nx = length(x); n1 = 0:19; % first signal
% Nx = 200;n = 0:Nx - 1; x = cos(pi*n/4);   % second signal
y = x(1:2:Nx - 1);              % downsampling with M = 2
X = fft(x);Y = fft(y);         % ffts of original and downsampled signals
L = length(X);w = 0:2*pi/L:2*pi - 2*pi/L;w1 = (w - pi)/pi; % frequency range
z = decimate(x,2,'fir');       % decimation with M = 2
Z = fft(z);                    % fft of decimated signal
%%%%%%%%%%%%%%%%%%%%%%%%%%%%
% interpolation
%%%%%%%%%%%%%%%%%%%%%%%%%%%%
s = interp(y,2);
```

As shown in Figure 10.4, the rectangular pulse is not band limited to $\pi/2$ since it has frequency components beyond $\pi/2$, while the sinusoid is band limited. The DTFT of the downsampled rectangular pulse (a narrower pulse) is not an expanded version of the DTFT of the pulse, while the DTFT of the downsampled sinusoid is an expanded version. The MATLAB function decimate uses an FIR low-pass filter to smooth out $x[n]$ to a frequency of $\pi/2$ before downsampling. In the case of the sinusoid, which satisfies the downsampling condition, the downsampling and the decimation provide the same results, but not for the rectangular pulse.

The original discrete-time signal can be recovered by interpolation. This procedure is composed of upsampling followed by low-pass filtering. The MATLAB function interp is used to that effect. If we use the downsampled signal as input to this function, we obtain slightly better results for the sinusoid than for the pulse when comparing the interpolated signal to the original signal. The results are shown in Figure 10.5. The error $s[n] - x[n]$ is shown also. The signal $s[n]$ is the interpolation of the downsampled signal $y[n]$. ∎

10.2.5 Parseval's Energy Result

Just like in the case of continuous-time signals, the energy or power of a discrete-time signal $x[n]$ can be equally computed in either the time or the frequency domain.

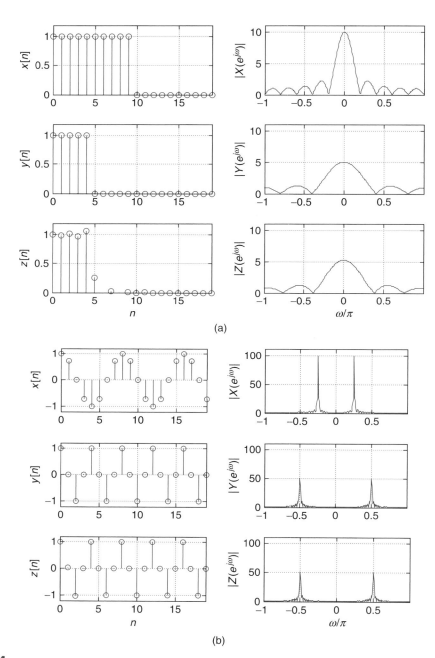

FIGURE 10.4

Downsampling of (a) non-band-limited and (b) band-limited discrete-time signals. The signals $x[n]$ correspond to the original signals, while $y[n]$ and $z[n]$ are their downsampled and decimated signals, respectively. The corresponding magnitude spectra are shown. Notice the difference between the downsampled and the decimated signals, they are identical when the signals are band-limited, slightly different otherwise.

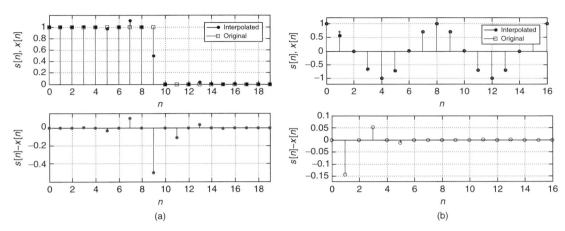

FIGURE 10.5

Interpolation of (a) non-band-limited and (b) bandlimited discrete-time signals. The interpolated signal is compared to the original signal, and the interpolation error is shown. The errors signals show that the original signal can be recovered almost exactly when the signal satisfies the bandlimiting condition, not otherwise.

If the DTFT of a finite-energy signal $x[n]$ is $X(e^{j\omega})$, we have that the energy of the signal is given by

$$E_x = \sum_{n=-\infty}^{\infty} |x[n]|^2 = \frac{1}{2\pi} \int_{-\pi}^{\pi} |X(e^{j\omega})|^2 d\omega \qquad (10.18)$$

The magnitude square $|X(e^{j\omega})|^2$ has the units of energy per radian, and so it is called an *energy density*. When $|X(e^{j\omega})|^2$ is plotted against frequency ω, the plot is called the *energy spectrum* of the signal, or how the energy of the signal is distributed over frequencies.

10.2.6 Time and Frequency Shifts

Shifting in time does not change the frequency content of a signal. Thus, the magnitude of the signal DTFT is not affected, only the phase is. Indeed, if $x[n]$ has a DTFT $X(e^{j\omega})$, then the DTFT of $x[n - N]$ for some integer N is

$$\mathcal{F}(x[n - N]) = \sum_{n} x[n - N]e^{-j\omega n}$$

$$= \sum_{m} x[m]e^{-j\omega(m+N)} = e^{-j\omega N}X(e^{j\omega})$$

If $x[n]$ has a DTFT

$$X(e^{j\omega}) = |X(e^{j\omega})|e^{j\theta(\omega)}$$

where $\theta(\omega)$ is the phase, the shifted signal $x_1[n] = x[n - N]$ has a DTFT of

$$X_1(e^{j\omega}) = X(e^{j\omega})e^{-j\omega N}$$

$$= |X(e^{j\omega})|e^{-j(\omega N - \theta(\omega))}$$

In a dual way, when we multiply a signal by a complex exponential $e^{j\omega_0 n}$ for some frequency ω_0, the spectrum of the signal is shifted in frequency. So if $x[n]$ has a DTFT $X(e^{j\omega})$, the modulated signal $x[n]e^{j\omega_0 n}$ has as DTFT $X(e^{j(\omega-\omega_0)})$. Indeed, the DTFT of $x_1[n] = x[n]e^{j\omega_0 n}$ is

$$X_1(e^{j\omega}) = \sum_n x_1[n]e^{-j\omega n} = \sum_n x[n]e^{-j(\omega-\omega_0)n} = X(e^{j(\omega-\omega_0)})$$

The following pairs illustrate the duality in time and frequency shifts: if the DTFT of $x[n]$ is $X(e^{j\omega})$ then

$$x[n-N] \Leftrightarrow X(e^{j\omega})e^{-j\omega N}$$

$$x[n]e^{j\omega_0 n} \Leftrightarrow X(e^{j(\omega-\omega_0)})$$

(10.19)

Remark *The signal $x[n]e^{j\omega_0 n}$ was called* modulated *because $x[n]$ modulates the complex exponential or discrete-time sinusoids. It can be written as*

$$x[n]\cos(\omega_0 n) + jx[n]\sin(\omega_0 n)$$

■ Example 10.7

The DTFT of $x[n] = \cos(\omega_0 n)$, $-\infty < n < \infty$, cannot be found from the Z-transform or from the sum defining the DTFT as $x[n]$ is not a finite-energy signal. Use the frequency-shift and the time-shift properties to find the DTFTs of $x[n] = \cos(\omega_0 n)$ and $y[n] = \sin(\omega_0 n)$.

Solution

Using Euler's identity we have that

$$x[n] = \cos(\omega_0 n) = \frac{e^{j\omega_0 n} + e^{-j\omega_0 n}}{2}$$

and so the DTFT of $x[n]$ is given by

$$X(e^{j\omega}) = \mathcal{F}[0.5e^{j\omega_0 n}] + \mathcal{F}[0.5e^{-j\omega_0 n}]$$

$$= \mathcal{F}[0.5]_{\omega-\omega_0} + \mathcal{F}[0.5]_{\omega+\omega_0}$$

$$= \pi[\delta(\omega-\omega_0) + \delta(\omega+\omega_0)]$$

Since

$$y[n] = \sin(\omega_0 n) = \cos(\omega_0 n - \pi/2) = \cos(\omega_0(n - \pi/(2\omega_0))) = x[n - \pi/(2\omega_0)]$$

we have that according to the time-shift property its DTFT is given by

$$Y(e^{j\omega}) = X(e^{j\omega})e^{-j\omega\pi/(2\omega_0)}$$

$$= \pi\left[\delta(\omega-\omega_0)e^{-j\omega\pi/(2\omega_0)} + \delta(\omega+\omega_0)e^{-j\omega\pi/(2\omega_0)}\right]$$

$$= \pi \left[\delta(\omega - \omega_0)e^{-j\pi/2} + \delta(\omega + \omega_0)e^{j\pi/2} \right]$$

$$= -j\pi \left[\delta(\omega - \omega_0) - \delta(\omega + \omega_0) \right]$$

Thus, the frequency content of the cosine and the sine is concentrated at the frequency ω_0. Although the sinusoids are infinite-energy signals they have finite power and their spectra can be measured with a spectrum analyzer, which displays how the power is distributed over the frequencies. ∎

10.2.7 Symmetry

When plotting or displaying the spectrum of a real-valued discrete-time signal it is important to know that it is only necessary to show the magnitude and the phase spectra for frequencies $[0 \ \pi]$, since the magnitude and the phase of $X(e^{j\omega})$ are even and odd functions of ω, respectively. This can be shown by considering a real-valued discrete-time signal $x[n]$, with inverse DTFT given by

$$x[n] = \frac{1}{2\pi} \int_{-\pi}^{\pi} X(e^{j\omega})e^{j\omega n}d\omega$$

and its complex conjugate is

$$x^*[n] = \frac{1}{2\pi} \int_{-\pi}^{\pi} X^*(e^{j\omega})e^{-j\omega n}d\omega = \frac{1}{2\pi} \int_{-\pi}^{\pi} X^*(e^{-j\omega'})e^{j\omega' n}d\omega'$$

Since $x[n] = x^*[n]$, as $x[n]$ is real, comparing the above integrals we have that

$$X(e^{j\omega}) = X^*(e^{-j\omega})$$

$$|X(e^{j\omega})|e^{j\theta(\omega)} = |X(e^{-j\omega})|e^{-j\theta(-\omega)}$$

$$\mathcal{R}e[X(e^{j\omega})] + j\mathcal{I}m[X(e^{j\omega})] = \mathcal{R}e[X(e^{-j\omega})] - j\mathcal{I}m[X(e^{-j\omega})]$$

or that the magnitude is an even function of ω—that is,

$$|X(e^{j\omega})| = |X(e^{-j\omega})| \tag{10.20}$$

and that the phase is an odd function of ω, or

$$\theta(\omega) = -\theta(-\omega) \tag{10.21}$$

Likewise, the real and the imaginary parts of $X(e^{j\omega})$ are also even and odd functions of ω:

$$\mathcal{R}e[X(e^{j\omega})] = \mathcal{R}e[X(e^{-j\omega})]$$

$$\mathcal{I}m[X(e^{j\omega})] = -\mathcal{I}m[X(e^{-j\omega})] \tag{10.22}$$

■ **Example 10.8**

For the signal $x[n] = \alpha^n u[n]$, $0 < \alpha < 1$, find the magnitude and the phase of its DTFT $X(e^{j\omega})$.

Solution

The DTFT of $x[n]$ is

$$X(e^{j\omega}) = \frac{1}{1 - \alpha z^{-1}} \Big|_{z=e^{j\omega}} = \frac{1}{1 - \alpha e^{-j\omega}}$$

since the Z-transform has a region of convergence $|z| > \alpha$ that includes the unit circle. Its magnitude is

$$|X(e^{j\omega})| = \frac{1}{\sqrt{(1 - \alpha \cos(\omega))^2 + \alpha^2 \sin^2(\omega)}}$$

which is an even function of ω given that $\cos(\omega) = \cos(-\omega)$ and $\sin^2(-\omega) = (-\sin(\omega))^2 = \sin^2(\omega)$. The phase is given by

$$\theta(\omega) = -\tan^{-1}\left[\frac{\alpha \sin(\omega)}{1 - \alpha \cos(\omega)}\right]$$

which is an odd function of ω. As functions of ω, the numerator is odd and the denominator is even, so that the argument of the inverse tangent is odd, which is in turn odd.   ■

■ **Example 10.9**

For a discrete-time signal

$$x[n] = \cos(\omega_0 n + \phi) \qquad -\pi \le \phi < \pi$$

determine how the magnitude and the phase responses of the DTFT $X(e^{j\omega})$ change with ϕ.

Solution

The signal $x[n]$ has a DTFT

$$X(e^{j\omega}) = \pi\left[e^{-j\phi}\delta(\omega - \omega_0) + e^{j\phi}\delta(\omega + \omega_0)\right]$$

Its magnitude is

$$|X(e^{j\omega})| = |X(e^{-j\omega})| = \pi[\delta(\omega - \omega_0) + \delta(\omega + \omega_0)]$$

for all values of ϕ. The phase of $X(e^{j\omega})$ is

$$\theta(\omega) = \phi\delta(\omega + \omega_0) - \phi\delta(\omega - \omega_0)$$

$$= \begin{cases} \phi & \omega = -\omega_0 \\ -\phi & \omega = \omega_0 \\ 0 & \text{otherwise} \end{cases}$$

In particular, if $\phi = 0$, the signal $x[n]$ is a cosine and its phase is zero. If $\phi = -\pi/2$, $x[n]$ is a sine and its phase is $\pi/2$ at $\omega = -\omega_0$ and $-\pi/2$ at $\omega = -\omega_0$. The DTFT of a sine is

$$X(e^{j\omega}) = \pi \left[\delta(\omega - \omega_0)e^{-j\pi/2} + \delta(\omega + \omega_0)e^{j\pi/2} \right]$$

The DTFTs of the cosine and the sine are only different in the phase. ∎

The symmetry property, like other properties, also applies to systems. If $h[n]$ is the impulse response of an LTI discrete-time system, and it is real valued, its DTFT is

$$H(e^{j\omega}) = \mathcal{Z}\left(h[n]\right)\big|_{z=e^{j\omega}} = H(z)\big|_{z=e^{j\omega}}$$

if the region of convergence of $H(z)$ includes the unit circle. As with the DTFT of a signal, the frequency response of the system, $H(e^{j\omega})$, has a magnitude that is an even function of ω, and a phase that is an odd function of ω. Thus, the *magnitude response* of the system is such that

$$|H(e^{j\omega})| = |H(e^{-j\omega})| \tag{10.23}$$

and the *phase response* is such that

$$\angle H(e^{j\omega}) = -\angle H(e^{-j\omega}) \tag{10.24}$$

According to these symmetries and that the frequency response is periodic, it is only necessary to give these responses in $[0, \pi]$ rather than in $(-\pi, \pi]$.

Computation of the Phase Spectrum

Computation of the phase using MATLAB is complicated by the following three issues:

- *Definition of the phase of a complex number*: Given a complex number $z = x + jy = |z|e^{j\theta}$, its phase θ is computed using the inverse tangent function

$$\theta = \tan^{-1}\left(\frac{y}{x}\right)$$

This computation is not well defined because the principal values of $\tan^{-1}$ are $[-\pi/2, \pi/2]$, while the phase can extend beyond those values. By adding the information of which quadrants x and y are in, the principal values can be extended to $[-\pi, \pi)$. When the phase is linear (i.e., $\theta = -N\omega$ for some integer N), using the extended principal values is not good enough.
- *Significance of magnitude when computing phase*: Given two complex numbers, $z_1 = 1 + j = \sqrt{2}e^{j\pi/4}$ and $z_2 = z_1 \times 10^{-16} = \sqrt{2} \times 10^{-16}e^{j\pi/4}$, they both have the same phase of $\pi/4$ but the magnitudes are very different, $|z_2| = 10^{-16}|z_1|$. For practical purposes, the magnitude of z_1 is more significant than that of z_2, which is very close to zero, so one could disregard the phase of z_2 with no effect on computations.
- *Noisy measurements*: Given that noise is ever present in actual measurements, even very small noise present in the signal can change the computation of phase.

Phase unwrapping

Another problem with phase computation has to do with the way the phase is displayed as a function of frequency with values within $[-\pi, \pi)$—the *wrapped phase*. If the DTFT of a signal is zero or infinite at some frequencies, the phase at those frequencies is not determined, as any value would be as good as any other because the magnitude is zero or infinity. On the other hand, if there are no zeros or poles on the unit circle to make the magnitude zero or infinite at some frequencies, the phase is continuous. However, because of the way the phase is computed and displayed with values between $-\pi$ to π, it seems discontinuous. These phase discontinuities are 2π wide, making the phase values right before and right after the discontinuity identical. Finding the frequencies where these discontinuities occur and patching the phase, it is possible to obtain the continuous phase. The process is called the *unwrapping of the phase*. The MATLAB function unwrap is used for this purpose.

■ Example 10.10

Consider a sinusoid $x[n] = \sin(\pi n/2)$ to which we add a Gaussian noise $\eta[n]$ generated by the MATLAB function randn, which theoretically can take any real value. Use the significance of the magnitude computed by MATLAB to estimate the phase.

Solution

The DTFT of $x[n]$ consists of two impulses, one at $\pi/2$ and the other at $-\pi/2$. Thus, the phase should be zero everywhere except at these frequencies. However, any amount of added noise (even in noiseless cases, given the computational precision of MATLAB) would give nonzero phases in places where it should be zero.

In this case, since we know that the magnitude spectrum is significant at the negative and the positive values of the sinusoid frequency, we use the magnitude as a mask to indicate where the phase computation can be considered important given the significance of the magnitude. The following script is used to illustrate the masking using the significance of the magnitude.

```
%%%%%%%%%%%%%%%%%%%%%%%%%%%
% Example 10.10---phase of sinusoid in noise
%%%%%%%%%%%%%%%%%%%%%%%%%%%%%
n = 0:99; x = sin(pi * n/2) + 0.1 * randn(1,100);    % sine plus noise
X = fftshift(fft(x)); % fft of signal
X1 = abs(X); theta = angle(X); % magnitude and phase
theta1 = theta. * X1/max(X1);    % masked phase
L = length(X); w = 0:2 * pi/L:2 * pi - 2 * pi/L;w1 = (w - pi)/pi; % frequency range
```

Using the mask, the noisy phase (Figure 10.6(b)) is converted into the phase of the sine, which occurs when the magnitude is significant (Figures 10.6(a) and 10.6(c)). ■

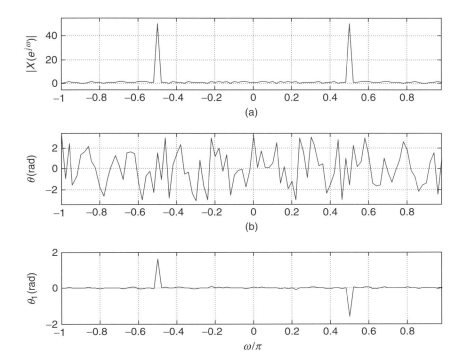

FIGURE 10.6

Phase spectrum of sinusoid in Gaussian noise using magnitude masking: (a) magnitude response, (b) wrapped phase and (c) unwrapped phase.

■ Example 10.11

Consider two FIR filters with impulse responses

$$h_1[n] = \sum_{k=0}^{9} \frac{1}{10}\delta[n-k]$$

$$h_2[n] = 0.5\delta[n-3] + 1.1\delta[n-4] + 0.5\delta[n-5]$$

Determine which of these filters has linear phase, and use the MATLAB function unwrap to find their unwrapped phase functions. Explain the results.

Solution

The transfer function of the filter with $h_1[n]$ is

$$H_1(z) = \frac{1}{10}\sum_{n=0}^{9} z^{-n} = 0.1\frac{1-z^{-10}}{1-z^{-1}} = 0.1\frac{z^{10}-1}{z^9(z-1)} = 0.1\frac{\prod_{k=1}^{9}(z - e^{j2\pi k/10})}{z^9}$$

Because this filter has nine zeros on the unit circle, its phase is not continuous and it cannot be unwrapped. The impulse response of the second filter is symmetric about $n = 4$; thus its phase is linear and continuous. Indeed, the transfer function of this filter is

$$H_2(z) = 0.5z^{-3} + 1.1z^{-4} + 0.5z^{-5} = z^{-4}(0.5z + 1.1 + 0.5z^{-1})$$

which gives the frequency response

$$H_2(e^{j\omega}) = e^{-j4\omega}(1.1 + \cos(\omega))$$

Since $1.1 + \cos(\omega) > 0$ for $-\pi \le \omega < \pi$, the phase is $\angle H_2(e^{j\omega}) = -4\omega$, which is a line through the origin with slope -4 (i.e., a linear phase)

The following script is used to compute the frequency responses of the two filters using fft, their wrapped phases using angle, and then unwrapping them using unwrap. Figure 10.7 displays the magnitude responses of the two filters, as well as their wrapped and unwrapped phases.

```
%%%%%%%%%%%%%%%%%%%%%%%%%
% Example 10.11---Phase unwrapping
%%%%%%%%%%%%%%%%%%%%%%%%%
h1 = (1/10) * ones(1,10); % fir filter 1
h2 = [ zeros(1,3) 0.5 1.1 0.5 zeros(1,3)];  % fir filter 2
H1 = fft(h1,256); % fft of h1
H2 = fft(h2,256); % fft of h2
H1m = abs(H1(1:128)); H1p = angle(H1(1:128)); % magnitude/phase of H1(z)
```

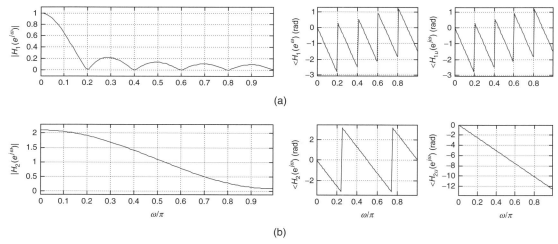

FIGURE 10.7

Unwrapping of the phase: (a) first filter magnitude response, and wrapped and unwrapped phase responses; (b) second filter magnitude response, and wrapped and unwrapped phase (linear phase) responses. Notice that the phase of the first filter cannot be unwrapped because the zeros of this filter are on the unit circle (see corresponding magnitude response).

```
H1up = unwrap(H1p); % unwrapped phase of H1(z)
H2m = abs(H2(1:128)); H2p = angle(H2(1:128)); % magnitude/phase of H2(z)
H2up = unwrap(H2p); % unwrapped phase of H1(z)
```

10.2.8 Convolution Sum

The computation of the convolution sum, just like the convolution integral in the continuous-time domain, is simplified in the Fourier domain.

If $h[n]$ is the impulse response of a stable LTI system, its output $y[n]$ can be computed by means of the convolution sum

$$y[n] = \sum_k x[k] \, h[n-k]$$

where $x[n]$ is the input. The Z-transform of $y[n]$ is the product

$$Y(z) = H(z)X(z) \qquad \text{ROC: } \mathcal{R}_Y = \mathcal{R}_H \cap \mathcal{R}_X$$

If the unit circle is included in $\mathcal{R}_Y$, then

$$Y(e^{j\omega}) = H(e^{j\omega})X(e^{j\omega}) \qquad \text{or}$$

$$|Y(e^{j\omega})| = |H(e^{j\omega})||X(e^{j\omega})|$$

$$\angle Y(e^{j\omega}) = \angle H(e^{j\omega}) + \angle X(e^{j\omega}) \tag{10.25}$$

Remarks

- *Since the system is stable, the ROC of $H(z)$ includes the unit circle, and so if the ROC of $X(z)$ includes the unit circle, the intersection of these regions will also include the unit circle.*
- *We will see later that it is still possible for $y[n]$ to have a DTFT when the input $x[n]$ does not have a Z-transform with a region of convergence including the unit circle, as when the input is periodic. In this case the output is also periodic. These signals are not finite energy, but finite power, and can be represented by DTFT containing analog delta functions.*

■ Example 10.12

Let $H(z)$ be the cascade of first-order systems with transfer functions

$$H_i(z) = K_i \frac{z - 1/\alpha_i}{z - \alpha_i^*} \qquad |z| > |\alpha_i|, \quad i = 1, \ldots, N-1$$

where $|\alpha_i| < 1$ and $K_i > 0$. Such a system is called an *all-pass system* because its magnitude response is a constant for all frequencies. If the DTFT of the filter input $x[n]$ is $X(e^{j\omega})$, determine the gains $\{K_i\}$ so that the magnitude of the DTFT of the output $y[n]$ of the system coincides with the magnitude of $X(e^{j\omega})$.

Solution

Notice that if $1/\alpha_i$ is a zero of $H_i(z)$, a pole at the conjugate reciprocal α_i^* exists. To show that the magnitude of $H_i(e^{j\omega})$ is a constant for all frequencies, consider the magnitude-squared function

$$|H_i(e^{j\omega})|^2 = H_i(e^{j\omega})H^*(e^{j\omega}) = K_i^2 \frac{(e^{j\omega} - 1/\alpha_i)(e^{-j\omega} - 1/\alpha_i^*)}{(e^{j\omega} - \alpha_i^*)(e^{-j\omega} - \alpha_i)}$$

$$= K_i^2 \frac{e^{j\omega}(e^{-j\omega} - \alpha_i)e^{-j\omega}(e^{j\omega} - \alpha_i^*)}{\alpha_i\alpha_i^*(e^{j\omega} - \alpha_i^*)(e^{-j\omega} - \alpha_i)} = \frac{K_i^2}{|\alpha_i|^2}$$

Thus, by letting $K_i = |\alpha_i|$, the above gives a unit magnitude. The cascade of the $H_i(z)$ gives a transfer function of

$$H(z) = \prod_i H_i(z) = \prod_i |\alpha_i| \frac{z - 1/\alpha_i}{z - \alpha_i}$$

so that

$$H(e^{j\omega}) = \prod_i H_i(e^{j\omega}) = \prod_i |\alpha_i| \frac{e^{j\omega} - 1/\alpha_i}{e^{j\omega} - \alpha_i}$$

$$|H(e^{j\omega})| = \prod_i |H_i(e^{j\omega})| = 1$$

$$\angle H(e^{j\omega}) = \sum_i \angle H_i(e^{j\omega})$$

which in turn gives

$$Y(e^{j\omega}) = |X(e^{j\omega})|e^{j(\angle X(e^{j\omega}) + \angle H(e^{j\omega}))}$$

so that the magnitude of the output coincides with that of the input; however, the phase of $Y(e^{j\omega})$ is the sum of the phases of $X(e^{j\omega})$ and $H(e^{j\omega})$. Thus, the all-pass system allows all frequency components in the input to appear at the output with no change in the magnitude spectrum but with a phase shift. ∎

10.3 FOURIER SERIES OF DISCRETE-TIME PERIODIC SIGNALS

Like in the continuous-time domain, we are interested in finding the response of an LTI system to a periodic signal. As in that case, we represent the periodic signal as a combination of complex exponentials and use the eigenfunction property of LTI systems to find the response.

Notice that we are proceeding in the reverse order we followed in the continuous-time case: We are considering the Fourier representation of periodic signals after that of aperiodic signals. Theoretically, there is no reason why this cannot be done, but practically it has the advantage of ending with a representation for discrete-time periodic signals that is discrete and periodic in frequency so it can be implemented in a computer. This is the basis of the so-called discrete Fourier transform (DFT), which is fundamental in digital signal processing (see Section 10.4). An algorithm called the Fast Fourier Transform (FFT) implements it very efficiently (see Chapter 12).

Before finding the representation of periodic discrete-time signals recall that:

■ A discrete-time signal $x[n]$ is periodic if there is a positive integer N such that $x[n + kN] = x[n]$ for any integer k. This value N is the smallest positive integer that satisfies this condition and it is called the period of $x[n]$. For the periodicity to hold we need that $x[n]$ be of infinite support (i.e., $x[n]$ must be defined in $-\infty < n < \infty$).

■ According to the eigenfunction property of discrete-time LTI systems, whenever the input to such systems is a complex exponential $Ae^{j(\omega_0 n + \theta)}$ the corresponding output in the steady state is

$$y[n] = Ae^{j(\omega_0 n + \theta)} H(e^{j\omega_0})$$

where $H(e^{j\omega_0})$ is the frequency response of the system at the input frequency ω_0. The advantage of this, as demonstrated in the continuous-time domain, is that if we are able to express the input signal as a linear combination of complex exponentials, then superposition gives us a linear combination of the responses to each exponential. Thus, if the input signal is of the form

$$x[n] = \sum_k A[k] e^{j\omega_k n}$$

then the output will be

$$y[n] = \sum_k A[k] e^{j\omega_k n} H(e^{j\omega_k})$$

This property is valid whether the frequency components of the input signal are harmonically related (when $x[n]$ is periodic) or not.

■ We showed before that a signal $x(t)$ that is periodic of period T_0 can be represented by its Fourier series

$$x(t) = \sum_{k=-\infty}^{\infty} \hat{X}[k] e^{j\frac{2\pi kt}{T_0}} \tag{10.26}$$

If we sample $x(t)$ using a sampling period $T_s = T_0/N$ where N is a positive integer, we get

$$x(nT_s) = \sum_{k=-\infty}^{\infty} \hat{X}[k] e^{j\frac{2\pi knT_s}{T_0}} = \sum_{k=-\infty}^{\infty} \hat{X}[k] e^{j\frac{2\pi kn}{N}}$$

The last summation repeats the frequencies between 0 to 2π. To avoid these repetitions, we let $k = m + rN$ where $0 \le m \le N - 1$ and $r = 0, \pm 1, \pm 2, \ldots$—that is, we divide the infinite support of k into an infinite number of finite segments of length N. We then have

$$x(nT_s) = \sum_{m=0}^{N-1} \sum_{r=-\infty}^{\infty} \hat{X}[m+rN]e^{j\frac{2\pi(m+rN)n}{N}}$$

$$= \sum_{m=0}^{N-1} \left[\sum_{r=-\infty}^{\infty} \hat{X}[m+rN] \right] e^{j\frac{2\pi mn}{N}}$$

$$= \sum_{m=0}^{N-1} X[m]e^{j\frac{2\pi mn}{N}}$$

This representation is in terms of complex exponentials with frequencies $2\pi m/N, m = 0, \ldots, N-1$, from 0 to $2\pi(N-1)/N$. It is this Fourier series representation that we will develop next.

Circular Representation of Discrete-Time Periodic Signals

Considering that the period N of a periodic signal $x[n]$ and the samples in a first period $x_1[n]$ completely characterize a periodic signal $x[n]$, a circular rather than a linear representation would more efficiently represent the signal. The circular representation is obtained by locating uniformly around a circle the values of the first period starting with $x[0]$ and putting in a clockwise direction the remaining terms $x[1], \ldots, x[N-1]$. Continuing in the clockwise direction are the values $x[N] = x[0], x[N+1] = x[1], \ldots, x[2N-1] = x[N-1]$, and so on. In general, any value $x[m]$ where m is represented as

$$m = kN + r$$

for integers k, the exact divisor of m by N, and the residue $0 \leq r < N$, equals one of the samples in the first period—that is,

$$x[m] = x[kN+r] = x[r]$$

This representation is called *circular* in contrast to the equivalent linear representation introduced before:

$$x[n] = \sum_{k=-\infty}^{\infty} x_1[n+kN]$$

which superposes shifted versions of the first period. The circular representation becomes very useful in the computation of the DFT, as we will see later in this chapter.

Figure 10.8 shows the circular and the linear representations of a periodic signal $x[n]$ of period $N = 4$.

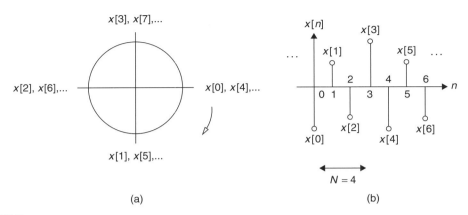

FIGURE 10.8

(a) Circular and (b) linear representation of a periodic discrete-time signal $x[n]$ of period $N = 4$. Notice how the circular representation shows the periodicity $x[0] = x[4], \ldots, x[3] = x[7], \ldots$ for positive as well as negative integers.

10.3.1 Complex Exponential Discrete Fourier Series

Consider the representation of a discrete-time signal $x[n]$ periodic of period N, using the orthogonal functions $\{\phi[k, n] = e^{j2\pi kn/N}\}$ for $n, k = 0, \ldots, N - 1$. Two important characteristics of these functions are:

■ The functions $\{\phi[k, n]\}$ are periodic with respect to k and n with period N. In fact,

$$\phi[k + \ell N, n] = e^{j\frac{2\pi(k+\ell N)n}{N}}$$

$$= e^{j\frac{2\pi kn}{N}} e^{j2\pi \ell n}$$

$$= e^{j\frac{2\pi kn}{N}}$$

where we used that $e^{j2\pi \ell n} = 1$. It can be equally shown that the functions $\{\phi(k, n)\}$ are periodic with respect to n with a period N.

■ The functions $\{\phi(k, n)\}$ are orthogonal with respect to n—that is,

$$\sum_{n=0}^{N-1} e^{j\frac{2\pi}{N} kn} (e^{j\frac{2\pi}{N} \ell n})^* = \begin{cases} N & \text{if } k - \ell = 0 \\ 0 & \text{if } k - \ell \neq 0 \end{cases}$$

and can be normalized by dividing them by $\sqrt{N}$. So $\{\phi[k, n]/\sqrt{N}\}$ are orthonormal functions.

These two properties will be used in obtaining the Fourier series representation of periodic discrete-time signals.

The Fourier series representation of a periodic signal $x[n]$ of period N is

$$x[n] = \sum_{k=k_0}^{k_0+N-1} X[k] e^{j\frac{2\pi}{N}kn} \tag{10.27}$$

where the Fourier series coefficients $\{X[k]\}$ are obtained from

$$X[k] = \frac{1}{N} \sum_{n=n_0}^{n_0+N-1} x[n] e^{-j\frac{2\pi}{N}kn} \tag{10.28}$$

The frequency $\omega_0 = 2\pi/N$ rad is the fundamental frequency, and k_0 and n_0 in Equations (10.27) and (10.28) are arbitrary integer values. The Fourier series coefficients $X[k]$, as functions of frequency $2\pi k/N$, are periodic of period N.

Remarks

- *The connection of the above two equations can be verified by using the orthonormality of the $\{\phi[k, n]/\sqrt{N}\}$ functions. In fact, if we multiply $x[n]$ by $e^{-j(2\pi/N)\ell n}$ and sum these values for n changing over a period, using Equation (10.27) we get:*

$$\sum_n x[n] e^{-j2\pi n\ell/N} = \sum_n \sum_k X[k] e^{j2\pi(k-\ell)n/N}$$

$$= \sum_k X[k] \sum_n e^{j2\pi(k-\ell)n/N} = NX[\ell]$$

  since $\sum_n e^{j2\pi(k-\ell)n/N}$ is zero, except when $k - \ell = 0$, in which case the sum is equal to N.
- *Both $x[n]$ and $X[k]$ are periodic with respect to n and k of the same period N, as can be easily shown using the periodicity of the functions $\{\phi(k, n)\}$. Consequently, the sum over k in the Fourier series and the sum over n in the Fourier coefficients are computed over any period of $x[n]$ and $X[k]$. Thus, the sum in the Fourier series can be computed in any period, or from $k = k_0$ to $k_0 + N - 1$ for any value of k_0. Likewise, the summation in the computation of the Fourier coefficients goes from $n = n_0$ to $n_0 + N - 1$, which is an arbitrary period for any integer value n_0.*
- *Notice that both $x[n]$ and $X[k]$ can be computed with a computer since the frequency is discrete and only sums are needed to implement them. We will use these characteristics in the practical computation of the Fourier transform of discrete-time signals, or DFT.*

■ Example 10.13

Find the Fourier series of a periodic signal

$$x[n] = 1 + \cos(2\pi n/4) + \sin(2\pi n/2) \qquad -\infty < n < \infty$$

Solution

The period of $x[n]$ is $N = 4$. Indeed,

$$x[n+4] = 1 + \cos(2\pi(n+4)/4) + \sin(2\pi(n+4)/2)$$
$$= 1 + \cos(2\pi n/4 + 2\pi) + \sin(2\pi n/2 + 4\pi) = x[n]$$

The frequencies in $x[n]$ are: a DC frequency, corresponding to the constant, and frequencies $\omega_0 = 2\pi/4 = \pi/2$ and $\omega_1 = 2\pi/2 = 2\omega_0$, corresponding to the cosine and the sine. No other frequencies are present in the signal. The fundamental frequency is $\omega_0 = 2\pi/N = \pi/2$, and the complex exponential Fourier series can be obtained directly from $x[n]$ using Euler's equation:

$$x[n] = 1 + 0.5(e^{j\pi n/2} + e^{-j\pi n/2}) - 0.5j(e^{j\pi n} - e^{-j\pi n})$$
$$= X[0] + X[1]e^{j\omega_0 n} + X[-1]e^{-j\omega_0 n} + X[2]e^{j2\omega_0 n} + X[-2]e^{-j2\omega_0 n} \qquad \omega_0 = \frac{\pi}{2}$$

so that the Fourier series coefficients are $X[0] = 1$, $X[1] = X^*[-1] = 0.5$, and $X[2] = X^*[-2] = -0.5j$. ∎

10.3.2 Connection with the Z-Transform

Recall that the Laplace transform was used to find the Fourier series coefficients. Likewise, for periodic discrete-time signals we will show that the Z-transform of a period of the signal, which always exists, can be connected with the Fourier series coefficients.

If $x_1[n] = x[n](u[n] - u[n-N])$ is a period of a periodic signal $x[n]$ of period N, its Z-transform

$$\mathcal{Z}(x_1[n]) = \sum_{n=0}^{N-1} x[n]z^{-n}$$

has the whole plane, except for the origin, as its region of convergence. The Fourier series coefficients of $x[n]$ are thus determined as

$$X[k] = \frac{1}{N} \sum_{n=0}^{N-1} x[n]e^{-j\frac{2\pi}{N}kn}$$
$$= \frac{1}{N}\mathcal{Z}(x_1[n]) \Big|_{z=e^{j\frac{2\pi}{N}k}} \qquad (10.29)$$

■ Example 10.14

Consider a discrete pulse $x[n]$ with a period $N = 20$, and $x_1[n] = u[n] - u[n-10]$ is the period between 0 and 19. Find the Fourier series of $x[n]$.

Solution

The Fourier series coefficients are found as ($\omega_0 = 2\pi/20$ rad)

$$X[k] = \frac{1}{20} \mathcal{Z}(x_1[n]) \Big|_{z=e^{j\frac{2\pi}{20}k}} = \frac{1}{20} \sum_{n=0}^{9} z^{-n} \Big|_{z=e^{j\frac{2\pi}{20}k}} = \frac{1}{20} \frac{1-z^{-10}}{1-z^{-1}} \Big|_{z=e^{j\frac{2\pi}{20}k}}$$

A close expression for $X[k]$ is obtained as follows:

$$X[k] = \frac{z^{-5}(z^5 - z^{-5})}{20z^{-0.5}(z^{0.5} - z^{-0.5})} \Big|_{z=e^{j\frac{2\pi}{20}k}}$$

$$= \frac{e^{-j\pi k/2} \sin(\pi k/2)}{20e^{-j\pi k/20} \sin(\pi k/20)}$$

$$= \frac{e^{-j9\pi k/20}}{20} \frac{\sin(\pi k/2)}{\sin(\pi k/20)} \qquad \blacksquare$$

10.3.3 DTFT of Periodic Signals

A discrete-time periodic signal $x[n]$ of period N with a Fourier series representation of

$$x[n] = \sum_{k} X[k]e^{j2\pi nk/N} \tag{10.30}$$

has a DTFT

$$X(e^{j\omega}) = \sum_{k} 2\pi X[k]\delta(\omega - 2\pi k/N) \qquad -\pi \le \omega < \pi \tag{10.31}$$

If we let $\mathcal{F}(.)$ indicate the DTFT, the DTFT of a periodic signal $x[n]$ is

$$X(e^{j\omega}) = \mathcal{F}(x[n]) = \mathcal{F}\left(\sum_{k} X[k]e^{j2\pi nk/N}\right) = \sum_{k} \mathcal{F}\left(X[k]e^{j2\pi nk/N}\right)$$

$$= \sum_{k} 2\pi X[k]\delta(\omega - 2\pi k/N) \qquad -\pi \le \omega < \pi$$

where $\delta(\omega)$ is the analog delta function since ω varies continuously.

■ Example 10.15

The periodic signal

$$\delta_M[n] = \sum_{m=-\infty}^{\infty} \delta[n - mM]$$

has a period M. Find its DTFT.

Solution

The DTFT of $\delta_M[n]$ is given by

$$\Delta_M(e^{j\omega}) = \sum_{m=-\infty}^{\infty} \mathcal{F}(\delta[n - mM]) = \sum_{m=-\infty}^{\infty} e^{-j\omega mM} \tag{10.32}$$

An equivalent result can be obtained if before we find the DTFT we find the Fourier series of $\delta_M[n]$. The coefficients of the Fourier series of $\delta_M[n]$ are

$$\frac{1}{M} \sum_{n=0}^{M-1} \delta_M[n] e^{-j2\pi nk/M} = \frac{1}{M} \sum_{n=0}^{M-1} \delta[n] e^{-j2\pi nk/M} = \frac{1}{M}$$

so that the Fourier series of $\delta_M[n]$ is

$$\delta_M[n] = \sum_{k=0}^{M-1} \frac{1}{M} e^{j2\pi nk/M}$$

and its DTFT is then given by

$$\Delta_M(e^{j\omega}) = \sum_{k=0}^{M-1} \mathcal{F}\left(\frac{1}{M} e^{j2\pi nk/M}\right) = \frac{2\pi}{M} \sum_{k=0}^{M-1} \delta\left(\omega - \frac{2\pi k}{M}\right) \qquad -\pi \leq \omega < \pi \tag{10.33}$$

Notice the equivalence of Equations (10.32) and (10.33).

Putting the pair $\delta_M[n]$ and $\Delta_M(e^{j\omega})$ together gives an interesting relation:

$$\sum_{m=-\infty}^{\infty} \delta[n - mM] \Leftrightarrow \frac{2\pi}{M} \sum_{k=0}^{M-1} \delta\left(\omega - \frac{2\pi k}{M}\right) \qquad \pi \leq \omega < \pi$$

Both terms are discrete in time and in frequency, and both are periodic. The period of $\delta_M[n]$ is M and the one of $\Delta(e^{j\omega})$ is $2\pi/M$. Furthermore, the DTFT of an impulse train in time is also an impulse train in frequency. However, it should be observed that the delta functions $\delta[n - mM]$ on the left term are discrete, while the ones on the right term, $\delta(\omega - 2\pi k/M)$, are continuous. ∎

Computation of the Fourier Series using MATLAB

Given that periodic signals only have discrete frequencies, and that the Fourier series coefficients are obtained using summations, the computation of the Fourier series can be implemented with a frequency discretized version of the DTFT, or the DFT, which can be efficiently computed using the FFT algorithm. To illustrate this using MATLAB, consider three different signals as indicated in the following script and displayed in Figure 10.9.

Each of the signals is periodic and we consider 10 periods to compute their FFTs, which provides the Fourier series coefficients (only 3 periods are displayed in Figure 10.9). A very important issue to remember is that one needs to input exactly one or more periods, and that the FFT length must be

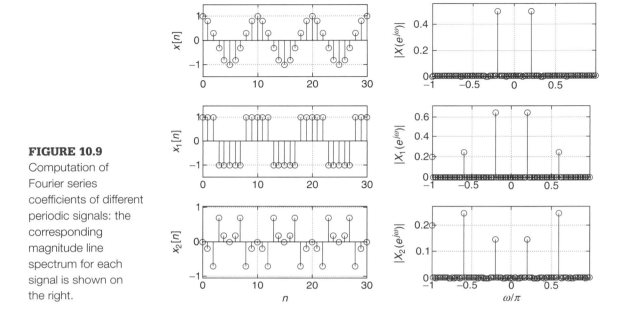

FIGURE 10.9

Computation of Fourier series coefficients of different periodic signals: the corresponding magnitude line spectrum for each signal is shown on the right.

that of a period or of multiples of a period. Notice that the MATLAB function sign is used to generate a periodic train of pulses from the cosine function. The need to divide by the number of periods used will be discussed later in section 10.4.3.

```
%%%%%%%%%%%%%%%%%%%%%%%%%%%
% Fourier series using FFT
%%%%%%%%%%%%%%%%%%%%%%%%%%%
N = 10; M = 10; N1 = M * N;n = 0:N1 - 1;
x = cos(2 * pi * n/N); % sinusoid
x1 = sign(x); % train of pulses
x2 = x - sign(x); % sinusoid minus train of pulses
X = fft(x)/M;X1 = fft(x1)/M;X2 = fft(x2)/M; % ffts of signals
X = X/N;X1 = X1/N;X2 = X2/N; % FS coefficients
```

10.3.4 Response of LTI Systems to Periodic Signals

Let $x[n]$, a periodic signal of period N, be the input of an LTI system with transfer function $H(z)$. If the Fourier series of $x[n]$ is

$$x[n] = \sum_{k=0}^{N-1} X[k]e^{j(k\omega_0)n} \qquad \omega_0 = \frac{2\pi}{N} \quad \text{fundamental frequency}$$

then according to the eigenfunction property of LTI systems, the output is also periodic of period N with Fourier series

$$y[n] = \sum_{k=0}^{N-1} X[k]H(e^{jk\omega_0})e^{jk\omega_0 n} \qquad \omega_0 = \frac{2\pi}{N} \text{ fundamental frequency}$$

and coefficients $Y[k] = X[k]H(e^{jk\omega_0})$. That is, it is affected by the frequency response of the system

$$H(e^{jk\omega_0}) = H(z)\Big|_{z=e^{jk\omega_0}}$$

at the harmonic frequencies $\{k\omega_0, k = 0, \ldots, N-1\}$.

Remarks

- *Although the input $x[n]$ and the output $y[n]$ of the LTI system are both periodic of the same period, the Fourier series coefficients of the output are affected by the frequency response $H(e^{jk\omega_0})$ of the system at each of the harmonic frequencies.*
- *A similar result is obtained by using the convolution property of the DTFT, so that if $X(e^{j\omega})$ is the DTFT of the periodic signal $x[n]$, then the DTFT of the output $y[n]$ is given by*

$$Y(e^{j\omega}) = X(e^{j\omega})H(e^{j\omega})$$

$$= \left[\sum_{k=0}^{N-1} 2\pi X[k]\delta(\omega - 2\pi k/N)\right]H(e^{j\omega})$$

$$= \sum_{k=0}^{N-1} 2\pi X[k]H(e^{j2\pi k/N})\delta(\omega - 2\pi k/N)$$

and letting $X[k]H(e^{j2\pi k/N}) = Y[k]$, we get the DTFT of the periodic output $y[n]$.

■ Example 10.16

Consider how to implement a crude spectral analyzer for discrete-time signals using MATLAB. Divide the discrete frequencies $[0\ \pi]$ in three bands: $[0\ 0.1\pi]$, $(0.1\pi\ 0.6\pi]$, and $(0.6\pi\ \pi]$, to obtain low-pass, band-pass, and high-pass components of the signal $x[n] = \text{sign}(\cos(0.2\pi n))$. Use the MATLAB function fir1 to design the three filters. Plot the original signal and its components in the three bands. Verify that the overall filter is an all-pass filter. Obtain the sum of the outputs of the filters, and explain how it relates to the original signal.

Solution

The script for this example uses several MATLAB functions that facilitate the filtering of signals, and for which you can get more information by using help.

The DTFT of $x[n]$ is found using the FFT algorithm as indicated before. The low-pass, band-pass, and high-pass filters are obtained using fir1, and the function filter allows us to obtain the corresponding outputs $y_1[n]$, $y_2[n]$, and $y_3[n]$. The frequency responses $\{H_i(e^{j\omega})\}$, $i = 1, 2, 3$ of the filters are found using the function freqz.

The three filters separate $x[n]$ into its low-, middle-, and high-band components from which we are able to obtain approximately the power of the signal in these three bands—that is, we have a crude spectral analyzer. Ideally, we would like the sum of the filters outputs to be equal to $x[n]$, with some delay, and so the sum of the frequency responses

$$H(e^{j\omega}) = H_1(e^{j\omega}) + H_2(e^{j\omega}) + H_3(e^{j\omega})$$

should be the frequency response of an all-pass filter. Indeed, that is the result shown in Figure 10.10, where we obtain the input signal as the output of the filter with transfer function $H(z)$, delayed 15 samples (which corresponds to half the order of the filters used).

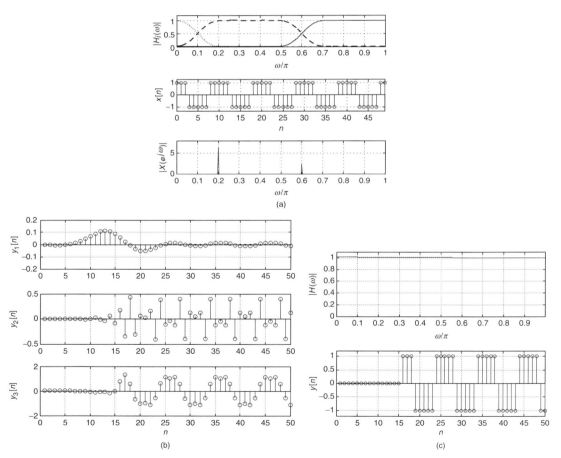

FIGURE 10.10

A crude spectral analyzer: (a) magnitude response of low-pass, band-pass, and high-pass filters; input signal and its magnitude spectrum; (b) outputs of filters; (c) overall magnitude response of the bank of filters, an all-pass filter, and overall response. Delay is due to linear phase of the bank of filters.

```
%%%%%%%%%%%%%%%%%%%%%%%%%%%%%%%%%%%%%%
% Example 10.16---Filtering of a periodic signal
%%%%%%%%%%%%%%%%%%%%%%%%%%%%%%%%%%%%%%
N = 500;n = 0: N - 1; x = cos(0.2 * pi * n); x = sign(x); % pulse signal
X = fft(x)/50; X = X(1:250); % approximate DTFT of signal using fft
L = 500; w1 = 0:2 * pi/L:pi - 2 * pi/L;w1 = w1/pi; % range of frequencies
h1 = fir1(30,0.1);  % low-pass filter
h2 = fir1(30,0.6,'high'); % high-pass filter
h3 = fir1(30,[0.1 0.6]); % band-pass filter
y1 = filter(h1,1,x); y2 = filter(h2,1,x); y3 = filter(h3,1,x);
y = y1 + y2 + y3; % outputs of filters
[H1,w] = freqz(h1,1); [H2,w] = freqz(h2,1); [H3,w] = freqz(h3,1);
H = H1 + H2 + H3; % frequency responses
```
■

10.3.5 Circular Shifting and Periodic Convolution
Circular Shifting

When a periodic signal $x[n]$ of period N is shifted by M samples the signal is still periodic. The circular representation provides the appropriate visualization of this shift, as it concentrates on the period displayed by the representation. Values are rotated circularly.

The Fourier series of the shifted signal $x_1[n] = x[n - M]$ is obtained from the Fourier series of $x[n]$ by replacing n by $n - M$ to get

$$x_1[n] = x[n - M] = \sum_k X[k]e^{j2\pi(n-M)k/N}$$

$$= \sum_k \left(X[k]e^{-j2\pi Mk/N}\right)e^{j2\pi nk/N}$$

so that the shifted signal and its Fourier series coefficients are related as

$$x[n - M] \Leftrightarrow X[k]e^{-j2\pi Mk/N} \tag{10.34}$$

It is important to consider what happens for different values of M. This shift can be represented as

$$M = mN + r, \qquad m = 0, \pm 1, \pm 2, \dots, \quad 0 \le r \le N - 1$$

and as such

$$e^{-j2\pi Mk/N} = e^{-j2\pi(mN+r)k/N} = e^{-j2\pi rk/N} \tag{10.35}$$

for any value of M, so that shifting by more than a period is equivalent to shifting by the residue r of dividing the shift M by N.

■ Example 10.17

To visualize the difference between a linear shift and a circular shift consider the periodic signal $x[n]$ of period $N = 4$ with a first period

$$x_1[n] = n \qquad n = 0, \ldots, 3$$

Plot $x[-n]$ and $x[n-1]$ as functions of n using the linear and the circular representations.

Solution

In the circular representation of $x[n]$, the samples $x[0]$, $x[1]$, $x[2]$, and $x[3]$ of the first period are located in a clockwise direction in the E(ast), S(outh), W(est), and N(orth) directions in the circle. Considering the E direction the origin of the representation, circular shifting is done analogously to linear shifting with respect to the origin. Delaying by M in the circular representation corresponds to shifting circularly, or rotating, M positions in the clockwise direction. Advancing by M corresponds to shifting circularly M positions in the counterclockwise direction. Reflection corresponds to reversing the order of the samples of $x_1[n]$ to E, N, W, and S (i.e., placing the entries of $x_1[n]$ counterclockwise starting with $x[0]$ in E).

The circular representation of $x[-n]$ starts with $x[0]$ in the first quadrant and then follows with $x[1]$, $x[2]$, and $x[3]$ in a counterclockwise direction (Figure 10.11). Looking at the circular representation in a clockwise direction we obtain the linear representation that as expected coincides with the reflection of $x[n]$.

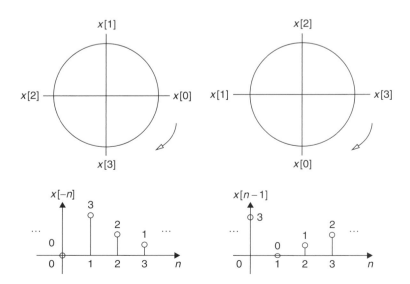

FIGURE 10.11
Circular representation of $x[-n]$ and $x[n-1]$.

For the circular representation of $x[n-1]$ we shift the term $x[0]$ from the E to the S in the clockwise direction and shift the others the same angle to get the circular representation of $x[n-1]$— a circular shift of one. Shifting linearly by one, we obtain the equivalent representation of $x[n-1]$. ∎

Periodic Convolution

Consider then the multiplication of two periodic signals $x[n]$ and $y[n]$ of the same period N. The product $v[n] = x[n]y[n]$ is also periodic of period N, and its Fourier series coefficients are

$$V[m] = \sum_{k=0}^{N-1} X[k]Y[m-k] \qquad 0 \le m \le N-1$$

as we will show next. That $v[n]$ is periodic of period N is clear. Its Fourier series is found by letting the fundamental frequency be $\omega_0 = \frac{2\pi}{N}$ and replacing the given Fourier coefficients,

$$v[n] = \sum_{m=0}^{N-1} V[m]e^{j\omega_0 nm} = \sum_{m=0}^{N-1}\sum_{k=0}^{N-1} X[k]Y[m-k]e^{j\omega_0 nm}$$

$$= \sum_{k=0}^{N-1} X[k]\left(\sum_{m=0}^{N-1} Y[m-k]e^{j\omega_0 n(m-k)}\right)e^{j\omega_0 kn} = \sum_{k=0}^{N-1} X[k]y[n]e^{j\omega_0 kn} = y[n]x[n]$$

Thus, we have that the Fourier series coefficients of the product of two periodic signals of the same period give

$$x[n]y[n] \Leftrightarrow \sum_{k=0}^{N-1} X[k]Y[m-k] \tag{10.36}$$

and by duality

$$\sum_{k=0}^{N-1} x[k]y[n-k] \Leftrightarrow NX[k]Y[k] \tag{10.37}$$

Although

$$\sum_{k=0}^{N-1} x[k]y[n-k] \quad \text{and} \quad \sum_{k=0}^{N-1} X[k]Y[m-k]$$

look like the convolution sums we had before, the periodicity of the sequences makes them different. These are called *periodic convolution sums*. Given the infinite support of periodic signals, the convolution sum of periodic signals does not exist—it would not be finite. The periodic convolution is done only for a period of the signals.

Remarks

- *As before, multiplication in one domain causes convolution in the other domain.*
- *In computing a periodic convolution we need to remember that: (1) the convolving sequences must have the same period, and (2) the Fourier series coefficients of a periodic signal share the same period with the signal.*

■ Example 10.18

To understand how the periodic convolution sum results, consider the product of two periodic signals $x[n]$ and $y[n]$ of period $N = 2$. Find the Fourier series of their product $v[n] = x[n]y[n]$.

Solution

The multiplication of the Fourier series

$$x[n] = X[0] + X[1]e^{j\omega_0 n}$$

$$y[n] = Y[0] + Y[1]e^{j\omega_0 n} \qquad \omega_0 = 2\pi/N = \pi$$

can be seen as a product of two polynomials in complex exponentials $\zeta[n] = e^{j\omega_0 n}$ so that

$$x[n]y[n] = (X[0] + X[1]\zeta[n])(Y[0] + Y[1]\zeta[n])$$

$$= X[0]Y[0] + (X[0]Y[1] + X[1]Y[0])\,\zeta[n] + X[1]Y[1]\zeta^2[n]$$

Now $\zeta^2[n] = e^{j2\omega_0 n} = e^{j2\pi n} = 1$ so that

$$x[n]y[n] = \underbrace{(X[0]Y[0] + X[1]Y[1])}_{V[0]} + \underbrace{(X[0]Y[1] + X[1]Y[0])}_{V[1]}\,e^{j\omega_0 n} = v[n]$$

after replacing $\zeta[n]$. When using the periodic convolution formula, we have

$$V[0] = \sum_{k=0}^{1} X[k]Y[-k] = X[0]Y[0] + X[1]Y[-1] = X[0]Y[0] + X[1]Y[2-1]$$

$$V[1] = \sum_{k=0}^{1} X[k]Y[1-k] = X[0]Y[1] + X[1]Y[0]$$

where in the upper equation we used the periodicity of $Y[k]$ so that $Y[-1] = Y[-1+N] = Y[-1+2] = Y[1]$. Thus, the multiplication of periodic signals can be seen as the product of polynomials in $\zeta[n] = e^{j\omega_0 n} = e^{j2\pi n/N}$ such that

$$\zeta^{pN+m}[n] = e^{j\frac{2\pi}{N}(pN+m)} = e^{j\frac{2\pi}{N}m} = \zeta^m[n] \qquad p = \pm 1, \pm 2, \ldots, \; 0 \le m \le N-1$$

which ensures that the resulting polynomial is always of order $N-1$.

Graphically, we proceed in a manner analogous to the convolution sum (see Figure 10.12) by representing $X[k]$ and $Y[m-k]$ circularly, and shifting $Y[m-k]$ clockwise by $m=0$ and $m=1$,

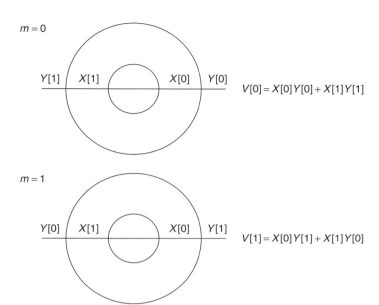

FIGURE 10.12

Periodic convolution of the Fourier series coefficients $X[k]$ and $Y[k]$.

while keeping $X[k]$ stationary. The circular representation of $X[k]$ is given by the internal circle with the values of $X[0]$ and $X[1]$ in the clockwise direction, while $Y[m-k]$ is represented in the outer circle with the two values of a period in the counterclockwise direction (corresponding to the reflection of the signal or $Y[-k]$ for $m=0$). Multiplying the values opposite to each other and adding them we get $V[0] = X[0]Y[0] + X[1]Y[1]$. If we shift the outer circle 180 degrees clockwise for $m=1$ and multiply the values opposite to each other and add their product, we get $V[1] = X[0]Y[1] + X[1]Y[0]$. There is no need for further shifting, as the results would coincide with the ones obtained before. The process is similar to the linear convolution but implemented circularly. ∎

For periodic signals $x[n]$ and $y[n]$ of period N, we have

(a) *Duality in time and frequency circular shifts:* The Fourier series coefficients of the signals on the left are the terms on the right:

$$x[n-M] \Leftrightarrow X[k]e^{-j2\pi Mk/N}$$

$$x[n]e^{j2\pi Mn/N} \Leftrightarrow X[k-M] \tag{10.38}$$

(b) *Duality in multiplication and periodic convolution sum:* The Fourier series coefficients of the signals on the left are the terms on the right:

$$z[n] = x[n]y[n] \Leftrightarrow Z[k] = \sum_{m=0}^{N-1} X[m]Y[k-m]$$

$$v[n] = \sum_{m=0}^{N-1} x[m]y[n-m] \Leftrightarrow V[k] = NX[k]Y[k] \tag{10.39}$$

■ Example 10.19

A periodic signal $x_1[n]$ of period $N = 4$ has a period

$$x_1[n] = \begin{cases} 1 & n = 0, 1 \\ 0 & n = 2, 3 \end{cases}$$

Suppose that we want to find the periodic convolution sum of $x[n]$ with itself—call it $v[n]$. Let then $y[n] = x[n-2]$, and find the periodic convolution sum of $x[n]$ and $y[n]$—call it $z[n]$. How does $v[n]$ compare with $z[n]$?

Solution

The circular representation of $x[n]$ is shown in Figure 10.13. To find the periodic convolution sum we consider a period $x_1[n]$ and represent the stationary signal by the internal circle, and the circularly shifted signal by the outside circle. Multiplying the values in each of the spokes and adding them we find the values of a period of $v[n]$, which is given by

$$v_1[n] = \begin{cases} 1 & n = 0 \\ 2 & n = 1 \\ 1 & n = 2 \\ 0 & n = 3 \end{cases}$$

Analytically, the Fourier series coefficients of $v[n]$ are $V[k] = N(X[k])^2 = 4(X[k])^2$. Using the Z-transform, $X_1(z) = 1 + z^{-1}$ so that $X_1^2(z) = 1 + 2z^{-1} + z^{-2}$ and

$$V[k] = 4\frac{X_1^2(z)}{4 \times 4}\Big|_{z=e^{j2\pi k/4}}$$

$$= \frac{1}{4}(1 + 2e^{-j2\pi k/4} + e^{-j2\pi 2k/4}) = \frac{1}{4}(1 + 2e^{-j\pi k/2} + e^{-j\pi k})$$

This can be verified by using the period obtained from the periodic convolution sum so that

$$V[k] = \frac{1}{N}\sum_{n=0}^{N-1} v[n]e^{-j2\pi nk/N} = \frac{1}{4}(1 + 2e^{-j2\pi k/4} + e^{-j2\pi 2k/4})$$

which equals the above expression.

Graphically, the periodic convolution of $x[n]$ and $y[n]$ is shown in Figure 10.13(c) where the stationary signal is chosen as $x[m]$, represented by the inner circle, and the circularly shifted signal is chosen as $y[n-m]$, represented by the outer circle. The result of the convolution is a periodic signal $z[n]$ of period

$$z_1[n] = \begin{cases} 1 & n = 0 \\ 0 & n = 1 \\ 1 & n = 2 \\ 2 & n = 3 \end{cases}$$

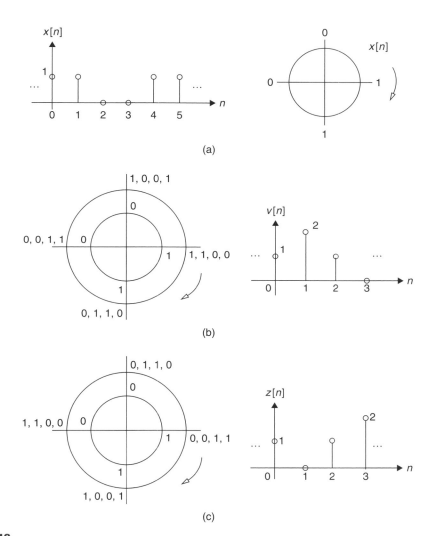

FIGURE 10.13

Periodic convolution sum of $x[n]$ with itself to get $v[n]$: (a) linear and circular representations of $x[n]$; (b) periodic convolution sum giving $v[n]$. (c) Circular representation of periodic convolution sum of $x[n]$ and $y[n] = x[n-2]$, the result is $z[n] = v[n-2]$.

As before, the Fourier series coefficients of $z[n]$ are given by

$$Z[k] = 4\frac{X_1(z)Y_1(z)}{4 \times 4}\Big|_{z=e^{j2\pi k/4}} = \frac{z^{-2} + 2z^{-3} + z^{-4}}{4}\Big|_{z=e^{j2\pi k/4}}$$

$$= \frac{1}{4}(e^{-j2\pi\ 2k/4} + 2e^{-j2\pi\ 3k/4} + e^{-j2\pi\ 4k/4}) = \frac{1}{4}(1 + e^{-j2\pi\ 2k/4} + 2e^{-j2\pi\ 3k/4})$$

which coincides with the $Z[k]$ obtained by the periodic convolution, which is given by

$$Z[k] = \frac{1}{N} \sum_{n=0}^{N-1} z[n]e^{-j2\pi nk/N} = \frac{1}{4}(1 + z^{-2} + 2z^{-3})|_{z=e^{j2\pi k/4}} = \frac{1}{4}(1 + e^{-j2\pi \; 2k/4} + 2e^{-j2\pi \; 3k/4})$$

■

10.4 DISCRETE FOURIER TRANSFORM

Recall that the direct and the inverse DTFTs corresponding to a discrete-time signal $x[n]$ are

$$X(e^{j\omega}) = \sum_{n} x[n]e^{-j\omega n} \qquad -\pi \le \omega < \pi$$

$$x[n] = \frac{1}{2\pi} \int_{-\pi}^{\pi} X(e^{j\omega})e^{j\omega n} d\omega$$

These equations have the following computational disadvantages:

■ The frequency ω varies continuously from $-\pi$ to π, and as such computing $X(e^{j\omega})$ needs to be done for an uncountable number of frequencies.
■ The inverse DTFT requires integration that cannot be implemented exactly in a computer.

To resolve these issues we consider the discrete Fourier transform or DFT (notice the name difference with respect to the DTFT), which is computed at discrete frequencies and its inverse does not require integration. Moreover, the DFT is efficiently implemented using an algorithm called the Fast Fourier Transform (FFT).

The development of the DFT is based on the representation of periodic discrete-time signals. Both the signal and the Fourier coefficients are periodic of the same period. Thus, the representation of discrete-time periodic signals is discrete in both time and frequency. We need then to consider how to extend aperiodic signals into periodic signals, with an appropriate period, to obtain their DFTs.

10.4.1 DFT of Periodic Discrete-Time Signals

A periodic signal $\tilde{x}[n]$ of period N is represented by N values in a period. Its discrete Fourier series is

$$\tilde{x}[n] = \sum_{k=0}^{N-1} \tilde{X}[k]e^{j\omega_0 nk} \qquad 0 \le n \le N-1 \tag{10.40}$$

where $\omega_0 = 2\pi/N$ is the fundamental frequency. The coefficients $\{\tilde{X}[k]\}$ correspond to harmonic frequencies $\{k\omega_0\}$ for $0 \le k \le N-1$, so that $\tilde{x}[n]$ has no frequency components at any other frequencies. Thus, $\tilde{x}[n]$ and $\tilde{X}[k]$ are both discrete and periodic of the same period N. Moreover, the Fourier series coefficients can be calculated using the Z-transform as

$$\tilde{X}[k] = \frac{1}{N} \mathcal{Z}[\tilde{x}_1[n]] \Big|_{z=e^{jk\omega_0}}$$

$$= \frac{1}{N} \sum_{n=0}^{N-1} \tilde{x}[n]e^{-j\omega_0 nk} \qquad 0 \le k \le N-1, \quad \omega_0 = 2\pi/N \qquad (10.41)$$

where $\tilde{x}_1[n] = \tilde{x}[n]W[n]$ is a period of $\tilde{x}[n]$ and $W[n]$ is a rectangular window—that is,

$$W[n] = u[n] - u[n-N] = \begin{cases} 1 & 0 \le n \le N-1 \\ 0 & \text{otherwise} \end{cases}$$

Thus, the periodic signal $\tilde{x}[n]$ can be expressed as

$$\tilde{x}[n] = \sum_{r=-\infty}^{\infty} \tilde{x}_1[n+rN] \qquad (10.42)$$

Although one could call Equation (10.41) the DFT of the periodic signal $\tilde{x}[n]$ and Equation (10.40) the corresponding inverse DFT, traditionally the DFT of $\tilde{x}[n]$ is $N\tilde{X}[k]$, or

$$X[k] = N\tilde{X}[k] = \sum_{n=0}^{N-1} \tilde{x}[n]e^{-j\omega_0 nk} \qquad 0 \le k \le N-1, \quad \omega_0 = 2\pi/N \qquad (10.43)$$

and the inverse DFT is

$$\tilde{x}[n] = \frac{1}{N} \sum_{k=0}^{N-1} X[k]e^{j\omega_0 nk} \qquad 0 \le n \le N-1 \qquad (10.44)$$

Equations (10.43) and (10.44) show that the representation of periodic signals is completely discrete: summations instead of integrals and discrete rather than continuous frequencies. Thus, the DFT and its inverse can be evaluated by computer.

Given a periodic signal $x[n]$ of period N, its DFT is given by

$$X[k] = \sum_{n=0}^{N-1} x[n]e^{-j2\pi nk/N} \qquad 0 \le k \le N-1 \qquad (10.45)$$

Its inverse DFT is

$$x[n] = \frac{1}{N} \sum_{k=0}^{N-1} X[k]e^{j2\pi nk/N} \qquad 0 \le n \le N-1 \qquad (10.46)$$

Both $X[k]$ and $x[n]$ are periodic of the same period N.

10.4.2 DFT of Aperiodic Discrete-Time Signals

We obtain the DFT of an aperiodic signal $y[n]$ by sampling its DTFT, $Y(e^{j\omega})$, in frequency. Suppose we choose $\{\omega_k = 2\pi k/L, \ k = 0, \ldots, L-1\}$ as the sampling frequencies, where an appropriate value for the integer $L > 0$ needs to be determined. Analogous to the sampling-in-time we did before, sampling-in-frequency generates a periodic signal in time:

$$\tilde{y}[n] = \sum_{r=-\infty}^{\infty} y[n + rL] \tag{10.47}$$

Now, if $y[n]$ is of finite length N, then when $L \geq N$ the periodic expansion $\tilde{y}[n]$ clearly displays a first period equal to the given signal $y[n]$ (with some zeros attached at the end when $L > N$). On the other hand, if the length $L < N$ the first period of $\tilde{y}[n]$ does not coincide with $y[n]$ because of superposition of shifted versions of it (this corresponds to *time aliasing*, the dual of frequency aliasing, which occurs in time sampling).

Assuming $y[n]$ is of finite length N and that $L \geq N$, as the dual of sampling in time we then have that

$$\tilde{y}[n] = \sum_{r=-\infty}^{\infty} y[n + rL] \ \Leftrightarrow \ Y[k] = Y(e^{j2\pi k/L}) = \sum_{n=0}^{N-1} y[n]e^{-j2\pi nk/L} \quad k = 0, \ldots, L-1 \tag{10.48}$$

The equation on the right is the DFT of $y[n]$. The inverse DFT is the Fourier series representation of $\tilde{y}[n]$ (normalized with respect to L) or its first period

$$y[n] = \frac{1}{L}\sum_{k=0}^{L-1} Y[k]e^{j2\pi nk/L} \qquad 0 \leq n \leq L-1 \tag{10.49}$$

where $Y[k] = Y(e^{j2\pi k/L})$.

Thus, instead of the frequency aliasing that sampling-in-time causes, we have time-aliasing whenever the length N of $y[n]$ is greater than the chosen L in the sampling-in-frequency. In practice, the generation of the periodic extension $\tilde{y}[n]$ is not needed—we just need to generate a period that either coincides with $y[n]$ when $L = N$, or when $L > N$ that coincides with $y[n]$ with a sequence of $L - N$ zeros attached to it (i.e., $y[n]$ is *padded with zeros*). To avoid time aliasing we do not consider choosing $L < N$.

If the signal $y[n]$ is a very long signal, in particular if $N \to \infty$, it does not make sense to compute its DFT, even if we could. Such a DFT would give the frequency content of the whole signal and since an infinite-length signal could have all types of frequencies its DFT would just give no valuable information. A possible approach to obtain, over time, the frequency content of a signal with a large time support is to window it and compute the DFT of each of these segments. Thus, when $y[n]$ is of infinite length, or its length is much larger than the desired or feasible length L, we use a window $W_L[n]$ of length L, and represent $y[n]$ as the superposition

$$y[n] = \sum_m y_m[n] \quad \text{where} \quad y_m[n] = y[n]W_L[n - mL] \tag{10.50}$$

Therefore, by the linearity of the DFT, we have the DFT of $y[n]$ is

$$Y[k] = \sum_m \text{DFT}(y_m[n]) = \sum_m Y_m[k] \qquad (10.51)$$

where each $Y_m[k]$ provides a frequency characterization of the windowed signal or the local frequency content of the signal. Practically, this would be more meaningful than finding the DFT of the whole signal. Now we have frequency information corresponding to segments of the signal and possibly evolving over time.

The DFT of an aperiodic signal $x[n]$ of finite length N is found as follows:

■ Choose an integer $L \geq N$ that is the length of the DFT to be the period of a periodic extension $\tilde{x}[n]$ having $x[n]$ as a period with padded zeros if necessary.
■ Find the normalized Fourier series representation of $\tilde{x}[n]$,

$$\tilde{x}[n] = \frac{1}{L} \sum_{k=0}^{L-1} \tilde{X}[k] e^{j2\pi nk/L} \qquad 0 \leq n \leq L-1 \qquad (10.52)$$

where

$$\tilde{X}[k] = \sum_{n=0}^{L-1} \tilde{x}[n] e^{-j2\pi nk/L} \qquad 0 \leq k \leq L-1 \qquad (10.53)$$

■ Then,
   ▪ $X[k] = \tilde{X}[k]$ for $0 \leq k \leq L-1$ is the DFT of $x[n]$.
   ▪ $x[n] = \tilde{x}[n]W[n]$ where $W[n] = u[n] - u[n-L]$ is a rectangular window of length N, is the IDFT of $X[k]$. The IDFT $x[n]$ is defined for $0 \leq n \leq L-1$.

10.4.3 Computation of the DFT via the FFT

Although we now have discrete frequencies and use only summations to compute the direct and the inverse DFT, there are still several issues that should be understood when computing these transforms. Assuming that the given signal is finite length, or it is made finite length by windowing, we have:

■ *Efficient computation with the FFT algorithm:* A very efficient computation of the DFT is done by means of the FFT algorithm, which takes advantage of some special characteristics of the DFT, as we will discuss in Chapter 12. It should be understood that the FFT is not another transformation but an algorithm to efficiently compute DFTs. For now, we will consider the FFT as a black box that for an input $x[n]$ (or $X[k]$) gives as output the DFT $X[k]$ (or IDFT $x[n]$).
■ *Causal aperiodic signals:* If the given signal $x[n]$ is causal of length N—that is, the samples

$$\{x[n], \ n = 0, 1, \ldots, N-1\}$$

are given—we can proceed to obtain $\{X[k], \ k = 0, 1, \ldots, N-1\}$ or the DFT of $x[n]$ by means of an FFT of length $L = N$. To compute an $L > N$ DFT we simply attach $L - N$ zeros at the end of the above sequence and obtain L values corresponding to the DFT of $x[n]$ of length L (why this could be seen as a better version of the DFT of $x[n]$ is discussed below in frequency resolution).

■ *Noncausal aperiodic signals:* When the given signal $x[n]$ is noncausal of length N—that is, the samples

$$\{x[n], \; n = -n_0, \ldots, 0, 1, \ldots, N - n_0 - 1\}$$

are given—we need to recall that a periodic extension of $x[n]$ or $\tilde{x}[n]$ was used to obtain its DFT. This means that we need to create a sequence of N values corresponding to the first period of $\tilde{x}[n]$—that is,

$$\underbrace{x[0]\, x[1]\; \cdots x[N - n_0 - 1]}_{\text{causal samples}}\; \underbrace{x[-n_0]\, x[-n_0 + 1] \cdots x[-1]}_{\text{noncausal samples}}$$

where as indicated the samples $x[-n_0]\, x[-n_0 + 1] \cdots x[-1]$ are the values that make $x[n]$ noncausal. If we wish to consider zeros after $x[N - n_0 - 1]$ to be part of the signal, so as to obtain a better DFT transform as we discuss later in frequency resolution, we simply attach zeros between the causal and noncausal components—that is,

$$\underbrace{x[0]\, x[1]\; \cdots x[N - n_0 - 1]}_{\text{causal samples}}\; 0\, 0 \; \cdots \; 0\, 0 \; \underbrace{x[-n_0]\, x[-n_0 + 1] \cdots x[-1]}_{\text{noncausal samples}}$$

to compute an $L > N$ DFT of the noncausal signal. The periodic extension $\tilde{x}[n]$ represented circularly instead of linearly would clearly show the above sequence.

■ *Periodic signals:* If the signal $x[n]$ is periodic of period N we will then choose $L = N$ (or a multiple of N) and calculate the DFT $X[k]$ by means of the FFT algorithm. If we use a multiple of the period (e.g., $L = MN$ for some integer $M > 0$), we need to divide the obtained DFT by the value M. For periodic signals we cannot choose L to be anything but a multiple of N as we are really computing the Fourier series of the signal. Likewise, no zeros can be attached to a period (or periods when $M > 1$) to improve the frequency resolution of its DFT—by attaching zeros to a period we distort the signal.

■ *Frequency resolution:* When the signal $x[n]$ is periodic of period N, the DFT values are normalized Fourier series coefficients of $x[n]$ that only exist for the harmonic frequencies $\{2\pi k/N\}$, as no frequency components exist for any other frequencies. On the other hand, when $x[n]$ is aperiodic, the number of possible frequencies depend on the length L chosen to compute its DFT. In either case, the frequencies at which we compute the DFT can be seen as frequencies around the unit circle in the z-plane. In both cases one would like to have a significant number of frequencies in the unit circle so as to visualize the frequency content of the signal well. The number of frequencies considered is related to the *frequency resolution* of the DFT of the signal.

  ▪ If the signal is aperiodic we can improve the frequency resolution of its DFT by increasing the number of samples in the signal without distorting the signal. This can be done by *padding the signal with zeros* (i.e., attaching zeros to the end of the signal). These zeros do not change the frequency content of the signal (they can be considered part of the aperiodic signal) but permit us to increase the available frequency components of the signal.

▨ On the other hand, the harmonic frequencies of a periodic signal of period N are fixed to $2\pi k/N$ for $0 \leq k < N$. In such a case we cannot pad the given period of the signal with an arbitrary number of zeros, because such zeros are not part of the periodic signal. As an alternative, to increase the frequency resolution of a periodic signal we consider more periods which give the same harmonic frequencies as for one period, but add zeros in between the harmonic frequencies when considering more than one period.

■ *Frequency scales:* When computing the DFT we obtain a sequence of complex values $X[k]$ for $k = 0, 1, \ldots, N - 1$ corresponding to an input signal $x[n]$ of length N. Since each of the k values corresponds to a discrete frequency $2\pi k/N$ the $k = 0, 1, \ldots, N - 1$ scale is converted into a discrete frequency scale $[0 \ 2\pi(N - 1)/N]$ (rad) (the last value is always smaller than 2π to keep the periodicity in frequency of $X[k]$) by multiplying the integer scale $\{0 \leq k \leq N - 1\}$ by $2\pi/N$. Subtracting π to this frequency scale we obtain discrete frequencies $[-\pi \ \pi - 2\pi/N]$ (rad) where again the last frequency does not coincide with π in order to keep the periodicity of 2π of the $X[k]$. We obtain a normalized discrete-frequency scale by dividing the above scale by π so as to obtain a nonunits normalized scale of $[0 \ 2(N - 1)/N]$ or $[-1 \ 1 - 2/N]$. Finally, if the signal is the result of sampling and we wish to display the analog frequency, we then use the relation where T_s is the sampling period and f_s is the sampling frequency:

$$\Omega = \frac{\omega}{T_s} = \omega f_s \ (\text{rad/sec}) \ \text{or} \ f = \frac{\omega}{2\pi T_s} = \frac{\omega f_s}{2\pi} \ (\text{Hz}) \tag{10.54}$$

giving analog scales $[0 \ \pi f_s]$ (rad/sec) and $[0 \ f_s/2]$ (Hz).

■ Example 10.20

Consider the DFT computation via the FFT of a causal signal $x[n] = (\sin(\pi n/32))(u[n] - u[n - 34])$ and of its advanced version $x[n + 16]$. To improve its frequency resolution compute FFTs of length $N = 512$. Explain the difference between computing the FFTs of the causal and the noncausal signals.

Solution

As indicated above, when computing the FFT of a causal signal the signal is simply inputted into the function. However, to improve the frequency resolution of the FFT we attach zeros to the signal. These zeros make it possible to have additional values of the frequency response of the signal, with no effect on the frequency content of the signal.

For the noncausal signal $x[n + 16]$, we need to recall that the DFTs of an aperiodic signal were computed by extending the signal into a periodic signal with an arbitrary period N, which exceeds the length of the signal. Thus, the periodic extension of $x[n + 16]$ can be obtained by creating an input vector consisting of $x[n]$, $n = 0, \ldots, 16$; $N - 33$ zeros (N being the length of the FFT and 33 the length of the signal) to improve the frequency resolution, and $x[n]$, $n = -16, \ldots, -1$.

In either case, the output of the FFT is available as an array of length $N = 512$ values. This array $X[k]$, $k = 0, \ldots, N - 1$ can be understood as values of the spectrum at frequencies $2\pi k/N$—that is, from 0 to $2\pi(N-1)/N$ rad (notice that this value is close to 2π but always smaller than 2π as needed to display a period of $X[k]$). We can change this scale into other frequency scales—for instance, if we wish a scale that considers positive as well as negative frequencies, to the above scale we subtract π, and if we wish a normalized scale $[-1\ 1)$, we simply divide the previous scale by π. When shifting to a $[-\pi\ \pi)$ or $[-1\ 1)$ frequency scale, the spectrum also needs to be shifted accordingly—this is done using the fftshift function. To understand this change recall that $X[k]$ is also periodic of period N.

The following script is used to compute the DFT of $x[n]$ and $x[n+16]$ given above. The results are shown in Figure 10.14.

```
%%%%%%%%%%%%%%%%%%%%%%%%%%%%%%%%%%%%%%%%%%%%%
% Example 10.20---FFFT computation of causal and
%                     noncausal signals
%%%%%%%%%%%%%%%%%%%%%%%%%%%%%%%%%%%%%%%%%%%%%
clear all; clf
N = 512;  % order of the FFT
n = 0:N - 1;
% causal signal
x = [ones(1,33) zeros(1,N - 33)]; x = x. * sin(pi * n/32); % zero-padding
X = fft(x); X = fftshift(X); % fft and its shifting to [-1 1] frequency scale
w = 2 * [0:N - 1]./N - 1; % normalized frequencies
n1 = [-9:40]; % time scale
% noncausal signal
xnc = [zeros(1,3) x(1:33) zeros(1,3)]; % noncausal signal
x = [x(17:33) zeros(1,N-33) x(1:16)]; % periodic extension and zero-padding
X = fft(x); X = fftshift(X);
n1 = [-19:19]; % time scale
```

▪ Example 10.21

Consider improving the frequency resolution of a periodic sampled signal

$$y(nT_s) = 4\cos(2\pi f_0 nT_s) - \cos(2\pi f_1 nT_s) \qquad f_0 = 100\,\text{Hz}, f_1 = 4f_0$$

where the sampling period is $T_s = 1/(3f_1)$ sec/sample.

Solution

In the case of a periodic signal, the frequency resolution of its FFT cannot be improved by attaching zeros. The length of the FFT must be the period or a multiple of the period of the signal. The following script illustrates how the FFT of the given periodic signal can be obtained by using 4 or 12 periods. As the number of periods increases the harmonic components appear in each case at exactly the same frequencies, and only zeros in between these fixed harmonic frequencies result from increasing the number of periods. However, the magnitude frequency response is increasing

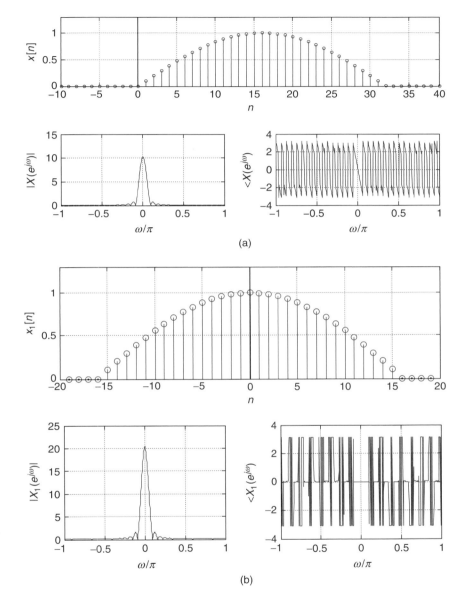

FIGURE 10.14

Computation of the FFT of (a) a causal signal and (b) a noncausal signal. Notice that as expected the magnitude responses are equal—only the phase responses change.

as the number of periods increases. Thus, we need to divide by the number of periods used in computing the FFT.

Since the signal is sampled, it is of interest to have the frequency scale of the FFTs in hertz, so we convert the discrete frequency ω (rad) into f (Hz) according to

$$f = \frac{\omega}{2\pi T_s} = \frac{\omega f_s}{2\pi}$$

where $f_s = 1/T_s$ is the sampling rate given in samples/second. The results are shown in Figure 10.15.

```
%%%%%%%%%%%%%%%%%%%%%%%%%%%%%%%%%%%%%%%%%%%%%%
% Example 10.21---Improving frequency resolution of FFFT of periodic signals
%%%%%%%%%%%%%%%%%%%%%%%%%%%%%%%%%%%%%%%%%%%%%%
f0 = 100; f1 = 4 * f0; % frequencies in Hz of signal
Ts = 1/(3 * f1);  %  sampling period
t = 0:Ts:4/f0;  % time for 4 periods
y = 4 * cos(2 * pi * f0 * t) - cos(2 * pi * f1 * t); % sampled signal (4 periods)
M = length(y);
Y = fft(y,M); Y = fftshift(Y)/4; % fft using 4 periods, shifting and normalizing
t1 = 0:Ts:12/f0; % time for 12 periods
y1 = 4 * cos(2 * pi * f0 * t1) - cos(2 * pi * f1 * t1); % sampled signal (12 periods)
Y1 = fft(y1);Y1 = fftshift(Y1)/12; % fft using 12 periods, shifting and normalizing
w = 2 * [0:M - 1]./M - 1;f = w/(2 * Ts); % frequency scale (4 periods)
N = length(y1);
w1 = 2 * [0:N - 1]./N - 1;f = w/(2 * Ts); % frequency scale (12 periods)
```
■

10.4.4 Linear and Circular Convolution Sums

The most important property of the DFT is the convolution property, which permits the computation of the linear convolution sum very efficiently by means of the FFT.

Consider the convolution sum that gives the output $y[n]$ of a discrete-time LTI system with impulse response $h[n]$ and input $x[n]$:

$$y[n] = \sum_m x[m]h[n - m]$$

In frequency, $y[n]$ is the inverse DTFT of the product

$$Y(e^{j\omega}) = X(e^{j\omega})H(e^{j\omega})$$

Assuming that $x[n]$ has a finite length M and that $h[n]$ has a finite length K, then $y[n]$ has a finite length $N = M + K - 1$. If we choose a period $L \geq N$ for the periodic extension $\tilde{y}[n]$ of $y[n]$, we would obtain the frequency-sampled periodic sequence

$$Y(e^{j\omega})|_{\omega=2\pi k/L} = X(e^{j\omega})H(e^{j\omega})|_{\omega=2\pi k/L}$$

or the DFT of $y[n]$ as the product of the DFTs of $x[n]$ and $h[n]$:

$$Y[k] = X[k]H[k] \quad \text{for} \quad k = 0, 1, \ldots, L - 1$$

We then obtain $y[n]$ as the inverse DFT of $Y[k]$. It should be noticed that the L-length DFT of $x[n]$ and of $h[n]$ requires that we pad $x[n]$ with $L - M$ zeros and $h[n]$ with $L - K$ zeros, so that both $X[k]$ and $H[k]$ have the same length L and can be multiplied at each k.

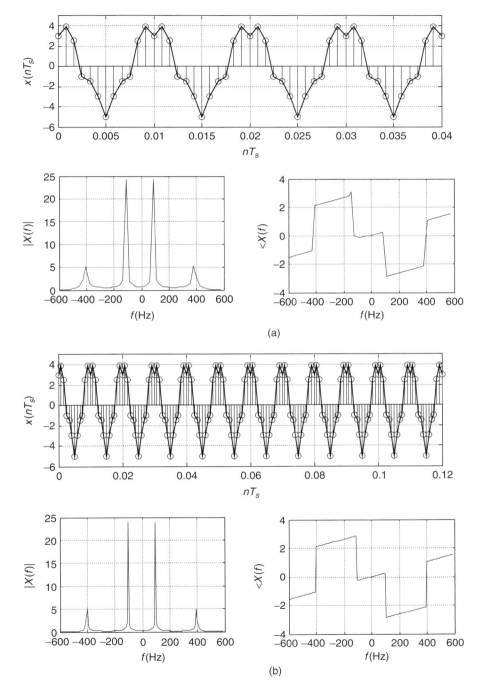

FIGURE 10.15

Computation of the FFT of a periodic signal using (a) 4 and (b) 12 periods to improve the frequency resolution of the FFT. Notice that both magnitude and phase responses look alike, but when we use 12 periods these spectra look sharper due to the increase in the number of frequency components added.

Given $x[n]$ and $h[n]$ of lengths M and K, the linear convolution sum $y[n]$ of length $N = M + K - 1$ can be found by following these three steps:

- Compute DFTs $X[k]$ and $H[k]$ of length $L \geq N$ for $x[n]$ and $h[n]$.
- Multiply them to get $Y[k] = X[k]H[k]$.
- Find the inverse DFT of $Y[k]$ of length L to obtain $y[n]$.

Although it seems computationally more expensive than performing the direct computation of the convolution sum, the above approach implemented with the FFT can be shown to be much more efficient.

The above procedure could be implemented by a *circular convolution sum* in the time domain, although in practice it is not done due to the efficiency of the implementation with FFTs. A circular convolution uses circular rather than linear representation of the signals being convolved. The *periodic convolution sum* introduced before is a circular convolution of fixed length—the period of the signals being convolved. When we use the DFT to compute the response of an LTI system the length of the circular convolution is given by the possible length of the linear convolution sum. Thus, if the system input is a finite sequence $x[n]$ of length M and the impulse response of the system $h[n]$ has a length K, then the output $y[n]$ is given by a linear convolution of length $M + K - 1$. The length $L \geq M + K - 1$ of the DFT $Y[k] = X[k]H[k]$ corresponds to a circular convolution of length L of the $x[n]$ and $h[n]$ padded with zeros so that both have length L. In such a case the circular and the linear convolutions coincide.

If $x[n]$ of length M is the input of an LTI system with impulse response $h[n]$ of length K, then

$$Y[k] = X[k]H[k] \Leftrightarrow y[n] = (x \otimes_L h)[n] \qquad (10.55)$$

where $X[k]$, $H[k]$, and $Y[k]$ are, respectively, DFTs of length L of the input, the impulse response, and the output of the LTI system, and $\otimes_L$ stands for the circular convolution of length L.
If L is chosen so that $L \geq M + K - 1$, the circular and the linear convolution sums coincide—that is,

$$y[n] = (x \otimes_L h)[n] = (x * h)[n] \qquad (10.56)$$

Remark *If we consider the periodic expansions of $x[n]$ and $h[n]$ with period $L = M + K - 1$, we can use their circular representations and implement the circular convolution as shown in Figure 10.16. Since the length of the linear convolution or convolution sum, $M + K - 1$, coincides with the length of the circular convolution, the two convolutions coincide. Given the efficiency of the FFT algorithm in computing the DFT, the convolution is typically done using the DFT as indicated above.*

■ Example 10.22

To illustrate the connection between the circular and the linear convolution, compute using MATLAB the circular convolution of a pulse signal $x[n] = u[n] - u[n-21]$ of length $N = 20$ with itself for different values of its length. Determine the length for which the circular convolution coincides with the linear convolution of $x[n]$ with itself.

FIGURE 10.16

Circular convolution of length $L = 8$ of $x[n]$ and $y[n]$. The signal $x[k]$ is stationary with a circular representation given by the inside circle, while $y[n - k]$ is represented by the outside circle and rotated in the clockwise direction. The shown circular convolution sum corresponds to $n = 0$.

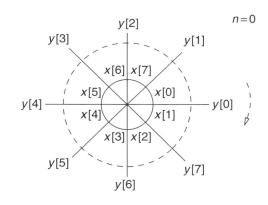

FIGURE 10.17

Circular versus linear convolutions: (a) Plot corresponds to linear convolution. (b) and (c) Plots are circular convolutions wih $L < 2N - 1$. (d) Plot is circular convolution with $L > 2N - 1$ coinciding with the linear convolution.

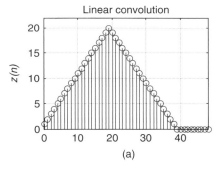
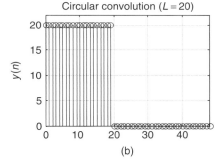
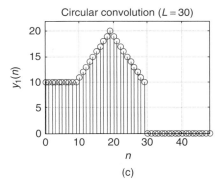
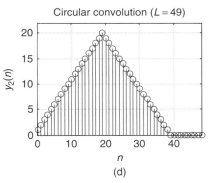

Solution

We know that the length of the linear convolution $z[n] = (x * x)[n]$ is $N + N - 1 = 39$. If we use the function circonv2 shown below to compute the circular convolution of $x[n]$ with itself with length $N < 2N - 1$, for instance $L = 20$ as shown in Figure 10.17(b), the result will not equal the linear convolution. Likewise, if the circular convolution is of length $N + 10 = 30 < 2N - 1$, only part of the result resembles the linear convolution (see Figure 10.17(c)). If we let the length of the circular convolution be $2N + 9 = 49 > 2N - 1$, the result is identical to the linear convolution (see Figure 10.17(d)). The script is given as follows.

```
%%%%%%%%%%%%%%%%%%%%%%%%%%%%%%%%%%%%%%%%%%%%
% Example 10.22---Linear and circular convolution
%%%%%%%%%%%%%%%%%%%%%%%%%%%%%%%%%%%%%%%%%%%%
clear all; clf
N = 20; x = ones(1,N);
% linear convolution
z = conv(x,x);z = [z zeros(1,10)];
% circular convolution
y = circonv2(x,x,N);
y1 = circonv2(x,x,N + 10);
y2 = circonv2(x,x,2 * N + 9);
Mz = length(z); My = length(y); My1 = length(y1);My2 = length(y2);
y = [y zeros(1,Mz - My)]; y1 = [y1 zeros(1,Mz - My1)]; y2 = [y2 zeros(1,Mz-My2)];
```

The function circonv2 has as inputs the signals to be convolved and the desired length of the circular convolution. It computes and multiplies the FFTs of the signals and then finds the inverse FFT to obtain the circular convolution. If the desired length of the circular convolution is larger than the length of each of the signals, the signals are padded with zeros to make them the length of the circular convolution. The following is the code for this function.

```
function xy = circonv2(x,y,N)
M = max(length(x),length(y))
if M>N
   disp('Increase N')
end
x = [x zeros(1,N - M)];
y = [y zeros(1,N - M)];
% circular convolution
X = fft(x,N); Y = fft(y,N); XY = X. * Y;
xy = real(ifft(XY,N));
```

■

■ Example 10.23

A significant advantage of using the FFT for computing the DFT is in filtering. Assume that the signal to filter consists of the MATLAB file "laughter.mat," multiplied by 5, to which a signal that continuously changes between -0.3 and 0.3 is added. We wish to recover the original "laughter.mat" signal using a filter. Use the MATLAB function fir1 to design the required filter.

Solution

Noticing that since the disturbance $0.3(-1)^n$ is a signal of frequency π, we need a low-pass filter with a wide bandwidth so as to get rid of the disturbance while trying to keep the frequency components of the desired signal. The following script is used to design the desired low-pass filter, and to implement the filtering. To compare the results obtained with the FFT we use the function conv to find the output of the filter in the time domain. The results are shown in Figure 10.18.

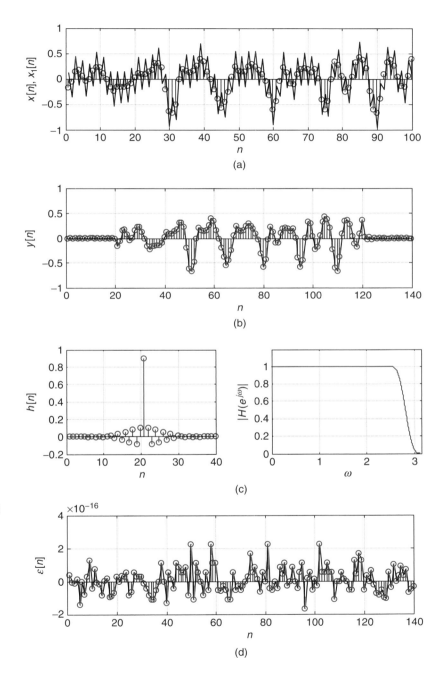

FIGURE 10.18

FIR filtering of disturbed signal. Comparison of results using `conv` and `fft` functions. (a) Actual and noisy signal (plotted as a continuous signal using black lines); (b) denoised signal (notice the delay caused by the FIR filter); (c) impulse response and magnitude response of low-pass FIR filter; and (d) error signal $\epsilon[n] = y[n] - y_1[n]$ between output from `conv`- and `fft`-based filtering.

```
%%%%%%%%%%%%%%%%%%%%%%%%%%%%%%%%%%%%%%%%%%%%%%%%
% Example 10.23---Filtering using convolution and FFT
%%%%%%%%%%%%%%%%%%%%%%%%%%%%%%%%%%%%%%%%%%%%%%%%

clear all; clf
N = 100; n = 0:N - 1;
load laughter
x = 5 * y(1:N)'; x1 = x + 0.3 * (-1).^n; % desired signal plus disturbance
h = fir1(40,0.9); [H,w] = freqz(h,1); %  low-pass FIR filter design
% filtering using convolution
y = conv(x,h); % convolution of signal and impulse response of FIR

% computing using FFT
M = length(x) + length(h) - 1; %  circular and linear convolutions equal
X = fft(x,M);
H = fft(h,M);
Y = X. * H;
y1 = ifft(Y); % output of filtering
```

■

10.5 WHAT HAVE WE ACCOMPLISHED? WHERE DO WE GO FROM HERE?

In this chapter we have considered the Fourier representation of discrete-time signals and systems. Just as with the Laplace and the Fourier transforms, in the continuous case there is a large class of discrete-time signals and impulse responses of systems for which we are able to find their discrete-time Fourier transform from their Z-transforms. For signals that are not absolutely summable, the time-frequency duality and other properties of the transform are used to find their DTFTs. Properties of the DTFT are very similar to those of the Z-transform. Although theoretically useful, the DTFT is computationally not feasible, due to the continuity of the frequency variable and to the integration required in the inverse transformation. It is the Fourier series of discrete-time signals that makes the Fourier representation computationally feasible. In Table 10.1, the DTFT of common signals and some DTFT properties are given.

The Fourier series coefficients constitute a periodic sequence of the same period as the signal; thus both are periodic. Moreover, the Fourier series and its coefficients are obtained as sums, and the frequency used is discretized. Thus, they can be obtained by computer. To take advantage of this, the spectrum of an aperiodic signal resulting from the DTFT is sampled so that in the time domain there is a periodic repetition of the original signal. For finite-support signals we can then obtain a periodic extension that gives the discrete Fourier transform or DFT. The significance of this result is that we have frequency representations of discrete-time signals that are computed algorithmically. Table 10.2 displays properties of the discrete Fourier series and of the Discrete Fourier Transform (DFT). What remains then is to take a look at the algorithm used for those computations or the fast Fourier transform (FFT). We will do that in Chapter 12, where we will show that this algorithm efficiently computes the DFT and makes the convolution sum a more feasible procedure.

Table 10.1 DTFT of Common Signals and DTFT Properties

Discrete-Time Fourier Transforms

	Discrete-Time Signal	DTFT $X(e^{j\omega})$, Periodic of Period 2π
1.	$\delta[n]$	$1, \quad -\pi \le \omega < \pi$
2.	A	$2\pi A \delta(\omega), \quad -\pi \le \omega < \pi$
3.	$e^{j\omega_0 n}$	$2\pi \delta(w - \omega_0), \quad -\pi \le \omega < \pi$
4.	$\alpha^n u[n], \quad |\alpha| < 1$	$\dfrac{1}{1-\alpha e^{-j\omega}}, \quad -\pi \le \omega < \pi$
5.	$n\alpha^n u[n], \quad |\alpha| < 1$	$\dfrac{\alpha e^{-j\omega}}{(1-\alpha e^{-j\omega})^2}, \quad -\pi \le \omega < \pi$
6.	$\cos(\omega_0 n) u[n]$	$\pi\left[\delta(\omega - \omega_0) + \delta(\omega + \omega_0)\right], \quad -\pi \le \omega < \pi$
7.	$\sin(\omega_0 n) u[n]$	$-j\pi\left[\delta(\omega - \omega_0) + \delta(\omega + \omega_0)\right], \quad -\pi \le \omega < \pi$
8.	$\alpha^{|n|}, \quad |\alpha| < 1$	$\dfrac{1-\alpha^2}{1-2\alpha \cos(\omega)+\alpha^2}, \quad -\pi \le \omega < \pi$
9.	$u[n + N/2] - u[n - N/2]$	$\dfrac{\sin(\omega(N+1)/2)}{\sin(\omega/2)}, \quad -\pi \le \omega < \pi$
10.	$\alpha^n \cos(\omega_0 n) u[n]$	$\dfrac{1-\alpha \cos(\omega_0)e^{-j\omega}}{1-2\alpha \cos(\omega_0)e^{-j\omega}+\alpha^2 e^{-2j\omega}}, \quad -\pi \le \omega < \pi$
11.	$\alpha^n \sin(\omega_0 n) u[n]$	$\dfrac{\alpha \sin(\omega_0)e^{-j\omega}}{1-2\alpha \cos(\omega_0)e^{-j\omega}+\alpha^2 e^{-2j\omega}}, \quad -\pi \le \omega < \pi$

Properties of the DTFT

Z-transform:	$x[n], X(z), |z| = 1 \in ROC$	$X(e^{j\omega}) = X(z)|_{z=e^{j\omega}}$
Periodicity:	$x[n]$	$X(e^{j\omega}) = X(e^{j(\omega+2\pi k)}), \quad k$ integer
Linearity:	$\alpha x[n] + \beta y[n]$	$\alpha X(e^{j\omega}) + \beta Y(e^{j\omega})$
Time-shift:	$x[n - N]$	$e^{-j\omega N} X(e^{j\omega})$
Frequency-shift:	$x[n]e^{j\omega_0 n}$	$X(e^{j(\omega-\omega_0)})$
Convolution:	$(x * y)[n]$	$X(e^{j\omega})Y(e^{j\omega})$
Multiplication:	$x[n]y[n]$	$\frac{1}{2\pi}\int_{-\pi}^{\pi} X(e^{j\theta})Y(e^{j(\omega-\theta)})d\theta$
Symmetry:	$x[n]$, real valued	$|X(e^{j\omega})|$, even function of ω
		$\angle X(e^{j\omega})$, odd function of ω
Parseval's relation:	$\sum_{n=\infty}^{\infty} |x[n]|^2 = \frac{1}{2\pi}\int_{-\pi}^{\pi} |X(e^{j\omega})|^2 d\omega$	

PROBLEMS

10.1. Eigenfunction property and frequency response—MATLAB

An IIR filter is characterized by the difference equation

$$y[n] = 0.5y[n - 1] + x[n] - 2x[n - 1] \qquad n \ge 0$$

where $x[n]$ is the input and $y[n]$ is the output of the filter. Let $H(z)$ be the transfer function of the filter.

Table 10.2 Properties of Discrete Fourier Series and Discrete Fourier Transform.

Fourier Series of Discrete-Time Periodic Signals		
	$x[n]$, periodic signal of period N	$X[k]$, periodic FS coefficients of period N
Z-transform	$x_1[n] = x[n](u[n] - u[n - N])$	$X[k] = \frac{1}{N} \mathcal{Z}(x_1[n]) \big\vert_{z=e^{j2\pi k/N}}$
DTFT	$x[n] = \sum_k X[k]e^{j2\pi nk/N}$	$X(e^{j\omega}) = \sum_k 2\pi X[k]\delta(\omega - 2\pi k/N)$
LTI response	Input: $x[n] = \sum_k X[k]e^{j2\pi nk/N}$	Output: $y[n] = \sum_k X[k]H(e^{jk\omega_0})e^{j2\pi nk/N}$
		$H(e^{j\omega})$, frequency response of system
Time shift (circular shift)	$x[n - M]$	$X[k]e^{-j2\pi kM/N}$
Modulation	$x[n]e^{j2\pi Mn/N}$	$X[k - M]$
Multiplication	$x[n]y[n]$	$\sum_{m=0}^{N-1} X[m]Y[k - m]$, periodic convolution
Periodic convolution	$\sum_{m=0}^{N-1} x[m]y[n - m]$	$NX[k]Y[n]$
Discrete Fourier Transform		
	$x[n]$, finite-length N aperiodic signal	$\tilde{x}[n]$, periodic extension of period $L \geq N$
	$\tilde{x}[n] = \frac{1}{N} \sum_{k=0}^{L-1} \tilde{X}[k]e^{j2\pi nk/L}$	$\tilde{X}[k] = \sum_{n=0}^{L-1} \tilde{x}[n]e^{-j2\pi nk/L}$
IDFT/DFT	$x[n] = \tilde{x}[n]W[k]$, $W[n] = u[n] - u[n - N]$	$X[k] = \tilde{X}[k]W[k]$, $W[k] = u[k] - u[k - N]$
Circular convolution	$(x \otimes_L y)[n]$	$X[k]Y[k]$
Circular and linear convolution	$(x \otimes_L y)[n] = (x * y)[n]$, $L \geq M + K - 1$	
	$M = $ length of $x[n]$, $K = $ length of $y[n]$	

(a) The given filter is LTI, and as such the eigenfunction property applies. Obtain the magnitude response $|H(e^{j\omega})|$ of the filter using the eigenfunction property.

(b) Compute the magnitude response $|H(e^{j\omega})|$ at discrete frequencies $\omega = 0$, $\pi/2$, and π radians. Show that the magnitude response is constant for $0 \leq \omega \leq \pi$, and as such this is an all-pass filter.

(c) Use the MATLAB function freqz to compute the frequency response (magnitude and phase) of this filter and to plot them.

(d) Determine the transfer function $H(z) = Y(z)/X(z)$. Find its pole and zero and indicate how they are related.

10.2. Frequency transformation of low-pass to high-pass filters—MATLAB

You have designed an IIR low-pass filter with an input–output relation given by the difference equation

$$y[n] = 0.5y[n - 1] + x[n] + x[n - 1] \qquad n \geq 0$$

where $x[n]$ is the input and $y[n]$ is the output. You are told that by changing the difference equation to

$$y[n] = -0.5y[n - 1] + x[n] - x[n - 1] \qquad n \geq 0$$

you obtain a high-pass filter.

(a) From the eigenfunction property find the frequency response of the two filters at $\omega = 0$, $\pi/2$, and π radians. Use the MATLAB functions freqz and abs to compute the magnitude responses of the two filters. Plot them to verify that the filters are low pass and high pass.

(b) Call $H_1(e^{j\omega})$ the frequency response of the first filter and $H_2(e^{j\omega})$ the frequency response of the second filter. Show that

$$H_2(e^{j\omega}) = H_1(e^{j(\pi-\omega)})$$

and relate the impulse response $h_2[n]$ to $h_1[n]$.

(c) Use the MATLAB function zplane to find and plot the poles and the zeros of the filters and determine the relation between the poles and the zeros of the two filters.

10.3. Computations from definition of DTFT and IDTFT

Consider the discrete-time signal $x[n] = 0.5^{|n|}$, and find its DTFT $X(e^{j\omega})$. From the direct and the inverse DTFT of $x[n]$:

(a) Determine the infinite sum

$$\sum_{k=-\infty}^{\infty} 0.5^{|n|}$$

(b) Find the integral

$$\int_{-\pi}^{\pi} X(e^{j\omega})d\omega$$

(c) Find the phase of $X(e^{j\omega})$.

(d) Determine the sum

$$\sum_{k=-\infty}^{\infty} (-1)^n 0.5^{|n|}$$

10.4. Frequency shift of FIR filters—MATLAB

Consider a moving-average FIR filter with an impulse response

$$h[n] = \frac{1}{3}(\delta[n] + \delta[n-1] + \delta[n-2])$$

Let $H(z)$ be the Z-transform of $h[n]$.

(a) Find the frequency response $H(e^{j\omega})$ of the FIR filter.

(b) Let the impulse response of a new filter be given by

$$h_1[n] = h[n]e^{j\pi n}$$

Use the eigenfunction property to find the frequency response $H_1(e^{j\omega})$ of the new FIR filter.

(c) Use the MATLAB functions freqz and abs to compute the magnitude response of the two filters. Plot them and determine the location of the poles and the zeros of the two filters. What type of filters are these?

10.5. Duality of DTFT

The DTFT of a discrete-time signal $x[n]$ is given as

$$X(e^{j\omega}) = e^{j\pi/4}\delta(\omega - 0.5\pi) + e^{-j\pi/4}\delta(\omega + 0.5\pi) - 2\pi e^{-j\pi/8}\delta(\omega - 0.71) - 2\pi e^{j\pi/8}\delta(\omega + 0.71)$$

(a) Is the signal $x[n]$ periodic? If so, indicate its period.

(b) Determine the signal $x[n]$, and verify your answer above.

10.6. Chirps for jamming—MATLAB

A chirp signal is a sinusoid of continuously changing frequency. Chirps are frequently used to jam communication transmissions. Consider the chirp

$$x[n] = \cos(\theta n^2)u[n] \qquad \theta = \frac{\pi}{2L} \quad 0 \leq n \leq L-1$$

(a) A measure of the frequency of the chirp is the so-called instantaneous frequency, which is defined as the derivative of the phase in the cosine—that is,

$$IF(n) = \frac{d\theta n^2}{dn}$$

Find the instantaneous frequency of the given chirp. Use MATLAB to plot $x[n]$ for $L = 256$.

(b) Let $L = 256$ and use MATLAB to compute the DTFT of $x[n]$ and to plot its magnitude. Indicate the range of discrete frequencies that would be jammed by the given chirp.

10.7. Time specifications for FIR filters—MATLAB

When designing discrete filters the specifications can be given in the time domain. One can think of converting the frequency-domain specifications into the time domain. Assume you wish to obtain a filter that approximates an ideal low-pass filter with a cut-off frequency $\omega_c = \pi/2$ and that has a linear phase $-N\omega$. Thus, the frequency response is

$$H(e^{j\omega}) = \begin{cases} 1e^{-jN\omega} & -\pi/2 \leq \omega \leq \pi/2 \\ 0 & -\pi \leq \omega < -\pi/2 \text{ and } \pi/2 < \omega \leq \pi \end{cases}$$

(a) Find the corresponding impulse response using the inverse DTFT of $H(e^{j\omega})$.

(b) If $N = 50$, plot $h[n]$ using the MATLAB function stem for $0 \leq n \leq 100$. Comment on the shape of the plot.

(c) Suppose we want a band-pass filter of center frequency $\omega_0 = \pi/2$. Use the above impulse response $h[n]$ to obtain the impulse response of the desired band-pass filter.

10.8. Z-transform and DTFT—MATLAB

Consider a discrete pulse $p[n] = u[n] - u[n-N]$.

(a) Use the definition of the DTFT to determine $P(e^{j\omega})$ and then use the Z-transform $P(z)$ of $p[n]$ to verify your result.

(b) For $N = 5$, 10, and 20, use the MATLAB function zplane to find the zeros of $P(z)$ and indicate at what frequencies $P(e^{j\omega})$ is zero. Verify your response using freqz.

(c) Suppose the impulse response of a filter is $h[n] = u[n] - u[n-4]$, and its input is

$$v[n] = \sum_{k=1}^{2} \cos(k\omega_0 n)$$

For what value of ω_0 is the steady-state response of the filter zero?

10.9. Downsampling and DTFT—MATLAB

Consider pulses $x_1[n] = u[n] - u[n-20]$ and $x_2[n] = u[n] - u[n-10]$, and their product $x[n] = x_1[n]x_2[n]$.

(a) Plot the three pulses. Could you say that $x[n]$ is a downsampled version of $x_1[n]$? What would be the downsampling rate? Find $X_1(e^{j\omega})$.

(b) Find directly the DTFT of $x[n]$ and compare it to $X_1(e^{j\omega/M})$ where M is the downsampling rate found above. If we downsample $x_1[n]$ to get $x[n]$, would the result be affected by aliasing? Use MATLAB to plot the magnitude DTFT of $x_1[n]$ and $x[n]$ to provide an answer.

10.10. Cascading of interpolators and decimators—MATLAB

Suppose you cascade an interpolator (an upsampler and a low-pass filter) and a decimator (a low-pass filter and a downsampler).

(a) If both the interpolator and the decimator have the same rate M, carefully draw a block diagram of the interpolator–decimator system.

(b) Suppose that the interpolator is of rate 3 and the decimator of rate 2. Carefully draw a block diagram of the interpolator–decimator system. What would be the equivalent of sampling the input of this system to obtain the same output?

(c) Use the MATLAB functions interp and decimate to process the first 100 samples of the test signal handel where the interpolator's rate is 3 and the decimator's is 2. How many samples does the output have?

10.11. MATLAB and phase computation

The computation of the phase of a complex number or function using MATLAB has some issues that you need to understand:

(a) The range of possible values of the inverse tangent needs to be extended to $[-\pi, \pi)$ depending on the quadrant the complex number is in. Consider the complex numbers $1 + j$, $-1 + j$, $-1 - j$, and $1 - j$, and represent each by a vector from the origin and consider what changes are needed when we use the formula to find the phase.

(b) The phase of a complex number is only significant if its magnitude is significant. Use MATLAB to compute the magnitudes and the phases of the complex numbers $x = 1 + j$ and $y = 10^{-6} + j10^{-6}$. How do the phases of these numbers compare? What about their magnitudes? Explain.

(c) If the function $X(e^{j\omega})$ is zero or infinite at a frequency ω_0, the phase is undetermined at that frequency and of no significance since the corresponding magnitude is zero or infinity. Let $X(z) = z - 1$, so that $X(e^{j\omega}) = e^{j\omega} - 1$. Can you determine the phase of $X(e^{j0})$? Explain. Likewise, if $X(z) = 1/(z - 1)$, what is the phase of $X(e^{j\omega})$ at $\omega = 0$?

(d) If the phase is linear (i.e., $\theta = -N\omega$), MATLAB will plot the values only between $[-\pi, \pi]$ and so the phase will not appear linear. Let $X(z) = z^{-4}$. Find the phase of $X(e^{j\omega})$—is it linear? Then use the MATLAB functions freqz and angle to compute the phase of $X(e^{j\omega})$—does it appear linear? Explain.

10.12. Linear phase and phase unwrapping—MATLAB

A DTFT $X(e^{j\omega})$ is said to have linear phase if its phase is a line through the origin of the frequency plane. Let

$$X(e^{j\omega}) = 2e^{-j4\omega} \qquad -\pi \le \omega < \pi$$

(a) Carefully plot the magnitude and the phase of $X(e^{j\omega})$. Is the phase linear?

(b) Use the MATLAB functions freqz and angle to compute the phase of $X(e^{j\omega})$ and then plot it. (Hint: Let $z = e^{j\omega}$ to be able to use freqz.) Does the phase computed by MATLAB appear linear? What are the maximum and minimum values of the phase, and how many radians separate the minimum from the maximum?

(c) Now, recalculate the phase, but after using angle use the function unwrapping in the resulting phase and plot it. Does the phase appear linear?

10.13. Linear phase and symmetry—MATLAB

Consider the signal $x[n] = A\delta[n] + u[n + 9] - u[n - 10]$.

(a) Carefully plot $x[n]$. Find the Z-transform $X(z)$ of $x[n]$ and from it $X(e^{j\omega})$, the DTFT of $x[n]$. Find the value of A so that the phase of $X(e^{j\omega})$ is zero. Use MATLAB to verify your results.

(b) Consider now $x_1[n] = x[n - 9]$, and use the value of A found before. Carefully plot $x_1[n]$ and find its DTFT using the Z-transform $X_1(z)$. Is its phase linear? Use MATLAB to verify your results. Use freqz, angle, and unwrap to compute the phase.

10.14. Sinusoidal form of DTFT

A triangular pulse is given by

$$t[n] = \begin{cases} 3+n & -2 \le n \le -1 \\ 3-n & 0 \le n \le 2 \\ 0 & \text{otherwise} \end{cases}$$

(a) The pulse can be written as

$$t[n] = \sum_{k=-\infty}^{\infty} A_k \delta[n-k]$$

Find the $\{A_k\}$ coefficients.

(b) Find a sinusoidal expression for the DTFT of $t[n]$—that is,

$$T(e^{j\omega}) = B_0 + \sum_{k=1}^{\infty} B_k \cos(k\omega)$$

Express the coefficients B_0 and B_k in terms of the A_k coefficients.

10.15. DTFT and Z-transform—MATLAB

Let $x[n] = r[n] - r[n-3] - u[n-3]$ where $r[n]$ is the ramp signal.

(a) Carefully plot $x[n]$ and find its Z-transform $X(z)$.

(b) If $y[n] = x[-n]$, give $Y(z)$ in terms of $X(z)$.

(c) Use the above results to find the DTFT of $x[n]$, $x[-n]$, and $x[n] + x[-n]$. Find the magnitude of each of these DTFTs and then use MATLAB to compute them and plot them.

10.16. Computations from DTFT definition

For simple signals it is possible to obtain some information on their DTFTs without computing them. Let

$$x[n] = \delta[n] + 2\delta[n-1] + 3\delta[n-2] + 2\delta[n-3] + \delta[n-4]$$

(a) Find $X(e^{j0})$ and $X(e^{j\pi})$ without computing the DTFT $X(e^{j\omega})$.

(b) Find

$$\int_{-\pi}^{\pi} |X(e^{j\omega})|^2 d\omega$$

(c) Find the phase of $X(e^{j\omega})$. Is it linear?

10.17. DTFT of even and odd functions

A signal

$$x[n] = 0.5^n u[n]$$

is neither even nor odd.

(a) Find the even $x_e[n]$ and the odd $x_o[n]$ components of $x[n]$, and carefully plot them.

(b) Find the Z-transforms of $x_e[n]$ and $x_o[n]$, and from them find the DFTs $X_e(e^{j\omega})$ and $X_o(e^{j\omega})$. Are they real or imaginary?

(c) Since $x[n] = x_e[n] + x_o[n]$ so that $X(e^{j\omega}) = X_e(e^{j\omega}) + X_o(e^{j\omega})$, how dos the real and the imaginary parts of $X(e^{j\omega})$ relate to $X_e(e^{j\omega})$ and $X_o(e^{j\omega})$? Explain.

(d) Use Parseval's result to obtain that $E_x = E_{xe} + E_{xo}$ i.e., the energy of the signal is the sum of the energies of its even and odd components.

10.18. Power spectral density

Consider an autocorrelation function

$$c[n] = 0.5^{|n|} \qquad -\infty < n < \infty$$

(a) Find the magnitude square of the DTFT $C(e^{j\omega})$ of $c[n]$, which is called the power spectral density.

(b) Find the Z-transform of $c[n]$ and determine where its poles and zeros are. Are there any zeros or poles on the unit circle?

(c) Find $C(e^{j0})$—that is, the dc value of the power spectral density. Determine the phase of $C(e^{j\omega})$—is it linear?

10.19. Convolution sum and product of polynomials

The convolution sum can be seen as a way to compute the coefficients of the product of polynomials. This is because

$$[x * y][n] \Leftrightarrow X(z)Y(z) \Leftrightarrow X(e^{j\omega})Y(e^{j\omega})$$

(a) Let $X(z) = 1 + 2z^{-1} + 3z^{-2}$ and $Y(z) = z^{-2} + 4z^{-3}$ if $x[n] = 1\delta[n] + 2\delta[n-1] + 3\delta[n-2]$ and $y[n] = 1\delta[n-2] + 4\delta[n-3]$ are sequences formed by the coefficients of the polynomials. Compute the convolution sum $[x * y][n]$ and compare it to the coefficients of the polynomial $Z(z) = X(z)Y(z)$, or $Z(e^{j\omega}) = X(e^{j\omega})Y(e^{j\omega})$.

(b) Suppose that the transfer function of a discrete-time system is

$$H(z) = \frac{W(z)}{V(z)} = 3z^2 + 2z + 2z^{-1} + 3z^{-2}$$

and that it is known that the input is $v[n] = u[n] - u[n-3]$. Use the connection between the product of the polynomials and the convolution sum to find the output $w[n]$ of the system.

10.20. Windowing and DTFT—MATLAB

A window $w[n]$ is used to consider the part of a signal we are interested in.

(a) Let $w[n] = u[n] - u[n-20]$ be a rectangular window of length 20. Let $x[n] = \sin(0.1\pi n)$. We are interested in a period of the infinite length $x[n]$, or $y[n] = x[n]w[n]$. Compute the DTFT of $y[n]$ and compare it with the DTFT of $x[n]$. Write a MATLAB script to compute $Y(e^{j\omega})$.

(b) Let $w_1[n] = (1 + \cos(2\pi n/11))(u[n+5] - un[n-5])$ be a raised-cosine window that is symmetric with respect to $n = 0$ (noncausal). Adapt the script in the previous part to find the DTFT of

$$z[n] = x[n]w_1[n]$$

where $x[n]$ is the sinusoid given above.

10.21. Z-transform and Fourier series—MATLAB

Let

$$x_1[n] = 0.5^n \qquad 0 \le n \le 9$$

be a period of a periodic signal $x[n]$.

(a) Use the Z-transform to compute the Fourier series coefficients.

(b) Use MATLAB to plot the magnitude and the phase line spectrum (i.e., $|X_k|$ and $\angle X_k$ versus frequency $-\pi \le \omega \le \pi$).

10.22. Linear equations and Fourier series—MATLAB

The Fourier series of a signal $x[n]$ and its coefficients X_k are both periodic of some value N, and as such can be written as

- $$x[n] = \sum_{k=0}^{N-1} X_k e^{j2\pi nk/N} \qquad 0 \le n \le N-1$$

- $$X_k = \frac{1}{N} \sum_{n=0}^{N-1} x[n] e^{-j2\pi nk/N} \qquad 0 \le k \le N-1$$

(a) To find the $x[n]$, $0 \le n \le N-1$ given X_k, $0 \le k \le N-1$, write a set of N linear equations. Indicate how you would find the $x[n]$ from the matrix equation.

(b) As you can see, there is a lot of duality in the Fourier series and its coefficients. If you consider the reverse problem in the previous part, how would you solve for X_k given the $x[n]$?

(c) Let $x[n] = n$ for $n = 0, 1, 2$, and 0 for $n = 3$ be a period of a periodic signal $x[n]$ of period $N = 4$. Use the above method to solve for the Fourier series coefficients X_k, $0 \le k \le 3$. Use MATLAB to find the inverse of the complex exponential matrix.

(d) Suppose that when computing the X_k for the $x[n]$ signal given above, you separate the sum into two sums, one for the even values of n (i.e., $n = 0, 2$) and the other for the odd values of n (i.e., $n = 1, 3$). Try to simplify the complex exponentials and write an equivalent matrix expression for the X_k.

10.23. Operations on Fourier series—MATLAB

A periodic signal $x[n]$ of period N can be represented by its Fourier series

$$x[n] = \sum_{k=0}^{N-1} X_k e^{j2\pi nk/N} \qquad 0 \le n \le N-1$$

If you consider this a representation of $x[n]$:

(a) Is $x_1[n] = x[n-3]$ periodic? If so, use the Fourier series of $x[n]$ to obtain the Fourier series coefficients of $x_1[n]$.

(b) Let $x_2[n] = x[n] - x[n-1]$ (i.e., the finite difference). Determine if $x_2[n]$ is periodic, and if so, find its Fourier series coefficients.

(c) If $x_3[n] = x[n](-1)^n$, is $x_3[n]$ periodic? If so, determine its Fourier series coefficients.

(d) Let $x_4[n] = sign[\cos(0.5\pi n)]$ where $sign(\xi)$ is a function that gives 1 when $\xi \ge 0$ and -1 when $\xi < 0$. Determine the Fourier coefficients of $x_4[n]$ if periodic.

(e) Use MATLAB to find the Fourier series coefficients for $x_i[n]$, $i = 1, 2, 3$, and 4 and to plot them as functions of k.

10.24. Fourier series of even and odd signals—MATLAB

Let $x[n]$ be an even signal and $y[n]$ be an odd signal.

(a) Determine whether the Fourier coefficients X_k and Y_k corresponding to $x[n]$ and $y[n]$ are complex, real, or imaginary.

(b) Consider $x[n] = \cos(2\pi n/N)$ and $y[n] = \sin(2\pi n/N)$ for $N = 3$ and $N = 4$. Use the above results to find the Fourier series coefficients for the two signals with the different periods.

(c) Use MATLAB to find the Fourier series coefficients of the above two signals with the different periods, and plot their magnitude and phase spectra.

10.25. Response of LTI systems to periodic signals—MATLAB

Suppose you get noisy periodic measurements

$$y[n] = (-1)^n x[n] + A\eta[n]$$

where $x[n]$ is the desired signal and $\eta[n]$ is a noise that varies from 0 to 1 at random.

(a) Let $A = 0$ and $x[n] = sign[\cos(0.1\pi n)]$. Determine how to process $y[n]$, indicating the type of filter to obtain an approximate version of $x[n]$. Consider the first 100 samples of $x[n]$ and use MATLAB to find the spectrum of $x[n]$ and $y[n]$ to show that the filter you recommend will do the job.

(b) Use the MATLAB function fir1 to generate the kind of filter you decided to use above and show that when filtering $y[n]$ for $A = 0$, you obtain the desired result.

(c) Consider the first 100 samples of the MATLAB file "handel.mat" a period of a signal that continuously replays these values over and over. Let $x[n]$ be the desired signal that results from this. Now let $A = 0.01$, and use the function rand to generate the noise, and come up with suggestions as to how to get rid of the effects of the multiplication by $(-1)^n$ and of the noise $\eta[n]$. Recover the desired signal $x[n]$.

10.26. DFT of an aperiodic and periodic signal—MATLAB
Consider a signal $x[n] = (-0.95)^n(u[n] - u[n-70])$.

(a) To compute the DFT of $x[n]$ we pad it with zeros so as to obtain a signal with length 2^γ, the larger but closest to the length of $x[n]$. Determine the value of γ and use the MATLAB function fft to compute the DFT $X[k]$ of the padded-with-zeros signal. Plot its magnitude and phase.

(b) Let now $x_1[n] = x[n-10]$. Compute its DFT $X_1[k]$ using the fft function. Pad $x_1[n]$ with zeros to compute the γ-length FFT where γ is the value obtained above.

(c) Consider $x[n]$ a period of a periodic signal of period $N = 70$. Compute its DFT using the fft algorithm and then plot its magnitude and phase. What is the length of the FFT? Can we pad with zeros the period to find its DFT?

10.27. Frequency resolution of DFT—MATLAB
When we pad an aperiodic signal with zeros, we are improving its frequency resolution—that is, the more zeros we attach to the original signal the better the frequency resolution, as we obtain the frequency representation at a larger number of frequencies around the unit circle.

(a) Consider an aperiodic signal $x[n] = u[n] - u[n-10]$, and compute its DFT by means of the fft function padding it with 10 and then 100 zeros. Plot the magnitude response using stem. Comment on the frequency resolution of the two DFTs.

(b) When the signal is periodic, one cannot pad a period with zeros. When computing the FFT in theory we generate a periodic signal of period L equal or larger than the length of the signal when the signal is aperiodic, but if the signal is periodic we must let L be the signal period or a multiple of it. Adding zeros to the period makes the signal different from the periodic signal. Consider $x[n] = \cos(\pi n/5)$, $-\infty < n < \infty$ as a periodic signal, and do the following:

- Consider exactly one period of $x[n]$ and compute the FFT of this sequence.
- Consider 10 periods of $x[n]$ and compute the FFT of this sequence.
- Consider attaching 10 zeros to one period and compute the FFT of the resulting sequence.

If we consider the first of these cases giving the correct DFT of $x[n]$, how many harmonic frequencies does it show. What happens when we consider the 10 periods? Are the harmonic frequencies the same as before? What are the values of the DFT in frequencies in between the harmonic frequencies? What happened to the magnitude at the original frequencies? Finally, does the last FFT relate at all to the first FFTs?

10.28. DFT and IIR filters—MATLAB
A definite advantage of the FFT is that it reduces considerably the computation in the convolution sum. Thus, if $x[n]$, $0 \le n \le N-1$ is the input of an FIR filter with impulse response $h[n]$, $0 \le n \le M-1$, their convolution sum $y[n] = [x * h][n]$ will be of length $M + N - 1$. Now if $X[k]$ and $H[k]$ are the DFTs (computed by the FFT) of $x[n]$ and $h[n]$, and if $Y[k] = X[k]H[k]$ is the DFT of the convolution sum of length bigger or equal to $M + N - 1$, then to be able to multiply the FFTs $X[k]$ and $H[k]$ they both should

be of the same length as $Y[k]$ (i.e., bigger or equal to $M + N - 1$). Consider what happens when the filter is an IIR that possibly has an impulse response of very large length. Let

$$y[n] - 1.755y[n - 1] + 0.81y[n - 2] = x[n] + 0.5x[n - 1]$$

be the difference equation representing an IIR filter with input $x[n]$ and output $y[n]$. Assume the initial conditions are zero, and the input is $x[n] = u[n] - u[n - 50]$. Use MATLAB to obtain your results.

(a) Compute using filter the first 40 values of the impulse response $h[n]$ and use them to approximate it (call it $\hat{h}[n]$). Compute the filter output $\hat{y}[n]$ using the FFT as indicated above. In this case, we are approximating the IIR filter by an FIR filter of length 40. Plot the input and the output. Use FFTs of length 128.

(b) Suppose now that we do not want to approximate $h[n]$, so consider the following procedure. Find the transfer function of the IIR filter, say $H(z) = B(z)/A(z)$, and if $X(z)$ is the Z-transform of the input, then

$$Y(z) = \frac{B(z)X(z)}{A(z)}$$

Compute as before the FFT for $x[n]$, of length 128 (call it $X[k]$) and compute the 128 length of the coefficients of $B(z)$ and $A(z)$ to obtain DFTs $B[k]$ and $A[k]$. Multiplying $X[k]$ by $B[k]$ and dividing by $A[k]$, all of length 128, results in a sequence of length 128 that should correspond to $Y[k]$, the DFT of $y[n]$. Compute the inverse FFT to get $y[n]$.

(c) Use filter to solve the difference equation and obtain $y[n]$ for $x[n] = u[n] - u[n - 50]$. If this is the exact solution, calculate the error with respect to the other responses in (a) and (b)?

10.29. **Circular and linear convolutions—MATLAB**

Consider the circular convolution of two signals $x[n] = n$, $0 \leq n \leq 3$, and $y[n] = 1$, $n = 0, 2$ and zero for $n = 1, 3$.

(a) Compute the convolution sum or linear convolution of $x[n]$ and $y[n]$. Do it graphically and verify your results by multiplying the DTFTs of $x[n]$ and $y[n]$.

(b) Use MATLAB to find the linear convolution. Plot $x[n]$, $y[n]$, and the linear convolution $z[n] = (x * y)[n]$.

(c) We wish to compute the circular convolution of $x[n]$ and $y[n]$ for different lengths $N = 4$, $N = 7$, and $N = 10$. Determine for which of these values does the circular and the linear convolutions coincide. Show the circular convolution for the three cases. Use MATLAB to verify your results.

(d) If we use the convolution property of the DFT verify your result in the above parts of this problem.

CHAPTER 11

Introduction to the Design of Discrete Filters

When in doubt, don't.
Benjamin Franklin (1706–1790)
Printer, inventor, scientist, and diplomat

11.1 INTRODUCTION

In this chapter we introduce the design of discrete filters. This material complements the theory of analog and discrete filtering presented in previous chapters, and in particular provides continuity to the introduction to analog filter design from Chapter 6.

Filtering is an important application of linear time-invariant (LTI) systems. According to the eigenfunction property of LTI systems (Figure 11.1) the steady-state response of a discrete-time LTI system to a sinusoidal input—with a certain frequency, magnitude, and phase—is also a sinusoid of the same frequency as the input, but with the magnitude and the phase affected by the response of the system at the input frequency. Since periodic as well as aperiodic signals have Fourier representations consisting of sinusoids of different frequencies, the frequency components of any signal can be modified by appropriately choosing the frequency response of an LTI system or filter. Filtering can thus be seen as a way to change the frequency content of an input signal.

The appropriate filter is specified using the spectral characterization of the input and the desired spectral characteristics of the output of the filter. Once the specifications of the filter are set, the problem becomes one of approximation, either by a ratio of polynomials or by a polynomial (if possible). After establishing that the filter resulting from the approximation satisfies the given specifications, it is then necessary to check its stability (if not guaranteed by the design method) in the case of the filter being a rational approximation, and if stable we need to figure out what would be the best possible way to implement the filter in either hardware or software. If not stable, we need either to repeat the approximation or to stabilize the filter before its implementation.

In the continuous-time domain, filters are obtained by means of rational approximation. In the discrete-time domain, there are two possible types of filters. The first is the result of rational

Signals and Systems Using MATLAB®. DOI: 10.1016/B978-0-12-374716-7.00015-6

FIGURE 11.1
Eigenfunction property of LTI systems.

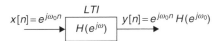

$$x[n]=e^{j\omega_0 n} \xrightarrow{\quad} \boxed{\begin{array}{c} LTI \\ H(e^{j\omega}) \end{array}} \xrightarrow{\quad} y[n]=e^{j\omega_0 n} H(e^{j\omega_0})$$

approximation—these filters are called recursive or infinite-impulse response (IIR) filters. The other is the nonrecursive or finite-impulse response (FIR) filters that result from a polynomial approximation. In the continuous-time domain, depending on their implementation, filters are either passive or active. Passive filters are implemented using resistors, capacitors, and inductors, while active filters are implemented with resistors, capacitors, and operational amplifiers. The implementation of discrete or digital filters is done by means of software or dedicated hardware.

As we will see, the discrete filter specifications can be in the frequency or in the time domain. For recursive or IIR filters, the specifications are typically given in the form of magnitude and phase specifications, while the specifications for nonrecursive or FIR filters can be in the time domain as a desired impulse response. The discrete filter design problem then consists in: Given the specifications of a filter we look for a polynomial or rational (ratio of polynomials) approximation to the specifications. The resulting filter should be realizable, which besides causality and stability requires that the filter coefficients be real valued.

The typical approach in filter design is to consider low-pass prototypes with normalized frequency and magnitude responses that may be transformed into other filters with the desired frequency response. Thus, a great deal of effort is put into the design of low-pass filters and into developing frequency transformations to map low-pass filters into other types of filters. Using cascade and parallel connections of filters also provides a way to obtain different types of filters.

There are different ways to obtain the rational approximation for discrete IIR filters—by transformation of analog filters, or by optimization methods that include stability as a constraint. We will see that the classical analog design methods (Butterworth, Chebyshev, Elliptic, etc.) can be used to design discrete filters by means of the bilinear transformation that maps the analog s-plane into the z-plane. Given that the FIR filters are unique to the discrete domain, the approximation procedures for FIR filters are unique to that domain.

The difference between discrete and digital filters is in the quantization and coding. For a discrete filter we assume that the input and the coefficients of the filter are represented with infinite precision—that is, using an infinite number of quantization levels—and thus no coding is performed. The coefficients of a digital filter are binary and the input and output are quantized and coded. Thus, quantization affects the performance of a digital filter, while it has no effect in discrete filters.

Considering continuous-to-discrete converters (CDCs) and discrete-to-continuous converters (DCCs) as simply samplers and reconstruction filters, it is possible to implement the filtering of band-limited analog signals using discrete filters (Figure 11.2). In such an application, an additional specification for the filter design is the sampling period. In this process it is crucial that the sampling period in the CDCs and DCCs be synchronized. In practice, filtering of analog signals is done using analog-to-digital (ADC) and digital-to-analog (DAC) together with digital filters.

FIGURE 11.2

Discrete filtering of analog signals using CDCs and DCCs. The CDC is simply a sampler while the DCC is a reconstruction filter.

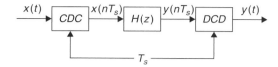

11.2 FREQUENCY-SELECTIVE DISCRETE FILTERS

The principle behind discrete filtering is easily understood by considering the response of an LTI system to sinusoids. If $H(z)$ is the transfer function of a discrete-time LTI system, and

$$x[n] = \sum_k A_k \cos(\omega_k n + \phi_k) \tag{11.1}$$

is the input of the system, according to the eigenfunction property of LTI systems the steady-state response of the system is

$$y_{ss}[n] = \sum_k A_k |H(e^{j\omega_k})| \cos(\omega_k n + \phi_k + \theta(\omega_k)) \tag{11.2}$$

where $|H(e^{j\omega_k})|$ and $\theta(\omega_k)$ are the magnitude and the phase of $H(e^{j\omega})$, which is the frequency response of the system, at the frequency ω_k. The frequency response is the transfer function computed on the unit circle (i.e., $H(e^{j\omega}) = H(z)|_{z=e^{j\omega}}$). It becomes clear from Equation (11.2) that by judiciously choosing the frequency response of the LTI system we can select the frequency components of the input we wish to have at the output, and attenuate or amplify their amplitudes or change their phases.

In general, for an input $x[n]$ with Z-transform $X(z)$, the Z-transform of the output of the filter is

$$Y(z) = H(z)X(z) \tag{11.3}$$

or on the unit circle (when $z = e^{j\omega}$),

$$Y(e^{j\omega}) = H(e^{j\omega})X(e^{j\omega}). \tag{11.4}$$

By selecting the frequency response $H(e^{j\omega})$ we allow some frequency components of $x[n]$ to appear in the output, and others to be filtered out.

Ideal frequency-selective filters, such as low-pass, high-pass, band-pass, and stopband filters, cannot be realized. They serve as prototypes for the actual filters.

11.2.1 Linear Phase

A filter changes the spectrum of its input in magnitude as well as in phase. Distortion in magnitude can be avoided by using an all-pass filter with unit magnitude for all frequencies. Phase distortion can be avoided by requiring the phase response of the filter to be linear. For instance, when transmitting a voice signal in a communication system it is important that the signals at the transmitter and at the receiver be ideally equal within a time delay and a constant attenuation factor. To achieve this, the transfer function of an ideal communication channel should equal that of an all-pass filter with a linear phase.

Indeed, if the output of the transmitter is a discretized baseband signal $x[n]$ and the recovered signal at the receiver is $\alpha x[n - N_0]$, for an attenuation factor α and a time delay N_0, ideally the channel is represented by a transfer function

$$H(z) = \frac{\mathcal{Z}(\alpha x[n - N_0])}{\mathcal{Z}(x[n])} = \alpha z^{-N_0} \qquad (11.5)$$

The constant gain of the all-pass filter permits all frequency components of the input to appear in the output. The linear phase simply delays the signal, which is a very tolerable distortion.

To appreciate the effect of linear phase, consider the filtering of a signal

$$x[n] = 1 + \cos(\omega_0 n) + \cos(\omega_1 n) \qquad \omega_1 = 2\omega_0 \qquad n \geq 0$$

using an all-pass filter with transfer function $H(z) = \alpha z^{-N_0}$. The magnitude response of this filter is α, and its phase is linear, as shown in Figure 11.3(a). The steady-state output of the all-pass filter is

$$y_{ss}[n] = 1H(e^{j0}) + |H(e^{j\omega_0})| \cos(\omega_0 n + \angle H(e^{j\omega_0})) + |H(e^{j\omega_1})| \cos(\omega_1 n + \angle H(e^{j\omega_1}))$$
$$= \alpha \left[1 + \cos(\omega_o(n - N_0)) + \cos(\omega_1(n - N_0)) \right] = \alpha x[n - N_0]$$

which is the input signal attenuated by α and delayed N_0 samples.

Suppose then that the all-pass filter has a phase function that is nonlinear, for instance, the one in Figure 11.3(b). The steady-state output would then be

$$y_{ss}[n] = 1H(e^{j0}) + |H(e^{j\omega_0})| \cos(\omega_0 n + \angle H(e^{j\omega_0})) + |H(e^{j\omega_1})| \cos(\omega_1 n + \angle H(e^{j\omega_1}))$$
$$= \alpha[1 + \cos(\omega_0(n - N_0)) + \cos(\omega_1(n - 0.5 N_0))] \neq \alpha x[n - N_0]$$

In the case of the linear phase each of the frequency components is delayed N_0 samples, and thus the output is just a delayed version of the input. On the other hand, in the case of a nonlinear phase the frequency component of frequency ω_1 is delayed less than the other two frequency components, creating distortion in the signal so that the output is not a delayed version of the input.

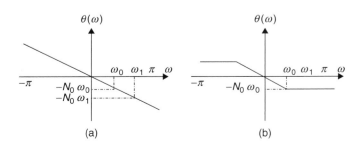

FIGURE 11.3

(a) Linear and (b) nonlinear phase.

(a)

(b)

Group Delay

A measure of linearity of the phase is obtained from the *group delay function,* which is defined as

$$\tau(\omega) = -\frac{d\theta(\omega)}{d\omega} \tag{11.6}$$

The group delay is constant when the phase is linear. Deviation of the group delay from a constant indicates the degree of nonlinearity of the phase. In the above cases, when the phase is linear (i.e., for $0 \leq \omega \leq \pi$),

$$\theta(\omega) = -N_0\omega \quad \Rightarrow \quad \tau(\omega) = N_0$$

and when the phase is nonlinear or

$$\theta(\omega) = \begin{cases} -N_0\omega & 0 < \omega \leq \omega_0 \\ -N_0\omega_0 & \omega_0 < \omega \leq \pi \end{cases}$$

for $0 \leq \omega \leq \pi$, then we have that the group delay is

$$\tau(\omega) = \begin{cases} N_0 & 0 < \omega \leq \omega_0 \\ 0 & \omega_0 < \omega \leq \pi \end{cases}$$

which is not constant.

11.2.2 IIR and FIR Discrete Filters

- A discrete filter with transfer function

$$H(z) = \frac{B(z)}{A(z)} = \frac{\sum_{m=0}^{M-1} b_m z^{-m}}{1 + \sum_{k=1}^{N-1} a_k z^{-k}} = \sum_{n=0}^{\infty} h[n]z^{-n} \tag{11.7}$$

  is called *infinite-impulse response* or *IIR* since its impulse response $h[n]$ typically has infinite length. It is also called *recursive* because if the input of the filter $H(z)$ is $x[n]$ and $y[n]$ is its output, the input–output relationship is given by the difference equation

$$y[n] = -\sum_{k=1}^{N-1} a_k y[n-k] + \sum_{m=0}^{M-1} b_m x[n-m] \tag{11.8}$$

  where the output recurs on previous outputs (i.e., the output is fed back).
- The transfer function of a *finite-impulse response* or *FIR* filter is

$$H(z) = B(z) = \sum_{m=0}^{M-1} b_m z^{-m} \tag{11.9}$$

  Its impulse response is $h[n] = b_n, n = 0, \ldots, M-1$, and zero elsewhere, thus of finite length. This filter is called *nonrecursive* given that the input–output relationship is given by

$$y[n] = \sum_{m=0}^{M} b_m x[n-m] = (b*x)[n] \tag{11.10}$$

  or the convolution sum of the filter coefficients (or impulse response) and the input.

For practical reasons, these filters are causal (i.e., $h[n] = 0$ for $n < 0$) and bounded-input bounded-output (BIBO) stable (i.e., all the poles of $H(z)$ must be *inside the unit circle*). This guarantees that the filter can be implemented and used in real-time processing, and that the output remains bounded when the input is bounded.

Remarks

- *Calling the IIR filters recursive is more appropriate. It is possible to have a filter with a rational transfer function that does not have an infinite-length impulse response. However, it is traditional to refer to these filters as IIR.*
- *When comparing the IIR and the FIR filters, neither has a definite advantage:*
  - *IIR filters are implemented more efficiently than FIR filters in terms of number of operations and required storage (having similar frequency responses, an IIR filter has fewer coefficients than an FIR filter).*
  - *The implementation of an IIR filter using the difference equation resulting from its transfer function is simple and computationally efficient, but FIR filters can be implemented using the Fast Fourier Transform (FFT) algorithm, which is computationally very efficient.*
  - *Since the transfer function of any FIR filter only has poles at the origin of the z-plane, FIR filters are always BIBO stable, but for an IIR filter we need to check that the poles of its transfer function (i.e., zeros of $A(z)$) are inside the unit circle if the design procedure does not guarantee stability.*
  - *FIR filters can be designed to have linear phase, while IIR filters usually have nonlinear phase, but approximately linear phase in the passband region.*

■ Example 11.1

The phase of IIR filters is always nonlinear. Although it is possible to design FIR filters with linear phase, not all FIR filters have linear phase. As we will see, symmetry conditions on the impulse response of FIR filters are needed to have linear phase. Consider the following two filters with input–output equations

$$(a)\ y[n] = 0.5y[n-1] + x[n]$$

$$(b)\ y[n] = \frac{1}{3}\left(x[n-1] + x[n] + x[n+1]\right)$$

where $x[n]$ is the input and $y[n]$ is the output. Use MATLAB to compute and plot the magnitude and the phase response of each of these filters.

Solution

The transfer functions of the given filters are:

$$(a)\ H_1(z) = \frac{1}{1 - 0.5z^{-1}}$$

$$(b)\ H_2(z) = \frac{1}{3}[z^{-1} + 1 + z] = \frac{1 + z + z^2}{3z} = \frac{(z - 1e^{j2.09})(z - 1e^{-j2.09})}{3z}$$

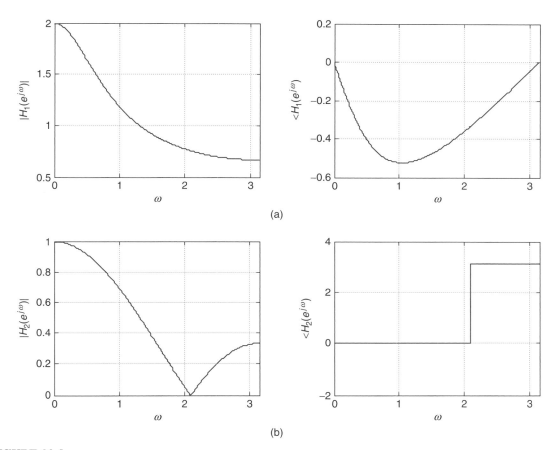

FIGURE 11.4

Magnitude and phase responses of an IIR filter with a transfer function of (a) $H_1(z) = 1/(1 - 0.5z^{-1})$, and of an FIR filter with a transfer function of (b) $H_2(z) = (z - 1e^{j2.09})(z - 1e^{-j2.09})/3z$. Notice the phase responses are nonlinear.

Thus, the first is an IIR filter and the second is an FIR filter (notice that this filter is noncausal as it requires future values of the input to compute the present output). The phase responses of these filters are clearly nonlinear. The transfer function $H_2(z)$ has zeros on the unit circle $z = 1e^{\pm j2.09}$, making the phase of this filter not continuous and so it cannot be unwrapped. Figure 11.4 shows the magnitude and the phase responses of the filters $H_1(z)$ and $H_2(z)$.   ■

■ Example 11.2

A simple model for the multipath effect in the channel of a wireless system is

$$y[n] = x[n] - \alpha x[n - N_0] \qquad \alpha = 0.8, \ N_0 = 11$$

That is, the output $y[n]$ is a combination of the input $x[n]$ and of a delayed and attenuated version of the input. Determine the transfer function of the filter that gives the above input–output equation. Use MATLAB to plot its magnitude and phase. If the phase is nonlinear, how would you recover the input $x[n]$ (which is the message)? Let the input be $x[n] = 2 + \cos(\pi n/4) + \cos(\pi n)$. In practice, the delay N_0 and the attenuation α are not known at the receiver and need to be estimated. What would happen if the delay is estimated to be 12 and the attenuation 0.79?

Solution

The transfer function of the filter with input $x[n]$ and output $y[n]$ is

$$H(z) = \frac{Y(z)}{X(z)} = 1 - 0.8z^{-11} = \frac{z^{11} - 0.8}{z^{11}}$$

with a pole of $z = 0$ of multiplicity 11, and zeros the roots of $z^{11} - 0.8 = 0$, or

$$z_k = (0.8)^{1/11} e^{j2\pi k/11} \qquad k = 0, \ldots, 10$$

Using the freqz function to plot its magnitude and phase responses (see Figure 11.5), we find that the phase is nonlinear, and as such the output of $H(z)$, $y[n]$, will not be a delayed version of the input. To recover the input, we use an *inverse filter* $G(z)$ such that cascaded with $H(z)$ the overall filter is an all-pass filter (i.e., $H(z)G(z) = 1$). Thus,

$$G(z) = \frac{z^{11}}{z^{11} - 0.8}$$

The poles and zeros and the magnitude and the phase responses of $H(z)$ are shown in Figure 11.5(a). The filters with transfer functions $H(z)$ and $G(z)$ are called *comb filters* given the shape of their magnitude responses.

If the delay is estimated to be 11 and the attenuation 0.8, the input signal $x[n]$ (the message) is recovered exactly; however, if we have slight variations on these values the message might not be recovered. When the delay is estimated to be 12 and the attenuation 0.79, the inverse filter is

$$G(z) = \frac{z^{12}}{z^{12} - 0.79}$$

having the poles, zeros, and magnitude and phase responses shown in Figure 11.5(b). In Figure 11.5(c), the effect of these changes are illustrated. The output of the inverse filter $z[n]$ does not resemble the sent signal $x[n]$. The signal $y[n]$ is the output of the channel with a transfer function $H(z)$. ∎

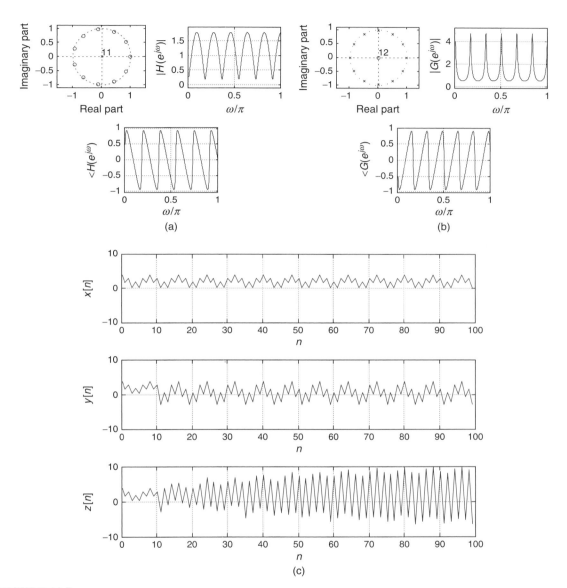

FIGURE 11.5

(a) Poles and zeros and frequency response of the FIR comb filter $H(z) = (z^{11} - 0.8)/z^{11}$, and (b) the estimated inverse IIR comb filter $G(z) = z^{12}/(z^{12} - 0.79)$. (c) The message $x[n] = 2 + \cos(\pi n/4) + \cos(\pi n)$, the output $y[n]$ of channel $H(z)$, and the output $z[n]$ of the estimated inverse filter $G(z)$.

11.3 FILTER SPECIFICATIONS

There are two ways to specify a discrete filter: in the frequency domain and in the time domain. The frequency-domain specification of the desired filter's magnitude and phase is more common in IIR filter design, while the time-domain specification in terms of the desired impulse response of the filter is more common in FIR filter design.

11.3.1 Frequency-Domain Specifications

Conventionally, when designing IIR filters a prototype low-pass filter is obtained first and then converted into the desired filter. The magnitude specifications of a discrete low-pass filter are given for frequencies $[0, \pi]$ due to the periodicity and the even characteristics of the magnitude. Typically, the phase is not specified but it is expected to be approximately linear.

For a low-pass filter, the desired magnitude $|H_d(e^{jw})|$ is to be close to unity in a passband frequency region, and close to zero in a stopband frequency region. A transition frequency region where the filter is not specified is needed. Thus, the magnitude specifications are displayed in Figure 11.6(a). The passband $[0, \omega_p]$ is the band of frequencies for which the attenuation specification is the smallest; the stopband $[\omega_{st}, \pi]$ is the band of frequencies where the attenuation specification is the greatest; and the transition band (ω_p, ω_{st}) is the frequency band where the filter is not specified. The frequencies ω_p and ω_{st} are called the passband and the stopband frequencies, respectively.

Loss Function

As for analog filters, the linear, or normal, scale specifications shown in Figure 11.6(a) do not give the sense of the attenuation, and thus the loss or log specification in decibels (dB) is preferred. The logarithmic scale also provides a greater resolution of the magnitude. Figure 11.7 shows the relation between a magnitude value $|G|$ and its corresponding loss value in decibels. Notice that the loss increases by 20 dB whenever the filter attenuates the input signal by a factor of 10^{-1}.

The magnitude specifications of a discrete low-pass filter in a linear scale are (Figure 11.6(a)):

$$\text{Passband:} \quad \delta_1 \leq |H(e^{j\omega})| \leq 1 \qquad 0 \leq \omega \leq \omega_p$$

$$\text{Stopband:} \quad 0 < |H(e^{j\omega})| \leq \delta_2 \qquad \omega_{st} \leq \omega \leq \pi \tag{11.11}$$

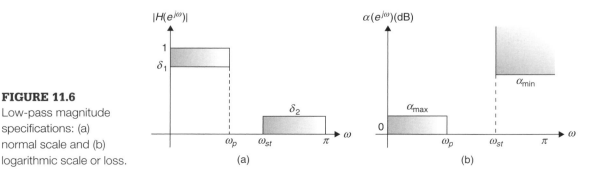

FIGURE 11.6
Low-pass magnitude specifications: (a) normal scale and (b) logarithmic scale or loss.

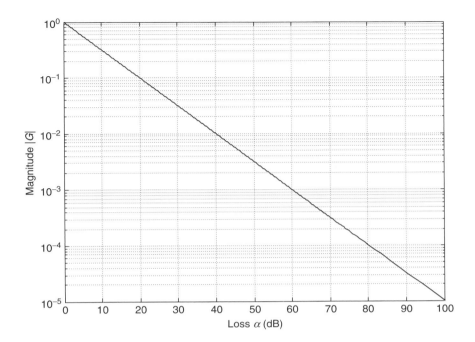

FIGURE 11.7

Loss α (dB) for normalized gain $|G|$.

Defining the *loss function* for a discrete filter as

$$\alpha(e^{j\omega}) = -10\log_{10}|H(e^{j\omega})|^2 = -20\log_{10}|H(e^{j\omega})| \quad \text{dB} \tag{11.12}$$

equivalent magnitude specifications for a discrete low-pass filter are (Figure 11.6(b)):

$$
\begin{aligned}
&\textit{Passband:} && 0 \leq \alpha(e^{j\omega}) \leq \alpha_{\max} && 0 \leq \omega \leq \omega_p \\
&\textit{Stopband:} && \alpha_{\min} \leq \alpha(e^{j\omega}) < \infty && \omega_{st} \leq \omega \leq \pi
\end{aligned} \tag{11.13}
$$

where $\alpha_{\max} = -20\log_{10}\delta_1$ and $\alpha_{\min} = -20\log_{10}\delta_2$ (which are positive since both δ_1 and δ_2 are positive and smaller than 1).

■ Example 11.3

Consider the following specifications for a low-pass filter:

$$
\begin{aligned}
0.9 \leq |H(e^{j\omega})| \leq 1.0 && 0 \leq \omega \leq \pi/2 \\
0 < |H(e^{j\omega})| \leq 0.1 && 3\pi/4 \leq \omega \leq \pi
\end{aligned}
$$

Determine the equivalent loss specifications.

Solution

The loss specifications are then

$$0 \leq \alpha(e^{j\omega}) \leq 0.92 \quad 0 \leq \omega \leq \pi/2$$
$$\alpha(e^{j\omega}) \geq 20 \quad\quad\quad 3\pi/4 \leq \omega \leq \pi$$

where $\alpha_{max} = 0.92$ dB and $\alpha_{min} = 20$ dB. These specifications indicate that in the passband the loss is small, or that the magnitude would change from 1 to

$$10^{-\alpha_{max}/20} = 10^{-0.92/20} = 0.9$$

while in the stopband we would like a large attenuation, at least α_{min}, or that the magnitude would have values smaller than

$$10^{-\alpha_{min}/20} = 0.1 \qquad\qquad\qquad ■$$

Remarks

- *The dB scale is an indicator of attenuation: If we have a unit magnitude the corresponding loss is 0 dB, and for every 20 dB in loss this magnitude is attenuated by 10^{-1}, so that when the loss is 100 dB the unit magnitude would be attenuated to 10^{-5}. The dB scale also has the physiological significance of being a measure of how humans detect levels of sound.*
- *Besides the physiological significance, the loss specifications have intuitive appeal. They indicate that in the passband, where minimal attenuation of the input signal is desired, the "loss" is minimal as it is constrained to be below a maximum loss of α_{max} dB. Likewise, in the stopband where maximal attenuation of the input signal is needed, the "loss" is set to be larger than α_{min} dB.*
- *When specifying a high-quality filter the α_{max} value should be small, the α_{min} value should be large, and the transition band should be as narrow as possible—that is, approximating as much as possible the frequency response of an ideal low-pass filter. The cost of this is a large order for the resulting filter, making the implementation expensive computationally and requiring large memory space.*

Magnitude Normalization

The specifications of the low-pass filter in Figure 11.6 are normalized in magnitude: The dc gain is assumed to be unity (or the dc loss is 0 dB), but there are many cases where that is not so. See Figure 11.8.

Not-normalized magnitude specifications: In general, the dc loss is different from 0 dB, so that the loss specifications are

$$\alpha_1 \leq \hat{\alpha}(e^{j\omega}) \leq \alpha_2 \quad 0 \leq \omega \leq \omega_p$$
$$\alpha_3 \leq \hat{\alpha}(e^{j\omega}) \quad\quad\quad \omega_{st} \leq \omega \leq \pi$$

Writing the above loss as

$$\hat{\alpha}(e^{j\omega}) = \alpha_1 + \alpha(e^{j\omega}) \qquad\qquad\qquad (11.14)$$

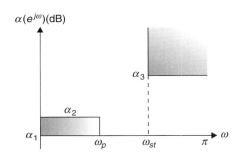

FIGURE 11.8
Not-normalized loss specifications for a low-pass filter.

the dc loss of α_1 is achieved by multiplying a magnitude-normalized filter by a constant K, such that

$$\hat{\alpha}(e^{j0}) = \alpha_1 = -20\log_{10} K \quad \text{or} \quad K = 10^{-\alpha_1/20} \tag{11.15}$$

The normalized specifications are then

$$0 \le \alpha(e^{j\omega}) \le \alpha_{max} \quad 0 \le \omega \le \omega_p$$

$$\alpha_{min} \le \alpha(e^{j\omega}) \quad \omega_{st} \le \omega \le \pi$$

where $\alpha_{max} = \alpha_2 - \alpha_1$ and $\alpha_{min} = \alpha_3 - \alpha_1$.

■ Example 11.4

Suppose the loss specifications of a low-pass filter are

$$10 \le \hat{\alpha}(e^{j\omega}) \le 11 \quad 0 \le \omega \le \frac{\pi}{2}$$

$$\hat{\alpha}(e^{j\omega}) \ge 50 \quad \frac{3\pi}{4} \le \omega \le \pi$$

Determine the loss specifications that can be used to design a magnitude-normalized filter, and the gain K to make it satisfy the above specifications.

Solution

If we let

$$\hat{\alpha}(e^{j\omega}) = 10 + \alpha(e^{j\omega})$$

the loss specifications for a normalized filter would be

$$0 \le \alpha(e^{j\omega}) \le 1 \quad 0 \le \omega \le \frac{\pi}{2}$$

$$\alpha(e^{j\omega}) \ge 40 \quad \frac{3\pi}{4} \le \omega \le \pi$$

Then the dc loss is $10\,\text{dB}$ and $\alpha_{\max} = 1$ and $\alpha_{\min} = 40\,\text{dB}$. Suppose that we design a filter $H(z)$ that satisfies the normalized filter specifications. If we let $\hat{H}(z) = KH(z)$ be the filter that satisfies the given loss specifications, at the dc frequency we must have that

$$-20\log_{10}|\hat{H}(e^{j0})| = -20\log_{10}K - 20\log_{10}|H(e^{j0})|$$

$$10 = -20\log_{10}K + 0$$

so that $K = 10^{-0.5} = 1/\sqrt{10}$.   ∎

Frequency Scales

Given that discrete filters can be used to process continuous-time as well as discrete-time signals, there are different equivalent ways in which the frequency of a discrete filter can be expressed (see Figure 11.9).

> In the discrete processing of continuous-time signals the sampling frequency (f_s in hertz or Ω_s in rad/sec) is known, and so we have the following possible scales:
>
> - The $f\,(\text{Hz})$ scale from 0 to $f_s/2$, the foldover or Nyquist frequency, that comes from the sampling theory.
> - The scale $\Omega = 2\pi f$ (rad/sec) where f is the previous scale, the frequency range is then from 0 to $\Omega_s/2$.
> - The discrete frequency scale $\omega = \Omega T_s$ (rad) ranging from 0 to π.
> - A normalized discrete-frequency scale ω/π (no units) ranging from 0 to 1.
>
> If the specifications are in the discrete domain, the scale is the ω (rad) or the normalized ω/π.

Remarks *Other scales are possible, but less used. One of these consists in dividing by the sampling frequency either in hertz or in rad/sec: The f/f_s (no units) scale goes from 0 to $1/2$, and so does the Ω/Ω_s (no units) scale. It is clear that when the specifications are given in any scale, it can be easily transformed into any other desired scale. If the filter is designed for use in the discrete domain only the scales in radians and the normalized ω/π are meaningful.*

11.3.2 Time-domain Specifications

Time-domain specifications consist in giving a desired impulse response $h_d[n]$. For instance, when designing a low-pass filter with cut-off frequency ω_c and linear phase $\phi(\omega) = -N\omega$, the desired

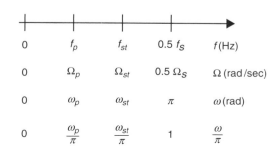

FIGURE 11.9
Frequency scales used in discrete filter design.

frequency response in $0 \le \omega \le \pi$ is

$$H_d(e^{j\omega}) = \begin{cases} 1e^{-j\omega N} & 0 \le \omega \le \omega_c \\ 0 & \omega_c < \omega \le \pi \end{cases}$$

The desired impulse response for this filter is then found from

$$h_d[n] = \frac{1}{2\pi} \int_{-\omega_c}^{\omega_c} 1e^{-j\omega N} e^{j\omega n} d\omega$$

The resulting $h_d[n]$ will be used as the desired impulse response to approximate.

■ Example 11.5

Consider an FIR filter with the following desired magnitude response in $0 \le \omega \le \pi$:

$$|H_d(e^{j\omega})| = \begin{cases} 1 & 0 \le \omega \le \frac{\pi}{4} \\ 0 & \text{elsewhere in } 0 \le \omega \le \pi \end{cases}$$

and zero phase. Find the desired impulse response $h_d[n]$ that we wish to approximate.

Solution

The desired impulse response is computed as follows:

$$h_d[n] = \frac{1}{2\pi} \int_{-\pi}^{\pi} H_d(e^{j\omega}) e^{j\omega n} d\omega = \frac{1}{2\pi} \int_{-\pi/4}^{\pi/4} e^{j\omega n} d\omega$$

$$= \begin{cases} \sin(\pi n/4)/\pi n & n \ne 0 \\ 0.25 & n = 0 \end{cases}$$

which corresponds to the impulse response of a noncausal system. As we will see later, windowing and shifting of $h_d[n]$ are needed to make it into a causal, finite-length filter. ■

11.4 IIR FILTER DESIGN

Two possible approaches in the design of IIR filters are:

- Using analog filter design methods and transformations between the s-plane and the z-plane.
- Using optimization techniques.

The first is a frequency transformation approach. Using a mapping between the analog and the discrete frequencies, we obtain the specifications for an analog filter from the discrete filter specifications. Applying well-known analog filter design methods, we then design the analog filter from the transformed specifications. The discrete filter is finally obtained by transforming the designed analog filter.

The optimal approach designs the filter directly, setting the rational approximation as a nonlinear optimization. The added flexibility of this approach is diminished by the need to ensure stability of the designed filter. Stability is guaranteed, on the other hand, in the transformation approach.

11.4.1 Transformation Design of IIR Discrete Filters

To take advantage of well-understood analog filter design, a common practice is to design discrete filters by means of analog filters and mappings of the s-plane into the z-plane. Two mappings used are:

- The sampling transformation $z = e^{sT_s}$.
- The bilinear transformation,

$$s = K \frac{1 - z^{-1}}{1 + z^{-1}}$$

Recall the transformation $z = e^{sT_s}$ was found when relating the Laplace transform of a sampled signal with its Z-transform. Using this transformation, we convert the analog impulse response $h_a(t)$ of an analog filter into the impulse response $h[n]$ of a discrete filter and obtain the corresponding transfer function. The resulting design procedure is called the *impulse-invariant method*. Advantages of this method are:

- It preserves the stability of the analog filter.
- Given the linear relation between the analog and the discrete frequencies the specifications for the discrete filter can be easily transformed into the specifications for the analog filter.

Its drawback is possible frequency aliasing. Sampling of the analog impulse response requires that the analog filter be band limited, which might not be possible to satisfy in all cases. Due to this we will concentrate on the approach based on the bilinear transformation.

Bilinear Transformation

The bilinear transformation results from the trapezoidal rule approximation of an integral. Suppose that $x(t)$ is the input and $y(t)$ is the output of an integrator with transfer function

$$H(s) = \frac{Y(s)}{X(s)} = \frac{1}{s} \tag{11.16}$$

Sampling the input and the output of this filter using a sampling period T_s, we have that the integral at time nT_s is

$$y(nT_s) = \int_{(n-1)T_s}^{nT_s} x(\tau)d\tau + y((n-1)T_s) \tag{11.17}$$

where $y((n-1)T_s)$ is the integral at time $(n-1)T_s$. Consider then the approximation of the integral. If T_s is very small, the integral between $(n-1)T_s$ and nT_s can be approximated by the area of a trapezoid with bases $x((n-1)T_s)$ and $x(nT_s)$ and height T_s (this is called the trapezoidal rule approximation of an integral):

$$y(nT_s) \approx \frac{[x(nT_s) + x((n-1)T_s)]\, T_s}{2} + y((n-1)T_s) \qquad (11.18)$$

with a Z-transform given by

$$Y(z) = \frac{T_s(1+z^{-1})}{2(1-z^{-1})} X(z)$$

The discrete transfer function is thus

$$H(z) = \frac{Y(z)}{X(z)} = \frac{T_s}{2} \frac{1+z^{-1}}{1-z^{-1}} \qquad (11.19)$$

which can be obtained directly from $H(s)$ by letting

$$s = \frac{2}{T_s} \frac{1-z^{-1}}{1+z^{-1}} \qquad (11.20)$$

The resulting transformation is linear in both numerator and denominator, and thus it is called the *bilinear transformation*. Thinking of the above transformation as a transformation from the z to the s variable, solving for the variable z in that equation, we obtain a transformation from the s to the z variable:

$$z = \frac{1 + (T_s/2)s}{1 - (T_s/2)s} \qquad (11.21)$$

The bilinear transformation:

$$z\text{- to } s\text{-plane:} \quad s = K\frac{1-z^{-1}}{1+z^{-1}} \qquad K = \frac{2}{T_s}$$

$$s\text{- to } z\text{-plane:} \quad z = \frac{1+s/K}{1-s/K} \qquad (11.22)$$

maps
- The $j\Omega$ axis in the s-plane into the unit circle in the z-plane.
- The open left-hand s-plane $\mathcal{Re}[s] < 0$ into the inside of the unit circle in the z-plane, or $|z| < 1$.
- The open right-hand s-plane $\mathcal{Re}[s] > 0$ into the outside of the unit circle in the z-plane, or $|z| > 1$.

Thus, as shown in Figure 11.10, for point A, $s = 0$ or the origin of the s-plane is mapped into $z = 1$ on the unit circle; for points B and B', $s = \pm j\infty$ are mapped into $z = -1$ on the unit circle; for point C, $s = -1$ is mapped into $z = (1 - 1/K)/(1 + 1/K) < 1$, which is inside the unit circle; and finally for point D, $s = 1$ is mapped into $z = (1 + 1/K)/(1 - 1/K) > 1$, which is located outside the unit circle.

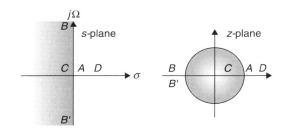

FIGURE 11.10

Bilinear transformation mapping of s-plane into z-plane.

In general, by letting $K = \frac{2}{T_s}$, $z = re^{j\omega}$ and $s = \sigma + j\Omega$ in Equation (11.21), we obtain

$$r = \sqrt{\frac{(1 + \sigma/K)^2 + (\Omega/K)^2}{(1 - \sigma/K)^2 + (\Omega/K)^2}}$$

$$\omega = \tan^{-1}\left(\frac{\Omega/K}{1 + \sigma/K}\right) + \tan^{-1}\left(\frac{\Omega/K}{1 - \sigma/K}\right) \tag{11.23}$$

From this we have that:

- In the $j\Omega$ axis of the s-plane (i.e., when $\sigma = 0$ and $-\infty < \Omega < \infty$), we obtain $r = 1$ and $-\pi \leq \omega < \pi$, which correspond to the unit circle of the z-plane.
- On the open left-hand s-plane, or equivalently when $\sigma < 0$ and $-\infty < \Omega < \infty$, we obtain $r < 1$ and $-\pi \leq \omega < \pi$, or the inside of the unit circle in the z-plane.
- Finally, on the open right-hand s-plane, or equivalently when $\sigma > 0$ and $-\infty < \Omega < \infty$, we obtain $r > 1$ and $-\pi \leq \omega < \pi$, or the outside of the unit circle in the z-plane.

The above transformation can be visualized by thinking of a giant who puts a nail in the origin of the s-plane and then grabs the plus and minus infinity extremes of the $j\Omega$ axis and pulls them together to make them agree into one point, getting a magnificent circle, keeping everything in the left plane inside, and keeping out the rest. If our giant lets go, we get back the original s-plane!

Remarks *The bilinear transformation maps the whole s-plane into the whole z-plane, differently from the transformation $z = e^{sT_s}$ that only maps a slab of the s-plane into the z-plane (see Chapter 9 on the Z-transform). Thus, a stable analog filter with poles in the open left-hand s-plane will generate a discrete filter that is also stable as it has poles inside the unit circle.*

Frequency Warping

A minor drawback of the bilinear transformation is the nonlinear relation between the analog and the discrete frequencies. Such a relation creates a warping that needs to be taken care of when specifying the analog filter using the discrete filter specifications.

The analog frequency Ω and the discrete frequency ω according to the bilinear transformation are related by

$$\Omega = K \tan(\omega/2) \tag{11.24}$$

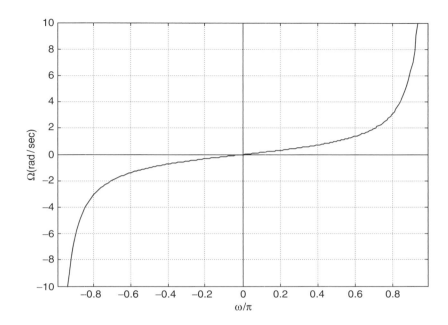

FIGURE 11.11

Relation between Ω and ω for $K = 1$.

which when plotted displays a linear relation around the low frequencies but it warps as we get into large frequencies (see Figure 11.11).

The relation between the frequencies is obtained by letting $\sigma = 0$ in the second equation in Equation (11.23). The linear relationship at low frequencies can be seen using the expansion of the tan(.) function

$$\Omega = K\left[\frac{\omega}{2} + \frac{\omega^3}{24} + \cdots\right] \approx \frac{\omega}{T_s}$$

for small values of ω or $\omega \approx \Omega T_s$. As frequency increases the effect of the terms beyond the first one makes the relation nonlinear. See Figure 11.11.

To compensate for the nonlinear relation between the frequencies, or the warping effect, the following steps to design a discrete filter are followed:

1. Using the frequency warping relation (Eq. 11.24) the specified discrete frequencies ω_p and ω_{st} are transformed into specified analog frequencies Ω_p and Ω_{st}. The magnitude specifications remain the same in the different bands—only the frequency is being transformed.
2. Using the specified analog frequencies and the discrete magnitude specifications, an analog filter $H_N(s)$ that satisfies these specifications is designed.
3. Applying the bilinear transformation to the designed filter $H_N(s)$, the discrete filter $H_N(z)$ that satisfies the discrete specifications is obtained.

11.4.2 Design of Butterworth Low-Pass Discrete Filters

Our aim in this section is to show how to design discrete low-pass filters based on the analog Butterworth low-pass filter design using the bilinear transformation as a frequency transformation.

Applying the warping relation between the continuous and the discrete frequencies

$$\Omega = K \tan(\omega/2) \tag{11.25}$$

to the magnitude-squared function of the Butterworth low-pass analog filter

$$\left| H_N(\Omega') \right|^2 = \frac{1}{1 + (\Omega')^{2N}} \qquad \Omega' = \frac{\Omega}{\Omega_{hp}}$$

gives the magnitude-squared function for the Butterworth low-pass discrete filter:

$$|H_N(e^{j\omega})|^2 = \frac{1}{1 + \left[\frac{\tan(0.5\omega)}{\tan(0.5\omega_{hp})} \right]^{2N}} \tag{11.26}$$

As a frequency transformation (no change to the loss specifications) we directly obtain the minimal order N and the half-power frequency bounds by replacing

$$\frac{\Omega_s}{\Omega_p} = \frac{\tan(\omega_{st}/2)}{\tan(\omega_p/2)} \tag{11.27}$$

in the corresponding formulas for N and Ω_{hp} of the analog filter, giving

$$N \geq \frac{\log_{10}[(10^{0.1\alpha_{min}} - 1)/(10^{0.1\alpha_{max}} - 1)]}{2 \log_{10}\left[\frac{\tan(\omega_{st}/2)}{\tan(\omega_p/2)} \right]}$$

$$2 \tan^{-1}\left[\frac{\tan(\omega_p/2)}{(10^{0.1\alpha_{max}} - 1)^{1/2N}} \right] \leq \omega_{hp} \leq 2 \tan^{-1}\left[\frac{\tan(\omega_{st}/2)}{(10^{0.1\alpha_{min}} - 1)^{1/2N}} \right] \tag{11.28}$$

The normalized half-power frequency $\Omega'_{hp} = 1$ in the continuous domain is mapped into the discrete half-power frequency ω_{hp}, giving the constant in the bilinear transformation

$$K_b = \frac{\Omega'}{\tan(0.5\omega)} \bigg|_{\Omega'=1, \omega=\omega_{hp}} = \frac{1}{\tan(0.5\omega_{hp})} \tag{11.29}$$

The bilinear transformation $s = K_b(1 - z^{-1})/(1 + z^{-1})$ is then used to convert the analog filter $H_N(s)$, satisfying the transformed specifications, into the desired discrete filter,

$$H_N(z) = H_N(s) \big|_{s=K_b(1-z^{-1})/(1+z^{-1})}$$

The basic idea of this design is to convert an analog frequency-normalized Butterworth magnitude-squared function into a discrete function using the relationship in Equation (11.25). To understand why this is an efficient approach consider the following issues that derive from the application of the bilinear transformation to the Butterworth design:

- Since the discrete magnitude specifications are not changed by the bilinear transformation, we only need to change the analog frequency term in the formulas obtained before for the Butterworth low-pass analog filter.
- It is important to recognize that when finding the minimal order N and the half-power relation the value of K is not used. This constant is only important in the final step where the analog filter is transformed into the discrete filter using the bilinear transformation.
- When considering that $K = 2/T_s$ depends on T_s, one might think that a small value for T_s improves the design, but that is not the case. Given that the analog frequency is related to the discrete frequency as

$$\Omega = \frac{2}{T_s} \tan\left(\frac{\omega}{2}\right) \qquad (11.30)$$

for a given value of ω if we choose a small value of T_s the specified analog frequency would increase, and if we choose a large value of T_s the analog frequency would decrease. In fact, in the above equation we can only choose either Ω or T_s. To avoid this ambiguity, we ignore the connection of K with T_s and concentrate on K.
- An appropriate value for K for the Butterworth design is obtained by connecting the normalized half-power frequency $\Omega'_{hp} = 1$ in the analog domain with the corresponding frequency ω_{hp} in the discrete-domain. This allows us to go from the discrete-domain specifications *directly* to the analog normalized frequency specifications. Thus, we map the normalized half-power frequency $\Omega'_{hp} = 1$ into the discrete half-power frequency ω_{hp}, by means of K_b.
- Once the analog filter $H_N(s)$ is obtained, using the bilinear transformation with the K_b we transform $H_N(s)$ into a discrete filter

$$H_N(z) = H_N(s) \Big|_{s=K_b \frac{z-1}{z+1}}$$

- The filter parameters (N, ω_{hp}) can also be obtained directly from the discrete loss function

$$\alpha(e^{j\omega}) = 10 \log \left[1 + (\tan(0.5\omega)/\tan(0.5\omega_{hp}))^{2N}\right] \qquad (11.31)$$

and the loss specifications

$$0 \le \alpha(e^{j\omega}) \le \alpha_{max} \qquad 0 \le \omega \le \omega_p$$

$$\alpha(e^{j\omega}) \ge \alpha_{min} \qquad \omega \ge \omega_{st}$$

just as we did in the continuous case. The results coincide with those where we replace the warping frequency relation.

■ Example 11.6

The analog signal

$$x(t) = \cos(40\pi t) + \cos(500\pi t)$$

is sampled using the Nyquist frequency and processed with a discrete filter $H(z)$ that is obtained from a second-order, high-pass analog filter

$$H(s) = \frac{s^2}{s^2 + \sqrt{2}s + 1}$$

The discrete-time output $y[n]$ is then converted into analog. Apply MATLAB's bilinear function to obtain the discrete filter with half-power frequencies $\omega_{hp} = \pi/2$. Use MATLAB to plot the poles and the zeros of the discrete filter in the z-plane and the corresponding magnitude and phase responses. Use the function plot to plot the sampled input and the filter output and consider these approximations to the analog signals. Change the frequency scale of the discrete filter into f in hertz and indicate what is the corresponding half-power frequency in hertz.

Solution

The coefficients of the numerator and denominator of the discrete filter are found from $H(s)$ using the MATLAB function bilinear. The input F_s in this function equals $K_b/2$ where K_b corresponds to the transformation of the discrete half-power frequency ω_{hp} into the normalized analog half-power frequency $\Omega_{hp} = 1$. The following script is used.

```
%%%%%%%%%%%%%%%%%
% Example 11.6
%%%%%%%%%%%%%%%%%
b = [1 0 0]; a = [1 sqrt(2) 1]; % coefficients of analog filter
whp = 0.5 * pi; % desired half-power frequency of discrete filter
Kb = 1/tan(whp/2); Fs = Kb/2; [num, den]=bilinear(b,a,Fs); % bilinear transformation
Ts = 1/500;  % sampling period
n = 0:499; x1 = cos(2 * pi * 20 * n * Ts)+cos(2 * pi * 250 * n * Ts);  % sampled signal
zplane(num, den) % poles/zeros of discrete filter
[H,w] = freqz(num,den); % frequency response of discrete filter
phi = unwrap(angle(H)); %  unwrapped phase of discrete filter
y = filter(num,den,x1); % output of discrete filter with input x1
```

We find the transfer function of the discrete filter to be

$$H(z) = \frac{0.2929(1 - z^{-1})^2}{1 + 0.1715z^{-2}}$$

The poles and the zeros of $H(z)$ can be found with the MATLAB function roots and plotted with zplane. The frequency response is obtained using freqz. To have the frequency scale in hertz we consider that $\omega = \Omega T_s$, letting $\Omega = 2\pi f$, then

$$f = \frac{\omega}{2\pi T_s} = \left(\frac{\omega}{\pi}\right)\left(\frac{f_s}{2}\right)$$

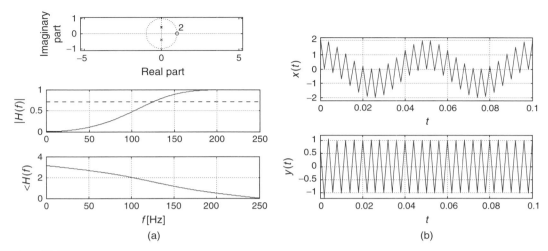

FIGURE 11.12

Bilinear transformation of a high-pass analog filter into a discrete filter with half-power frequencies $\omega_{hp} = \pi/2$ or $f_{hp} = 125$ Hz. (a) Poles and zeros and magnitude and phase responses of the discrete filter. (b) The analog input and output obtained using the MATLAB function `plot` to interpolate the sampled signal $x(nT_s)$ and the output of the discrete filter $y(nT_s)$ into $x(t)$ and $y(t)$.

so we multiply the normalized discrete frequency ω/π by $f_s/2 = 250$, resulting in a maximum frequency of 250 Hz. The half-power frequency in hertz is thus 125 Hz. The magnitude and the phase responses of $H(z)$ are shown in Figure 11.12. Notice that phase is approximately linear in the passband, despite the fact that no phase specifications are considered.

Since the maximum frequency of $x(t)$ is 250 Hz we choose $T_s = 1/500$. As a high-pass filter, when we input $x(nT_s)$ into $H(z)$ its low-frequency component $\cos(40\pi n T_s)$ is attenuated. The input and corresponding output of the filter are shown in Figure 11.12. ■

Remarks

- *The design is simplified by giving a desired half-power frequency ω_{hp}, as we then only need to calculate the order of the filter by using the stopband constraint. In this case, by setting $\alpha_{max} = 3$ dB and $\omega_p = \omega_{hp}$ one could also use the upper equation in Equation (11.28) to find N.*
- *A very important consequence of using the bilinear transformation is that the resulting transfer function $H_N(z)$ is guaranteed to be BIBO stable. This transformation maps the poles of a stable filter $H_N(s)$ in the left-hand s-plane into poles inside the unit circle corresponding to $H_N(z)$, making the discrete filter stable.*
- *The bilinear transformation creates a pole and a zero in the z-plane for each pole in the s-plane. Analytic calculation of the poles of $H_N(z)$ is not as important as in the analog case. The MATLAB function zplane can be used to plot its poles and zeros, and the function roots can be used to find the values of the poles and zeros.*

■ Applying the bilinear transformation by hand to filters of order higher than 2 is cumbersome. When doing so, $H_N(s)$ should be expressed as a product or sum of first- and second-order transfer functions before applying the bilinear transformation to each. That is, we express $H_N(s)$ as

$$H_N(s) = \prod_i H_{N_i}(s) \quad or$$

$$= \sum_\ell H_{N_\ell}(s)$$

where $H_{Ni}(s)$ or $H_{N\ell}(s)$ are first- or second-order functions with real coefficients. Applying the bilinear transformation to each of the $H_{Ni}(s)$ or $H_{N\ell}(s)$ components to obtain $H_{Ni}(z)$ and $H_{N\ell}(z)$, the discrete filter becomes

$$H_N(z) = \prod_i H_{N_i}(z) \quad or$$

$$= \sum_\ell H_{N_\ell}(z)$$

■ Since the resulting filter has normalized magnitude, a specified dc gain can be attained by multiplying $H_N(z)$ by a constant value G so that $|GH(e^{j0})|$ equals the desired dc gain.

■ Example 11.7

The specifications of a low-pass discrete filter are:

$$\omega_p = 0.47\pi \text{ (rad)} \qquad \alpha_{max} = 2 \text{ dB}$$
$$\omega_{st} = 0.6\pi \text{ (rad)} \qquad \alpha_{min} = 9 \text{ dB}$$
$$\alpha(e^{j0}) = 0 \text{ dB}$$

Use MATLAB to design a discrete low-pass Butterwoth filter by means of the bilinear transformation.

Solution

Since the frequency specifications are in radians, we use these values directly in the MATLAB function buttord, which for inputs α_{max}, α_{min}, ω_p/π, and ω_{st}/π, provides the minimal order N and the half-power frequency ω_{hp}/π of the filter. Notice that the input and the output frequencies are normalized (i.e., divided by π). With the outputs of buttord as inputs of the function butter we obtain the coefficients of the numerator and the denominator of the designed filter $H(z) = B(z)/A(z)$. The function roots is used to find the poles and the zeros of $H(z)$, while zplane plots them. The magnitude and the phase responses are found using freqz aided by the functions abs, angle, and unwrap. Notice that butter obtains the normalized analog filter and transforms it using the bilinear transformation. The script used is as follows.

```
%%%%%%%%%%%%%%%%%
% Example 11.7
%%%%%%%%%%%%%%%%%
% LP Butterworth
alphamax = 2; alphamin = 9; % loss specifications
wp = 0.47;ws = 0.6; % passband and stopband frequencies
[N,wh] = buttord(wp,ws,alphamax,alphamin) % minimal order, half-power frequency
[b,a] = butter(N,wh); % coefficients of designed filter
[H,w] = freqz(b,a);w = w/pi;N = length(H); % frequency response
spec1 = alphamax * ones(1,N); spec2 = alphamin * ones(1,N); % specification lines
hpf = 3.01 * ones(1,N); % half-power frequency line
disp('poles') % display poles
roots(a)
disp('zeros') % display zeros
roots(b)
alpha = -20 * log10(abs(H));   % loss in dB
```

The results of the design are shown in Figure 11.13. The order of the designed filter is $N = 3$ and the half-power frequency is $\omega_{hp} = 0.499\pi$. The poles are along the imaginary axis of the z-plane ($K_b = 1$) and there are three zeros at $z = -1$. The transfer function of the designed filter is

$$H(z) = \frac{0.166 + 0.497z^{-1} + 0.497z^{-2} + 0.166z^{-3}}{1 - 0.006z^{-1} + 0.333z^{-2} - 0.001z^{-3}}$$

Finally, to verify that the specifications are satisfied we plot the loss function $\alpha(e^{j\omega})$ along with three horizontal lines corresponding to $\alpha_{\max} = 2$ dB, 3 dB for the half-power frequency, and $\alpha_{\min} = 9$ dB.

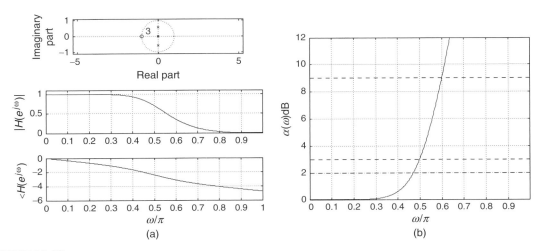

FIGURE 11.13

Design of low-pass Butterworth filter using MATLAB: (a) poles and zeros and magnitude and phase responses, and (b) verification of specifications using loss function $\alpha(\omega)$.

The crossings of these lines with the filter loss function indicate that at the normalized frequency of 0.6 the loss is 9 as desired, and that at the normalized frequency 0.47 the loss is less than 2 dB, so that the normalized half-power frequency is about 0.5. ∎

■ Example 11.8

In this example we consider designing a Butterworth low-pass discrete filter for processing an analog signal. The filter specifications are:

$$f_p = 2250 \text{ Hz} \text{ passband frequency}$$
$$f_{st} = 2700 \text{ Hz} \text{ stopband frequency}$$
$$f_s = 9000 \text{ Hz} \text{ sampling frequency}$$
$$\alpha_1 = -18 \text{ dB} \text{ dc loss}$$
$$\alpha_2 = -15 \text{ dB} \text{ loss in passband}$$
$$\alpha_3 = -9 \text{ dB} \text{ loss in stopband}$$

Solution

The specifications are not normalized (see Figure 11.14). Normalizing them, we have that:

$$\alpha(e^{j0}) = -18 \text{ dB}$$
$$\alpha_{\max} = \alpha_2 - \alpha_1 = 3 \text{ dB}$$
$$\alpha_{\min} = \alpha_3 - \alpha_1 = 9 \text{ dB}$$

and

$$\omega_p = \frac{2\pi f_{hp}}{f_s} = 0.5\pi$$
$$\omega_{st} = \frac{2\pi f_{st}}{f_s} = 0.6\pi$$

Note that $\omega_p = \omega_{hp}$ since the difference in the losses at dc and at ω_p is 3 dB.

FIGURE 11.14

Loss specifications for a discrete low-pass filter for processing an analog signal.

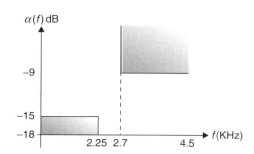

Since the sample period is

$$T_s = 1/f_s = (1/9) \times 10^{-3} \text{ sec/sample} \quad \Rightarrow \quad K_b = \cot(\pi f_{hp} T_s) = 1$$

Given that the half-power frequency is known, only the minimal order of the filter is needed. The loss function for the Butterworth filter is then

$$\alpha(e^{j\omega}) = 10 \log_{10} \left(1 + \left[\frac{\tan(0.5\omega)}{\tan(0.5\omega_{hp})} \right]^{2N} \right) = 10 \log_{10}(1 + (\tan(0.5\omega))^{2N})$$

since $0.5\omega_{hp} = \pi/4$. For $\omega = \omega_{st}$, letting $\alpha(e^{j\omega_{st}}) = \alpha_{min}$, solving for N we get

$$N = \frac{\log_{10}(10^{0.1\alpha_{min}} - 1)}{2 \log_{10}(\tan(0.5\omega_{st}))}$$

Replacing α_{min} and ω_{st} we obtain that the minimal order is $N = 4$. The MATLAB script used in the design is as follows.

```
%%%%%%%%%%%%%%%%%%%%%%%%%%%%%%%%%%%
% Example 11.8---filtering of analog signal
%%%%%%%%%%%%%%%%%%%%%%%%%%%%%%%%%%%
wh = 0.5*pi; ws = 0.6*pi; alphamin = 9; Fs = 9000; % filter specifications
N = log10((10^(0.1*alphamin)-1))/(2*log10(tan(ws/2)/tan(wh/2)));N = ceil(N)
[b,a] = butter(N,wh/pi);
[H,w] = freqz(b,a);w = w/pi;N = length(H);f = w*Fs/2;
alpha0 = -18;
G = 10^(-alpha0/20);H = H*G;
spec2 = alpha0 + alphamin*ones(1,N);
hpf = alpha0 + 3.01*ones(1,N);
disp('poles'); p = roots(a)
disp('zeros'); z = roots(b)
alpha = -20*log10(abs(H));
```

The transfer function of the discrete filter is found as

$$H(z) = H_1(z)H_2(z)$$

where each $H_i(z)$, $i = 1, 2$ is the result of applying the bilinear transformation to $H_i(s)$, $i = 1, 2$ formed by pairs of complex-conjugate poles of the analog low-pass Butterworth. To ensure that the loss is -18 dB at $\omega = 0$, we included a gain G in the numerator so that

$$H'(z) = GH(z) = \frac{G(z+1)^4}{10.61(z^2 + 0.45)(z^2 + 0.04)}$$

satisfies the loss specifications. Notice that when $K_b = 1$, the poles are imaginary. The dc gain G of the filter is found from the dc loss $\alpha(e^{j0}) = -18 = -20 \log_{10} G$ as $G = 7.94$.

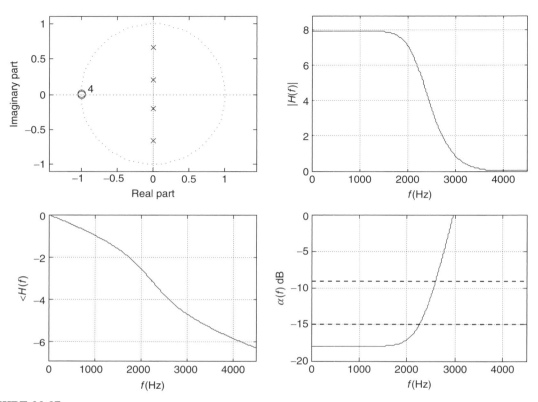

FIGURE 11.15
Low-pass filter for filtering of an analog signal.

Since this filter is used to filter an analog signal the frequency scale of the magnitude and the phase responses of the filter is given in hertz (see Figure 11.15). To verify that the specifications are satisfied the loss function is plotted and compared with the losses corresponding to f_{hp} and f_{st}. The loss at $f_{hp} = 2250\,\text{Hz}$ coincides with the dc loss plus 3 dB, and the loss at $f_{st} = 2700\,\text{Hz}$ is above the specified value. ∎

11.4.3 Design of Chebyshev Low-Pass Discrete Filters
The design of Chebyshev low-pass filters is very similar to the design of Butterworth low-pass filters.

The constant K_c of the bilinear transform for the Chebyshev filter is calculated by transforming the normalized pass frequency $\Omega'_p = 1$ into the discrete frequency ω_p:

$$K_c = \frac{1}{\tan(0.5\omega_p)}$$

(11.32)

Replacing

$$\Omega' = \frac{\Omega}{\Omega_p} = K_c \tan(0.5\omega) = \frac{\tan(0.5\omega)}{\tan(0.5\omega_p)} \qquad (11.33)$$

into the magnitude-squared function for the Chebyshev analog filter it yields the magnitude-squared function of the discrete Chebyshev low-pass filter,

$$|H_N(e^{j\omega})|^2 = \frac{1}{1 + \varepsilon^2 C_N^2(\tan(0.5\omega)/\tan(0.5\omega_p))} \qquad (11.34)$$

where $C(.)$ are the Chebyshev polynomials of the first kind encountered before in the analog design. The ripple parameter remains the same as in the analog design (since it does not depend on frequency):

$$\varepsilon = (10^{0.1\alpha\,\max} - 1)^{1/2} \qquad (11.35)$$

while using the warping relation between the continuous and discrete frequencies gives that the minimal order of the filter is

$$N \geq \frac{\cosh^{-1}[(10^{0.1\alpha\,\min} - 1)/(10^{0.1\alpha\,\max} - 1)]^{1/2}}{\cosh^{-1}[\tan(0.5\omega_{st})/\tan(0.5\omega_p)]} \qquad (11.36)$$

and that the half-power frequency can be found to be

$$\omega_{hp} = 2\tan^{-1}\left[\tan(0.5\omega_p)\cosh\left(\frac{1}{N}\cosh^{-1}\left(\frac{1}{\varepsilon}\right)\right)\right] \qquad (11.37)$$

After calculating these parameters, the transfer function of the Chebyshev discrete filter is found by transforming the Chebyshev analog filter of order N into a discrete filter using the bilinear transformation

$$H_N(z) = H_N(s)|_{s=K_c(1-z^{-1})/(1+z^{-1})} \qquad (11.38)$$

Remarks

■ *Just as with the Butterworth filter, the equations for the filter parameters (N, ω_{hp}) can be obtained from the analog formulas by substituting*

$$\frac{\Omega_{st}}{\Omega_p} = K_c \tan(0.5\omega_{st}) = \frac{\tan(0.5\omega_{st})}{\tan(0.5\omega_p)}$$

■ *The filter parameters $(N, \omega_{hp}, \varepsilon)$ can also be found from the loss function, obtained from the discrete Chebyshev squared magnitude,*

$$\alpha(e^{j\omega}) = 10\log_{10}\left[1 + \varepsilon^2 C_N^2\left(\frac{\tan(0.5\omega)}{\tan(0.5\omega_p)}\right)\right] \qquad (11.39)$$

This is done by following a similar approach to the one in the analog case.

- *Like in the discrete Butterworth, for Chebyshev filters the dc gain (i.e., gain at $\omega = 0$) can be set to any desired value by allowing a constant gain G in the numerator such that*

$$|H_N(e^{j0})| = |H_N(1)| = G\frac{|N(1)|}{|D(1)|} = \text{desired gain} \tag{11.40}$$

- *MATLAB provides two functions to design Chebyshev filters. The function* cheby1 *is for designing the filters covered in this section, while* cheby2 *is to design filters with a flat response in the passband and with ripples in the stopband. The order of the filter is found using* cheb1ord *and* cheb2ord. *The functions* cheby1 *and* cheby2 *will give the filter coefficients.*

■ Example 11.9

Consider the design of two low-pass Chebyshev filters. The specifications for the first filter are:

$$\alpha(e^{j0}) = 0 \text{ dB}$$

$$\omega_p = 0.47\pi \text{ rad} \qquad \alpha_{\max} = 2 \text{ dB}$$

$$\omega_{st} = 0.6\pi \text{ rad} \qquad \alpha_{\min} = 9 \text{ dB}$$

For the second filter, let $\omega_p = 0.48\pi$ rad and keep the other specifications. Determine the half-power frequency of the two filters. Use MATLAB for the design.

Solution

We obtained in Example 11.7 a third-order Butterworth low-pass filter that satisfies the specifications of the first filter. According to the results in this example a second-order Chebyshev filter satisfies the same specifications. It is always so that a Chebyshev filter satisfies the same specifications as a Butterworth with a lower order. For the second filter we narrow the transition band by 0.01π radians, and so the order increases by one. The following is the script for the design of the two filters.

```
%%%%%%%%%%%%%%%%%%%%%%%%%%%
% Example 11.9---LP Chebyshev
%%%%%%%%%%%%%%%%%%%%%%%%%%%%%%
alphamax = 2;  alphamin = 9; % loss specs
figure(1)
for i = 1:2,
  wp = 0.47 + (i-1) * 0.01; ws = 0.6; % normalized frequency specs
  [N,wn] = cheb1ord(wp,ws,alphamax,alphamin)
  [b,a] = cheby1(N,alphamax,wn);
  wp = wp * pi;
  % magnitude and phase
  [H,w] = freqz(b,a); w = w/pi; M = length(H);H = H/H(1);
  % to verify specs
  spec0 = zeros(1,M); spec1 = alphamax * ones(1,M) * (-1)^(N+1);
  spec2 = alphamin * ones(1,M);
  alpha = -20 * log10(abs(H));
```

```
hpf = (3.01 + alpha(1)) * ones(1,M);
% epsilon and half-power frequency
epsi = sqrt(10^(0.1 * alphamax)-1);
whp = 2 * atan(tan(0.5 * wp) * cosh(acosh(sqrt(10^(0.1 * 3.01)-1)/epsi)/N));
whp = whp/pi
% plotting
subplot(221); zplane(b,a)
subplot(222)
plot(w,abs(H)); grid; axis([0 max(w) 0 1.1 * max(abs(H))])
subplot(223)
plot(w,unwrap(angle(H)));grid;
subplot(224)
plot(w,alpha);
hold on; plot(w,spec0,'r'); hold on; plot(w,spec1,'r')
hold on; plot(w,hpf,'k'); hold on
plot(w,spec2,'r'); grid; axis([0 max(w) 1.1 * min(alpha) 1.1 * (alpha(1) + 3)]);
hold off
figure(2)
end
```

The transfer function of the first filter is

$$H_1(z) = \frac{0.224 + 0.449z^{-1} + 0.224z^{-2}}{1 - 0.264z^{-1} + 0.394z^{-2}}$$

and its half-power frequency is $\omega_{hp} = 0.493\pi$ rad. The second-order filter has a transfer function of

$$H_2(z) = \frac{0.094 + 0.283z^{-1} + 0.283z^{-2} + 0.094z^{-3}}{1 - 0.691z^{-1} + 0.774z^{-2} - 0.327z^{-3}}$$

and a half-power frequency of $\omega_{hp} = 0.4902\pi$. The poles and the zeros as well as the magnitude and the phase responses of the two filters are shown in Figure 11.16. Notice the difference in the gain (or losses) in the passband of the two filters. In order for the dc gain to be unity, the gain in the even-order filter reaches values above 1, while the odd-order filter does the opposite.

The cut-off frequency given as output by cheb1ord and that cheby1 uses is ω_p. Since the half power is not given by cheb1ord, the half-power frequency of the filter is calculated using the minimal order N, the ripple factor ε, and the passband frequency ω_p in Equation (11.37). See the script. ∎

■ Example 11.10

Consider the following specifications of a filter that will be used to filter an acoustic signal:

$$\text{dc gain} = 10$$

$$\text{Half-power frequency: } f_{hp} = 4 \text{ KHz}$$

$$\text{Band-stop frequency: } f_{st} = 5 \text{ Khz, } \alpha_{\min} = 60 \text{ dB}$$

$$\text{Sampling frequency: } f_s = 20 \text{ KHz}$$

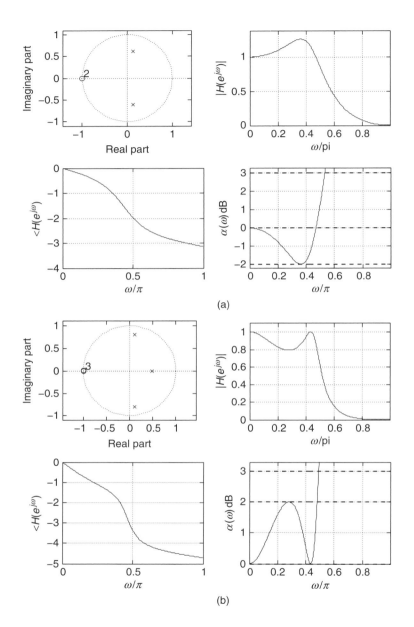

FIGURE 11.16

Two Chebyshev filters with different transition bands: (a) even-order filter for $\omega_p = 0.47\pi$, and (b) odd-order filter for $\omega_p = 0.48\pi$ (narrower transition band).

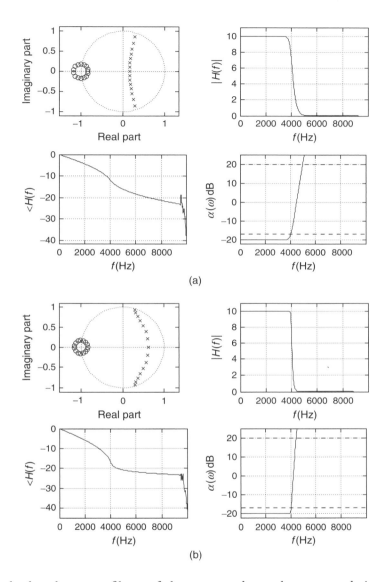

FIGURE 11.17

Equal-order ($N = 15$) (a) Butterworth and (b) Chebyshev filters for filtering of acoustic signal.

Design a Butterworth and a Chebyshev low-pass filters of the same order and compare their frequency responses.

Solution

The specifications of the discrete filter are:

$$\text{dc gain} = 10 \Rightarrow \alpha(e^{j0}) = -20 \, \text{dB}$$

$$\text{Half-power frequency: } \omega_{hp} = 2\pi f_{hp}(1/f_s) = 0.4\pi \, \text{rad}$$

$$\text{Band-stop frequency: } \omega_{st} = 2\pi f_{st}(1/f_s) = 0.5\pi \, \text{rad}$$

When designing the Butterworth filter we only need to find the minimal order N given that the half-power frequency is specified. We find that $N = 15$ satisfies the specifications. Using this value with the given discrete half-power frequency the function butter gives the coefficients of the filter. See Figure 11.17 for the results.

The design of the Chebyshev filter with order $N = 15$ and half-power frequency $\omega_{hp} = 0.4\pi$ cannot be done directly with the function cheby1, as we do not have the passband frequency ω_p. Using the equation to compute the half-power frequency, we solve for ω_p after we give a value to α_{max} (arbitrarily chosen to be 0.001 dB), which allows us to compute the ripple factor ε. See the script part corresponding to the Chebyshev design. The function cheby1 with inputs N, α_{max}, and ω_p gives the coefficients of the designed filter. Using the coefficients we plot the poles and the zeros, the magnitude and the phase responses, and the loss function, as shown in Figure 11.17. According to the loss function plots the Chebyshev filter displays a sharper response in the transition band than the Butterworth filter, as expected.

```
% Butterworth/Chebyshev filters for analog signal
wh = 0.4 * pi;ws = 0.5 * pi; alphamin = 40;Fs = 20000;
% Butterworth
N = log10((10^(0.1 * alphamin)-1))/(2 * log10(tan(ws/2)/tan(wh/2))); N = ceil(N);
% [b,a] = butter(N,wh/pi); % to get Butterworth filter get rid of ' % '
% Chebyshev
alphamax = 0.001;
epsi = sqrt(10^(0.1 * alphamax)-1);
% computation of wp for Chebyshev design
wp = 2 * atan(tan(0.5 * wh)/(cosh(acosh(sqrt(10^(0.1 * 3.01)-1)/epsi)/N))); wp = wp/pi;
[b,a] = cheby1(N,alphamax,wp);
% magnitude and phase
[H,w] = freqz(b,a);w = w/pi;M = length(H);f = w * Fs/2;
alpha0 = -20;H = H * 10;
% to verify specs
spec2 = alpha0 + alphamin * ones(1,M);
hpf = alpha0 + 3.01 * ones(1,M);
alpha = -20 * log10(abs(H));
Ha = unwrap(angle(H));
```

11.4.4 Rational Frequency Transformations

As indicated before, the conventional approach to filter design is to obtain first a prototype low-pass filter and then to transform it into different types of filters by means of frequency transformations. The magnitude specifications remain unchanged.

When using analog filters to design IIR discrete filters the frequency transformation could be done in two ways:

■ Transform a prototype low-pass analog filter into a desired analog filter, which in turn is converted into the desired discrete filter using the bilinear or other transformation.

- Design a prototype low-pass discrete filter and then transform it into the desired discrete filter [1,55].

The first approach has the advantage that the analog frequency transformations (See Chapter 6) are available and well understood. Its drawback appears when applying the bilinear transformation as it may cause undesirable warping in the higher frequencies. So the second approach will be used.

Given a prototype low-pass filter $H_{\ell p}(Z)$, we wish to transform it into a desired filter $H(z)$, which is typically another low-pass, band-pass, high-pass, or stopband filter. The transformation

$$G(z^{-1}) = Z^{-1} \tag{11.41}$$

should preserve the rationality and the stability of the low-pass prototype. Accordingly, $G(z^{-1})$ should

- Be rational to preserve the rationality.
- Map the inside of the unit circle in the Z-plane into the inside of the unit circle in the z-plane to preserve stability.
- Map the unit circle $|Z| = 1$ into the unit circe $|z| = 1$.

If $Z = Re^{j\theta}$ and $z = re^{j\omega}$, the third condition on $G(z^{-1})$ corresponds to

$$G(e^{-j\omega}) = |G(e^{-j\omega})|e^{j\angle(G(e^{-j\omega}))} = \underbrace{1\, e^{-j\theta}}_{\text{unit circle in Z-plane}} \tag{11.42}$$

indicating that the frequency transformation $G(z^{-1})$ has the characteristics of an all-pass filter, with magnitude $|G(e^{-j\omega})| = 1$ and phase $\angle G(e^{-j\omega}) = -\theta$.

Using the general form of the transfer function of an all-pass filter (ratio of two equal-order polynomials with poles and zeros being the conjugate inverse of each other), we obtain the general form of the rational transformation as

$$Z^{-1} = G(z^{-1}) = K \prod_k \frac{z^{-1} - \alpha_k}{1 - \alpha_k^* z^{-1}} \tag{11.43}$$

where $|\alpha_k| < 1$ and K is ± 1. The values of K and $\{\alpha_k\}$ are obtained from the prototype and the desired filters.

Low-Pass to Low-Pass Transformation

We wish to obtain the transformation $Z^{-1} = G(z^{-1})$ to convert a prototype low-pass filter into a different low-pass filter. The all-pass transformation should be able to expand or contract the frequency support of the prototype low-pass filter but keep its order. Thus, it should be a ratio of linear transformations,

$$Z^{-1} = K \frac{z^{-1} - \alpha}{1 - \alpha z^{-1}} \tag{11.44}$$

for some parameters K and α. Since the zero frequency in the Z-plane is to be mapped into the zero frequency in the z-plane, we let $Z = z = 1$ in the transformation to get $K = 1$. To obtain α, we let

$Z = e^{j\theta}$ and $z = e^{j\omega}$ in Equation (11.44) to obtain

$$e^{-j\theta} = \frac{e^{-j\omega} - \alpha}{1 - \alpha e^{-j\omega}} \tag{11.45}$$

The value of α, in Equation (11.44) that maps the cut-off frequency θ_p of the prototype into the desired cut-off frequency ω_d (see Figure 11.18(a)), is found from Equation (11.45) as follows. First, we have that

$$\alpha = \frac{e^{-j\omega} - e^{-j\theta}}{1 - e^{-j(\theta+\omega)}} = \frac{e^{-j\omega} - e^{-j\theta}}{2je^{-j0.5(\theta+\omega)} \sin((\theta+\omega)/2)} = \frac{\sin((\theta-\omega)/2)}{\sin((\theta+\omega)/2)}$$

and then replacing θ and ω by θ_p and ω_d gives

$$\alpha = \frac{\sin((\theta_p - \omega_d)/2)}{\sin((\theta_p + \omega_d)/2)} \tag{11.46}$$

Notice that if the prototype filter coincides with the desired filter (i.e., $\theta_p = \omega_d$), then $\alpha = 0$, and the transformation is $Z^{-1} = z^{-1}$. For different values of α between 0 and 1 the transformation shrinks the

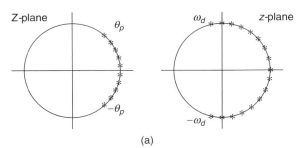

(a)

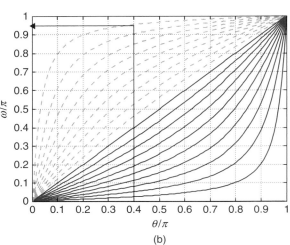

FIGURE 11.18
(a) Frequency transformation from a prototype low-pass filter with cut-off frequency θ_p into a low-pass filter with desired cut-off frequency ω_d.
(b) Mapping of θ into ω frequencies in low-pass to low-pass frequency transformation: the continuous lines correspond to $0 < \alpha \leq 1$, while the dashed lines correspond to values $-1 \leq \alpha \leq 0$. The arrow shows the transformation of $\theta_p = 0.4\pi$ into $\omega_d \approx 0.95\pi$ when $\alpha = -0.9$.

(b)

support of the prototype low-pass filter, and conversely for $-1 \leq \alpha < 0$ the transformation expands the support of the prototype. (In Figure 11.18 the frequencies θ and ω are normalized to values between 0 and 1—that is, both are divided by π.)

Remarks

- *The low-pass to low-pass (LP-LP) transformation then consists in:*
  - *Given θ_p and ω_d, find the corresponding α value using Equation (11.46).*
  - *Use the found α in the transformation (Eq. 11.44) with $K = 1$.*
- *Even in the simple case of a low-pass to low-pass transformation, the relation between the frequencies θ and ω is highly nonlinear. Indeed, solving for $e^{j\omega}$ in the transformation (Eq. 11.45), we get*

$$
e^{-j\omega} = \left(\frac{e^{-j\theta} + \alpha}{1 + \alpha e^{-j\theta}} \right) \left(\frac{1 + \alpha e^{j\theta}}{1 + \alpha e^{j\theta}} \right) = \frac{e^{-j\theta} + 2\alpha + \alpha^2 e^{j\theta}}{1 + 2\alpha \cos(\theta) + \alpha^2}
$$

$$
= \underbrace{\frac{2\alpha + (1 + \alpha^2)\cos(\theta)}{1 + 2\alpha \cos(\theta) + \alpha^2}}_{A} - j\underbrace{\frac{(1 - \alpha^2)\sin(\theta)}{1 + 2\alpha \cos(\theta) + \alpha^2}}_{B}
$$

and since $e^{-j\omega} = \cos(\omega) - j\sin(\omega)$, comparing the two sides we find that $\cos(\omega) = A$ and $\sin(\omega) = B$ so that $\tan(\omega) = B/A$, and thus

$$
\omega = \tan^{-1}\left[\frac{(1 - \alpha^2)\sin(\theta)}{2\alpha + (1 + \alpha^2)\cos(\theta)} \right]
$$

which when plotted for different values of α gives Figure 11.18(b). These curves clearly show the mapping of a frequency θ_p into ω_d and the value of α needed to perform the correct transformation.

Low-Pass to High-Pass Transformation

The duality between low-pass and high-pass filters indicates that this transformation, like the LP-LP, should be linear in both numerator and denominator. Also notice that the prototype low-pass filter can be transformed into a high-pass filter with the same bandwidth, by changing Z^{-1} into $-Z^{-1}$. Indeed, complex poles or zeros $R_1 e^{\pm j\theta_1}$ of the low-pass filter are mapped into $-R_1 e^{\pm j\theta_1} = R_1 e^{j(\pi \pm \theta_1)}$ corresponding to a high-pass filter. For instance, a low-pass filter

$$
H(Z) = \frac{Z + 1}{Z - 0.5}
$$

with a zero at -1 and a pole at 0.5 becomes

$$
H_1(Z) = \frac{-Z + 1}{-Z - 0.5} = \frac{Z - 1}{Z + 0.5}
$$

with a zero at 1 and a pole at -0.5, which makes it a high-pass filter.

The low-pass to high-pass (LP-HP) transformation is then

$$
Z^{-1} = -\frac{z^{-1} - \alpha}{1 - \alpha z^{-1}}
$$

and to obtain α we replace θ_p by $\pi - \theta_p$ in Equation (11.46) to get:

$$\alpha = \frac{\sin(-(\theta_p + \omega_d)/2 + \pi/2)}{\sin((\theta_p - \omega_d)/2 + \pi/2)}$$

$$= \frac{-\sin((\theta_p + \omega_d)/2 - \pi/2)}{\sin((\theta_p - \omega_d)/2 + \pi/2)} = \frac{\cos((\theta_p + \omega_d)/2)}{\cos((\theta_p - \omega_d)/2)} \tag{11.47}$$

As before, θ_p is the cut-off frequency of the prototype low-pass filter and ω_d is the desired cut-off frequency of the high-pass filter.

When the low-pass and the high-pass filters have the same bandwidth, $\omega_d = \pi - \theta_p$, we have that $\theta_p + \omega_d = \pi$ and so $\alpha = 0$ giving as a transformation $Z^{-1} = -z^{-1}$, which transforms the low-pass prototype into a high-pass filter, both of the same bandwidth.

Low-Pass to Band-Pass and Band-Stop Transformations

By being linear in both the numerator and the denominator, the LP-LP and LP-HP transformations preserve the number of poles and zeros of the prototype filter. To transform a low-pass filter into a band-pass or into a band-stop filter, the number of poles and zeros must be doubled. For instance, if the prototype is a first-order low-pass filter (with real-valued poles and zeros) we need a quadratic, rather than a linear, transformation in both numerator and denominator to obtain band-pass or band-stop filters from the low-pass filter since band-pass or band-stop filters cannot be first-order filters.

The low-pass to band-pass (LP-BP) transformation is

$$Z^{-1} = -\frac{z^{-2} - bz^{-1} + c}{cz^{-2} - bz^{-1} + 1} \tag{11.48}$$

while the low-pass to band-stop (LP-BS) transformation is

$$Z^{-1} = \frac{z^{-2} - (b/k)z^{-1} - c}{-cz^{-2} - (b/k)z^{-1} + 1} \tag{11.49}$$

where

$$b = 2\alpha k/(k+1)$$

$$c = (k-1)/(k+1)$$

and

$$\alpha = \frac{\cos((\omega_{d2} + \omega_{d1})/2)}{\cos((\omega_{d2} - \omega_{d1})/2)}$$

$$k = \cot((\omega_{d2} - \omega_{d1})/2)\tan(\theta_p/2)$$

The frequencies ω_{d1} and ω_{d2} are the desired lower and higher cut-off frequencies in the band-pass and band-stop filters.

11.4.5 General IIR Filter Design with MATLAB

The following function buttercheby1 can be used to design low-pass, high-pass, band-pass, and stop-band Butterworth as well as Chebyshev filters. One important thing to remember when designing band-pass and band-stop filters is that the order of the low-pass prototype is half that of the desired filter.

```
function [b,a,H,w] = buttercheby1(lp_order,wn,BC,type)
%
%  Design of frequency discriminating filters
%  using Butterworth and Chebyshev methods, the bilinear transformation and
%  frequency transformations
%
%  lp_order : order of low-pass filter prototype
%  wn : vector containing the cut-off normalized frequency(ies)
%  (entries must be normalized)
%  BC: Butterworth (0) or Chebyshev1 (1)
%  type : type of filter desired
%     1 = low-pass
%     2 = high-pass
%     3 = band-pass
%     4 = stopband
%  [b,a] : numerator, denominator coefficients of designed filter
%  [H,w] : frequency response, frequency range
%  USE:
%  [b,a,H,w] = buttercheby1(lp_order,wn,BC,type)
 if BC == 0; % Butterworth filter
   if type == 1
   [b,a] = butter(lp_order,wn);     % lowpas
   elseif type == 2
   [b,a] = butter(lp_order,wn,'high');  % high-pass
   elseif type == 3
   [b,a] = butter(lp_order,wn);     % band-pass
   else
   [b,a] = butter(lp_order,wn,'stop'); % stopband
   end
[H,w] = freqz(b,a,256);
else % Chebyshev1 filter
R = 0.01;
if type == 1,
   [b,a] = cheby1(lp_order,R,wn);      % lowpas
   elseif type == 2,
   [b,a] = cheby1(lp_order,R,wn,'high');  % high-pass
   elseif type == 3,
   [b,a] = cheby1(lp_order,R,wn);      % band-pass
   else
```

```
    [b,a] = cheby1(lp_order,R,wn,'stop');  % stopband
end
[H,w] = freqz(b,a,256);
end
```

To illustrate the design of filters other than low-pass filters, consider the design of a Butterworth and Chebyshev band-pass and band-stop filter of order $N = 30$ and half-power frequencies $[0.4\pi, \ 0.6\pi]$. The following script is the design, which is shown in Figure 11.19.

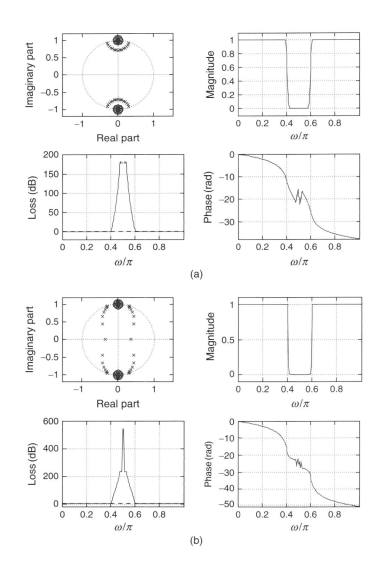

FIGURE 11.19

(a) Band-Stop Butterworth and (b) Chebyshev filters: (clockwise for each side from top left) poles and zeros, magnitude, phase frequency responses, and loss.

```
%%%%%%%%%%%%%%%%
% band-stop Butterworth
%%%%%%%%%%%%%%%%
figure(1)
[b1,a1] = buttercheby1(15,[0.4 0.6],0,4)
%%%%%%%%%%%%%%%%
% band-stop Chebyshev
%%%%%%%%%%%%%%%%
figure(2)
[b2,a2] = buttercheby1(15,[0.4 0.6],1,4)
```

There are other filters that can be designed with MATLAB, following a procedure similar to the previous cases. For instance, to design a band-pass elliptic filter with cut-off frequencies $[0.45\pi, 0.55\pi]$ of order 20 and with loss specifications of 0.1 and 40 dB in the passband and the stopband, we use the command shown below. Likewise, to design a high-pass filter using the cheby2 function we specify the order 10, the loss in the stopband, and the cut-off frequency 0.55π and indicate it is a high-pass filter. The results are shown in Figure 11.20.

```
%%%%%%%%%%%%%%%%
% Elliptic and Cheby2
%%%%%%%%%%%%%%%%
[b1,a1] = ellip(10,0.1,40,[0.45 0.55]);
[b2,a2] = cheby2(10,40, 0.55,'high');
```

11.5 FIR FILTER DESIGN

The design of FIR filters is typically discrete. The specification of FIR filters is usually given in the time domain rather than in the frequency domain. FIR filters have three definite advantages: (1) stability, (2) possible linear phase, and (3) efficient implementation. Indeed, the poles of an FIR filter are at the origin of the z-plane; thus FIR filters are stable. An FIR filter can be designed to have linear phase, and since the input–output equation of an FIR filter is equivalent to a convolution sum, FIR filters are implemented using the Fast Fourier Transform (FFT). A minor disadvantage is the storage required—typically FIR filters have a large number of coefficients.

■ Example 11.11

A moving-average filter has an impulse response

$$h[n] = \frac{1}{M} \quad 0 \le n \le M - 1$$

and zero otherwise. The transfer function of this filter is

$$H(z) = \sum_{n=0}^{M-1} \frac{1}{M} z^{-n} = \frac{1}{M} \frac{z^M - 1}{z^{M-1}(z - 1)}$$

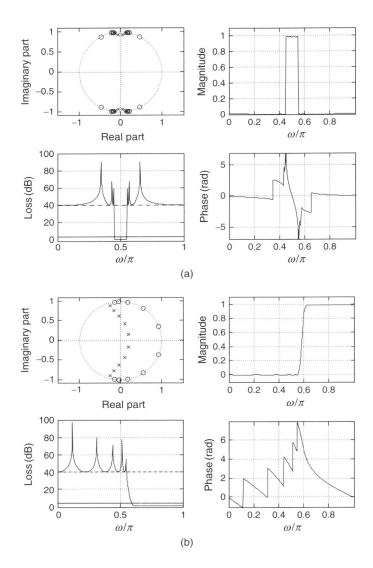

FIGURE 11.20

(a) Elliptic band-pass filter and (b) high-pass filter using cheby2: (clockwise for each side from top left) poles and zeros, magnitude, phase frequency responses, and loss.

Consider the stability of this filter, and determine if the phase of this filter is linear and what type of filter it is.

Solution

The impulse response $h[n]$ is absolutely summable given its finite length M; thus the filter is BIBO stable. Indeed, the apparent pole at $z = 1$, which would make the filter unstable, is canceled by a zero also at $z = 1$ (notice that $H(1)$ is $0/0$, according to the final expression above, indicating that a pole and a zero at $z = 1$ exist, but also from the sum $H(1) = 1$, so there are no poles at $z = 1$).

The remaining $M - 1$ poles of this filter are at the origin of the z-plane, making the filter stable. The zeros of $H(z)$ are complex numbers that make the numerator zero (i.e., $z^M - 1 = 0$, or $z_k = e^{j2\pi/M}$, for $k = 0, \ldots, M - 1$). When $k = 0$ the zero is 1, which cancels the pole at 1, so that

$$H(z) = \frac{(z - 1) \prod_{k=1}^{M-1}(z - e^{j2\pi/M})}{Mz^{M-1}(z - 1)} = \frac{\prod_{k=1}^{M-1}(z - e^{j2\pi/M})}{Mz^{M-1}}$$

To convince yourself of the pole–zero cancellation let $M = 3$, for which

$$H(z) = \frac{1}{3}\frac{z^3 - 1}{z^2(z - 1)} = \frac{1}{3}\frac{(z^2 + z + 1)(z - 1)}{z^2(z - 1)} = \frac{z^2 + z + 1}{3z^2}$$

showing the pole–zero cancellation.

Since the zeros of the filter are on the unit circle, the phase of this filter is not linear. Although the filter is considered a low-pass filter, it is of very poor quality in terms of its magnitude response. ∎

11.5.1 Window Design Method

The usual filter specifications of magnitude and linear phase can be translated into a time-domain specification (i.e., a desired impulse response) by means of the discrete-time Fourier transform. In this section, we will show how to design FIR filters using this specification with the *window method*. You will see that this is a trial-and-error method, as there is no measure of how close the designed filter is to the desired response, and that using different windows we obtain different results.

Let $H_d(e^{j\omega})$ be the desired frequency response of an ideal discrete low-pass filter. Assume that the phase of $H_d(e^{j\omega})$ is zero. The desired impulse response is given by the inverse discrete-time Fourier transform

$$h_d[n] = \frac{1}{2\pi} \int_{-\pi}^{\pi} H_d(e^{j\omega})e^{j\omega n} \, d\omega \tag{11.50}$$

which is noncausal and of infinite length. The filter

$$H_d(z) = \sum_{n=-\infty}^{\infty} h_d[n]z^{-n} \tag{11.51}$$

is thus not an FIR filter. To obtain an FIR filter that approximates $H_d(e^{j\omega})$ we need to window the impulse response $h_d[n]$ to get a finite length, and then delay the resulting windowed impulse response to achieve causality.

For an odd integer N, define

$$h_W[n] = h_d[n]w[n] = \begin{cases} h_d[n] & -(N-1)/2 \le n \le (N-1)/2 \\ 0 & \text{elsewhere} \end{cases}$$

where $w[n]$ is a *rectangular window*,

$$w[n] = \begin{cases} 1 & -(N-1)/2 \le n \le (N-1)/2 \\ 0 & \text{otherwise} \end{cases} \tag{11.52}$$

that causes the truncation of $h_d[n]$. The windowed impulse response $h_W[n]$ has a discrete-time Fourier transform of

$$H_W(e^{j\omega}) = \sum_{n=-(N-1)/2}^{(N-1)/2} h_W[n]e^{-j\omega n}$$

For a large value of N we have that $H_W(e^{j\omega})$ must be a good approximation of $H_d(e^{j\omega})$—that is,

$$|H_W(e^{j\omega})| \approx |H_d(e^{j\omega})|$$
$$\angle H_W(e^{j\omega}) = \angle H_d(e^{j\omega}) = 0$$

It is not clear how the value of N should be chosen—this is what we meant by this design is a trial-and-error method.

To make $H_W(z)$ a causal filter, we shift to the right the impulse response $h_W[n]$ by $(N-1)/2$ (assume N is chosen to be an odd number so that this division is an integer) samples to obtain

$$\hat{H}(z) = H_W(z)z^{-(N-1)/2} = \sum_{m=-(N-1)/2}^{(N-1)/2} h_W[m]z^{-(m+(N-1)/2)}$$

$$= \sum_{n=0}^{N-1} h_d[n-(N-1)/2]w[n-(N-1)/2]z^{-n}$$

after letting $n = m + (N-1)/2$. For a large value of N, we have

$$|\hat{H}(e^{j\omega})| = |H_W(e^{j\omega})e^{-j\omega(N-1)/2}| = |H_W(e^{j\omega})| \approx |H_d(e^{j\omega})|$$

$$\angle\hat{H}(e^{j\omega}) = \angle H_W(e^{j\omega}) - \frac{N-1}{2}\omega = -\frac{N-1}{2}\omega \qquad (11.53)$$

since $\angle H_W(e^{j\omega}) = \angle H_d(e^{j\omega}) = 0$. That is, the magnitude response of the FIR filter $\hat{H}(z)$ is approximately (depending on the value of N) the desired response and its phase response is linear. These results can be generalized as follows.

■ If the desired low-pass frequency response has a magnitude

$$|H_d(e^{j\omega})| = \begin{cases} 1 & -\omega_c \le \omega \le \omega_c \\ 0 & \text{otherwise} \end{cases} \qquad (11.54)$$

and a linear phase

$$\theta(\omega) = -\omega(N-1)/2$$

the corresponding impulse response is given by

$$h_d[n] = \begin{cases} \sin(\omega_c(n-\alpha))/(\pi(n-\alpha)) & n \ne \alpha \\ \omega_c/\pi & n = \alpha \end{cases} \qquad (11.55)$$

where $\alpha = (N-1)/2$. Using a window $w[n]$ of length N and centered at $(N-1)/2$, the windowed impulse response is $h[n] = h_d[n]w[n]$, and the designed FIR filter is

$$H(z) = \sum_{n=0}^{N-1} h[n]z^{-n}$$

- The design using windows is a trial-and-error procedure. Different trade-offs can be obtained by using various windows and various lengths of the windows.
- The symmetry of the impulse response $h[n]$ with respect to $(N-1)/2$, independent of whether this is an integer or not, guarantees the linear phase of the filter.

11.5.2 Window Functions

In the previous section, the windowed impulse response $h_W[n]$ was written as

$$h_W[n] = h_d[n]w[n]$$

where

$$w[n] = \begin{cases} 1 & -(N-1)/2 \le n \le (N-1)/2 \\ 0 & \text{otherwise} \end{cases} \tag{11.56}$$

is a rectangular window of length N. If we wish $H_W(e^{j\omega}) = H_d(e^{j\omega})$, we would need a rectangular window of infinite length so that the impulse responses $h_W[n] = h_d[n]$ (i.e., no windowing). This ideal rectangular window has a discrete-time Fourier transform

$$W(e^{j\omega}) = 2\pi \delta(\omega) \qquad -\pi \le \omega < \pi \tag{11.57}$$

Since $h_W[n] = w[n]h_d[n]$, then $H_W(e^{j\omega})$ is the convolution of $H_d(e^{j\omega})$ and $W(e^{j\omega})$ in the frequency domain—that is,

$$H_W(e^{j\omega}) = \frac{1}{2\pi} \int_{-\pi}^{\pi} H_d(e^{j\theta})W(e^{j(\omega-\theta)})d\theta$$

$$= \int_{-\pi}^{\pi} H_d(e^{j\theta})\delta(\omega-\theta)d\theta = H_d(e^{j\omega})$$

Thus, for $N \to \infty$, the result of this convolution is $H_d(e^{j\omega})$, but if N is finite the convolution in the frequency domain would give a distorted version of $H_d(e^{j\omega})$. Thus, to obtain a good approximation of $H_d(e^{j\omega})$ using a finite window $w[n]$ the window must have a spectrum approximating that of the ideal rectangular window. That is, an impulse in frequency in $-\pi \le \omega < \pi$ as in Equation (11.57) with most of its energy concentrated in the low frequencies. The smoothness of the window makes this possible.

Examples of windows that are smoother than the rectangular window are:

Triangular or Barlett window: $w[n] = \begin{cases} 1 - \frac{2|n|}{N-1} & -(N-1)/2 \le n \le (N-1)/2 \\ 0 & \text{otherwise} \end{cases}$ (11.58)

Hamming window: $w[n] = \begin{cases} 0.54 + 0.46\cos(2\pi n/(N-1)) & -(N-1)/2 \le n \le (N-1)/2 \\ 0 & \text{otherwise} \end{cases}$ (11.59)

Kaiser window: This window has a parameter β that can be adjusted. It is given by

$$w[n] = \begin{cases} \frac{I_0\left(\beta\sqrt{1-(n/(N-1))^2}\right)}{I_0(\beta)} & -(N-1)/2 \le n \le (N-1)/2 \\ 0 & \text{otherwise} \end{cases}$$ (11.60)

where $I_0(x)$ is the zero-order Bessel function of the first kind, which can be computed by the series

$$I_0(x) = 1 + \sum_{k=1}^{\infty}\left(\frac{(0.5x)^k}{k!}\right)^2$$ (11.61)

When $\beta = 0$ the Kaiser window coincides with a rectangular window, since $I_0(0) = 1$. As β increases the window becomes smoother.

The above definitions are for windows symmetric with respect to the origin. Figures 11.21 and 11.22 show the causal rectangular, Barlett, Hamming, and Kaiser windows, and their magnitude spectra.

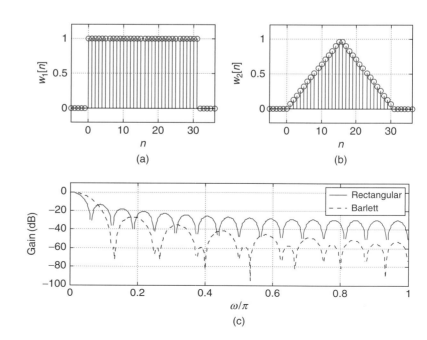

FIGURE 11.21

(a) Rectangular and (b) Barlett causal windows and (c) their spectra.

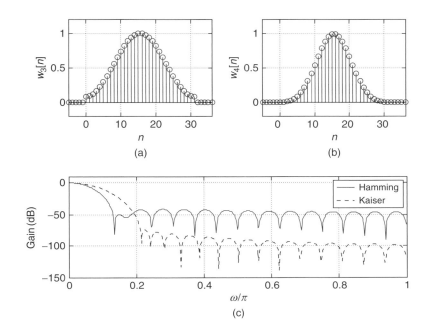

FIGURE 11.22

(a) Hamming and (b) Kaiser causal windows and (c) their spectra.

Given that the sidelobes for the Kaiser window have the largest loss, the Kaiser window is considered the best of these four, followed by the Hamming, the Bartlett, and the rectangular windows. Notice that the width of the first lobe is the widest for the Kaiser and the narrowest for the rectangular, as this width depends on the smoothness of the window.

Remarks *The linear phase is a result of the symmetry of the impulse response of the designed filter. It can be shown that if the impulse response h[n] of the FIR filter is even or odd symmetric with respect to the sample (N − 1)/2 (whether this is a integer or not) the filter has a linear phase.*

■ Example 11.12

Design a low-pass FIR filter with $N = 21$ to be used in filtering analog signals and that approximates the following ideal frequency response:

$$H_d(e^{j\omega}) = \begin{cases} 1 & 0 \le f \le 125 \text{ Hz} \\ 0 & \text{elsewhere in } 0 \le f \le f_s/2 \end{cases}$$

where $\omega = 2\pi f/f_s$ and $f_s = 1000$ Hz is the sampling rate. Use first a rectangular window, and then a Hamming window. Compare the designed filters.

Solution

The discrete frequency response is given by

$$H_d(e^{j\omega}) = \begin{cases} 1 & 0 \le \omega \le \pi/4 \text{ rad} \\ 0 & \text{elsewhere in } 0 \le \omega \le \pi \end{cases}$$

The desired impulse response is thus

$$h_d[n] = \frac{1}{2\pi} \int_{-\pi}^{\pi} H_d(e^{j\omega}) e^{j\omega n} d\omega = \frac{1}{2\pi} \int_{-\pi/4}^{\pi/4} e^{j\omega n} d\omega$$

$$= \begin{cases} \sin(\pi n/4)/(\pi n) & n \neq 0 \\ 0.25 & n = 0 \end{cases}$$

Using a rectangular window, the FIR filter is then of the form (the delay is $(N-1)/2 = 10$)

$$\hat{H}(z) = H_W(z) z^{-10} = \sum_{n=0}^{20} h_d[n-10] z^{-n}$$

$$= 0.25 z^{-10} + \sum_{n=0, n \neq 10}^{20} \frac{\sin(\pi(n-10)/4)}{\pi(n-10)} z^{-n}$$

The magnitude and the phase responses of this filter are shown in Figure 11.23 when we use a rectangular and a Hamming window.

The magnitude and the phase responses of the filter designed using the Hamming window is much improved over the one obtained using the rectangular window. Notice that the second lobe in the stopband for the Hamming window design is at about -50 dB while for the rectangular window design it is at about -20 dB, a significant difference. In both cases, the phase response is linear in the passband of the filter, corresponding to the impulse response $h[n]$ being symmetric with respect to the $(N-1)/2 = 10$ sample. ∎

■ Example 11.13

Design a high-pass filter of order 14 and a cut-off frequency of 0.2π using the Kaiser window. Use MATLAB.

Solution

Let $h_{lp}[n]$ be the impulse response of an ideal low-pass filter:

$$H_{lp}(e^{j\omega}) = \begin{cases} 1 & -\omega_c \leq \omega \leq \omega_c \\ 0 & \text{otherwise in } [-\pi, \pi) \end{cases}$$

According to the modulation property of the DTFT, we have that

$$2h_{lp}[n] \cos(\omega_0 n) \quad \Leftrightarrow \quad H_{lp}(e^{j(\omega+\omega_0)}) + H_{lp}(e^{j(\omega-\omega_0)})$$

If we let $\omega_0 = \pi$, then the right term gives a high-pass filter, and so $h_{hp}[n] = 2h_{lp}[n] \cos(\pi n) = 2(-1)^n h_{lp}[n]$ is the desired impulse response of the high-pass filter. The following script shows how to use the function fir to design the high-pass filter.

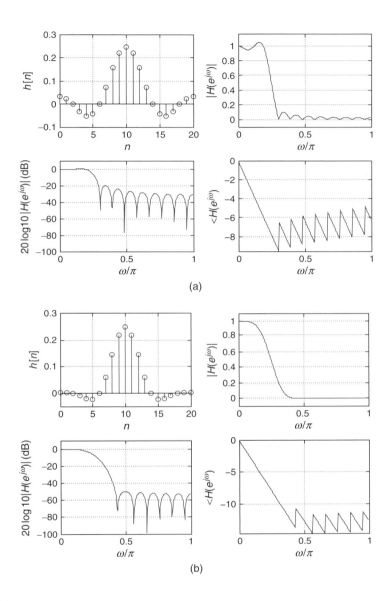

FIGURE 11.23

Low-pass FIR filters using (a) rectangular and (b) Hamming windows.

```
%%%%%%%%%%%%%%%%%
% Example 11.13---FIR filter from 'fir'
%%%%%%%%%%%%%%%%%%
M = 14;wc = 0.2;wo = 1;wind = 4;
[b] = fir(M,wc,wo,wind);
[H,w] = freqz(b,1,256);
```

The results are shown in Figure 11.24. Notice the symmetry of the impulse response with respect to $M/2 = 7$ gives a linear phase in the passband of the high-pass filter. The second lobe of the gain in dB is about -50 dB.

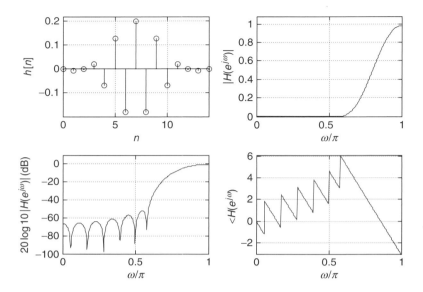

FIGURE 11.24

High-pass FIR filter design using Kaiser window.

The function fir can be used to design low-pass, high-pass, and band-pass FIR filters using differ-ent types of windows. When designing high-pass and band-pass FIRs, fir first designs a prototype low-pass filter and then uses the modulation property to shift it in frequency to a desired center frequency.

```
function [b] = fir(N,wc,wo,wind)
%
% FIR filter design using window method and
% frequency modulation
%
%  N : order of the FIR filter
%  wc : normalized cutoff frequency (between 0 and 1)
%  of low-pass prototype
%  wo : normalized center frequency (between 0 and 1)
%  of high-pass, bandpass filters
%  wind : type of window function
%    1 : rectangular
%    2 : hanning
%    3 : hamming
%    4 : kaiser
%  [b] : coefficients of designed filter
%
% USE:
%  [b] = fir(N,wc,wo,wind)
%
```

```
n = 0:N;
if wind == 1
window = boxcar(N + 1);
disp(' ***** RECTANGULAR WINDOW  *****')
elseif wind == 2
window = hanning(N + 1);
disp(' *****HANNING WINDOW *****')
elseif wind == 3
window = hamming(N + 1);
disp(' *****   HAMMING WINDOW *****')
else
window = kaiser(N + 1,4.55);
disp(' *****   KAISER WINDOW *****')
end
% calculation of ideal impulse response
den = pi * (n - N/2);
num = sin(wc*den);
% if N even, this prevents 0/0
if fix(N/2) == N/2,
num(N/2 + 1) = wc;
den(N/2 + 1) = 1;
end
b = (num./den). * window';
% frequency shifting
[H,w] = freqz(b,1,256); %% low-pass
if wo > 0 & wo < 1,
b = 2 * b.*cos(wo * pi * (n - N/2))/H(1);
elseif wo == 0,
   b = b/abs(H(1));
elseif wo == 1;
   b = b. * cos(wo * pi * (n - N/2));
end
```

MATLAB provides the function fir1 to design FIR filters with the window method. As expected, the results are identical for either fir1 and fir. The reason for writing fir is to simplify the code and to show how the modulation property can be used in the design of filters different from low-pass filters.   ∎

11.6 REALIZATION OF DISCRETE FILTERS

The realization of a discrete filter can be done in hardware or in software. In either case, the implementation of the transfer function $H(z)$ of a discrete filter requires delays, adders, and constant multipliers as actual hardware or as symbolic components. Figure 11.25 depicts the operation of each of these components as block diagrams.

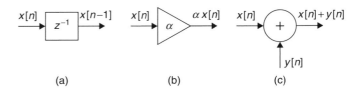

FIGURE 11.25
Block diagrams of different components used to realize discrete filters: (a) delay, (b) constant multiplier, and (c) adder.

In choosing a structure over another to realize a filter, two factors to consider are:

- *Computational complexity*, which relates to the number of operations (mainly multiplications and additions), but more importantly to the number of delays used. The aim is to reduce to a minimum the number of delays in the structure.
- *Quantization effects* or the representation of filter parameters using finite-length registers. The aim is to minimize quantization effects on parameters and on operations.

We will consider here the computational complexity of the structures seeking to obtain minimal realizations—that is, to optimize the number of delays used. The quantization effects are not considered.

11.6.1 Realization of IIR Filters

The structures commonly used to realize IIR filters are:

- Direct forms I and II
- Cascade
- Parallel

The direct forms represent the difference equation resulting from the transfer function of the IIR filter while attempting to minimize the number of delays. The cascade and parallel structures are based on the product or sum of first- and second-order filters to express the filter transfer function, which are in turn implemented using a direct form.

Direct Form Realizations

Given the transfer function of an IIR filter

$$H(z) = \frac{Y(z)}{X(z)} = \frac{\sum_{k=0}^{M-1} b_k z^{-k}}{1 + \sum_{k=1}^{N-1} a_k z^{-k}} \tag{11.62}$$

where $Y(z)$ and $X(z)$ are the Z-transforms of the output $y[n]$ and the input $x[n]$, the input–output relationship is given by the difference equation

$$y[n] = -\sum_{k=1}^{N-1} a_k y[n-k] + \sum_{k=0}^{M-1} b_k x[n-k] \tag{11.63}$$

The direct forms attempt to realize this equation with no more than $N-1$ delays.

Direct Form I

The *direct form I* is the implementation of the above difference Equation (11.63) as is, by means of delays, constant multipliers, and adders. Assuming the input $x[n]$ is available, then $M - 1$ delays are needed to generate the delayed inputs $\{x[n - k]\}$ for $k = 1, \ldots, M - 1$. Likewise, the output components require additional $N - 1$ delays. Thus, a direct form I realization requires $M + N - 2$ delays for an $(N - 1)$th-order difference equation. In terms of number of delays, the direct form I is the least efficient realization.

■ Example 11.14

Use the direct form I to realize the transfer function

$$H(z) = \frac{1 + 1.5z^{-1}}{1 + 0.1z^{-1}}$$

of a discrete filter.

Solution

The transfer function corresponds to a system with a first-order difference equation

$$y[n] = x[n] + 1.5x[n - 1] - 0.1y[n - 1]$$

so $M = N = 2$ and this equation can be realized as shown in Figure 11.26 with $M + N - 2 = 2$ delays.

The difference equation, and thus the transfer function, for this filter can be easily obtained from the realization. The above realization is nonminimal since it uses two delays to represent a first-order system.

The output of the above realization is seen to be

$$y[n] = -0.1y[n - 1] + x[n] + 1.5x[n - 1]$$

so that the transfer function is

$$H(z) = \frac{Y(z)}{X(z)} = \frac{1 + 1.5z^{-1}}{1 + 0.1z^{-1}}$$

by letting a delay be represented by z^{-1} in the z-domain. ■

FIGURE 11.26
Direct form I realization of
$H(z) = (1 + 1.5z^{-1})/(1 + 0.1z^{-1})$.

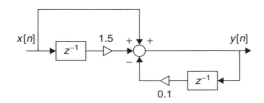

Remarks

- *In general, given a direct form I realization one can easily obtain the difference equation and consequently the transfer function of the filter from it.*
- *The* minimal realization *of qth-order discrete filter must use q delays. The direct form I is only capable of providing these minimal realizations for* all-pole *filters (i.e., when the numerator in Equation (11.62) is a constant), otherwise we need to use the direct form II to to obtain minimal realizations. If the transfer function has only poles, then it is possible to obtain a minimal realization with direct form I. Indeed, if*

$$H(z) = \frac{Y(z)}{X(z)} = \frac{b_0}{1 + \sum_{k=1}^{N-1} a_k z^{-k}} \tag{11.64}$$

where $Y(z)$ and $X(z)$ are the Z-transforms of the output $y[n]$ and the input $x[n]$, the input–output relationship is given by the difference equation

$$y[n] = -\sum_{k=1}^{N-1} a_k y[n-k] + b_0 x[n] \tag{11.65}$$

which only requires $N - 1$ delays for the output, and none for the input. This is a minimal realization of $H(z)$ as only $N - 1$ delays are needed.

Direct Form II
If the polynomials

$$B(z) = \sum_{k=0}^{M-1} b_k z^{-k} \quad \text{and} \quad A(z) = \sum_{k=0}^{N-1} a_k z^{-k} \qquad M \leq N$$

represent the numerator and denominator of the transfer function $H(z)$ of the filter we wish to realize, we have

$$H(z) = \frac{Y(z)}{X(z)} = \frac{B(z)}{A(z)}$$

where $X(z)$ and $Y(z)$ correspond to the Z-transforms of the input and of the output of the filter. We then have that

$$Y(z) = H(z)X(z) = B(z)\left[\frac{X(z)}{A(z)}\right] \tag{11.66}$$

Defining an output $w[n]$ with $W(z) = X(z)/A(z)$, corresponding to the second term in the last equation, we obtain an all-pole filter with transfer function

$$\frac{W(z)}{X(z)} = \frac{1}{A(z)} \tag{11.67}$$

The output $y[n]$ is then obtained as the inverse of

$$Y(z) = B(z)W(z) \tag{11.68}$$

By realizing the all-pole filter given in Equation (11.67), and using its output $w[n]$ in the realization of Equation (11.68), we minimize the number of delays used. The realization of Equation (11.68) does not require new delays, as the delayed $w[n]$'s are already available from the realization of the all-pole filter. Thus, the number of delays used corresponds to the order of the denominator $A(z)$, which is the order of the filter.

■ Example 11.15

Consider the same transfer function as in Example 11.14 to obtain a direct form II realization of it.

Solution

We have that

$$W(z) = \frac{X(z)}{1 + 0.1z^{-1}}$$

gives

$$w[n] = x[n] - 0.1w[n-1]$$

Now, according to Equation (11.68),

$$Y(z) = (1 + 1.5z^{-1})W(z)$$

which corresponds to the difference equation

$$y[n] = w[n] + 1.5w[n-1] \tag{11.69}$$

Notice that in Equation (11.69) $w[n]$ and $w[n-1]$ are already available, and thus there is no need for new delays in this step. The direct form II realization using only one delay is shown in Figure 11.27.

Obtaining the transfer function from the direct form II realization is not as obvious as from the direct form I. In this case, we need to obtain the transfer function corresponding to the all-pole filter first and use it to obtain the overall transfer function. From the above realization, we have

$$w[n] = x[n] - 0.1w[n-1]$$
$$y[n] = w[n] + 1.5w[n-1]$$

FIGURE 11.27

Direct form II realization of $H(z) = (1 + 1.5z^{-1})/(1 + 0.1z^{-1})$. This is a minimal realization of $H(z)$, which is a first-order system, as only one delay is used.

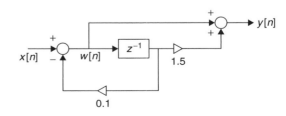

If we replace the first equation into the second we obtain an expression containing $w[n]$ and $w[n-2]$ and $x[n]$ so that we cannot express $y[n]$ directly in terms of the input. Instead, consider the Z-transforms of the above equation,

$$(1 + 0.1z^{-1})W(z) = X(z)$$

$$Y(z) = (1 + 1.5z^{-1})W(z) \tag{11.70}$$

Thus, we obtain from the top equation in Equation (11.70):

$$W(z) = \frac{X(z)}{1 + 0.1z^{-1}}$$

which when replaced in the bottom equation in Equation (11.70) gives

$$Y(z) = \frac{X(z)(1 + 1.5z^{-1})}{1 + 0.1z^{-1}}$$

giving the transfer function $H(z)$. ∎

Remarks

- *Direct form II is more advantageous than direct form I because of the consequence of using fewer delays. We will use direct form II to realize first- and second-order modules in the cascade and parallel realizations.*
- *The cascade and the parallel realizations will connect first- and second-order systems to realize a given transfer function. General direct form II realizations for a first and second-order filter with respective transfer functions*

$$H_1(z) = \frac{b_0 + b_1 z^{-1}}{1 + a_1 z^{-1}} \tag{11.71}$$

$$H_2(z) = \frac{b_0 + b_1 z^{-1} + b_2 z^{-2}}{1 + a_1 z^{-1} + a_2 z^{-2}} \tag{11.72}$$

are given in Figure 11.28. The coefficients of the above transfer functions are real. The realization of $H_1(z)$ is obtained by getting rid of the lower part of the realization (i.e., getting rid of the constant multipliers

FIGURE 11.28
Direct form II realization of first- and second-order filters (for the first-order filters let $a_2 = b_2 = 0$ thus eliminating the constant multipliers and the lower delay).

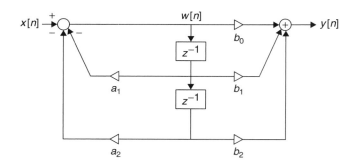

because $a_2 = b_2 - 0$ and the lower delay because it is not needed once these constant multiplier are deleted).

■ *Obtaining the transfer function from a direct form II realization is not as obvious as it is from a direct form I realization. For a given realization direct form II, we need to use the auxiliary variable $w[n]$ to obtain the transfer function from the realization. Instead of using the time-domain equations one should use the Z-transform equations to obtain the overall transfer function. For instance, consider the direct form II realization in Figure 11.28 for the second-order system. From the realization we obtain*

$$w[n] = x[n] - a_1 w[n-1] - a_2 w[n-2] \Rightarrow (1 + a_1 z^{-1} + a_2 z^{-2}) W(z) = X(z)$$

$$y[n] = b_0 w[n] + b_1 w[n-1] + b_2 w[n-2] \Rightarrow Y(z) = (b_0 + b_1 z^{-1} + b_2 z^{-2}) W(z)$$

Solving for $W(z)$ in the top equation and replacing it in the bottom equation will give $H_2(z)$.

Cascade Realization

The *cascade realization* is obtained by representing the given transfer function $H(z) = B(z)/A(z)$ as a product of first- and second-order filters $H_i(z)$ with real coefficients:

$$H(z) = \prod_i H_i(z) \tag{11.73}$$

Each transfer function $H_i(z)$ is realized by direct form II and cascaded. Different from the analog case, this cascade realization is not constrained by loading.

■ Example 11.16

Obtain a cascade realization of the filter with transfer function

$$H(z) = \frac{3 + 3.6z^{-1} + 0.6z^{-2}}{1 + 0.1z^{-1} - 0.2z^{-2}}$$

Solution

The poles of $H(z)$ are $z = -0.5$ and $z = 0.4$, and the zeros are $z = -1$ and $z = -0.2$, all of which are real. One way of obtaining the cascade realization is to express $H(z)$ as

$$H(z) = \left[\frac{3(1 + z^{-1})}{1 + 0.5z^{-1}} \right] \left[\frac{1 + 0.2z^{-1}}{1 - 0.4z^{-1}} \right]$$

If we let

$$H_1(z) = \frac{3(1 + z^{-1})}{1 + 0.5z^{-1}}$$

$$H_2(z) = \frac{1 + 0.2z^{-1}}{1 - 0.4z^{-1}}$$

Realizing $H_1(z)$ and $H_2(z)$ separately and then cascading them we obtain the realization for $H(z)$ shown in Figure 11.29.

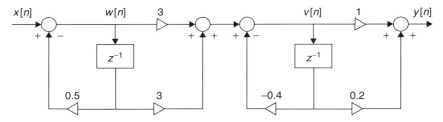

FIGURE 11.29

Cascade realization of $H(z) = (3 + 3.6z^{-1} + 0.6z^{-2})/(1 + 0.1z^{-1} - 0.2z^{-2})$.

It is also possible to express $H(z)$ as

$$H(z) = \underbrace{\left[\frac{1 + 0.2z^{-1}}{1 + 0.5z^{-1}}\right]}_{\hat{H}_1(z)} \underbrace{\left[\frac{3(1 + z^{-1})}{1 - 0.4z^{-1}}\right]}_{\hat{H}_2(z)}$$

which would give a different but equivalent realization of $H(z)$.

Since loading is not applicable, the product of the transfer functions always gives the overall transfer function. As LTI systems these realizations can be cascaded in different orders with the same result. ∎

■ Example 11.17

Obtain a cascade realization of

$$H(z) = \frac{1 + 1.2z^{-1} + 0.2z^{-2}}{1 - 0.4z^{-1} + z^{-2} - 0.4z^{-3}}$$

Solution

The zeros of $H(z)$ are $z = -1$ and $z = -0.2$, while its poles are $z = \pm j$ and $z = 0.4$. We can thus rewrite $H(z)$ as the following two equivalent expressions:

$$H(z) = \left[\frac{1 + z^{-1}}{1 + z^{-2}}\right] \left[\frac{1 + 0.2z^{-1}}{1 - 0.4z^{-1}}\right]$$

$$= \left[\frac{1 + 0.2z^{-1}}{1 + z^{-2}}\right] \left[\frac{1 + z^{-1}}{1 - 0.4z^{-1}}\right]$$

where the complex-conjugate poles give the denominator of the first filter. Realizing each of these components and cascading in any order would give different but equivalent representation of $H(z)$. Figure 11.30 shows the realization of the top form of $H(z)$. ∎

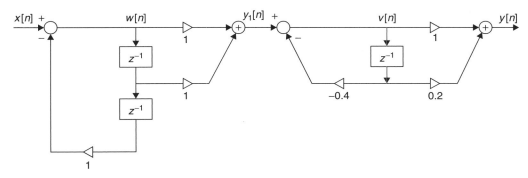

FIGURE 11.30
Cascade realization of $H(z) = [(1 + z^{-1})/(1 + z^{-2})] \; [(1 + 0.2z^{-1})/(1 - 0.4z^{-1})]$.

Parallel Realization

In this case the given transfer function $H(z)$ is represented as a partial fraction expansion,

$$H(z) = \frac{B(z)}{A(z)} = C + \sum_{i=1}^{r} H_i(z) \tag{11.74}$$

where C is a constant and the r filters $H_i(z)$ are first- or second-order systems with real coefficients that are implemented with the direct form II.

The constant C in the expansion is needed when the numerator (in positive powers of z) is of larger or equal order than the denominator. If the numerator is of larger order than the denominator, the filter is noncausal. To illustrate this, consider a first-order filter with a transfer function where the numerator is of second order (in terms of positive powers of z)

$$H(z) = \frac{Y(z)}{X(z)} = \frac{b_0 z^2 + b_1 z + b_2}{z + a_1} = \frac{b_0 z + b_1 + b_2 z^{-1}}{1 + a_1 z^{-1}}$$

The difference equation representing this system is

$$y[n] = -a_1 y[n - 1] + b_0 x[n + 1] + b_1 x[n] + b_2 x[n - 1]$$

requiring a future input $x[n + 1]$ to compute the present $y[n]$ (i.e., corresponding to a noncausal filter).

The cascade and parallel realizations are shown in Figure 11.31.

■ Example 11.18

Let

$$H(z) = \frac{3 + 3.6z^{-1} + 0.6z^{-2}}{1 + 0.1z^{-1} - 0.2z^{-2}} = \frac{3z^2 + 3.6z + 0.6}{z^2 + 0.1z - 0.2}$$

Obtain a parallel realization.

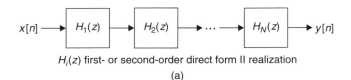

$H_i(z)$ first- or second-order direct form II realization

(a)

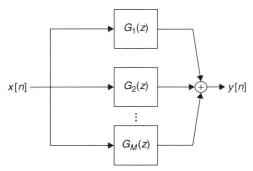

$G_i(z)$ first- or second-order direct form II realization

(b)

FIGURE 11.31

(a) Cascade and (b) parallel realizations of IIR filters.

Solution

The transfer function $H(z)$ is not proper rational, in either positive or negative powers of z, and the poles are $z = -0.5$ and $z = 0.4$. Thus, the transfer function can be expanded as

$$H(z) = A_1 + \frac{A_2}{1 + 0.5z^{-1}} + \frac{A_3}{1 - 0.4z^{-1}}$$

In this case we need A_1 because the numerator, in positive as well as in negative powers of z, is of the same order as the denominator. We then have

$$A_1 = H(z)|_{z=0} = -3$$

$$A_2 = H(z)(1 + 0.5z^{-1})|_{z^{-1}=-2} = -1$$

$$A_3 = H(z)(1 - 0.4z^{-1})|_{z^{-1}=2.5} = 7$$

Letting

$$H_1(z) = \frac{-1}{1 + 0.5z^{-1}}$$

$$H_2(z) = \frac{7}{1 - 0.4z^{-1}}$$

we obtain the parallel realization for $H(z)$ shown in Figure 11.32.   ∎

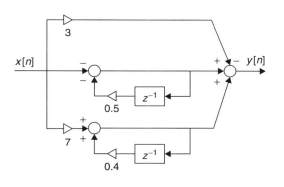

FIGURE 11.32
Parallel realization for $H(z) = (3 + 3.6z^{-1} + 0.6z^{-2})/(1 + 0.1z^{-1} - 0.2z^{-2})$.

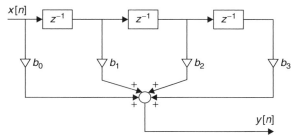

FIGURE 11.33
Direct form realization of FIR filter.

11.6.2 Realization of FIR Filters

The realization of FIR filters can be done using direct and cascade forms. Since these filters are nonrecursive, there are no different direct forms and there is no way to implement FIR filters in parallel.

The direct realization of an FIR filter consists in realizing the input–output equation using delays, constant multipliers, and summers. For instance, if the transfer function of an FIR filter is given by

$$H(z) = \sum_{k=0}^{M} b_k z^{-k} \qquad (11.75)$$

the Z-transform of the filter output can be written as

$$Y(z) = H(z)X(z)$$

where $X(z)$ is the Z-transform of the filter input. In the time domain we have

$$y[n] = \sum_{k=0}^{M} b_k x(n - k)$$

which can be realized as shown in Figure 11.33 in the case of $M = 3$.

Notice that M is the number of delays needed and that there are $M + 1$ taps, which has given the name of *tapped filters* to FIR filters realized this way.

The cascade realization of an FIR filter is based on the representation of $H(z)$ in Equation (11.75) as a cascade of first- and second-order filters—that is, we let

$$H(z) = \prod_{i=1}^{r} H_i(z)$$

where

$$H_i(z) = b_{oi} + b_{1i}z^{-1} \quad \text{or}$$
$$H_i(z) = b_{oi} + b_{1i}z^{-1} + b_{2i}z^{-2}$$

■ Example 11.19

Provide the cascade realization of an FIR filter with transfer function

$$H(z) = 1 + 3z^{-1} + 3z^{-2} + z^{-3}$$

Solution

The transfer function is factored as

$$H(z) = (1 + 2z^{-1} + z^{-2})(1 + z^{-1})$$

which can be realized as the cascade of two FIR filters,

$$y_1[n] = x[n] + x[n-1]$$
$$y[n] = y_1[n] + 2y_1[n-1] + y_1[n-2]$$

which are realized as shown in Figure 11.34.

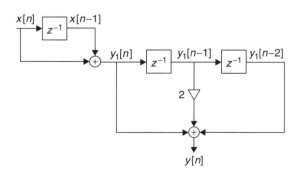

FIGURE 11.34
Cascade realization of FIR filter.

11.7 WHAT HAVE WE ACCOMPLISHED? WHERE DO WE GO FROM HERE?

In Chapter 6 and in this chapter you have been introduced to the most important application of linear time-invariant systems: filtering. The design and realization of analog and discrete filters gathers many practical issues in signals and systems. If you pursue this topic, you will see the significance, for instance, of passive and active elements, feedback and operational amplifiers, reactance functions, and frequency transformation in analog filtering. The design and realization of discrete filters brings together interesting topics such as quantization error and its effect on the filters, optimization methods for filter design, stabilization of unstable filters, etc. If you pursue filtering deeper, you will find that there is a lot more on filter design than what we have provided you in this chapter. A lot more. Also remember that MATLAB has a large number of tools to design and implement filters.

PROBLEMS

11.1. FIR filters: causality and phase—MATLAB

A three-point moving-average filter is of the form:

$$y[n] = \beta \left(\alpha x[n-1] + x[n] + \alpha x[n+1] \right)$$

where α and β are constants, and $x[n]$ is the input and $y[n]$ is the output of the filter.

(a) Determine the transfer function $H(z) = Y(z)/X(z)$ of the filter, and from it determine the frequency response $H(e^{j\omega})$ of the filter in terms of α and β.

(b) Find the values of α and β so that the dc gain of the filter is unity, and the filter has a zero phase. For $\alpha = 0.5$ and the corresponding value of β, sketch $H(e^{j\omega})$ and find the poles and zeros of $H(z)$ and plot them in the z-plane. Verify your results using MATLAB.

(c) Suppose we let $v[n] = y[n-1]$ be the output of a second filter. Is this filter causal? Find its transfer function $G(z) = V(z)/X(z)$. Use MATLAB to compute the unwrapped phase of $G(z)$ and to plot the poles and zeros of $G(z)$ and explain the relation between $G(z)$ and $H(z)$.

11.2. FIR and IIR filters: causality and zero phase—MATLAB

Let the filter $H(z)$ be the cascade of a causal filter with transfer function $G(z)$ and an anti-causal filter with transfer function $G(z^{-1})$, so that

$$H(z) = G(z)G(z^{-1})$$

(a) Suppose that $G(z)$ is an FIR filter with transfer function

$$G(z) = \frac{1}{3}(1 + 2z^{-1} + z^{-2})$$

Find the frequency response $H(e^{j\omega})$ and determine its phase.

(b) Determine the impulse response of the filter $H(z)$. Is $H(z)$ a causal filter? If not, would delaying its impulse response make it causal? Explain. What would be the transfer function of the causal filter?

(c) Use MATLAB to verify the unwrapped phase of $H(z)$ you obtained analytically, and to plot the poles and zeros of $H(z)$.

(d) How would you use the MATLAB function conv to find the impulse response of $H(z)$?

(e) Suppose then that $G(z) = 1/(1 - 0.5z^{-1})$. Find the filter $H(z) = G(z)G(z^{-1})$. Is this filter zero phase? If so, where are its poles and zeros? If you think of filter $H(z)$ as causal, is it BIBO stable?

11.3. FIR and IIR filters: symmetry of impulse response and linear phase—MATLAB

Consider two FIR filters with transfer functions

$$H_1(z) = 0.5 + 0.5z^{-1} + 2.2z^{-2} + 0.5z^{-3} + 0.5z^{-4}$$

$$H_2(z) = -0.5 - 0.5z^{-1} + 0.5z^{-3} + 0.5z^{-4}$$

(a) Find the impulse responses $h_1[n]$ and $h_2[n]$ corresponding to $H_1(z)$ and $H_2(z)$. Plot them carefully and determine the sample value for which these impulse responses are even or odd.

(b) Show that $G(z) = z^2 H_1(z)$ is zero phase, and from it determine the phase of the filter $H_1(z)$. Use MATLAB to find the unwrapped phase of $H_1(z)$ and confirm your analytic results.

(c) Find the phase of $H_2(z)$ by finding the phase of $F(z) = z^2 H_2(z)$. Use MATLAB to find the unwrapped phase of $H_2(z)$. Is it linear?

(d) If $H(z)$ is an IIR filter, according to the above arguments could it be possible for it to have linear phase? Explain.

11.4. Effect of phase on filtering—MATLAB

Consider two filters with transfer functions

$$H_1(z) = z^{-100}$$

$$H_2(z) = \left(0.5 \frac{1 - 2z^{-1}}{1 - 0.5z^{-1}}\right)^{10}$$

(a) Verify that the magnitude of these two filters is unity, but that they have different phases. Find analytically the phase of $H_1(z)$ and use MATLAB to find the unwrapped phase of $H_2(z)$ and to plot it.

(b) Consider the MATLAB signal "handel.mat" (a short piece of the "Messiah" by composer George Handel). Use the MATLAB function filter to filter it with the two given filters. Listen to the output and plot them and compare them. What is the difference (look at first 200 samples of the outputs from the two filters)?

(c) Can you recover the original signal by advancing either of the outputs? Explain.

11.5. Butterworth versus Chebyshev specifications

A Butterworth low-pass discrete filter of order N has been designed to satisfy the following specifications:

$$\text{Sampling period } T_s = 100 \ \mu\text{sec}$$

$$\alpha_{max} = 0.7 \text{ dB for } 0 \le f \le f_p = 1000 \text{ Hz}$$

$$\alpha_{min} = 10 \text{ dB for } f_{st} = 1200 \le f \le f_s/2 \text{ Hz}$$

What should be the new value of the stopband frequency f_{st} so that an Nth-order Chebyshev low-pass filter satisfies the design specifications for T_s, α_{max}, α_{min}, and f_p.

11.6. Bilinear transformation and pole location—MATLAB

Find the poles of the discrete filter obtained by applying the bilinear transformation with $K = 1$ to an analog second-order Butterworth low-pass filter. Determine the half-power frequency ω_{hp} of the resulting discrete filter. Use the MATLAB function bilinear to verify your results.

11.7. Warping effect of the bilinear transformation—MATLAB

The nonlinear relation between the discrete frequency ω (rad) and the continuous frequency Ω (rad/sec) in the bilinear transformation causes warping in the high frequencies. To see this consider the following:

(a) Use MATLAB to design a Butterworth analog band-pass filter of order $N = 12$ and with half-power frequencies $\Omega_1 = 10$ and $\Omega_2 = 20$ (rad/sec). Use the MATLAB function bilinear with $K = 1$ to transform the resulting filter into a discrete filter. Plot the magnitude and the phase of the discrete filter.

(b) Increase the order of the filter to $N = 14$ and keep the other specifications the same. Design an analog band-pass filter and use again bilinear with $K = 1$ to transform the analog filter into a discrete filter. Plot the magnitude and the phase of the discrete filter. Explain your results.

11.8. Warping effect of the bilinear transformation—MATLAB

The warping effect of the bilinear transformation also affects the phase of the transformed filter. Consider a filter with transfer function $G(s) = e^{-5s}$.

(a) Find the transformed discrete frequencies ω (rad) corresponding to $0 \leq \Omega \leq 20$ (rad/sec) using a bilinear transformation with $K = 1$. Plot Ω versus ω.

(b) Discretize the continuous frequencies $0 \leq \Omega \leq 20$ (rad/sec) to compute values of $G(j\Omega)$ and use MATLAB functions to plot the phase of $G(j\Omega)$.

(c) Find the function

$$H(e^{j\omega}) = G(j\Omega)|_{\Omega=\tan(\omega/2)}$$

and plot its unwrapped phase using MATLAB for the discrete frequencies corresponding to the analog frequencies to $0 \leq \Omega \leq 20$ (rad/sec).

11.9. Discrete Butterworth filter for analog processing—MATLAB

Design a Butterworth low-pass discrete filter that satisfies the following specifications:

$$0 \leq \alpha(e^{j\omega}) \leq 3 \text{ dB} \qquad \text{for } 0 \leq f \leq 25 \text{ Hz}$$

$$\alpha(e^{j\omega}) \geq 38 \text{ db} \qquad \text{for } 50 \leq f \leq Fs/2 \text{ Hz}$$

The sampling frequency is $F_s = 2000$ Hz. Express the transfer function $H(z)$ of the designed filter as a cascade of filters. Use first the design formulas and then use MATLAB to confirm your results. Show that the designed filter satisfies the specifications, plotting the loss function of the designed filter.

11.10. All-pass IIR filter—MATLAB

Consider an all-pass analog filter

$$G(s) = \frac{s^4 - 4s^3 + 8s^2 - 8s + 4}{s^4 + 4s^3 + 8s^2 + 8s + 4}$$

(a) Use MATLAB functions to plot the magnitude and the phase responses of $G(s)$. Indicate whether the phase is linear.

(b) A discrete filter $H(z)$ is obtained from $G(s)$ by the bilinear transformation. By trial and error, find the value of K in the bilinear transformation so that the poles and zeros of $H(z)$ are on the imaginary axis of the z-plane. Use MATLAB functions to do the bilinear transformation and to plot the magnitude and unwrapped phase of $H(z)$ and its poles. Is it an all-pass filter? If so, why?

(c) Let the input to the filter $H(z)$ be $x[n] = \sin(0.2\pi n)$, $0 \leq n < 100$, and the corresponding output be $y[n]$. Use MATLAB functions to compute and plot $y[n]$. From these results would you say that the phase of $H(z)$ is approximately linear? Why or why not?

11.11. Butterworth filtering of analog signal—MATLAB

We wish to design a discrete Butterworth filter that can be used in filtering a continuous-time signal. The frequency components of interest in this signal are between 0 and 1 KHz, so we would like the filter to have a maximum passband attenuation of 3 dB within that band. The undesirable components of the input signal occur beyond 2 KHz, and we would like to attenuate them by at least 10 dB. The maximum frequency present in the input signal is 5 KHz. Finally, we would like the dc gain of the filter to be 10. Choose the Nyquist sampling frequency to process the input signal. Use MATLAB to design the filter. Give the transfer function of the filter, plot its poles and zeros and its magnitude and unwrapped phase response using an analog frequency scale in KHz.

11.12. Butterworth versus Chebyshev filtering—MATLAB

If we wish to preserve low-frequency components of the input, a low-pass Butterworth filter could perform better than a Chebyshev filter. MATLAB provides a second Chebyshev filter function cheby2 that has a flat response in the passband and a rippled one in the stopband. Let the signal to be filtered be the first 100 samples from MATLAB's "train" signal. To this signal add some Gaussian noise to be generated by randn, multiply it by 0.1, and add it to the 100 samples of the train signal. Design three discrete filters, each of order 20, and a half frequency (for Butterworth butter) and passband frequency (for the Chebyshev filters) of $\omega_n = 0.5$. For the design with cheby1 let the maximum passband attenuation be 0.01 dB, and for the design with cheby2 let the minimum stopband attenuation be 60 dB. Obtain the three filters and use them to filter the noisy "train" signal.

Using MATLAB plot the following for each of the three filters:

(a) Using the fft function compute the DFT of the original signal, the noisy signal, and the noise, and plot their magnitudes. Is the cut-off frequency of the filters adequate to get rid of the noise? Explain.

(b) Compute and plot the magnitude and the unwrapped phase and the poles and the zeros for each of the three filters. Comment on the differences in the magnitude responses.

(c) Use the filter function to obtain the output of each of the filters, and plot the original noiseless signal and the filtered signals. Compare them.

11.13. Butterworth, Chebyshev, and elliptic filters—MATLAB

The gain specifications of a filter are:

$$-0.1 \leq 20 \log_{10} |H(e^{j\omega})| \leq 0 \ \ (\text{dB}) \qquad 0 \leq \omega \leq 0.2\pi$$

$$20 \log_{10} |H(e^{j\omega})| \leq -60 \ \ (\text{dB}) \qquad 0.3\pi \leq \omega \leq \pi$$

(a) Find the loss specifications for this filter.

(b) Design using MATLAB a Butterworth, a Chebyshev (using cheby1), and an elliptic filter. Plot in one plot the magnitude response of the three filters, and compare them and indicate which gives the lowest order.

11.14. Notch and all-pass filters—MATLAB

Notch filters are a family of filters that include the all-pass filter. For the filter

$$H(z) = K \frac{(1 - \alpha_1 z^{-1})(1 + \alpha_2 z^{-1})}{(1 - 0.5z^{-1})(1 + 0.5z^{-1})}$$

(a) Determine the values of α_1, α_2, and K that would make $H(z)$ an all-pass filter of unit magnitude. Use MATLAB to compute and plot the magnitude response of $H(z)$ using the obtained values for α and K. Plot the poles and the zeros of this filter.

(b) If we would like the filter $H(z)$ to be a notch filter of unit gain at $\omega = \pi/2$ rad, determine the values of α and K to achieve this and then determine where the notch(es) are. Use MATLAB functions to verify that the filter is a notch filter, and to plot the poles and the zeros.

(c) Place the zeros of $H(z)$ at positions between the zeros for the all-pass and the notch filters, and use MATLAB to plot the magnitude responses. Each of these filters must have unit gain at $\omega = \pi/2$ rad. Explain the connection between the all-pass and the notch filters.

(d) Suppose we use the transformation $z^{-1} = jZ^{-1}$ to obtain a filter $H(Z)$. Repeat the above part of the problem for $H(Z)$. Where are the notches of this new filter. What would be the difference between the all-pass filters $H(z)$ and $H(Z)$?

11.15. IIR comb filters—MATLAB

Consider a filter with transfer function

$$H(z) = K \frac{1 + z^{-4}}{1 + (1/16)z^{-4}}$$

(a) Find the gain K so that this filter has a unit dc gain. Use then MATLAB to find and plot the magnitude response of $H(z)$ and its poles and zeros. Indicate why it is called a comb filter.

(b) Use MATLAB to find the phase response of the filter $H(z)$. Why is it that the phase seems to be wrapped and it cannot be unwrapped by MATLAB?

(c) Suppose you wish to obtain an IIR comb filter that is sharper around the notches of $H(z)$ and flatter in between notches. Implement such a filter using the function butter to obtain two notch filters of order 10 and appropriate cut-off frequencies. Decide how to connect the two filters. Plot the magnitude and the phase of the resulting filter and its poles and zeros.

11.16. Three-band discrete spectrum analyzer—MATLAB

To design a three-band discrete spectrum analyzer for speech signals, we need to design a low-pass, a band-pass, and a high-pass filter. Let the sampling frequency be $F_s = 10$ KHz. Consider the three bands, in KHz, to be $[0 \ F_s/4]$, $(F_s/4 \ 3F_s/8]$, and $(3F_s/8 \ F_s/2]$. Let all the filters be of order $N = 4$, and choose the cut-off frequencies so that the sum of the three filters is approximately an all-pass filter of unit gain.

11.17. FIR filter design with different windows

Design a low-pass FIR digital filter with $N = 21$. The desired response of the filter is

$$|H_d(e^{j\omega T})| = \begin{cases} 1 & 0 \le f \le 250 \text{ Hz} \\ 0 & \text{elsewhere in } 0 \le f \le (f_s/2) \end{cases}$$

where $\omega = 2\pi f/f_s$ and the phase is zero for all frequencies. The sampling frequency is $f_s = 2000$ Hz.

(a) Use a rectangular window in your design. Plot the magnitude and the phase of the designed filter.

(b) Use a triangular window in the design and compare the magnitude and the phase plots of this filter with those obtained in (a).

11.18. FIR filter design—MATLAB

Design an FIR low-pass filter with a cut-off frequency of $\pi/3$ and lengths $N = 21$ and then $N = 81$:

(a) Using a rectangular window.

(b) Use MATLAB to design the filter using the rectangular, Hamming, and Kaiser windows, and compare the magnitude of the resulting filters.

11.19. Modulation property transformation for IIR filters—MATLAB

The modulation-based frequency transformation is applicable to IIR filters. It is obvious in the case of FIR filters, but requires a few more steps in the case of IIR filters. In fact, if we have that the transfer function of the prototype IIR low-pass filter is $H(z) = B(z)/A(z)$, with impulse response $h[n]$, let the transformed filter be $\hat{H}(z) = \mathcal{Z}(2h[n] \cos(\omega_o n))$ for some frequency ω_0.

(a) Find the transfer function $\hat{H}(z)$ in terms of $H(z)$.

(b) Consider an IIR low-pass filter

$$H(z) = \frac{1}{1 - 0.5z^{-1}}$$

If $\omega_0 = \pi/2$, determine $\hat{H}(z)$.

(c) How would you obtain a high-pass filter from $H(z)$ given in the previous item? Use MATLAB to plot the resulting filters here and in the past item.

11.20. Down-sampling transformations—MATLAB

Consider down sampling the impulse response $h[n]$ of a filter with transfer function

$$H(z) = \frac{1}{1 - 0.5z^{-1}}$$

(a) Use MATLAB to plot $h[n]$ and the down sampled impulse response $g[n] = h[2n]$.
(b) Plot the magnitude responses corresponding to $h[n]$ and $g[n]$ and comment on the effect of the down sampling.

11.21. Modulation property transformation—MATLAB

Consider a moving-average, low-pass, FIR filter,

$$H(z) = \frac{1 + z^{-1} + z^{-2}}{3}$$

(a) Use the modulation property to convert the given filter into a high-pass filter.
(b) Use MATLAB to plot the magnitude responses of the low-pass and the high-pass filters.

11.22. Implementation of IIR rational transformation—MATLAB

Use MATLAB to design a Butterworth second-order low-pass discrete filter $H(Z)$ with half-power frequency $\theta_{hp} = \pi/2$ and a dc gain of 1. Consider this low-pass filter a prototype that can be used to obtain other filters. Implement using MATLAB the frequency transformations $Z^{-1} = N(z)/D(z)$ using the convolution property to multiply polynomials to obtain:
(a) A high-pass filter with a half-power frequency $\omega_{hp} = \pi/3$ from the low-pass filter.
(b) A band-pass filter with $\omega_1 = \pi/2$ and $\omega_2 = 3\pi/4$ from the low-pass filter.
(c) Plot the magnitude of the low-pass, high-pass, and band-pass filters.
Give the corresponding transfer functions for the low-pass as well as the high-pass and the band-pass filters.

11.23. Parallel connection of IIR filters—MATLAB

Use MATLAB to design a Butterworth second-order low-pass discrete filter with half-power frequency $\theta_{hp} = \pi/2$ and a dc gain of 1; call it $H(z)$. Use this filter as a prototype to obtain a filter composed of a parallel combination of the following filters:
(a) Assume that we upsample by $L = 2$ the impulse response $h(n)$ of $H(z)$ to get a new filter $H_1(z) = H(z^2)$. Determine $H_1(z)$ and plot its magnitude using MATLAB.
(b) Assume then that we shift $H(z)$ by $\pi/2$ to get a band-pass filter $H_2(z)$. Find the transfer function of $H_2(z)$ from $H(z)$ and then plot its magnitude.
(c) If the filters $H_1(z)$ and $H_2(z)$ are connected in parallel, what is the overall transfer function $G(z)$ of the parallel connection? Plot the magnitude response corresponding to $G(z)$.

11.24. Realization of IIR filters

Consider the following transfer function:

$$H(z) = \frac{2(z - 1)(z^2 + \sqrt{2}\, z + 1)}{(z + 0.5)(z^2 - 0.9z + 0.81)}$$

(a) Develop a cascade realization of $H(z)$ using first- and second-order sections. Use direct form II to realize each of the sections.
(b) Develop a parallel realization of $H(z)$ by considering first- and second-order sections, each realized using direct form II.

11.25. Realization of IIR filters
  Given the realization in Figure 11.35:

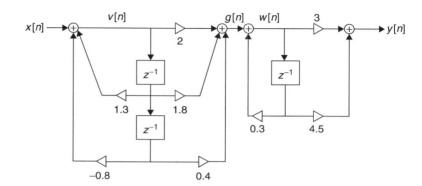

FIGURE 11.35

Problem – 11.25: IIR realization.

(a) Obtain the difference equations relating $g[n]$ to $x[n]$ and $g[n]$ to $y[n]$.
(b) Obtain the transfer function $H(z) = Y(z)/X(z)$ for this filter.

CHAPTER 12

Applications of Discrete-Time Signals and Systems

Nullius in verba
(Take nobody's word for it).
Motto of the Royal Society,
Britain's 350-year-old science fraternity

12.1 INTRODUCTION

In this chapter we will present applications of the theory of discrete-time signals and systems to three important areas: digital signal processing, digital control, and digital communications. The material in this chapter is meant to be a more motivational than detailed presentation. We encourage our readers to look for the details in excellent textbooks in these three areas [54, 22, 16].

Given the advances in digital technologies and computers, processing of signals is being done mostly digitally. Early results in sampling, analog-to-digital conversion, and the fast computation of the output of linear systems using the Fast Fourier Transform (FFT) made it possible for digital signal processing to become a technical area on its own. (The first books in this area [63, 54] come from the mid-1970s.) Although the origins of the FFT have been traced back to the German mathematician Gauss in the early 1800s, the modern theory of the algorithm comes from the 1960s. It should be understood that the FFT is not yet another transform, but an efficient algorithm to compute the discrete Fourier transform (DFT), which we covered in Chapter 10.

Analog classic control systems can be implemented digitally using analog-to-digital and digital-to-analog converters and computers to implement the control laws. The theory of sampled data shows the connection between the Laplace and the Z-transform. The difficulty in the analysis of these systems is the mixing of continuous- and discrete-time signals.

Digital communication systems provide a more efficient way to communicate information than analog communication systems, but they are more demanding in terms of bandwidth. As

Signals and Systems Using MATLAB®. DOI: 10.1016/B978-0-12-374716-7.00016-8

indicated before, digital communications began with the introduction of pulse code modulation (PCM). Telephony and radio using baseband and band-pass signals have converged into wireless communications. Many of the principles of analog communications have remained, but its implementation has changed from analog to digital with slightly different objectives. Efficient use of the radio spectrum and efficient processing have become the objectives of modern wireless communication systems such as spread spectrum and orthogonal frequency-division multiplexing, which we introduce here.

12.2 APPLICATION TO DIGITAL SIGNAL PROCESSING

In many applications, such as speech processing or acoustics, one would like to digitally process analog signals. In practice, this is possible by converting the analog signals into binary signals using an analog-to-digital converter (ADC), and if the output is desired in analog form a digital-to-analog converter (DAC) is used to convert the binary signal into a continuous-time signal. Ideally, if no quantization is considered and if the discrete-time signal is converted into an analog signal by sinc interpolation the system can be visualized as in Figure 12.1.

Viewing the whole system as a black box with an analog signal $x(t)$ as input, and giving as output also an analog signal $y(t)$, the processing can be seen as a continuous-time system with a transfer function $G(s)$. Under the assumption of no quantization, the discrete-time signal $x[n]$ is obtained by sampling $x(t)$ using a sampling period determined by the Nyquist sampling condition. Likewise, considering the transformation of a discrete-time (or sampled signal) $y[n]$ into a continuous-time signal $y(t)$ by means of the sinc interpolation, the ideal DAC is an analog low-pass filter that interpolates the discrete-time samples to obtain an analog signal. Finally, the discrete-time signal $x[n]$ is processed by a discrete-time system with transfer function $H(z)$, which depends on the desired transfer function $G(s)$.

Thus, one can process discrete- or continuous-time signals using discrete systems. A great deal of the computational cost of this processing is due to the convolution sum used to obtain the output of the discrete system. That is where the significance of the FFT algorithm lies. Although the DFT allows us to simplify the convolution to a multiplication, it is the FFT that as an algorithm provides a very efficient implementation of this process. In the next section, we will introduce you to the FFT and provide some of the basics of this algorithm for you to understand its efficiency.

FIGURE 12.1

Discrete processing of analog signals using an ideal ADC and DAC. $G(s)$ is the transfer function of the overall system, while $H(z)$ is the transfer function of the discrete-time system.

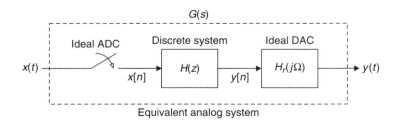

12.2.1 Fast Fourier Transform

Comparing the equations for the DFT and the inverse DFT, or IDFT

$$X[k] = \sum_{n=0}^{N-1} x[n] W_N^{kn} \qquad k = 0, \ldots, N-1 \tag{12.1}$$

$$x[n] = \frac{1}{N} \sum_{k=0}^{N-1} X[k] W_N^{-kn} \qquad n = 0, \ldots, N-1 \tag{12.2}$$

where $W_N = e^{-j2\pi/N}$, one sees a great deal of duality (more so if both the DFT and the IDFT had the term $1/\sqrt{N}$ instead of only the $1/N$ in the IDFT). Since $X[k]$ is complex, one can also see that one algorithm could be used for both the direct and the inverse DFTs if we assume $x[n]$ to be complex.

Two issues used to assess the complexity of an algorithm are:

- *Total number of additions and multiplications:* Typically, the complexity of a computational algorithm is assessed by determining the number of additions and multiplications it requires. The direct calculation of $X[k]$ using Equation (12.1) for $k = 0, \ldots, N-1$ requires $N \times N$ complex multiplications, and $N \times (N-1)$ complex additions. Computing the number of real multiplications and real divisions needed, it is found that the total number of these operations is of the order of N^2.
- *Storage:* Besides the number of computations, the required storage is an issue of interest.

Radix 2 FFT Algorithm

In the following, we consider the basic algorithm for the FFT. We assume that the FFT length is $N = 2^\gamma$ for an integer $\gamma > 1$. Excellent references on the DFT and the FFT are Briggs and Henson [10] and Brigham [11]

The FFT algorithm:

- Uses the fundamental principle of "divide and conquer": Dividing a problem into smaller problems with similar structure, the original problem can be successfully solved by solving each of the smaller problems.
- Takes advantage of periodicity and symmetry properties of W_N^{nk}:
  - **(a)** *Periodicity:* W_N^{nk} is periodic of period N with respect to n, and with respect to k—that is,

$$W_N^{nk} = \begin{cases} W_N^{(n+N)k} \\ W_N^{n(k+N)} \end{cases}$$

  - **(b)** *Symmetry:* The conjugate of W_N^{nk} is such that

$$\left[W_N^{nk} \right]^* = W_N^{(N-n)k} = W_N^{n(N-k)}$$

Decimation-in-Time Algorithm

Applying the divide-and-conquer principle, we express $X[k]$ as

$$X[k] = \sum_{n=0}^{N-1} x[n]W_N^{kn} \qquad k = 0, \ldots, N-1$$

$$= \sum_{n=0}^{N/2-1} \left[x[2n]W_N^{k(2n)} + x[2n+1]W_N^{k(2n+1)} \right]$$

That is, we gather the samples with even arguments separately from those with odd arguments.

From the definition of W_N^{nk} we have that

$$W_N^{k(2n)} = e^{-j2\pi(2kn)/N} = e^{-j2\pi kn/(N/2)} = W_{N/2}^{kn}$$

$$W_N^{k(2n+1)} = W_N^k W_{N/2}^{kn}$$

which allows us to write

$$X[k] = \sum_{n=0}^{N/2-1} x[2n]W_{N/2}^{kn} + W_N^k \sum_{n=0}^{N/2-1} x[2n+1]W_{N/2}^{kn}$$

$$= Y[k] + W_N^k Z[k] \qquad k = 0, \ldots, N-1 \tag{12.3}$$

where $Y[k]$ and $Z[k]$ are DFTs of length $N/2$ of the even-numbered sequence $\{x[2n]\}$ and of the odd-numbered sequence $\{x[2n+1]\}$, respectively.

Although it is clear how to compute the values of $X[k]$ for $k = 0, \ldots, (N/2) - 1$ as

$$X[k] = Y[k] + W_N^k Z[k] \qquad k = 0, \ldots, (N/2) - 1 \tag{12.4}$$

it is not clear how to proceed for $k \geq N/2$. The $N/2$ periodicity of $Y[k]$ and $Z[k]$ allows us to find those values:

$$X[k + N/2] = Y[k + N/2] + W_N^{k+N/2} Z[k + N/2]$$

$$= Y[k] - W_N^k Z[k] \qquad k = 0, \ldots, N/2 - 1 \tag{12.5}$$

where besides the periodicity of $Y[k]$ and $Z[k]$, we used

$$W_N^{k+N/2} = e^{-j2\pi[k+N/2]/N} = e^{-j2\pi k/N}e^{-j\pi} = -W_N^k$$

Writing Equations (12.4) and (12.5) in a matrix form we have

$$\mathbf{X}_N = \begin{bmatrix} \mathbf{I}_{N/2} & \mathbf{\Omega}_{N/2} \\ \mathbf{I}_{N/2} & -\mathbf{\Omega}_{N/2} \end{bmatrix} \begin{bmatrix} \mathbf{Y}_{N/2} \\ \mathbf{Z}_{N/2} \end{bmatrix} = \mathbf{A}_1 \begin{bmatrix} \mathbf{Y}_{N/2} \\ \mathbf{Z}_{N/2} \end{bmatrix} \tag{12.6}$$

where $\mathbf{I}_{N/2}$ is a unit matrix and $\Omega_{N/2}$ is a diagonal matrix with entries $\{W_N^k, k = 0, \ldots, N/2 - 1\}$; both matrices have dimension $N/2 \times N/2$. The vectors $\mathbf{X}_N$, $\mathbf{Y}_{N/2}$, and $\mathbf{Z}_{N/2}$ contain the coefficients of $x[n]$, $y[n]$, and $z[n]$.

Repeating the above computation for $Y[k]$ and $Z[k]$ we can express it in a similar matrix form until we reduce the process to 2×2 matrices. While performing these computations, the ordering of $x[n]$ is changed. This scrambling of $x[n]$ is obtained by a permutation matrix $\mathbf{P}_N$ (with 1 and 0 entries indicating the resulting ordering of the $x[n]$ samples). If $N = 2^\gamma$, the $\mathbf{X}_N$ vector, containing the DFT terms $X[k]$, is obtained as the product of γ matrices $\mathbf{A}_i$ and the permutation matrix. That is,

$$\mathbf{X}_N = \left[\prod_{i=1}^{\gamma} \mathbf{A}_i \right] \mathbf{P}_N \mathbf{x} \qquad \mathbf{x} = [x[0], \ldots, x[N-1]]^T \qquad (12.7)$$

where T stands for transpose. Given the large number of 1s and 0s in the $\{\mathbf{A}_i\}$ and the $\mathbf{P}_N$ matrices, the number of additions and multiplications is much lower than those in the original formulas. The number of operations is found to be of the order of $N\log_2 N = \gamma N$, which is much smaller than the original number of order N^2. For instance, if $N = 2^{10} = 1024$, the number of additions and multiplications for the computation of the DFT from its original formula is $N^2 = 2^{20} = 1,048,576$, while the FFT computation requires $N\log_2 N = 1024 \times 10 = 10,240$—that is, the FFT requires about 1% of the number of operations required by the original formula for the DFT.

■ Example 12.1

Consider the decimation-in-time FFT algorithm for $N = 4$. Give the equations to compute the four DFT values $X[k]$, $k = 0, \ldots, 3$ in matrix form.

Solution

If we compute the DFT of $x[n]$ directly we have that

$$X[k] = \sum_{n=0}^{3} x[n] W_4^{nk} \qquad k = 0, \ldots, 3$$

which can be rewritten in the matrix form as

$$
\begin{bmatrix} X[0] \\ X[1] \\ X[2] \\ X[3] \end{bmatrix} =
\begin{bmatrix}
1 & 1 & 1 & 1 \\
1 & W_4^1 & W_4^2 & W_4^3 \\
1 & W_4^2 & 1 & W_4^2 \\
1 & W_4^3 & W_4^2 & W_4^1
\end{bmatrix}
\begin{bmatrix} x[0] \\ x[1] \\ x[2] \\ x[3] \end{bmatrix}
$$

where we used

$$W_4^4 = W_4^{4+0} = e^{-j2\pi 0/4} = W_4^0 = 1$$
$$W_4^6 = W_4^{4+2} = e^{-j2\pi 2/4} = W_4^2$$
$$W_4^9 = W_4^{4+4+1} = e^{-j2\pi 1/4} = W_4^1$$

which requires 16 multiplications (8 if multiplications by 1 are not counted) and 12 additions. Thus, either 28 or 20, if multiplications by 1 are not counted, multiplications and additions are required. Since the entries are complex, these are complex additions and multiplications. A complex addition requires 2 real additions, and a complex multiplication requires 4 real multiplications and 3 real additions. Indeed, for two complex numbers $z = a + jb$ and $v = c + jd$, $z + v = (a + c) + j(b + c)$ and $zv = (ac - bd + j(bc + ad))$. Thus, the total number of real multiplications is 16×4 and real additions is $12 \times 2 + 16 \times 3$ for a total of 136 operations.

Separating the even- and the odd-numbered entries of $x[n]$, we have

$$X[k] = \sum_{n=0}^{1} x[2n]W_2^{kn} + W_4^k \sum_{n=0}^{1} x[2n + 1]W_2^{kn}$$

$$= Y[k] + W_4^k Z[k] \qquad k = 0, \ldots, 3$$

which can be written as

$$X[k] = Y[k] + W_4^k Z[k]$$

$$X[k + 2] = Y[k] - W_4^k Z[k] \qquad k = 0, 1$$

In matrix form the above equations can be written as

$$
\begin{bmatrix} X[0] \\ X[1] \\ \cdots \\ X[2] \\ X[3] \end{bmatrix}
=
\begin{bmatrix} 1 & 0 & \vdots & 1 & 0 \\ 0 & 1 & \vdots & 0 & W_4^1 \\ \cdots & \cdots & \cdots & \cdots & \cdots \\ 1 & 0 & \vdots & -1 & 0 \\ 0 & 1 & \vdots & 0 & -W_4^1 \end{bmatrix}
\begin{bmatrix} Y[0] \\ Y[1] \\ \cdots \\ Z[0] \\ Z[1] \end{bmatrix}
= \mathbf{A}_1
\begin{bmatrix} Y[0] \\ Y[1] \\ Z[0] \\ Z[1] \end{bmatrix}
$$

which is in the form indicated by Equation (12.6).

Now we have that

$$Y[k] = \sum_{n=0}^{1} x[2n]W_2^{kn} = x[0]W_2^0 + x[2]W_2^k$$

$$Z[k] = \sum_{n=0}^{1} x[2n + 1]W_2^{kn} = x[1]W_2^0 + x[3]W_2^k \qquad k = 0, 1$$

which in matrix form is

$$
\begin{bmatrix} Y[0] \\ Y[1] \\ \cdots \\ Z[0] \\ Z[1] \end{bmatrix} = \begin{bmatrix} 1 & 1 & \vdots & 0 & 0 \\ 1 & -1 & \vdots & 0 & 0 \\ \cdots & \cdots & \cdots & \cdots & \cdots \\ 0 & 0 & \vdots & 1 & 1 \\ 0 & 0 & \vdots & 1 & -1 \end{bmatrix} \begin{bmatrix} x[0] \\ x[2] \\ \cdots \\ x[1] \\ x[3] \end{bmatrix} = \mathbf{A}_2 \begin{bmatrix} x[0] \\ x[2] \\ x[1] \\ x[3] \end{bmatrix}
$$

where we replaced $W_2^1 = e^{-j2\pi/2} = -1$. Notice the change in the ordering of the $\{x[n]\}$.

Reordering the $x[n]$ entries we have

$$
\begin{bmatrix} x[0] \\ x[2] \\ x[1] \\ x[3] \end{bmatrix} = \begin{bmatrix} 1 & 0 & 0 & 0 \\ 0 & 0 & 1 & 0 \\ 0 & 1 & 0 & 0 \\ 0 & 0 & 0 & 1 \end{bmatrix} \begin{bmatrix} x[0] \\ x[1] \\ x[2] \\ x[3] \end{bmatrix} = \mathbf{P}_4 \begin{bmatrix} x[0] \\ x[2] \\ x[1] \\ x[3] \end{bmatrix}
$$

which finally gives

$$
\begin{bmatrix} X[0] \\ X[1] \\ X[2] \\ X[3] \end{bmatrix} = \mathbf{A}_1 \, \mathbf{A}_2 \, \mathbf{P}_4 \begin{bmatrix} x[0] \\ x[1] \\ x[2] \\ x[3] \end{bmatrix}
$$

The count of multiplications is now much lower given the number of 1s and 0s. The complex additions and multiplications are now 10 (2 complex multiplications and 8 complex additions) if we do not count multiplications by 1 or -1. Half of what it was before! ∎

12.2.2 Computation of the Inverse DFT

The FFT algorithm can be used to compute the inverse DFT without any changes in the algorithm. Assuming the input $x[n]$ is complex ($x[n]$ being real is a special case), the complex conjugate of the inverse DFT equation, multiplied by N, is

$$
Nx^*[n] = \sum_{k=0}^{N-1} X^*[k] W^{nk} \tag{12.8}
$$

Ignoring that the right term is in the frequency domain, we recognize it as the DFT of a sequence $\{X^*[k]\}$ and it can be computed using the FFT algorithm discussed before. The desired $x[n]$ is thus obtained by computing the complex conjugate of Equation (12.8) and dividing it by N. As a result, the same algorithm, with the above modification, can be used to compute both the direct and the inverse DFTs.

Remark *In the FFT algorithm the 2N memory allocations for the complex input (one allocation for the real part and another for the imaginary part of the input) are the same ones used for the output. Each step uses* *the same locations. Since X[k] is typically complex, to have identical allocation with the output, the input sequence x[n] is assumed to be complex. If x[n] is real, it is possible to transform it into a complex sequence and use properties of the DFT to obtain X[k].*

12.2.3 General Approach of FFT Algorithms

Although there are many algorithmic approaches to the FFT, the initial idea was to represent the finite one-dimensional signal $x[n]$ as a two-dimensional array. This can be done by representing the length N of $x[n]$ as the product of smaller integers, provided N is not prime. If N is prime, the DFT computation is done with the conventional formula, as the FFT does not provide any simplification. However, in that case we could attach zeros to the signal (if the signal is not periodic) to increase its length to a nonprime number. This factorization approach has historic significance as it was the technique used by Cooley and Tukey, the authors of the FFT.

Suppose N can be factored as $N = pq$; then the frequency and time indices k and n in the direct DFT can be written as

$$k = k_1 p + k_0 \quad \text{for} \quad k_0 = 0, \ldots, p - 1, \ k_1 = 0, \ldots, q - 1$$

$$n = n_1 q + n_0 \quad \text{for} \quad n_0 = 0, \ldots, q - 1, \ n_1 = 0, \ldots, p - 1$$

The values of k range from 0 (when $k_0 = k_1 = 0$) to $N - 1$ (when $k_1 = q - 1$ and $k_0 = p - 1$ then $k = k_1 p + k_0 = (q - 1)p + (p - 1) = qp - 1 = N - 1$). Likewise for n. The direct DFT

$$X[k] = \sum_{n=0}^{N-1} x[n] W^{nk}$$

can be written to reflect the dependence on the new indices as

$$X[k_0, k_1] = \sum_{n_0=0}^{q-1} \sum_{n_1=0}^{p-1} x[n_0, n_1] W_N^{(n_1 q + n_0)(k_1 p + k_0)} \tag{12.9}$$

giving a two-dimensional array.

The decimation-in-time FFT presented before may be viewed in this framework by letting $q = 2$ and $p = N/2$, where as before $N = 2^\gamma$ is even. We then have using $W_N^{2n_1(k_1 p + k_0)} = W_{N/2}^{n_1(k_1 p + k_0)}$,

$$X[k_0, k_1] = \sum_{n_0=0}^{1} W_N^{n_0(k_1 p + k_0)} \sum_{n_1=0}^{N/2-1} x[n_0, n_1] W_{N/2}^{n_1(k_1 p + k_0)}$$

$$= \sum_{n_1=0}^{N/2-1} x[0, n_1] W_{N/2}^{n_1(k_1 p + k_0)} + W_N^{(k_1 p + k_0)} \sum_{n_1=0}^{N/2-1} x[1, n_1] W_{N/2}^{n_1(k_1 p + k_0)}$$

$$= Y[k] + W_N^{(k_1 p + k_0)} Z[k] \qquad k_0 = 0, \ldots, N/2 - 1, \ k_1 = 0, 1 \tag{12.10}$$

where when $n_0 = 0$ the even terms $(x[0, n_1] = x[qn_1] = x[2n_1])$ in the input are being transformed, while when $n_0 = 1$ the odd terms $(x[1, n_1] = x[qn_1 + 1] = x[2n_1 + 1])$ of the input are being transformed. Since $k = k_1 p + k_0$ the final equation is $Y[k] + W_N^k Z[k]$, which we obtained in the decimation-in-time approach (see Eq. 12.3).

Factoring $N = 2 \times N/2$ corresponds to one step of the decimation-in-time method. If we factor $N/2$ as $N/2 = (2)(N/4)$, we would obtain the second step in the decimation-in-time algorithm. If $N = 2^\gamma$, this process is repeated γ times or until the length is 2.

Remark *A dual of the decimation-in-time FFT algorithm is the decimation-in-frequency method.*

The Modern FFT

A paper by James Cooley, an IBM researcher, and Professor John Tukey from Princeton University [15] describing an algorithm for the machine calculation of complex Fourier series appeared in *Mathematics of Computation* in 1965. Cooley, a mathematician, and Tukey, a statistician, had in fact developed an efficient algorithm to compute the discrete Fourier transform (DFT), which will be called the FFT. Their result was a turning point in digital signal processing: The proposed algorithm was able to compute the DFT of a sequence of length N using $N \log N$ arithmetic operations, much smaller than the N^2 operations that had blocked the practical use of the DFT. As Cooley indicated in his paper "How the FFT Gained Acceptance" [14], his interest in the problem came from a suggestion from Tukey on letting N be a composite number, which would allow a reduction in the number of operations of the DFT computation.

The FFT algorithm was a great achievement for which the authors received deserved recognition, but also benefited the new digital signal processing area, and motivated further research on the FFT. But as in many areas of research, Cooley and Tukey were not the only ones who had developed an algorithm of this class. Many other researchers before them had developed similar procedures. In particular, Danielson and Lanczos, in a paper published in the *Journal of the Franklin Institute* in 1942 [19], proposed an algorithm that came very close to Cooley and Tukey's results. Danielson and Lanczos showed that a DFT of length N could be represented as a sum of two $N/2$ DFTs proceeding recursively with the condition that $N = 2^\gamma$. Interestingly, they mention that (remember this was in 1942!):

Adopting these improvements the approximation time for Fourier analysis are: 10 minutes for 8 coefficients, 25 minutes for 16 coefficients, 60 minutes for 32 coefficients, and 140 minutes for 64 coefficients.

■ Example 12.2

Consider computing the FFT of a signal of length $N = 2^3 = 8$ using the decimation-in-time algorithm.

Solution

Letting $N = qp = 2 \times 4$, we then have that

$$X[k_0, k_1] = \sum_{n_0=0}^{1} W_8^{n_0(k_1 p + k_0)} \sum_{n_1=0}^{N/2-1} x[n_0, n_1] W_4^{n_1(k_1 p + k_0)}$$

$$= \sum_{n_1=0}^{3} x[0, n_1] W_4^{n_1(k_1 p + k_0)} + W_8^{(k_1 p + k_0)} \sum_{n_1=0}^{3} x[1, n_1] W_4^{n_1(k_1 p + k_0)}$$

where

$$k = 4k_1 + k_0 \quad \text{for } k_0 = 0, \dots, 3, \; k_1 = 0, 1$$
$$n = 2n_1 + n_0 \quad \text{for } n_0 = 0, 1, \; n_1 = 0, \dots, 3$$

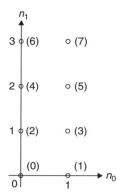

FIGURE 12.2

Lattice for $n_0 = 0, 1$ and $n_1 = 0, \ldots, 3$ (the values in parentheses are the indices of the samples). Notice the ordering in the two columns.

Figure 12.2 displays a lattice for n_0 and n_1 and the indices of the samples are in parentheses. By replacing $k = 4k_1 + k_0$, we get the first step of the decimation-in-time.

If we let $y[n] = x[2n]$ and $z[n] = x[2n + 1]$, $n = 0, \ldots, 3$, we can then repeat the above procedure by factoring $4 = pq = 2 \times 2$ and expressing $Y[k]$ and $Z[k]$ as we did for $X[k]$. Thus, we have

$$Y[k_0, k_1] = \sum_{n_0=0}^{1} W_4^{n_0(k_1 p + k_0)} \sum_{n_1=0}^{1} y[n_0, n_1] W_2^{n_1(k_1 p + k_0)}$$

$$= \sum_{n_1=0}^{1} y[0, n_1] W_2^{n_1(k_1 p + k_0)} + W_4^{(k_1 p + k_0)} \sum_{n_1=0}^{1} y[1, n_1] W_2^{n_1(k_1 p + k_0)}$$

and

$$Z[k_0, k_1] = \sum_{n_1=0}^{1} z[0, n_1] W_2^{n_1(k_1 p + k_0)} + W_4^{(k_1 p + k_0)} \sum_{n_1=0}^{1} z[1, n_1] W_2^{n_1(k_1 p + k_0)}$$

where now

$$k = 2k_1 + k_0 \quad \text{for} \ k_0 = 0, 1, \ k_1 = 0, 1$$
$$n = 2n_1 + n_0 \quad \text{for} \ n_0 = 0, 1, \ n_1 = 0, 1$$

If we replace $k = 2k_1 + k_0$, we obtain

$$Y[k] = I[k] + W_4^k G[k]$$
$$Z[k] = H[k] + W_4^k F[k]$$

where

$$I[k] = \sum_{n_1=0}^{1} y[0, n_1] W_2^{n_1 k} = \sum_{n_1=0}^{1} y[2n_1] W_2^{n_1 k} = \sum_{n_1=0}^{1} x[4n_1] W_2^{n_1 k}$$

$$G[k] = \sum_{n_1=0}^{1} y[1, n_1] W_2^{n_1 k} = \sum_{n_1=0}^{1} y[2n_1 + 1] W_2^{n_1 k} = \sum_{n_1=0}^{1} x[4n_1 + 2] W_2^{n_1 k}$$

Likewise,

$$H[k] = \sum_{n_1=0}^{1} z[0, n_1] W_2^{n_1 k} = \sum_{n_1=0}^{1} z[2n_1] W_2^{n_1 k} = \sum_{n_1=0}^{1} x[4n_1 + 1] W_2^{n_1 k}$$

$$F[k] = \sum_{n_1=0}^{1} z[1, n_1] W_2^{n_1 k} = \sum_{n_1=0}^{1} z[2n_1 + 1] W_2^{n_1 k} = \sum_{n_1=0}^{1} x[4n_1 + 3] W_2^{n_1 k}$$

■

■ Example 12.3

In this example we wish to compare the efficiency of the FFT algorithm with that of our algorithm dft.m that computes the DFT using its definition. Consider the computation of the FFT and the DFT of a signal consisting of ones of increasing lengths $N = 2^r$, $r = 8, \ldots, 11$, or 256 to 2048.

Solution

To compare the algorithms we use the following script. The MATLAB function cputime measures the time it takes for each of the algorithms to compute the DFT of the sequence of ones.

```
%%%%%%%%%%%%%%
% example 12.3
%  fft vs dft
%%%%%%%%%%%%%%
clf; clear all
time = zeros(1,4); time1 = time;
for r = 8:11,$$
   N(r) = 2^r;
   i = r - 7;
   t = cputime;
   fft(ones(1,N(r)),N(r));
   time(i) = cputime - t;
   t = cputime;
   dft(ones(N(r),1),N(r));
   time1(i) = cputime - t;
end

   %%%%%%%%%%%
   % function dft
   %%%%%%%%%%%
   function X = dft(x,N)
   n = 0:N - 1;
```

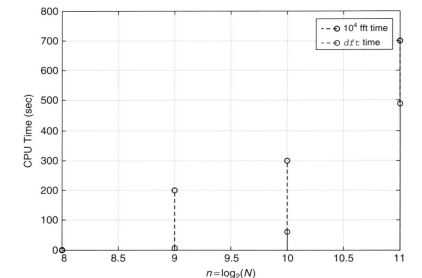

FIGURE 12.3

CPU times for the fft and the dft functions used in computing the DFT of sequences of ones of lengths $N = 256$ to 2048 (corresponding to $n = 8, \ldots 11$). The CPU time for the FFT is multiplied by 10^4.

```
W = ones(1,N);
for k = 1:N - 1,
    W = [W; exp(-j * 2 * pi * n * k/N)];
end
X = W * x;
```

The results of the comparison are shown in Figure 12.3. Notice that to make the Computer Processing Unit (CPU) time for the FFT comparable with that of the dft algorithm, it is multiplied by 10^4, illustrating how much faster the FFT is compared to the computation of the DFT from its definition. ■

■ Example 12.4

The convolution sum is computationally very expensive. Compare the CPU time used by the MATLAB function conv, which is used to compute the convolution sum in the time domain, with the CPU time used by an implementation of the convolution sum in frequency using the FFT. Recall that the frequency implementation requires computing the DFT of the signals being convolved, their multiplication, and finally the computation of the IDFT to the final convolution result.

Solution

To illustrate the efficiency in computation provided by the FFT in computing the convolution sum we compare the CPU times used by the conv function and the implementation of the convolution sum using the FFT. As indicated before, the convolution of two signals $x[n]$, and $y[n]$ of lengths N and M is obtained in the frequency domain by following these three steps:

- Compute the DFTs $X[k]$, $Y[k]$ of $x[n]$, and $y[n]$ of length $M + N - 1$.
- Multiply these complex DFTs to get $X[k]Y[k] = U[k]$.
- Compute the IDFT of $U[k]$ corresponding to the convolution $x[n] * y[n]$.

Implementing the DFT and the IDFT with the FFT algorithm it can be shown that the computational complexity of the above three steps is much smaller than that of computing the convolution sum directly using the conv function.

To demonstrate the efficiency of the FFT implementation we consider the convolution of a signal, for increasing lengths, with itself. The signal is a sequence of ones of increasing length of 1000 to 10,000 samples. The CPU times used by the functions conv and the FFT three-step procedure are measured and compared for each of the lengths. The CPU time used by conv is divided by 10 to be able to plot it with the CPU of the FFT-based procedure shown in the following script. The results are shown in Figure 12.4.

```
%%%%%%%%%%%%%
% example 12.4
%  conv vs fft
%%%%%%%%%%%%%%
time1 = zeros(1,10);time2 = time1;
for i = 1:10,
    NN = 1000 * i;
    x = ones(1,NN);
```

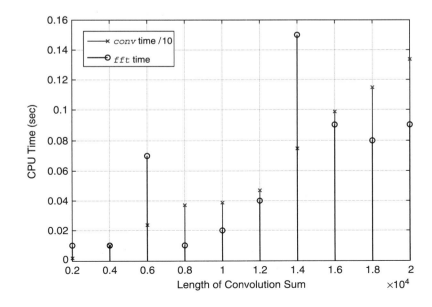

FIGURE 12.4

CPU times for the fft and the conv functions when computing the convolution of sequences of ones of lengths $N = 1000$ to 10,000. The CPU time used by conv is divided by 10.

```
M = 2 * NN-1;
t0 = cputime;
y = conv(x,x); % convolution using conv
time1(i) = cputime-t0;
t1 = cputime;
X = fft(x,M); X = fft(x,M); Y = X. * X; y1 = ifft(Y); %  convolution using fft
time2(i) = cputime-t1
sum(y-y1) % check conv and fft results coincide
pause % check for small difference
end
```

Gauss and the FFT

Going back to the sources used by the FFT researchers it was discovered that many well-known mathematicians had developed similar algorithms for different values of N. But that an algorithm similar to the modern FFT had been developed and used by Carl Gauss, the German mathematician, probably in 1805, predating even Fourier's work on harmonic analysis in 1807, was an interesting discovery—although not surprising [31]. Gauss has been called the "prince of mathematicians" for his prodigious work in so many areas of mathematics, and for the dedication to his work. His motto was *Pauca sed matura* (few, but ripe); he would not disclose any of his work until he was very satisfied with it. Moreover, as it was customary in his time, his treatises were written in Latin using a difficult mathematical notation, which made his results not known or understood by modern researchers. Gauss's treatise describing the algorithm was not published in his lifetime, but appeared later in his collected works. He, however, deserves the paternity of the FFT algorithm.

The developments leading to the FFT, as indicated by Cooley [14], point out two important concepts in numerical analysis (the first of which applies to research in other areas): (1) the *divide-and-conquer approach*—that is, it pays to break a problem into smaller pieces of the same structure; and (2) the *asymptotic behavior of the number of operations*. Cooley's final recommendations in his paper are worth serious consideration by researchers in technical areas:

- Prompt publication of significant achievements is essential.
- Review of old literature can be rewarding.
- Communication among mathematicians, numerical analysts, and workers in a wide range of applications can be fruitful.
- Do not publish papers in neoclassic Latin.

12.3 APPLICATION TO SAMPLED-DATA AND DIGITAL CONTROL SYSTEMS

Most control systems being used today use computers and ADCs and DACs. Control systems where continuous- and discrete-time signals appear are called **sample-data systems**. The analysis of these systems is more complicated than that of either continuous- or discrete-time systems, given the mixed signals in the system. In the following analysis we will ignore the effect of the quantizer and the coder, so that we are not considering digital control systems, but rather sampled-data or discrete control systems. Understanding the effects of sampling and the conversion of signals from continuous to

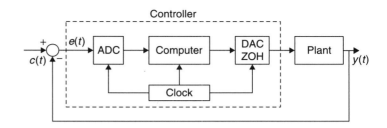

FIGURE 12.5

Digital implementation of a
continuous-feedback system. ZOH
stands for zero-order hold.

discrete and back from discrete to continuous is very important in obtaining a discrete-time system
from a sampled-data control system.

Consider the relation between a continuous control system and its implementation using a computer
as shown in Figure 12.5. In the continuous-feedback system, the controller responds to an error signal
$e(t)$, which is the difference between a reference input signal $c(t)$ and the system output $y(t)$, attempting to change the dynamics of an analog plant. A digital realization of this continuous-feedback
system typically requires that the error signal be converted into a digital signal by means of an ADC
before being fed to a computer implementing the controller (e.g., a PID controller). A DAC with a
zero-order hold that is synchronously connected and has the same sampling period as the previous
ADC is used to generate a signal that will act on the plant. The output of the plant $y(t)$ is fed back and
compared with the command signal $c(t)$ to obtain the error $e(t)$. The digital controller is composed
of the ADC, the computer, and the DAC with the zero-order hold, all of which are synchronized by a
common clock.

Why are sampled-data and digital control systems needed? In part, because many systems are inherently discrete—for example, a radar tracking system scans to convert azimuth and elevation into
sampled data. But in general we have that:

■ A continuous control system operates in real time, and the amplitude of its signals are allowed to
   take any possible value, but its elements are susceptible to degradation with time, and the system
   is sensitive to noise and difficult to change since it is hardwired.
■ Digital components are less susceptible to aging, environmental variations, and noise. A digital
   controller can be modified easily by changing software without changing the hardware. However,
   computational speed and resolution (word length) are limitations of digital controllers that can
   cause instabilities.

Remarks

■ *In the following development you need to remember:*
   1. *The Laplace transform of an ideally sampled signal,*

$$x_s(t) = \sum_n x(nT_s)\delta(t - nT_s)$$

where $x(nT_s) = x(t)|_{t=nT_s}$ are the sample values of the continuous-time signal $x(t)$,

$$X_s(s) = \mathcal{L}[x_s(t)] = \sum_n x(nT_s)e^{-snT_s}$$

2. *The Laplace transform $X_s(s)$ is related to the Z-transform of the discrete-time signal $x(nT_s)$ by letting $z = e^{sT_s}$.*

■ *Recall that the ideal sampler is a time-varying system and that the quantizer is a nonlinear system; thus sampled-data and digital control systems are time varying and time-varying nonlinear, respectively. Therefore, the complexity of their analyses.*

12.3.1 Open-Loop Sampled-Data System

Consider the system shown in Figure 12.6. Assume the discrete-time signal $x(nT_s)$ coming from a computer is used to drive an analog plant with a transfer function $G(s)$. To change the state of the plant, $x(nT_s)$ is converted into a continuous signal that holds the value of the sample for the duration of the sample period T_s. This can be implemented using a DAC with a *zero-order hold* (ZOH), which holds the value of $x(nT_s)$ until the next sample arrives at $(n + 1)T_s$. Furthermore, to allow the output signal to be possibly processed by a computer, assume the output of the plant $y(t)$ is also sampled to get $y(nT_s)$. We are interested in the transfer function that relates the discrete input $x(nT_s)$ to the discrete output $y(nT_s)$ where T_s is the sampling period chosen according to the maximum frequency present in the analog input $x(t)$.

As we saw in Chapter 7, the transfer function of a zero-order hold is

$$H_{zoh}(s) = \frac{1 - e^{-sT_s}}{s} \tag{12.11}$$

which corresponds to an impulse response

$$h_{zoh}(t) = u(t) - u(t - T_s) \tag{12.12}$$

or a pulse of duration T_s and unit amplitude. If the sampled signal is written as

$$x_s(t) = \sum_n x(nT_s)\delta(t - nT_s) \tag{12.13}$$

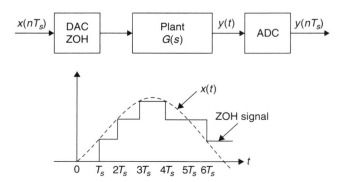

FIGURE 12.6
Open-loop sampled-data system for an analog plant $G(s)$. The output of the DAC with a ZOH is illustrated.

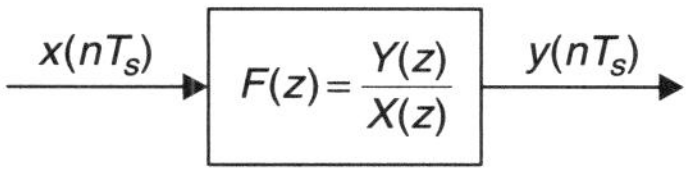

FIGURE 12.7

Equivalent discrete-time system of the open-loop sampled-data system.

$$\xrightarrow{x(nT_s)} \boxed{F(z) = \dfrac{Y(z)}{X(z)}} \xrightarrow{y(nT_s)}$$

(i.e., a sequence of impulses at times $\{nT_s\}$ with amplitude the sampled values $x(nT_s)$), then the output of the DAC with ZOH is

$$v(t) = [x_s * h_{zoh}](t) = \sum_n x(nT_s)h_{zoh}(t - nT_s) \tag{12.14}$$

or a piecewise constant signal (see Figure 12.6). Putting together the transfer function of the ZOH with that of the plant so that $F(s) = H_{zoh}(s)G(s)$, we have that $Y(s) = F(s)X_s(s)$.

If we let $f(t) = \mathcal{L}^{-1}[F(s)]$, then the output of the plant is given by the convolution integral as

$$y(t) = [x_s * f](t) = \sum_n x(nT_s)[\delta * f](t - nT_s) = \sum_n x(nT_s)f(t - nT_s)$$

which is the convolution sum of the discrete input and the sampled-impulse response of the plant combined with that of the ZOH. For $Y(z) = \mathcal{Z}[y(nT_s)]$ and $X(z) = \mathcal{Z}[x(nT_s)]$, we have that when we sample $y(t)$, then

$$y(kT_s) = y(t)|_{t=kT_s} = \sum_n x(nT_s)f(kT_s - nT_s)$$

The transfer function of the discrete system is

$$F(z) = \mathcal{Z}[f(nT_s)] = \frac{Y(z)}{X(z)} \tag{12.15}$$

which can be obtained by sampling the inverse Laplace transform $f(t) = \mathcal{L}^{-1}[F(s)]$ and then computing the Z-transform of $f(nT_s)$. We have thus obtained the equivalent discrete-time system to the sampled-data system shown in Figure 12.7.

■ Example 12.5

Consider the open-loop sampled-data system shown in Figure 12.6, where the DAC with ZOH is synchronized with an ADC, which is just an ideal sampler. Let $T_s = 1$ sec/sample be the sampling period. If the transfer function of the plant is

$$G(s) = \frac{1}{(s + 1)(s + 2)}$$

find the transfer function $F(z) = Y(z)/X(z)$.

Solution

The combined transfer function of the ZOH and the plant is

$$F(s) = \frac{G(s)(1 - e^{-s})}{s} = \frac{G(s)}{s} - \frac{G(s)e^{-s}}{s}$$

so that if we find the inverse Laplace transform of $\hat{G}(s) = G(s)/s$, call it $\hat{g}(t)$, then

$$f(t) = \hat{g}(t) - \hat{g}(t - 1)$$

The inverse of $\hat{G}(s) = G(s)/s$ is obtained by partial fraction expansion

$$\hat{G}(s) = \frac{G(s)}{s} = \frac{1}{s(s + 1)(s + 2)} = \frac{A}{s} + \frac{B}{s + 1} + \frac{C}{s + 2}$$

$$= \frac{0.5}{s} - \frac{1}{s + 1} + \frac{0.5}{s + 2}$$

so that

$$\hat{g}(t) = [0.5 - e^{-t} + 0.5e^{-2t}]u(t)$$

Sampling $f(t) = \hat{g}(t) - \hat{g}(t - 1)$ with a sampling period $T_s = 1$ gives

$$f(n) = \hat{g}(n) - \hat{g}(n - 1)$$

where $\hat{g}(n) = [0.5 - e^{-n} + 0.5e^{-2n}]u(n)$. The Z-transform of $f(n)$ is then the transfer function

$$F(z) = \frac{Y(z)}{X(z)} = \hat{G}(z)(1 - z^{-1})$$

$$= (1 - z^{-1}) \left(\frac{0.5}{1 - z^{-1}} - \frac{1}{1 - e^{-1}z^{-1}} + \frac{0.5}{1 - e^{-2}z^{-1}} \right)$$

$$= 0.5 - \frac{(1 - z^{-1})}{1 - e^{-1}z^{-1}} + \frac{0.5(1 - z^{-1})}{1 - e^{-2}z^{-1}}$$ ■

12.3.2 Closed-Loop Sampled-Data System

Consider the feedback system shown in Figure 12.8 where for simplicity we assume $H(s) = 1$ (i.e., no feedback sensor). An equivalent block diagram is obtained by moving back the sampler at the input. Consider finding the transfer function of the sampled input command $c_s(t)$ and the sampled output of the system $y_s(t)$. The above open-loop development can be used to find the transfer function of the feedback system.

The sampled error signal is

$$e_s(t) = c_s(t) - y_s(t)$$

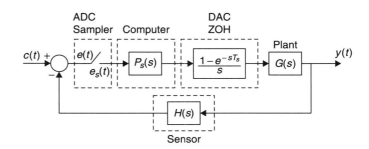

FIGURE 12.8
Closed-loop sampled-data control system.

with Laplace transform

$$E_s(s) = C_s(s) - Y_s(s) \qquad (12.16)$$

The function $P_s(s)$ corresponds to the discretization of an analog controller, such as a PID controller. The Laplace transform of the output of the computer is then

$$M_s(s) = P_s(s)E_s(s) = \sum_n m(nT_s)e^{-snT_s} \qquad (12.17)$$

or the Laplace transform of a sampled signal. On the other hand, the DAC with ZOH and the plant have together a transfer function

$$\hat{G}(s) = \frac{(1 - e^{-sT_s})G(s)}{s}$$

Thus, the Laplace transform of the output of the plant is

$$M_s(s)\hat{G}(s) = \sum_n m(nT_s)\left[\hat{G}(s)e^{-snT_s}\right] \qquad (12.18)$$

Using the time-shifting property, the inverse Laplace transform of the above equation is

$$\sum_n m(nT_s)\hat{g}(t - nT_s)$$

which when sampled at $t = kT_s$ gives the convolution sum

$$\sum_n m(nT_s)\hat{g}(kT_s - nT_s) = y_s(nT_s) \qquad (12.19)$$

so that $M(z)\hat{G}(z) = Y(z)$.

Letting $z = e^{sT_s}$ in Equation (12.16), we obtain $E(z) = C(z) - Y(z)$, and replacing it in Equation (12.17) gives

$$M(z) = P(z)E(z) = P(z)[C(z) - Y(z)]$$

We thus have

$$\underbrace{P(z)[C(z) - Y(z)]}_{M(z)} \hat{G}(z) = Y(z)$$

from which we obtain

$$Y(z) = \frac{P(z)C(z)\hat{G}(z)}{1 + P(z)\hat{G}(z)}$$

Calling $F(z) = P(z)\hat{G}(z)$ (the feed-forward transfer function consisting of the discretized analog controller and the ZOH and the plant), we get

$$\frac{Y(z)}{C(z)} = \frac{F(z)}{1 + F(z)} \tag{12.20}$$

or the transfer function of the data-sampled system. Notice that this equation looks like the equation of a continuous-feedback system.

Remarks

■ *In the equivalent discrete-time system obtained above, the information of the output of the open-loop or the closed-loop systems in between the sampling instants is not available; only the samples $y(nT_s)$ are. This is also indicated by the use of the Z-transform.*

■ *Depending on the location of the sampler, there are some sampled-data control systems for which we cannot find a transfer function. This is due to the time-variant nature of the system.*

■ Example 12.6

Suppose we wish to have a data-sampled system like the one shown in Figure 12.8 that simulates the effects of an integral analog controller. Let the plant be a first-order system,

$$G(s) = \frac{1}{s + 1}$$

Let the sampling period be $T_s = 1$. Determine $P(z)$ and find the discrete transfer function of the sampled-data system when $H(s) = 1$.

Solution

If $e(t)$ is the input of an integrator and $v(t)$ its output, letting $t = nT_s$ and approximating the integral by a sum we have that

$$v(nT_s) = \sum_{k=0}^{n} e(kT_s)T_s = \sum_{k=0}^{n-1} e(kT_s)T_s + e(nT_s)T_s$$
$$= v(nT_s - T_s) + T_s e(nT_s)$$

After replacing $T_s = 1$ it becomes $v[n] = v[n-1] + e[n]$, so the transfer function of the integrator is

$$P(z) = \frac{V(z)}{E(z)} = \frac{1}{1 - z^{-1}}$$

The transfer function of the DAC with ZOH and the plant is

$$\hat{G}(s) = \frac{(1 - e^{-sT_s})G(s)}{s} = \frac{(1 - e^{-s})}{s(s+1)}$$

If we let $D(s) = 1/s(s+1)$, then $\hat{g}(t) = d(t) - d(t-1)$. Expanding $D(s)$ as

$$D(s) = \frac{1}{s(s+1)} = \frac{A}{s} + \frac{B}{s+1} = \frac{1}{s} - \frac{1}{s+1}$$

so that $d(t) = u(t) - e^{-t}u(t)$ sampled, gives

$$d(nT_s) = u(nT_s) - e^{-nT_s}u(nT_s)$$

which has a Z-transform

$$D(z) = \frac{1}{1 - z^{-1}} - \frac{1}{1 - e^{-1}z^{-1}} \quad \Rightarrow \quad \hat{G}(z) = (1 - z^{-1})D(z)$$

The transfer function is then

$$\frac{Y(z)}{C(z)} = \frac{P(z)(1 - z^{-1})D(z)}{1 + P(z)(1 - z^{-1})D(z)}$$

$$= \frac{D(z)}{1 + D(z)}$$

since $P(z)(1 - z^{-1}) = 1$. ∎

12.4 APPLICATION TO DIGITAL COMMUNICATIONS

Although over the years the principles of communications have remained the same, their implementation has changed considerably. Analog communications transitioned into digital communications, while telephony and radio have coalesced into wireless communications. The scarcity of radio spectrum changed the original focus on bandwidth and energy efficiency into more efficient utilization of the available spectrum by sharing it, and by transmitting different types of data together. Wireless communications has allowed the growth of cellular telephony, personal communication systems, and wireless local area networks.

Modern digital communications was initiated with the concept of pulse code modulation, which allowed the transmission of binary signals. PCM is a practical implementation of sampling, quantization, and coding, or analog-to-digital conversion, of an analog message into a digital message. Using the sample representation of a message, the idea of mixing several messages—possibly of different types—developed into time-division multiplexing (TDM), which is the dual of frequency-division multiplexing (FDM). In TDM, samples from different messages are interspersed into one message that can be quantized, coded, and transmitted and then separated into the different messages at the receiver.

As multiplexing techniques, FDM and TDM became the basis for similar techniques used in wireless communications. Typically, three forms of sharing the available radio spectrum are: frequency-division multiple access (FDMA) where each user is assigned a band of frequencies all of the time; time-division multiple access (TDMA) where a user is given access to the available bandwidth for a limited time; and code-division multiple axis (CDMA) where users share the available spectrum without interfering with each other thanks to a unique code given to each user.

In this section we will introduce some of these techniques avoiding technical details, which we leave to excellent books in communications, telecommunications, and wireless communications. As you will learn, digital communications has a number of advantages over analog communications:

- The cost of digital circuits decreases as digital technology becomes more prevalent.
- Data from voice and video can be merged with computer data and transmitted over a common transmission system.
- Digital signals can be denoised easier than analog signals, and errors in digital communications can be corrected using special coding.

However, digital communication systems require a much larger bandwidth than analog communication systems, and quantization noise is introduced when transforming an analog signal into a digital signal.

12.4.1 Pulse Code Modulation

PCM can be seen as an implementation of analog-to-digital conversion of analog signals providing a serial bit stream. This means that sampling applied to a continuous-time message gives a pulse amplitude–modulated (PAM) signal that is then quantized and assigned a binary code to differentiate the different quantization levels. If each of the digital words has b binary digits, there are 2^b levels, and to each a different code word is assigned. An example of a code commonly used is the Gray code where only one bit changes from one quantization level to another. The most significant bit can be thought to correspond to the sign of the signal (1 for positive values, and 0 for negative values) and the others differentiate each level.

PCM is widely used in digital communications given that:

- It is realized with inexpensive digital circuitry.
- It allows merging and transmission of data from different sources (audio, video, computers, etc.) by means of time-division multiplexing, which we will see next.
- PCM signals can be easily regenerated by repeaters in long-distance digital telephony.

Despite all of these advantages, you need to remember that because of the sampling and the quantization the process used to obtain PCM signals is neither linear nor time invariant, and as such

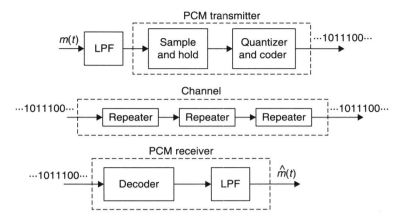

FIGURE 12.9
PCM system: transmitter, channel, and receiver.

its analysis is complicated. Figure 12.9 shows a transmitter, a channel, and a receiver of a PCM system.

The main disadvantage of PCM is that its bandwidth is wider than that of the analog message it represents. This is due to the rectangular pulses in the signal. If we represent the PCM signal $s(t)$ as

$$s(t) = \sum_{n=0}^{N-1} a_n \varphi(t - n\tau_s)$$

where $\varphi(t)$ is a function and τ_s is the duration of a symbol, the spectrum of $s(t)$ will be

$$S(\Omega) = \sum_{n=0}^{N} a_n \mathcal{F}[\varphi(t - n\tau_s)]$$

$$= \sum_{n=0}^{N} a_n \phi(\Omega) e^{-jn\tau_s \Omega}$$

Suppose that the function is a sinc function having an infinite time support so that $s(t)$ also has an infinite time support. If

$$\varphi(t) = \frac{\sin(\pi t/\tau_s)}{\pi t}$$

it has a band-limited spectrum,

$$\phi(\Omega) = u(\Omega + \pi/\tau_s) - u(\Omega - \pi/\tau_s)$$

so that the spectrum of the signal is also band limited. On the other hand, by duality if we use a rectangular pulse as the function $\varphi(t)$, its spectrum will spread over all frequencies, making the spectrum of the signal of very large bandwidth.

■ Example 12.7

Suppose you have a binary signal 01001001, with a duration of 8 units of time, and wish to represent it using rectangular pulses and sinc functions. Consider the bandwidth of each of the representations.

Solution

Using pulses $\varphi(t)$, the digital signal can be expressed as

$$s(t) = \sum_{n=0}^{7} b_n \varphi(t - n\tau_s)$$

where b_n are the binary digits of the digital signal (i.e., $b_0 = 0$, $b_1 = 1$, $b_2 = 0$, $b_3 = 0$, $b_4 = 1$, $b_5 = 0$, $b_6 = 0$, $b_7 = 1$, and $\tau_s = 1$). Thus,

$$s(t) = \varphi(t - 1) + \varphi(t - 4) + \varphi(t - 7)$$

and the spectrum of $s(t)$ is

$$
\begin{aligned}
S(\Omega) &= \phi(\Omega)(e^{-j\Omega} + e^{-j4\Omega} + e^{-j7\Omega}) \\
&= \phi(\Omega)e^{-j4\Omega}(e^{j3\Omega} + 1 + e^{-j3\Omega}) \\
&= \phi(\Omega)e^{-j4\Omega}(1 + 2\cos(3\Omega))
\end{aligned}
$$

so that

$$|S(\Omega)| = |\phi(\Omega)||1 + 2\cos(3\Omega)|$$

If the pulses are rectangular,

$$\varphi(t) = u(t) - u(t - 1)$$

the PCM signal will have an infinite-support spectrum because the pulse is of finite support. On the other hand, if we use sinc functions,

$$\varphi(t) = \frac{\sin(\pi t/\tau_s)}{\pi t}$$

its time support is infinite but its frequency support is finite (i.e., the sinc function is band limited). In which case, the spectrum of the PCM signal is also of finite support.

If this digital signal is transmitted and received without any distortion, at the receiver we can use the orthogonality of the $\varphi(t)$ signals or sample the received signal at nT_s to obtain the b_n. Clearly, each of these pulses has disadvantages—the advantage of having a finite support in the time or in the frequency becomes a disadvantage in the other domain. ■

Baseband and Band-Pass Communication Systems

A baseband signal can be transmitted over a pair of wires (like in a telephone), coaxial cables, or optical fibers. But a baseband signal cannot be transmitted over a radio link or a satellite because this would require a large antenna to radiate the low-frequency spectrum of the signal. Thus, the signal spectrum must be shifted to a higher frequency by modulating a carrier by the baseband signal. This can be done by amplitude and by angle modulation (frequency and phase). Several forms are possible.

■ Example 12.8

Suppose the binary signal 01001101 is to be transmitted over a radio link using AM and FM modulation. Discuss the different band-pass signals obtained.

Solution

The binary message can be represented as a sequence of pulses with different amplitudes. For instance, we could represent the binary digit 1 by a pulse of constant amplitude, and the binary 0 is represented by switching off the pulse (see the corresponding modulating signal $m_1(t)$ in Figures 12.10(a) and 12.10(b)). Another possible representation would be to represent the binary digit 1 with a positive pulse of constant amplitude, and 0 with the negative of the pulse used for 1 (see the corresponding modulating signal $m_2(t)$ in Figures 12.10(c) and 12.10(d)).

In AM modulation, if we use $m_1(t)$ to modulate a sinusoidal carrier $\cos(\Omega_0 t)$ we obtain the amplitude-shift keying (ASK) signal shown in Figure 12.10(a). On the other hand, when using $m_2(t)$ to modulate the same carrier we obtain a phase-shift keying (PSK) signal shown in Figure 12.10(c). In this case, the phase of the carrier is shifted 180° as the pulse changes from positive to negative.

Using FM modulation, the symbol 0 is transmitted using a windowed cosine of some frequency Ω_{c0} and the symbol 1 is transmitted with a windowed cosine of frequency Ω_{c1} resulting in frequency-shift keying (FSK). The data are transmitted by varying the frequency. In this case it is possible to get the same modulated signals for both $m_1(t)$ and $m_2(t)$. The modulated signals are shown in Figure 12.10(b and d).

The ASK, PSK, and FSK are also known as BASK, BPSK, and BFSK, respectively, by adding the word "binary" (B) to the corresponding amplitude-, phase-, and frequency-shift keying. ■

12.4.2 Time-Division Multiplexing

In a telephone system, multiplexing enables multiple conversations to be carried across a single shared circuit. The first multiplexing system used was frequency-division multiplexing (FDM), which we covered in Chapter 6. In FDM an available bandwidth is divided among different users. In the case of voice communications, each user is allocated a bandwidth of 4 KHz, which provides good fidelity. In FDM, a user could use the allocated frequencies all of the time, but the user could not go outside the allocated band of frequencies.

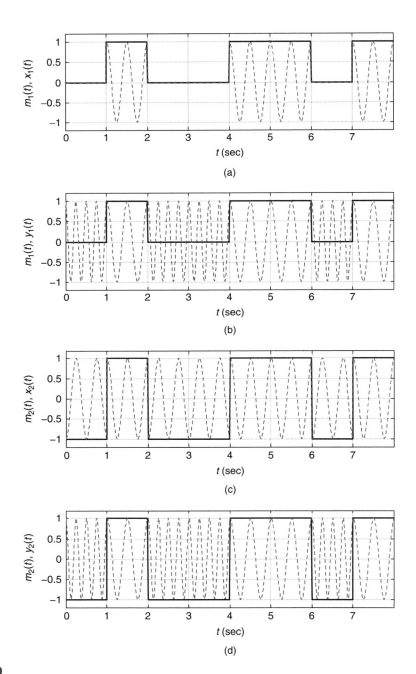

FIGURE 12.10

Pulse signals (continuous line) $m_1(t)$ and $m_2(t)$ corresponding to binary sequence 01001101. (a, c), ASK signals $x_1(t) = m_1(t)\cos(4\pi t)$ and $x_2(t) = m_2(t)\cos(4\pi t)$ in dashed lines. (b, d) In dashed lines the FSK signals $y_1(t)$ and $y_2(t)$ equal $\cos(4\pi t)$ when $m_1(t)$, $m_2(t)$ are 1 and $\cos(8\pi t)$ when $m_2(t)$ is 0 or -1.

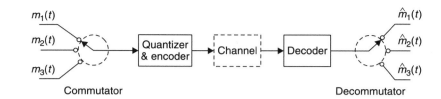

FIGURE 12.11

TDM system: transmitter, channel, and receiver.

Pulse-modulated signals have large bandwidths, and as such, when transmitted together they overlap in frequency, interfering with each other. However, these signals only provide information at each of the sampling times, so that one could insert in between these times other samples that will be separated at the receiver. This is the principle of *time-division multiplexing* (TDM), where pulses from different signals are interspersed into one signal and converted into a PCM signal and transmitted. See Figure 12.11. At the receiver, the signal is changed back into the pulse-modulated signal and separated into the number of signals interspersed at the input. Repeaters placed between the transmitter and the receiver regenerate a clean binary signal from a noisy binary signal along the way. The noisy signal coming into the repeater is thresholded to known binary levels and resent. A large part of the cost of a transmission facility is due to these repeaters that are placed about every mile along the line.

TDM allows the transmission of different types of data, and mixture of analog and digital using different multiplexing techniques. Not to lose information, the switch at the receiver needs to be synchronized with the switch at the transmitter. *Frame synchronization* consists in sending a synchronizing signal for each frame. An additional channel is allocated for this purpose. To accommodate more users, the width of the pulses used for each user needs to become narrower, which increases the overall bandwidth of the multiplexed signal.

12.4.3 Spread Spectrum and Orthogonal Frequency-Division Multiplexing

The objective of TDM is to put several users or different types of data together sharing the same bandwidth at different times. Likewise, FDM users share part of the available bandwidth all the time. TDM and FDM are examples of how to use bandwidth in an efficient way. In other situations, like in quadrature-amplitude modulation (QAM), the objective is to send two messages over the same bandwidth using the orthogonality of the carriers to recover them. In spread spectrum, the objective is to use the orthogonality of the carriers associated with different users to share the available spectrum, while spreading the message in frequency so that it occupies a bandwidth much larger than that of the message. On the other hand, orthogonal frequency-division multiplexing (OFDM) is a multicarrier system where the carriers are orthogonal.

Sharing the radio spectrum among users, or multiple access, is a basic strategy of wireless communication systems. Basic modalities are derived from FDM, TDM, and spread spectrum. In FDMA the spectrum is shared by assigning specific channels to users, permanently or temporarily. TDMA allows access to all of the available spectrum, but each user is assigned a time interval in which to access it. CDMA uses spread spectrum, where a user's message is spread or encrypted over the available spectrum using a code to differentiate the different users. The objective of these three techniques is to maximize the radio spectrum utilization.

Spread Spectrum—A Famous Actress Idea

Not surprisingly, the first mention of the use of frequency hopping, a form of spread spectrum, for secure communications came from a patent by Nikolas Tesla in 1903. As you recall, Tesla is the world-renowned Serbian-American inventor, and physicist, and mechanical and electrical engineer who pioneered amplitude modulation.

The most celebrated invention of frequency-hopping was, however, that of Hedy Lamarr and George Antheil, who in 1942 received a U.S. patent for their "secret communications system," in which they used a piano-roll for frequency-hopping. This was during World War II, and their idea was to stop the enemy from detecting or jamming radio-guided torpedoes. To avoid the jammer, in frequency-hopping spread spectrum the transmitter changes in a quasi-random way the center frequency of the transmitted signal. Hedy Lamarr (1913–2000) was an Austrian-American actress and communications technology innovator, while George Antheil (1900–1959) was an American composer and pianist. Their patent was never applied, and it would be many years before the technology was actually deployed. Ms. Lamarr conceived the idea of hopping from frequency to frequency just as a piano player plays the same notes, but in different octaves. Their concept eventually provided the basis for the CDMA airlink, which Qualcomm commercialized in 1995. Today, CDMA and its core principles provide the backbone for wireless communications, thanks to the creative vision of an extraordinary woman [70, 74, 62].

Spread Spectrum

A spread-spectrum system is one in which the transmitted signal is spread over a wide frequency band, much wider than the bandwidth required to transmit the message. Such a system would take a baseband voice signal with a bandwidth of a few kilohertz and spread it to a band of many megahertz. Two types of spread-spectrum systems are:

- *Direct-sequence system:* A digital code sequence with a bit rate higher than the message is used to obtain the modulated signal.
- *Frequency-hopping system:* The carrier frequency is shifted in discrete increments in a pattern dictated by a code sequence. We will not consider this here.

Direct-Sequence Spread-Spectrum

Suppose the message $m(t)$ we wish to transmit is a polar binary signal, and that a spreading code $c(t)$, also in polar binary form, is modulated by the message to obtain the modulated baseband signal

$$x(t) = m(t)c(t) \tag{12.21}$$

The sequence $c(t)$ is pseudo-random, unpredictable to an outsider, but that can be generated deterministically. Each user is assigned uniquely one of these sequences—that is, the spreading codes assigned to two users are not related at all. Moreover, the bit rate of $c(t)$ is much higher than that of the message. As in many other modulation systems, the modulated baseband signal $x(t)$ has a much higher rate than the message, and as such its spectrum is much wider than that of the message that is already wide given that it is a sequence of pulses. This can also be seen by considering that $x(t)$ as the product of $m(t)$, and $c(t)$, its spectrum is the convolution of the spectrum of $m(t)$ with the spectrum of $c(t)$ with a bandwidth equal to the sum of the bandwidths of these spectra.

When transmitting over a radio link the baseband signal $x(t)$ modulates an analog carrier to obtain the transmitting signal $s(t)$. At the receiver, if no interference occurred in the transmission, the received signal $r(t) = s(t)$, and after demodulation using the analog carrier frequency, the spread signal $x(t)$ is obtained. If we multiply it by $c(t)$ we get

$$x(t)c(t) = c^2(t)m(t) = m(t) \tag{12.22}$$

since $c^2(t) = 1$ for all t. See Figure 12.12.

Two significant advantages of direct-sequence spread spectrum are:

- *Robustness to noise and jammers:* The above detection or despreading is idealized. The received signal will have interferences due to channel noise, interference from other users, and even, in military applications, intentional jamming. Jamming attempts to corrupt the sent message by adding to it either a narrowband or a wideband signal. If at the receiver, the spread signal contains additive noise $\eta(t)$ and a jammer $j(t)$, it is demodulated by the BPSK system. The received baseband signal is

$$\hat{r}(t) = x(t) + \hat{\eta}(t) + \hat{j}(t) \tag{12.23}$$

where the noise and the jammer have been affected by the demodulator.
Multiplying it by $c(t)$ gives

$$\hat{r}(t)c(t) = m(t) + \hat{\eta}(t)c(t) + \hat{j}(t)c(t) \tag{12.24}$$

or the desired message with a spread noise and jammer. Thus, the transmitted signal is resistant to interferences by spreading them over all frequencies.

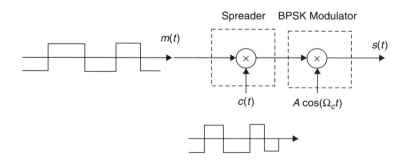

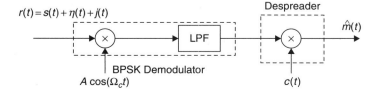

FIGURE 12.12
Direct-sequence
spread-spectrum system.

- *Robustness to interference from other users:* Assuming no noise or jammer, if the received baseband signal comes from two users—that is,

$$\hat{r}(t) = m_1(t)c_1(t) + m_2(t)c_2(t) \tag{12.25}$$

where the codes $c_1(t)$ and $c_2(t)$ are the corresponding codes for the two users, and $m_1(t)$ and $m_2(t)$ their messages. At the receiver of user 1, despreading using code $c_1(t)$ we get

$$\hat{r}(t)c_1(t) = m_1(t)c_1^2(t) + m_2(t)c_2(t)c_1(t) \approx m_1(t) \tag{12.26}$$

since the codes are generated so that $c_1^2(t) = 1$ and $c_1(t)$ and $c_2(t)$ are not correlated. Thus, we detect the message corresponding to user 1. The same happens when there is interference from more than one user.

Simulation of direct sequence spread spectrum. In this simulation we consider that the message is randomly generated and that the spreading code is also randomly generated (our code does not have the same characteristics as the one used to generate the code for spread-spectrum systems). To generate the train of pulses for the message and the code we use filters of different length (recall the spreading code changes more frequently than the message). The spreading makes the transmitting signal have a wider spectrum than that of the message (see Figure 12.13).

The binary transmitting signal modulates a sinusoidal carrier of frequency 100 Hz. Assuming the communication channel does not change the transmitted signal and perfect synchronization at the analog receiver is possible, the despread signal coincides with the sent message. In practice, the effects of multipath in the channel, noise, and possible jamming would not make this possible.

```
%%%%%%%%%%%%%%%%
% Simulation of
% spread spectrum
%%%%%%%%%%%%%%%%%
clear all; clf
% message
m1 = rand(1,12)>0.9;m1 = (m1-0.5) * 2;
m = zeros(1,00);
m(1:9:100) = m1
h = ones(1,9);
m = filter(h,1,m);
% spreading code
c1 = rand(1,25)>0.5;c1 = (c1-0.5) * 2;
c = zeros(1,100);
c(1:4:100) = c1;
h1 = ones(1,4);
c = filter(h1,1,c);
Ts = 0.0001; t = [0:99] * Ts;
s = m. * c;
figure(1)
```

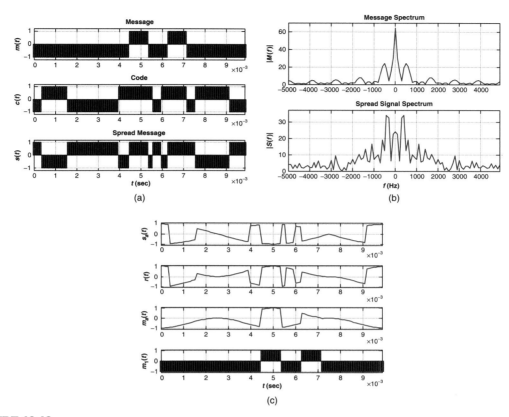

FIGURE 12.13

Simulation of direct-sequence spread-spectrum communication. (a) Displays from top to bottom the message, the code, and the spread signal. (b) Displays the spectrum of the message and of the spread signal (notice it is wider than that of the message). (c) Displays the band-pass signals sent and received (assumed equal), the despread analog, and the binary message.

```
subplot(311)
bar(t,m); axis([0 max(t) -1.2 1.2]);grid; ylabel('m(t)')
subplot(312)
bar(t,c); axis([0 max(t) -1.2 1.2]);grid; ylabel('c(t)')
subplot(313)
bar(t,s); axis([0 max(t) -1.2 1.2]);grid; ylabel('s(t)'); xlabel('t (sec)')
% spectrum of message and spread signal
M = fftshift(abs(fft(m)));
S = fftshift(abs(fft(s)));
N = length(M);K = [0:N-1];w = 2 * K * pi/N-pi; f = w/(2 * pi * Ts);
figure(2)
subplot(211)
```

```
plot(f,M);grid; axis([min(f) max(f) 0 1.1 * max(M)]); ylabel('—M(f)—')
subplot(212)
plot(f,S); grid; axis([min(f) max(f) 0 1.1  * max(S)]);ylabel('—S(f)—'); xlabel('f (Hz)')
% analog modulation and demodulation
s = s. * cos(200 * pi * t);
r = s. * cos(200 * pi * t);
% despreading
mm = r. * c;
for k = 1:length(mm);
if mm(k) > 0
   m2(k) = 1;
else
   m2(k) = -1;
end
end
figure(3)
subplot(411)
plot(t,s); axis([0 max(t) 1.1 * min(s) 1.1 * max(s)]);grid; ylabel('s_a(t)')
subplot(412)
plot(t,r); axis([0 max(t) 1.1 * min(r) 1.1 * max(r)]);grid; ylabel('r(t)')
subplot(413)
plot(t,mm); axis([0 max(t) 1.1 * min(mm) 1.1 * max(mm)]);grid; ylabel('m_a(t)')
subplot(414)
bar(t,m2); axis([0 max(t) -1.2 1.2]);axis([0 max(t) 1.1 * min(mm) 1.1 * max(mm)])
grid;ylabel('\m(t)'); xlabel('t (sec)')
```

Orthogonal Frequency-Division Multiplexing

OFDM is a multicarrier modulation technique where the available bandwidth is divided into narrowband subchannels. It is used for high data-rate transmission over mobile wireless channels [27, 60, 4].

If $\{d_k, \ k = 0, \ldots, N-1\}$ are symbols to be transmitted, the OFDM-modulated signal is

$$s(t) = \sum_{m=-\infty}^{\infty} \sum_{k=0}^{N-1} d_k e^{j2\pi f_k t} p(t - mT) \tag{12.27}$$

where T is the symbol duration, $f_k = f_0 + k\Delta f$ for a subchannel bandwidth $\Delta f = 1/T$ with initial frequency f_0, and $p(t) = u(t) - u(t - T)$. Thus, the carriers are conventional complex exponentials. Considering a baseband transmission, at the receiver the orthogonality of these exponentials in $[0, T]$ allows us to recover the symbols. Indeed, assuming that no interference is introduced by the transmission channel (i.e., the received signal $r(t) = s(t)$), multiplying $r(t)$ by the conjugate of the exponential carrier and smoothing the result we obtain for $k = 0, \ldots, N-1$, and $mT \le t \le (m+1)T$

(where $p(t - mT) = 1$),

$$\frac{1}{T} \int_{mT}^{(m+1)T} r(t)e^{-j2\pi f_k t} dt = \frac{1}{T} \int_{mT}^{(m+1)T} \sum_{\ell=0}^{N-1} d_\ell e^{j2\pi f_\ell t} e^{-j2\pi f_k t} dt$$

$$= \sum_{\ell=0}^{N-1} d_\ell \frac{1}{T} \int_{mT}^{(m+1)T} e^{-j2\pi (f_k - f_\ell)t} dt$$

$$= \sum_{\ell=0}^{N-1} d_\ell \delta[k - \ell] = d_k$$

for any $-\infty < m < \infty$, and where we let $f_k - f_\ell = (k - \ell)\Delta f = (k - \ell)/T$.

OFDM Implementation with FFT

If the modulated signal $s(t)$, $0 \le t \le T$, in Equation (12.27) is sampled at $t = nT/N$, we obtain for a frame the inverse DFT

$$s[n] = \sum_{k=0}^{N-1} d_k e^{j2\pi f_k nT/N} = \sum_{k=0}^{N-1} d_k e^{j2\pi kn/N} \qquad 0 \le n \le N - 1 \qquad (12.28)$$

where $2\pi f_k T/N = 2\pi k/N$ are the discrete frequencies in radians. At the receiver, with no interferences present, the symbols $\{d_k\}$ are obtained by computing the DFT of the baseband received signal. Given that the inverse and the direct DFT can be efficiently implemented by the FFT, the OFDM is a very efficient technique that is used in wireless local area networks (WLANs) and digital audio broadcasting (DAB).

Figure 12.14 gives a general description of the transmitter and receiver in an OFDM system: The high-speed data in binary form coming into the system are transformed from serial to parallel and fed into an IFFT block giving as output the transmitting signal that is sent to the channel. The received signal is then fed into an FFT block providing estimates of the sent symbols that are finally put in serial form.

FIGURE 12.14
Discrete model of baseband OFDM. The blocks S/P and P/S convert a serial into a parallel stream and a parallel to serial, respectively.

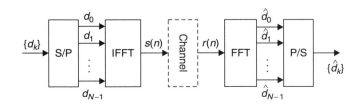

12.5 WHAT HAVE WE ACCOMPLISHED? WHERE DO WE GO FROM HERE?

In this chapter we have seen how the theoretical results presented in the third part of the book relate to digital signal processing, digital control, and digital communications. The Fast Fourier Transform made possible the establishment and significant growth of digital signal processing as a technical area. The next step for you could be to get into more depth in the theory and applications of digital signal processing, preferably including some theory of random variables and processes, toward statistical signal processing, speech, and image processing. We have shown you also the connection of the discrete-time signals and systems with digital control and communications. Deeper understanding of these areas would be an interesting next step. You have come a long way, but there is more to learn.

Useful Formulas

Trigonometric Relations

Reciprocal

$$\csc(\theta) = \frac{1}{\sin(\theta)}$$

$$\sec(\theta) = \frac{1}{\cos(\theta)}$$

$$\cot(\theta) = \frac{1}{\tan(\theta)}$$

Pythagorean Identity

$$\sin^2(\theta) + \cos^2(\theta) = 1$$

Sum and Difference of Angles

$$\sin(\theta \pm \phi) = \sin(\theta)\cos(\phi) \pm \cos(\theta)\sin(\phi)$$

$$\sin(2\theta) = 2\sin(\theta)\cos(\theta)$$

$$\cos(\theta \pm \phi) = \cos(\theta)\cos(\phi) \mp \sin(\theta)\sin(\phi)$$

$$\cos(2\theta) = \cos^2(\theta) - \sin^2(\theta)$$

Multiple Angle

$$\sin(n\theta) = 2\sin((n-1)\theta)\cos(\theta) - \sin((n-2)\theta)$$

$$\cos(n\theta) = 2\cos((n-1)\theta)\cos(\theta) - \cos((n-2)\theta)$$

Signals and Systems Using MATLAB®. DOI: 10.1016/B978-0-12-374716-7.00017-x

Products

$$\sin(\theta)\sin(\phi) = \frac{1}{2}[\cos(\theta - \phi) - \cos(\theta + \phi)]$$

$$\cos(\theta)\cos(\phi) = \frac{1}{2}[\cos(\theta - \phi) + \cos(\theta + \phi)]$$

$$\sin(\theta)\cos(\phi) = \frac{1}{2}[\sin(\theta + \phi) + \sin(\theta - \phi)]$$

$$\cos(\theta)\sin(\phi) = \frac{1}{2}[\sin(\theta + \phi) - \sin(\theta - \phi)]$$

Euler's Identity

$$e^{j\theta} = \cos(\theta) + j\sin(\theta) \qquad j = \sqrt{-1}$$

$$\cos(\theta) = \frac{e^{j\theta} + e^{-j\theta}}{2} \qquad \sin(\theta) = \frac{e^{j\theta} - e^{-j\theta}}{2j}$$

$$\tan(\theta) = -j\left[\frac{e^{j\theta} - e^{-j\theta}}{e^{j\theta} + e^{-j\theta}}\right]$$

Hyperbolic Trigonometry Relations

$$\textit{Hyperbolic cosine:} \quad \cosh(\alpha) = \frac{1}{2}(e^{\alpha} + e^{-\alpha})$$

$$\textit{Hyperbolic sine:} \quad \sinh(\alpha) = \frac{1}{2}(e^{\alpha} - e^{-\alpha})$$

$$\cosh^2(\alpha) - \sinh^2(\alpha) = 1$$

Calculus

Derivatives (u, v functions of x; α, β constants)

$$\frac{duv}{dx} = u\frac{dv}{dx} + v\frac{du}{dx}$$

$$\frac{du^n}{dx} = nu^{n-1}\frac{du}{dx}$$

Integrals

$$\int \phi(\gamma)dx = \int \frac{\phi(\gamma)}{\gamma'}d\gamma, \text{ where } \gamma' = \frac{d\gamma}{dx}$$

$$\int u\,dv = uv - \int v\,du$$

$$\int x^n dx = \frac{x^{n+1}}{n+1} \qquad n \neq -1, \text{ integer}$$

$$\int x^{-1}dx = \log(x)$$

$$\int e^{ax}dx = \frac{e^{ax}}{a} \qquad a \neq 0$$

$$\int xe^{ax}dx = \frac{e^{ax}}{a^2}(ax-1)$$

$$\int \sin(ax)dx = -\frac{1}{a}\cos(ax)$$

$$\int \cos(ax)dx = \frac{1}{a}\sin(ax)$$

$$\int \frac{\sin(x)}{x}dx = \sum_{n=0}^{\infty}(-1)^n \frac{x^{2n+1}}{(2n+1)(2n+1)!} \qquad \text{integral of sinc function}$$

$$\int_0^{\infty} \frac{\sin(x)}{x}dx = \int_0^{\infty} \left[\frac{\sin(x)}{x}\right]^2 dx = \frac{\pi}{2}$$

Bibliography

[1] A. Antoniou. *Digital Filters*. New York: McGraw-Hill, 1979.

[2] E. T. Bell. *Men of Mathematics*. New York: Simon and Schuster, 1965.

[3] J. Belrose. Fessenden and the early history of radio science. *http://www.ieee.ca/millennium/radio/radio_radioscientist.html*, accessed 2010.

[4] J. Bingham. Multicarrier modulation for data transmission: An idea whose time has come. *IEEE Communications Magazine*, May 1990: 5–14.

[5] N. K. Bose. *Digital Filters*. Salem, MA: Elsevier, 1985.

[6] J. L. Bourjaily. *http://www-personal.umich.edu/~jbourj/money.htm*, accessed 2010.

[7] C. Boyer. *A History of Mathematics*. New York: Wiley, 1991.

[8] R. Bracewell. *The Fourier Transform and Its Application*. Boston: McGraw-Hill, 2000.

[9] M. Brain. How CDs work. *http://electronics.howstuffworks.com/cd7.htm*, accessed 2010.

[10] W. Briggs and V. Henson. *The DFT*. Philadelphia: Society for Industrial and Applied Mathematics, 1995.

[11] O. Brigham. *The Fast Fourier Transform and Its Applications*. Englewood Cliffs, NJ: Prentice-Hall, 1988.

[12] D. Cheng. *Analysis of Linear Systems*. Reading, MA: Addison-Wesley, 1959.

[13] L. Chua, C. Desoer, and E. Kuh. *Linear and Nonlinear Circuits*. New York: McGraw-Hill, 1987.

[14] J. Cooley. How the FFT gained acceptance. In *A History of Scientific Computing*, edited by S. Nash. New York: Association for Computing Machinery, Inc. Press, 1990, pp. 133–140.

[15] J. W. Cooley and J. W. Tukey. An algorithm for the machine calculation of complex Fourier series. *Mathematics of Computation*, 19, April 1965: 297.

[16] L. W. Couch. *Digital and Analog Communication Systems*. Upper Saddle River, NJ: Pearson/Prentice-Hall, 2007.

[17] E. Craig. *Laplace and Fourier Transforms for Electrical Engineers*. New York: Holt, Rinehart, and Winston, 1966.

[18] E. Cunningham. *Digital Filtering*. Boston: Houghton Mifflin, 1992.

[19] G. Danielson and C. Lanczos. Some improvements in practical Fourier analysis and their applications to X-ray scattering from liquids. *J. Franklin Institute*, 1942: 365–380.

[20] C. Desoer and E. Kuh. *Basic Circuit Theory*. New York: McGraw-Hill, 1969.

[21] R. Dorf and R. Bishop. *Modern Control Systems*. Upper Saddle River, NJ: Prentice-Hall, 2005.

[22] G. Franklin, J. Powell, and M. Workman. *Digital Control of Dynamic Systems*. Reading, MA: Addison-Wesley, 1998.

Signals and Systems Using MATLAB®. DOI: 10.1016/B978-0-12-374716-7.00025-9

[23] G. Johnson. Claude Shannon, Mathematician, Dies at 84. New York Times, February 27, 2001.

[24] R. Gabel and R. Roberts. *Signals and Linear Systems*. New York: Wiley, 1987.

[25] S. Goldberg. *Introduction to Difference Equations*. New York: Dover, 1958.

[26] R. Graham, D. Knuth, and O. Patashnik. *Concrete Mathematics: A Foundation for Computer Science*. Reading, MA: Addison-Wesley, 1994.

[27] S. Haykin and M. Moher. *Communication Systems*. Hoboken, NJ: Wiley, 2009.

[28] S. Haykin and M. Moher. *Modern Wireless Communications*. Upper Saddle River, NJ: Pearson/Prentice-Hall, 2005.

[29] S. Haykin and M. Moher. *Introduction to Analog and Digital Communications*. Hoboken, NJ: Wiley, 2007.

[30] S. Haykin and B. Van Veen. *Signals and Systems*. New York: Wiley, 2003.

[31] M. Heideman, D. Johnson, and S. Burrus. Gauss and the history of the FFT. *IEEE Acoustics, Speech and Signal Processing (ASSP) Magazine*, vol. 1, pp. 14–21, Oct. 1984.

[32] K. Howell. *Principles of Fourier Analysis*. Boca Raton, FL: Chapman & Hall, CRC Press 2001.

[33] IEEE. Nyquist biography. *http://www.ieee.org/web/aboutus/history_center/biography/nyquist.html*, accessed 2010.

[34] Intel. Microprocessor quick reference guide. *http://www.intel.com/pressroom/kits/quickreffam.htm*, accessed 2010.

[35] Intel. Moore's law. *http://www.intel.com/technology/mooreslaw/index.htm*, accessed 2010.

[36] L. Jackson. *Signals, Systems, and Transforms*. Reading, MA: Addison-Wesley, 1991.

[37] E. I. Jury. *Theory and Application of the Z-Transform Method*. New York: Wiley, 1964.

[38] E. Kamen and B. Heck. *Fundamentals of Signals and Systems*. Upper Saddle River, NJ: Pearson/Prentice-Hall, 2007.

[39] R. Keyes. Moore's law today. *IEEE Circuits and Systems Magazine*, 8, 2008: 53–54.

[40] B. P. Lathi. *Modern Digital and Analog Communication Systems*. New York: Oxford University Press, 1998.

[41] B. P. Lathi. *Linear Systems and Signals*. New York: Oxford University Press, 2002.

[42] E. Lee and P. Varaiya. *Structure and Interpretation of Signals and Systems*. Boston: Addison-Wesley, 2003.

[43] W. Lehr, F. Merino, and S. Gillet. Software radio: Implications for wireless services, industry structure, and public policy. *http://itc.mit.edu*, accessed 2002.

[44] D. Luke. The origins of the sampling theorem. *IEEE Communications Magazine*, April 1999:106–108.

[45] M. Bellis. The invention of radio. *http://inventors.about.com/od/rstartinventions/a/radio.htm*, accessed 2010.

[46] D. McGillem and G. Cooper. *Continuous and Discrete Signals and System Analysis*. New York: Holt, Rinehart, and Wiston, 1984.

[47] Z. A. Melzak. *Companion to Concrete Mathematics*. New York: Wiley, 1973.

[48] S. Mitra. *Digital Signal Processing*. New York: McGraw-Hill, 2006.

[49] C. Moler. Cleve's Corner—The Origins of MATLAB. *http://www.mathworks.com/company/newsletters/news_notes/clevescorner/dec04.html*, accessed 2010.

[50] National. Op-amp history. *http://www.analog.com/library/analogdialogue/.../39.../web_chh_final.pdf*, accessed 2008.

[51] F. Nebeker. *Signal Processing—The Emergence of a Discipline, 1948 to 1998*. New Brunswick, NJ: IEEE History Center, 1998. (This book was especially published for the 50th anniversary of the creation of the IEEE Signal Processing Society in 1998).

[52] MIT news. MIT Professor Claude Shannon dies; was founder of digital communications. *http://web.mit.edu/newsoffice/2001/shannon.html*, accessed 2009.

[53] K. Ogata. *Modern Control Engineering*. Upper Saddle River, NJ: Prentice-Hall, 1997.

[54] A. Oppenheim and R. Schafer. *Digital Signal Processing*. Englewood Cliffs, NJ: Prentice-Hall, 1975.

[55] A. Oppenheim and R. Schafer. *Discrete-Time Signal Processing*. Upper Saddle River, NJ: Prentice-Hall, 2010.

[56] A. Oppenheim and A. Willsky. *Signals and Systems*. Upper Saddle River, NJ: Prentice-Hall, 1997.

[57] A. Papoulis. *Signal Analysis*. New York: McGraw-Hill, 1977.

[58] PBS.org. Who invented radio? *http://www.pbs.org/tesla/ll/ll_whoradio.html*, accessed 2010.

[59] C. Phillips, J. Parr, and E. Riskin. *Signals, Systems and Transforms*. Upper Saddle River, NJ: Pearson/Prentice-Hall, 2003.

[60] R. Prasad and S. Hara. Overview of multi-carrier CDMA. IEEE Communications Magazine, vol. 35, pp. 126–133, Dec. 1997.

[61] J. Proakis and M. Salehi. *Communication Systems Engineering*. Upper Saddle River, NJ: Prentice-Hall, 2002.

[62] Qualcomm. Who we are: History. *http://www.qualcomm.com/who_we_are/history.html*, accessed 2010.

[63] L. Rabiner and B. Gold. *Theory and Application of Digital Signal Processing*. Englewood Cliffs, NJ: Prentice-Hall, 1975.

[64] GNU Radio. The GNU software radio. *http://www.gnu.org/software/gnuradio/*, accessed 2008.

[65] Jaycar Electronics Reference Data Sheet. Understanding decibels. *http://www.jaycar.com.au/images_uploaded/decibels.pdf*, accessed 2009.

[66] D. Slepian. On bandwidth. *Proceeding of the IEEE*, 64, March 1976. 292–300.

[67] S. Smith. *The Scientist and Engineer's Guide to Digital Signal Processing (http://www.dspguide.com)*. California Technical Publishing, San Diego, CA, 1997.

[68] S. Soliman and M. Srinath. *Continuous and Discrete Signals and Systems*. Upper Saddle River, NJ: Prentice-Hall, 1998.

[69] A. Stanoyevitch. *Introduction to Numerical Ordinary and Partial Differential Equations Using MATLAB* . New York: Wiley, 2005.

[70] Statemaster.com. Spread spectrum. *http://www.statemaster.com/encyclopedia/Spread-spectrum*, accessed 2009.

[71] H. Stern and S. Mahmoud. *Communication Systems—Analysis and Design*. Upper Saddle River, NJ: Pearson/Prentice-Hall, 2004.

[72] M. Van Valkenburg. *Analog Filter Design*. New York: Oxford University Press, 1982.

[73] Wikipedia. Euler's identity. *http://en.wikipedia.org/wiki/Euler's_identity*, accessed 2009.

[74] Wikipedia. Hedy Lamarr. *http://en.wikipedia.org/wiki/Hedy_Lamarr*, accessed 2009.

[75] Wikipedia. Nikola Tesla. *http://en.wikipedia.org/wiki/Nikola_Tesla*, accessed 2009.

[76] Wikipedia. Oliver Heaviside. *http://en.wikipedia.org/wiki/Oliver_Heaviside*, accessed 2009.

[77] Wikipedia. Field-programable gate array. *http://en.wikipedia.org/wiki/Field-programmable_gate_array*, accessed 2008.

[78] M. R. Williams. *A History of Computing Technologies*. New York: Wiley/IEEE Computer Society Press, 1997.

Index